Oceanographic and Marine Environmental Studies around the Arabian Peninsula

Oceanographic and Marine Environmental Studies around the Arabian Peninsula presents studies on a range of topics related to the marine environment of the Red Sea and Arabian (Persian) Gulf.

This book contains invited and peer-reviewed chapters from diverse researchers active in their respective fields. The chapters offer new data and include comprehensive lists of references. Some of the main topics included in the book are pollution from heavy metals and petroleum, hydro-environmental characteristics of the seas, conservation of marine ecosystems, the risk of climate change on the Red Sea region, and the mangrove environment.

With new developments occurring in the coastal regions in recent decades, the book will be not only a helpful resource to researchers but also a valuable reference for anyone curious about managing the marine and littoral environment of these two unique seas.

Oceanographic and Marine Environmental Studies around the Arabian Peninsula

Edited by Najeeb M.A. Rasul and Ian C.F. Stewart

CRC Press
Taylor & Francis Group
Boca Raton London New York

CRC Press is an imprint of the
Taylor & Francis Group, an **informa** business
A BALKEMA BOOK

First published 2025
by CRC Press/Balkema
4 Park Square, Milton Park, Abingdon, Oxon, OX14 4RN

and by CRC Press/Balkema
2385 NW Executive Center Drive, Suite 320, Boca Raton FL 33431

CRC Press/Balkema is an imprint of the Taylor & Francis Group, an informa business

British Library Cataloguing-in-Publication Data
A catalogue record for this book is available from the British Library

Library of Congress Cataloging-in-Publication Data
Names: Rasul, Najeeb M. A., editor. | Stewart, Ian C. F. (Ian Charles Ferguson), editor.
Title: Oceanographic and marine environmental studies around the Arabian Peninsula /
edited by Najeeb M.A. Rasul and Ian C.F. Stewart.
Description: Boca Raton, FL : Taylor & Francis, 2024. | Includes bibliographical
references and index.
Identifiers: LCCN 2023058878 (print) | LCCN 2023058879 (ebook) |
ISBN 9781032342948 (hardback) | ISBN 9781032342955 (paperback) |
ISBN 9781003321408 (ebook)
Subjects: LCSH: Oceanography—Arabian Sea. | Hydrography—Arabian Sea. |
Chemical oceanography—Arabian Sea. | Marine debris—
Arabian Sea. | Ecosystem health—Arabian Sea.
Classification: LCC GC731 .O44 2024 (print) | LCC GC731 (ebook) |
DDC 551.46/153—dc23/eng/20240220
LC record available at https://lccn.loc.gov/2023058878
LC ebook record available at https://lccn.loc.gov/2023058879

ISBN: 978-1-032-34294-8 (hbk)
ISBN: 978-1-032-34295-5 (pbk)
ISBN: 978-1-003-32140-8 (ebk)

DOI: 10.1201/9781003321408

Typeset in Times
by Apex CoVantage, LLC

Contents

Preface

Due to their semi-enclosed nature, the Red Sea and Arabian (or Persian) Gulf have unique environments that are very susceptible to human activities. Large areas are relatively shallow, and their location within an arid region means that they also experience high evaporation rates that, together with limited contact with the Indian Ocean, result in high temperatures and salinities that have an effect on the pollution levels. Although the region, including the Arabian Sea to the south of the Peninsula, has hosted major marine trade routes for several thousand years, it is only in recent times that pollution has become a major concern. The region is host to major hydrocarbon deposits, both onshore and offshore, that together with rapidly expanding industrialisation and population growth have resulted in extensive development of the area. All this has had a far-reaching impact on the marine and terrestrial ecology of the region and on its human inhabitants, including significant pollution of the marine environment in various forms, which is one of the main areas with which this book is concerned.

A wide range of topics are examined in this volume, from the hydrodynamics and chemistry of the Red Sea and the Gulf to the shallow-water and littoral environments and associated pollution. Five chapters deal with the hydrodynamics and chemistry of seawater, with five chapters on heavy metals and a further two chapters on marine litter in the Red Sea. The ongoing problem of hydrocarbon pollution in the Gulf is discussed in three chapters. Ten chapters deal with marine and coastal ecosystems, including coastal mangroves. Changes in climate over time and their effects are discussed in a further chapter, and a final chapter discusses possible further research. This volume is not intended to be a comprehensive survey of the various subjects, but rather presents recent studies that will be of interest to researchers in the area. The chapters aim to highlight some of the current thinking and summaries of recent research in each field, including useful reference lists for further study.

The contributions of the many technical referees to improving the contents of the chapters are greatly appreciated.

Editor biographies

Najeeb M.A. Rasul obtained a PhD in geological oceanography from the University of Wales, Swansea, UK. He was with the Faculty of Marine Science, King Abdulaziz University, Jeddah, before joining the Saudi Geological Survey, where he was a technical adviser until 2022 and head of the Center for Marine Geology from 2003 to 2010. He has held key research positions at the Geological and Geophysical Research Systems in Canada, the Challenger Division for Seafloor Processes at the Southampton Oceanography Centre, currently NOC, Southampton, and Environment Canada. He has conducted research cruises with European Union research centres, including the Institute of Marine Sciences (ISMAR), Italy, to the Red Sea and the IOC-UNESCO (TTR programme) in the Atlantic Ocean and Mediterranean Sea. He was a EUROMARGINS and Saudi scientific coordinator. He has also been involved in research projects as chief scientist and principal investigator in the Western Basin of Lake Erie, the Indian and Atlantic Oceans, the Arabian Sea and, currently, the Red Sea. His research interests have concentrated on both shallow and deepwater sedimentary processes, depositional mechanisms and environmental aspects.

Ian C.F. Stewart received a PhD in seismology from the University of Adelaide, South Australia. He is an independent geophysical consultant based in Adelaide and has worked for numerous mining and oil companies in Australia, the Middle East and elsewhere. From 1974 until 1982, he was with the Department of Earth Sciences at Memorial University of Newfoundland, Canada, where he taught geophysics. He then carried out research for Saudi Aramco in the exploration department in Dhahran, Saudi Arabia, from 1982 until 1996 and was in charge of the development and application of potential field methods throughout the Kingdom. Later, he was involved in geophysical exploration in Australia. He was the Geophysical Technical Adviser to the Saudi Geological Survey (SGS), Jeddah, between 2003 and 2013, applying geophysical methods for mineral and groundwater exploration in the Kingdom, as well as carrying out earthquake studies. He has published numerous papers in addition to reports for the SGS, Saudi Aramco and other companies.

Contributors

Qamar Abbas
Department of Biology
College of Science
University of Bahrain
Sakhir Campus
Bahrain

Mohamed Abdelkareem
Geology Department
Faculty of Science
South Valley University
Qena, Egypt
and
Remote Sensing Laboratory
South Valley University
Qena, Egypt

Khalil H. Abdulla
Supreme Council for Environment
Manama, Bahrain

Nada M. Al-Dewail
Department of Environmental Sciences
Faculty of Environmental Sciences and Marine Biology
Hadhramout University
Yemen

Moahd K. Alghuson
Industrial Engineering Department
Faculty of Engineering
University of Tabuk
Tabuk, Saudi Arabia

Sama Al-Maarofi
Director of Administration and Program Development
New Brunswick Aboriginal Peoples Council
Fredericton, NB
Canada

Bandar Almur
Department of Environmental Science
King Abdulaziz University
Jeddah, Saudi Arabia

Awatif Alnair
National Center for Research
Khartoum, Sudan

Abdullah S. Alnasser
Department of Civil Engineering
College of Engineering
Qassim University
Unaizah, Saudi Arabia
and
Civil and Environmental Engineering
Southern Methodist University
Dallas, Texas, USA

Yousef Alosairi
Environment and Life Sciences Research Center
Kuwait Institute for Scientific Research
Kuwait City, Kuwait

Hamid Al-Saad
College of Marine Science
University of Basrah
Basrah, Iraq

Abeer Al-Saleh
Crises Decision Support Program
Environmental & Life Science Research Center
Kuwait Institute for Scientific Research
Kuwait City, Kuwait

Nabil Al-Shwafi
Faculty of Petroleum and Natural Resources
Sana'a University
Yemen

Hanin Alzubaidy
Biological and Environmental Science and
Engineering Division
King Abdullah University of Science and Technology
Thuwal, Saudi Arabia

Arafat T. Amer
Department of Biology
Faculty of Education – Radfan
University of Aden
Yemen

Neila Annabi-Trabelsi
Laboratoire Biodiversité Marine et Environnement
Université de Sfax
Sfax, Tunisia

André Antunes
State Key Laboratory of Lunar and Planetary Sciences
Macau University of Science and Technology
Macau SAR, China
and
China National Space Administration
Macau Center for Space Exploration and Science
Macau SAR, China

Mohamed Asharaf
Center for Environment & Marine Studies
King Fahd University of Petroleum & Minerals
Dhahran, Saudi Arabia

Dimitra Atri
Center for Space Science
New York University Abu Dhabi
Abu Dhabi, UAE

Habib Ayadi
Laboratoire Biodiversité Marine et Environnement
Université de Sfax
Sfax, Tunisia

Azad Baazeem
Biological and Environmental Science and Engineering Division
King Abdullah University of Science and Technology
Thuwal, Saudi Arabia

Abdulla Ba-Issa
Chemistry Department
Faculty of Science
Sana'a University
Yemen

Abdulrahman Ben-Hasan
Marine Science Department
College of Science
Kuwait University
Kuwait City, Kuwait

Ali A. DouAbul
Marine Science Center
University of Basrah
Basrah, Iraq

Mariam A. Eldai
Department of Coastal Management
Faculty of Marine Science and Fisheries
Red Sea University
Port Sudan, Sudan

Khalid M. Elfatih
Department of Coastal Management
Faculty of Marine Science and Fisheries
Red Sea University
Port Sudan, Sudan

El Hag A. El Hag
UNESCO
Chair for Oceanography
Khartoum, Sudan

Jinoy Gopalan
Center for Environment & Marine Studies
King Fahd University of Petroleum & Minerals
Dhahran, Saudi Arabia

Wassim Guermazi
Laboratoire Biodiversité Marine et Environnement
Université de Sfax
Sfax, Tunisia

Riyadh F. Halawani
Department of Environment
Faculty of Environmental Sciences
King Abdulaziz University
Jeddah, Saudi Arabia

Amna A. Hamid
National Center for Research
Khartoum, Sudan

Haiat S. Hassan
Department of Coastal Management
Faculty of Marine Science and Fisheries
Red Sea University
Port Sudan, Sudan

Yang He
State Key Laboratory of Lunar and Planetary Sciences
Macau University of Science and Technology
Macau SAR, China
and
China National Space Administration
Macau Center for Space Exploration and Science
Macau SAR, China

Eldirdery A.A. Ibrahim
Department of Coastal Management
Faculty of Marine Science and Fisheries
Red Sea University
Port Sudan, Sudan

Khadega I. Idrees
Department of Coastal Management
Faculty of Marine Science and Fisheries
Red Sea University
Port Sudan, Sudan

Thadickal V. Joydas
Center for Environment & Marine Studies
King Fahd University of Petroleum & Minerals
Dhahran, Saudi Arabia

Qusaie Karam
Crises Decision Support Program
Environmental & Life Sciences Research Center
Kuwait Institute for Scientific Research
Kuwait City, Kuwait

Song J. Kim
Department of Biological Sciences
College of Natural Sciences
Kongju National University
Gongju, South Korea

Vigneshwaran Krishnamoorthi
Center for Space Science
New York University Abu Dhabi
Abu Dhabi, UAE

Vincent Leignel
Laboratoire BIOSSE
Le Mans Université
France

Reynaldo Lindo
Center for Environment & Marine Studies
King Fahd University of Petroleum & Minerals
Dhahran, Saudi Arabia

Hashem Madkour
Marine and Environmental Geology
National Institute of Oceanography and Fisheries
Red Sea Branch
Hurghada, Egypt

Karuppasamy Manikandan
Center for Environment & Marine Studies
King Fahd University of Petroleum & Minerals
Dhahran, Saudi Arabia

Abbas M. Mansour
Geology Department
Faculty of Science
South Valley University
Qena, Egypt

Maged Marghany
Global Geoinformation Sdn. Bhd.
MasaKiara
Jln Dato Soliman
Kuala Lumpur, Malaysia

Hisham M. Nagi
Department of Environmental Sciences
Faculty of Petroleum and Natural Resources
Sana'a University
Yemen

Humood A. Naser
Department of Biology
College of Science
University of Bahrain
Sakhir, Bahrain

Dirar H. Nasr
Department of Marine Biology
Institute of Marine Researches
Red Sea University
Port Sudan, Sudan

Nahid A.R. Osman
Department of Coastal Management
Faculty of Marine Science and Fisheries
Red Sea University
Port Sudan, Sudan

Andrew N. Quicksall
Department of Civil and Environmental Engineering
Southern Methodist University
Dallas, Texas, USA

Mohammed Qurban
National Center for Wildlife
Ministry of Environment, Water and Agriculture
Riyadh, Kingdom of Saudi Arabia

Syed M. Rahman
Applied Research Center for Environment and Marine Studies
Research Institute
King Fahd University of Petroleum and Minerals
Dhahran, Saudi Arabia

Hope Rasmussen
Department of Civil and Environmental Engineering
Southern Methodist University
Dallas, Texas, USA

Najeeb M.A. Rasul
Geological and Geophysical Research Systems
Mississauga
Ontario, Canada

Hussain Raza
Department of Biological Sciences
College of Natural Sciences
Kongju National University
Gongju, South Korea

Amira Rekik
Laboratoire Biodiversité Marine et Environnement
Université de Sfax
Sfax, Tunisia

Afkar A. Rahama
Faculty of Applied Sciences
Red Sea University
Port Sudan, Sudan

Alexandre S. Rosado
Biological and Environmental Science and Engineering Division
King Abdullah University of Science and Technology
Thuwal, Saudi Arabia

Ahmed I. Rushdi
ETAL
Corvallis
Oregon, USA

Reshma Sahito
Department of Zoology
University of Sindh
Jamshoro, Pakistan

Nader A. Salman
Al-Manara College for Medical Sciences
Misan, Iraq

Júnia Schultz
Biological and Environmental Science and Engineering Division
King Abdullah University of Science and Technology
Thuwal, Saudi Arabia

Soha Shabaka
National Institute of Oceanography and Fisheries
NIOF
Egypt

Ahmed M. Shawky
EcoPro Training Center
Ismailia, Egypt

Fatima M. Shediwah
Department of Biology
Faculty of Science
University of Sana'a
Yemen

Marta F. Simões
State Key Laboratory of Lunar and Planetary Sciences
Macau University of Science and Technology
Macau SAR, China
and
China National Space Administration
Macau Center for Space Exploration and Science
Macau SAR, China

Bernd R. T. Simoneit (deceased)
Department of Chemistry
College of Sciences
Oregon State University
Corvallis, Oregon, USA

Ian C.F. Stewart
Stewart Geophysical Consultants Pty. Ltd.
Adelaide
South Australia, Australia

Awatif A. Sugga
Department of Marine Biology
Institute of Marine Researches
Red Sea University
Port Sudan, Sudan

Osman M. Tiah
Department of Coastal Management
Faculty of Marine Science and Fisheries
Red Sea University
Port Sudan, Sudan

Surya P. Tiwari
Applied Research Center for Environment and Marine Studies
Research Institute
King Fahd University of Petroleum and Minerals
Dhahran, Saudi Arabia

Peter J. Vine
Earth and Ocean Sciences
University of Galway
Ireland

Myra E. Wilson
Civil and Environmental Engineering
Southern Methodist University
Dallas, Texas, USA

Mohamed Youssef
Department of Geology and Geophysics
College of Science
King Saud University
Riyadh, Saudi Arabia
and
Department of Geology
Faculty of Science
South Valley University
Qena, Egypt

Sumaya A. Zakieldeen
Khartoum University
Khartoum, Sudan

Khalid M. Zubier
Marine Physics Department
Faculty of Marine Sciences
King Abdulaziz University
Jeddah, Saudi Arabia

Introduction to oceanographic and marine environmental studies around the Arabian Peninsula

1

Najeeb M.A. Rasul, Ian C.F. Stewart and Peter J. Vine

1.1 INTRODUCTION

Due to their contrasting geological settings, the Red Sea and Arabian (Persian) Gulf (PAG) have quite different morphologies. The Red Sea is about 2,000 km long, with a surface area of 458,620 km^2 and a maximum width of 355 km, and is a young ocean basin formed by the splitting of the Arabian-Nubian shield and the resulting ongoing separation of the African and Arabian plates. It has a narrow axial trough, the spreading centre, with depths greater than 1,000 m that underlies about 15% of the Red Sea, while depths of up to about 3,000 m occur in the "Deeps"; the shelves are much shallower, with about 40% of the Red Sea area having depths of less than 100 m (Figure 1.1). Some details of the Red Sea morphology, oceanography and environment are given by Rasul and Stewart (2015, 2019, and the chapters and many references therein) and will not be repeated here.

The Arabian Plate to the east of the Red Sea is fairly stable geologically; at its eastern side, it is overlaid by the Arabian Gulf, which covers about 251,00 km^2, is 989 km long and is quite shallow, with a maximum depth of over 90 m and an average depth of only 50 m. The deepest sections lie toward the east of the Gulf due to subduction of the Arabian plate under the Eurasian plate, forming the Zagros Fold Belt (Figure 1.2). At the northwestern end of the Gulf, the Euphrates and the Tigris Rivers converge in the delta of the Shatt al-Arab. The maximum width of the Gulf is 340 km, and at its narrowest point (56 km wide) at the Strait of Hormuz, the Gulf joins the Gulf of Oman and Indian Ocean.

While there are morphological differences between the Red Sea and the Gulf, they are both semi-enclosed with limited contact with the Indian Ocean. They are located within an arid region, and the environment is affected by high evaporation rates that result in high temperatures and salinities. The Gulf in particular, as well as parts of the Red Sea (such as the Gulf of Suez), are host to major hydrocarbon deposits and production facilities, and there is extensive shipping activity along both waterways. Considerable expansion of the population and urbanisation along the coasts of both the Red Sea and Gulf in recent decades have also had far-reaching impacts on the marine and terrestrial life of the region and the ecosystem in general.

The present volume contains chapters on a wide range of topics concerning the oceanography and marine environment of these two unique regions and follows on from the earlier volumes edited by Rasul and Stewart (2015, 2019) that dealt exclusively with the Red Sea, whereas here, chapters are also included on topics related to the Arabian Gulf (PAG).

1.2 THIS VOLUME

Brief summaries of the various topics covered in this volume are presented in the following sections.

1.2.1 Hydrodynamics

A study of the Northern Arabian Gulf (NAG) highlights the key hydrodynamic features of this region, particularly in relation to dissolved and particulate pollutants. Circulation patterns including mixing and transport affect the role of hydrodynamics whilst previous and ongoing activities have adversely impacted the coastal environments, including poor water quality, hypoxia, frequent harmful blooms and turbid waters. Moreover, climate change, including global warming and the reduction of precipitation and river flows, has altered the hydrodynamic regime. It has also affected various ecological components, particularly near the estuarine zones. In the same context, dust storms at the northern edge of the Gulf have increased in strength and frequency.

DOI: 10.1201/9781003321408-1

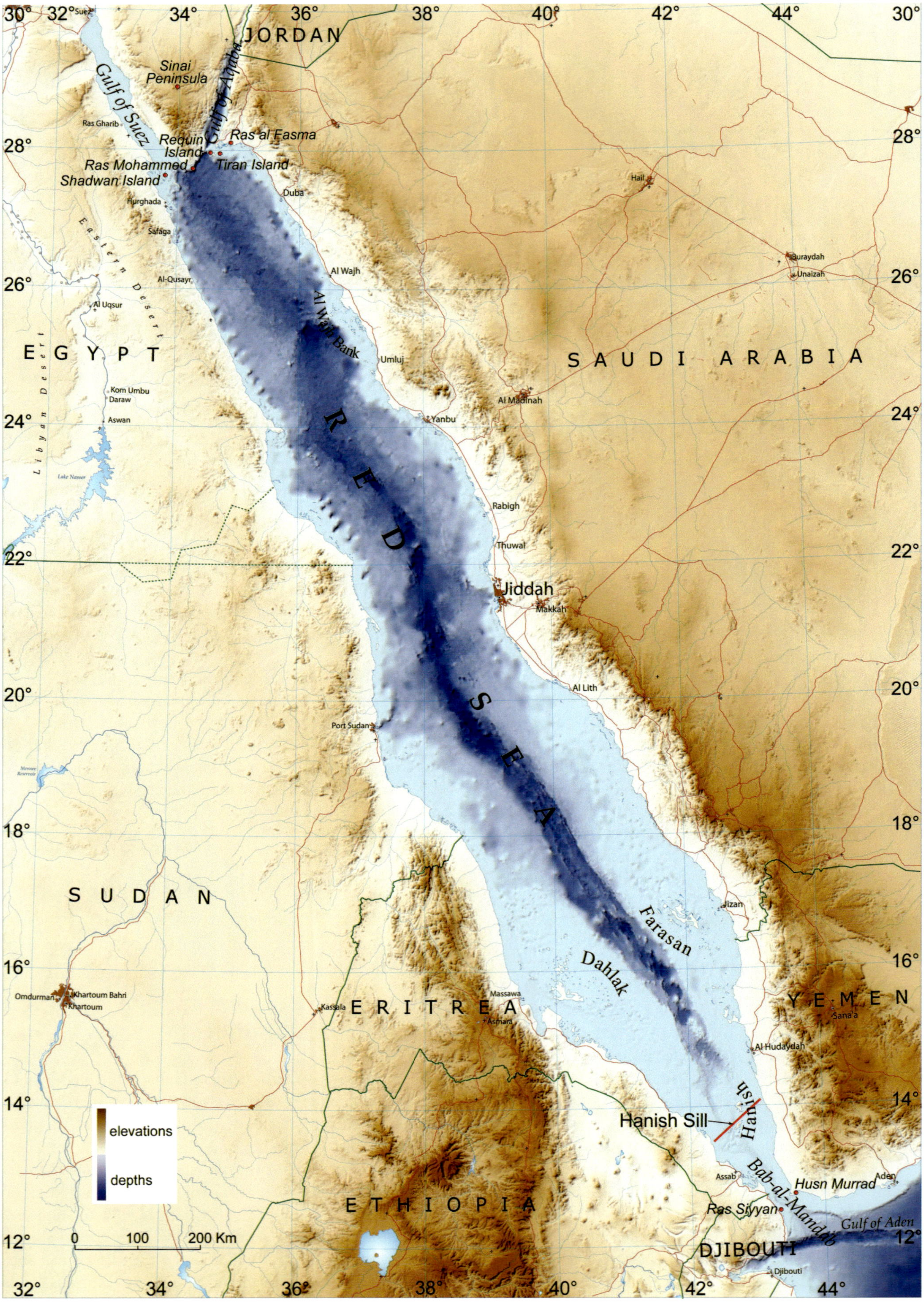

FIGURE 1.1 Map of the Red Sea area; darker colours indicate greater depths or higher elevations.

Source: Rasul et al. 2015

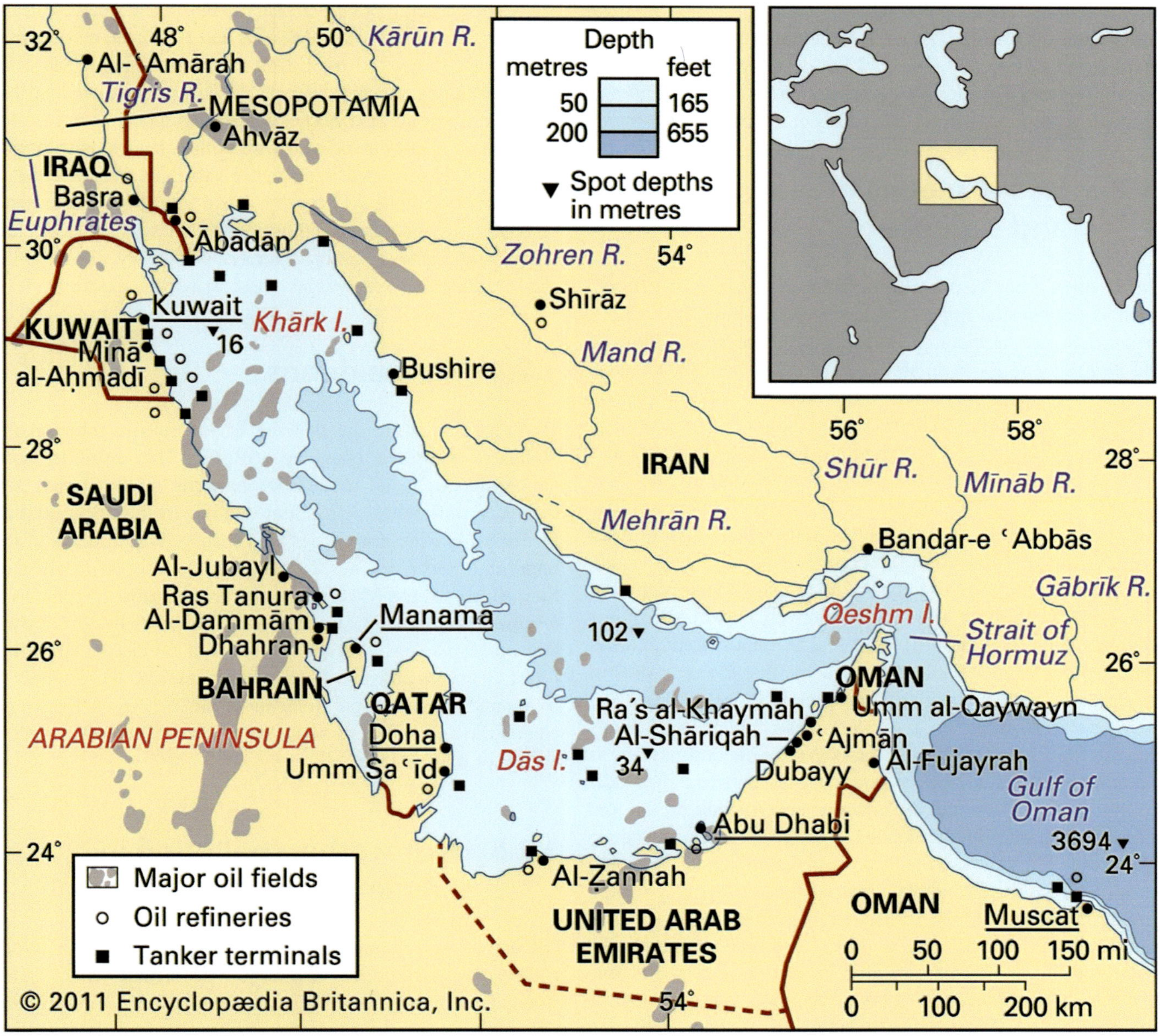

FIGURE 1.2 Map of the Arabian (Persian) Gulf.

Source: Encyclopædia Britannica 2023

ALOS PALSAR radar data is frequently acquired over the coral reef Farasan Islands in the Red Sea. Automatic turbulence clustering involves two advanced quantum image processing algorithms: 1) a quantum multi-objective evolutionary algorithm (QMEA) and 2) a quantum non-dominant sort and elitism (QNSGA-II). The ALOS PALSAR radar image textures and backscatters are initially encoded in qubit chromosomes. The results predict the structural morphology of turbulent flow, and QNSGA-II is a promising tool for the automatic detection of turbulent flow in satellite radar data, as shown in the Farasan data.

Wave conditions in the Red Sea vary both spatially and temporally under the influence of the varying wind field. A recent study utilized ERA5 hourly wave reanalysis data extending over a period of 72 years. Variability in terms of exceedance levels, annual and seasonal trends, return values and energy resource assessments were in general alignment with the previous studies, indicating that the ERA5 wave data are suitable for studying wave variability in the Red Sea. Accordingly, the utilisation of the other ERA5 wave parameters in more widespread investigations of the wave variability in the Red Sea is highly encouraged.

1.2.2 Chemical oceanography

The spectrally resolved water-leaving radiance, also known as ocean colour, is essential for comprehending the distribution and dynamics of phytoplankton (i.e., chlorophyll-a pigment) in seawater, since chlorophyll-a absorbs light at blue and red wavelengths and reflects green wavelengths. Remote sensing using satellite data allows continuous measurement of optically active substances in the water column, including chlorophyll-a, colour-dissolved organic matter and suspended sediments. Physical, chemical and biological factors cause the spatial and temporal changes in the ocean colour of the Arabian Gulf, including changes in nutrients, temperature, light, mixing, circulation and pollution. The information can be used to identify and address environmental issues such as ocean pollution and the impacts of climate change.

The most acute consequences of eutrophication of fresh water and seawater are hypoxia, a decrease in biodiversity and harmful algal blooms. The Red Sea is a semi-closed system where the amplification of pollution (chemical, physical and biological) is an important environmental factor.

1.2.3 Heavy metals in seawater and sediments

Information from Yemen covers the levels, variation and possible environmental impacts of heavy metals in coastal waters, sediments and the biota of Yemen, mainly the coastal cities of Al-Mukalla and Aden in the Gulf of Aden and Al-Hodeidah in the Red Sea. The selected heavy metals include manganese (Mn), iron (Fe), cobalt (Co), nickel (Ni), copper (Cu), zinc (Zn), cadmium (Cd) and lead (Pb). Examination of the available data showed that the concentrations of Fe, Co, Ni, Zn, Cd and Pb in Al-Mukalla are higher than their levels in Aden and Al-Hodeidah seawaters. Mn and Cu are higher in Al-Hodeidah than in Aden and Al-Mukalla seawaters. The concentrations of heavy metals in sediments from Al-Mukalla, Aden and Al-Hodeidah vary, whereas Mn, Fe, Co, Zn and Pb are relatively high in Aden while Ni and Cu are high in Al-Hodeidah. The values of the biogeochemical parameters and indices indicate that the sediments of the three cities are moderately contaminated by only Cd and Pb.

Lead (Pb) pollution is a worldwide public health issue. Identifying Pb inputs from anthropogenic pollution as well as natural geological sources is important for understanding how to limit inputs of Pb into the environment. A study on the Red Sea coastline near Jeddah looked at sediment cores from three sites, revealing significant pollutant and natural inputs, including dust-blown sediments based on regional wind patterns spreading highly radiogenic natural Pb isotopic signatures. Fine particles are transported to the region in dust storms and carried into the sediment via runoff from sporadic intense flood systems. This causes the downtown site to accumulate higher concentrations of anthropogenic Pb, with Pb contributions temporally increasing from lower to upper depths, likely due to changes in precipitation resulting from climate change and disturbances to the soil from urbanization. This study gives important context to the Pb inputs regionally into the Red Sea area off the Arabian Peninsula.

Lead ratios in Red Sea sediments from the Jeddah region were assessed from 40 sediment samples, indicating a mixed natural and gasoline-derived background across the region, with the lowest levels in the central latitudes of the study region, focused on the Jeddah urban area and increasing both to the north and south of that area. This research provides geochemical fingerprints of both a legacy-impacted background and a modern anthropogenic source of Pb pollution in the Saudi Arabian coastal areas of the Red Sea.

Sources of lead in mangrove sediments have been studied off three major industrial sites (i.e., Yanbu, Jeddah, and the Farasan Islands). Trends and Pb ratios were examined relative to the literature to identify likely sources for each location. Results show that despite broad-based lead contamination from gasoline, Jeddah and the Farasan Islands have relatively low comparable Pb concentrations, showing that the gasoline source has a significant impact on the isotopic signature while it has little impact on total loading.

Concentrations of Fe, Mn, Zn, Cu, Pb, Ni and Cd, were determined in shells of *Strombus tricornis* and associated marine sediments from 15 sites on the Egyptian coast of the Red Sea, with higher values occurring in association with continental contributions such as iron, manganese, copper and nickel across wadis, while heavy metals associated with pollution such as cadmium and lead were recorded with higher values in areas associated with human activities. Generally, metal variations reflect natural conditions and human activity. Nevertheless, there is no clear relationship between the concentrations of heavy metals in the shells of *Strombus tricornis* and those in associated sediments.

1.2.4 Marine debris

Two chapters study the distribution of marine litter on the Sudanese Red Sea coastline (SRSC). This exhibits similar composition and abundance patterns on the beach and in beach sediments. A gradual decline in the accumulation rate of all marine litter types was observed. Sediments of the upper and lower beaches differed significantly in densities of their marine litter. Meso-plastics were the most abundant (in number) plastic litter size category, followed by the macro- and mega-plastics. Further studies that precisely determine types, densities and size distribution of marine plastic litter in living and non-living components of SRSC ecosystems would enable the authorities to better control and lessen the adverse effects of marine litter.

1.2.5 Hydrocarbon pollution in sediments and seawater

Oil enters the marine environment through multiple routes such as oil exploration discharges, accidental tanker collisions and natural oil seepages from the seabed. The Regional Organization for the Protection of the Marine Environment (ROPME) sea area in the Arabian Gulf is susceptible to oil spillages with resulting toxicity-induced malformations on marine organisms such as sea bream *Sparidentex hasta* and orange-spotted grouper *Epinephelus coioides*, including inward curvature of the spine, kyphosis-hunchback, and scoliosis-lateral bending of the spine.

The distribution, characteristics, sources and environmental impacts of hydrocarbons in the Shatt Al-Arab River and the northern part of the Persian/Arabian Gulf (PAG) coast are discussed. The total oil-related sources were dominant and estimated to be >80% of total hydrocarbons (81% in the river, 80% in the estuary and 84% in the coastal zone). The presence and inputs of high amounts of petroleum hydrocarbons to the freshwater and marine ecosystems of the region will have short- and long-term ecological impacts, including toxicity to benthic organisms that depend on the bottom water and the sediments as spawning and nursery areas, together with associated fishes.

New SAR sensors, such as Sentinel-1, providing affordable data could be a valuable basis for guiding oil and gas exploration, raising the question of what role microwave remote-sensing technology can play in resolving oil and gas exploration performance. A new technique is described for automatically detecting oil and gas exploration activities and their consequences in SAR data, such as TerraSAR-X sequences. Toward that goal, quantum interferometry is used, which is based on 4-D phase unwrapping (the Marghany 4-D quantized holographic interferometry algorithm). The hologram interferometry technique can infer 4-D from 3-D phase unwrapping of SAR holographic interferometry and accurately determine oil and gas levels in storage tanks.

1.2.6 Conservation and the marine ecosystem

The unique location and habitats of Sanganeb atoll in the Sudanese Red Sea have attracted wide attention, including efforts by local government and institutions to declare it a marine national park. Such efforts focused on legal protection, public education, environmental monitoring and scientific research. Among the Sanganeb Marine National Park's (SMNP) unique features is its location in the northwest Indian Ocean biogeographic region, which, at that time, had no other World Heritage sites.

Current data on marine fisheries and aquaculture is fundamental in establishing sustainable management plans. Gulf-wide, total aquatic food production recently surpassed one million tonnes, with fisheries and aquaculture contributing almost equally to the total production. Around three-quarters of the total fisheries' yield comes from gillnets, traps, lines, shrimp trawls and encircling nets. Most of the production consists of medium (30–89 cm) to large (>= 90 cm) sizes of demersal, reef-associated and pelagic fishes. Some of the most common management regulations include, for example, gear restrictions and marine protected areas, yet problems pertinent to fisheries management (e.g., weak enforcement) and habitat degradation are undermining fisheries' sustainability. Addressing these problems represents a crucial step toward sustainable fisheries.

1.2.7 Mangroves and the coastal ecosystem

Extensive coastal developments involving dredging are carried out in the PAG with consequent impacts on the marine environment, including loss and degradation of coastal and marine habitats such as mangrove swamps, seagrass beds, coral reefs, sabkha flats and productive mudflats. Effective management to avoid or minimize the impacts of dredging and reclamation may include prevention, planning, mitigation, ecosystem restoration and compensation measures. An ecosystem-based management approach can serve as an important platform for holistic integration of the various efforts of management to deliver the economic, environmental and social benefits of dredging and reclamation activities.

Mangroves provide numerous benefits to humans and wildlife, including carbon sequestration, protection against coastal erosion and habitats for various species. The extent of mangroves in the Saudi Arabian Gulf was mapped using Sentinel-2 satellite images that showed that the total area of mangroves increased from 6.04 km^2 in 2018 to 6.86 km^2 in 2022. The mapping provides a baseline for monitoring the status of mangrove ecosystems in eastern Saudi Arabia and shows the importance of remote sensing and GIS techniques for monitoring coastal ecosystems at regional and global scales.

The Egyptian Red Sea coast is known for its highly productive and diversified marine and coastal ecosystems, which include coral reefs, seagrass beds, mangrove forests, beaches, elevated coral reefs and coastal dunes. These valuable resources are being impacted by an increasing number of anthropogenic activities and natural events posing various threats. The Red Sea coast and shallows are under high risk of flash flood hazards, which cause erosion of the coastal area and destruction of the marine ecosystem.

Red Sea mangrove ecosystems and the mangrove microbiome in general are discussed, including the abundance, distribution and phenotype of the mangroves inhabiting the coasts and islands of the Red Sea. A review of the microbiome's role in plant interactions and mangroves in particular is provided, along with a discussion of the current approaches to exploring and understanding the mangrove microbiome.

The Red Sea is a hotbed of biological diversity, and nowhere is this of greater importance to humankind than at the interface between land and sea – the mangrove zone. Not only are these enigmatic plants nurseries for fish and protection against coastal erosion, but they are also a valuable source of medicines widely used in traditional cures for a variety of human ailments, including cancer, tumours and inflammatory illnesses, together with fungal, bacterial, viral and diabetic conditions. Used in traditional medicine for millennia, extracts from mangrove species such as *Avicennia marina* and *Rhizophora mucronata* exhibit strong therapeutic values.

Recent field observations confirm that two mangrove species still exist in Sudan – namely, *Avicennia marina* and *Rhizophora mucronata* – and factors affecting growth and survival of seedlings of *A. marina* are discussed. The mangrove spatial extent increased from 1984 to 1990 and decreased sharply during the period from 1995 to 2000, possibly due to severe droughts in 1997 and 1998. Between 2000 and 2010, a large increase in the mangrove area was observed, which could be attributed to increased rainfall conditions after the droughts.

Coastal sabkhas are productive ecosystems accommodating considerable diversity. Common plant families reported from Red Sea coastal sabkhas include Asteraceae, Poaceae, Juncaceae, Portulaceae, Amaranthaceae, Palmae, Juncaceae, Cyperaceae, Chenopodiaceae, Leguminosae, Zygophyllaceae and Plumbaginaceae. While the two bacterial families Salinibacteraceae and Rubidibacteraceae dominate the bacterial community in sabkha biofilms, the archaeal community is dominated by Halobacteriaceae. However, knowledge of seaweed, biogeography and genetic diversity is incomplete. The central Red Sea has the highest diversity of seagrass (10 to 12 seagrass species) compared to the northern and the southern Red Sea. Twelve species of seagrasses were recorded: *Cymodocea rotundata*, *Cymodocea serrulata*, *Halodule uninervis*, *Halodule pinifolia*, *Syringodium isoetifolium*, *Thalassodendron ciliatum*, *Enhalus acoroides*, *Halophila ovalis*, *Halophila decipiens*, *Halophila minor*, *Halophila stipulacea* and *Thalassia hemprichii*.

The effect of grazing by dugongs on seagrass in Wadi El-Gemal National Park in Egypt is described. Different levels of grazing intensity were observed in order to understand the competition between and recovery of seagrass species and were compared to a control area with no grazing. In the no-grazing experiment, the abundance of *H. stipulacea* increased over the monitoring period while that of *H. ovalis* decreased. In the case of low grazing, *H. stipulacea* colonised the "feeding trails" first, and at the end of the experiment, *H. stipulacea* and *C. rotundata* had recovered to pre-treatment levels. At high grazing, *H. ovalis* showed significantly greater recovery than *H. stipulacea* or *C. rotundata*. Intensive grazing by dugongs changes the relative abundance of different species of seagrasses, and the density of feeding trails has a role in controlling the seagrass composition.

Coastal systems are particularly sensitive to three key drivers related to climate change: sea level, ocean temperature and ocean acidity. The effects of climate change on the southern Sudanese Red Sea coast were assessed through mapping and monitoring land cover, using remote sensing and GIS techniques. These include the effects of sea-level rise, its impacts on potable water reservoirs in the Tokar Delta and on key habitats, and the inundation expected by 2030 and 2100 is estimated.

1.2.8 Climate variability and its effects

Climate change is likely to exacerbate the situation as the number of flood events increases. These events have the capability to cause loss of a sizable portion of the coast and settlement/tourism areas, infrastructure and human activities. In addition to climate change impacts, the earthquake activity along the Red Sea also represents a risk in some tourism areas. Integration of remote sensing, GIS and field data is significant for natural resources management and monitoring ecosystems in order to adapt to climate change.

1.2.9 New developments

The unique setting of the Arabian Peninsula leads to a high number of extreme saline environments, particularly associated with the coasts of the Arabian Gulf (PAG), the Red Sea and sabkhas further inland, as well as the borders of the Arabian tectonic plate (deep-sea brines). These locations remain underexplored yet provide exciting new opportunities at the forefront of microbiology research, ranging from biotechnology to astrobiology. An overview of these environments, their microbial communities and some of the first results from new research are discussed.

REFERENCES

Encyclopædia Britannica. 2023. *Arabian (Persian) Gulf Map*. www.britannica.com/place/Persian-Gulf#/media/1/452764/1998. Accessed 30 April 2023.

Rasul, N.M.A., and I.C.F. Stewart (eds.). 2015. *The Red Sea: The Formation, Morphology, Oceanography and Environment of a Young Ocean Basin*. Berlin and Heidelberg: Springer Earth System Sciences, 633 p. http://doi.org/10.1007/978-3-662-45201-1.

Rasul, N.M.A., and I.C.F. Stewart (eds.). 2019. *Oceanographic and Biological Aspects of the Red Sea*. Cham, Switzerland: Springer Nature, 550 p. http://doi.org/10.1007/978-3-319-99417-8.

Rasul, N.M.A., Stewart, I.C.F., and Z.A. Nawab. 2015. Introduction to the Red Sea: Its origin, structure and environment. In: *The Red Sea: The Formation, Morphology, Oceanography and Environment of a Young Ocean Basin*. Eds. N.M.A. Rasul and I.C.F. Stewart, 1–28. Berlin and Heidelberg: Springer Earth System Sciences. http://doi.org/10.1007/978-3662-45201-1_1.

PART 1

Hydrodynamics

Hydro-environmental characteristics of the Northern Arabian Gulf

2

Yousef Alosairi

2.1 PHYSICAL CHARACTERISTICS OF THE NORTHERN ARABIAN GULF

The Gulf is a comparatively large northwest-southeast-oriented basin in the Northern Hemisphere that extends 980 km along its main axis (Figure 2.1). At the northern end of the basin, the Gulf has a width of 218 km between Kuwait and Iran, where the region is characterized as shallow water near the shoreline. During low tides, large mudflats are exposed, particularly during low-spring conditions. The mudflats support many ecological species, including birds, and are a nursery ground for shrimps etc. (Al-Zaidan et al. 2006). Several tidal creeks and wide channels uniquely characterize the northern edge of the Gulf. Tidal creeks also exist in Boubyan Island (Kuwait), the largest island in the NAG (863 km^2). The tidal creeks vary in width, up to several metres, and are occasionally flooded while tidal channels are much larger, up to several hundred metres, and are mostly wet.

The tidal creeks are stable morphologically due to the warm saline nature of the region (Alosairi et al. 2019). However, the narrow creeks are characterized by relatively deeper water (2 m in some cases) than other shallow estuarine systems, where the water is about 36 m deep. On the banks of the creeks, a thin salt crust overlies the sand formed during surge events while the excessive heat evaporates the water, and the salt remains. Therefore, tidal creeks are a source of saline water during the surge and spring tidal conditions, when they are mostly flooded. The tidal channels such as Khor Abdullah and Al Subiya (Kuwait) and Khor Mahshahr (Iran) are much larger and deeper than the tidal creeks (4–8 m). Shoals and sand bars are found in these channels, composed of different sediment types, including cohesive sediments. The shoals are submerged most of the time and are aligned with the flow regime of the channels. They vary in size and correspondingly play a role in the flow divergence and mixing in the channels.

Offshore, the bathymetric characteristics of the NAG are mainly described as shallow, with a mean depth of 36 m. Across the basin, the bathymetry takes the form of an asymmetric bowl, relatively deep near Iran (>20 m) and shallow near Kuwait and Saudi Arabia (Figure 2.1). There are several islands, including coral islands, in the NAG region. Some islands are used for oil transportation while oil rigs are present across the whole region.

2.2 DRIVING FORCES: TIDES, WIND AND DENSITY

Tidal propagation in shallow semi-enclosed waters is affected by factors like basin geometry, friction and Coriolis forces. Given the hydrographical setting of the Gulf, bounded by the Strait of Hormuz, and the influence of Coriolis forcing, the resonant period of the tides is nearly 27 hours. This results in two amphidromic points of semi-diurnal constituents, one at the northwest end of the basin and the other in the southeast end, and a single amphidromic point of diurnal constituents in the central part near Bahrain (Hoseini and Soltanpour 2022; Madah and Gharbi 2022).

For demonstration purposes, a sample of measurements at the NAG is illustrated in Figure 2.2. A clear mixed semi-diurnal tide dominates the NAG with a tidal range that exceeds 4 m during spring conditions, while during neap tides, the tidal range reduces to about 2.3 m. It is worth noting from the datasets that, in some instances, the strong winds affect the water levels. This is evident from the irregularities in the water level patterns (Figure 2.2).

The general range of the salinity in the NAG region is typically around 40–41 ppt and well mixed up to a depth of 30 m (Reynolds 1993; Alosairi and Pokavanich 2017a). Local variations in salinity exist at point sources such as rivers and desalination plants, usually found near the shore. The salinity may drop to as low as 36 ppt in the estuarine regions

DOI: 10.1201/9781003321408-3

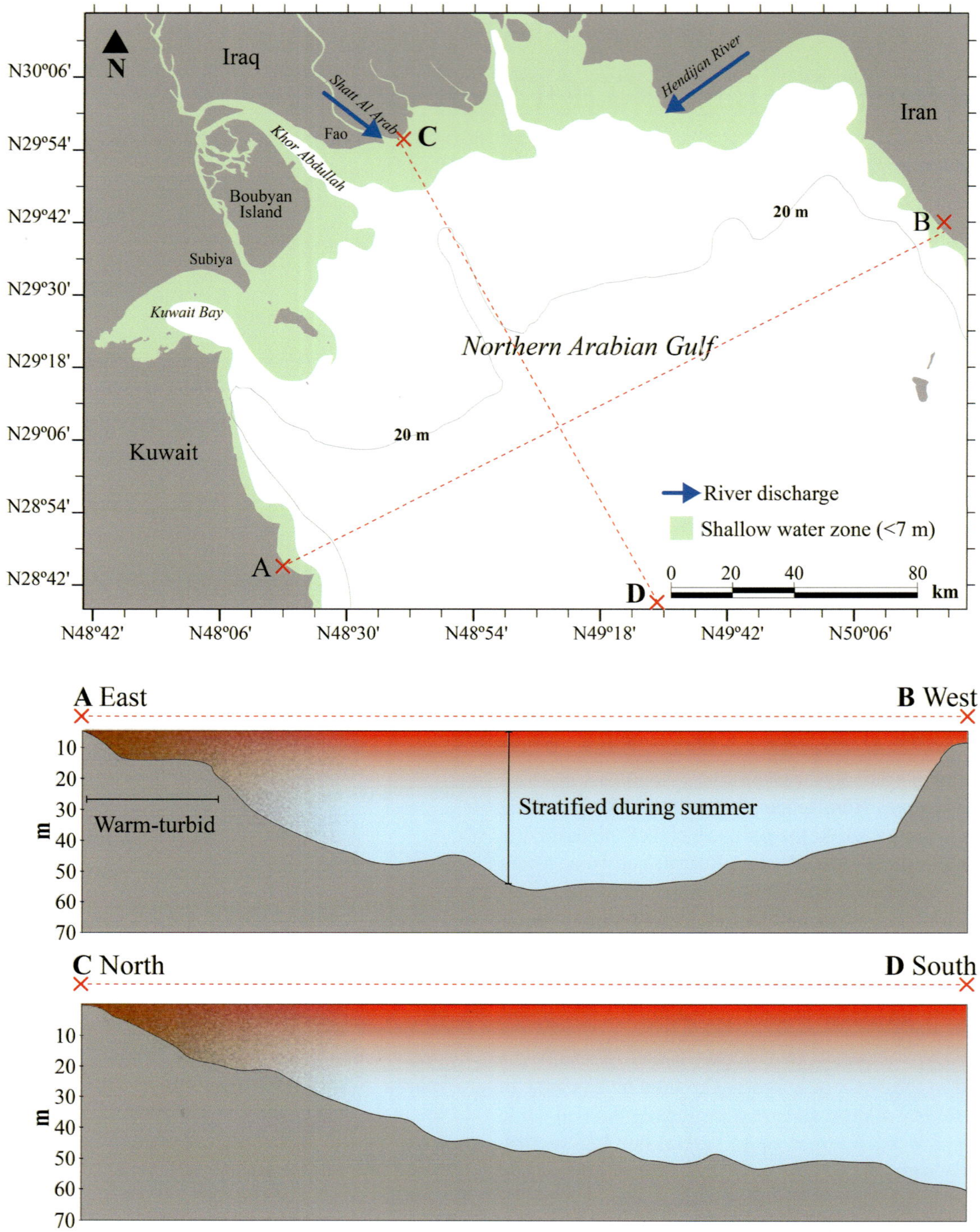

FIGURE 2.1 NAG general characteristics including the shallow coastal zones and river discharges (top panel), cross-section across the NAG (middle panel), and cross-section along the NAG (lower panel).

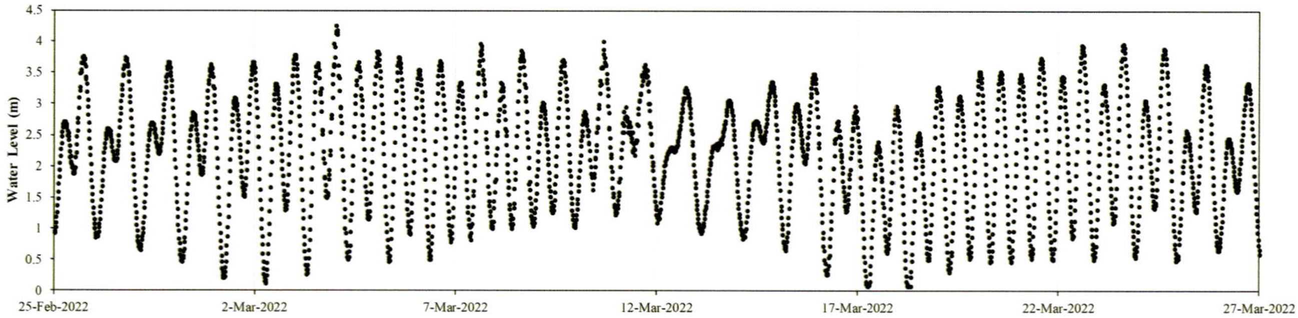

FIGURE 2.2 Sample of water level measurements in NAG.

during high freshwater discharge seasons. In contrast, in the waters receiving the desalination plant discharges, the salinity exceeds 46 ppt. The point sources impose horizontal density gradients extending from the near to far fields (several hundred metres). At water depths greater than 30 m, stratification becomes more apparent during summer. The thermocline development primarily responds to surface heating, which has increased over the past years (Figure 2.1). In 2020, an extreme sea surface temperature in Kuwait Bay (KB) reached 36.7°C (Alosairi et al. 2020a). The records were found to be the highest globally in 2020, which provided evidence of global warming. The impact of global warming on the extent of the thermocline has not yet been investigated. It would significantly affect the density-driven currents and vertical exchange of elements such as nutrients and trace metals.

The surface salinity is considerably lower due to freshwater inflow close to the Shatt Al-Arab mouth; this flow extends to the nearby channels, including Khor Abdullah. It is worth noting that the freshwater discharges from Shatt Al-Arab are not systematic, mainly due to human intervention. However, they follow seasonal trends in some years (Table 2.1). During extreme events, the flow from the Shatt Al Arab can be significant and may reduce the salinity at the northwest of the Gulf to as low as 36 ppt (Alosairi et al. 2019). Large volumes of freshwater discharge from the river and runoffs are directly linked to extreme rainfall events that occur over short periods (1–3 days). These events have been seen to reduce the salinity levels for almost all Kuwait's territorial waters. The extent of the extreme events seems to have increased over the past years, which aligns well with the disruption of rain patterns globally. However, this aspect has not been shown to be direct evidence of the impact of climate change.

Human interventions in the upstream regions of the Shatt Al Arab have significantly reduced the discharge (see Table 2.1). The major activities include the desiccation of the Mesopotamian Marshlands, dam constructions along the Tigris and Euphrates Rivers, and the shifting of the river channels. Therefore, the estuarine field is confined to the discharge near shore, extending a few kilometres. As a result, the horizontal gradients of the density have been greatly reduced over the past 30 years. These horizontal gradients, low near the Shatt Al Arab at 996 kg/m^3 and high in southern Kuwait at 1023 kg/m^3, imposed gravitational circulations that covered the full extent of the Kuwait coast (Alosairi and Pokavanich 2017b). The large estuarine fields would generally prompt seaward transport farther than the current conditions.

TABLE 2.1 River discharge in m^3 at the NAG, Shatt Al Arab discharge before and after human interventions and Hendijan River, Iran.

PERIOD	*PRE-REGULATED SHATT AL-ARAB*	*REGULATED SHATT AL-ARAB*	*HENDIJAN RIVER*
Jan	911	220	40
Feb	999	240	54
Mar	1546	270	47
Apr	1320	270	52
May	1299	300	56
Jun	1484	290	52
Jul	2433	270	48
Aug	2514	240	45
Sep	1515	220	44
Oct	823	180	39
Nov	688	180	34
Dec	688	190	39
Annual	**2053**	**239**	**46**

Source: Alosairi et al. 2018b

The wind in the NAG is characterized by the land-sea breeze temporarily overruled by the northwest wind known as the Shamal, eastern winds known as Nashi, southeast winds known as Sharqi (SE) or the southwest winds known as Kous. The Shamal wind is the most common in the region, blowing 63% of the time during summer and 43% of the time for the remainder of the year (Alosairi and Pokavanich 2017a) (Figure 2.3). The southeasterly winds occur for 19% of the time, but they produce the highest waves in the NAG because of the fetch length. Once the Kous timescales are sufficient, they may impose a positive surge in the northwest of the Gulf. The wave propagates into various northern channels, including Khor Abdullah, Khor Al Sabiya and the Shatt Al Arab, where it can intrude into the river channel due to the low discharge. Progressively, the wave dissipates as it reaches enclosed regions such as Warba Island and KB.

Most of the time, the residual currents are weak over the whole Gulf as their value does not exceed 2 cm (Alosairi and Pokavanich 2017a; Pous et al. 2012). A barotropic circulation dominates the general circulation in the Gulf from April to July. The circulation breaks down into minor features as the wind strengthens, and stratification diminishes due to winter cooling (Pous et al. 2012). Wind plays a crucial role in mixing shallow regions and enclosed areas like KB (Alosairi et al. 2018a), where environmental issues could arise during calm conditions like fishkill incidents and algal blooms (Alosairi and Alsulaiman 2019). Winds also play a crucial role in transporting organic and inorganic elements near the surface, including trace metals, microplastics (Alosairi et al. 2020b) and suspended sediments. Therefore, the dominant Shamal enhances the transport of various materials from the north toward the central axis of the Gulf in a transient mode. One of the key features of the NAG is the counterclockwise circulation that acts as a supplier to the western coast of the Gulf (Figure 2.4). The counterclockwise currents are enhanced by the strength of the Shamal winds (Figure 2.3).

2.3 MIXING AND TRANSPORT PATTERNS

Coastal communities in the Gulf region rely on the dispersion and transport of effluents once they enter coastal waters so that, once diluted, the biological breakdown renders them harmless or transports them away to offshore zones. Three distinct sets of processes combine to determine whether a specific domain has the assimilative capacity to accept additional effluent loading and can transport material from one location to another. First, the near and intermediate dispersal rates must

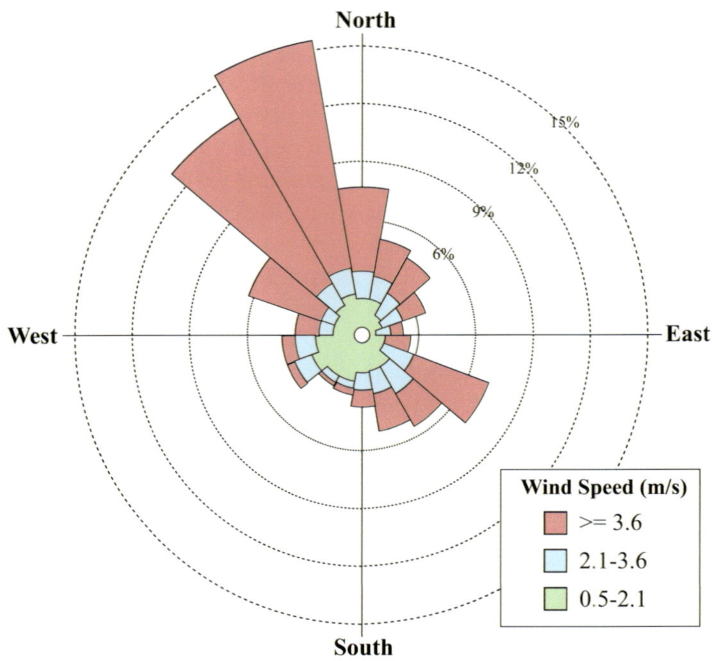

FIGURE 2.3 Annual wind conditions in the NAG.

Source: Reconstructed from Alosairi and Pokavanich (2017a)

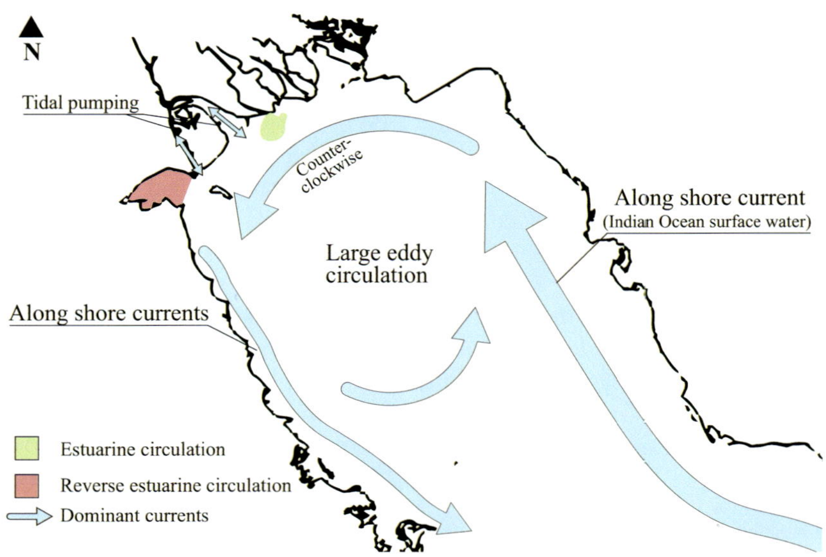

FIGURE 2.4 General transport patterns in the NAG.

Source: Reconstructed from the numerical modeling study in Alosairi and Pokavanich (2017a).

be sufficiently fast to dilute a given pollutant to a sufficiently low concentration that, when added to the background concentration, does not harm the environment and the functions of the ecosystem (Imberger et al. 2007). Second, the transport must effectively remove the by-products of the pollutant breakdown processes so there is no long-term build-up, particularly in the coastal zones. Third, the ecosystem must have the biochemical capacity to ensure the breakdown of those pollutants that may harm the rest of the ecosystem (Alosairi et al. 2011).

The mixing, dispersion and transport scale may vary from several hundred metres to kilometres, depending on the hydrodynamic processes and the meteorological effects. Based on recent studies (Alosairi et al. 2011), various scales of mixing and transport exist in the Gulf that will either deal with human pollutants, such as discharges from outfalls, or work to deliver nutrients from the ocean into the Gulf on large temporal and spatial scales (Figure 2.4).

The transport and mixing features in the Gulf region were studied previously, utilizing limited field measurements and mostly numerical models (Alosairi et al. 2015). At the near fields, the tides play a significant role in the mixing and dispersion of effluents (Alosairi et al. 2015). This is due to the region's shallowness, in which the basin's horizontal scale, several hundred kilometres, greatly exceeds the vertical scale of only 2–60 m. On the Arabian Peninsula, the action of the tides is much more significant than that on the Iranian side, where a typical pollutant discharge from an outfall would be highly dispersed within the tidal cycle induced by the tidal shear effects. Tidal asymmetries in the channels, such as Khor Abdullah, control the transport mode to a great extent (Figure 2.4).

Farther offshore, the role of gravitational forces, induced by the variations in temperature and salinity, takes an additional role in mixing and transport. In such cases, vertical exchange appears, although it is minimal. A gravitational circulation exists in the NAG. In rare conditions of estuarine regimes, estuarine and reverse-estuarine conditions exist in the NAG. Locally, near the river discharge, such as the Shatt Al Arab, an estuarine circulation dominates during high freshwater discharge (Alosairi and Pokavanich 2017b) (Figure 2.4). The low-density water overflows the dense water seaward, while the seawater moves toward land from the lower layers. The freshwater discharges highly govern the spatial scales of the estuarine fields and horizontal transport toward the offshore.

In contrast, reverse estuarine fields also exist in enclosed coastal areas such as KB (Figure 2.4). The KB is a hypersaline water body in which the salinity at the inner zone is higher than in the seaward areas. The density variations impose a reverse estuarine circulation at the main axis of the water body. Near surface, the water flows landward, while near the bed, it is seaward. The circulations are particularly enhanced by the desalination plants and the extensive evaporation in the shallow zones, estimated to be 1.4 to 5.0 m/yr (Reynolds 1993). It is worth noting that the residual circulations in the reverse estuarine fields can reach 20 cm (Alosairi et al. 2018a).

Along the Arabian coast, winds enhance transport near the surface. The alignment of the basin towards the Shamal and the shallow hydrography allow for alongshore transport toward the south. The winds impose shear forces near the surface and bed within the first several kilometres from the shoreline. The shear forces lead to large mixed fields and horizontal dispersion, particularly during summer (Figure 2.4).

2.4 WATER QUALITY

In general, water quality measurements in the NAG region are scarce compared to the physical parameters and are limited to the nearshore zones. Hence, a comprehensive explanation of various nutrients, phytoplankton and zooplankton cycles is difficult to achieve and requires further research. Therefore, recent studies conducted near the shore have provided insight into the impacts associated with human activities (Devlin et al. 2015; Ahmed et al. 2022; Fawzi and Mahdi 2014). In general, the literature indicates a continuous decline in the water quality in the NAG over recent decades (Alosairi et al. 2018b; Devlin et al. 2015). The key factors affecting the region's quality can be categorized as local, regional and global.

2.4.1 Algal blooms

The impacts related to human activities are mainly local, such as raw sewage and desalination plant discharges, solid waste (plastic) and mega waterfront development (e.g., causeway) (Figure 2.5). The untreated sewage discharges along the coastline of Kuwait have been shown to elevate the nutrient level two-fold at the near field of the discharge (Alosairi et al. 2015). In some areas in KB, extreme levels of nutrients were found to exceed 15.0 mg/l ammonia, 0.2 mg/l nitrites, 1.0 mg/l nitrate and 4.0 mg/l phosphorus (Alosairi et al. 2015). These measurements were obtained directly from the outfalls designed for stormwater. It is worth noting that discharges from sewage are not continuous, nor do they have the same levels of concentration since they operate sporadically. However, the trickling mode of discharge that has lasted for years and is ongoing has impacted low dynamic zones by accumulating dissolved and particulate matter within the receiving bodies.

The organic and inorganic material adsorbed to the cohesive sediments in the shallow area imposes a continuous threat to the ecological system during warm seasons (Beg et al. 2003). The fluxes from the seabed to the overlying water result in an imbalance, mostly enrichment, of nutrients in the system. During low dynamic conditions, neap tides and low winds, the horizontal dispersion and vertical mixing are minimal, favourable conditions for algal growth (Figure 2.5) and, consequently, consumption of dissolved oxygen. These circumstances are usually found before fishkill events in KB (Alosairi and Alsulaiman 2020). Over the past decade, the fishkill events have been shown to be cyclic, occurring in summer specifically and, in some cases, on large scale. Several components play a role in these events, including the sediment-water interactions in the coastal zone in the inhibited areas. The sediments of the shallow zones (Figure 2.1) have the highest interactions with the overlying water due to the shallow depths. In other words, the shallow zone absorbs much of the anthropogenic pollutant and acts as a potential source of different materials given the right circumstances. The interaction of marine sediments with the overlying water in the NAG area has not been evaluated comprehensively yet.

2.4.2 Marine plastics

Microplastics in the marine system have been increasing globally since early 1970 (Oliveira and Almeida 2019). In a similar trend, microplastics levels are elevated in the northwest of the Gulf (Al-Salem et al. 2020). The presence of microplastics in the marine system results in the deterioration of the water quality. It jeopardizes the well-being of multiple species as it can be mistaken for prey by smaller animals before entering the marine food chain (Chatterjee and Sharma 2019). A recent study conducted by Al-Salem et al. (2020) detected microplastic fragments in the guts of several commercial species in the northwest of the Gulf, which threatens the ecosystem and may affect the health of humans. It is worth noting that the impacts associated with plastics, whether micro or macro, are of local, regional and global impact. Plastics have a life span of several years in the marine system, and they gain buoyancy while decaying to lower densities. Given the dominant currents, both factors enable the particles to travel farther (see Figure 2.6, revised from Alosairi et al. 2020b).

Alosairi et al. (2020b) have recently studied the microplastics transport and accumulation in the northwest of the Gulf. The study was conducted using a numerical model in a three-dimensional manner. It concluded that the enclosed

FIGURE 2.5 Discharge from an outfall associated with an algal bloom (top left panel), large algal bloom in Kuwait bay (top right panel), algal bloom in the southern coast of Kuwait Bay (lower panel).

areas are the perfect trap for macroplastics, which eventually degrade into fibres and smaller fragments of plastic. The microplastics can float or oscillate in the water column until they reach the offshore zones of the Gulf. The transport time scale of microplastics and the dominant transport regime are depicted in Figure 2.6. Alosairi et al. (2020b) have tested the dominant wind effects on the transport scales of microplastics using numerical scenarios (Figure 2.6). The results showed that even during strong southeast winds (Kous), which oppose the dominant circulation pattern, the transport had a southward dominant flow but was slower than under Shamal conditions.

2.4.3 Shatt Al Arab impacts

Regionally, the Shatt Al Arab River discharge plays a significant role in the water quality, particularly near the estuarine fields (Alosairi et al. 2018b; Fawzi and Mahdi 2014). The Shatt Al Arab discharge controls the nutrient cycle in the northwest to a large extent; however, the levels have been interrupted. Over the past 50 to 60 years, a significant shift has occurred in the nutrient-rich freshwater discharge levels due to human interventions in the hydraulic regime in the Tigris and Euphrates Rivers and the associated wetlands (Alosairi et al. 2018b). The key activities in this aspect are the desiccation of the Mesopotamian marshlands and the construction of the dams along the catchment of the river (Al-Yamani et al. 2007; Saeed et al. 1999). Both have weakened the discharge volumes and quality (Saeed et al. 1999). Alosairi and Pokavanich (2017b) have reconstructed the estuarine fields and the distribution area associated with the historical Shatt Al Arab discharges. The study concluded that the Shatt Al Arab nutrients and trace metals during high discharges supplied almost the full length of Kuwait's coastline, more than 100 km, including biologically rich mudflats in KB.

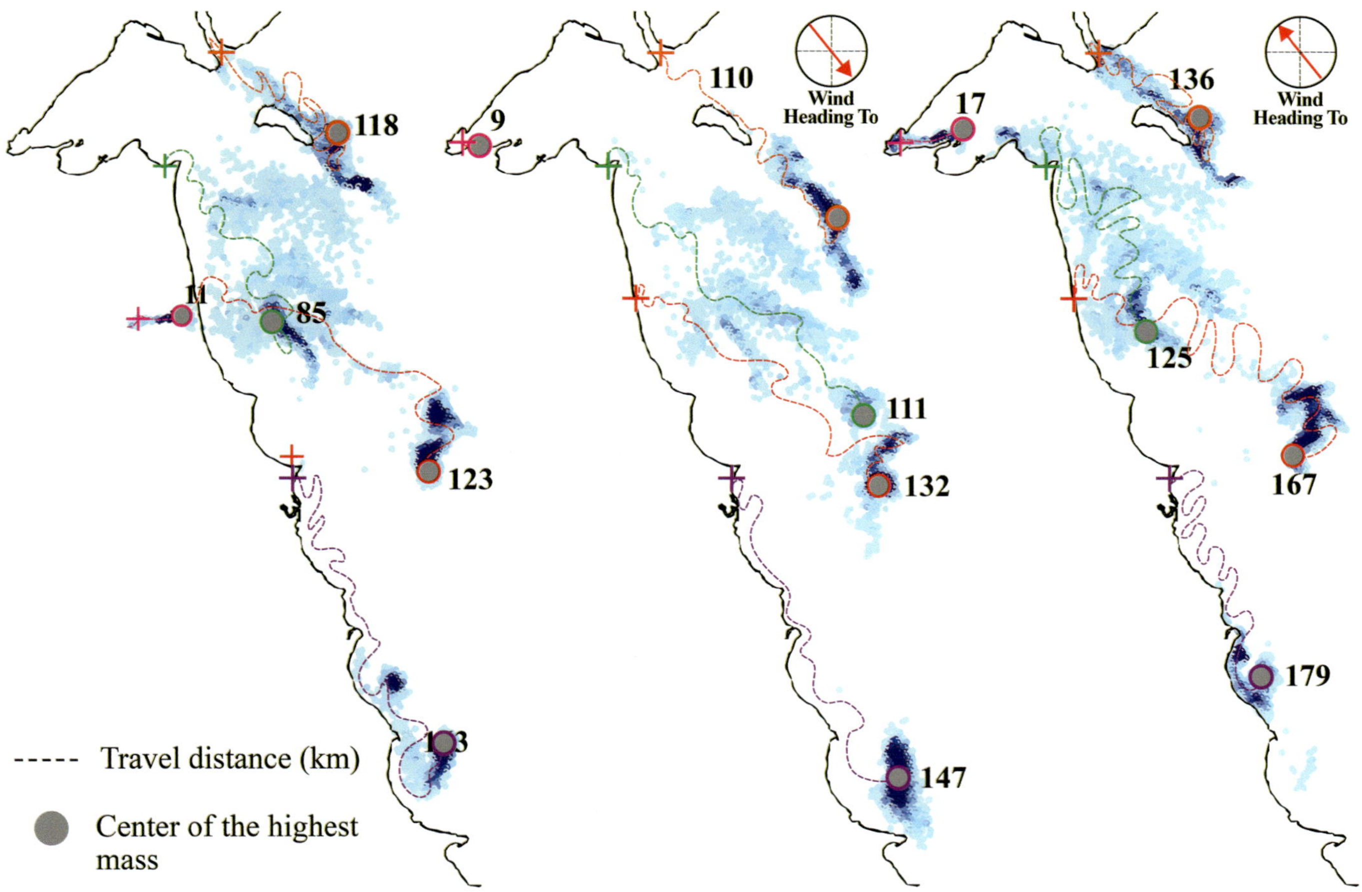

FIGURE 2.6 Patterns of microplastics in the NAG 15 days after release from different coastal zones during various wind conditions: varying winds (left panel), Shamal winds (middle panel), Kous wind (right panel) (+ denotes the release location).

Source: Alosairi et al. 2020b

The phytoplankton community across northeastern Kuwait's waters was found to be rather diverse, with 108 taxa identified and strongly diatom dominated, which is typical for these areas (Polikarpov et al. 2009). The phytoplankton community around Boubyan Island was similar through time in terms of species richness and composition of the dominant species. Still, it differed significantly in species diversity. This is mainly associated with the reductions in river discharge that shifted the phytoplanktonic structure. The taxonomic composition of microzooplankton in the Boubyan area of the Gulf has become poorer in the last ten years. Nine species of tintinnids seem to have disappeared from the Boubyan fauna, and one endemic species, *Leprotintinnus Boubyanicus*, also vanished (Polikarpov et al. 2009). The composition of the tintinnids community studied near the estuarine fields (low salinities) in 2004–2005 was significantly different from the open regions of the Gulf (high salinities); in 2015–2016, the composition of these areas had become similar (Alosairi et al. 2018b). That said, the transport associated with estuarine fields seems to be highly affected and confined to the river mouth of the Shatt Al Arab.

2.4.4 Hypoxia

Fish kills in warm regions are traditionally linked to hypoxia, resulting in low dissolved oxygen (DO) concentrations of less than 2 mg/l. Without sufficient DO levels, the ecosystem would be under stress, and some species would die, depending on their tolerance (Salacinska et al. 2010). However, the environmental circumstances, including the dynamics regime, leading to such events vary noticeably. For example, the vertical transport of toxic material in Lake Averno, Italy, led to a fishkill event (Caliro et al. 2008). In a similar context, up-welling and down-welling played a key role in the death of pelagic fish in the Neuse River estuary in the United States due to high fluxes of organic matter (Reynolds-Fleming and Luettich 2004).

In a similar trend to coasts worldwide (Diaz and Rosenberg 2001), the eutrophic status of the Gulf is of major concern due to anthropogenic activities and sea surface temperature rises. The transport regime in the NAG is transboundary (Alosairi and Pokavanich 2017b). In other words, the current in the Iranian waters eventually reaches Iraq, then Kuwait and, from there, Saudi Arabia. Therefore, the activities that lead to organic loading will affect the DO levels and elevate the chances of harmful algal blooms and fishkill events.

A recent study by Alosairi and Alsulaiman (2019) indicated that the fishkill events within the coastal zone of KB directly relate to the basin's hydrographical and dynamic features. Waste loads lead to the accumulation of particulate matter in a low dynamic zone and saturate the bed sediment, as indicated in Figure 2.7. During low wind conditions and neap tides, this period is favourable for sediment fluxes into the water column, in which a vast amount of organic content is released with high consumption of DO (Figure 2.7). The depletion of DO leads to fishkill events, particularly during the summer seasons in KB and other shallow enclosed regions in a similar

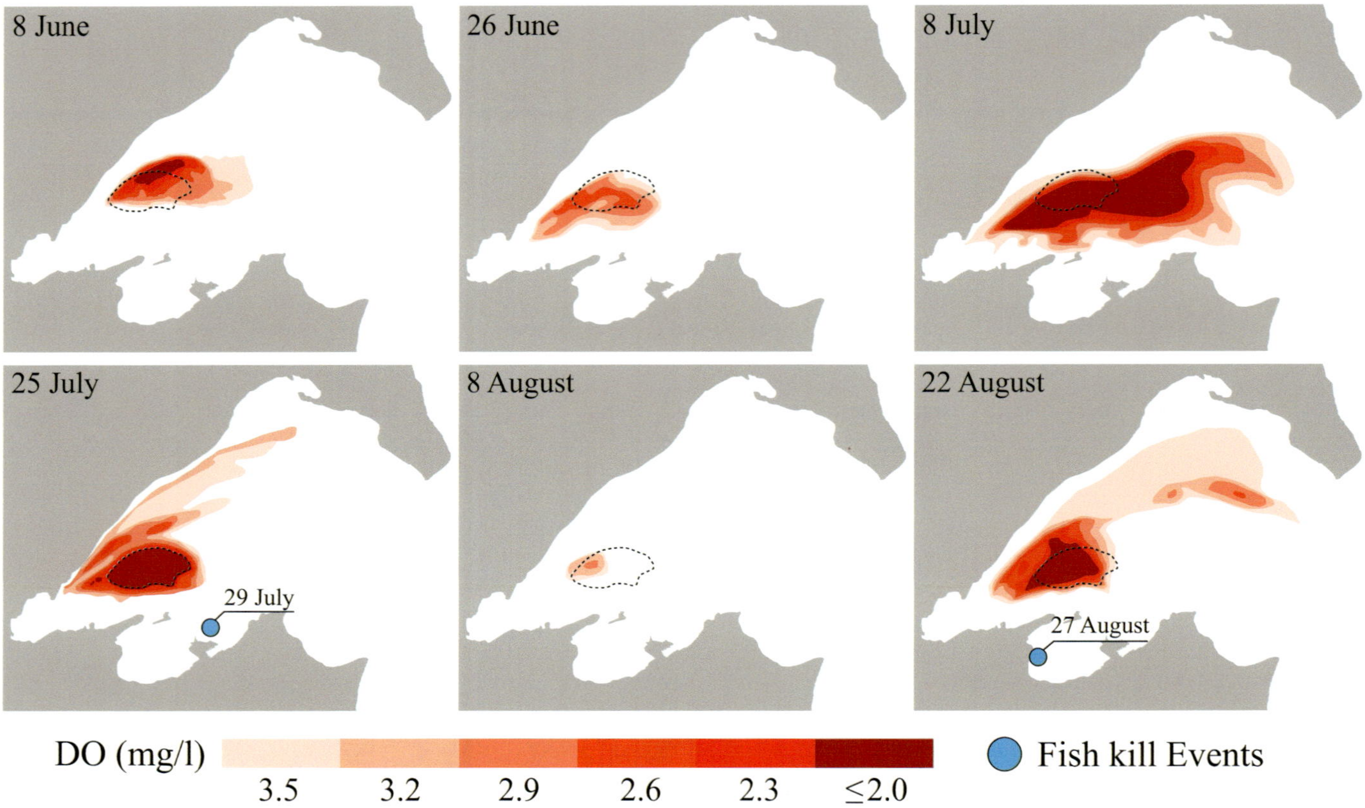

FIGURE 2.7 Hypoxic parcels in KB during the summer season: top panel computes DO for the near-surface and bed layers in the middle of KB; lower panels are the hypoxic parcels during neap tides for the near-bed layer.

Source: Alosairi and Alsulaiman (2019)

fashion. Therefore, it is important to note that the impact of waste loads in the coastal zone does not necessarily affect the alongshore areas only. Enclosed basins host organic content in the seabed, which could be transported to the offshore low dynamic zones by the residual current. A classic example of such conditions is KB, in which the reverse estuarine circulations are responsible for the transport and exchange between the coastal zone and the NAG.

Hypoxic conditions re-occur in the same vicinity in a given water body; however, the extent of the hypoxic parcels from one event to another varies considerably (Figure 2.7). This is due to the combined tidal, wind and density-driven forces that occur at different time scales and affect the derivation of organic material. In addition, the biological interactions in the ecological cycle vary considerably. The varying background concentration and nutrient transport from boundaries, light penetration and fluxes from sediment all result in competitive behaviour amongst the species. The blooming of species and the resulting consumption of DO would therefore vary given these circumstances (Figure 2.7).

In general, hypoxia and associated fish kills are closely associated with density gradients and specific dynamic settings that decelerate oxygen diffusion from a higher concentration layer to a lower concentration layer in the coastal zones (Alosairi and Alsulaiman 2019). However, fish kills may also occur without stratification in a highly eutrophic system due to harmful algal blooms at large scales. Hypoxia and harmful algal blooms may coexist during each fish-kill event in the NAG. Furthermore, hypoxia may also be introduced by the advection of low DO water parcels from neighbouring regions during some exceptional and infrequent situations.

2.5 SEDIMENTS

Sediments in the NAG are distinctive and play a large role in limiting the growth of different species by inhibiting light penetration. Previous studies categorize the input of the sediments into the NAG into two types, aeolian and fluvial (Milliman and Meade 1983; Al-Ghadban and El-Sammak 2005). The contribution of each source to the sediment budget in the estuary varies considerably since the former is driven by air that covers large areas and the latter by river discharge.

The Arabian Peninsula is one of nine major dust source regions worldwide (Tanaka and Chiba 2006; World Bank 2019). The Gulf is located at the global dust belt's centre, stretching from the Sahara Desert in the west through the Middle East to the Gobi Desert of China and Mongolia in the east (Middleton 2019). Hence, the supplies of various sediment particles are enormous. The NAG region is affected by several dust storm events from different locations and extents (Al-Hemoud et al. 2022). The west and Mesopotamian Iraq, the Iraq/Syria border and northern and southern Saudi Arabia are key sources of major dust events in the Gulf and the NAG (Al-Hemoud et al. 2022). Fine sediment is delivered by the dominant Shamal winds to the NAG most of the time (Yu et al. 2016). Al-Hemoud et al. (2020) noted that in some cases, the

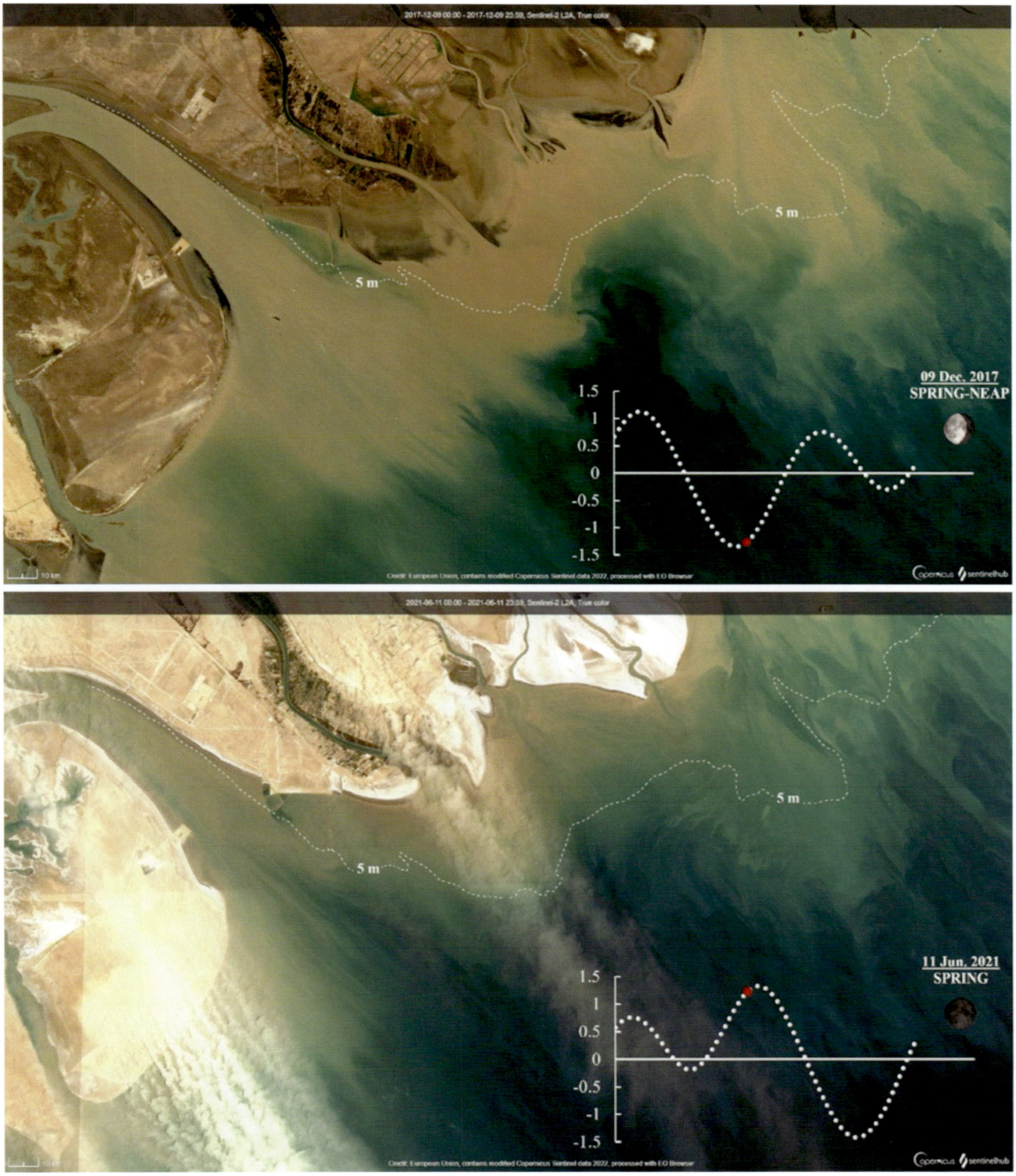

FIGURE 2.8 Turbid water at the northwest of the Arabian Gulf during different tidal phases; the top panel indicates the extent of the turbid water, including the plumes from the Shatt Al Arab, and the lower panel indicates the sediment supplies from Boubyan Island and Algasba near the Shatt Al Arab.

Source: https://apps.sentinel-hub.com/

suspended dust was sustained for 48 hours extreme post-event while it crossed over three countries in the region. During the wind event, it may be suspended in the air as high as 1,000 m above the land (Foda et al. 1985). Climate change, desertification and human activities have recently elevated the scale and frequency of dust storms in the region (Al-Hemoud et al. 2020). Locally, Boubyan and Warba Islands, the nearby depressions, sabkhas and dry marshes are major local sources of dust to the NAG (Al-Dousari and Al-Awadhi 2012) (Figure 2.8).

The effects of the dust storms on the marine sediments can extend to more than 200 km from the edge of the NAG, given the hydrodynamic features of the area (Alosairi and Pokavanich

2017a; Alosairi et al. 2018b; Al-Ghadban and El-Sammak 2005). This is due to the small sediment fraction, such as silt and clay, that dominates the type of sediment entering the marine system. Given the shallowness of the NAG (see the bathymetric line of 5 m in Figure 2.8), the tidal action imposes a large shear stress even during neap tides, which keeps the sediment particles always in suspension with travel to far distances. The siltation of the sediments in the NAG occurs at low dynamic periods and areas. Foda et al. (1985) estimated that the average siltation of the aeolian sediments (small fractions) for the first 200 km from the coast is 0.8 mm/yr in the NAG. However, the average yearly siltation rate may be 3–4 mm/yr near shore. In semi-enclosed zones, such as KB, the average yearly amount of dust was 265 tonnes/km^2, equivalent to 0.17 mm/yr (Al-Dousari and Al-Awadhi 2012).

River systems are commonly associated with the suspended solids of fine and coarse material and organic matter that supply estuaries (Figure 2.8, top panel). At the northwest of the Gulf, the Shatt Al Arab acts as the main supplier of fluvial sediment in the region (Alosairi et al. 2018b). Milliman and Meade (1983) estimated that the fluvial sediment from the Shatt Al Arab ranged between 53 and 105 million tonnes/yr during the 1960s. This rate is equivalent to 0.85 to 1.7 g/l of sediment concentration associated with the river discharges. Recently, the sediment load of the Shatt Al Arab has most likely decreased significantly for two reasons. First is the reduced discharge flow rate, as explained in Alosairi and Pokavanich (2017b). The second is the sedimentation in numerous reservoirs and marshes of Mesopotamia (Baltzer and Purser 1990). Considering the recent estimates of the river discharge given by Alosairi and Pokavanich (2017b), the sediment load today can be equated to 1% of that given by Milliman and Meade (1983). Hence, one to several million tonnes/yr is the best estimate for the present-day annual discharge of fine sediments to the NAG via the Shatt Al Arab.

Concerning the sediment fate and transport in the NAG, the horizontal dynamics are much greater than the vertical and dominate the transport mode. The semi-diurnal tides and the strong wind, which act on the shallow water, impose large shear fields sufficient to keep the fine suspended particles within the water column. Therefore, the waters within the first 60 km from the coast of the NAG are always turbid year round (Figure 2.8). The shift in the turbid fields is found post–dust storms and strong wind events. Although sediment behaviour in the water column differs from aeolian-based pollutants, the hydrodynamic controls are alike. The counterclockwise current delivers much sediment toward the Arabian coast along Kuwait and Saudi Arabia (Figure 2.4). Highly turbid water is also observed in the enclosed zones, particularly in Khor Al Subiya and Khor Abdullah near Boubyan Island. This is mainly due to the sediments being recycled in the enclosed region while residual current controls are minimal. In addition, the trapping and residence time is much higher in these areas, given the enclosed hydrographic nature and the local supply of sediments. The aeolian dust and sand drift from land into the marine water is the major contributor to the NAG compared to the Shatt Al Arab fluvial input (Figure 2.8).

The fluvial input of sediment from the Shatt Al Arab is seen to be local and rarely extends beyond the estuarine fields in the NAG (Figure 2.8). The distribution of the fluvial sediments is controlled by both the tides and the residual estuarine currents. Due to the low discharge, the sediments are adjacent to the coastline of the Shatt Al Arab and within 1 to 6 km from the coast. The estuarine turbidity maximum (ETM) has not been studied previously for the Shatt Al Arab discharge; however, it seems that it occurs during low dynamic conditions and high discharges. It is worth noting that high discharges were reported in 2018, but these occurred during extreme rain events and are non-cyclic (Alosairi et al. 2019). That said, the discharge from extreme events allows the sediment plumes to extend further offshore, replicating the estuarine fields that impose a seaward transport near the surface.

2.6 CONCLUSION

The NAG exhibits a unique environmental system compared to the central and southern parts of the Gulf. The hydrographical setting enables a very dynamic system which is mostly driven by tides and winds. Therefore, the NAG is well-mixed water in the vertical, and minimal gradients are found in the horizontal. Some variations could be found during interseason periods in the deep zones which impose density-driven currents. Near the shallow coastal zones, human activities lead to low water quality conditions. In addition, the shallow nature allows for sediment-water interaction, leading to major issues such as red tides and depletion of dissolved oxygen in enclosed areas. Suspended sediments at the NAG are mostly aeolian supply and characterized by fine particles; in COMPARISON, the fluvial input are very minimal.

REFERENCES

Ahmed, A., Madhusoodhanan, R., Yamamoto, T., Fernandes, L., Al-Said, T., Nithyanandan, M., Thuslim, F., Al-Zakri, W., and F. Al-Yamani. 2022. Analysis of phytoplankton variations and community structure in Kuwait Bay, Northwestern Arabian Gulf. *J Sea Res* 180: 102163. http://doi.org/10.1016/j.seares.2022.102163.

Al-Dousari, A.M., and J. Al-Awadhi. 2012. Dust fallout in northern Kuwait, major sources and characteristics. *Kuwait J Sci* 39, no. 2A: 171–187.

Al-Ghadban, A.N., and A. El-Sammak. 2005. Sources, distribution and composition of the suspended sediments, Kuwait Bay, Northern Arabian Gulf. *J Arid Environ* 60, no. 4: 647–661.

Al-Hemoud, A., Al-Dashti, H., Al-Saleh, A., Petrov, P., Malek, M., Elhamoud, E., Al-Khafaji, S., Li, J., Koutrakis, P., Doronzo, D., and N. Middleton. 2022. Dust storm 'hot spots' and transport pathways affecting the Arabian Peninsula. *J Atmos Sol-Terr Phys* 238–239: 105932. http://doi.org/10.1016/j.jastp.2022.105932.

Al-Hemoud, A., Al-Dousari, A., Al-Dashti, H., Petrov, P., Al-Saleh, A., Al-Khafaji, S., Behbehani, W., Li, J., and P. Koutrakis. 2020. Sand and dust storm trajectories from Iraq Mesopotamian flood plain to Kuwait. *Sci Total Environ* 710: 136291. http://doi.org/10.1016/j.scitotenv.2019.136291.

Alosairi, Y., and N. Alsulaiman. 2019. Hydro-environmental processes governing the formation of hypoxic parcels in an inverse estuarine water body: Model validation and discussion. *Mar Pollut Bull* 144: 92–104.

Alosairi, Y., and N. Alsulaiman. 2020. Hydro-environmental processes governing the formation of hypoxic parcels in an inverse estuarine water body: Assessment of physical controls. *Mar Pollut Bull* 157: 111311. http://doi.org/10.1016/j.marpolbul.2020.111311.

Alosairi, Y., Alsulaiman, N., Petrov, P., and Q. Karam. 2019. Responses of salinity and chlorophyll-a to extreme rainfall events in the northwest Arabian Gulf: Emphasis on Shatt Al-Arab. *Mar Pollut Bull* 149: 110564. http://doi.org/10.1016/j.marpolbul.2019.110564.

Alosairi, Y., Alsulaiman, N., Rashed, A., and D. Al-Houti. 2020a. World record extreme sea surface temperatures in the northwestern Arabian/Persian Gulf verified by in situ measurements. *Mar Pollut Bull* 161: 111766. http://doi.org/10.1016/j.marpolbul.2020.111766.

Alosairi, Y., Al-Salem, S.M., and A. Al Ragum. 2020b. Three-dimensional numerical modelling of transport, fate and distribution of microplastics in the northwestern Arabian/Persian Gulf. *Mar Pollut Bull* 161: 111723. http://doi.org/10.1016/j.marpolbul.2020.111723.

Alosairi, Y., Al Sulaiman, N., Neelamani, S., Polikarpov, I., Kwarteng, A., Small, C., and A. Al Handal. 2018b. The effect of developments in Mesopotamian marshland on Boubyan and Warba Islands. *Kuwait Institute for Scientific Research, Report EC060K, Report No. 14765*. www.kisr.edu.kw.

Alosairi, Y., Imberger, J., and R.A. Falconer. 2011. Mixing and flushing in the Persian Gulf (Arabian Gulf). *J Geophys Res Oceans* 116, no. C3: C03029. http://doi.org/10.1029/2010JC006769.

Alosairi, Y., and T. Pokavanich. 2017a. Seasonal circulation assessments of the northern Arabian/Persian Gulf. *Mar Pollut Bull* 116, nos. 1–2: 270–290.

Alosairi, Y., and T. Pokavanich. 2017b. Residence and transport time scales associated with Shatt Al-Arab discharges under various hydrological conditions estimated using a numerical model. *Mar Pollut Bull* 118, nos. 1–2: 85–92.

Alosairi, Y., Pokavanich, T., Al-Salem, K., Taqi, A., Al-Said, T., Al-Hashem, A., Al-Rubaiaan, E., Karam, Q., Al-Yaegoub, A., Al-Dousari, N., and S. Al-Awadhi. 2015. The optimal and sustainable extension of the emergency outfalls along Kuwait coastline. *Kuwait Institute for Scientific Research, Report No. 12782*. www.kisr.edu.kw.

Alosairi, Y., Pokavanich, T., and N. Alsulaiman. 2018a. Three-dimensional hydrodynamic modelling study of reverse estuarine circulation: Kuwait Bay. *Mar Pollut Bull* 127: 82–96.

Al-Salem, S.M., Uddin, S., and B. Lyons. 2020. Evidence of microplastics (MP) in gut content of major consumed marine fish species in the State of Kuwait (of the Arabian/Persian Gulf). *Mar Pollut Bull* 154: 111052. http://doi.org/10.1016/j.marpolbul.2020.111052.

Al-Yamani, F.Y., Bishop, J.M., Al-Rifaie, K., and W. Ismail. 2007. The effects of the river diversion, Mesopotamian marsh drainage and restoration, and river damming on the marine environment of the northwestern Arabian Gulf. *Aquat Ecosyst Health Manag* 10, no. 3: 277–289.

Al-Zaidan, A.S.Y., Kennedy, H., Jones, D.A., and S.Y. Al-Mohanna. 2006. Role of microbial mats in Sulaibikhat Bay (Kuwait) mudflat food webs: Evidence from δ13C analysis. *Mar Ecol Progr Ser* 308: 27–36.

Baltzer, F., and B.H. Purser. 1990. Modern alluvial fan and deltaic sedimentation in a foreland tectonic setting: The lower Mesopotamian plain and the Arabian Gulf. *Sediment Geol* 67, nos. 3–4: 175–197.

Beg, M.U., Saeed, T., Al-Muzaini, S., Beg, K.R., and M. Al-Bahloul. 2003. Distribution of petroleum hydrocarbon in sediment from coastal area receiving industrial effluents in Kuwait. *Ecotoxicol Environ Saf* 54, no. 1: 47–55.

Caliro, S., Chiodini, G., Izzo, G., Minopoli, C., Signorini, A., Avino, R., and D. Granieri. 2008. Geochemical and biochemical evidence of lake overturn and fish kill at Lake Averno, Italy. *J Volcanol Geotherm Res* 178, no. 2: 305–316.

Chatterjee, S., and S. Sharma. 2019. Microplastics in our oceans and marine health. *Field Actions Sci Rep J Field Act, Special Issue* 19: 54–61.

Devlin, M.J., Massoud, M.S., Hamid, S.A., Al-Zaidan, A., Al-Sarawi, H., Al-Enezi, M., Al-Ghofran, L., Smith, A.J., Barry, J., Stentiford, G.D., and S. Morris. 2015. Changes in the water quality conditions of Kuwait's marine waters: Long term impacts of nutrient enrichment. *Mar Pollut Bull* 100, no. 2: 607–620.

Diaz, R.J., and R. Rosenberg. 2001. Overview of anthropogenically-induced hypoxic effects on marine benthic fauna. In: *Coastal Hypoxia: Consequences for Living Resources and Ecosystems*, vol. 58. Eds. N.N. Rabalais and R.E. Turner, 129–145. Am Geophys Union. http://doi.org/10.1029/CE058p0129.

Fawzi, M.N., and A.B. Mahdi. 2014. Iraq's inland water quality and their impact on the north-western Arabian Gulf. *Marsh Bull* 9, no. 1: 1–22.

Foda, M.A., Khalaf, F.I., and A.S. Al-Kadi. 1985. Estimation of dust fallout rates in the northern Arabian Gulf. *Sedimentology* 32, no. 4: 595–603.

Hoseini, S.M., and M. Soltanpour. 2022. Numerical study of influencing factors on tidal wave propagation in the Persian Gulf. *Ocean Sci J* 57, no. 4: 642–671.

Imberger, J., Mamouni, E.A.D., Anderson, J., Ng, M.L., Nicol, S., and A. Veale. 2007. The index of sustainable functionality: A new adaptive, multicriteria measurement of sustainability – application to Western Australia. *Int J Environ Sustain Dev* 6, no. 3: 323–355.

Madah, F., and S.H. Gharbi. 2022. Numerical simulation of tidal hydrodynamics in the Arabian Gulf. *Oceanologia* 64, no. 2: 327–345.

Middleton, N. 2019. Variability and trends in dust storm frequency on decadal timescales: Climatic drivers and human impacts. *Geosciences* 9, no. 6: 261. http://doi.org/10.3390/geosciences9060261.

Milliman, J.D., and R.H. Meade. 1983. World-wide delivery of river sediment to the oceans. *J Geol* 91, no. 1: 1–21.

Oliveira, M., and M. Almeida. 2019. The why and how of micro (nano) plastic research. *Trends Analytical Chem* 114: 196–201.

Polikarpov, I., Al-Yamani, F., and M. Saburova. 2009. Space-time variability of phytoplankton structure and diversity in the north-western part of the Arabian Gulf (Kuwait). *BioRisk* 3: 86–96.

Pous, S., Carton, X., and P. Lazure. 2012. A process study of the tidal circulation in the Persian Gulf. *Open J Marine Sci* 2, no. 4: 131–140.

Reynolds, R.M. 1993. Physical oceanography of the Gulf, Strait of Hormuz, and the Gulf of Oman – Results from the Mt Mitchell expedition. *Mar Pollut Bull* 27: 35–59.

Reynolds-Fleming, J.V., and R.A. Luettich. 2004. Wind-driven lateral variability in a partially mixed estuary. *Estuarine Coastal Shelf Sci* 60, no. 3: 395–407.

Saeed, T., Al-Ghadban, A.N., Al-Shemmari, H., Al-Mutairi, M., and H. Al-Hashash. 1999. Preliminary assessment of the impact of draining of Iraqi marshes on Kuwait's northern marine environment. Part II. Sediment associated pollutants. *Water Sci Technol* 40, no. 7: 89–98.

Salacinska, K., El Serafy, G.Y., Los, F.J., and A. Blauw. 2010. Sensitivity analysis of the two dimensional application of the Generic Ecological Model (GEM) to algal bloom prediction in the North Sea. *Ecological Modelling* 221, no. 2: 178–190.

Tanaka, T.Y., and M. Chiba. 2006. A numerical study of the contributions of dust source regions to the global dust budget. *Glob Planet Change* 52, nos. 1–4: 88–104.

World Bank. 2019. *Sand and Dust Storms in the Middle East and North Africa (MENA) Region: Sources, Costs, and Solutions*. Washington, DC: World Bank.

Yu, Y., Notaro, M., Kalashnikova, O.V., and M.J. Garay. 2016. Climatology of summer Shamal wind in the Middle East. *J Geophys Res Atmos* 121, no. 1: 289–305.

Quantum multi-objective algorithm and radar satellite data for automatic detection of nonlinear turbulent flow in the Red Sea

3

Maged Marghany

3.1 INTRODUCTION

The phase between laminar and irregular flow is turbulence. It occurs when a fluid's velocity exceeds a specific critical value, causing the flow to become unpredictable and chaotic. In natural phenomena like rivers, oceans and atmospheric airflows as well as engineering applications like fluid transport systems and aircraft design, turbulence is frequently seen. In fluid dynamics, turbulence poses a significant challenge because it can increase drag, cause energy loss and damage structures. To comprehend its effects and create methods for turbulence control in various applications, researchers are still researching the phenomenon.

Standard mathematical techniques are required for a turbulent flow investigation to comprehend its substantial properties and mechanisms. Due to its extensive use in a variety of scientific domains, such as mechanical engineering, fluid dynamics, aerospace, physics and hydrodynamics, turbulent science is highly valued in practical terms. Principally, a turbulent pattern exhibits chaotic behavior in both time and space and occupies an extraordinary level of spatial freedom (Barenghi et al. 2014). Spatial-temporal chaos is a term that has been used to describe fluid instability. The rapid flow of any fluid through an obstacle or an airfoil causes turbulence in the boundary layers and forms a turbulent wake, both of which effectively increase the drag the flow exerts on the obstruction. This is referred to as turbulent flow. Since there are nonlinearities in the equations of motion, turbulent flow is by definition unstable in this scenario.

Pioneering researchers concur that a turbulent flow is adequately regulated by the complex interaction of mass and momentum (Chorin 1968; Doering and Gibbon 1995; Temam 2001; Constantin and Foias 2020). The complex Navier-Stokes equations were used by experienced researchers in this plausible scenario to precisely quantify the extraordinary complexity of turbulent flow. The Navier-Stokes formulae cannot be solved analytically due to significant uncertainties, and the possible solutions to these challenging equations are robustly shown to have a sufficient discrepancy. This area of mathematics has been explored and debated in great depth over the past centuries, and it remains an ongoing discussion even today. In this sense, tracking the spatial fluctuation of turbulent flow in the ocean or an estuary could be possible by exploiting remote-sensing techniques.

3.1.1 Can ocean turbulence be quantized?

In the context of quantum fluids, quantum turbulence refers to the turbulent motion of quantum fluid, like a superfluid, which is characterized by quantized vorticity, superfluidity and, at finite temperatures, a two-fluid pattern and, notably, quantum front boundaries. Ocean flow systems are properly viewed from this modern perspective as superfluids. Therefore, three characteristics are involved in the quantification of ocean flow: 1) it reveals two-fluid sorts at nonzero temperatures or when thermocline layers exist; 2) it can flow freely, without the dissipative consequence of viscous forces, particularly in tsunami wave propagation; and 3) its constrained rotation is restricted to discrete vortex lines (Donnelly and Swanson 1986; Barenghi et al. 2014; Tsubota et al. 2017). Nonetheless, unlike vortices in ordinary ocean flow, which are constant and can vary in size, shape and strength, eddies in aqueous fluids are periodic. Superfluidity and quantized vorticity are remarkable macroscopic-length scale-specific indices of quantum ocean flow. In this concept, the identical vortices have a continuous size. This

DOI: 10.1201/9781003321408-4

is yet another intriguing feature of superfluids, which distinguishes them from random eddies in a conventional fluid and arises from quantum mechanics, the implications of which are observable on a massive scale at low temperatures. Quantum turbulence, therefore, is an entanglement of these quantized vortices, culminating in a clear geometry of water instability that is significantly more convenient to model than regular turbulence, in which various potential interactions of the eddies rapidly cause the issue to become too complex to predict what would take place in the ocean (Vinen and Niemela 2002). Comprehensively, quantum turbulence describes the turbulent flow of quantum fluids (chaotic motion of fluids at high speeds).

In the Gulf of Mexico loop current, ocean water particles spin up and down in an uneven rhythm, resulting in chaotic movement on a large scale. In this sense, the Gulf loop current can have a significant impact on weather patterns and ocean ecosystems as it can transport warm water and nutrients to different areas of the Gulf. Understanding the behaviour of the loop current is crucial for predicting and managing these effects. This phenomenon is known as eddies, and they can have a significant impact on the ocean's circulation and marine life. Scientists are studying these eddies to better understand their behaviour and potential effects. In this understanding, the significant question is now: Why should loop currents exhibit quantum turbulence? Quantum turbulence is a phenomenon that occurs in superfluids and Bose-Einstein condensates, which are systems with zero viscosity. Loop currents in these systems can exhibit quantum turbulence due to the presence of vortices, which are topological defects in the fluid flow.

3.2 IMAGINARY MECHANISMS FOR TURBULENCE IN SAR DATA

In this perspective, turbulence appears to be relatively common in the part of the atmosphere adjacent to the earth's surface. It is propelled by microscale gradients in velocity, temperature and moisture, and its responsibility is to extinguish the equivalent physical gradients that sustain it. As a result, the microscale instability created by the differences in these gradients is eliminated, allowing for a return to normal atmospheric conditions. This process of self-elimination is an example of a thermodynamic process called convective collapse.

In this regard, this reduction invariably occurs as a potential consequence of turbulent energy fluxes, without which buoyancy, adequate moisture and considerable momentum would be transported by molecular diffusion between the earth's complex surface and the prevailing atmosphere. In contrast to turbulent diffusion, its scales remain three to six orders of magnitude less. Large-scale movements would henceforth be dominated by turbulent flux. The key question is: How does the SAR sensor image ocean turbulence?

When a SAR sensor illuminates the ocean surface at moderate incident angles between 20° and 60°, the prevailing ratio of backscattered energy is created by ocean surface roughness on the scale of the radar wavelength projected onto the ocean surface – the "Bragg" wavelength – as shown in the previous studies of Tomiyasu (1978), Alpers et al. (1981), Jain (1981) and Marghany (2021). The dominant wind propagation initiates slight waves in its longitudinal direction. In this regard, the intensity backscattered from the turbulent surface is greatest when the local wind is in the radar's line of sight. As well as the wind blowing away from the microwave radar antenna, there is an equivalent but somewhat smaller local extreme in the backscattered energy. Once the radar's direct line of sight is perpendicular to the prevailing wind direction, the backscattered energy value is the lowest. SAR would thus detect the key constraints that atmospheric and oceanic phenomena typically impose on the wave spectrum at centimetre scales. In the meantime, precipitation can occasionally interfere with the radar signal, making the dominant atmospheric conditions evident to SAR. Conversely, areas of rain can emerge in SAR images as an incoherent pattern that typically signifies constrained cell dynamics.

Spaceborne SAR data is distinguished by its 10–100 m resolution and a swath width of the order of 100–1,000 km. Therefore, spaceborne SAR successfully and efficiently provides an extensive view of the roughness patterns caused by sea-surface tension. Macroscale, mesoscale and microscale meteorological events generate visible traces, from which one can infer the presence of a corresponding dynamic and thermodynamic environment (e.g., statically unstable versus statically stable, baroclinic versus barotropic) (Di Martino et al. 2015). Consequently, patches flow continuously along the prevailing wind direction at the continuous surface or approximately a few degrees clockwise in the leading case of barotropic or weakly baroclinic convective marine atmospheric boundary layers detectable in SAR images (Macklin and Cordey 1991; Kanevsky 2008; Rieu et al. 2021).

In this view, buoyancy-driven/shear-organized rolls are spiral circulations triggered by thermodynamic instability in a vertical wind shear environment. As wind shear increases for a specified quantity of upwelling in the marine atmospheric boundary layer, a province of randomly organized extended speckles transforms into a distinct field of linearly arranged elongated speckles. Ascending and descending circulation zones, therefore, cause identical uniform development and a significant decline in sea surface roughness. The produced SAR thickness pattern accurately depicts the frequent appearance of irregular dark and bright speckled lines. The specific orientation of these rollers' surface tracks, and hence their SAR footprint, is usually performed using matching methods or techniques (Sikora et al. 2006; Di Martino et al. 2015; Kanevsky 2008; Marghany 2021).

3.3 CAN A QUANTUM ALGORITHM AUTOMATICALLY DETECT TURBULENT FLOWS IN SAR IMAGES?

The unique challenge of segmenting turbulent flows properly using partition clustering is considered a combinatorial optimization task. Nevertheless, global optimizing techniques like conventional genetic algorithms (CGAs) are frequently time consuming, invariably have sluggish convergence rates and are prone to be stuck in a local optimum. Since an intelligent machine can exclusively operate in one independent state at a time, conventional algorithms ensure only one completed phase at a time. According to quantum physics, the constituent particles of turbulent flow and ocean waves in nature are unfixed but can occupy multiple phases at the same time, a phenomenon known as superposition. Once those particles are distributed in SAR images, they decompose to a particular state.

In this understanding, fuzziness, coherent superposition, interference and implicit parallelism are notable examples of quantum computing's local roots in quantum mechanics. These benefits result in a greater variety and consistency between the investigation and usage of standard evolutionary algorithms. Several leading scientists have exploited quantum codes in evolutionary computation. For an excellent example, Han and Kim (2004) progressively developed a quantum-inspired evolutionary algorithm for combinatorial optimization based on quantum computing perception and coding. Jiao et al. (2008) investigated a quantum-inspired immune clonal technique for global optimization in this possible scenario.

The defects of single-factor clustering techniques, as well as potentially fundamental errors in the image processing segmentation of visible brightness and darkness patches, typically dominate the automatic detection of turbulent boundary flow in the SAR image. Can a specific multi-criterion, quantitative genetic algorithm reliably detect turbulence currents based on bottom structures including coral reefs?

3.4 QUANTUM IMAGE PROCESSING

The key question is now: What is the difference between quantum image processing and classic ones? The key difference involves a qubit. In quantum image processing, a qubit is exploited instead of a bit. Utilizing Hilbert space would construct a set of different states, which can be described effectively by a wave function (Ψ). Therefore, a qubit exists in two-dimensional complex vector space (Williams 2011). Put differently, a vector space is an abstract mathematical structure formed by a set of vectors which can be added together and multiplied (scaled) by scalar numbers. Therefore, if $\vec{v}_1, \vec{v}_2, ..., \vec{v}_m$ are vectors in a complex vector space, then a linear combination is of the form $c_1\vec{v}_1 + c_2 v_2 + ...c_m v_m$. In this perspective view, the linear superposition with different quantum states can be mathematically expressed as:

$$|\Psi>=\alpha|0\rangle+\beta|1\rangle \tag{3.1}$$

$$|\dot{\alpha}^2|+|\beta^2|=1 \tag{3.2}$$

How do radar signals relate to quantum states in both equations 1 and 2? Radar systems' sensitivity and resolution can be improved using quantum states. Radar signals can be measured more precisely using quantum entanglement, enabling more precise target detection and tracking. The quantum state equations can distinguish between strongly scattered signals due to rough turbulence on the sea surface and weak ones that are scattered from a smooth sea surface state. This is because quantum states can provide more information about the radar signal, allowing for better differentiation between different types of scattering. Additionally, quantum sensors can be used to detect very weak radar signals that would otherwise be undetectable with traditional radar systems.

Equations 1 and 2 demonstrate that a two-dimensional complex vector α and β with a length of 1 and $|0\rangle$and$|1\rangle$ present the binary orthonormal origins for the dual-dimensional vector space (Kaye et al. 2007; Kanevsky 2008; Williams 2011). Consequently, equation 2 reveals that the qubit proves a linear superposition of the base states through fluctuating the values of α and β with rotation angles θ and ϕ. This procedure can be explored mathematically based on the Bloch sphere (Figure 3.1) as follows:

$$|\Psi\rangle = 0.5\cos\theta|0\rangle + e^{0.5(i\phi)\sin\theta}|1\rangle \tag{3.3}$$

Equation 3 reveals that the Bloch sphere can be explored to present the qubit in three dimensions as a vector $\vec{r}$ from the origin to a point on the 3-D sphere with a radius of one.

According to this perspective, the Bloch sphere can greatly assist us in distinguishing between qubit and bit. In this sense, the qubit spins up and down and superpositions between $|0\rangle$and$|1\rangle$in the same time in less than a second (Figure 3.2). Nevertheless, the conventional bit is either 0 or 1, which leads to a longer time of computation than a qubit (Hey 1999).

The remaining question is now: How can a qubit explore multi-objective evolution to create a novel quantum multi-objective evolution (QME) for automatic detection of turbulent patterns in synthetic aperture radar (SAR) images?

3.5 QUANTUM MULTI-OBJECTIVE EVOLUTIONARY ALGORITHM (QMEA)

According to this perspective, when the projection on a particular basis is recognized, the quantum system is considered to be collapsed. The measures and decoherence are other names for wave function $|\Psi\rangle$ collapsing. In other words, it should be $|\Psi\rangle = \alpha|0\rangle$ in the particular instance of the projection of $|\Psi\rangle$ on the $|0\rangle$ state. Consequently, on the state $|0\rangle$, the probability of the qubit collapsing would approach $|\dot{\alpha}^2|$. Accordingly, a novel qubit chromosome representation is adopted in quantum multi-objective evolution (QME) on the assumption and core principles

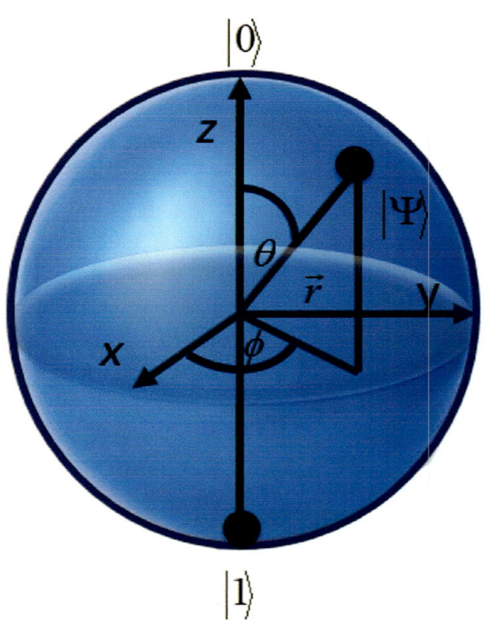

FIGURE 3.1 3-D illustration of Bloch sphere.

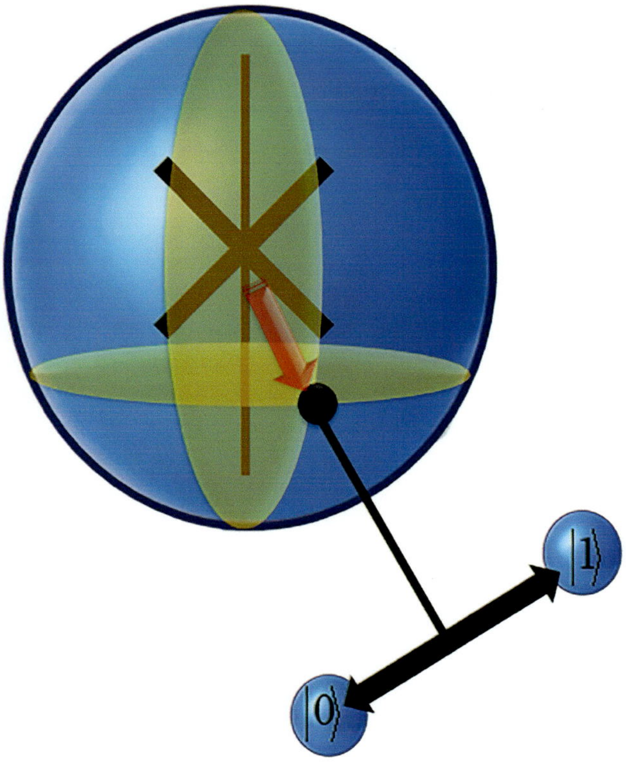

FIGURE 3.2 Sketch of a qubit.

of quantum computing, including such qubits and linear superposition (Kaye et al. 2007). The representation's distinguishing feature is its ability to represent any linear superposition of solutions. A quantum bit (qubit) is the smallest component of data that can be stored in a dual-state quantum computer. It can be in either the "1" or "0" states or any superposition of both $|0\rangle$and$|1\rangle$. Equation 3 can be utilized to represent the state of a qubit. In this view, let us consider P as the turbulent flow pattern in SAR data, which has the wave function:

$$|\Psi(P)\rangle = \sum_{m\in P} \Psi(P)|P\rangle \tag{3.4}$$

The pattern-sorting process is depicted by m in equation 4, which includes the pattern P (Hey 1999; Wang et al. 2005; Williams 2011; Bardin et al. 2020). Accordingly, m patterns of the turbulence flow features are assumed as a training set D in the SAR data, which is described as: $D = \{\Psi(P)\}$.

According to this perspective, the coherence state of the SAR features $|\Psi_{SAR}\rangle$ is mathematically described as follows:

$$|\Psi_{SAR}\rangle = |x_1x_2\text{.......}, x_n, R_1R_2, \text{....}, R_{n-1}, C_1C_2\rangle \tag{3.5}$$

where n is the total number of the SAR coherence features $x_1x_2\text{.......}, x_n$, which are stored in patterns P. The controlled register patterns are depicted by $R_1R_2, \text{....}, R_{n-1}$ accompanied by C_1C_2.

Equation 5 indicates the restoring P patterns as $x_1x_2\text{.......}, x_n$, with the total number n. Also, $R_1R_2, \text{....}, R_{n-1}$ as well as C_1C_2 are controlled register patterns. The Grover algorithm, therefore, involves three operations, which deliver the coherent state of the SAR features $|\psi_{SAR}\rangle$. In this understanding, exploring Grover algorithms, n qubits are vital to signify P patterns, as well as $n+1$ qubits as demanded in controlling qubits in QME algorithms. That delivers the entirety of $2n+1$ qubits, which are imposed to differentiate m patterns. Therefore, the Grover algorithm also provides the quantity of $2n+1$ encodes in embedding the QME network algorithms (Hey 1999; Wang et al. 2005; Williams 2011; Marghany 2021). Consequently, the M matrix generation is the first operation, and it is cast as follows:

$$M = \begin{bmatrix} 1 & 0 & 0 & 0 \\ 0 & 1 & 0 & 0 \\ 0 & 0 & \sqrt{\frac{p-1}{p}} & \frac{\Psi(m)}{\sqrt{p}} \\ 0 & 0 & \frac{\Psi(m)}{\sqrt{p}} & \sqrt{\frac{p-1}{p}} \end{bmatrix} \tag{3.6}$$

The main question is now how to achieve M. In the circumstance of $1 \le P \le m$, and 1 <= P<=m, M can be accurately achieved. The flip transformation Ψ^0, which is identified by the matrix transformation, is then cast as the second giving operation:

$$\Psi^0 = \begin{bmatrix} \Psi & 0 \\ 0 & I \end{bmatrix} \tag{3.7}$$

In such cases of dual qubits, one for the turbulent flow and the other for the surrounding environment, one of them must regulate the flip transformation on the other, and equation 7 is exploited. In other words, let us assume that the SAR image $\Psi^0|\text{turbulent}\rangle$ represents the state of the turbulent flow, and $|\Psi \ge \beta|0\rangle + \alpha||1\rangle$ represents the state of the surrounding environment (Bardin et al. 2020; Cuomo et al. 2020; Marghany 2021). Using the M matrix, the pattern state detection of turbulent flow is given by:

$$\Psi^0|\text{turbulent}\rangle\langle non-\text{turbulent}| = \begin{bmatrix} 0 & 1 & 0 & 0 & 1 & \beta \\ 1 & 0 & 0 & 0 & 1 & \alpha \\ 0 & 0 & 0 & 1 & 0 & \beta \\ 0 & 0 & 1 & 0 & 0 & \alpha \end{bmatrix} \tag{3.8}$$

Therefore, achieving the third operation requires a three-qubit operation, which is a function of transformation Ψ^{00}(Bardin et al. 2020; Cuomo et al. 2020; Hassija et al. 2020). In this perspective, a three-qubit operation can be recognized as $|\Psi_1 \ge 1|0\rangle$, $|\Psi_2 \ge 1|1\rangle$, and $|\Psi_3 \ge \beta|0\rangle + \alpha|1\rangle$, in which the turbulent flow levels (i.e., strong, medium, and light) are encoded, respectively, in the SAR data. Therefore, a three-qubit operation can be identified as:

$$\Psi^{00}|\Psi_1\Psi_2\Psi_3\rangle = \begin{bmatrix} \alpha \\ \beta \\ 0 \\ 0 \\ 0 \\ 0 \\ 0 \\ 0 \end{bmatrix} \tag{3.9}$$

where equation 9 represents the different levels of the turbulent pattern flows $|\Psi_1\rangle, |\Psi_2\rangle$ and $|\Psi_3\rangle$, respectively, in the SAR image.

In line with Wang et al. (2005), evolutionary computing with qubit representation has a better characteristic of population diversity than other representations since it can represent a linear superposition of state probabilities. In binary representation, at least eight strings – (000), (001), (010), (011), (100), (101), (110), and (111) – are required, whereas the single qubit shown earlier is enough to depict eight states (Figure 3.3). In this view, the quantum evolutionary algorithm (QMEA) and quantum genetic algorithms (QGA) operate efficiently on probability amplitudes of fundamental quantum states, whereas in the latter two (QEA) active crossovers and identified mutations are properly used to maintain population diversity (Kim et al. 2006).

The initial stage of QMEA is to generate accurate individual qubit populations of the turbulence flow levels in an SAR image, which can be signified as $|Q(t)\rangle = \{|q_1^t\rangle, |q_2^t\rangle, \ldots., |q_n^t\rangle\}$ (Figure 3.4). In this sense, let us consider that $|q_1^t\rangle = \begin{Bmatrix} |\alpha_{i1}^t\rangle, |\alpha_{i2}^t\rangle, \ldots\ldots, |\alpha_{im}^t\rangle \\ |\beta_{i1}^t\rangle, |\beta_{i2}^t\rangle, \ldots\ldots\ldots, |\beta_{im}^t\rangle \end{Bmatrix}$ where $(i = 1, 2, \ldots.., n)$ is a qubit individual, n is the size of the population, and m is the span of the individual.

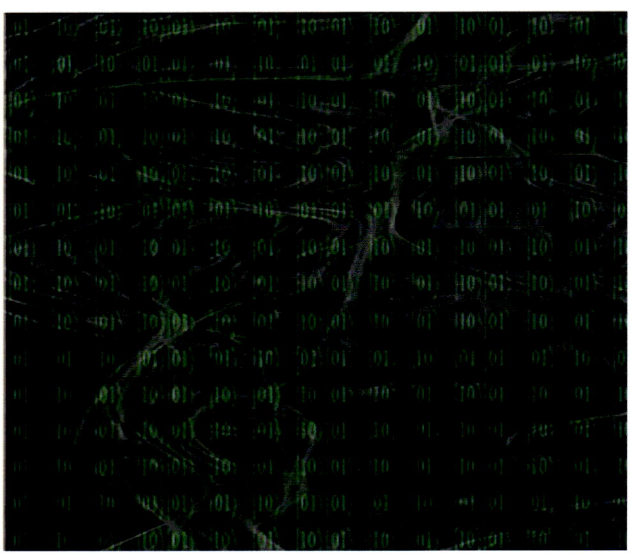

FIGURE 3.3 Qubits of turbulent generation in SAR image.

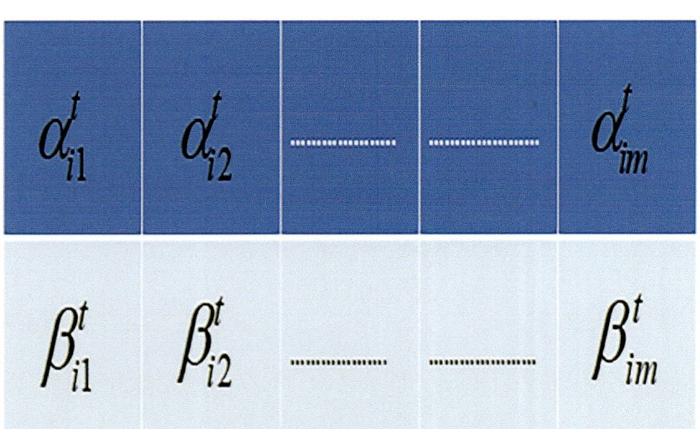

FIGURE 3.4 An array of one-qubit q_1^t populations.

3.6 POPULATION PATTERN OF QUANTUM TURBULENT FLOW CREATION

In light of contemporary viewpoints, let us consider the constraint of the random qubit generation of the turbulence flow as $Q|N_O\rangle \in |0,1\rangle$. In this, the random number $Q|N_O\rangle$ is confirmed as a turbulent flow if $Q|N_O\rangle > |\alpha_i^t|^2$ in which the equivalent qubit equals 1 and is dissimilar when q_n^t is equivalent to 0, and turbulence does not exist in the SAR data. Consequently, the clustering centre point of the turbulence is depicted by qubits $|1\rangle$ individually (Balicki et al. 2010; Balicki 2014; Marghany 2021).

3.6.1 Quantum fitness

The qubit chromosome form needs to be coded with the boundaries of the turbulence footprint. Every qubit in the qubit chromosome $|q_m\rangle$ in this issue corresponds to a coefficient in the north-order surface fitting polynomial, which is given by:

$$f|i,j\rangle_q = |q_0\rangle + |q_1 i\rangle + |q_2 j\rangle + |q_3 i^2\rangle + |q_4 ij\rangle + |q_5 j^2\rangle + \ldots + |q_m j^n\rangle \tag{3.10}$$

The connectedness metric is then employed to assess fitness. It is conceivable to mathematically express the level of the qubit chromosomes that are signed from adjacent pixels in the SAR image and clustered as the turbulent flow as follows:

$$conn|q^t\rangle = \sum_{i=1}^{m}\left(\sum_{j=1}^{L}|q_{i,j}^t\rangle\right) \tag{3.11}$$

Equation 11 explains how to distinguish turbulent clusters in the environment of several qubit chromosomes L of m size. In this particular circumstance, each 10 qubit chromosome for every pixel closest to pixel i and its adjoining j-th is distinguished as the turbulent flows. In contrast, the absence of turbulent occurrence in SAR images and features belonging to the surrounding sea surface is recognized as well $|q_{i,j}^t\rangle = |j^{-1}\rangle$.

3.6.2 Quantum mutation

The quantum mutation is carried out through the quantum rotation gate $G(\theta)$ with the assistance of the existing outstanding individual qubit, which behaves as a mutation operator in QEA (Figure 3.5):

$$G(\theta) = \begin{bmatrix} \cos\theta & -\sin(\theta) \\ \sin\theta & \cos\theta \end{bmatrix} \tag{3.12}$$

Equation 12 demonstrates the quantum rotation gate for searching in the quantum pixel in a SAR image (Figure 3.6) as a function of rotation angle, which is cast as:

$$\theta = k \times f|\alpha_1, \beta_1\rangle \tag{3.13}$$

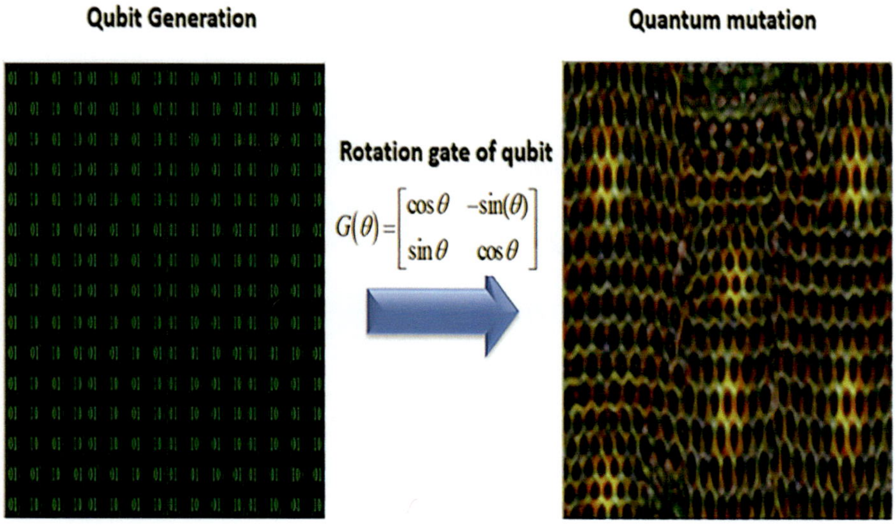

FIGURE 3.5 Quantum mutation of SAR image for turbulent detection.

FIGURE 3.6 Quantum rotation gate for turbulent detection in SAR data.

TABLE 3.1 Lookup table of $f|\alpha_i,\beta_i\rangle$.

		$f\lvert\alpha_i,\beta_i\rangle$	
$\lvert\alpha_1\rvert*\lvert\beta_1\rvert>0$	$\lvert\alpha_2\rvert*\lvert\beta_2\rvert>0$	arctan ($\lvert\beta_1\rvert/\lvert\alpha_1\rvert$) *(BEST SOLUTION)*	arctan ($\lvert\beta_2\rvert/\lvert\alpha_2\rvert$) *(CURRENT SOLUTION)*
True	True	+1	−1
True	False	−1	+1
False	True	−1	+1
False	False	+1	−1

In this case, the value of the coefficient k affects how quickly convergence arises. The number k must be selected wisely. The search grid of the algorithm would be large if k is too large, which would cause the solutions to diverge or prematurely converge to a local optimum. Conversely, if k is too small, the search grid would also be small, which would induce the algorithm to potentially become stuck. Additionally, $f|\alpha_i,\beta_i\rangle$ maintains the direction of convergence of the search to a global optimum (Wang et al. 2005). The quantum gate determines whether the clustering of turbulence detection is a false alarm or a true clustering. Figure 3.6 reveals how a quantum rotation gate can be applied to exhibit the turbulent detection as bright patches for the initial moment. The look-up table for determining which quantum chromosome of turbulence detection is best is presented in Table 3.1.

The rules in the opposite direction can be provided by:

$$|q_i^{t+1}\rangle = G(t)\times|q_j^t\rangle \tag{3.14}$$

In this sense, the probability amplitude of the qubit is signified by an individual population at *t-th* creation as t is the evolutionary initiation at the quantum gate $G(t)$ (Beige et al. 2000; Wang et al. 2005; Balicki 2014; Cuomo et al. 2020).

3.7 PARETO OPTIMAL SOLUTION FOR QUANTUM NON-DOMINATED SORT AND ELITISM (QNSGA-II)

In short, the second version of the *non-dominated sorting genetic algorithm* introduced by Professor Kalyanmoy Deb for solving *non-convex* and *non-smooth* single and multi-objective optimization problems is then developed by Marghany (2021) to deliver QNSGA-II. Let us assume that the qubit chromosome is $Q|q_1^t\rangle,\ldots\ldots\ldots\ldots\ldots,Q|q_N^t\rangle$ for population size number N. In this regard, each solution m would be compared with every other solution in the population to find if it is dominated. Therefore, $S_{Q|q_1^t\rangle}$ is considered as every quantum multi-objective solution that would be compared to other solutions to resolve the level of domination: i.e., initialize $S_{Q|q_1^t\rangle}=\Psi\left(Q|q_1^t\rangle\right)$ (Deb et al. 2000; Deb 2001). In this sense, the main features of QNSGA-II involve: 1) a sorting non-dominated procedure of qubit chromosome $Q|q_N^t\rangle$ where all the individuals are sorted according to the level of non-domination; 2) it implements elitism, which stores all non-dominated solutions, hence enhancing convergence properties; 3) it adopts a suitable automatic mechanics based on the crowding distance to guarantee diversity and spread of solutions; and 4) constraints are implemented using a modified definition of dominance without the use of penalty functions (Deb 2001). Toward this end, the best classification technique of Deb et al. (2000) can be examined in turbulence automatic detection as follows:

In main population *P*, every individual qubit $Q|q_N^t\rangle$ is validated:

Let F_1 be the first front,
Set $Q|q_1\rangle = 1$.
Individuals dominate $Q|q_1\rangle$ then $Q|q_1\rangle \in F_1$,
Set first front set $F_1\left(Q|q_1\rangle\right)$,
Update by adding $Q|q_1\rangle$ to front 1,
$F_1\left(Q|q_1\rangle\right) = F_1\left(Q|q_N^t\rangle\right) \cup Q|q_1\rangle$,
Then $F_i\left(Q|q_N^t\rangle\right) \neq \Psi\left(Q|q_N^t\rangle\right)$,

Hence set $Q(q_1)_{rank} = i + |1\rangle$. Update the set $Q|q_i^t\rangle$ with individual *m*, i.e., $Q|q_i^t\rangle = Q|q_i^t\rangle \cup m$.
– increment the front of one.
– For turbulence, now the set $Q|q_i^t\rangle$ is the next front,
$F_i(Q|q_i^t\rangle) = Q|q_i^t\rangle$.
Return
End

Consequently, the quantum probability of the maximum sorting of the qubit chromosome is $P_{\max}(Q|q_i^t\rangle)$, which is also corresponds to quantum qubit chromosomes of probability amplitude of turbulence footprint in SAR data and should comply with:

1 The probability of turbulent flow pixels is signified as $(P_{\max}(Q|q_i^t\rangle)$: the discrepancy of maximum probability of the quantum qubit chromosomes containing the turbulence quantum $P_{\max}(Q|q_i^t\rangle) = \max\left\{P_1(Q|q_1^t\rangle),\ldots\ldots,P_k(Q|q_k^t\rangle)\right\}$, which designates the probability existence of the turbulence in Pareto front $j,\ \forall j = 1,2,\ldots,k$.
2 Each row and column in the SAR image involves the sum of the quantum probability of turbulence qubit chromosomes: $(\sum P_i(Q|q_i^t\rangle)_i)$.

According to the example, Pareto optimal solutions are operated to hold the perceptiveness of turbulence variety and its encompassing circumstances in SAR images. Therefore, computing the Pareto front grade of convergence would be a function of the following, giving the Euclid distance:

$$P_f\left(P_i(Q|q_i^t\rangle)\right) = \sum_{\omega=1}^{\omega} \sqrt{\sum_{m=1}^{N} \min_{u=1,U}\left(P_{um}(Q(|q_{um}^t\rangle) - |a_{\omega m}\rangle\right)^2} \tag{3.15}$$

Therefore, equation 15 is applied to compute the Pareto front for automatic detection of the turbulence flows in SAR data based on the QMEA (Deb et al. 2000; Deb 2001; Marghany 2021). In this understanding, *um* represents the number of Pareto front points, and *N* represents the number of optimized criteria for the automatic detection of turbulent flow in SAR data (Figure 3.7). Indeed, they are completely excluded from the possibility of finding quantum qubit chromosomes in turbulence flow. Needless to say, in multi-objective optimization problems, the Pareto front is an accumulation of ideal solutions that are utilized to identify trade-offs between various objectives rather than to evaluate accuracy. Researchers can choose the best solution based on their priorities by having a solid understanding of the Pareto front.

3.8 RESULTS AND DISCUSSION

The Advanced Land Observing Satellite (ALOS), which was launched on 24 January 2006 by the Japan Aerospace and Exploration Agency (JAXA), and operated until 12 May 2011 provides the SAR data. ALOS data used here was acquired on 22 March 2007 around the Farasan Islands in Saudi Arabia's Red Sea coastal waters (Figures 3.8 and 3.9). More precisely,

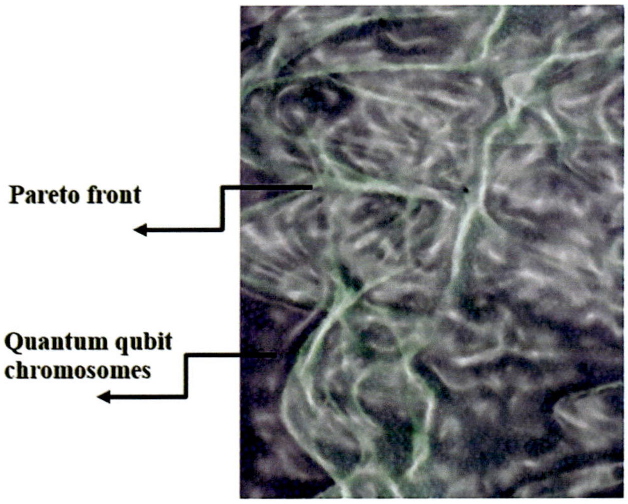

FIGURE 3.7 Pareto front curves engendered by quantum qubit chromosomes in SAR data.

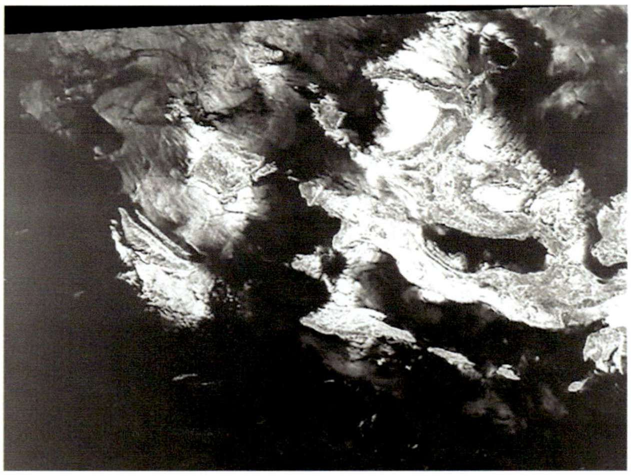

FIGURE 3.8 ALOS data acquired in this study.

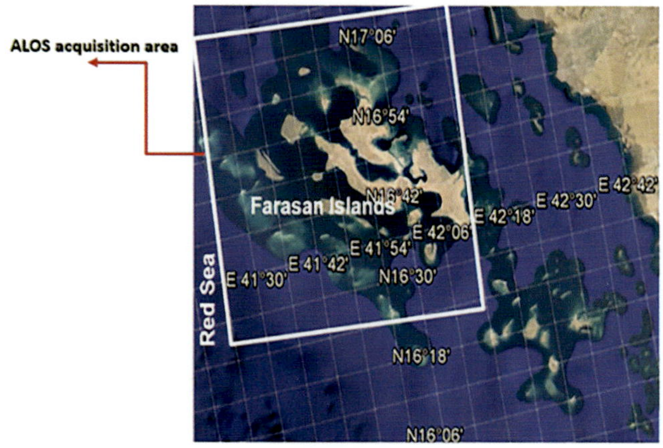

FIGURE 3.9 Google Earth map of the Farasan Islands, Red Sea and ALOS acquisition area (white box).

the white box shown in Figure 3.9 represents the ALOS data acquisition coverage area in the Red Sea.

ALOS (which carried PALSAR) was put into a sun-synchronous orbit and circumnavigated the earth every 100 minutes, or 14 times throughout the day. Every 46 days, ALOS PALSAR made a return to the initial course (repeat cycle). At the equator, the inter-orbital distance was roughly 59.7 kilometres. The PALSAR was an active microwave sensor that could be operated day or night and was unaffected by the weather. The synthetic aperture radar (SAR) on Japan's first Earth observation satellite served as its foundation (JERS-1). The PALSAR was a right-looking SAR that operated in the L-band and also had HH polarization and a cross-track aiming range of 8° to 60°.

The coral reef–induced turbulent flow is investigated here using ALOS PALSAR data in the fine mode. The greatest ground resolution for fine mode data, which became 7 m, is one of the highest for a SAR (in contrast, the JERS-1 SAR had an approximately 18-metre resolution). With a look angle of 35°, a swath width of 70 km, a noise equivalent backscattering coefficient of -25 dB and a signal-to-ambiguity (S/A) ratio of 25 dB, the fine resolution mode provides a spatial resolution of 10 m in both range and azimuth for HH or VV with swath width of 70 km. However, the spatial resolution for full polarization HH+HV+VV+VH is 30 m with a swath width of 30 km.

The normalized radar backscatter cross-section NRSC variation in the ALOS PALSAR image is demonstrated in Figure 3.10. Over the Farasan Islands, the greatest NRSC of -5dB occurs. The Farasan Islands contain sandy coasts, mangroves and algae flats that reach their maximum extent near the coastline (Andréfouët et al. 2002) while reefs are only found offshore (Al-Anbaawy 1993), which contributes to the maximum peak of NRSC. The Farasan Islands' coral reefs can generate an inconsistent pattern of water flow because of friction (Abu-Zied et al. 2011), which forces sea surface currents to modulate surface waves and create a very rough sea surface that produces the greatest SAR backscatter.

The ALOS PALSAR fine mode instrument has a reduced noise floor range of -21dB, allowing for improved sea surface imaging. In particular, for the HH backscatter from turbulent flow cover, the instrument noise floor is only reached at the extreme edge of the swath. This could also aid in determining the radar cross-section of water utilizing L-band radar, when look-alikes such as alga blooms, low wind zones and high concentrations of chlorophyll are dampening the surface waves. Regarding parameter settings, certain quantities are used, including quantum population, encoding length and maximum iteration number. To guarantee a precise consequence, we set N equal to 100, encoding length equal to 20, and iteration number equal to 100 (Marghany 2013). Figure 3.11 depicts the results achieved using QMEA (Marghany 2014a, 2014b). In this sense, QMEA can automatically distinguish between Bragg scattering and non-Bragg scattering (Marghany 2021). The dynamic variation between Bragg and non-Bragg scattering is precisely calculated. Bragg scattering tends to capture the uneven turbulent flow between the Farasan Islands (Marghany 2015). The chaotic structure of the turbulent flow is distinctly twisted by QME in this regard (Figure 3.11). This

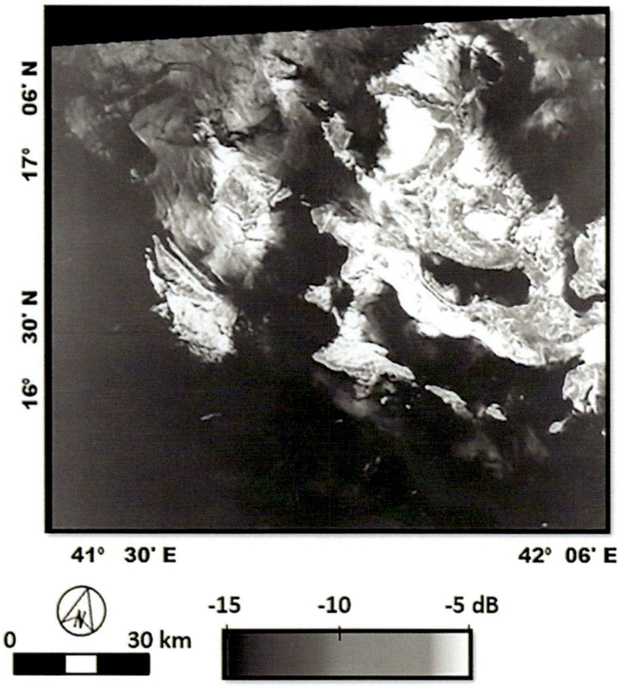

FIGURE 3.10 Normalized radar backscatter cross-section (NRSC) variation in ALOS SAR image.

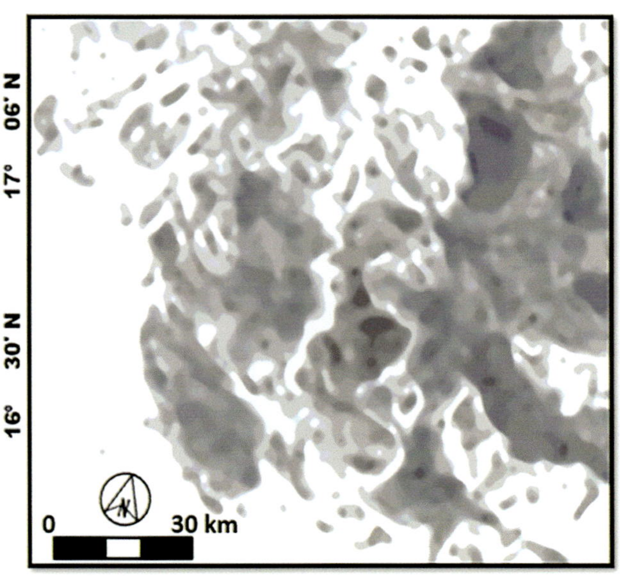

FIGURE 3.11 The turbulent flow's chaotic pattern by QMEA.

result agreed with previous work by Marghany and Hakami (2019), which used optical satellite data to track turbulent eddy flow. Indeed, PALSAR's unique characteristics would enable it to capture images of turbulence in areas where other sensors cannot. This would improve the accuracy of automatic detection systems and provide valuable data for weather forecasting and aviation safety (Singh et al. 2023).

Additionally, QNSGA-II can provide a variety of clusters in ALOS, such as spinning mesoscale eddies that envelop the sea surface (Figure 3.12). As well as the iteration number increased to 100, N remains at 80 individual qubits, the encoding length is set to 20, and a crisp chaotic turbulent morphological flow is perfectly detected in ALOS data (Figure 3.12).

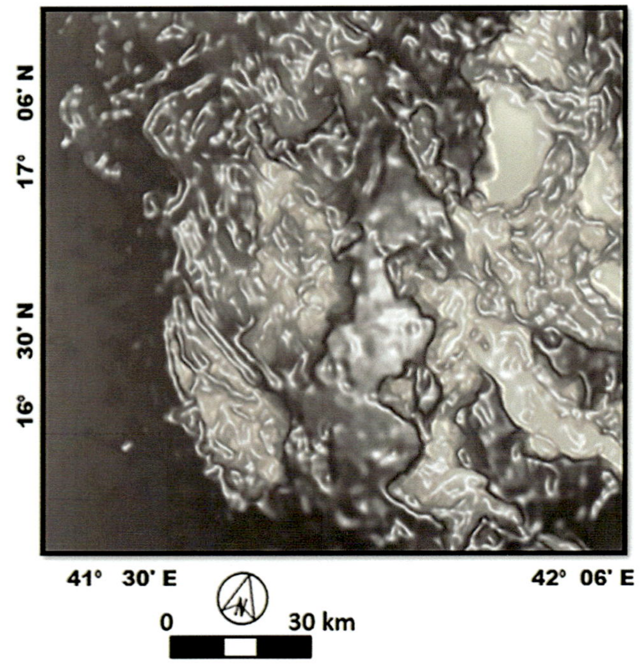

FIGURE 3.12 Automatic turbulent morphology flow discrimination in ALOS PALSAR data using QNSGA-II.

Precise clusters of the turbulent morphological flow are obtained by increasing the number of variables. In other words, these clusters of turbulent morphology flow have demonstrated the range of turbulent morphology flow characteristics, from strong to mild flow (Figure 3.13). In actuality, the water flow, which has a maximum magnitude of 1.5 m s^{-1}, creates erratic patterns as it strikes the Farasan Islands. Marghany and Hakami (2019) claim that the molecular diffusion that occurs at the coral-water interface results in a fluid viscosity that reduces turbulent motions and vertical transport. As noted by Marghany (2021), the synthesis or consumption of dissolved elements by the coral surface results in a steep gradient that forms a diffusive boundary layer.

In essence, ALOS PALSAR detects variations in the roughness of a turbulent flow's surface and, for vigorous streams, variations in the conductivity of its surface layer. Because the vast majority of the radar energy that strikes a smooth surface is refracted away from the radar antenna, ALOS PALSAR "seems to see" substantial turbulence as a rough (radar-bright) patch against the smoother (radar-dark) ocean surface. The antenna was able to distinguish between strong and weak flows for ocean turbulence using a radar system because of the great sensitivity of the fine mode and other aspects. It is preferred to use HH polarization since its considerable tilt makes it susceptible to scattering from turbulent sea surfaces. Strong, moderate and weak turbulent level flows are all analyzed quantitatively by QNSGA-II (Figure 3.13).

In 100 iteration runs, QNSGA-II can therefore obtain the best solutions more frequently and with the highest clustering efficiency. The QNSGA-II runs stably and optimally. The consistent result of QNSGA-II, which shows a greater number of distinct clusters with turbulent flows tied to coral reefs, seems instantly to be precisely classified (Figure 3.14).

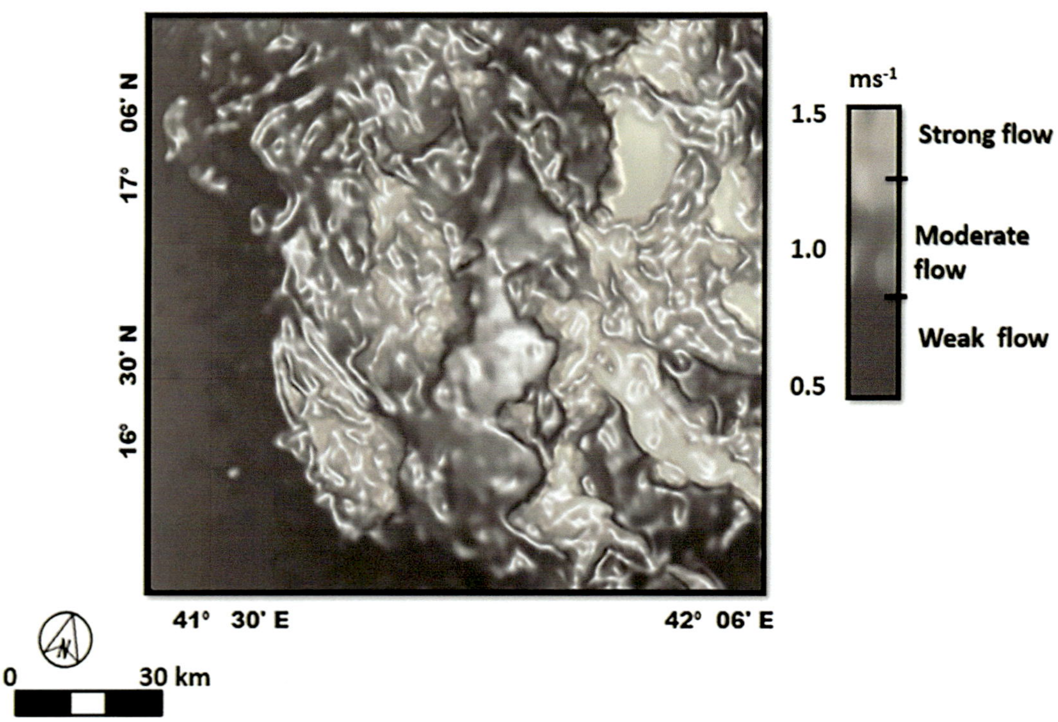

FIGURE 3.13 Clustering of turbulent patterns exploiting QNSGA-II.

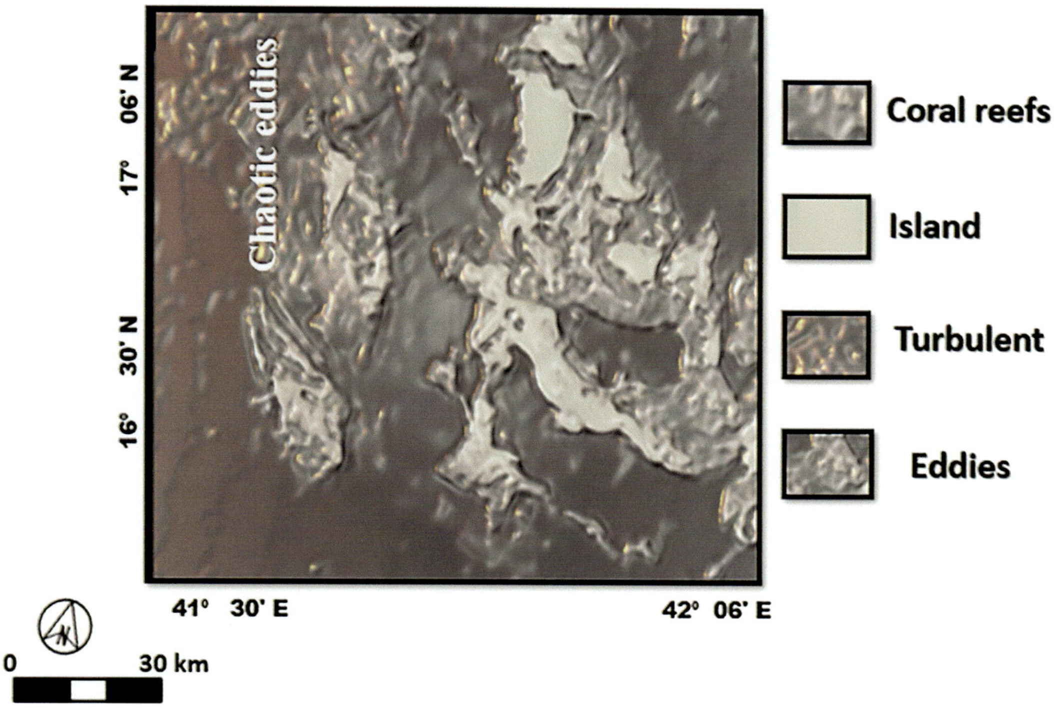

FIGURE 3.14 QNSGA-II automatically detects turbulent flow caused by coral reefs.

The QNSGA-II algorithm reveals the possible interaction between water currents and coral reefs. Additionally, QNSGA-II is capable of reliably distinguishing between the island, coral reefs and turbulence flow. It is possible to accurately confirm the profound influence of boundary movement on the generation of turbulent flow intensity, such as strong flow patterns and weak or gentle flow, with QNSGA-II (Figure 3.14). Properly compared to the irregular surface of the flowing sea, the turbulent flow typically retains a more marked degree of coherence.

According to this perspective, irregular reefs are a common feature at shallow water depths between 5 and 20 m along the Farasan Islands. In this sense, the main surface current flow with a maximum velocity of 1 m s^{-1} along the Farasan Islands coastal waters modifies the tide. Therefore, the coral reefs, sand bars, low islands and shoals can cause turbulent flow with a maximum velocity of 1.5 m s^{-1} when a tidal current passes through. This turbulent flow can cause the development of chaotic eddies in shallow water areas and thus has a significant effect on the hydrodynamics of the Farasan Islands' coastal waters. Consequently, irregular reefs in shallow water can significantly alter the hydrodynamic patterns and cause chaotic eddies (Figure 3.14). This finding agrees with the study of Marghany and Hakami (2019).

Needless to say, for instance, one such sensor is the Sentinel-1 SAR, which operates at the C-band and can provide high-resolution images of ocean turbulence patterns. Other sensors are the RADARSAT-2 SAR and TerraSAR-X, which operate at the C-band and X-band, respectively, and can provide information on sea turbulence patterns included in oil spills and ship wake turbulence detection (Belcher 2008).

Figures 3.15 and 3.16 demonstrate the non-dominated practical solutions of various algorithms. It appears that QNSGA-II (Figure 3.15) is significantly more efficient than Q-MEA (Figure 3.16). Q-MEA is often held at the established centre of the Pareto front. In this unique case, Q-MEA

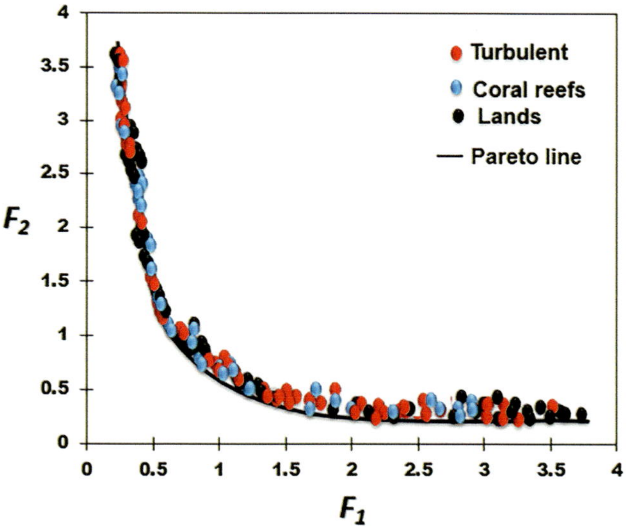

FIGURE 3.15 Non-dominated solutions delivered by QNSGA-II.

emphasizes one critical component of the Pareto front. When searching for the most effective solution for turbulent flow intensity changes, coral reefs, local land or surrounding sea footprint boundary levels in SAR data, QNSGA-II accurately preserved a high degree of diversity in the available solutions. As a consequence, the QNSGA-II can distribute its weak population diversity along the obtained front more accurately than the Q-MEA. This could lead to more efficient optimization of the Pareto front as the QNSGA-II is capable of finding more accurate solutions that are closer to the optimal trade-off between objectives.

In addition, QNSGA-II significantly increases the core characteristics of the non-dominated set, which is often indicated by the stable population in multi-objective issues including SAR textures. QNSGA-II is a speculative approach that

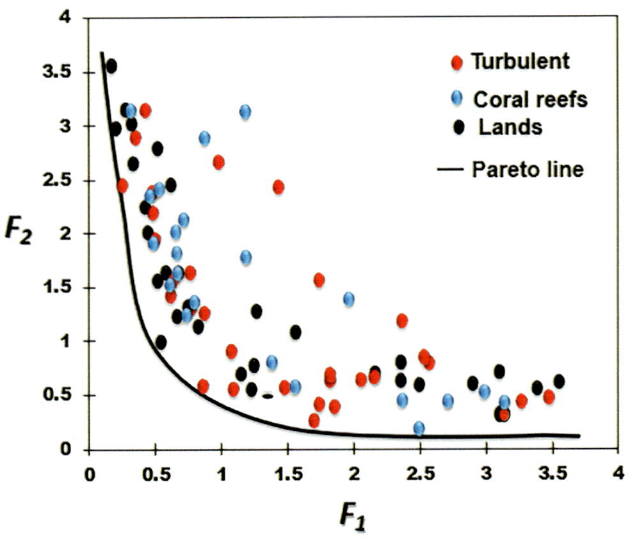

FIGURE 3.16 Non-dominated solutions provided by Q-MEA.

involves proper utilization of the key concepts and guiding principles of quantum computing, such as significant uncertainty, coherent superposition and constructive interference. Indeed, QNSGA-II accurately locates the comprehensive solutions of automatic detection of turbulent flow magnitudes, coral reefs, land and adjacent sea footprint bounds that are adjacent to the Pareto-optimal front (Figure 3.15). The accuracy is shown by the Pareto front of 3.7 in both dimensions of the front one (F_1) and front two (F_2) axes. This finding validates the speculations of Deb et al. (2000) and Deb (2001).

In other words, QNSGA-II gradually improves computing proximity to the Pareto-optimal front while preserving diversity integral by retaining quantum evolutionary algorithm compensations (QMEA). By taking advantage of the benefits of local optimization, these solutions can provide a more accurate representation of the problem while minimizing computational complexity. It is important to keep in mind that the Pareto front only represents a set of optimal solutions and may not necessarily include the most accurate solution. Therefore, it is crucial to evaluate the accuracy of each solution individually before making a final decision.

In appropriate comparison to conventional evolutionary algorithmic programmes, the used QNSGA-II algorithm adequately incorporates both local and global evaluation, which optionally allows it to typically determine the optimal situation with remarkable precision. As a potential consequence, QNSGA-II can be accurately exploited to satisfactorily solve optimization problems and to preemptively detect multiple access code splitting (Vinen and Niemela 2002; Wang et al. 2005; Marghany 2013, 2015). In this unique perspective, QNSGA-II may be combined with a genetic algorithm to positively enhance great precision (Williams 2011; Marghany 2013; Marghany and Hakami 2019). Eventually, advanced SAR image segmentation and the digital process will routinely require quantum computing algorithms like QNSGA-II.

3.9 CONCLUSIONS

This chapter convincingly demonstrates a novel technique for the automatic detection of the level of turbulent flow using SAR data like ALOS PALSAR. Toward this end, the quantum multi-objective algorithm is developed and examined in a SAR image which was acquired over the Farasan Islands in the Red Sea. In short, SAR backscatter and texture information is instantly turned into qubit chromosomes. In conclusion, quantum multi-objective evolution (QMEA) employs a novel qubit chromosomal depiction based on the modern notion and conceptual frameworks of quantum computing, including qubits and nonlinear superposition. The quantum multi-objective evolutionary (QMEA) can efficiently deliver a distinguishable classification map of turbulent flow without distinct boundaries and edges.

In this practical sense, quantum non-dominated sort and elitism (QNSGA-II) are properly used to boost the efficiency of QMEA for automatic detection of the particular cause of turbulent flow in the Farasan Islands in ALOS PALSAR data. The efficient Pareto front algorithm is implemented to accurately determine the most effective optimal solution for the automatic detection of turbulence and surrounding features. The boundaries of the turbulent footprint are more accurately and automatically clustered and identified than QMEA. The surrounding sea features coral reefs and the local land, and these are perfectly distinguished from the surrounding gradient variation of the turbulent flows.

Therefore, the QNSGA-II can also better detect the varying level of turbulence currents such as strong, moderate and smooth than QMEA. In addition, QNSGA-II can show that the coral reefs are the leading cause of turbulent flow around the Farasan Islands. This accuracy is well shown by the Pareto front of value 3.7 in both dimensions of the front one (F_1) and front two (F_2) axes. In conclusion, QNSGA-II represents a promising tool for the automatic detection of turbulent flow in shallow coastal island waters such as the Farasan Islands. This approach can be further extended to other coastal regions for efficient monitoring and management of water resources. However, the effectiveness of QNSGA-II may be influenced by the availability and quality of input data.

REFERENCES

Abu-Zied, R.H., Bantan, R.A., Basaham, A.S., El Mamoney, M.H., and H.A. Al-Washmi. 2011. Composition, distribution, and taphonomy of nearshore benthic foraminifera of the Farasan Islands, southern Red Sea, Saudi Arabia. *J Foraminiferal Res* 41, no. 4: 349–362.

Al-Anbaawy, M.I.H. 1993. Reefal facies and diagenesis of Pleistocene sediments in Kamaran Island, southern Red Sea, Republic of Yemen. *Egypt J Geol* 37, no. 2: 137–151.

Alpers, W.R., Ross, D.B., and C.L. Rufenach. 1981. On the detectability of ocean surface waves by real and synthetic aperture radar. *J Geophys Res Oceans* 86, no. C7: 6481–6498.

Andréfouët, S., Mumby, P., McField, M., Hu, C., and F. Muller-Karger. 2002. Revisiting coral reef connectivity. *Coral Reefs* 21, no. 1: 43–48.

Balicki, J.M. 2014. Quantum-inspired multi-objective evolutionary algorithms for decision making: Analyzing the state-of-the-art. In: *Advances in Applied and Pure Mathematics*. Ed. J. Balicki, 383–389. Proc 2nd International Conference on Mathematical, Computational and Statistical Sciences MCSS'14, Gdańsk, Poland, May 15–17, 2014. WSEAS Press, Attica.

Balicki, J.M., Balicka, H.T., Masiejczyk, J., and A. Zacniewski. 2010. Multi-criterion decision making in distributed systems by quantum evolutionary algorithms. In: *Proc 2nd European Conference on Computer Science*, Puerto de la Cruz, Spain, pp. 328–333.

Bardin, J.C., Sank, D., Naaman, O., and E. Jeffrey. 2020. Quantum computing: An introduction for microwave engineers. *IEEE Microw Mag* 21, no. 8: 24–44.

Barenghi, C.F., Skrbek, L., and K.R. Sreenivasan. 2014. Introduction to quantum turbulence. *Proc Natl Acad Sci* 111, supplement 1: 4647–4652.

Beige, A., Braun, D., Tregenna, B., and P.L. Knight. 2000. Quantum computing using dissipation to remain in a decoherence-free subspace. *Phys Rev Lett* 85, no. 8: 1762.

Belcher, D.P. 2008. Sidelobe prediction in transionospheric SAR imaging radar from the ionospheric turbulence strength CkL. In: *Proc 2008 Int Conf on Radar*, IEEE, New York City, pp. 54–59.

Chorin, A.J. 1968. Numerical solution of the Navier-Stokes equations. *Math Comput* 22, no. 104: 745–762.

Constantin, P., and C. Foias. 2020. *Navier-Stokes Equations*. Chicago: University of Chicago Press.

Cuomo, D., Caleffi, M., and A.S. Cacciapuoti. 2020. Towards a distributed quantum computing ecosystem. *IET Quantum Comm* 1, no. 1: 3–8.

Deb, K. 2001. Nonlinear goal programming using multi-objective genetic algorithms. *J Oper Res Soc* 52, no. 3: 291–302.

Deb, K., Agrawal, S., Pratap, A., and T. Meyarivan. 2000. A fast elitist non-dominated sorting genetic algorithm for multi-objective optimization: NSGA-II. In: *International Conference on Parallel Problem Solving from Nature*, 849–858. Berlin and Heidelberg: Springer.

Di Martino, G., Iodice, A., Riccio, D., and G. Ruello. 2015. Ocean monitoring with SAR: An overview. In: *OCEANS 2015-Genova*, 18–21 May 2015, Genova: Institute of Electrical and Electronics Engineers (IEEE), pp. 1–5.

Doering, C.R., and J.D. Gibbon. 1995. *Applied Analysis of the Navier-Stokes Equations (No. 12)*. Cambridge: Cambridge University Press.

Donnelly, R.J., and C.E. Swanson. 1986. Quantum turbulence. *J Fluid Mech* 173: 387–429.

Han, K.H., and J.H. Kim. 2004. Quantum-inspired evolutionary algorithms with a new termination criterion, H/sub/spl epsi//gate, and two-phase scheme. *IEEE Trans Evol Comput* 8, no. 2: 156–169.

Hassija, V., Chamola, V., Saxena, V., Chanana, V., Parashari, P., Mumtaz, S., and M. Guizani. 2020. Present landscape of quantum computing. *IET Quantum Commun* 1, no. 2: 42–48.

Hey, T. 1999. Quantum computing: An introduction. *Comput Control Eng J* 10, no. 3: 105–112.

Jain, A.T.U.L. 1981. SAR imaging of ocean waves: Theory. *IEEE J Ocean Eng* 6, no. 4: 130–139.

Jiao, L., Li, Y., Gong, M., and X. Zhang. 2008. Quantum-inspired immune clonal algorithm for global optimization. *IEEE Trans Syst Man Cybern B (Cybernetics)* 38, no. 5: 1234–1253.

Kanevsky, M.B. 2008. *Radar Imaging of the Ocean Waves*. Amsterdam: Elsevier.

Kaye, P., Laflamme, R., and M.A. Mosca. 2007. *Introduction to Quantum Computing*. Oxford: Oxford University Press.

Kim, Y., Kim, J.H., and K.H. Han. 2006. Quantum-inspired multiobjective evolutionary algorithm for multiobjective 0/1 knapsack problems. In: *2006 IEEE Int Conf on Evolutionary Computation*, 16–21 July 2006, Vancouver: Institute of Electrical and Electronics Engineers (IEEE), pp. 2601–2606.

Macklin, J.T., and R.A. Cordey. 1991. Seasat SAR observations of ocean waves. *Int J Remote Sens* 12, no. 8: 1723–1740.

Marghany, M. 2013. Genetic algorithm for oil spill automatic detection from MultiSAR satellite data. In: *Proc 34th Asian Conference on Remote Sensing 2013, Bali, Indonesia, October 20–24, 2013*, Asian Association on Remote Sensing, Tokyo, pp. SC03671–SC03677.

Marghany, M. 2014a. Utilization of a genetic algorithm for the automatic detection of oil spill from RADARSAT-2 SAR satellite data. *Mar Pollut Bull* 89: 20–29.

Marghany, M. 2014b. Multi-objective evolutionary algorithm for oil spill detection from COSMO-SkyMed satellite. In: *Computational Science and Its Applications – ICCSA 2014*. Eds. B. Murgante, S. Misra, M. Carlini, C.M. Torre, H.-Q. Nguyen, D. Taniar, B.O. Apduhan and O. Gervasi, 355–371. Amsterdam: Springer.

Marghany, M. 2015. Flock 1 data multi-objective evolutionary algorithm for turbulent flow detection. In: *CD of 36th Asian Conf Remote Sensing: Fostering Resilient Growth in Asia, ACRS 2015*, Quezon City, Manila, Philippines, 24 October 2015–28 October 2015, pp. 1–6.

Marghany, M. 2021. *Nonlinear Ocean Dynamics: Synthetic Aperture Radar*. Amsterdam: Elsevier.

Marghany, M., and M. Hakami. 2019. Automatic detection of coral reef induced turbulent boundary flow in the Red Sea from Flock-1 satellite data. In: *Oceanographic and Biological Aspects of the Red Sea*. Eds. N.M.A. Rasul and I.C.F. Stewart, 105–122. Cham, Switzerland: Springer Oceanography, Springer Nature.

Rieu, P., Moreau, T., Cadier, E., Raynal, M., Clerc, S., Donlon, C., and C. Maraldi. 2021. Exploiting the Sentinel-3 tandem phase dataset and azimuth oversampling to better characterize the sensitivity of SAR altimeter sea surface height to long ocean waves. *Adv Space Res* 67, no. 1: 253–265.

Sikora, T.D., Young, G.S., Beal, R.C., Monaldo, F.M., and P.W. Vachon. 2006. Applications of synthetic aperture radar in marine meteorology. *Atmos Ocean Int* 2: 83–105.

Singh, A., Kushwaha, S.K.P., Nandy, S., Padalia, H., Ghosh, S., Srivastava, A., and N. Kumari. 2023. Aboveground forest biomass estimation by the integration of TLS and ALOS PALSAR data using machine learning. *Remote Sens* 15, no. 4: 1143.

Temam, R. 2001. *Navier-Stokes Equations: Theory and Numerical Analysis*, vol. 343. New York City: American Mathematical Soc.

Tomiyasu, K. 1978. Tutorial review of synthetic-aperture radar (SAR) with applications to imaging of the ocean surface. *Proc IEEE* 66, no. 5: 563–583.

Tsubota, M., Fujimoto, K., and S. Yui. 2017. Numerical studies of quantum turbulence. *J Low Temp Phys* 188, no. 5: 119–189.

Vinen, W.F., and J.J. Niemela. 2002. Quantum turbulence. *J Low Temp Phys* 128, no. 5: 167–231.

Wang, L., Tang, F., and H. Wu. 2005. Hybrid genetic algorithm based on quantum computing for numerical optimization and parameter estimation. *Appl Math Comput* 171, no. 2: 1141–1156.

Williams, C.P. 2011. Quantum cryptography. In: *Explorations in Quantum Computing*, 507–563. London: Springer.

Wind wave variability in the Red Sea

4

Khalid M. Zubier

4.1 INTRODUCTION

Nearly one-third of world's maritime trading passes through the Red Sea, with energy supplies flowing annually through the Suez Canal and Bab El-Mandeb Strait accounting for 20–25% of the total seaborne-traded energy commodities (crude oil, refined petroleum products and liquefied natural gas). Huge entertainment cruise ships have recently started sailing between major Red Sea ports (Saudi, Egyptian and Jordanian). In the last few years, the Kingdom of Saudi Arabia has announced and initiated three Red Sea coastal giga projects (NEOM, AMALA and the Red Sea Project), which are now under rapid development. Projects pertaining to the development of the coastal area of Jeddah city centre were recently announced. Major desalination and electricity-generating plants are located on the coastal Red Sea areas along with major Saudi Arabian seaports.

The energetic wave conditions that occur in the Red Sea threaten the safety of the extensive navigation in the Red Sea as well as the safety of the coastal structures, including the giga projects and their coastal environment. For many years, the Red Sea has been lacking in terms of wave information, partially due to the spatially and temporally limited wave measurements. However, during the last decade, a considerable amount of research has been conducted on the Red Sea wave conditions and their variability utilizing long-term hindcasted wind and wave fields and observations from satellite altimeters, as well as limited wave measurements.

In this chapter, available wave measurements are discussed in section 2, while section 3 incorporates a comprehensive review of the literature relating to the Red Sea wave conditions and their variability. The review will be followed, in section 4, by an analytical study of wave conditions' variability in the Red Sea using reanalysis wave data that has been recently released by the European Centre for Medium-Range Weather Forecasts (ECMWF) under the title ECMWF Reanalysis Version 5 (ERA5) (Hersbach et al. 2020). Concluding remarks along with a few recommendations are given in section 5.

4.2 WAVE MEASUREMENTS

The Red Sea is extremely lacking in terms of conventional wave measurements. This statement is supported by the fact that an interested researcher will not be able to find conventional wave measurements in the Red Sea either in real time or for any time within the last five years. Based on our knowledge, all the documented conventional wave measurements made in the Red Sea through the last two decades are limited to its eastern-central part (Figure 4.1).

Using wave and tide gauges, King Abdulaziz University (KAU) collected short-term wave measurements off Jeddah for two months during the summer of 1991 (Abdelrahman 1995), two months (more than a decade later) during the summer of 2005 just outside Obhur Creek (WTR) and two months during the summer of 2006 near Al-Gahaz (DWR) (Zubier et al. 2007). KAU, in collaboration with Ifm-Geomar, University of Kiel and the German Technical Cooperation Agency (GTZ) managed to deploy a Met/Ocean buoy near the Jeddah shore (approximately 11 km from the shoreline at a water depth of 50 m), which was used to collect wave data from early summer 2012 to early spring 2013. In addition, a wave rider buoy (WRB) was moored near Saroom (nearly 10 km from Jeddah Sea Port at a water depth of 20 m) that collected wave measurements for nearly six months between late November 2012 and mid-May 2013 (Mayerle et al. 2016).

Through collaboration with Woods Hole Oceanographic Institution (WHOI), King Abdullah University of Science and Technology (KAUST) managed to deploy a Met/Ocean buoy just off Thuwal (at a water depth of 694 m) and collected wave measurements from October 2008 to May 2010 (Ralston et al. 2013). As part of a study conducted at King Abdullah Economic City (KAEC), waves were measured from December 2006 to February 2008 using two buoys (Buoy-M and Buoy-N), which were deployed at the coastal area of KAEC (Langodan et al. 2020).

These relatively short-term wave measurements have been utilized in the investigations of physical and statistical characteristics of the waves in the eastern-central Red Sea as well as for the validation/verification of wave models applied in the

DOI: 10.1201/9781003321408-5

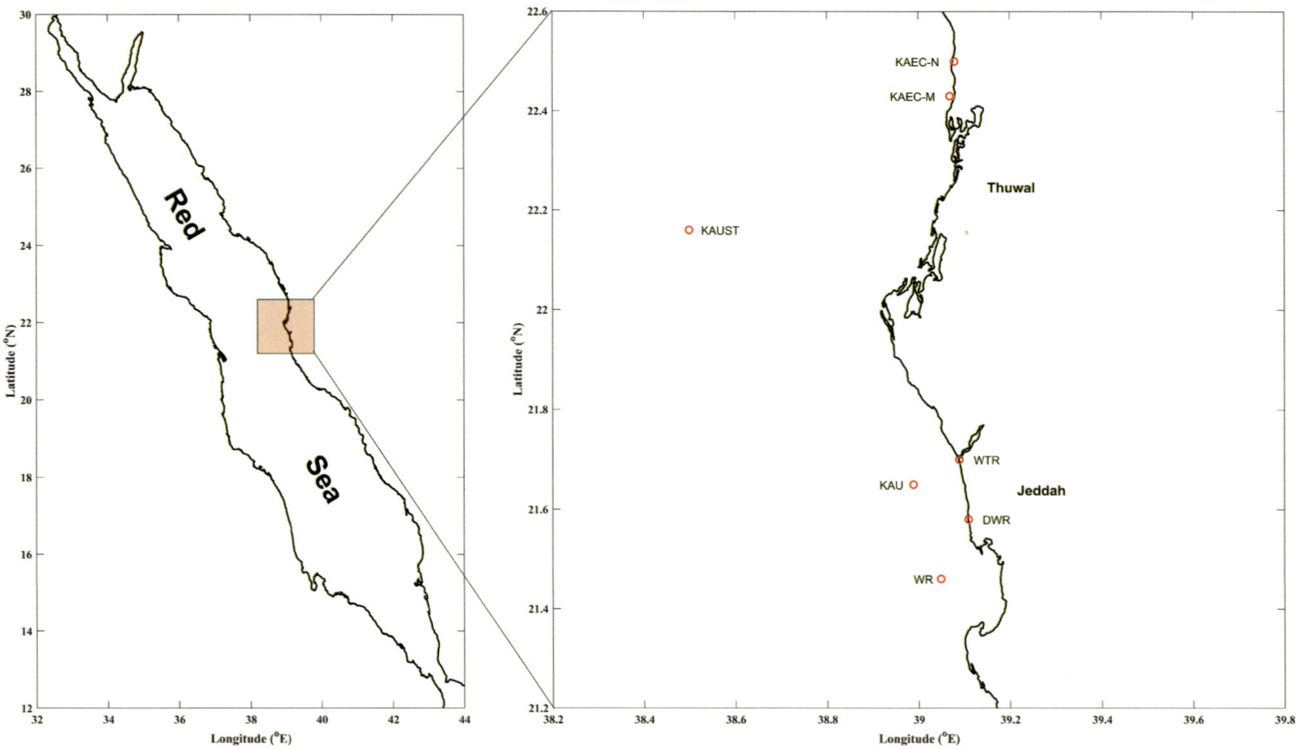

FIGURE 4.1 Red Sea (left panel) and Eastern Central Red Sea (right panel), including Jeddah and Thuwal coastlines.

Red Sea. However, it is necessary to point out that neither local university (KAU and KAUST) was able to keep its Met/Ocean buoys operational for longer periods due to the limitations in their collaborations with the other international institutions and, consequently, the lack of adequate resources for their operation and maintenance.

Although considered to be very accurate, conventional wave measurements in the Red Sea were very limited both spatially and temporally. Satellite-based wave measurements provide a much more extended spatial coverage (along track), but their temporal coverage is still very limited because it takes a few days for the repeated along-track measurements. Some of the studies discussed in section 3 elaborate on the validation/verification of Red Sea numerical wave models using along-track altimeter wave measurements of different satellite missions, including ERS-1, ERS-2, TOPEX, POSEIDON, JASON-1, JASON-2 and ENVISAT. None of these studies, however, has utilized wave measurements based on satellite-obtained synthetic aperture radar (SAR) images, which have a high spatial resolution and can provide detailed information on the wave fields, including 2-D wave spectra and integral wave parameters.

4.3 PREVIOUS STUDIES

4.3.1 Numerical wave modeling

With the lack of conventional wave measurements and the spatial and temporal limitations of both conventional and satellite-based wave measurements, numerical wave modeling constituted the most practical and cost-effective way to obtain reliable wave data in the Red Sea. Studies pertaining to wave modeling in the Red Sea depended on either Wave Modeling (WAM) (Wamdi Group 1988), Wave Watch III (WW3) (Tolman and the WAVEWATCH III Development Group 2014) or Simulation of Waves Nearshore (SWAN) (Booij et al. 1999), which are third-generation spectral wave models based on wave action density balance equations. The performance of each of these phase-averaging models in simulating the wave conditions in the Red Sea has been determined through the comparisons of model results with in-situ and/or satellite measurements. Nevertheless, the performance of these models depends on many factors including their setup (e.g., the computational time step size, the computational grid resolution, the accounted-for physical mechanisms and their formulations), accuracy and spatial and temporal resolutions of wind fields and the accuracy and resolution of the bathymetry.

4.3.1.1 WAM model

Saad (2010) used the EMCWF-ERA-Interim reanalysis wind field to force the WAM model to simulate the wave conditions in the Red Sea during 2007. The comparisons between the WAM model results and the remotely obtained wave height measurements by three satellite altimetry missions (ENVISAT, JASON and GFO) clearly showed underestimation of wave conditions. This underestimation was attributed to the low-resolution modeling scheme and forcing wind field.

4.3.1.2 WW3 model

Langodan et al. (2014) used the WW3 model to simulate the observed wave conditions over a period of a full calendar year (2009). They forced the wave model with five different wind fields having different spatial and temporal resolutions (NOAA-NCEP, ECMWF, ECMWF-Enhanced, WRF-Final and WRF-Local). The wave model results were compared with KAUST buoy observations and altimeter data obtained through the ENVISAT, JASON-1 and JASON-2 satellite missions. The comparisons showed that the best model results were obtained using WRF-Local and ECMWF-Enhanced forcing wind fields. Based on further investigations, it was concluded that due to the unusual wind and wave patterns in the Red Sea, the wave model source functions may not properly replicate the local wave fields and therefore need to be updated. Accordingly, Langodan et al. (2015) modified the source functions and achieved a satisfactory fit through this. Further, Langodan et al. (2016a) carried out WW3 simulations in the Red Sea for a period of one year (2009) using two WRF wind fields, with (WRF-VARFC) and without (WRF-CTL) data assimilation of observations. The validation using KAUST buoy data and altimeter data (ERS-2 ENVISAT, JASON-1 and JASON-2) clearly showed considerable improvements in the quality of the wave model outputs when using wind fields with assimilation (WRF-VARFC). Further investigations revealed that during periods of increased wind intensity (e.g., due to mountain gaps), an accurate spatial structure of winds is significantly important to properly simulate the waves associated with those events in the Red Sea. Using WRF-generated reanalysis high-resolution wind fields, Langodan et al. (2016b, 2017b) forced WW3 to simulate the wave conditions in the Red Sea. Model validations using KAUST buoy observations and altimeter-derived measurements allowed for demonstrating the model relevance for simulating the Red Sea wave conditions. Later, Langodan et al. (2018) improved the wind hindcasting by producing high-resolution wind fields using WRF-ARW with data assimilation. These generated wind fields have been used to force the wave model WW3. Comparison of wave model results with observations obtained by KAUST buoy and satellite altimetry (ENVISAT, JASON-1, JASON-2, ERS-1, ERS-2, CRYOSAT-2 and ALTIKA) revealed good agreement between the model results and the observations.

Apart from ECMWF and WRF winds, researchers also utilized CFSR winds for the prediction of waves in the Red Sea. Shanas et al. (2017a) applied the WW3 model to simulate Red Sea wave conditions using CFSR and ERA-Interim as forcing wind fields. Comparing the model results with KAUST buoy observations revealed a slightly better wave model performance when forced using the CFSR wind field than when forced using the ERA-Interim wind field. Shanas et al. (2017b) and Aboobacker et al. (2017) also used CFSR wind to force WW3 to simulate Red Sea wave conditions. To determine the model performance, they compared the model results with KAUST buoy observations and satellite-based wave measurements. The obtained reasonable agreement for the wave heights among altimeters, buoy and the model indicated the validity of the wave model (WW3) simulation of the wave conditions in the Red Sea. More recently, Alsaaq and Shamji (2022) used ERA5 wind to force the wave model (WW3) to simulate Red Sea wave conditions. For model validation, they compared the model results at the KAUST buoy location with the actual buoy observations. Their model validation statistics (i.e., bias, root mean squared error, scatter index and correlation coefficient) were almost equal to those obtained by Shanas et al. (2017b), despite the fact that the two studies used totally different wind forcings.

3.1.3 SWAN model

Zubier et al. (2009) used SWAN to numerically simulate the wave conditions in the Red Sea for several days in the summer of 2006. The relatively coarse-grid Navy Operational Global Atmospheric Prediction System (NOGAPS) winds were used to force the numerical model. The wave model outputs were compared with conventional wave measurements collected by Zubier et al. (2007), as well as remotely obtained wave measurements (ENVISAT, JASON-1). Their results showed fair agreement between the model outputs and the conventionally and remotely obtained wave measurements, with some discrepancies that were attributed to the accuracy as well as the low spatial and temporal resolutions of the forcing wind fields. The SWAN model was also used by Gharbi (2012) to simulate the wave conditions in the Red Sea from February through May 2010, applying three different wind fields: NOGAPS, Global Forecast System-3-Analysis (GFS3) and Global Forecast System-4-Analysis (GFS4). Through comparisons of model outputs and wave observations obtained by the KAUST buoy and along 36 ENVISAT tracks, it was clearly shown that the simulation forced by the relatively higher spatial resolution GFS4 wind fields outperformed the two simulations forced by the other two wind fields of lower spatial resolution (GFS3 and NOGAPS).

Ralston et al. (2013) used the SWAN wave model coupled with the Advance Circulation Model (ADCIRC) to simulate the Red Sea wave conditions over a period of almost two years between 2007 and 2009. With the option of an unstructured computational grid, the model was forced with two wind fields having different spatial resolutions, which were generated using the weather research and forecasting (WRF) model. The wave model performance was evaluated by comparing its results with observations from the buoy (KAUST) and five different satellite altimetry missions. Their results clearly showed that when forced with the coarser resolution wind field (30 km grid size), the wave model underestimated the wave heights in the central Red Sea while the model with 10 km wind resolution performed well. This indicated that the spatial resolution of the forcing wind fields has significant impact on the quality of the model results.

Fery et al. (2012) tested the sensitivity of a SWAN-based Red Sea wave model through a two-month (March–April 2010) simulation of the wave conditions. The comparisons of the model results with the wave observations by the KAUST buoy revealed the model's high sensitivity not only to the input wind fields but also to the wave growth formulations, including wave energy dissipation by white-capping. Later, Fery et al. (2015) used the SWAN model (implemented in Delft3d software by Deltares) to simulate the Red Sea wave conditions from February 2010 through January 2013. The model forcing wind fields were obtained from the reanalysis cycle of the global meteorological model of the German Weather Service (DWD), which has a range of spatial resolutions between 20 km (after 2012) and 60 km (before 2008). In order to test the model's sensitivity to wave growth by wind formulation (based on the preliminary investigation), two separate model simulations were made, each with a different wave growth formulation. For model validation, they used observations obtained from the KAU wave rider, KAU buoy and KAUST buoy, as well as altimeter wave measurements from the ENVISAT and

JASON-2 satellites. The comparison of model outputs with the observations revealed a significant difference in terms of wave mean directions at the KAUST buoy location that was attributed to the inaccuracy of the forcing wind fields. The results also showed that better agreement between the model results and the observations at the KAUST buoy location is obtained when the wave growth formulation by JANSSEN (Janssen 1991) is used. However, at the KAU buoy and wave rider buoy locations, better agreement is obtained in terms of wave height when the KOMEN wave growth formulation (Komen et al. 1984) is used. The wave period is better predicted using the JANNSEN formulation. These results suggested that there are regional consequences for different formulations depending on the fetch, bathymetry and available wind field, and the model sensitivity must be tested before carrying out the actual hindcasting.

Albarakati and Aboobacker (2018) used two SWAN model setups to simulate the wave conditions in the Red Sea over a period of 14 months (June 2005–August 2006). The two model setups differ in terms of the spectra cut-off frequency and frequency discretization, such that one of the setups is exactly similar to those of the measured spectra used in the validation while the other is not. For model validation, they used parametric and spectral wave data collected near shore by KAU at two locations during the summers of 2005 and 2006 using two wave gauges (Zubier et al. 2007). The two model simulations were forced by the same hourly wind fields obtained from the climate forecast system reanalysis (CFSR). In terms of wave parameters, their results showed that model-obtained significant wave heights were slightly overestimated during low wave conditions and underestimated during high wave conditions while mean wave periods were slightly overestimated during high wave conditions. They suggested that these inconsistencies were attributable to the differences between observed and model-obtained wave spectra, which, in turn, might be related to the low cut-off frequency of measured spectra and apparent irregularities in the input coastal bathymetry data of low resolution. In terms of spectra, the comparison between the model outputs and the observations showed small inconsistencies in the high-frequency range that were attributed to the coarse spatial resolution of the forcing wind field.

Zubier and Gharbi (2019) studied the sensitivity of a SWAN-based Red Sea wave model to the spatial and temporal resolutions of the forcing wind fields. The reanalysis wind fields obtained from the global forecast system (GFS4), originally at a spatial-resolution of 0.5° and a temporal-resolution of three hours, were used to create three additional wind fields of different spatial and temporal resolutions (0.5°, 6 hrs; 1°, 3 hrs; 1°, 6 hrs). The Red Sea wave model was forced by the four wind fields of different resolutions to simulate the wave conditions in the Red Sea during spring 2010, and the results were compared with KAUST buoy measurements and ENVISAT observations. Comparisons showed that when forced with coarser spatial resolution winds, the wave model significantly underestimated the higher wave heights. Such an underestimation was not detected when forcing the model with winds having lower temporal resolution, indicating that, in terms of wave parameter (significant wave height), the Red Sea wave model is more sensitive to spatial resolution and less sensitive to temporal resolution of the forcing winds. Unlike these results, the results in terms of frequency and directional wave spectra showed that model outputs are significantly affected not only by the spatial resolution but also by the temporal resolution of the wind fields.

Beyramzade et al. (2019) used SWAN, with two different model setups and three different forcing wind fields, to simulate the Red Sea wave conditions over the period of one year (2010). The two model setups, which differed only in terms of the applied wave growth by wind formulation (KOMEN vs. JANSSEN), were forced by cross-calibrated multi-platform (CCMP), ERA-interim, and ERA5 wind fields. The results were compared with wave observations obtained using the KAUST buoy and different satellites (CRYOSAT, ERS-2, ENVISAT, JASON-1 and JASON-2). The comparisons showed that optimal agreement was obtained when ERA5 winds were used with the JANSSEN wave growth formulation. More recently, the coupled models (SWAN and ADCIRC) were used by Langodan et al. (2020) to simulate the wave conditions in the central Red Sea with the option of unstructured grid and WRF wind forcing for a period of 30 years. The model validation was carried out using nearly three months of wave observation by the two KAEC buoys. The results indicated that while the wave model showed reasonable performance, a few events were not very well predicted due to the rapid changes in wave heights.

Most of these studies, whether using SWAN, WW3 or WAM, have clearly shown that the quality of the outputs of these numerical wave models depends mainly on the quality of the forcing wind field. Nevertheless, these numerical wave modeling studies have added a great deal of knowledge regarding waves in the Red Sea, including their physical characteristics, spatial and temporal variabilities and potential as energy resources. In the following sections, more details are given regarding the major findings of the previous studies.

4.3.2 Wave variabilities

Abdelrahman (1993) suggested seasonal variability in the wave parameters in the eastern-central Red Sea, which was derived by applying an empirical formula using one-year wind measurements from a nearby land-based weather station. Numerical wave modeling, introduced in the previous section, allowed for conducting long-term wave hindcasting studies for the entire Red Sea. The advanced computational resources represented by the high-performance computers (HPCs) at both KAUST and KAU gave the researchers at the two universities great opportunities to widen their research in this field. The Red Sea wave hindcasting results were utilized in investigating the wave variability for the whole Red Sea as well as for selected nearshore locations, and conducting extreme value analysis at these locations. Assessment of wave energy potentials for the entire Red Sea and in selected regions became possible through the utilization of these wave hindcasting results.

Through the application of a calibrated numerical wave model, Fery et al. (2012) were able to assess the seasonality of the regional wave characteristics in the Red Sea that were attributed to the seasonality of the regional wind patterns. Their results suggested that the central Red Sea experiences southeast-propagating higher and longer waves during the spring season while during the fall season, it experiences smaller and shorter opposing waves propagating from both the southeast and the northwest.

Based on the results of Red Sea wave simulations over a period of two years, Ralston et al. (2013) attributed the seasonal variabilities to the monsoonal winds. According to their results, the NW winds that blow over the entire Red Sea during summer generate smaller waves than in winter. On the other hand, the SE monsoon wind that blows over the southern part of the Red Sea during winter can generate waves with mean significant waves that exceed 2 m in height, while the waves in the northern part of the Red Sea remain smaller due to the weak NW wind that blows over the northern part. Furthermore, the seasonal variability in the waves at the central part of the Red Sea are mainly related to the mountain-gap effect that is well pronounced during the summer. Mountain wind jets from the westerly located Tokar Gap blow eastward over the central part of the Red Sea and generate high waves in this region, with maximum wave heights that reach up to 14 m, while the other regions of the Red Sea are relatively calm. The northern part of the Red Sea also experiences larger-than-usual waves during fall and winter due to the effects of smaller mountain gaps that are located along the northeast coast. According to Langodan et al. (2015), the two opposing wave systems that are forced by opposite winds in the northern and southern Red Sea and converge at the central region create a challenge for numerical modeling and, consequently, for any model-based wave climatology study. Langodan et al. (2016b) pointed out the consistency between the wave height and wind variabilities in the Red Sea. The results of their 18-year wave hindcast by a validated numerical wave model indicated that, during summer, the northwesterly winds generate energetic wave conditions in the northern Red Sea while the southeasterly wind generates energetic wave conditions in the southern Red Sea during winter. They suggested that these energetic waves in the northern and southern Red Sea might sometimes have the potential to progress beyond the central Red Sea.

The long-term variabilities of waves in the Red Sea have been studied by Shanas et al. (2017b) by simulating wave conditions over a 32-year period using a validated wave model. Their results showed a negative trend in the significant wave height in the northern Red Sea during summer that was attributed to the reduction in intensity of extreme events during this period. They have attributed the significant positive trend found in the southern Red Sea during winter and pre-summer to waves that are generated in the central Red Sea by extreme wind events during this period and propagate southward as swells. They have also attributed the negative trend in the significant wave height found in the central Red Sea during summer to the reduction in the intensity of mountain-gap jets. Generally, their seasonal trend analysis results indicated that the wave height trends in the Red Sea are mainly influenced by regional factors, including the monsoon and mountain-gaps, that vary seasonally. In terms of climatic indices, both the North Atlantic Oscillation (NAO) and East Atlantic/Western Russia (EAWR) indices correlated positively with the wave heights in the northern Red Sea and negatively in the southern Red Sea while, for El Nino, the correlation with the wave heights was exactly the opposite. The central Red Sea area was found to act as a barrier between the NAO and El Nino-Southern Oscillation (ENSO) influential regions, suggesting possible contributions to climatic indices by the mountain-gap winds and the wind convergence. However, the authors pointed out that further investigations are necessary to understand and quantify the regional contribution to the global climate.

Based on a 30-year hindcast, the two-part study of the Red Sea Climatology – Part 1: Winds (Langodan et al. 2017a) and Part 2: Waves (Langodan et al. 2017b) – identified four main wind systems in the Red Sea (each with a distinctive pattern) that characterized the corresponding wave climatology. The prominent wave system that occurs year round in the northeast consists of southward propagating high waves (with decreasing heights) that are associated with southward-propagating relatively cold air entering from the Mediterranean Sea. From October through April, this system is opposed by the northward-propagating wave system associated with monsoonal winds in the south. Like the second system, the third wave system is also seasonal because it occurs only from July through September, and it is associated with the wind that blows eastward through the Tokar Gap and generates the highest waves in the Red Sea. The fourth wave system is non-seasonal, and it occurs near the northeastern coast in association with the narrow valley wind jets that blow westward. The decreasing trend in wave heights of the Red Sea that was identified in the study was stronger in the first system, milder in the second system and slightly present in the third system. To further investigate the wave trends at the different wave systems, Langodan et al. (2018) used wave spectra partitioning for better definition of the characteristics and trends of each wave system among the four that were identified, again through a validated model simulation of waves for 30 years in the Red Sea. One of the major findings of the study was that the first wave system is present throughout the year (although at variable levels) while the other three systems are distinctly seasonal, with the second and the fourth systems occurring in the winter and the third system occurring in the summer. The other major finding was that, in terms of both total energy and number of occurrences, the most significant decreasing trend is associated with the first system while a milder decreasing trend is found for the second system, and no marked decreasing trend is found for the third and fourth wave systems.

Analysis of KAUST buoy wave measurements made by Shanas et al. (2017a) indicated a multi-directional field that consists of locally generated waves and the waves propagated from distant northern areas (swells), whose interaction leads to superimposition on diurnal and monthly cycles. On a daily basis, these multi-directional waves were found in the early morning and from afternoon to midnight. Results also showed that the periods from March through June and from August through November are dominated by the superimposition of those multi-directional waves. Through the utilization of one year of simulated wave conditions by a validated numerical wave model, the analysis was extended to cover the whole Red Sea. Spatially, it was found that only 28% of the Red Sea area is dominated by superimposed waves while 58% of the area is dominated by non-superimposed waves, and 14% of the area is characterized by the coexistence of both superimposed and non-superimposed waves. Temporally, the percentage of the area of the Red Sea that was dominated by superimposed waves was found to vary between 14% (in March) and 36% (in May) while the percentage of the area with coexistence varies between 13% (in June–August) and 19% (in January and February). The superimposition of locally generated waves and the swells propagating from the northern Red Sea was also identified along the Jeddah coast by Shanas et al. (2018) through the analysis of measured waves, including 2-D spectra, that were collected during two consecutive summer seasons (Zubier et al. 2007). Albarakati and Aboobacker (2018) extended this investigation further by simulating the wave conditions off Jeddah over a period of one year using a validated numerical wave model. Their results allowed for the identification of multi-directional swells that propagate to the Jeddah nearshore area from the northern and southern Red Sea. Their analysis of diurnal variability showed that in the early morning, swells are dominant while locally generated waves are absent, whereas during the evening hours, superimposition of the swells and the wind seas occurs. The results also indicated the winter dominance of swells over locally generated waves in the Jeddah nearshore area.

Shamji et al. (2020) analyzed the wave climate around Farasan Island (southern Red Sea) using a 32-year wave hindcast generated by a validated wave model. Their results indicated a contrasting behaviour of seasonal waves at its north and south, such that the northern side of the island experiences a relatively higher contribution of the northerly waves while the southern side experiences a relatively higher contribution of the southerly waves. Their estimation of seasonal return values showed that the wave height return value for winter is higher than for summer due to the effect of the strong monsoon-driven SSE wind that blows during winter. Langodan et al. (2020) used a 30-year wave hindcast generated by a validated wave model to determine the wave climate in the nearshore area of KAEC (central Red Sea). Their results showed that extreme waves occur in this region more often from February through June than in other months of the year. Using a generalized Pareto distribution (GPD) with peak over threshold (POT) method, the highest estimated return value of wave height offshore KAEC was 4.36 m, for a 100-year return period. Alsaaq and Shamji (2022) investigated the wave climate in the nearshore area of Jeddah using a 39-year wave hindcast generated by a validated wave model. Their results indicated that, in comparison to other seasons, the Jeddah nearshore area gets exposed to slightly higher waves during winter. The seasonal trend investigation showed that Jeddah nearshore extreme wave conditions have a positive trend during the winter, pre-summer and summer seasons and a significant negative trend during the pre-winter season. Using the GPD-POT model, their highest estimated wave height return value offshore Jeddah was 3.44 m over a 100-year period.

Langodan et al. (2016b) utilized an 18-year wave hindcast generated by a validated wave model to assess the wave energy potential in the Red Sea. Their results suggested inhomogeneous distribution of wave energy with the concentration of energy at specific areas that have specific atmospheric conditions. Their monthly variability investigation showed that higher mean wave power occurs in the southern Red Sea from November through April while higher mean wave power occurs in the northern Red Sea from February through September. The central Red Sea experiences higher mean wave power from June through August. Based on their analysis of the wave energy variation, they concluded that the Red Sea has no potential sites from which wave energy can be harvested because the amount of wave energy in the Red Sea is not sufficient for harvesting using the currently available wave energy converters. However, they suggested further investigation should be carried out in the future using a multi-decadal hindcast with a higher spatial resolution and considering the efficiency of a hybrid wind-wave energy system. The same conclusion and recommendation were reached by Alkhayyat et al. (2020), whose study was more focused on the northwestern Red Sea. Their results suggested two potential sites for installing a wave energy converter (WEC) in the Gulf of Aqaba that could provide efficient power only during the month of September. Aboobacker et al. (2017) conducted a wave energy resource assessment in the Red Sea using a 32-year wave hindcast carried out by a validated wave model. Their analysis of the estimated wave power on monthly, seasonal and an interannual basis showed significant spatial and seasonal variations in wave power. Their results also indicated that the deep waters of the Red Sea have a distinct amount of wave power that varies seasonally from higher in the winter to lower in the summer. The highest winter mean wave power has been found along the coasts of Yemen, Egypt, Sudan and Eritrea, while the highest summer mean wave power was found at the coast of Saudi Arabia. The apparent spatial and temporal variabilities of the wave power in the Red Sea were attributed to different prevailing wind systems, including mountain-gap winds and sea breezes. Interannual variabilities in mean wave power in the Red Sea were found to be mainly influenced by the ENSO and Indian Ocean dipole (IOD) climatic modes. Analysis of the interannual trends of the mean wave power showed a positive trend at 12 selected locations with an average increase of mean wave power of 20% over the 32-year period.

4.3.3 Wave forecasting

Most of the studies discussed here have used wave hindcasting obtained using validated numerical wave models to study the variability in the Red Sea wave conditions. However, their findings are essential for developing and improving operational wave forecasting systems for the Red Sea. In an early preliminary study, Zubier et al. (2009) developed an experimental SWAN-based wave forecasting system for the Red Sea. Later, Mayerle et al. (2016) developed a SWAN-based operational wave forecasting component as a part of a coastal information system for the Red Sea. This system, which was developed through scientific collaboration between KAU, Ifm-Geomar, the University of Kiel and GTZ, became operational during the last few months of the collaboration and continued working for several months afterward before it stopped working for technical reasons. At KAUST, Hoteit et al. (2021) have developed an integrated high-resolution data-driven regional analysis and prediction modeling system for the Red Sea that contains 1 km–50 m resolution wave models. Development and validation of long-term Red Sea regional atmospheric-oceanic wave reanalysis and forecasting capacities are among their achievements.

Two recent studies by Zubier (2020) and Miky et al. (2021) have used a totally different approach to wave forecasting in the central Red Sea that is based on artificial neural networks (ANNs). Zubier (2020) used the nonlinear auto-regressive network with exogenous inputs (NARX) to develop two models that provide wave forecasting over a forecasting horizon that extends from 3 to 24 hours. The two NARX-based models differ only by the inclusion of wind and wave directional difference as an input for one model (B) and not for the other (A). Through comparison with KAUST buoy actual wave observations, the results showed that the two models have predicted significant wave heights with a decreasing performance over the forecasting horizon; however, the higher performance shown by model (B) indicated the importance of inclusion of the directional difference input term. Miky et al. (2021) developed a new ANN-based approach (RNNARX) based on the integration through a sequential system of recurrent neural networks (RNN) and NARX that was applied to forecasting the wave conditions in the central Red Sea over a forecasting horizon that extended for five days. The performance comparison showed that RNNARX outperformed RNN and NARX in wave height predictions for high lead times, indicating that RNNARX can be suitable for making five-day wave predictions.

4.4 CURRENT STUDY

In this section, the relatively new ERA5 wave reanalysis data have been utilized in a brief analysis to determine some of the Red Sea wave statistics. Since September 2019, the European Centre for Medium-Range Weather Forecast (ECMWF) has replaced its older ERA-interim reanalysis with the new ERA5 (www.ecmwf.int/en/forecasts/datasets/reanalysis-datasets/era5) that provides hourly estimates of a large number of atmospheric and oceanic variables over an extended period of 72 years (1950–2021). In two recent Red Sea wave studies, ERA5 wind analysis data have been utilized as forcing wind fields for the SWAN wave model (Beyramzade et al. 2019) and for the WW3 wave model (Alsaaq and Shamji 2022). It is worth mentioning here that the ERA5 wind reanalysis data is provided with a higher spatial resolution (0.25° x 0.25°) than the ERA5 wave reanalysis data (0.5° x 0.5°). However, in this section, the ERA5 wave reanalysis data was used to investigate the variability of the Red Sea wave conditions.

4.4.1 Data validation

Figure 4.2 shows the 167 Red Sea grid points for which ERA5 provided hourly wave reanalysis data over the 72-year period. For validation purposes, KAUST buoy measurements from October 2008 through September 2010 were compared with the ERA5 data at the nearest grid point to the buoy location. The comparisons shown in Figures 4.3 and 4.4 were made in terms of significant wave height, mean wave period and peak wave period. In terms of significant wave heights, Figure 4.3 shows a remarkable agreement in the pattern between ERA5 wave data and KAUST observed waves. Both Figures 4.3 and 4.4a show some underestimation by ERA5, mainly at the higher wave height range. The statistics given in Table 4.1 indicate a slightly better agreement between ERA5 wave data and KAUST buoy wave observations than those obtained between KAUST buoy observations and numerical wave model results in the study by Alsaaq and Shamji (2022) using ERA5 forcing wind fields and the study by Shanas et al. (2017b) using CFSR forcing wind fields. The comparisons between ERA5 data and actual wave measurements by the KAUST buoy, shown in Figure 4.4, also indicate a better agreement in terms of the mean wave periods than in terms of the peak wave periods. The validation has been extended further by comparing significant wave heights obtained from ERA5 with those obtained from 14 satellite altimetry missions (CRYOSAT, ENVISAT,

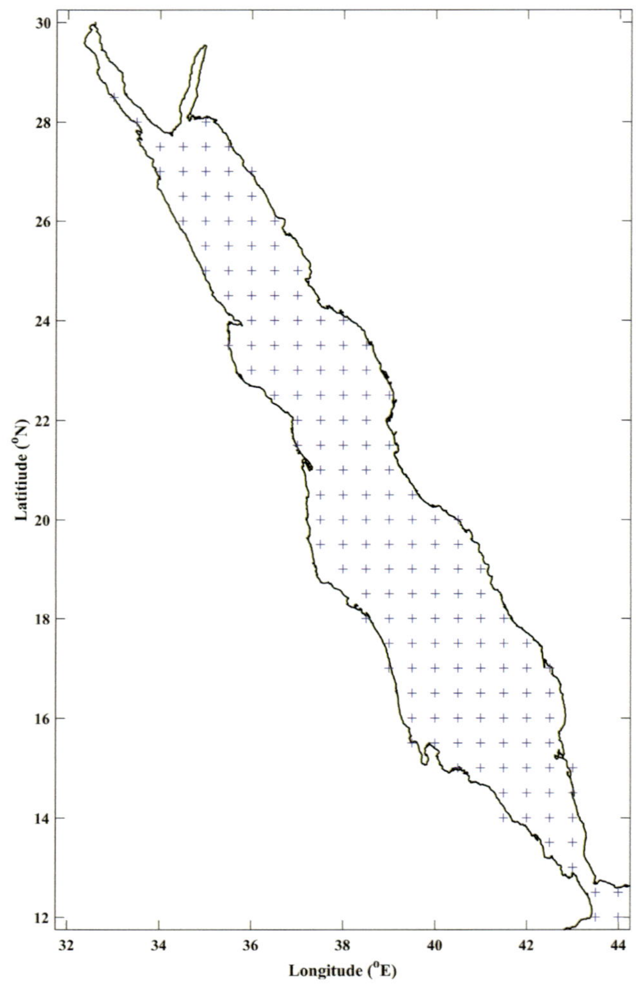

FIGURE 4.2 ERA5 wave grid points in the Red Sea.

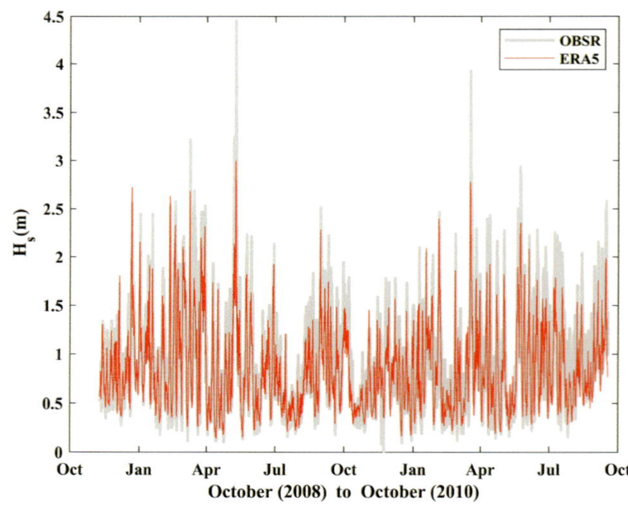

FIGURE 4.3 Comparison between ERA5 significant wave height and KAUST buoy significant wave height.

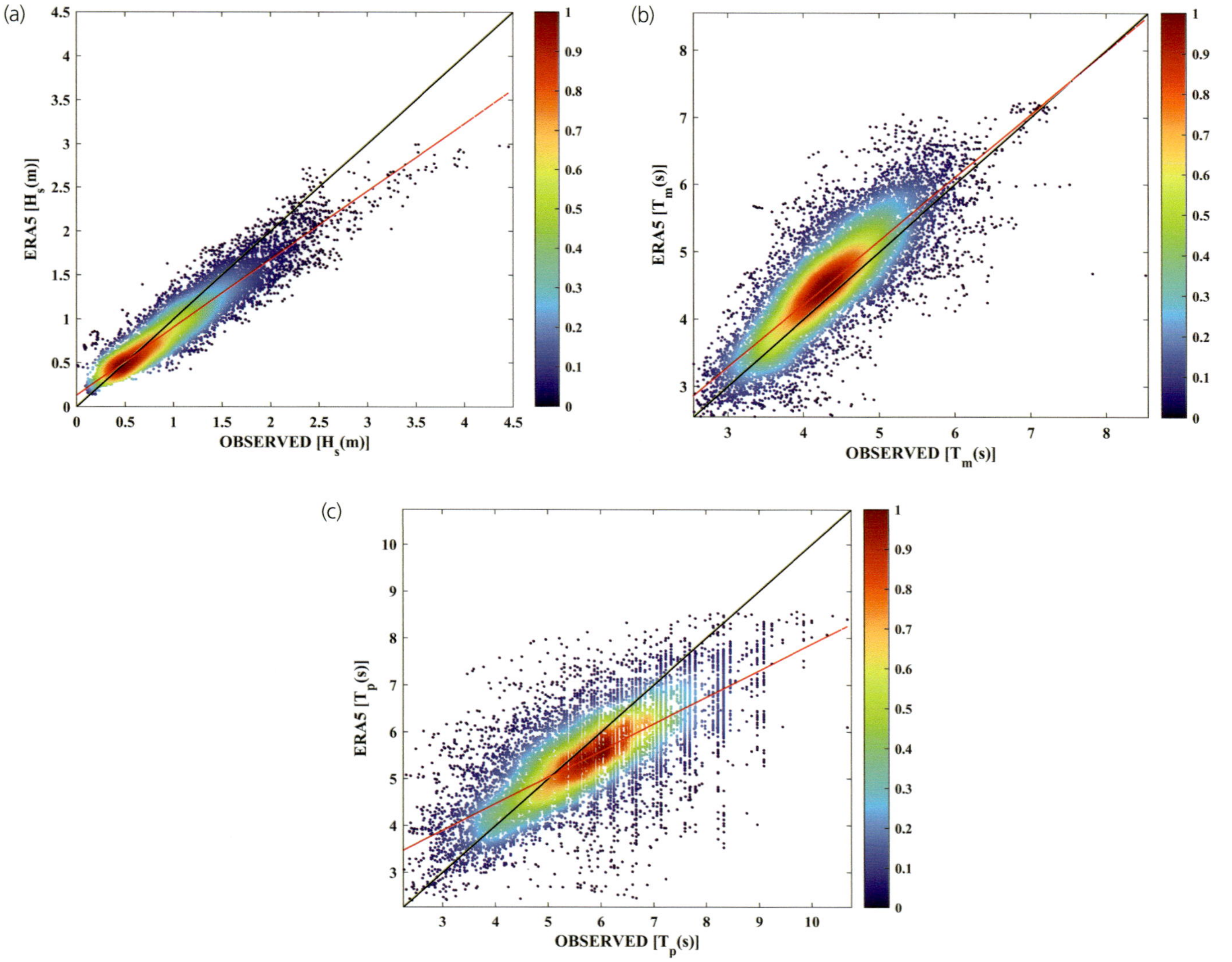

FIGURE 4.4 Scatter plots of ERA5 vs. observed significant wave heights (a), mean wave periods (b) and peak wave periods (c). The black line represents 45° line (ERA5 = observed), the red line is obtained through linear regression fitting, and the colour bar scale represents normalized intensity of scatter points.

TABLE 4.1 Error statistics of ERA5 wave parameters vs. observed wave parameters by KAUST buoy. Also given are the statistics estimated by Shanas et al. (2017b) and Alsaaq and Shamji (2022).

	STATISTICS			
WAVE PARAMETERS	*BIAS*	*CORRELATION COEFFICIENT*	*ROOT MEAN SQUARE ERROR*	*SCATTER INDEX*
H_s (m) [This Chapter]	−0.08	0.94	0.22	0.20
H_s (m) (Shanas et al. 2017b)	−0.08	0.91	0.26	0.24
H_s (m) (Alsaaq and Shamji 2022)	0.08	0.90	0.25	0.25
T_m (s)	0.20	0.83	0.48	0.11
T_p (s)	−0.31	0.75	0.90	0.16

ERS-1, ERS-2, GEOSAT, GFO, HYS, JASON-1, JASON-2, JASON-3, SARAL, SENTINEL-3A, SENTINEL-3B and TOPEX), which were made available by the Integrated Marine Observing System (IMOS) through the Australian Ocean Data Network (AODN) portal (https://imos.org.au/facilities/aodn). The comparisons in Figure 4.5 show consistent overestimations of lower wave heights by ERA5 and underestimations of higher wave heights among all 14 missions. The statistics for the comparisons pertaining to each satellite altimetry mission as well as the combined satellite data (also shown in Figure 4.5) are given in Table 4.2. The range of the correlation coefficient, between 0.69 and 0.87, indicates a reasonable agreement in the patterns between the ERA5 and the satellite-obtained wave heights, despite the apparent under- and overestimations shown in Figure 4.5. In general, the validation procedures carried

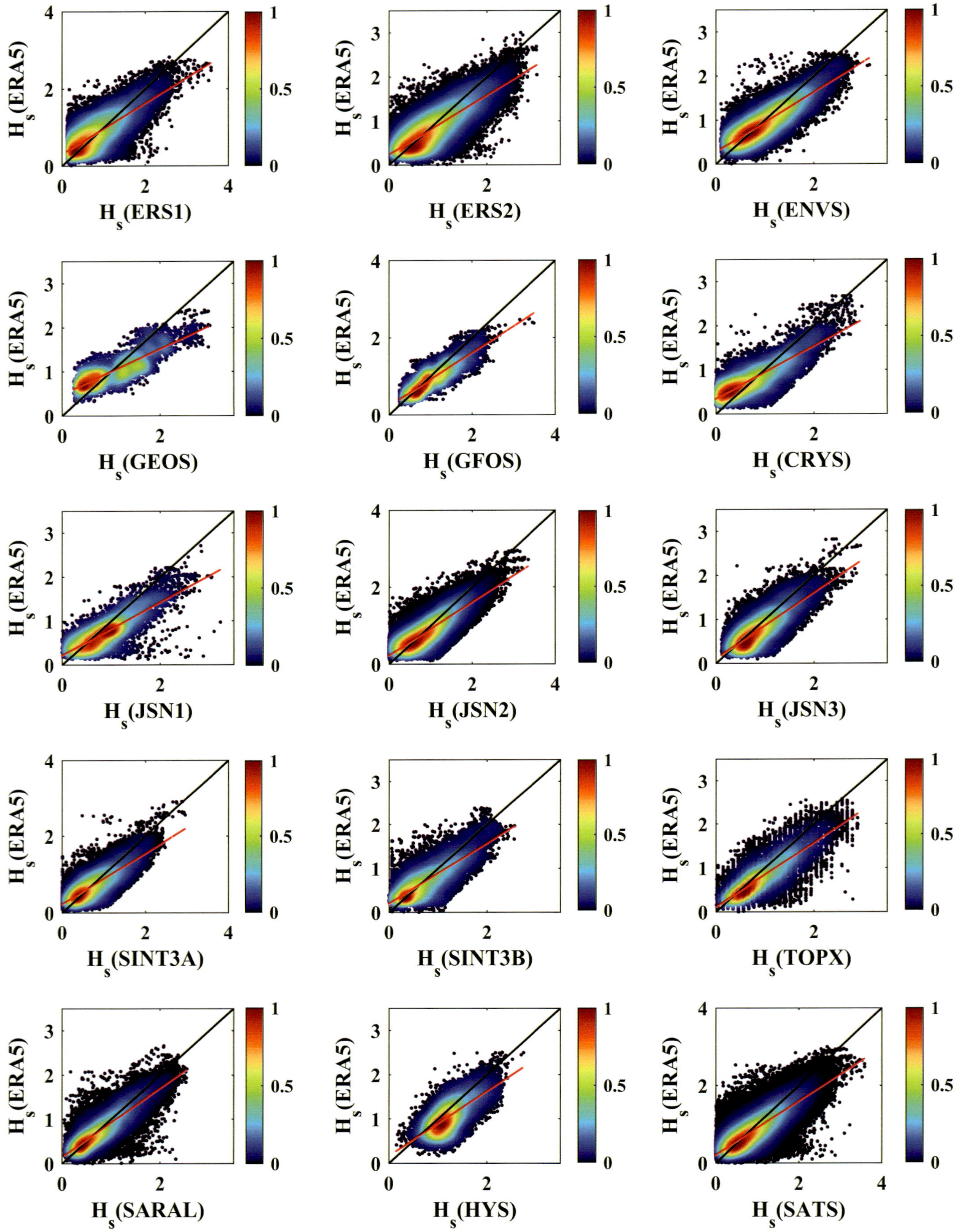

FIGURE 4.5 Scatter plots of ERA5 significant wave heights vs. remotely obtained significant wave heights by 14 satellite altimetry missions (ERS-1, ERS-2, ENVISAT, GEOSAT, GFO, CRYOSAT, JASON-1, JASON-2, JASON-3, SENTINEL-3A, SENTINEL-3B, TOPEX, SARAL and HYS) separately and altogether (SATS, in the bottom right panel). The black line represents 45° line (ERA5 = observed), the red line is obtained through linear regression fitting, and the colour bar scale represents normalized intensity of scatter points.

TABLE 4.2 Error statistics of ERA5 significant wave heights (m) vs. remotely obtained significant wave heights (m) by 14 satellite altimetry missions separately and all together.

SATELLITES	*BIAS*	*CORRELATION COEFFICIENT*	*ROOT MEAN SQUARED ERROR*	*SCATTER INDEX*
ERS-1	−0.0001	0.78	0.36	0.41
ERS-2	0.07	0.78	0.33	0.36
ENVISAT	0.07	0.84	0.29	0.28
GEOSAT	0.05	0.82	0.35	0.33
GFO	0.12	0.86	0.26	0.24
CRYOSAT	−0.02	0.82	0.32	0.41
JASON-1	0.14	0.83	0.31	0.34
JASON-2	0.07	0.81	0.29	0.32
JASON-3	0.13	0.79	0.29	0.31
SENTINEL-3A	0.05	0.80	0.28	0.36
SENTINEL-3B	0.05	0.80	0.29	0.37
TOPEX	0.15	0.87	0.29	0.31
SARAL	0.04	0.86	0.23	0.29
HYS	0.17	0.69	0.32	0.26
All together	0.08	0.81	0.30	0.33

TABLE 4.3 Monthly spatially averaged 99th percentile significant wave heights (m).

MONTH	*JAN*	*FEB*	*MAR*	*APR*	*MAY*	*JUN*	*JUL*	*AUG*	*SEP*	*OCT*	*NOV*	*DEC*
H_s (m)	1.913	1.927	1.906	1.736	1.501	1.553	1.421	1.355	1.342	1.294	1.533	1.732

out using actual and remotely obtained wave measurements gave satisfactory results, considering the fact that the ERA5 grid points do not necessarily coincide with the points of measurements along the satellite tracks, and the nearest ERA5 grid point (used in the comparison) to the KAUST buoy location is more than 18.5 km away.

4.4.2 Wave variabilities

The monthly 99th percentile (extreme) significant wave heights obtained for each grid point over the 72-year period were spatially averaged (Table 4.3). The maximum spatially averaged value of 1.927 m was obtained for February (winter) while the minimum spatially averaged value of 1.294 m was obtained for October (fall). The monthly exceedances of the 99th percentile (extreme) significant wave heights above the maximum spatially averaged value (1.927 m) were estimated for each grid point and are shown in Figure 4.6. From January through April (winter and post-winter), the plots show three distinct exceedances areas in the northern, central and southern Red Sea. Previous studies have shown and discussed this pattern, which is associated with the strong NW winds in the northern Red Sea, the even stronger SE winds in the southern Red Sea and the wind convergence at the central Red Sea. In Figure 4.6, the exceedance magnitudes experience significant drops all over the Red Sea in May, which is marked by the disappearance of the exceedance area in the southern Red Sea due to the monsoon-associated termination of the SE wind in this area. The absence of the exceedance in the southern Red Sea continues through summer and post-summer (June–September) before it returns in October and continues to develop through November and December (pre-winter and early winter). Figure 4.6 also shows that during summer (July and August), the only area showing a significantly high exceedance level is the central Red Sea, indicating the effect of the strong mountain-gap winds that blow from the west over the central Red Sea area during the summer season.

Trend analysis, through estimating Sen's slope and testing of significance (Sen 1968; Kendall 1975; Hess et al. 2001), was also conducted on the 72-year ERA5 Red Sea wave data. Figure 4.7 shows the spatial distribution of the annual trend of the 99th percentile significant wave heights for the entire Red Sea. However, the trend analysis shows that the statistically significant trends (at the 95% confidence level) are limited to the central Red Sea as well as the northeastern and southwestern nearshore areas. The positive trend in the central Red Sea indicates a possible increase in the role of the westward-blowing strong mountain-gap winds on the waves in this area over the years. The influence of the westward-blowing mountain-gap winds in the northeastern Red Sea seems to be increasing as well, based on the positive trend in that area. Figure 4.8 shows the spatial distribution of the monthly trends of the 99th percentile wave heights for the 72-year period. The trend analysis shows that the statistical significance of the monthly trends varies spatially and temporally; however, only statistically significant trends (at the 95% confidence level) were considered. In the central Red Sea, the 99th percentile significant wave heights show an increasing (positive) trend during February, while a decreasing (negative) trend is shown in August at the southern part of the central Red Sea. The 99th percentile significant wave heights in the northern Red Sea have two different phases: negative in June and positive in December. For the southern Red Sea, the trends are exactly the opposite of those

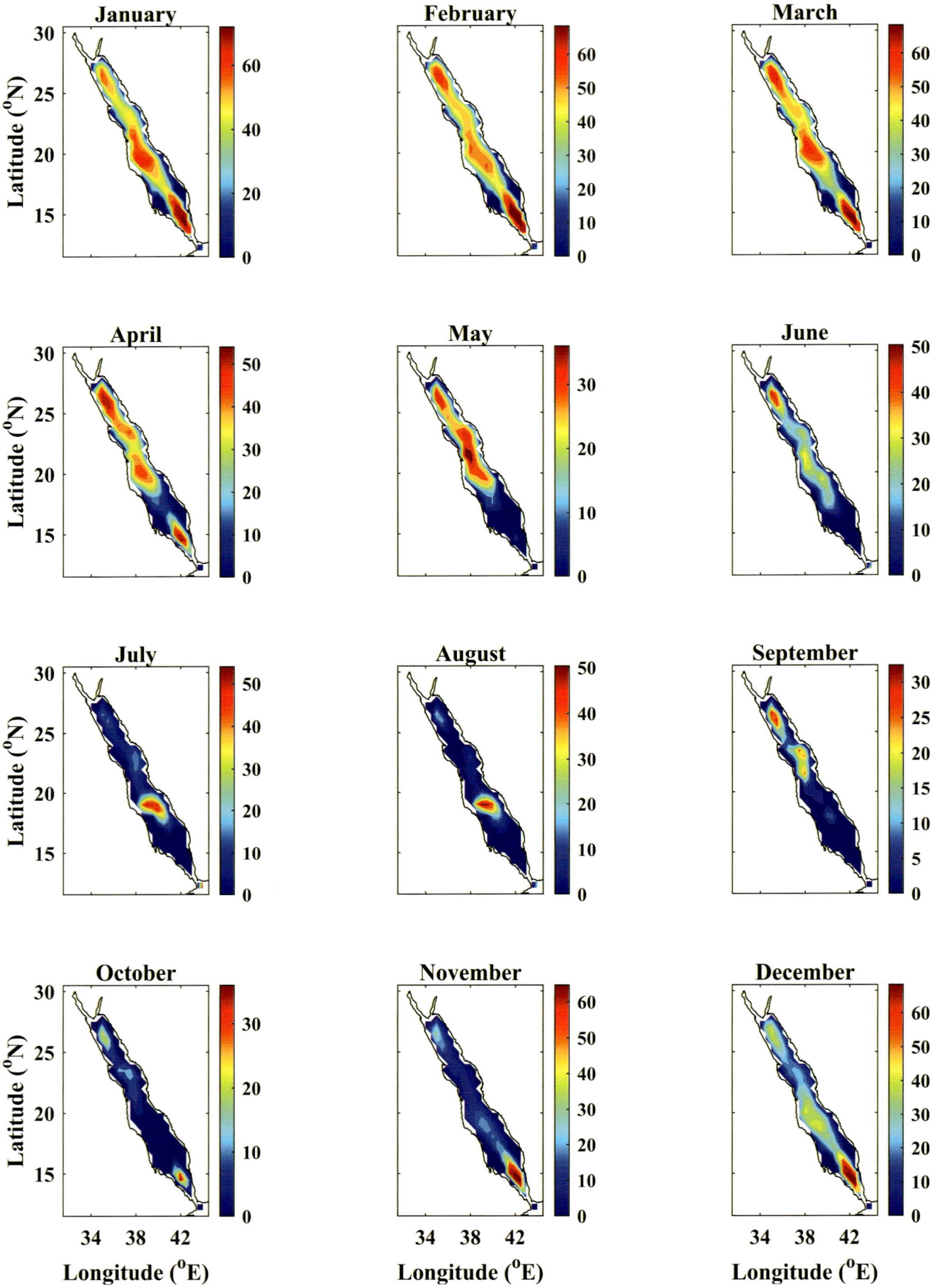

FIGURE 4.6 Spatio-temporal variability of exceedances of 99th percentile significant wave heights. Colour bar scale represents number of exceedances.

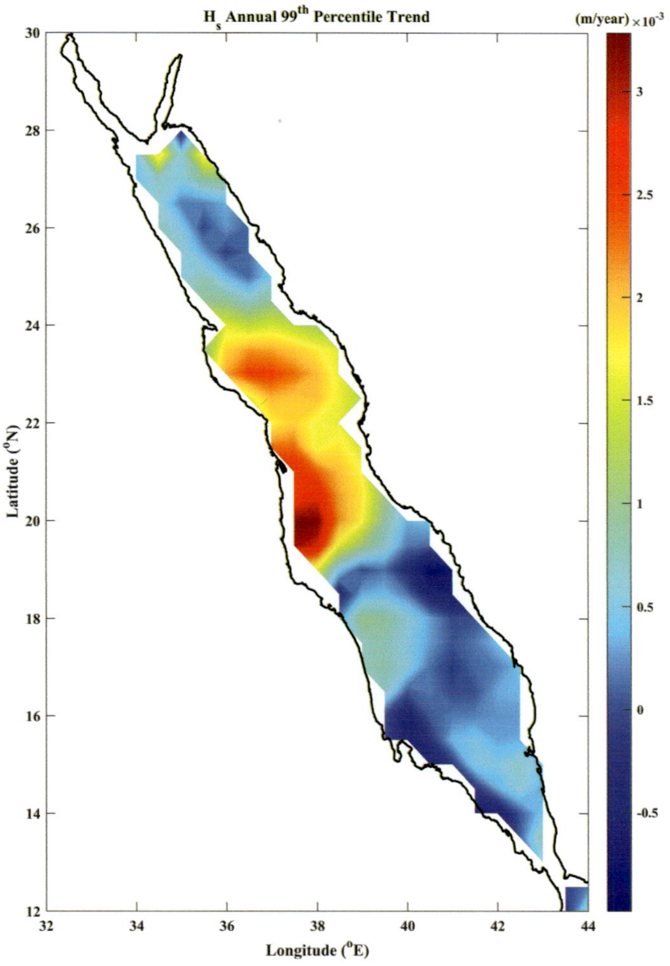

FIGURE 4.7 Spatial variability of annual trend of 99th percentile significant wave heights (m/year).

in the northern Red Sea in June and December. These results indicate that the wave height annual and seasonal trends in the Red Sea are mainly controlled by the spatially and temporally varying winds in this region.

The GPD-POT method (Coles 2001) was applied to conduct an extreme value analysis on the ERA5 significant wave height data using the "ismev" R-package (Stephenson 2018). The threshold selection algorithm within the "ercv" R-package (del Castillo and Padilla 2016) was applied to determine the optimum threshold value for every single grid point based on its 72-year hourly wave height record. For verification purposes, Figure 4.9 shows diagnostic plots for the GPD maximum-likelihood (ML) fit (based on the optimum threshold) for the nearest grid point to Jeddah City. Both the probability and quantile plots show remarkable agreement between the empirical and the model values. Similar agreements are seen in the return level plot between the actual data and the estimated return levels and also in the density plot between the fitted density and the histogram of the data values above the selected threshold. Such agreements clearly indicate a highly satisfactory fitting using the applied GPD-POT method. The estimated return levels for the 10-, 20-, 50- and 100-year return periods for the nearest grid point to Jeddah City are shown in Table 4.4, as well as the return levels for the same return periods estimated by Alsaaq and Shamji (2022) at a nearby point using the same method. Although their study was based on wave data obtained using a 39-year hindcasting by the WW3 model that was forced with the high-resolution ERA5 wind field, the agreement between the estimated return levels is quite notable. The spatial variation in the 100-year return level in the Red Sea, shown in Figure 4.10, indicates maximum return levels (that exceed 4 m in height) between the latitudes 18° and 20° N that can be attributed to the mountain-gap effect discussed earlier. Figure 4.10 also shows two distinct areas of low return values at both sides of the Red Sea around 16° N, where the return level drops to less than 1.5 m closer to the shoreline.

To estimate the mean wave power in the Red Sea, both ERA5 significant wave heights and ERA5 peak wave periods for the 72-year period were utilized. Figure 4.11 shows the spatial variation in the mean wave power for the entire 72-year period, indicating a distinctively higher mean wave power at the southern part of the Red Sea and two areas of relatively high mean wave power at the central and northern parts of the Red Sea. The spatio-temporal variability in the mean wave power, given in Figure 4.12, indicates that during winter, the high mean wave powers are limited to the southern part of the Red Sea, whereas the high mean wave powers in the central and northern Red Sea start developing in the late-winter/early-spring period. The high mean wave powers in the Red Sea central and northern areas reach their maximum in the early summer, when the high mean wave power in the southern part totally disappears. For most of the summer period, the high wave powers are confined to the central part of the Red Sea, due to the strong summer wind jets blowing through the westerly located mountain gap. This distinctive high mean wave power in the central Red Sea disappears totally in the month of September, during which the northern Red Sea experiences its highest mean wave power due to the strong westward-blowing wind jets. The disappearance of high mean power in the southern part lasts through the whole summer and early fall before it starts developing again in the mid-fall as winter starts approaching.

4.5 CONCLUDING REMARKS

Navigational safety and protection of coastal structures require adequate wave data that can be analyzed to resolve wave conditions and their spatial and temporal variabilities. The conventional wave measurements in the Red Sea are very limited, both spatially and temporally. The satellite-obtained wave measurements, although providing more spatial coverage, are temporally limited due to the long time span between repeated tracks. However, this spatially and/or temporally limited data has been extensively utilized in the validation/verification procedures of numerical wave models, which constitute the most practical and cost-effective methods to obtain the required wave information in the Red Sea. The accuracy of the outputs of these models depends mainly on the accuracy and the resolutions of the inputs, including the forcing wind fields and the bathymetry (in shallow waters). Wave hindcasting obtained through application of these validated wave models allowed the study of wave variability in the Red Sea over periods that

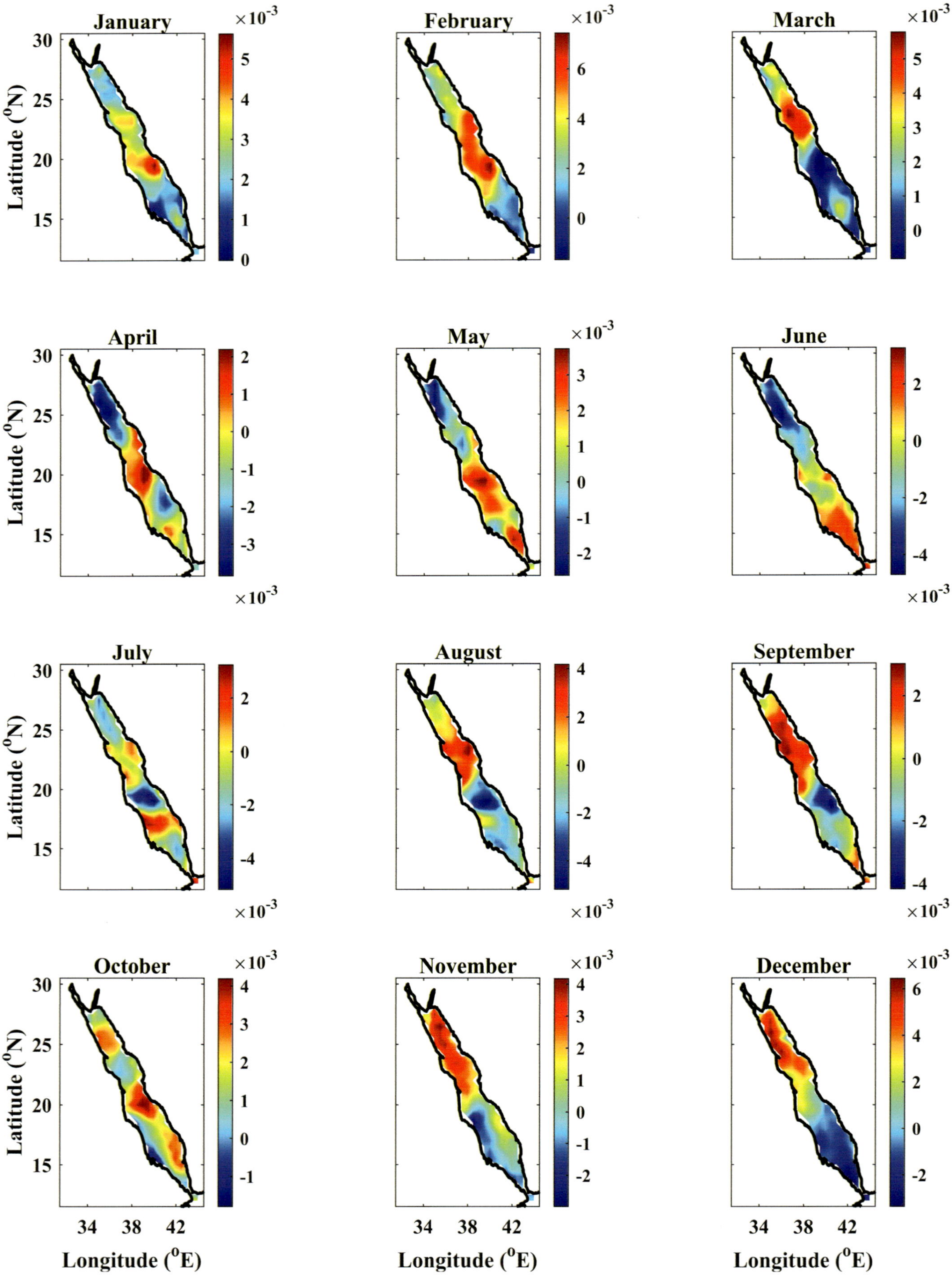

FIGURE 4.8 Spatio-temporal variability of trends of 99th percentile significant wave heights (m/year).

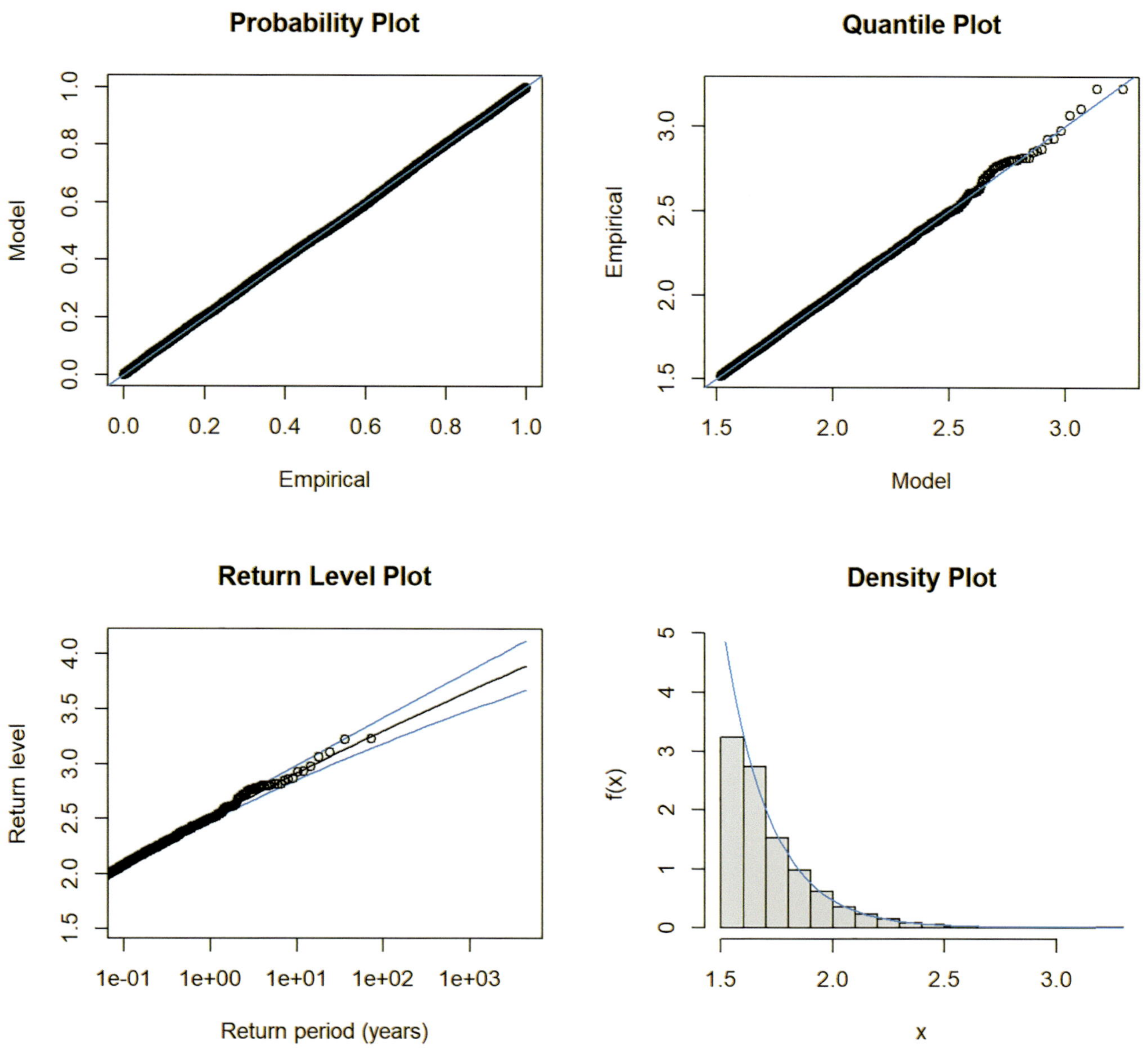

FIGURE 4.9 Diagnostic plots for GPD fit obtained with optimum threshold for nearest grid point to Jeddah City.

TABLE 4.4 Estimated return values for different return periods for the nearest grid point to Jeddah City based on ERA5 wave data and those estimated by Alsaaq and Shamji (2022) at a nearby point.

RETURN PERIODS (YEARS)	*RETURN VALUES (M)*	
	THIS CHAPTER	*ALSAAQ AND SHAMJI (2022)*
10	2.96	2.99
20	3.07	3.11
50	3.19	3.19
100	3.30	3.27

ranged from a few hours to several decades. In this chapter, the longest available wave hindcasting data (ERA5 72-year hourly wave data) has been utilized to study the variabilities of the wave conditions in the Red Sea in terms of exceedance levels, annual and seasonal trends, return values and energy resource assessments. The results obtained through these methods of analysis are in good agreement with the results of the previous studies that have been comprehensively reviewed in this chapter. Such agreements indicate the suitability of using the ERA5 wave data to study the variability of the wave conditions in the Red Sea. Accordingly, the utilization of the other ERA5-provided wave parameters in more widespread investigations of the wave variability in the Red Sea is highly encouraged.

Wave-hindcasting-based investigations, such as the one conducted in this chapter and in the reviewed previous studies, provide the wave variability information necessary for long-term planning of coastal construction and navigational routes. Nevertheless, more accurate long-term planning should also take into account the potential impact of climate change on wind variability and, consequently, on wave variability. The essential wave variability information for near-future (one to five days) planning can only be provided through operational wave forecasting. Even though the two widely used numerical wave models (SWAN and WW3) that have been employed to simulate the varying wave conditions

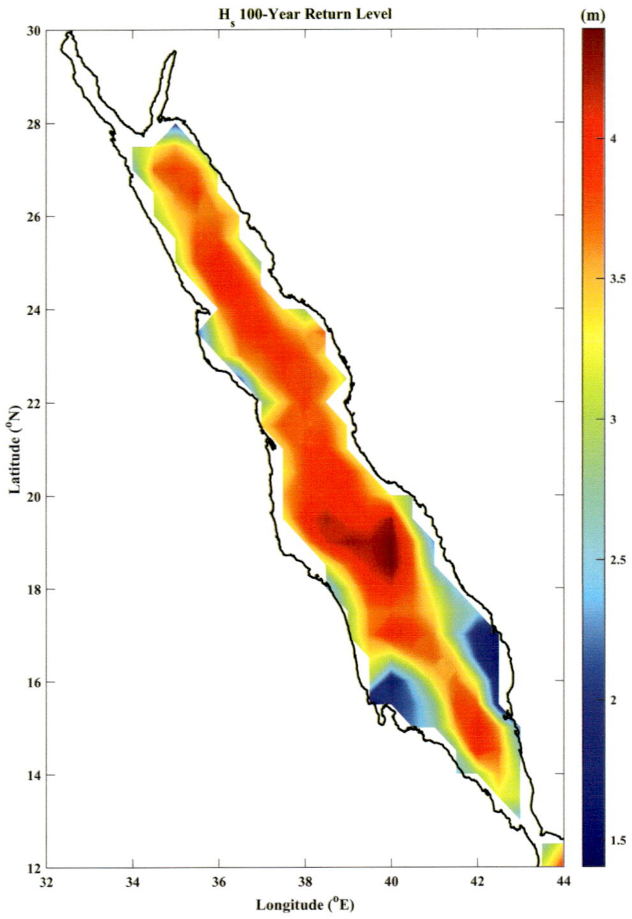

FIGURE 4.10 Spatial variability of significant wave height – 100-year return level.

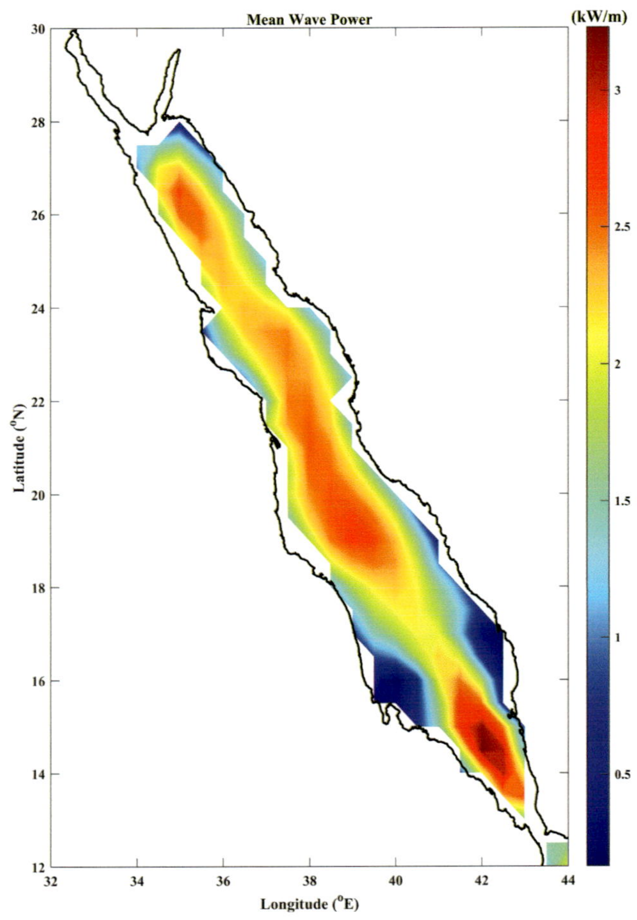

FIGURE 4.11 Spatial variability of mean wave power (kW/m).

in the Red Sea were extensively validated, the utilization of these models in forecasting near-future waves in the Red Sea is very limited. This is partially due to the limited availability of accurate and reliable forecasted wind fields to force the numerical wave models in a forecasting mode. Such wind fields must be in spatial and temporal resolutions that are high enough for resolving the short-distance and short-time wave variabilities in the Red Sea. The other reason pertains to the limited availability of bathymetric data for the coastal and nearshore areas with sufficiently high spatial resolution to accurately simulate the wave variability due to the complex bathymetry in those regions.

For further investigations of the Red Sea wave variability, continuous research and development are required in order to further improve the quality of atmospheric model–predicted wind fields over the Red Sea in both the spatial and temporal domains so small-scale variabilities (e.g., those due to mountain gaps) and short-term variabilities (e.g., those due to extreme wind events) can be resolved. Furthermore, continuous improvement in the quality of the conventional and satellite-based wave measurements will impact the quality of the model predictions of the spatially and temporally varying winds and waves in the Red Sea through: 1) utilization of the high-quality data for more extensive validation/verification of wind and wave models and 2) utilization of the high-quality data in data assimilation schemes by wind and wave models. Moreover, effective collaboration and data sharing between the local universities (including their relevant departments, colleges and research centres) and the relevant governmental sectors (including the National Center for Meteorology, the General Authority for Survey and Geospatial Information, the Saudi Geological Survey and the Saudi Ports Authority) would be very beneficial for reliable hindcasting and forecasting of both the long- and short-term wave variabilities in the Red Sea.

ACKNOWLEDGEMENTS

The author of this chapter would like to express his sincere thanks and appreciation to Dr. Najeeb M.A. Rasul and Dr. Ian C.F. Stewart for the invitation to contribute to this valuable book. The author would like to thank King Abdulaziz University (KAU) for all kinds of support he received over the years. Many thanks are also due to those researchers at both KAU and King Abdullah University of Science and Technology (KAUST), including faculty members and former postdoctoral and postgraduate students, for their valuable contributions to the knowledge concerning waves and their variability in the Red Sea.

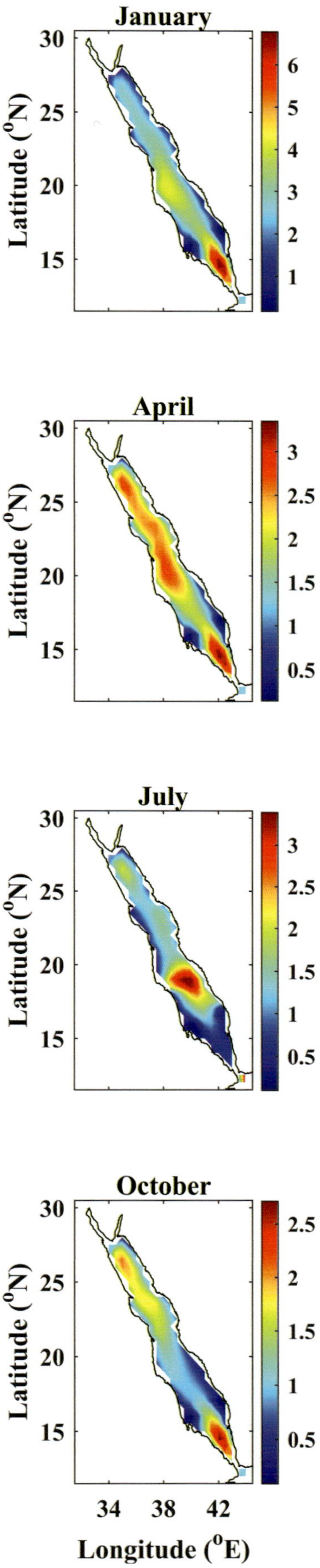

FIGURE 4.12 Spatio-temporal variability of the mean wave power (kW/m).

REFERENCES

Abdelrahman, S.M. 1993. Hindcasting wave climatology off Jeddah coast. *J KAU Mar Sci* 4: 3–16.

Abdelrahman, S.M. 1995. Statistical wave parameters offshore Jeddah coast. *J KAU Mar Sci* 6: 25–37.

Aboobacker, V.M., Shanas, P.R., Alsaafani, M.A., and A.M. Albarakati. 2017. Wave energy resource assessment for Red Sea. *Renew Energy* 114, no. A: 46–58.

Albarakati, A.M.A., and V.M. Aboobacker. 2018. Wave transformation in the nearshore waters of Jeddah, west coast of Saudi Arabia. *Ocean Eng* 163: 599–608.

Alkhayyat, M., Brahimi, T., Langodan, S., and I. Hoteit. 2020. Wave energy in the Red Sea region perspectives and analysis. In: *2020 6th IEEE International Energy Conference (ENERGYCon)*. http://doi.org/10.1109/energycon48941.2020.9236479.

Alsaaq, F., and V.R. Shamji. 2022. Extreme wind wave climate off Jeddah coast, the Red Sea. *J Mar Sci Eng* 10: 748.

Beyramzade, M., Siadatmousavi, S.M., and M.M. Nik. 2019. Skill assessment of SWAN model in the Red Sea using different wind data. *Reg Stud Mar Sci* 30: 100714.

Booij, N., Ris, R.C., and L.H. Holthuijsen. 1999. A third-generation wave model for coastal regions 1 – model description and validation. *J Geophys Res* 104: 7649–7666.

Coles, S.G. 2001. *An Introduction to Statistical Modeling of Extreme Values*. London: Springer-Verlag.

del Castillo, J., and M. Padilla. 2016. Modeling extreme values by the residual coefficient of variation. *SORT Stat Oper Res Trans* 40, no. 2: 303–320.

Fery, N., Al-Subhi, A.M., Zubier, K.M., and G. Bruss. 2015. Evaluation of the sea state near Jeddah based on recent observations and model results. *J Oper Oceanogr* 8: 1–10.

Fery, N., Bruss, G., Al-Subhi, A., and R. Mayerle. 2012. Numerical study of wind-generated waves in the Red Sea. In: *Proc 4th International Conference Coastlab12*. Ghent, Belgium, pp. 446–455.

Gharbi, S. 2012. *Numerical Wave Simulation in the Red Sea*. M.Sc. Thesis, King Abdulaziz University, Saudi Arabia.

Hersbach, H., Bell, B., and P. Berrisford. 2020. The ERA5 global reanalysis. *Q J R Meteorol Soc* 146: 1999–2049. http://doi.org/10.1002/qj.3803.

Hess, A., Iyer, H., and W. Malm. 2001. Linear trend analysis: A comparison of methods. *Atmos Environ* 35: 5211–5222.

Hoteit, I., Abualnaja, Y., Afzal, S., Ait-El-Fquih, B., Akylas, T., Antony, C., et al. 2021. Towards an end-to-end analysis and prediction system for weather, climate, and marine applications in the Red Sea. *Bull Am Meteorol Soc* 102, no. 1: E99–E122. http://doi.org/10.1175/bamsd-19-0005.1.

Janssen, P.A. 1991. Quasi-linear theory of wind-wave generation applied to wave forecasting. *J Phys Oceanogr* 21: 1631–1642.

Kendall, M.G. 1975. *Rank Correlation Methods*, 4th ed. London: Charles Griffin.

Komen, G.J., Hasselmann, S., and K. Hasselmann. 1984. On the existence of a fully developed wind-sea spectrum. *J Phys Oceanogr* 14: 1271–1285.

Langodan, S., Antony, C., Shanas, P.R., Dasari, H.P., Abualnaja, Y., Knio, O., and I. Hoteit. 2020. Wave modelling of a reef-sheltered coastal zone in the Red Sea. *Ocean Eng* 207: 107378.

Langodan, S., Cavaleri, L., Pomaro, A., Portilla, J., Abualnaja, Y., and I. Hoteit. 2018. Unraveling climatic wind and wave trends in the Red Sea using wave spectra partitioning. *J Clim* 31, no. 5: 1881–1895.

Langodan, S., Cavaleri, L., Pomaro, A., Viswanadhapalli, Y., Bertotti, L., and I. Hoteit. 2017b. The climatology of the Red Sea – part 2: The waves. *Int J Climatol* 37: 4518–4528.

Langodan, S., Cavaleri, L., Viswanadhapalli, Y., Pomaro, A., Bertotti, L., and I. Hoteit. 2017a. The climatology of the Red Sea – part 1: The wind. *Int J Climatol* 37: 4509–4517.

Langodan, S., Cavaleri, L., Viswanadhapalli, Y., and I. Hoteit. 2014. The Red Sea: A natural laboratory for wind and wave modeling. *J Phys Oceanogr* 44: 3139–3159.

Langodan, S., Cavaleri, L., Viswanadhapalli, Y., and I. Hoteit. 2015. Wind-wave source functions in opposing seas. *J Geophys Res Ocean* 120: 6751–6768.

Langodan, S., Viswanadhapalli, Y., Dasari, H.P., Kino, O., and I. Hoteit. 2016b. A high-resolution assessment of wind and wave energy potentials in the Red Sea. *Appl Energ* 181: 244–255.

Langodan, S., Viswanadhapalli, Y., and I. Hoteit. 2016a. The impact of atmospheric data assimilation on wave simulations in the Red Sea. *Ocean Eng* 116: 200–215.

Mayerle, R., Al-Subhi, A., Fernández Jaramillo, J., Salama, A., Burss, G., Zubier, K., Runte, K., Turki, A., Hesse, K., Jastania, H., Ladwig, N., and M. Muddaris. 2016. Development of a coastal information system for the management of Jeddah coastal waters in Saudi Arabia. *Comput Geosci* 89: 71–78.

Miky, Y., Kaloop, M.R., Elnabwy, M.T., Baik, A., and A. Alshouny. 2021. A recurrent-cascade-neural network – nonlinear autoregressive networks with exogenous inputs (NARX) approach for long-term time-series prediction of wave height based on wave characteristics measurements. *Ocean Eng* 204: 109958.

Ralston, D., Jiang, H., and J. Farrar. 2013. Waves in the Red Sea: Response to monsoonal and mountain gap winds. *Cont Shelf Res* 65: 1–13.

Saad, A. 2010. *Wave and Wind Conditions in the Red Sea – A Numerical Study Using a Third-Generation Wave Model*. M.Sc. Thesis, Geophysical Institute, University of Bergen, Bergen, Norway.

Sen, P.K. 1968. Estimates of the regression coefficient based on Kendall's Tau. *J Am Stat Assoc* 63: 1379–1389.

Shamji, V.R., Aboobacker, V.M., and T.C. Vineesh. 2020. Extreme value analysis of wave climate around Farasan Islands, southern Red Sea. *Ocean Eng* 207: 107395.

Shanas, P.R., Aboobacker, V.M., Albarakati, A.M.A., and K.M. Zubier. 2017a. Superimposed wind waves in the Red Sea. *Ocean Eng* 138: 9–22.

Shanas, P.R., Aboobacker, V.M., Albarakati, A.M.A., and K.M. Zubier. 2017b. Climate-driven variability of wind-waves in the Red Sea. *Ocean Model* 119: 105–117.

Shanas, P.R., Aboobacker, V.M., Zubier, K.M., and A.M.A. Albarakati. 2018. Spectral wave characteristics along the central coast of eastern Red Sea. *Arab J Geosci* 11: 90.

Stephenson, A.G. 2018. *ISMEV: An Introduction to Statistical Modeling of Extreme Values*. https://cran.r-project.org/web/packages/ismev/ismev.pdf. R package version 1.42.

Tolman, H.L., and the WAVEWATCH III Development Group. 2014. User manual and system documentation of WAVEWATCH III version 4.18. *Technical Note 316, NOAA/NWS/NCEP/MMAB, 282 pp. + Appendices*. https://polar.ncep.noaa.gov/waves/wavewatch/manual.v4.18.pdf.

Wamdi Group. 1988. The WAM model – A third generation ocean wave prediction model. *J Phys Oceanogr* 18: 1775–1810.

Zubier, K.M. 2020. Using an artificial neural network for wave height forecasting in the Red Sea. *Indian J Mar Sci* 49, no. 2: 184–191.

Zubier, K.M., Abualnaja, Y.O., and A.M. Al-Subhi. 2009. Development of an operational wave prediction system for the Red Sea: Experimental phase. *Paper presented at the Offshore Arabia Conference*, Dubai, United Arab Emirates.

Zubier, K.M., Albarakati, A., and M. Rady. 2007. Numerical simulation of waves at Jeddah near-shore region. *Final Report (Project 262/425), Deanship of Scientific Research*, King Abdulaziz University, Jeddah, Saudi Arabia.

Zubier, K.M., and S.H. Gharbi. 2019. Sensitivity of a Red Sea numerical wave model to spatial and temporal resolution of forcing wind field. *Indian J Mar Sci* 48, no. 4: 566–575.

PART 2

Chemical oceanography

Remote sensing of ocean colour in the Arabian Gulf

5

Surya P. Tiwari and Syed M. Rahman

5.1 INTRODUCTION

The Arabian Gulf is a shallow, semi-enclosed sea that connects to the Indian Ocean through the Strait of Hormuz. It is located in a subtropical, hyper-arid zone. The Arabian Gulf has a significant seasonal change in temperature, ranging from 11°C in the winter to 36°C in the summer, and it is a highly saline environment, with salinity surpassing 39 ppt in most areas (Sheppard et al. 2010). Salinity in Gulf embayments, such as the Gulf of Salwah, can even reach up to 70 ppt (Sheppard et al. 2010). This is due to poor freshwater inflow from rivers and rainfall, along with high rates of evaporation that lead to a net evaporation loss of 1.6 m year^{-1} (Sheppard et al. 2010). Critically, these extremes are currently increasing even further due to climate change, and the sea surface temperature is expected to increase by up to 4°C by the end of this century (Noori et al. 2019). To detect changes in the phytoplankton due to an increase in temperature, extensive knowledge of the current status of the phytoplankton community is required. Phytoplankton abundance depends on the availability of sunlight and nutrients, but it can also be influenced by other factors such as water temperature, salinity, bathymetry, winds and the number of marine creatures that graze on it.

Tracking changes in global marine plant biomass is fundamental to ocean biology. Most of the plant biomass in the ocean comprises microscopic photosynthetic organisms (phytoplankton) (Board et al. 2011). Satellite observation of ocean colour radiometry involves detection of spectral variations in the water-leaving radiance (or remote-sensing reflectance, sr^{-1}), which is the sunlight backscattered out of the ocean after interaction with water and its constituents at several spectral wavebands, from the visible to near infrared regions. In the open ocean, the signal is primarily dependent on phytoplankton, which contain photosynthetic pigments – that is, chlorophyll-a (chl-a) – and an assemblage of other pigments. Ocean colour remote sensing is thus the only satellite remote-sensing technique that opens a window to understanding ocean colour biology.

There is a linkage between sea surface temperature (SST), photosynthetically active radiation (PAR) and ocean colour biology as these variables can affect the productivity and distribution of phytoplankton, which, in turn, can influence the concentration of pigments such as chlorophyll-a and the ocean colour. SST can influence phytoplankton metabolism and growth rate while PAR can influence photosynthesis and productivity.

The ocean colour of the Arabian Gulf is influenced by a combination of physical, chemical and biological factors which can vary spatially and temporally. Some of the main factors that can affect the ocean colour in the Arabian Gulf include nutrient availability, temperature, light availability, mixing and circulation and pollution. For example, if nutrient concentrations in the water increase over time, this could lead to increased phytoplankton productivity and higher concentrations of pigments such as chlorophyll-a, which could result in a change in ocean colour. Similarly, if temperature or light availability increases over time, this could also lead to increased phytoplankton productivity and a change in ocean colour. Changes in mixing and circulation patterns in the water column could also affect the distribution and concentration of nutrients, temperature and light, which could influence ocean colour. Finally, increases in pollution levels could lead to changes in the concentration of pigments and other substances in the water, which could also affect ocean colour. Understanding the factors that influence the ocean colour biology of the Arabian Gulf is important for monitoring the health of the marine environment and for predicting and mitigating the impacts of environmental changes. Phytoplankton research in the Gulf has been the topic of numerous studies since the second half of the 20th century (Dorgham 2013). However, most of these were localized studies in Kuwaiti waters (Subba Rao and Al-Yamani 1998; Al-Yamani et al. 2006; Polikarpov et al. 2009) or Iranian waters (Hulburt et al. 1981); a few studies in Saudi Arabian waters focused on the area around Al-Jubail (Abdul Azis et al. 2003; Chandy et al. 1991). In-situ bio-geo-optical measurements are only available in comparatively few locations in the Arabian Gulf, which are expensive, and not for all times. Ocean colour data play a critical role in observation systems for monitoring coastal eutrophication, harmful algal blooms and sediment plumes.

This chapter describes the uses of ocean colour observations from MODIS-Aqua to investigate the remote sensing of ocean colour in the Arabian Gulf. It is often difficult to obtain accurate ocean colour observations in the coastal

DOI: 10.1201/9781003321408-7

shallow regions at a regional scale due a lack of local validation of ocean colour sensors. Therefore, it is possible that the measured values of ocean colour parameters could show a low agreement with satellite-based observations in coastal regions of the Arabian Gulf.

5.2 OBJECTIVES

The objective of this research is to provide data on the key ocean colour parameters, which are important for understanding ocean colour remote sensing in the Arabian Gulf. It is well understood that a variety of physical, chemical and biological factors could possibly influence ocean colour. These changes manifest as an increase or decrease in the chlorophyll-a concentrations, suspended particles, coloured dissolved organic matter and the scattering properties of the water. Advancements in remote-sensing technology, algorithm development and data processing methodologies have made it possible to derive these parameters with greater accuracy at higher spatial and temporal resolutions. Finally, the chapter will enhance knowledge and promote the effective use of ocean colour data for studying and managing marine environments in the Arabian Gulf.

5.3 MATERIALS AND METHODS

The following subsections describe some of the common approaches used to derive ocean colour products from satellite remote-sensing data. From July 2002 to December 2022, the monthly level-3 moderate resolution imaging spectroradiometer (MODIS) Aqua standard mapped products of remote-sensing reflectance at 443 nm (R_{rs}(443)), chlorophyll-a (Chl-a), sea surface temperature (SST; 4 μ), absorption by gelbstoff and detritus at 443 nm (a_{dg}(443)), and photosynthetically active radiation (PAR) with a 4 km spatial resolution were downloaded from NASA's Ocean Colour data portal (http://oceandata.sci.gsfc.nasa.gov). The data were utilized to create monthly, seasonal (summer-autumn-winter-spring) and annual mean ocean colour maps of the study area to understand the changes in ocean colour in the Arabian Gulf. Due to space constraints, the annual average maps for Chl-a, SST, PAR and a_{dg}(443) are not presented in this chapter. However, annual variations in the ocean colour-derived parameters are described in the results and discussion section.

5.3.1 Remote-sensing reflectance

The spectral distribution of visible solar radiation reflected from below the ocean surface and passing through the sea-air interface is the primary quantity to be determined by ocean colour sensors. The key to extracting any ocean colour parameters from the satellite data is the spectrally resolved water-leaving radiance, commonly known as ocean colour. The following expression can be used to compute the reflectance R_{rs} from the ocean colour sensors:

$$R_{rs} = (L_w(\lambda) / (F_0 f_s \cos(\theta_s) t_{ds}) f_b(\lambda) f_\lambda \quad (5.1)$$

where:

$L_w(\lambda)$ = water-leaving radiance
F_0 = extraterrestrial solar irradiance
f_s = adjustment of F_0 for variation in Earth-sun distance
f_b = bidirectional reflectance correction
f_λ = correction for out-of-band response
$t_{ds}(\lambda)$ = the transmittance of diffuse radiation through the atmosphere in the viewing path from sun to surface

5.3.2 Chlorophyll-a from satellite remote sensing

Chlorophyll-a (Chl-a) is a pigment found in phytoplankton, which are microscopic plants that live in the oceans. It is an important indicator of phytoplankton productivity and is often used as a measure of the health of the marine environment. In ocean colour remote sensing, there are several algorithms for the retrieval of chlorophyll from remote-sensing reflectance; for example, the concentration of Chl-a is derived using empirical ocean colour algorithms that rely on R_{rs} band ratios in the blue-green region (O'Reilly et al. 1998; O'Reilly and Werdell 2019). A multi-band ratio (MBR) algorithm uses R_{rs} at multiple wavelengths (depending on the sensor) in the blue-green region. The following equation shows the basic mathematical formulation of the MBR algorithm (O'Reilly et al. 1998; O'Reilly and Werdell 2019):

$$\log_{10}(Chl-a) = a_0 + \sum_{i=1}^{4} a_i \left(log10 \left(\max \left(\frac{R_{rs}(\lambda_{blue})}{R_{rs}(\lambda_{green})} \right) \right) \right)^i \quad (5.2)$$

where a_0 to a_4 are sensor-dependent coefficients, $R_{rs}(\lambda_{blue})$ is the maximum of R_{rs} in the blue region: e.g., 443, 490, and 510 nm for the OC4 algorithm and $R_{rs}(\lambda_{green})$ is at a single wavelength in 547 to 565 nm. NASA's operational algorithm to derive Chl-a from various sensors combines the MBR with the colour index (CI) (Hu et al. 2019, 2012) algorithm in the following form:

$$CI = R_{rs}(\lambda_{green}) - \left[R_{rs}(\lambda_{blue}) + \left(\frac{\lambda_{green} - \lambda_{blue}}{\lambda_{red} - \lambda_{blue}} \right) * \left(R_{rs}(\lambda_{red}) - R_{rs}(\lambda_{blue}) \right) \right] \quad (5.3)$$

with $\lambda_{blue}, \lambda_{green}$ and λ_{red} close to 443, 555 and 670 nm, respectively. The CI algorithm demonstrated higher accuracy in derived Chl-a concentrations in oligotrophic waters than OCx algorithms. For chlorophyll retrievals below 0.15 mg m^{-3}, the CI algorithm is used.

For chlorophyll retrievals above 0.2 mg m^{-3}, the MBR algorithm is used. In between these values, the CI and MBR algorithms are blended using a weighted approach. More details about CI and switching between CI and MBR are presented by Hu et al. (2012).

5.3.3 Sea surface temperature

The sea surface temperature (SST) (^{0}C) measures the infrared radiation emitted by the top layer of the ocean and is an important factor in determining the climate and weather patterns in a region. For MODIS, the SST4 algorithm utilizing the 3.9 and 4μm spectral bands is used for the SST product. MODIS was the first and only sensor to date to use short wave infrared (SWIR) channels solely to determine SST.

$$SST4 = a_{ij0} + a_{ij1} * T_{3.9\mu m} + a_{ij2}\left(T_{3.9\mu m} - T_{4.0\mu m}\right) + a_{ij3}\left(sec\theta_{sza} - 1\right) + a_{ij4}\left(mirror\right) + a_{ij5}\left(\theta^*_{sza}\right) + a_{ij6}\left(\theta^2_{sza}\right) \quad (4)$$

where "mirror" corresponds to mirror side number (0 or 1), θ^*_{sza} is SZA (solar zenith angle) made negative for pixels in the first half of the scan line and $T_{3.9,4\mu m}$ correspond to brightness temperature at 3.959 and 4.050 μm, in ^{0}C, respectively. The latest version of the MODIS SST product (i.e., R2019) uses a cloud mask generated based on a cloud classification algorithm implementing alternating decision trees (ADtree) The SST products correspond to the skin temperature of the ocean, a ~1-mm-thick surface thermal skin layer.

5.3.4 Coloured dissolved organic matter

Coloured dissolved organic matter (CDOM), also known as gelbstoff and detritus, is the optically measurable component of dissolved organic matter in water. It occurs naturally in aquatic environments and is primarily leached from decaying detritus and organic matter. It strongly absorbs short-wavelength light ranging from ultraviolet to blue, whereas pure water absorbs longer wavelength red light (Tiwari and Shanmugam 2011). The generalized inherent optical property (GIOP) model is used to derive the absorption coefficient of non-algal material plus CDOM (a_{dg}) at 443 nm in m^{-1} (Werdell et al. 2013).

5.3.5 Photosynthetically Active Radiation

Photosynthetically active radiation (PAR) (Einstein m^{-2} day^{-1}) is the portion of the electromagnetic spectrum that is used by plants for photosynthesis. PAR is defined as the available solar quantum energy flux in 400 to 700 nm at the ocean surface. The generally used PAR product in primary productivity (PP) models is the PAR reaching the earth's surface over a 24-hour period and is different from instantaneous PAR.

As the PAR algorithm models the atmosphere into layers, clouds and clear regions in satellite pixels need not be carried out. Briefly, the solar flux reaching the ocean surface is computed as:

$$E = \frac{E_{clear}\left(1 - A\right)}{\left(1 - A_s\right)} \quad (5)$$

where A_s and A correspond to the albedo of the surface and cloud-surface system. E_{clear} is the solar flux.

5.3.6 Processing of ocean colour satellite sensor data

The past and present sensors launched on various satellites for ocean colour applications differ in design, calibration and data produced in terms of instantaneous field of view or spatial resolution, spectral resolution, swath and temporal resolution (Table 5.1).

At present, a two-step process is implemented in processing satellite imagery to generate spectral remote-sensing reflectance, which is subsequently used to derive the biogeochemical parameters and inherent optical properties (IOPs). The first term is called atmospheric correction, which converts top-of-the-atmosphere radiance to exact normalized water-leaving radiance and its equivalent spectral R_{rs} (IOCCG 2010; Gordon 2021; Mobley et al. 2016). For a measured water-leaving radiance $\left(L_w\right)$, the effect of solar zenith angle, atmospheric attenuation and Earth-sun distance are removed in the calculation of $R_{rs}\left(\theta_v, \phi_{azi}\right)$, where θ_v and ϕ_{azi} correspond to viewing and azimuth angles.

Level 1 radiance products can be atmospherically corrected using the "l2gen" tool in the SeaWiFS data analysis system (SeaDAS) openly available from NASA. The default atmospheric correction algorithm implemented in NASA's SeaDAS processing system is an iterative scheme that calculates R_{rs} in the near infra-red (NIR) using bio-optical empirical models (Bailey et al. 2010; Mobley et al. 2016).

TABLE 5.1 Spatial and spectral resolution of remote sensors widely used in ocean colour studies: coastal zone colour scanner (CZCS), moderate resolution imaging spectroradiometer (MODIS), sea-viewing wide field-of-view sensor (SeaWiFS) and visible infrared imaging radiometer suite (VIIRS).

SENSOR/SATELLITE	*OCEAN COLOUR PARAMETER*	*DATA AVAILABILITY*	*NUMBER OF SPECTRAL CHANNELS WITH RANGE*	*SPATIAL RESOLUTION AND SWATH*
CZCS/Nimbus-7*	*Chl-a*	1978–1986	6 in 443–1020 nm	825 m; 1556 km
MODIS-Aqua	*Chl-a*, PAR, SST,adg_443	2002–**	36 in 405–14,385 nm	250/500/1000 m; 2330 km
SeaWiFS/ObrView-2*	*Chl-a*, PAR, SST,adg_443	1997–2010	8 in 402–885 nm	1100 m; 2806 km
VIIRS/Suomi VIIRS/JPSS-1	*Chl-a*, SST, adg_443	2011–** 2017–**	22 in 402–11,800 nm	375/750; 3000 km

*- past sensors; **- till date

5.4 RESULTS AND DISCUSSION

5.4.1 Monthly mean of $R_{rs}(443)$

Figure 5.1 shows the monthly mean of the $R_{rs}(443)$ for the Arabian Gulf. It was observed from the spatial map of ocean colour that the satellite-based reflectance values ranged from 0.0001 to 0.0200 with a mean value of 0.0045 sr^{-1}. The northern part of the Arabian Gulf shows fairly high values of reflectance compared to the southern part. The central parts of the Gulf show low values of R_{rs} during the winter months. Most of the high values of reflectance are seen in the coastal areas due to shallow water depths with relatively high scattering by Chl-a and suspended sediment particles.

5.4.2 Monthly mean of Chl-a, SST, PAR and $a_{dg}(443)$

Figures 5.2a to 5.2d present the MODIS-derived monthly distribution of mean of Chl-a, SST, $a_{dg}(443)$ and PAR in the Arabian Gulf. The results revealed that the Chl-a values (Figure 5.2a) are comparatively low in the Arabian Gulf's central regions, while higher values were observed along the coastlines from north to south. From November to March, the relatively elevated levels of Chl-a (> 0.5 mg m^{-3}) extended to the eastern parts of the Gulf toward the Sea of Oman while SST decreased from the northwest to the east. During the period of May to October, however, the average Chl-a concentration in the eastern regions decreased significantly while the average concentration (0.5 mg m^{-3}) increased in the central and coastal regions.

5.4.3 Seasonal mean of Chl-a, SST, PAR and $a_{dg}(443)$

The spatio-temporal seasonal variations of Chl-a, SST, PAR and $a_{dg}(443)$ in the Arabian Gulf were investigated, using satellite-derived products from the MODIS-Aqua sensor over a period of 20 years (2002–2023). In this study, winter, spring, summer and autumn were defined as December to February, March to May, June to August and September to November, respectively.

Knowledge of the seasonal variation of Chl-a is useful for assessing the seasonal patterns of primary productivity in the Arabian Gulf. It is clear from the temporal variations plot that the Chl-a and $a_{dg}(443)$ are lower in the Arabian Gulf, with an increase in SST and PAR (Figures 5.3a to 5.3d). Figure 5.3a shows the Chl-a elevated patterns during winter and autumn. On the other hand, during spring and summer, the Chl-a concentration is observed to be low compared to other seasons in the central parts of the Arabian Gulf.

The spatial patterns of Chl-a, SST and PAR across four seasons are shown in Figures 5.3a, 5.3b, and 5.3c, respectively. Clear spatial and temporal variations of the Chl-a trend can be seen in the Arabian Gulf. The seasonal mean plot of the Arabian Gulf exhibits high SSTs (30–35°C) during the summer and is distributed evenly across the whole Gulf. During the wintertime, SST can be seen to be below 20°C, and the temperature difference between the summer and winter is approximately 10–15°C over the Gulf. In offshore waters, the increasing trend of Chl-a in the Arabian Gulf during the winter and autumn might be attributed to the increase in nutrient content. The water column condition (stratified and mixed) was likely an important physical factor that affects the seasonal pattern of Chl-a in the Gulf. Figures 5.3b and 5.3c show the highest SST and PAR during the summer, and the lowest SST and PAR are observed during the winter. The SST and PAR show similar patterns during the winter, spring, summer and autumn. It is clear from the seasonal plots that there is a strong seasonal pattern for the Chl-a, with maximum concentrations during winter and autumn and minima during spring and summer. However, in summer, Chl-a concentrations were relatively high in the coastal waters of the Gulf compared to the offshore waters. In particular, the coastal region of the Gulf of Salwah experienced higher Chl-a than the offshore waters of the Arabian Gulf during the spring and summer. The satellite remote-sensing data revealed that warming and high light intensities could decrease the productivity of the Arabian Gulf. However, during summer and autumn, the Chl-a trend is high in the coastal waters of the Gulf of Salwah. Therefore, in general, the Gulf of Salwah, where the Chl-a concentration is high throughout the year, is the most productive part of the Arabian Gulf across four seasons. The factors and mechanisms that contribute to high biological productivity in the Gulf of Salwah are still unexplored and require detailed study. However, numerous physical (salinity, temperature, currents, light, circulations, mixing, winds, tides, monsoons) and atmospheric factors (dust events) could play important roles in ocean colour biology of the Arabian Gulf. Additionally, wind-tide-thermohaline circulations, Indian Ocean monsoons and anthropogenic activity could significantly impact the seasonal pattern of Chl-a. SST and PAR could be key factors in explaining the variability of Chl-a in seasonal and inter-annual patterns.

In general, the Arabian Gulf tends to have relatively high Chl-a concentrations in the spring and summer and relatively low concentrations in the autumn and winter. The SST in the Arabian Gulf tends to be highest in the summer and lowest in the winter while PAR tends to be highest in the summer and lowest in the winter. However, these patterns can vary significantly depending on the location in the Gulf and the specific year or time period being considered. Overall, the combination of these factors can lead to relatively high Chl-a concentrations in the spring and summer in the Arabian Gulf.

Annual averages of satellite-derived Chl-a, SST, PAR and $a_{dg}(443)$ from 2002 to 2023 show that the distribution patterns of Chl-a and $a_{dg}(443)$ are elevated along the coasts of the Arabian Gulf and comparatively depleted in the centre of

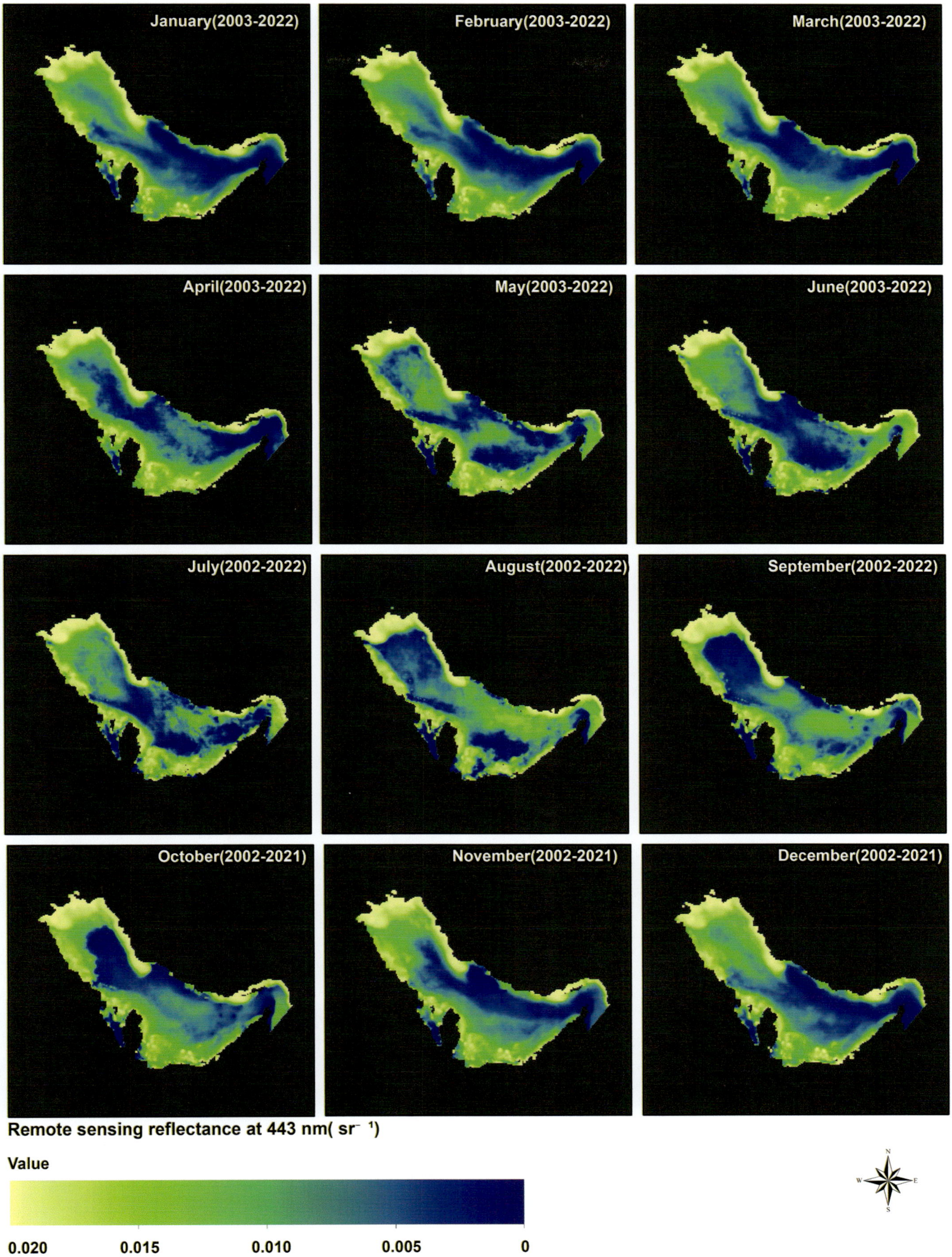

FIGURE 5.1 MODIS-Aqua monthly mean of remote-sensing reflectance at 443 nm in the Arabian Gulf.

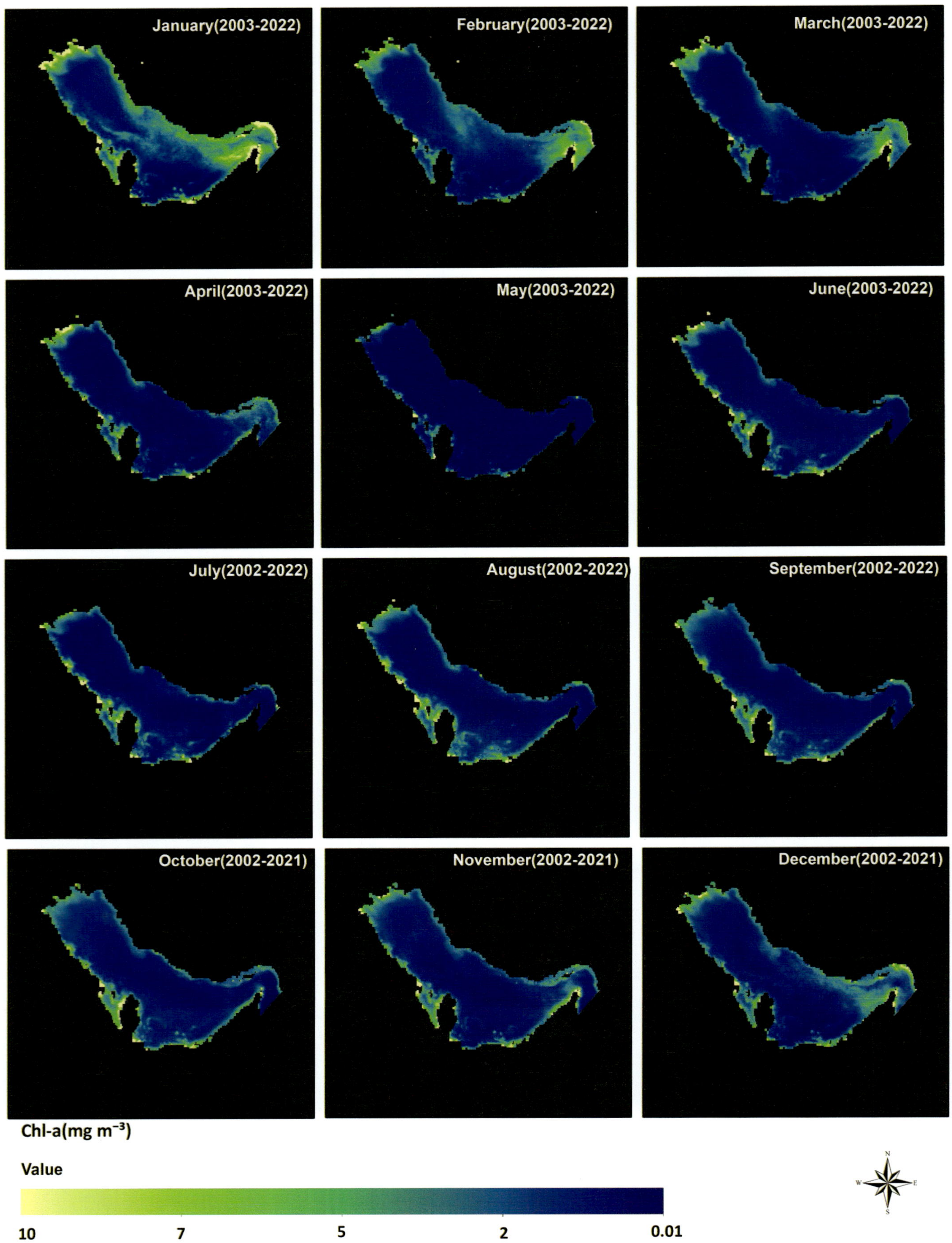

FIGURE 5.2A MODIS-Aqua monthly mean of Chl-a in the Arabian Gulf.

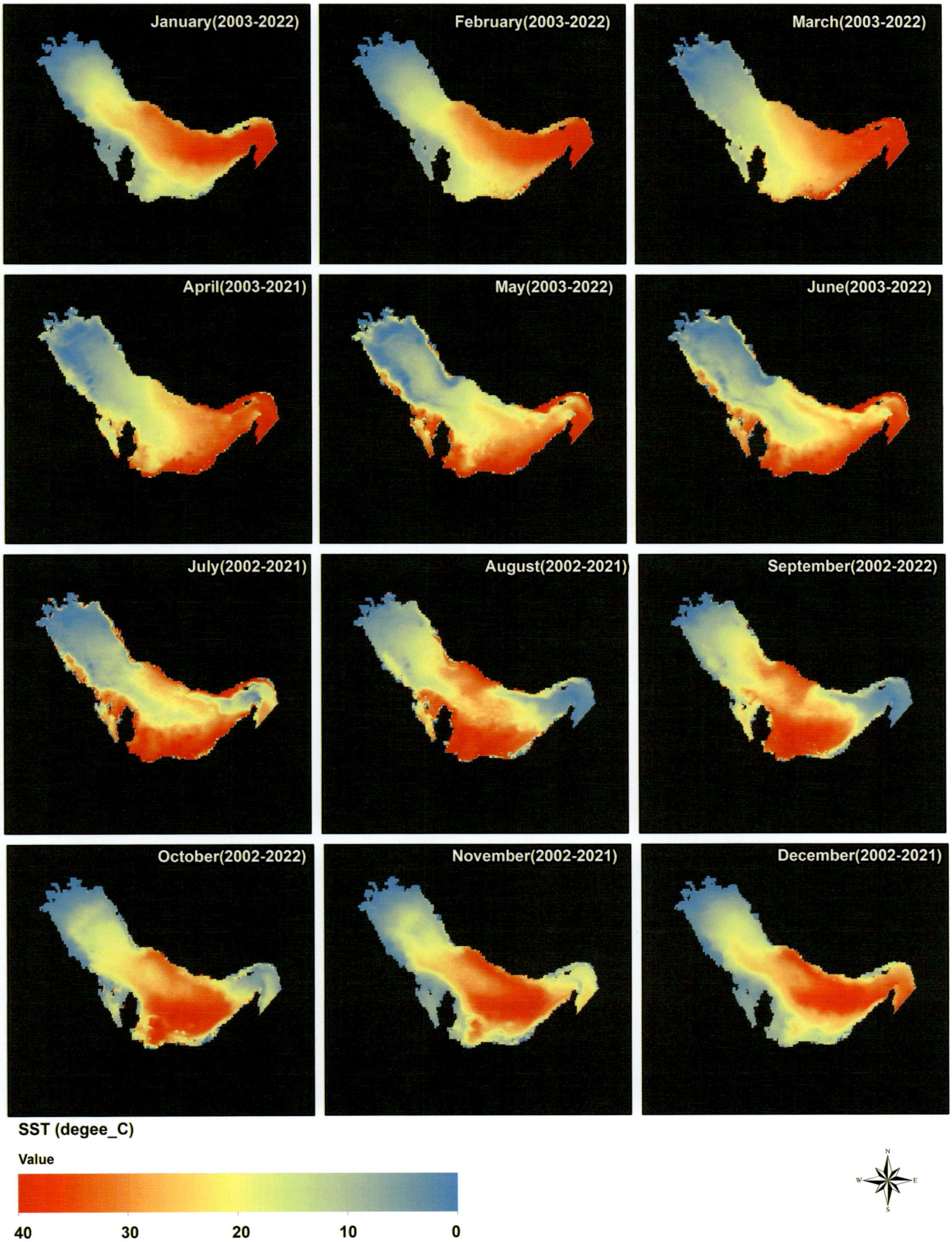

FIGURE 5.2B MODIS-Aqua monthly mean of SST in the Arabian Gulf.

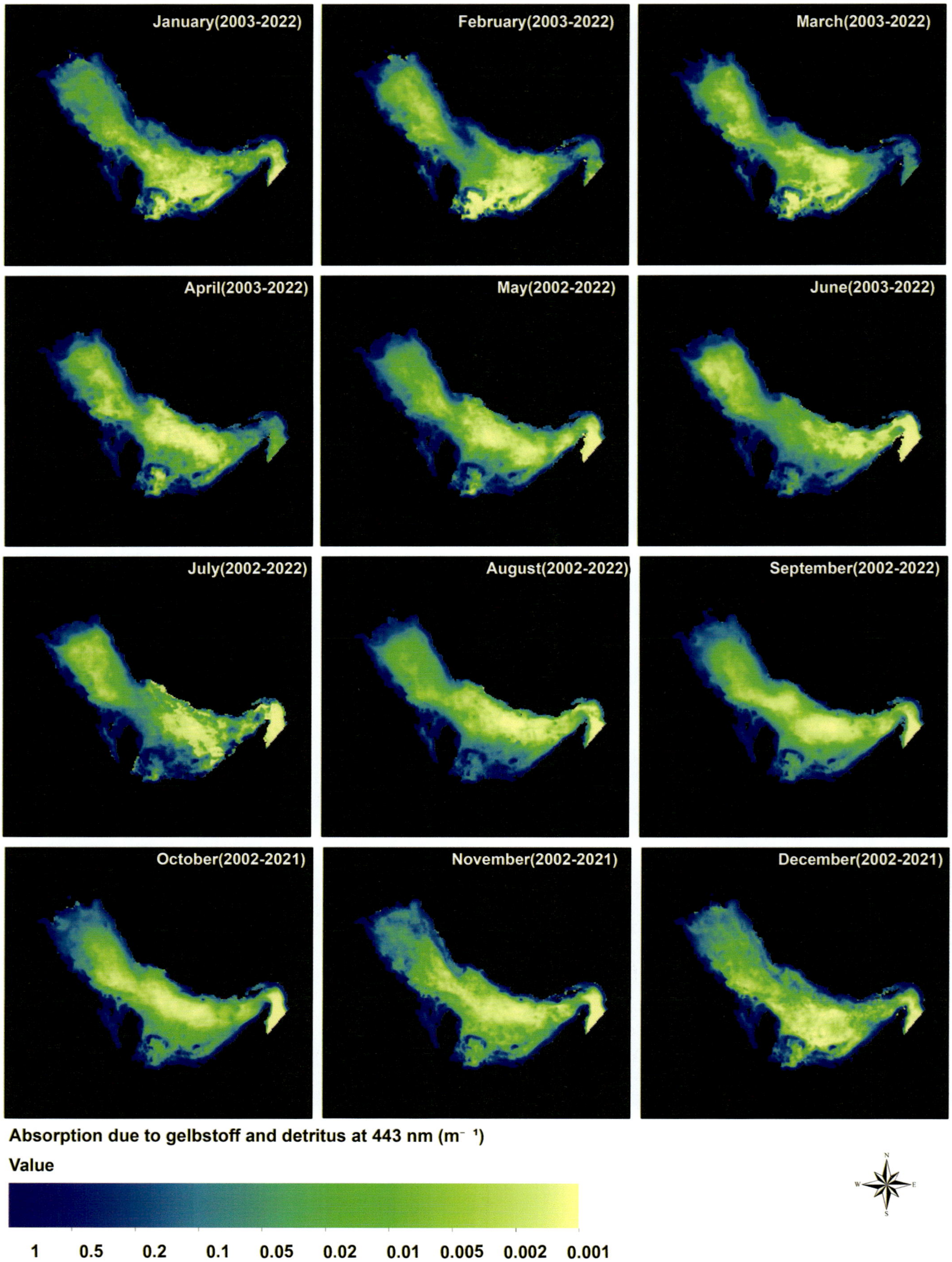

FIGURE 5.2C MODIS-Aqua monthly mean of $a_{dg}(443)$ in the Arabian Gulf.

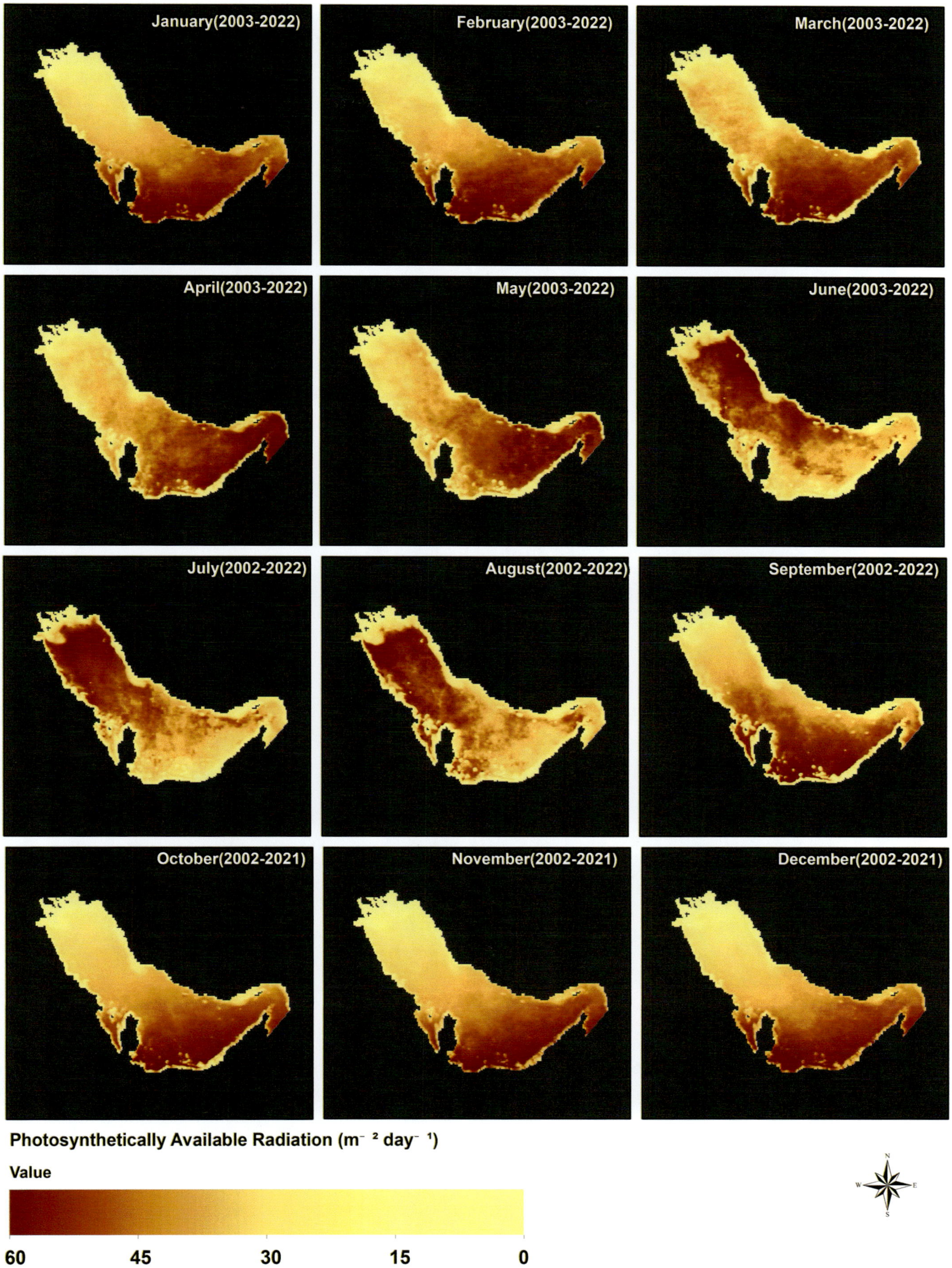

FIGURE 5.2D MODIS-Aqua monthly mean of PAR in the Arabian Gulf.

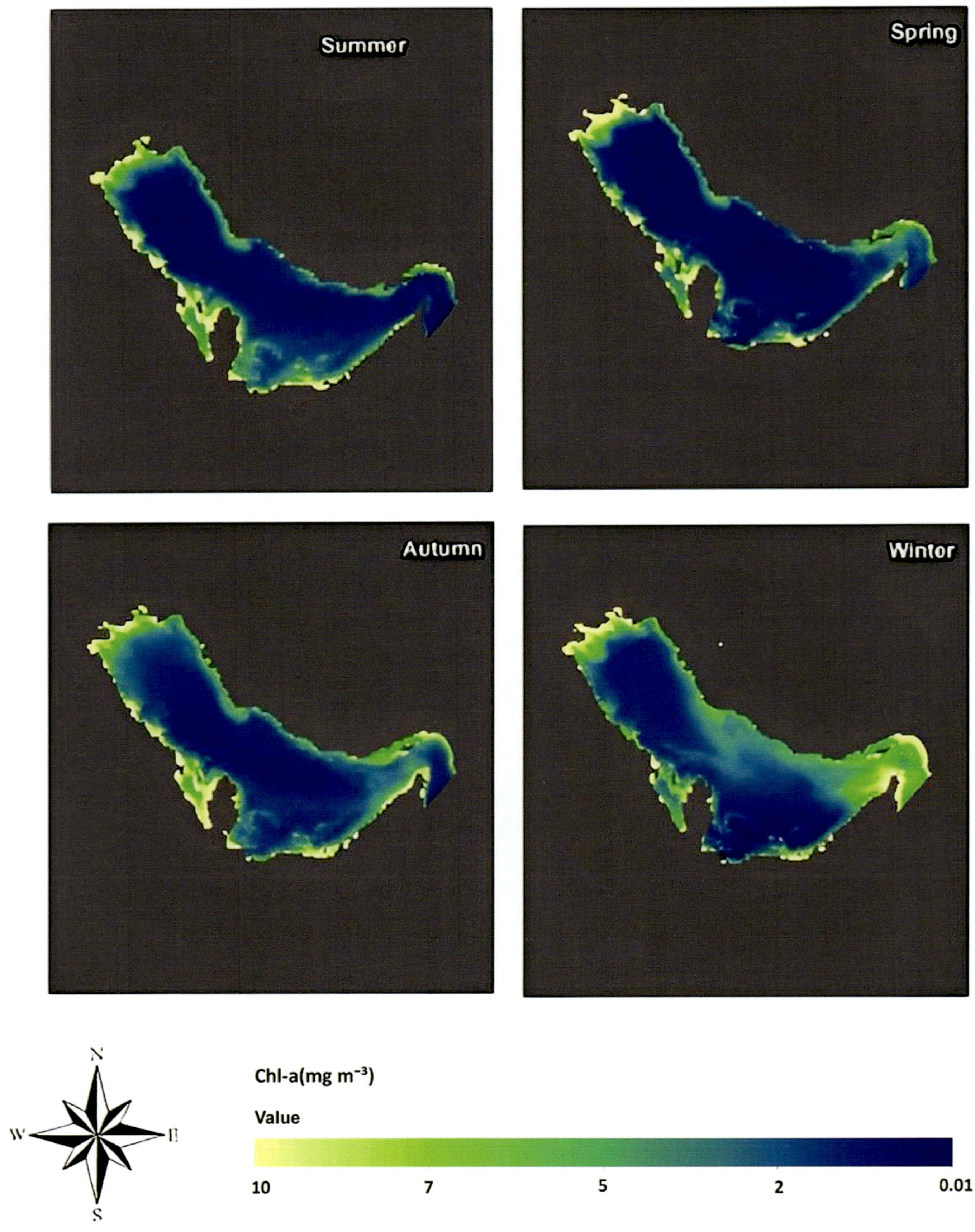

FIGURE 5.3A MODIS-Aqua seasonal mean of Chl-a in the Arabian Gulf.

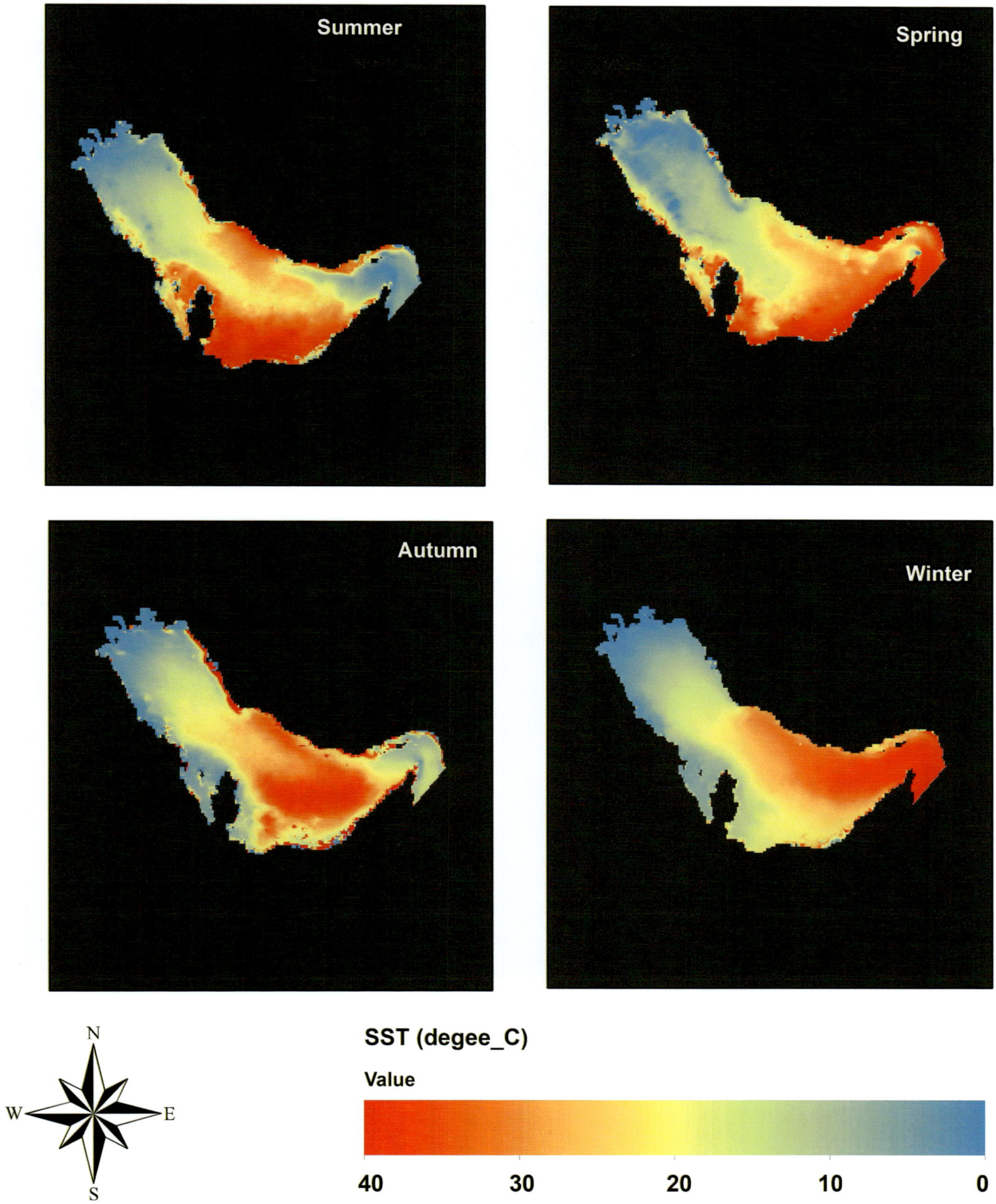

FIGURE 5.3B MODIS-Aqua seasonal mean of SST in the Arabian Gulf.

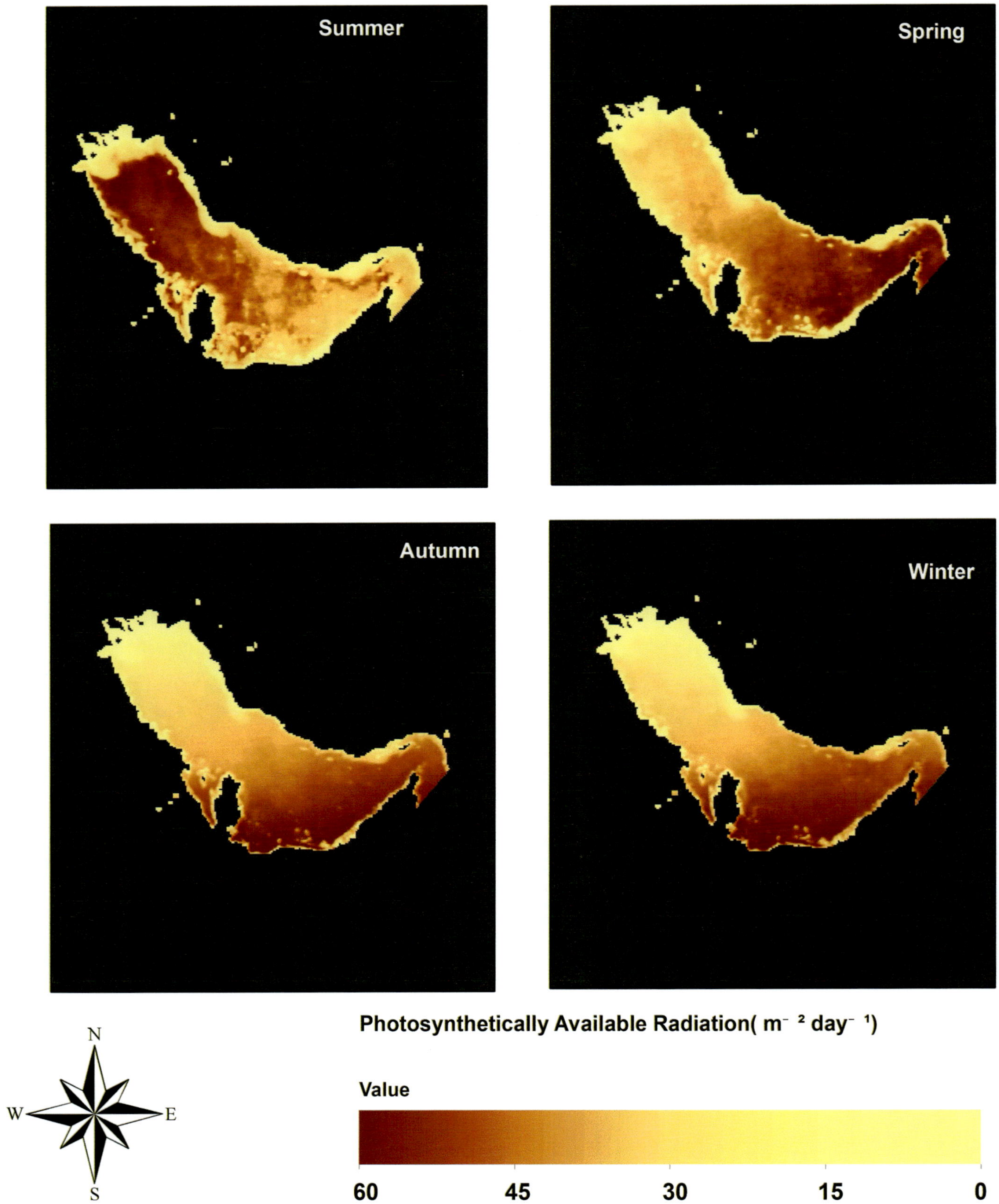

FIGURE 5.3C MODIS-Aqua seasonal mean of PAR in the Arabian Gulf.

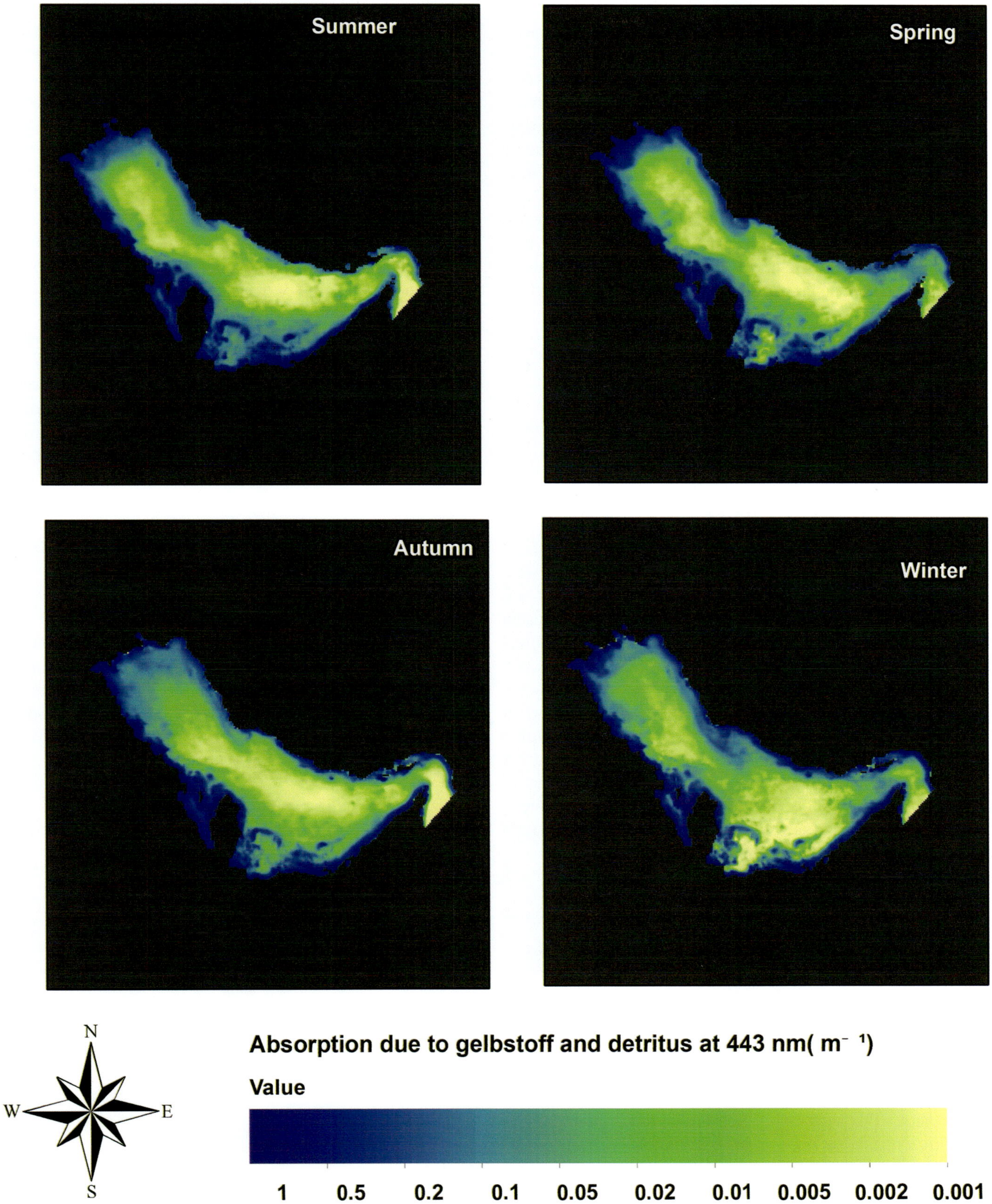

FIGURE 5.3D MODIS-Aqua seasonal mean of $a_{dg}(443)$ in the Arabian Gulf.

the offshore areas. The seasonal average for the summer and autumn was similar to the spatial variations of the annual average of SST.

In summary, the stationary levels of SST and Chl-a concentrations were stable in the study area, irrespective of the monthly, seasonal or annual cycles. This analysis revealed that the mean Chl-a concentration decreased from the coastal area to offshore (i.e., the centre) of the Arabian Gulf, and its value ranged from 0.01 to 0.5 mg m^{-3}. On the other hand, the average SST decreased from central offshore parts toward the head (northwest) and coasts, and the average SST decreased from west to east (i.e., entrance at the Sea of Oman) from summer to early autumn and increased from late autumn to spring (Moradi and Kabiri 2015).

5.5 CONCLUSIONS

We studied the distribution of ocean colour parameters such as R_{rs}(443), Chl-a, SST, PAR and a_{dg}(443) in the Arabian Gulf. These ocean colour parameters are the optically active constituents and are used to describe ocean colour biology. Spatial distribution maps for the Arabian Gulf were created using MODIS-Aqua ocean colour data products. These maps were examined to understand the ocean colour biology of the Gulf. In general, the winter peak of Chl-a concentrations was located mainly in some coastal areas while the summer peak was found in the southern and northwest regions. In conclusion, ocean colour remote sensing is a valuable tool for monitoring the aquatic environments of the Arabian Gulf, but it is important to carefully consider the unique optical properties of the Gulf's waters and the availability of in-situ measurements when using this technique. With continued development and refinement of algorithms and validation methods, ocean colour remote sensing has the potential to provide valuable insights into the health and productivity of the Gulf's marine ecosystems.

5.6 FUTURE RESEARCH DIRECTIONS AND OUTLOOK

This chapter has used long-term satellite observation to explain the ocean colour distribution pattern in the Arabian Gulf. As already discussed in this chapter, retrieval of Chl-a concentration from satellite sensors is based on the blue, green and red bands information for open ocean and coastal waters. Thus, the accuracy of the Chl-a estimates becomes lower for the water dominated by other constituents such as total suspended solids (TSS) and coloured dissolved organic matter. Consequently, there is considerable scope for improving the satellite retrieval of the ocean colour parameters using in-situ optical measurements from various platforms, requiring continuing efforts as follows:

1 A monitoring system for phytoplankton blooms in the Arabian Gulf would be highly advantageous; however, current ocean colour products cannot adequately reproduce Chl-a concentration in coastal parts of the Gulf. Therefore, accurate models need to be developed for mapping the phytoplankton distribution in the Gulf.
2 Independent global and regional studies with multiple types of measurement methods including Argo, LiDAR, buoy, underway flowthrough, and mooring of parameters (R_{rs}, CDOM, TSS and Chl-a) need to be conducted to assess the accuracy of the remote-sensing-derived parameters.
3 Integration of data from multiple sources with machine learning approaches could further improve the estimates of ocean colour parameters.

Overall, addressing the aforementioned factors is crucial for improving the accuracy and reliability of ocean colour data products in the Arabian Gulf. This can be achieved through the use of high-quality measured data and improved parameterizations and the development of regional optical models that are appropriately scaled and validated in the Arabian Gulf.

ACKNOWLEDGEMENTS

We would like to thank the King Fahd University of Petroleum & Minerals, Dhahran, Saudi Arabia, for the overall research support. We thank Mr. Omer Rehman Reshi for making available the satellite images. The authors would also like to thank the NASA GSFC Ocean Biology Processing Group (OBPG) for the ocean colour satellite remote-sensing data.

REFERENCES

Abdul Azis, P.K., Al-Tisan, I.A., Daili, M.A., Green, T.N., Dalvi, A.G.I., and M.A. Javeed. 2003. Chlorophyll and plankton of the Gulf coastal waters of Saudi Arabia bordering a desalination plant. *Desalination* 154, no. 3: 291–302. http://doi.org/10.1016/S0011-9164(03)80044-9.

Al-Yamani, F.Y., Subba Rao, D.V., Mharzi, A., Ismail, W., and K. Al-Rifaie. 2006. Primary production off Kuwait, an arid zone environment, Arabian Gulf. *Int J Oceans Oceanogr* 1, no. 1: 27–49.

Bailey, S.W., Franz, B.A., and P.J. Werdell. 2010. Estimation of near-infrared water-leaving reflectance for satellite ocean colour data processing. *Opt Express* 18, no. 7: 7521. http://doi.org/10.1364/oe.18.007521.

Board, S.S., Board, O.S., and National Research Council. 2011. *Assessing the Requirements for Sustained Ocean Colour Research and Operations*. Washington, DC: National Academies Press.

Chandy, J.P., Al-Tisan, I., Munshi, H.A., and H. Abd el Reheim. 1991. Marine phytoplankton: A study on seanal abuance and distribution in Jubail. *Technical Report No. SWCC (RDC)*. https://www.semanticscholar.org/paper/MARINE-PHYTOPLANKTON-%3A-A-STUDY-ON-SEASONAL-AND-IN-l-Chandy-Al-Tisan/fa586b89af78e7c63367584dbae7a61baef5063c.

Dorgham, M.M. 2013. Plankton research in the ROPME Sea Area, achievements and gaps. *Int J Environ Res* 7, no. 3: 767–778. www.sid.ir/en/VEWSSID/J_pdf/108220130324.pdf.

Gordon, H.R. 2021. Evolution of ocean colour atmospheric correction: 1970–2005. *Remote Sens* 13, no. 24: 1970–2005. http://doi.org/10.3390/rs13245051.

Hu, C., Feng, L., Lee, Z., Franz, B.A., Bailey, S.W., Werdell, P.J., and C.W. Proctor. 2019. Improving satellite global chlorophyll a data products through algorithm refinement and data recovery. *J Geophys Res Oceans* 124, no. 3: 1524–1543. http://doi.org/10.1029/2019JC014941.

Hu, C., Lee, Z., and B. Franz. 2012. Chlorophyll a algorithms for oligotrophic oceans: A novel approach based on three-band reflectance difference. *J Geophys Res Oceans* 117, no. 1: 1–25. http://doi.org/10.1029/2011JC007395.

Hulburt, E.M., Mhmoodian, F., Russell, M., Stalcup, F., Lalezary, S., and P. Amirhor. 1981. Attributes of the plankton flora at Bushehr, Iran. *Hydrobiologia* 79: 51–63.

IOCCG. 2010. *Atmospheric Correction for Remotely-Sensed Ocean-Colour Products: Vol. No. 10*. Ed. M. Wang. IOCCG. http://doi.org/10.25607/OBP-101.

Mobley, C.D., Werdell, J., Franz, B., Ahmad, Z., and S. Bailey. 2016. Atmospheric correction for satellite ocean colour radiometry. *NASA Ocean Biology Processing Group*, 73 p.

Moradi, M., and K. Kabiri. 2015. Spatio-temporal variability of SST and Chlorophyll-a from MODIS data in the Persian Gulf. *Mar Pollut Bull* 98, nos. 1–2: 14–25.

Noori, R., Tian, F., Berndtsson, R., Abbasi, M.R., Naseh, M.V., Modabberi, A., Soltani, A., and B. Kløve. 2019. Recent and future trends in sea surface temperature across the Persian Gulf and Gulf of Oman. *PLoS ONE* 14, no. 2: e0212790. http://doi.org/10.1371/journal.pone.0212790.

O'Reilly, J.E., Maritorena, S., Mitchell, B.G., Siegel, D.A., Carder, K.L., Garver, S.A., Kahru, M., and C.R. McClain. 1998. Ocean colour chlorophyll algorithms for SeaWiFS. *J Geophys Res* 103, no. C11: 24937–24953. http://doi.org/10.1029/98JC02160.

O'Reilly, J.E., and P.J. Werdell. 2019. Chlorophyll algorithms for ocean colour sensors – OC4, OC5 & OC6. *Remote Sens Environ* 229: 32–47. http://doi.org/10.1016/j.rse.2019.04.021.

Polikarpov, I., Al-Yamani, F., and M. Saburova. 2009. Space-time variability of phytoplankton structure and diversity in the north-western part of the Arabian Gulf (Kuwait). *BioRisk* 3: 86–96. http://doi.org/10.3897/biorisk.3.8.

Sheppard, C., Al-Husiani, M., Al-Jamali, F., Al-Yamani, F., Baldwin, R., Bishop, J., Benzoni, F., Dutrieux, E., Dulvy, N.K., Durvasula, S.R.V., Jones, D.A., Loughland, R., Medio, D., Nithyanandan, M., Pilling, G.M., Polikarpov, I., Price, A.R.G., Purkis, S., Riegl, B., Saburova, M., Namin, K., Taylor, O., Wilson, S., and K. Zainal. 2010. The Gulf: A young sea in decline. *Mar Pollut Bull* 60, no. 1: 13–38. http://doi.org/10.1016/J.MARPOLBUL.2009.10.017.

Subba Rao, D.V., and F. Al-Yamani. 1998. Phytoplankton ecology in the waters between Shatt Al-Arab and Straits of Hormuz, Arabian Gulf: A review. *Plankton Biol Ecol* 45, no. 2: 101–116.

Tiwari, S.P., and P. Shanmugam. 2011. An optical model for the remote sensing of coloured dissolved organic matter in coastal/ocean waters. *Estuarine Coastal Shelf Sci* 93, no. 4: 396–402.

Werdell, P.J., Franz, B.A., Bailey, S.W., Feldman, G.C., Boss, E., Brando, V.E., Dowell, M., Hirata, T., Lavender, S.J., Lee, Z., Loisel, H., Maritorena, S., Mélin, F., Moore, T.S., Smyth, T.J., Antoine, D., Devred, E., D'Andon, O.H.F., and A. Mangin. 2013. Generalized ocean colour inversion model for retrieving marine inherent optical properties. *Appl Opt* 52: 2019–2037. http://doi.org/10.1364/AO.52.002019.

Eutrophication in the Red Sea
Causes and effects

6

Neila Annabi-Trabelsi, Wassim Guermazi, Amira Rekik, Habib Ayadi and Vincent Leignel

6.1 THE RED SEA: MAIN CHARACTERISTICS

The Red Sea is a semi-enclosed narrow tropical sea separating northern Africa from the Arabian Peninsula (Western Asia), extending from 12.5°N to about 30°N over a distance of about 2250 km with an average width of 280 km (Acker et al. 2008). At the northern end, it separates into the Gulf of Aqaba and the Gulf of Suez, which have been connected to the Mediterranean Sea via the Suez Canal (Figure 6.1) since 1868. At the southern end, it has been connected to the Indian Ocean (Gulf of Aden and the Arabian Sea) through the Bab el Mandeb passage for about three million years (Edwards 1987).

The horizontal and vertical circulation of the Red Sea is governed by seasonally varying atmospheric forcing, which is characterized by strong winds and freshwater fluxes (Papadopoulos et al. 2013; Yao et al. 2014). The main wind patterns are a southward flow during summer along the basin axis and a reversal of wind direction in the southern part of the basin in response to the winter monsoon (Langodan et al. 2014). Inflowing surface and intermediate waters from the Gulf of Aden are gradually altered by evaporation and cooling, with the density increasing northward. This forms an eastern boundary current that appears to manifest between 16° and 21° N (Yao et al. 2014). This eastern boundary current episodically routes patches of warmer, less saline water rich in nutrients and plankton biomass from the south toward the central Red Sea (Acker et al. 2008).

The Red Sea is a spatially heterogeneous ecosystem based on gradients from north to south in salinity (35–41 psu), temperature (21–34°C) and primary productivity estimated by Chlorophyll *a* concentration (0.1–5.0 mg m^{-3}) (Sofianos and Johns 2003; Raitsos et al. 2013; DiBattista et al. 2016). The Red Sea is highly oligotrophic, particularly in the subtropical central and northern areas (Kürten et al. 2014; Roth et al. 2018). However, anthropogenic pressures on the Red Sea are constantly increasing (Carvalho et al. 2019; Karcher et al. 2020). The countries with coastlines represent sources of diversified pollution released into the Red Sea as illustrated by poorly or untreated sewage discharge from the cities and tourist centres (Gerges 2002; Ziegler et al. 2016). Gladstone et al. (2013) identified the direct impacts of coastal and marine tourism, mariculture, urbanization and land use practices on the environment in the Red Sea, including the Gulf of Aqaba and Gulf of Suez, such as local pollution, resource depletion, habitat loss and conversion and habitat and wildlife disturbances. Eutrophication is one of the greatest threats to the Red Sea (Loya 2004).

The Red Sea contains extensive areas of coral reefs (Figure 6.1), supporting high levels of diversity with numerous endemic species. There are 346 Scleractinian coral species (Anthozoa), 635 species of Polychaeta, 211 Echinodermata, 79 Tunicata, 231 Crustacea, 137 Mollusca and 1071 shallow fish species in the only coral reef environment of the Red Sea (DiBattista et al. 2016). The Red Sea is also rich in other marine habitats, such as mangroves, islands and shoals, as well as coastal lagoons, which contribute to the high levels of diversity (Carvalho et al. 2019). This probably corresponds to more than 3,500 species in general with biodiversity higher than that of comparable enclosed seas (e.g., the Baltic, the Black and the Caspian Seas), where the number of species never passes 1,000.

6.2 DEFINITION OF EUTROPHICATION

Eutrophication is defined as an increase in the rate of supply of organic matter to an ecosystem. (Vollenweider 1992; Nixon 1995, 2009). Nixon (1995) noted that increases in the supply of organic matter to coastal ecosystems have various causes, the most common being excess inputs of labile, inorganic nitrogen and phosphorus. The European Union defines cultural eutrophication as the enrichment of water by nutrients, especially compounds of nitrogen and phosphorus, causing an accelerated growth of algae and higher forms of plant life to produce an undesirable disturbance to the balance of organisms present in the water and to the quality of the water concerned (European Commission 1991). The OSPAR definition pointed out that eutrophication may result in a range of undesirable disturbances in the marine ecosystem,

DOI: 10.1201/9781003321408-8

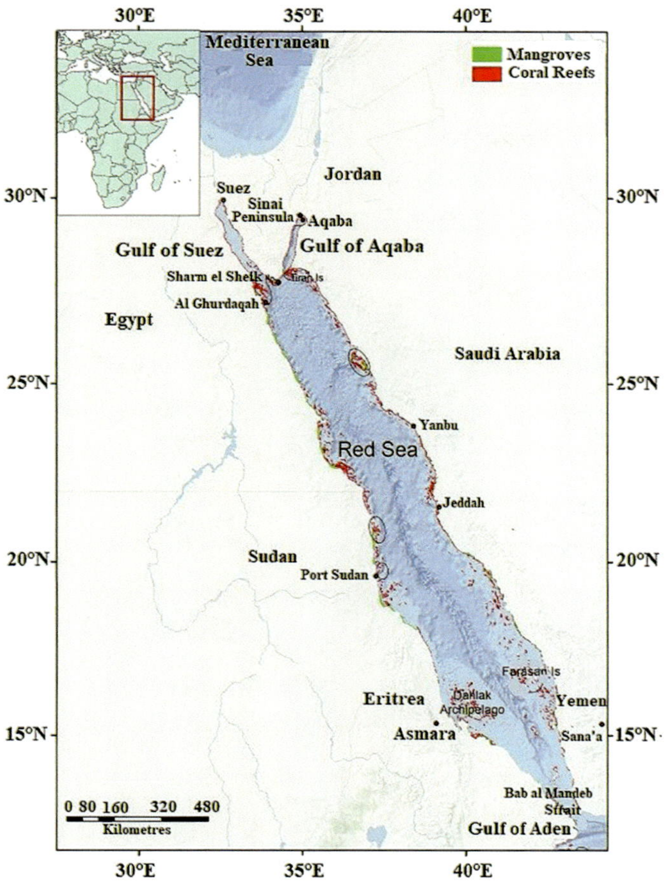

FIGURE 6.1 Map of the Red Sea showing the distribution of mangroves and coral reefs.

Source: Carvalho et al. (2019)

including a shift in the composition of the flora and fauna which affects habitats and biodiversity, and the depletion of oxygen, causing the death of fish and other species (OSPAR 1997). Coastal marine eutrophication has been recognized as a problem of "disturbance" of the marine ecosystems since the 1960s (Postma 1966) and one of the greatest threats to the health of coastal estuarine and marine ecosystems worldwide (Malone and Newton 2020).

The common procedure of OSPAR includes identifying the main parameters according to their causal relationships involved in eutrophication processes (Figure 6.2). In this illustration, all aspects of nutrient enrichment (nutrient inputs, concentrations and ratios) as well as possible direct effects (e.g., increased levels of nuisance and/or toxic phytoplankton species, shifts and/or losses of submerged aquatic vegetation) and indirect effects (e.g., oxygen deficiency, changes and/or death of benthos, death of fish, algal toxins) are covered. While the immediate effect of eutrophication is increased primary production via phytoplankton or macrophytes (McGlathery et al. 2007), it leads to changes in the energy flow of aquatic food webs (Turner 2001; van der Lee et al. 2021). According to Selman et al. (2008), the initial impacts of nutrient increases are the excessive growth of phytoplankton and macroalgae. These, in turn, can lead to other impacts such as:

1 Loss of subaquatic vegetation as excessive phytoplankton and macroalgae growth reduces light penetration (Rabalais 2002; Adams et al. 2020).
2 Change in species composition and biomass of the benthic (bottom-dwelling) aquatic community, eventually leading to reduced species diversity and the dominance of gelatinous organisms such as jellyfish (Richardson et al. 2009; Dorgham 2014).
3 Coral reef damage as increased nutrient levels favour algal growth in films over coral larvae (Hallock 2001). Coral growth is inhibited because the algae outcompete coral larvae for available surfaces on which to grow.
4 A shift in phytoplankton species composition, creating favourable conditions for the development of nuisance, toxic or otherwise harmful algal blooms (Cloern 2001; Wurtsbaugh et al. 2019; Brush et al. 2020).
5 Low dissolved oxygen and formation of hypoxic or "dead" zones (oxygen-depleted waters), particularly on the bottom, which, in turn can lead to ecosystem collapse (Wurtsbaugh et al. 2019; Brush et al. 2020).

Selman et al. (2008) identified 415 eutrophic and hypoxic coastal areas, with 169 documented hypoxic areas, 233 areas of concern and 13 systems in recovery worldwide (Figure 6.3). Three of these areas located in the Red Sea, at the level of the Gulf of Aqaba, the Egyptian coast and the Yemen coast, are described as hypoxic.

6.3 THE BEGINNING OF THE EUTROPHICATION PROCESS: RELEASE OF NUTRIENTS IN THE RED SEA

As mentioned in the definition of eutrophication, the major drivers are inorganic molecules containing nitrogen and phosphorus (Nixon 1995; Elser et al. 2007; Henderson 2015). The nutrient sources of the Red Sea include the intrusion of Indian Ocean water, aerosol deposition, atmospheric N_2 fixation by algae and urban runoff (Sofianos and Johns 2007; Acker et al. 2008; Aberle et al. 2010; Wankel et al. 2010; Raitsos et al. 2013). Due to the winter monsoon, nutrient concentrations gradually increase from north to south and toward the coasts (Sofianos and Johns 2007). In fact, during the period from November to March, the NE winter monsoon generates a current from the Gulf of Aden into the southern part of the Red Sea (Halim 1984).

Additional nutrient inputs to the Red Sea come from industrial and urban wastewater (Basaham et al. 2009; Al-Farawati 2010; Kürten et al. 2014; Peña-García et al. 2014) and from aquaculture (Loya et al. 2004; Kürten et al. 2015), inducing localized eutrophication. Moreover, the discharge of insufficiently treated sewage was shown to raise nearshore nitrogen availability (Sawall et al. 2014). In fact, regional eutrophication was recorded for the developed urban areas at the northern tip of the Gulf of Aqaba (Loya and Kramarsky-Winter 2003; Naumann et al. 2015; Hall et al. 2018). The Gulf of Suez is also exposed to different sources of pollution, including agricultural, industrial and domestic wastes; oil exploration; and

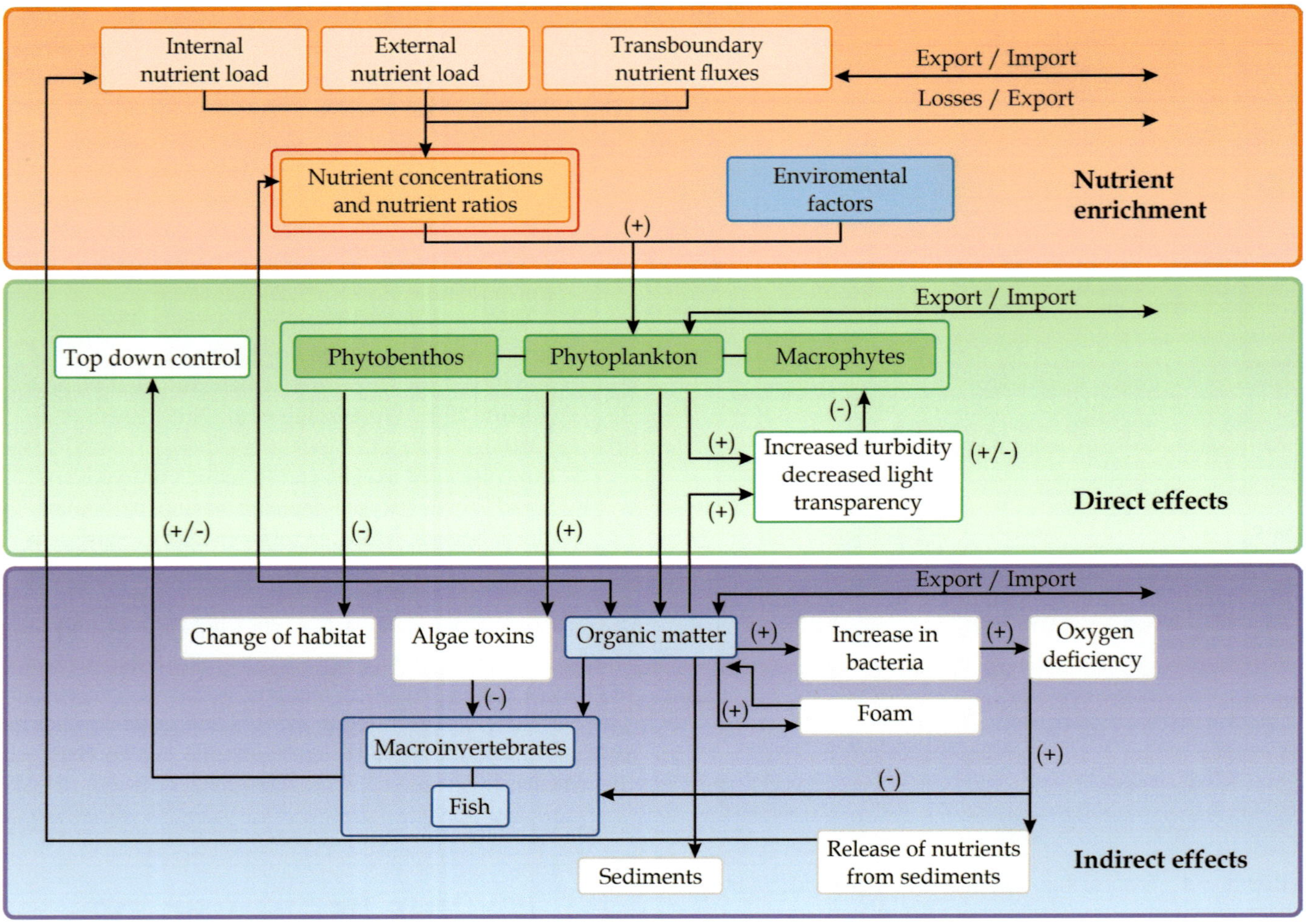

FIGURE 6.2 Conceptual diagram of eutrophication. The (+) and (-) symbols respectively indicate a positive (increase, reinforcement) and negative (decrease, inhibition) effect on the components of the ecosystem.

Source: Claussen et al. (2009)

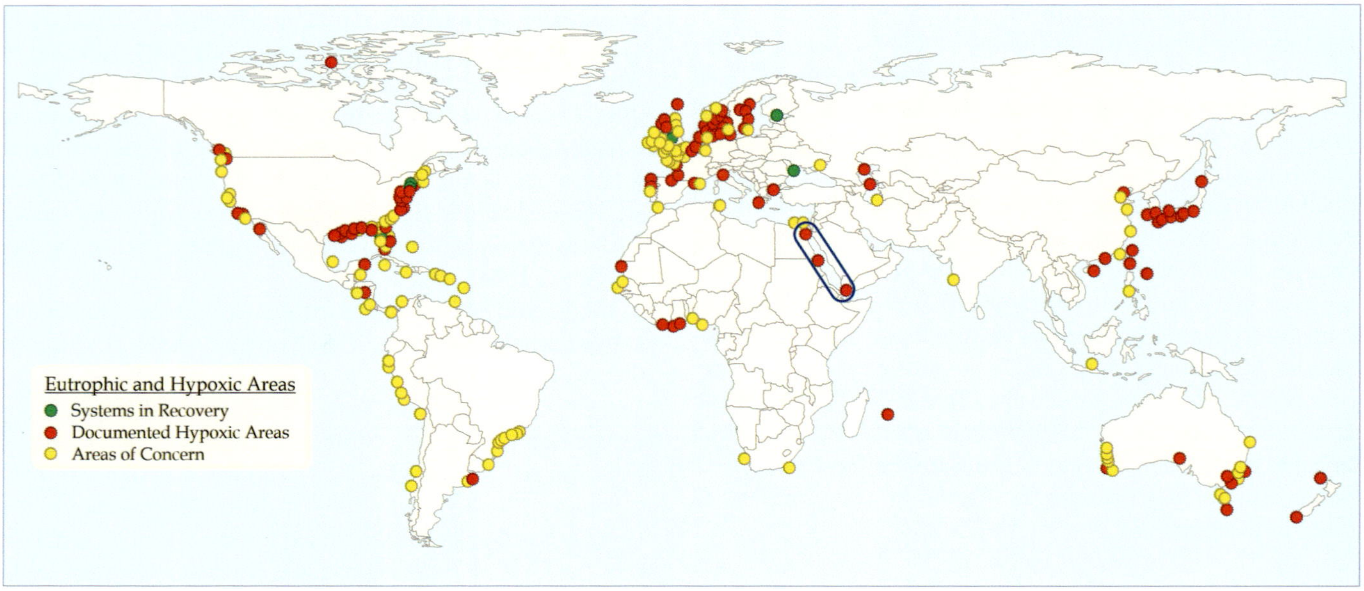

FIGURE 6.3 Eutrophic and hypoxic coastal systems worldwide.

Source: Selman et al. (2008)

ship ballasts (Nassar et al. 2015). The Suez Canal, being a main international maritime shipping route, experiences heavy traffic with illegal oil discharges (Abou El-Magd et al. 2020). Oil pollution harms the marine ecosystem and creates pressure on the coastal socioeconomic activities in this area (Abou El-Magd et al. 2020). This pollution increases nutrient concentrations in the Gulf water, which eventually accelerates eutrophication (Nassar et al. 2015).

In the polluted coastal water of Jeddah, high levels of nutrient loadings especially showed evident signs of eutrophication (Al-Harbi and Khomayis 2010): nitrates, nitrites, ammonium and phosphates reach at surface layers 62.31, 13.98, 103.98 and 58.31 μg L^{-1}, respectively, leading to high chlorophyll-*a* values between 1.84 and 242.2 mg L^{-1}. According to Peña-García et al. (2014), daily loads of total nitrogen (TN) and phosphorus (TP) from Jeddah into the Red Sea amount to 6,564 kg and 2241 kg, respectively, comprising 83% dissolved inorganic nitrogen and 33% dissolved phosphate. For example, Al Shabab and Al Arbaeen (Jeddah City) are coastal inlets that have been, for decades, the site of the dumping of about 100,000 m^3/day of semi-treated sewage and became a source of annoyance, particularly due to the emanation of bad odours (El Sayed et al. 2015). The pollution load, calculated using concentrations at zero salinity and nominal hydraulic flow of the effluents, amounted to 2,737 and 285 kg/day of inorganic nitrogen and phosphorus, respectively (El Sayed et al. 2015). Potential algal mass production that may result from the total assimilation of this nutrient influx is equivalent to 23–30 tonnes/day, which is several times higher than the direct organic matter input from the sewage effluents (El Sayed et al. 2015). The analysis of the Trophic State Index (TSI) using a logarithmic chlorophyll *a* concentration (Chl-a), Secchi depth (SD) and total phosphorus (TP) (Carlson 1977) demonstrated the eutrophic and even the progressed hypereutrophic condition in the coastal area situated around 140 km north of Jeddah (Touliabah and Elbassat 2017).

6.4 HARMFUL MICROALGAE BLOOMS IN THE RED SEA

The most studied aspect of marine eutrophication is the intense proliferation of some phytoplanktonic species in coastal areas (coloured waters). Subsequently, such blooms can cause severe environmental problems either directly, through toxin production, or indirectly, through their high biomass degradation (Stumpf and Tomlinson 2007). In fact, harmful algal blooms can cause fish kills, human illness through shellfish poisoning and the death of marine mammals and shore birds (Anderson et al. 2002). Five harmful algal blooms were recorded in the Red Sea during the last three decades, from which three blooms were formed by Dinophyta (*Noctiluca scintillans*, *Pyrodinuium bahamense* and *Protoperidinium quinquecorne*), one by Raphidiophyta (*Heterosigma akashiwo*) and one by Cyanobacteria (*Trichodesmium erythraeum*) (Mohamed 2018).

An intense bloom of *Noctiluca scintillans* was first noticed in the Red Sea off the Al-Shuqayq region, southwestern Saudi Arabia, in February 2004, with a concentration of 3 × 10 cells L^{-1} (Mohamed and Mesaad 2007). This heterotrophic dinoflagellate species is one of the most important and abundant red tide organisms and is distributed worldwide (Harrison et al. 2011). A *Noctiluca scintillans* bloom is indirectly linked to increasing eutrophication by an increase in prey abundance (Mohamed and Mesaad 2007). *Noctiluca scintillans* is known to have no photosynthetic pigments and to feed on other phytoplankton, small zooplankton and their eggs (Suthers et al. 2019). It is particularly abundant in high-productivity areas where diatoms dominate since they are its preferred food source (Turkoglu 2013). A bloom of the harmful dinoflagellate *Pyrodinium bahamense* (8×10^4 cells L^{-1}) in a lagoon north of Jeddah (Saudi Arabia) was recorded in November 2013 (Banguera-Hinestroza et al. 2016). A saxitoxin ELISA immunoassay of cultures and water samples confirmed the toxicity of the strain found in the Red Sea (Banguera-Hinestroza et al. 2016). Besides dinoflagellate blooms, the southern Red Sea coasts of Saudi Arabia witnessed a large bloom of the harmful raphidophyte *Heterosigma akashiwo* during May and June 2010 (Mohamed and Al-Shehri 2012). The *Protoperidinium quinquecorne* bloom (1.4×10^7 cells L^{-1}) observed on 6 June 2012 along the Al Hodeidah coast in the Yemeni Red Sea was accompanied by a massive die-off of sardines (*Sardinella sp.*) (Alkawri et al. 2016). On the Egyptian coast, many potentially harmful and bloom-forming species were recorded in the main body of the Red Sea, the Gulf of Aqaba and the Gulf of Suez. In the main body of the Red Sea, potentially harmful dinoflagellates of the genera *Ceratium*, *Protoperidinium*, *Dinophysis* and *Gonyaulax* were recorded in considerable numbers ($1–2 \times 10^3$ cells L^{-1}) during spring 2008 at the Hurghada and Sharm El-Sheikh coasts (Madkour et al. 2010). The Egyptian Red Sea coast showed a peak of the toxic cyanobacterium *Trichodesmium erythraeum* (3×10^3 individual L^{-1}) in the summer of 2008 (Madkour et al. 2010). Similarly, in the coastal waters of the Gulf of Aqaba, a bloom of the potentially toxic cyanobacteria *Trichodesmium* thiebautii and *Trichodesmium erythraeum* with > 10^6 colonies m^{-3} was observed during the fall of 1997 (Post et al. 2002).

6.5 IMPACT OF EUTROPHICATION ON CORAL REEFS

Coral reefs are the most abundant shallow water ecosystems in the Red Sea, harbouring a high species diversity and habitat complexity over large environmental gradients (DeVantier et al. 2000; DiBattista et al. 2016). They provide a variety of ecosystem services, including seafood, coastal protection, biogeochemical processes, recreational possibilities and tourism (Moberg and Folke 1999). Coral reefs are threatened by global (e.g., ocean acidification, ocean warming) and local (e.g., coral disease outbreaks, overfishing, eutrophication) anthropogenic stressors (Hughes et al. 2003; Fabricius 2005; Hoegh-Guldberg et al. 2007; Ali 2009; Sully et al. 2022). While Gulf of Aqaba corals may be resilient to ocean acidification and warming (Osman et al. 2017; Roik et al. 2017), eutrophication and overfishing are among the two most important factors affecting coral reefs that typically grow in nutrient-poor waters (Naumann et al. 2015). Eutrophication strongly affects coral reefs by way of nutrient enrichment, light loss from turbidity and the smothering and alteration of surface properties from sedimentation (Fabricius 2011). These changes lead to changes in trophic structures and reduced coral recruitment and diversity (Fabricius 2011; Duprey et al. 2016).

On the northern tip of the Gulf of Aqaba, coastal eutrophication caused a loss of up to 76% of the hard coral cover at Eilat between 1986 and 2000 (Loya 2004) and a decline of 12% in coral cover at the coastline of Dahab from 2010 to 2013 (Naumann et al. 2015). The rate of death of colonies of the coral *Stylophora pistillata* was found to be four to five times as great in the polluted area of the Gulf of Aqaba as in a control area (Walker and Ormond 1982). A rise in nutrient concentrations would severely impede the Gulf of Aqaba reef's capacity for growth and significantly alter the bacterial community structure and functionality of the coral holobiont (Hall et al. 2018). Hall et al. (2018) suggested that eutrophication may compromise the resilience of *Stylophora pistillata* to ocean acidification and warming conditions in the Gulf of Aqaba. Silverman et al. (2007) have assessed an apparent effect of elevated levels of dissolved nutrients on the calcification of an entire coral reef in the Gulf of Aqaba. Nutrients released from the fish farms in the Gulf of Aqaba have adverse effects on the successful production of larvae by *Stylophora pistillata* by causing failure of oocyte maturation, fertilization,= and ensuing larval development (Loya et al. 2004). Hall et al. (2018) reported a shift in the coral microbial community in eutrophication conditions that may, in turn, affect the coral health. In addition, increases in dissolved inorganic and organic nutrients may stimulate bacterial production and activity that can be harmful to corals and other marine organisms (Kline et al. 2006). For example, coral-associated microbe abundances increase under eutrophic conditions (Sawall et al. 2012), which favours those microbes that benefit from nutrient-rich waters (Jessen et al. 2013). In the Red Sea, coral *Acropora hemprichii*, microbes of the genus *Nautella* and *Defluvibacter*, particularly benefitted from inorganic nutrient enrichment (Jessen et al. 2013). Mohamed et al. (2012) studied nine sites along the northern Egyptian Red Sea, concluding that enhanced local anthropogenic stresses are causes of the initiation and the persistence of coral diseases in the studied reefs. In contrast, other benthic groups in coral reefs, such as turf and macroalgae, benefit from increased nutrient availability in many cases (Costa et al. 2000), particularly in combination with reduced herbivory. Hence, shifts from coral- to algal-dominated reefs occur in the Red Sea (Bahartan et al. 2010; Naumann et al. 2015).

6.6 IMPACT OF EUTROPHICATION ON MANGROVES

Mangroves are woody trees and shrubs that grow in unique intertidal environments and cover nearly 60–70% of the earth's tropical and subtropical coastal areas (Ramsey and Jensen 1996; Sohaib et al. 2023). Along the Red Sea, mangroves do not grow as luxuriously as most other tropical mangroves due to difficult conditions in their distribution range, including no permanent freshwater inputs, salinities over 40 psu, sea surface temperatures over 31°C in summer and the warming of the sea (Mandura et al. 1988; Raitsos et al. 2011). Mangroves are found in the most threatened ecosystems on Earth; about one-third of their global area has been lost since World War II (Alongi 2002). This decline continues at an annual rate of about 2.1% (Valiela et al. 2001). However, the area of Red Sea mangroves (135 km^2) has increased by 12% over the past 40 years (Almahasheer et al. 2016). In fact, losses of Red Sea mangroves, mostly due to coastal development, have been compensated by afforestation projects (Almahasheer et al. 2016). These reports identify aquaculture and urban development as the main drivers of mangrove decline (Webber et al. 2016).

Mangroves, localized along the coast of the Red Sea, are represented mainly by three species (*Avicennia marina*, *Rhizophora mucronata* and *Bruguiera gymnorhiza*) (Mandura and Khafaji 1994). *Avicennia marina* is the dominant species in the Red Sea (Mandura 1997). Urban expansion and different types of anthropogenic activities along the Red Sea coast of Saudi Arabia are among the major causes of the deterioration of mangrove habitats and the decline of mangrove plants (Sohaib et al. 2023). Moreover, Arshad et al. (2018) showed that sewage discharge with high concentrations of nutrients leads to the death of pneumatophores, which have been shown to affect mangrove productivity. The death of pneumatophores decreases the aeration area, which apparently affects the respiration rate of the root system, nutrient uptake and plant growth, consequently leading to retarded growth of mangroves (Mandura 1997). The mangrove stand of *Avicennia marina* located at the South Corniche of Jeddah City receives about 100 000 m^3 of sewage with high nutrient concentrations per day (Mandura 1997). The effect of sewage discharge on the South Corniche mangrove stand is summarised by Mandura (1997) as follows: it 1) leads to stunted mangrove plants, 2) causes death to pneumatophores, 3) induces pneumatophore abnormal growth and 4) reduces the areas of pneumatophore distribution, root respiration and nutrient uptake.

6.7 CONCLUSION

Eutrophication is an issue that requires greater attention from governments and society in general. Implementation of tools to help assess the effectiveness of alternative scenarios for reducing nutrient inputs is especially needed to estimate coastal impact and target appropriate policy responses in the Red Sea. We suggest strong management efforts for monitoring, assessing and regulating measures to offset the negative trends in the sustainability of coral reefs and mangroves in the Red Sea coastal regions. Dead corals reefs should be progressively restored by planting new nursery-grown corals on old or artificial reef materials.

In order to successfully implement efficient measures to mitigate nutrient pollution in the Jeddah Bay area, Peña-García et al. (2014) suggested first closing down anthropogenic nutrient inputs to the lagoons and redirecting wastewater to sewage works equipped with tertiary treatment, and, as much as possible, treated water should be reused for industrial processing and irrigation purposes, and then rejected through extended multiport diffuser pipes toward the deeper outer water body exhibiting sufficient cross-cutting currents to allow for rapid dilution.

REFERENCES

Aberle, N., Hansen, T., Boettger-Schnack, R., et al. 2010. Differential routing of 'new' nitrogen toward higher trophic levels within the marine food web of the Gulf of Aqaba, northern Red Sea. *Mar Biol* 157: 157–169. https://doi.org/10.1007/s00227-009-1306-y.

Abou El-Magd, I., Zakzouk, M., Abdulaziz, A.M., and E.M. Ali. 2020. The potentiality of operational mapping of oil pollution in the Mediterranean Sea near the entrance of the Suez Canal using Sentinel-1 SAR data. *Remote Sens* 12: 1352. http://doi.org/10.3390/rs12081352.

Acker, J., Leptoukh, G., Shen, S., Zhu, T., and S. Kempler. 2008. Remotely-sensed chlorophyll a observations of the northern Red Sea indicate seasonal variability and influence of coastal reefs. *J Marine Systems* 69, nos. 3–4: 191–204. http://doi.org/10.1016/j.jmarsys.2005.12.006.

Adams, J.B., Taljaard, S., Van Niekerk, L., and D.A. Lemley. 2020. Nutrient enrichment as a threat to the ecological resilience and health of South African microtidal estuaries. *Afr J Aquat Sci* 45, nos. 1–2: 23–40. http://doi.org/10.2989/16085914.2019.1677212.

Al-Farawati, R. 2010. Environmental conditions of the coastal waters of southern Corniche, Jeddah, eastern Red Sea: Physico-chemical approach. *Aust J Basic Appl Sci* 4, no. 8: 3324–3337.

Al-Harbi, S.M., and H.S. Khomayis. 2010. Eutrophication and chlorophyll-a in a severely polluted coastal water of Jeddah, Red Sea. *J King Abdulaziz Univ* 21, no. 2: 15.

Ali, M. 2009. Coral bleaching: Processes, extent, and implications: A review. *Mesopot J Mar Sci* 24, no. 1: 35–42.

Alkawri, A., Al Areeki, M., and K. Alsharaby. 2016. The first recorded bloom of *Protoperidinium quinquecorne* and its link to a massive fish kill in Yemeni coastal waters, southern Red Sea. *Plankton Benthos Res* 11, no. 2: 75–78.

Almahasheer, H., Aljowair, A., Duarte, C.M., and X. Irigoien. 2016. Decadal stability of Red Sea mangroves. *Estuar Coast Shelf Sci* 169: 164–172. http://doi.org/10.1016/j.ecss.2015.11.027.

Alongi, D.M. 2002. Present state and future of the world's mangrove forests. *Environ Conserv* 29, no. 3: 331–349. http://doi.org/10.1017/S0376892902000231.

Anderson, D.M., Glibert, P.M., and J.M. Burkholder. 2002. Harmful algal blooms and eutrophication: Nutrient sources, composition, and consequences. *Estuaries* 25: 704–726. http://doi.org/10.1007/BF02804901.

Arshad, M., Alrumman, S.A., and E.M. Eid. 2018. Evaluation of carbon sequestration in the sediment of polluted and non-polluted locations of mangroves. *Fundam Appl Limnol* 192, no. 1: 53–64.

Bahartan, K., Zibdah, M., Ahmed, Y., Israel, A., Brickner, I., and A. Abelson. 2010. Macroalgae in the coral reefs of Eilat (Gulf of Aqaba, Red Sea) as a possible indicator of reef degradation. *Mar Pollut Bull* 60, no. 5: 759–764.

Banguera-Hinestroza, E., Eikrem, W., Mansour, H., Solberg, I., Cúrdia, J., Holtermann, K., Edvardsen, B., and S. Kaartvedt. 2016. Seasonality and toxin production of Pyrodinium bahamense in a Red Sea lagoon. *Harmful Algae* 55: 163–171. http://doi.org/10.1016/j.hal.2016.03.002.

Basaham, A., Rifaat, A., El-Mamoney, M., and M. El-Sayed. 2009. Re-evaluation of the impact of sewage disposal on coastal sediments of the Southern Corniche, Jeddah, Saudi Arabia. *J King Abdulaziz Univ Mar Sci* 20, no. 1: 109–126.

Brush, M.J., Giani, M., Totti, C., Testa, J.M., Faganeli, J., Ogrinc, N., Kemp, W.M., and S.F. Umani. 2020. Eutrophication, harmful algae, oxygen depletion, and acidification. In: *Coastal Ecosystems in Transition*. Eds. T.C. Malone, A. Malej and J. Faganeli. American Geophysical Union, Geophysical Monograph 256. http://doi.org/10.1002/9781119543626.ch5.

Carlson, R.E. 1977. A trophic state index for lakes 1. *Limnol Oceanogr.* 22, no. 2: 361–369.

Carvalho, S., Kürten, B., Krokos, G., Hoteit, I., and J. Ellis. 2019. The Red Sea. In: *World Seas: An Environmental Evaluation*. Ed. C. Sheppard, 49–74. Amsterdam: Elsevier.

Claussen, U., Zevenboom, W., Brockmann, U., Topcu, D., and P. Bot. 2009. Assessment of the eutrophication status of transitional, coastal and marine waters within OSPAR. In: *Eutrophication in Coastal Ecosystems: Towards Better Understanding and Management Strategies*. Selected Papers from the Second International Symposium on Research and Management of Eutrophication in Coastal Ecosystems, 20–23 June 2006, Nyborg, Denmark, 49–58. Dordrecht: Springer. http://doi.org/10.1007/978-90-481-3385-7_5.

Cloern, J.E. 2001. Our evolving conceptual model of the coastal eutrophication problem. *Mar Ecol Prog Ser* 210: 223–253. http://doi.org/10.3354/meps210223.

Costa, O.S., Leão, Z.M.D.A.N., Nimmo, M., and M.J. Attrill. 2000. Nutrification impacts on coral reefs from northern Bahia, Brazil. In: *Island, Ocean and Deep-Sea Biology. Proc 34th European Marine Biology Symposium, Ponta Delgada (Azores), Portugal, 13–17 September 1999*, 307–315. Dordrecht: Springer.

DeVantier, L., Turak, E., Al-Shaikh, K., and G. De'ath. 2000. Coral communities of the central-northern Saudi Arabian Red Sea. *Fauna of Arabia* 18: 23–66.

DiBattista, J.D., Roberts, M.B., Bouwmeester, J., Bowen, B.W., Coker, D.J., Lozano-Cortés, D.F., et al. 2016. A review of contemporary patterns of endemism for shallow water reef fauna in the Red Sea. *J Biogeogr* 43, no. 3: 423–439. http://doi.org/10.1111/jbi.12649. http://doi.org/10.1016/0025-326X(82)90492-1.

Dorgham, M. 2014. Effects of eutrophication. In: *Eutrophication: Causes, Consequences and Control*. Eds. A. Ansari and S. Gill. Dordrecht: Springer. http://doi.org/10.1007/978-94-007-7814-6_3.

Duprey, N.N., Yasuhara, M., and D.M. Baker. 2016. Reefs of tomorrow: Eutrophication reduces coral biodiversity in an urbanized seascape. *Glob Chang Biol* 22, no. 11: 3550–3565. http://doi.org/10.1111/gcb.13432.

Edwards, F.J. 1987. Climate and oceanography. In: *Red Sea*. Eds. A.J. Edwards and S.M. Head, 45–69. Oxford: Pergamon Press.

El Sayed, M.A., Al Farawati, R.K., El Maradny, A.A., et al. 2015. Environmental status and nutrients and dissolved organic carbon budget of two Saudi Arabian Red Sea coastal inlets: A snapshot statement. *Environ Earth Sci* 74: 7755–7767. http://doi.org/10.1007/s12665-013-2557-y.

Elser, J.J., Bracken, M.E., Cleland, E.E., Gruner, D.S., Harpole, W.S., Hillebrand, H., et al. 2007. Global analysis of nitrogen and phosphorus limitation of primary producers in freshwater, marine and terrestrial ecosystems. *Ecol Lett* 10, no. 12: 1135–1142. http://doi.org/10.1111/j.1461-0248.2007.01113.x.

European Commission. 1991. *Council Directive 91/271/EEC of 21 May 1991 Concerning Urban Waste Water Treatment. O.J. L135, 30.5.* European Commission, Brussels.

Fabricius, K.E. 2005. Effects of terrestrial runoff on the ecology of corals and coral reefs: Review and synthesis. *Mar Pollut Bull* 50, no. 2: 125–146. http://doi.org/10.1016/j.marpolbul.2004.11.028.

Fabricius, K.E. 2011. Factors determining the resilience of coral reefs to eutrophication: A review and conceptual model. In: *Coral Reefs: An Ecosystem in Transition*. Eds. Z. Dubinsky and N. Stambler, 493–505. Dordrecht: Springer. http://doi.org/10.1007/978-94-007-0114-4_28.

Gerges, M.A. 2002. The Red Sea and Gulf of Aden action plan – facing the challenges of an ocean gateway. *Ocean Coast Manag* 45, nos. 11–12: 885–903. http://doi.org/10.1016/S0964-5691(02)00112-6.

Gladstone, W., Curley, B., and M.R. Shokri. 2013. Environmental impacts of tourism in the Gulf and the Red Sea. *Mar Pollut Bull* 72, no. 2: 375–388. http://doi.org/10.1016/j.marpolbul.2012.09.017.

Halim, Y. 1984. Plankton of the Red Sea and the Arabian Gulf. *Deep-Sea Res* 31: 969–982.

Hall, E.R., Muller, E.M., Goulet, T., Bellworthy, J., Ritchie, K.B., and M. Fine. 2018. Eutrophication may compromise the resilience of the Red Sea coral *Stylophora pistillata* to global change. *Mar Pollut Bull* 131: 701–711. http://doi.org/10.1016/j.marpolbul.2018.04.067.

Hallock, P. 2001. Coral reefs, carbonate sediments, nutrients, and global change. In: *The History and Sedimentology of Ancient Reef Systems. Topics in Geobiology*, vol. 17. Ed. G.D. Stanley. Boston, MA: Springer. http://doi.org/10.1007/978-1-4615-1219-6_11.

Harrison, P.J., Furuya, K., Glibert, P.M., et al. 2011. Geographical distribution of red and green Noctiluca scintillans. *Chin J Ocean Limnol* 29: 807–831. http://doi.org/10.1007/s00343-011-0510-z.

Henderson, A.D. 2015. Eutrophication. In: *Life Cycle Impact Assessment. LCA Compendium – The Complete World of Life Cycle Assessment*. Eds. M. Hauschild and M. Huijbregts. Dordrecht: Springer. http://doi.org/10.1007/978-94-017-9744-3_10.

Hoegh-Guldberg, O., Mumby, P.J., Hooten, A.J., Steneck, R.S., Greenfield, P., et al. 2007. Coral reefs under rapid climate change and ocean acidification. *Science* 318: 1737–1742.

Hughes, T., Baird, A., Bellwood, D., Card, M., Connolly, S., et al. 2003. Climate change, human impacts, and the resilience of coral reefs. *Science* 301: 929–933.

Jessen, C., Villa Lizcano, J.F., Bayer, T., Roder, C., Aranda, M., Wild, C., and C.R. Voolstra. 2013. In-situ effects of eutrophication and overfishing on physiology and bacterial diversity of the Red Sea coral Acropora hemprichii. *PLoS ONE* 8, no. 4: e62091. http://doi.org/10.1371/journal.pone.0062091.

Karcher, D.B., Roth, F., Carvalho, S., El-Khaled, Y.C., Tilstra, A., Kürten, B., Struck, U., Jones, B.H., and C. Wild. 2020. Nitrogen eutrophication particularly promotes turf algae in coral reefs of the central Red Sea. *Peer J* 8: e8737. http://doi.org/10.7717/peerj.8737.

Kline, D.I., Kuntz, N.M., Breitbart, M., Knowlton, N., and F. Rohwer. 2006. Role of elevated organic carbon levels and microbial activity in coral mortality. *Mar Ecol Prog Ser* 314: 119–125. http://doi.org/10.3354/meps314119.

Kürten, B., Al-Aidaroos, A.M., Struck, U., Khomayis, H.S., Gharbawi, W.Y., and U. Sommer. 2014. Influence of environmental gradients on C and N stable isotope ratios in coral reef biota of the Red Sea, Saudi Arabia. *J Sea Res* 85: 379–394. http://doi.org/10.1016/j.seares.2013.07.008.

Kürten, B., Khomayis, H.S., Devassy, R., Audritz, S., Sommer, U., Struck, U., El-Sherbiny, M.M., and A.M. Al-Aidaroos. 2015. Ecohydrographic constraints on biodiversity and distribution of phytoplankton and zooplankton in coral reefs of the Red Sea, Saudi Arabia. *Mar Ecol* 36, no. 4: 1195–1214.

Langodan, S., Cavaleri, L., Viswanadhapalli, Y., and I. Hoteit. 2014. The Red Sea: A natural laboratory for wind and wave modeling. *J Phys Oceanogr*, 44: 3139–3159.

Loya, Y. 2004. The coral reefs of Eilat – past, present and future: three decades of coral community structure studies. In: *Coral Health and Disease*. Eds. E. Rosenberg and Y. Loya, 1–34. Berlin and Heidelberg: Springer.

Loya, Y., and E. Kramarsky-Winter. 2003. In situ eutrophication caused by fish farms in the northern Gulf of Eilat (Aqaba) is beneficial for its coral reefs: A critique. *Mar Ecol Prog Ser* 261: 299–303.

Loya, Y., Lubinevsky, H., Rosenfeld, M., and E. Kramarsky-Winter. 2004. Nutrient enrichment caused by in situ fish farms at Eilat, Red Sea is detrimental to coral reproduction. *Mar Pollut Bull* 49, no. 4: 344–353. http://doi.org/10.1016/j.marpolbul.2004.06.011.

Madkour, F., El-Sherbiny, M., and M. Aamer. 2010. Phytoplankton population along certain Egyptian coastal regions of the Red Sea. *Egypt J Aquat Biol Fish* 14, no. 2: 95–109. http://doi.org/10.21608/EJABF.2010.2064.

Malone, T.C., and A. Newton. 2020. The globalization of cultural eutrophication in the coastal ocean: Causes and consequences. *Front Mar Sci* 7: 670. http://doi.org/10.3389/fmars.2020.00670.

Mandura, A.S. 1997. Mangrove stand under sewage pollution stress: Red Sea. *Mangroves Salt Marshes* 1: 255–262. http://doi.org/10.1023/A:1009927605517.

Mandura, A.S., and A.K. Khafaji. 1994. An aerial survey aided ecological remarks on mangrove forests of Farasan Archipelago (Red Sea). *International Conference on Desert Development (25–30 July 1994)*, Mexico, Published by International Desert Development Commission, pp. 16–24.

Mandura, A.S., Khafaji, A.K., and S.M. Saifullah. 1988. Ecology of a mangrove stand of a central Red Sea coast area: Ras Hatiba (Saudi Arabia). *Proc Saudi Biol Soc* 11, no. 1: 85–112.

McGlathery, K.J., Sundbäck, K., and I.C. Anderson. 2007. Eutrophication in shallow coastal bays and lagoons: The role of plants in the coastal filter. *Mar Ecol Prog Ser* 348: 1–18. http://doi.org/10.3354/meps07132.

Moberg, F., and C. Folke. 1999. Ecological goods and services of coral reef ecosystems. *Ecol Econ* 29, no. 2: 215–233. http://doi.org/10.1016/S0921-8009(99)00009-9.

Mohamed, A.R., Ali, A.H.A.M., and H.A. Abdel-Salam. 2012. Status of coral reef health in the northern Red Sea, Egypt. *Proc 12th International Coral Reef Symp*, Cairns, Australia, pp. 9–13.

Mohamed, Z.A. 2018. Potentially harmful microalgae and algal blooms in the Red Sea: Current knowledge and research needs. *Mar Environ Res* 140: 234–242. http://doi.org/10.1016/j.marenvres.2018.06.019.

Mohamed, Z.A., and A.M. Al-Shehri. 2012. The link between shrimp farm runoff and blooms of toxic *Heterosigma akashiwo* in Red Sea coastal waters. *Oceanologia* 54, no. 2: 287–309. http://doi.org/10.5697/oc.54-2.287.

Mohamed, Z.A., and I. Mesaad. 2007. First report on Noctiluca scintillans blooms in the Red Sea off the coasts of Saudi Arabia: Consequences of eutrophication. *Oceanologia* 49, no. 3: 337–351.

Nassar, M.Z., El-Din, N.G.S., and S.M. Gharib. 2015. Phytoplankton variability in relation to some environmental factors in the eastern coast of Suez Gulf, Egypt. *Environ Monit Assess* 187: 1–18. http://doi.org/10.1007/s10661-015-4874-y.

Naumann, M.S., Bednarz, V.N., Ferse, S.C.A., et al. 2015. Monitoring of coastal coral reefs near Dahab (Gulf of Aqaba, Red Sea) indicates local eutrophication as potential cause for change in benthic communities. *Environ Monit Assess* 187: 44. http://doi.org/10.1007/s10661-014-4257-9.

Nixon, S.W. 1995. Coastal marine eutrophication: A definition, social causes, and future concerns. *Ophelia* 41, no. 1: 199–219. http://doi.org/10.1080/00785236.1995.10422044.

Nixon, S.W. 2009. Eutrophication and the macroscope. *Hydrobiologia* 629: 5–19. http://doi.org/10.1007/s10750-009-9759-z.

Osman, E.O., Smith, D.J., Ziegler, M., Kurten, B., Conrad, C., El-Haddad, K.M., Voolstra, C.R., and D.J. Suggett. 2017. Thermal refugia against coral bleaching throughout the northern Red Sea. *Glob Change Biol* 24, no. 2: e474–e484. http://doi.org/10.1111/gcb.13895.

OSPAR. 1997. *OSPAR Agreement 1997–17*. The Common Procedure for the Identification of the Eutrophication Status of the OSPAR Maritime Area.

Papadopoulos, V.P., Abualnaja, Y., Josey, S.A., Bower, A., Raitsos, D.E., Kontoyiannis, H., et al. 2013. Atmospheric forcing of the winter air – sea heat fluxes over the northern Red Sea. *J Clim* 26: 1685–1701.

Peña-García, D., Ladwig, N., Turki, A.J., and M.S. Mudarris. 2014. Input and dispersion of nutrients from the Jeddah Metropolitan Area, Red Sea. *Mar Pollut Bull* 80, no. 1–2: 41–51. http://doi.org/10.1016/j.marpolbul.2014.01.052.

Post, A.F., Dedej, Z., Gottlieb, R., Li, H., Thomas, D.N., El-Absawi, M., El-Naggar, A., El-Gharabawi, M., and U. Sommer. 2002. Spatial and temporal distribution of Trichodesmium spp. in the stratified Gulf of Aqaba, Red Sea. *Mar Ecol Prog Ser* 239: 241–250. http://doi.org/10.3354/meps239241.

Postma, H. 1966. The cycle of nitrogen in the Wadden Sea and adjacent areas. *Neth J Sea Res* 3, no. 2: 186–221.

Rabalais, N.N. 2002. Nitrogen in aquatic ecosystems. *AMBIO: J Human Environ* 31, no. 2: 102–112. http://doi.org/10.1579/0044-7447-31.2.102.

Raitsos, D.E., Hoteit, I., Prihartato, P.K., Chronis, T., Triantafyllou, G., and Y. Abualnaja. 2011. Abrupt warming of the Red Sea. *Geophys Res Lett* 38: L14601. http://doi.org/10.1029/2011GL047984.

Raitsos, D.E., Pradhan, Y., Brewin, R.J., Stenchikov, G., and I. Hoteit. 2013. Remote sensing the phytoplankton seasonal succession of the Red Sea. *PLoS ONE* 8, no. 6: e64909. http://doi.org/10.1371/journal.pone.0064909.

Ramsey, E.W., and J.R. Jensen. 1996. Remote sensing of mangrove wetlands: Relating canopy spectra to site-specific data. *Photogramm Eng Remote Sens* 62: 939–948.

Richardson, A.J., Bakun, A., Hays, G.C., and M.J. Gibbons. 2009. The jellyfish joyride: Causes, consequences and management responses to a more gelatinous future. *Trends Ecol Evol* 24, no. 6: 312–322. http://doi.org/10.1016/j.tree.2009.01.010.

Roik, A., Rothig, T., Pogoreutz, C., and C.R. Voolstra. 2017. Coral reef carbonate budgets and ecological drivers in the naturally high temperature and high alkalinity environment of the Red Sea. *bioRxiv*. https://doi.org/10.1101/203885.

Roth, F., Saalmann, F., Thomson, T., Coker, D.J., Villalobos, R., Jones, B.H., Wild, C., and S. Carvalho. 2018. Coral reef degradation affects the potential for reef recovery after disturbance. *Mar Environ Res* 142: 48–58. http://doi.org/10.1016/j.marenvres.2018.09.022.

Sawall, Y., Al-Sofyani, A., Kürten, B., Al-Aidaroos, A., Hoang, X.B., Marimuthu, N., Khomayis, H., Sommer, U., Gharbawi, W., and M. Wahl. 2014. Coral communities, in contrast to fish communities, maintain a high assembly similarity along the large latitudinal gradient along the Saudi Red Sea coast. *J Ecosystem Ecography* s4: 3.

Sawall, Y., Richter, C., and A. Ramette. 2012. Effects of eutrophication, seasonality and macrofouling on the diversity of bacterial biofilms in equatorial coral reefs. *PLoS ONE* 7, no. 7: e39951. http://doi.org/10.1371/journal.pone.0039951.

Selman, M., Greenhalgh, S., Diaz, R., and Z. Sugg. 2008. *Eutrophication and Hypoxia in Coastal Areas: A Global Assessment of the State of Knowledge*. Washington, DC: World Resources Institute, 6 p.

Silverman, J., Lazar, B., and J. Erez. 2007. Effect of aragonite saturation, temperature, and nutrients on the community calcification rate of a coral reef. *J Geophys Res* 112: C05004. http://doi.org/10.1029/2006JC003770.

Sofianos, S.S., and W.E. Johns. 2003. An oceanic general circulation model (OGCM) investigation of the Red Sea circulation: 2. Three-dimensional circulation in the Red Sea. *J Geophys Res Oceans* 108, no. C3. http://doi.org/10.1029/2001JC001185.

Sofianos, S.S., and W.E. Johns. 2007. Observations of the summer Red Sea circulation. *J Geophys Res* 112: C06025. http://doi.org/10.1029/2006JC003886.

Sohaib, M., Al-Barakah, F.N., Migdadi, H.M., Alyousif, M., and I. Ahmed. 2023. Ecological assessment of physico-chemical properties in mangrove environments along the Arabian Gulf and the Red Sea coasts of Saudi Arabia. *Egypt J Aquat Res* 49, no. 1: 9–16.

Stumpf, R.P., and M.C. Tomlinson. 2007. Remote sensing of harmful algal blooms. In: *Remote Sensing of Coastal Aquatic Environments. Remote Sensing and Digital Image Processing*, vol. 7. Eds. R.L. Miller, C.E. Del Castillo and B.A. McKee. Dordrecht: Springer. http://doi.org/10.1007/978-1-4020-3100-7_12.

Sully, S., Hodgson, G., and R. van Woesik. 2022. Present and future bright and dark spots for coral reefs through climate change. *Glob Change Biol* 28, no. 15: 4509–4522. http://doi.org/10.1111/gcb.16083.

Suthers, I.M., Richardson, A.J., and D. Rissik. 2019. The importance of plankton. In: *Plankton: A Guide to Their Ecology and Monitoring for Water Quality*. Eds. I. Suthers, D. Rissik and A.J. Richardson, 1–19. Clayton, Victoria: CSIRO Publishing.

Touliabah, H.E., and R.A. Elbassat. 2017. Ecological study of the Rabigh Lagoon, eastern site of the Red Sea, Saudi Arabia with special reference to Eutrophication Index. *J Mar Sci Res Dev* 7: 242. http://doi.org/10.4172/2155-9910.1000242.

Turkoglu, M. 2013. Red tides of the dinoflagellate *Noctiluca scintillans* associated with eutrophication in the Sea of Marmara (the Dardanelles, Turkey). *Oceanologia* 55, no. 3: 709–732. http://doi.org/10.5697/oc.55-3.709.

Turner, R.E. 2001. Some effects of eutrophication on pelagic and demersal marine food webs. In: *Coastal Hypoxia: Consequences for Living Resources and Ecosystems*, vol. 58. Eds. N.N. Rabalais and R.E. Turner, 371–398. American Geophysical Union. http://doi.org/10.1029/CE058p0371.

Valiela, I., Bowen, J.L., and J.K. York. 2001. Mangrove forests: One of the World's threatened major tropical environments: At least 35% of the area of mangrove forests has been lost in the past two decades, losses that exceed those for tropical rain forests and coral reefs, two other well-known threatened environments. *Biosci* 51, no. 10: 807–815. http://doi.org/10.1641/0006-3568(2001)051[0807:MFOOTW]2.0.CO;2.

Van der Lee, G.H., Vonk, J.A., Verdonschot, R.C., Kraak, M.H., Verdonschot, P.F., and J. Huisman. 2021. Eutrophication induces shifts in the trophic position of invertebrates in aquatic food webs. *Ecology* 102, no. 3: e03275. http://doi.org/10.1002/ecy.3275.

Vollenweider, R.A. 1992. Coastal marine eutrophication: principles and control. In: *Marine Coastal Eutrophication*. Eds. R.A. Vollenweider, R. Marchetti and R. Viviani, 1–20. Elsevier. http://doi.org/10.1016/B978-0-444-89990-3.50011-0.

Walker, D.I., and R.F.G. Ormond. 1982. Coral death from sewage and phosphate pollution at Aqaba, Red Sea. *Mar Pollut Bull* 13, no. 1: 21–25.

Wankel, S.D., Chen, Y., Kendall, C., Post, A.F., and A. Paytan. 2010. Sources of aerosol nitrate to the Gulf of Aqaba: Evidence from δ15N and δ18O of nitrate and trace metal chemistry. *Mar Chem* 120, no. 1–4: 90–99. http://doi.org/10.1016/j.marchem.2009.01.013.

Webber, M., Calumpong, H., Ferreira, F., Granek, E., Green, S., Ruwa, R., and M. Soares. 2016. Mangroves. In: *The First Global Integrated Marine Assessment. World Ocean Assessment I*, 877–886. Geneva: Cambridge University Press.

Wurtsbaugh, W.A., Paerl, H.W., and W.K. Dodds. 2019. Nutrients, eutrophication and harmful algal blooms along the freshwater to marine continuum. *WIREs Water* 6: e1373. http://doi.org/10.1002/wat2.1373.

Yao, F., Hoteit, I., Pratt, L.J., Bower, A.S., Köhl, A., Gopalakrishnan, G., and D. Rivas. 2014. Seasonal overturning circulation in the Red Sea: 2. Winter circulation. *J Geophys Res Oceans* 119, no 4: 2263–2289.

Ziegler, M., Roik, A., Porter, A., Zubier, K., Mudarris, M.S., Ormond, R., and C.R. Voolstra. 2016. Coral microbial community dynamics in response to anthropogenic impacts near a major city in the central Red Sea. *Mar Pollut Bull* 105, no. 2: 629–640. http://doi.org/10.1016/j.marpolbul.2015.12.045.

PART 3

Heavy metals in seawater and sediments

Heavy metal levels in waters, sediments and biota from the Red Sea and Gulf of Aden coasts of Yemen

7

Ahmed I. Rushdi, Nabil Al-Shwafi, Abdulla Ba-Issa, Hamid Al-Saad, Nada M. Al-Dewail, Hisham M. Nagi, Arafat T. Amer and Fatima M. Shediwah

7.1 INTRODUCTION

The coastline of Yemen is about 2252 kilometres long (Nagi 2021), extending from Midi on the Red Sea to Ras Darbat Ali on the Arabian Sea (Figure 7.1). Different critical coastal habitats, which are considered to be important sources of nutrients in the region, are observed along the coast of Yemen. They include mangroves (IUCN 1987a, 1987b; Rushdi et al. 1994; Rouphael et al. 1998; Krupp et al. 2006), seaweed (IUCN 1987a, 1987b; Rushdi et al. 1994), coral reefs (IUCN 1987a, 1987b; Rushdi et al. 1994), seagrass beds (IUCN 1987a, 1987b; Rushdi et al. 1994; Rouphael et al. 1998), salt marshes (IUCN 1987a, 1987b; Rushdi et al. 1994) and sandy and rocky ecosystems (IUCN 1987a, 1987b; Dekker and de Ceuninck van Capelle 1994; Rushdi et al. 1994; Krupp et al. 2006). Human activities, including coastal industrial and unplanned development, haphazard fishing, hunting and tourism, have a major impact on the coastal and marine environments of Yemen (Smith et al. 1981; Walker and Ormond 1982; Haskoning 1991; Law 1991; Rushdi et al. 1991, 1994; Nagi and Abubakr 2013; Saleh et al. 2018; Nomaan et al. 2021; Al-Qadasy et al. 2018; Al-Shwafi et al. 2021). Oil-related pollution is recognized as the most serious problem in coastal environments of the region, particularly those essential habitats (Dick 1987; Rushdi et al. 1994; DouAbul et al. 1997; DouAbul and Al-Shiwafi 1998). There are different sources of chemicals in marine environments, including natural and anthropogenic sources (Rushdi et al. 2019, Rushdi 2021). Atmospheric particulate matter transported by wind and riverine inputs contributes to the levels of these chemical substances in coastal regions (Rushdi et al. 2019). Dust storms are common episodes on the Arabian Peninsula, especially during summertime, and can introduce a variety of materials that impact the regional seas (Kalenderski et al. 2013; Prakash et al. 2015). Dust loadings can be increased by human activities and widen the extent of the dust sources (Ashbaugh et al. 2003; Chen et al. 2018; Li et al. 2021).

Heavy metals are metallic elements with a relatively high density compared to water (Fergusson 1990). Other sources of heavy metals in the environment include natural inputs from biogeochemical processes (Bradl 2002; He et al. 2005). However, while some of these metals are important micronutrients for many floras and fauna, they can be damaging when they occur in high concentrations (He et al. 2005; Tchounwou et al. 2012). In the marine environment, heavy metals are persistent and non-degradable contaminants and bioaccumulate in suspended particulate matter and sediments, which are eventually released into the aqueous solution, causing adverse health effects in associated organisms at certain concentrations (Xu et al. 2017). They are toxic contaminants in aquatic environments when they exceed certain threshold levels (Zulkifli et al. 2010; Arfaeinia et al. 2019; Yunus et al. 2020). Thus, understanding the levels of heavy metals in the marine environments of Yemen, including neighbouring countries, is important for identifying the current conditions of the ecosystems and suggesting feasible actions to overcome the increasing ecological pressures.

In this chapter, we gathered the existing published and some unpublished data, including those obtained by the co-authors, on heavy metals in seawater, sediments and biota from the coastal environment of Yemen (at Al-Mukalla, Aden in the Gulf of Aden and Al-Hodeidah in the Red Sea) in order to examine their concentration levels and variations and compare the results with levels in other coastal environment in the Red Sea as well as in other regions of the world. We aim to assess the contamination and pollution conditions of the study area.

DOI: 10.1201/9781003321408-10

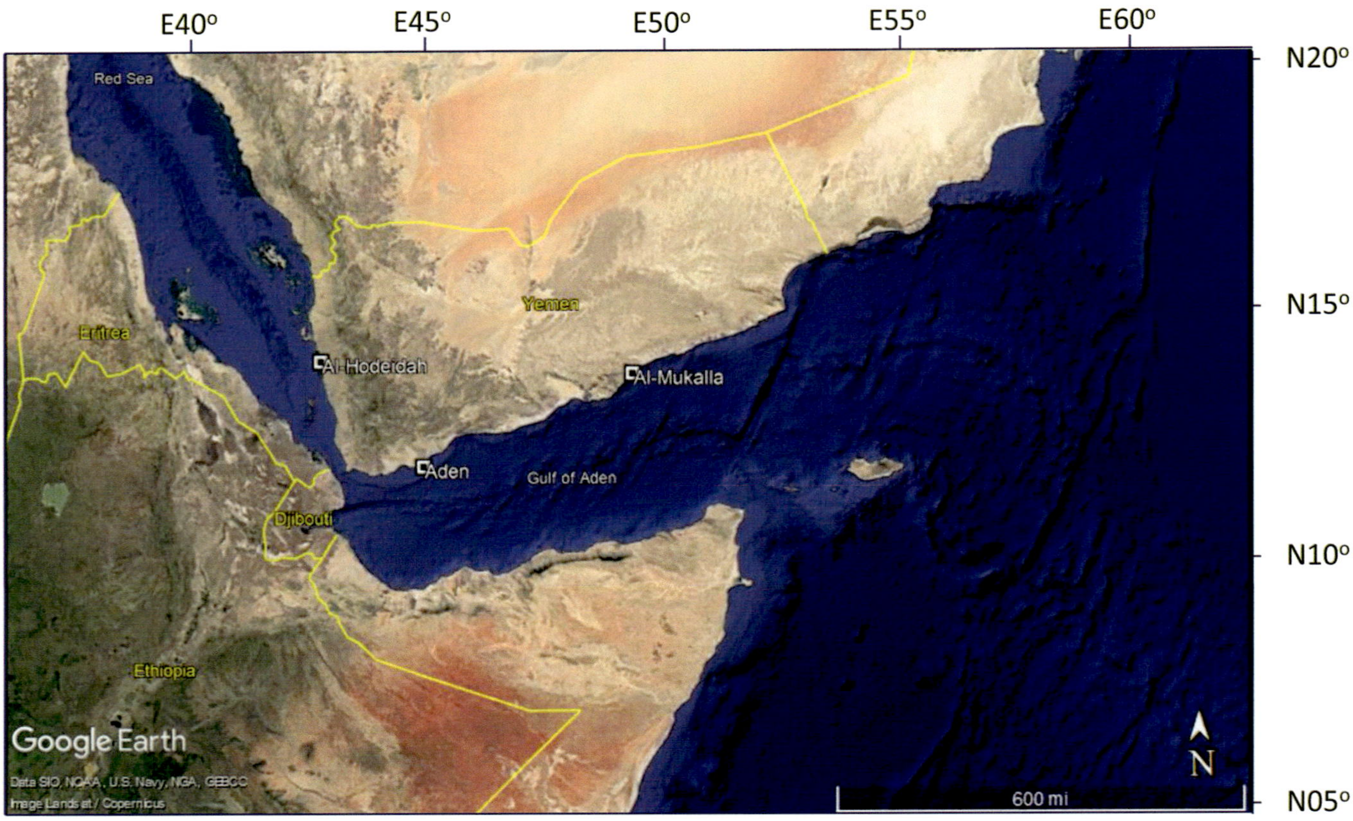

FIGURE 7.1 Map showing the locations of the data collected to investigate heavy metals in seawater, sediments and biota.

Source: Google Earth

7.2 METHODS AND MATERIALS

We collected data from published articles and reports on heavy metal concentrations in seawater, sediments and biota around the coastal cities of Al-Mukalla, Aden and Al-Hodeidah of Yemen. The heavy metal data include manganese (Mn), iron (Fe), cobalt (Co), nickel (Ni), copper (Cu), zinc (Zn), cadmium (Cd) and lead (Pb). We analyzed the data published by various authors, including unpublished results obtained by the co-authors, to evaluate the status of these heavy metals in the coastal seawaters, sediments, and biota (algae, fishes and molluscs) from the aforementioned coastal cities. To assess the pollution and contamination conditions of the sediments for these heavy metals, we calculated the following biogeochemical parameters and indices:

Contamination factor (CF) Turekian and Wedepohl 1961; Taylor 1964):

$$CF = \frac{C_x}{C_B} \quad (7.1)$$

Contamination degree (CD) (Hakanson 1980):

$$CD = \sum CF_i \quad (7.2)$$

Geo-accumulation index (I_{geo}) (Müller 1969):

$$I_{geo} = \left(\frac{C_x}{1.5 \cdot C_B}\right) \quad (7.3)$$

Pollution load index (PLI) (Tomlinson et al. 1980):

$$PLI = \sqrt[n]{\left(CF_1 + CF_2 + \ldots + CF_n\right)} \quad (7.4)$$

where C_x is the concentration of the metal x in the sediments, C_B is the background concentration of the same metal (Yu et al. 2011; Rahman and Ishiga 2012), CF_i is the contamination factor for element i and 1.5 is the correction matrix of the background to reduce the possible lithogenic variation effects.

7.3 RESULTS AND DISCUSSION

7.3.1 Seawater

The concentrations of heavy metals in the coastal seawaters of Yemen vary considerably among Al-Mukalla and Aden in the Gulf of Aden and Al-Hodeidah in the Red Sea (Hebba et al. 2004; Al-Qadasy et al. 2017, 2018; Al-Shwafi et al. 2021). The heavy metals include manganese (Mn), iron (Fe), cobalt (Co), nickel (Ni), copper (Cu), zinc (Zn), cadmium (Cd) and lead (Pb). Only Cd and Pb concentrations have been reported for seawater from Aden (Table 7.1 and Figure 7.2). Note that the concentrations are in mg L^{-1} for seawater and µg g^{-1} for sediments. The concentration of Mn varies from 0.20 mg L^{-1} to 0.30 mg L^{-1}, with an average value of 0.25±0.05 mg L^{-1}, in seawater from Al-Mukalla and from 0.0001 mg L^{-1} to 0.008 mg L^{-1}, with an average of 0.0019±0.0025 mg L^{-1}, in Al-Hodeidah. The concentration of Fe in seawater from Al-Mukalla ranges

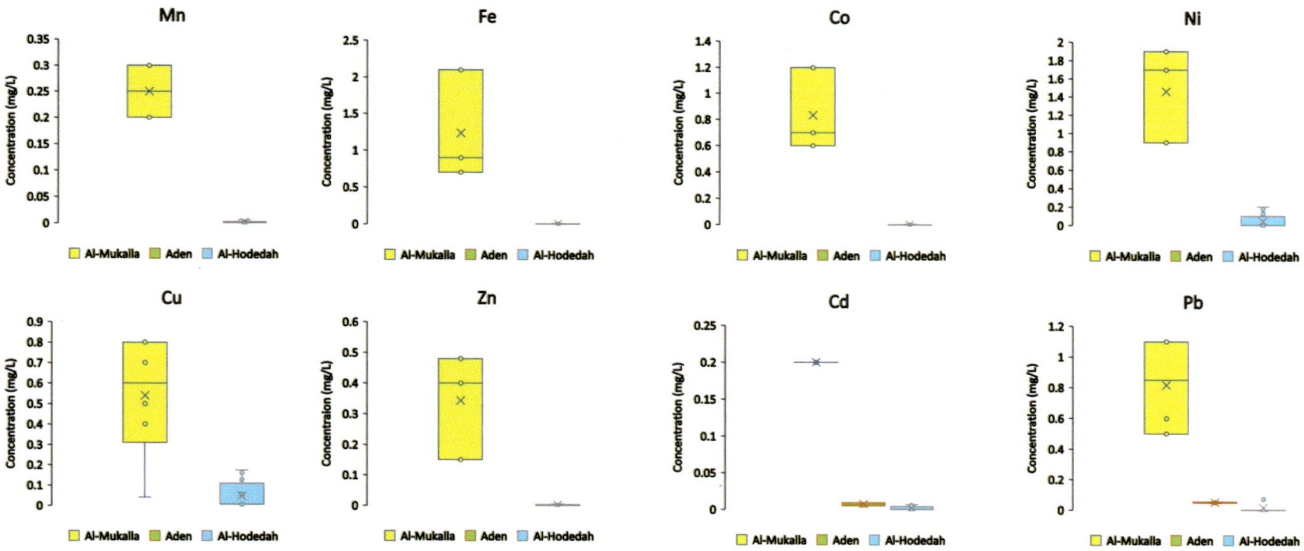

FIGURE 7.2 Box plots showing the concentrations of various heavy metals in seawater from Al-Mukalla, Aden (Gulf of Aden) and Al-Hodeidah (Red Sea). (Mean values shown by X.)

TABLE 7.1 Concentrations (mg L^{-1}) of heavy metals in the coastal seawater of Al-Mukalla, Aden (Gulf of Aden) and Al-Hodeidah (Red Sea), Yemen.

	AL-MUKALLA		*ADEN*		*AL-HODEIDAH*		
	AVG.	*SD*	*AVG.*	*SD*	*AVG.*	*SD*	*REFERENCE*
Mn	0.25	0.05	ND	ND	0.0019	0.0025	(1), (2)
Fe	1.23	0.62	ND	ND	0.0038	0.0067	(1), (2)
Co	0.83	0.26	ND	ND	0.0019	0.0025	(1), (2)
Ni	1.46	0.46	ND	ND	0.0498	0.0742	(1), (2)
Cu	0.54	0.27	ND	ND	0.0474	0.0639	(1), (2)
Zn	0.34	0.14	ND	ND	0.0022	0.0026	(1), (2)
Cd	0.2	0	0.0075	0.0021	0.0007	0.0009	(1), (2), (3)
Pb	0.82	0.29	0.0503	0.0036	0.0166	0.0317	(1), (2), (3)

(1) Al-Shwafi et al. (2017), (2) Hebba et al. (2004), (3) Al-Qadasy et al. (2017)

from 0.70 mg L^{-1} to 2.10 mg L^{-1}, with an average of 1.23±0.62 mg L^{-1}, and in Al-Hodeidah ranges from 0.0001 mg L^{-1} to 0.0224 mg L^{-1}, with an average of 0.0038±0.0067 mg L^{-1}. The Co concentrations are 0.60–1.2 mg L^{-1} (average 0.83±0.26 mg L^{-1}) in Al-Mukalla and 0.0001–0.0080 mg L^{-1} (average 0.0019±0.0025 mg L^{-1}) in Al-Hodeidah. The concentrations of Ni in Al-Mukalla range from 0.90 mg L^{-1} to 1.9 mg L^{-1}, with an average of 1.46±0.46 mg L^{-1}, and in Al-Hodeidah range from 0.004 mg L^{-1} mg L^{-1} to 0.207 mg L^{-1}, with an average of 0.050±0.074 mg L^{-1}. The Cu concentrations vary from 0.04 mg L^{-1} to 0.80 mg L^{-1} (average 0.54±0.27 mg L^{-1}) in Al-Mukalla and from 0.01 mg L^{-1} mg L^{-1} to 0.17 mg L^{-1} (average 0.05±0.06 mg L^{-1}) in Al-Hodeidah. The concentrations of Zn range from 0.15 mg L^{-1} to 0.48 mg L^{-1} (average 0.34±0.14 mg L^{-1}) in Al-Mukalla and from 0.000 mg L^{-1} to 0.010 mg. L^{-1} (average 0.002±0.003 mg L^{-1}) in Al-Hodeidah. The Cd concentrations range from 0.10 mg L^{-1} to 0.20 mg L^{-1} (average 0.16±0.05 mg L^{-1}) in Al-Mukalla, from 0.0001 mg L^{-1} to 0.009 mg L^{-1} (average 0.008±0.002 mg L^{-1}) in Aden and from 0.000 mg L^{-1} to 0.007 mg L^{-1} (average 0.0007±0.0009 mg L^{-1}) in Al-Hodeidah. The concentrations of Pb differ from 0.50 mg L^{-1} to 1.10 mg L^{-1} (average 0.82±0.29 mg L^{-1}) in Al-Mukalla, from 0.045 mg L^{-1} to 0.055 mg L^{-1} (average 0.050±0.004 mg L^{-1}) in Aden and from 0.000 mg L^{-1} to 0.087 mg L^{-1} (average 0.0166±0.032 mg L^{-1}) in Al-Hodeidah.

The levels of heavy metals in the coastal waters of Yemen are relatively higher than the values reported in the Red Sea (Abu-Hilal 1987; Saad and Fahmy 1994; Abd El-Wahab et al. 2005) and other coastal regions (Wang et al. 2012). For example, Mn, Fe, Co and Ni concentrations are high on the Yemen coast compared to the coastal waters of Jinzhou Zhahjiang Bay in China (Zhang et al. 2018), Alkhobar (Alharbi et al. 2017) and Dammam in Saudi Arabia (Mahboob et al. 2022), the Gulf of Aqaba (Al-Taani et al. 2014) and the Gulf of Chabahar (Oman Sea) (Bazzi 2014). The concentrations of these heavy metals are lower than the levels measured in seawater from Suva, Fiji (Arikibe and Prasad 2020). They are comparatively higher in Al-Mukalla relative to the average concentrations measured in open ocean seawater (e.g., Cd and Co = 0.00001–0.00005 mg L^{-1}, Mn = ~ 0.0002 mg L^{-1}, Cr = 0.0003 mg L^{-1}, Pb = 0.000002–0.000003 mg L^{-1}, Fe = 0.00003–0.00194 mg L^{-1}, Cu = 0.002–0.003 mg L^{-1}, Ni = 0.0002 mg L^{-1}, Zn = 0.003–0.005 mg L^{-1}, V = 0.0015 mg L^{-1}) (Gerlach 1976; Whitfield 1979; Clark 1989). The elevated concentrations of heavy metals in the coastal waters of Al-Mukalla are a warning sign of possible contamination as a result of corrosion and leaching of the paint from boats, sewage inputs or solid waste dumping. More studies are needed to investigate the sources of these heavy metals in the coastal waters of Al-Mukalla.

7.3.2 Sediment

The concentrations of heavy metals in the coastal sediments of Yemen vary between the Gulf of Aden and the Red Sea, as shown in Table 7.2 and Figure 7.3. The concentrations of Mn range from 23.8 μg g^{-1} to 112.0 μg g^{-1} (average 62.7±23.7 μg g^{-1}) in Al-Mukalla area, 138.4 μg g^{-1} to 658.9 μg g^{-1} (average 365.4±115.2 μg g^{-1}) in Aden and from 9.2 μg g^{-1} to 26.7 μg g^{-1} (average 16.4±6.1 μg g^{-1}) in the vicinity of Al-Hodeidah. The concentrations of Fe in Al-Mukalla range from 448 μg g^{-1} to 2950 μg g^{-1} (average 1476±1290 μg g^{-1}), in Aden from 40 μg g^{-1} to 3663 μg g^{-1} (average 1527±1455 μg g^{-1}) and in Al-Hodeidah

TABLE 7.2 Concentrations (µg g^{-1}) of heavy metals in sediments from the coastal water of Al-Mukalla, Aden (Gulf of Aden) and Al-Hodeidah (Red Sea), Yemen.

	AL-MUKALLA		*ADEN*		*AL-HODEIDAH*		
	AVG.	*SD*	*AVG.*	*SD*	*AVG.*	*SD*	*REFERENCE*
Mn	62.6	23.7	365.4	115.2	16.4	6.1	(1–6)
Fe	1475.6	607.6	1,527.3	1,454.6	172.3	38.1	(1–6)
Co	12.8	6.0	24.4	5.5	14.4	6.4	(1–6)
Ni	12.9	4.8	36.0	7.2	57.3	31.0	(1–6)
Cu	20.4	14.2	16.1	4.0	61.0	13.8	(1–6)
Zn	55.6	18.7	130.3	79.4	11.5	4.5	(1–6)
Cd	1.5	0.5	1.9	1.1	3.0	1.5	(1–6)
Pb	10.3	4.8	78.0	25.7	5.6	0.3	(1–6)

(1) Mostafa et al. (2009), (2) Nasr et al. (2006), (3) Al Hagibi et al. (2018), (4) Al-Qadasy et al. (2017), (5) Saleh et al. (2018), (6) Wasel and Al-Edresy (2016)

from 99.7 µg g^{-1} to 199.8 µg g^{-1} (average 172.3±83.1 µg g^{-1}). The Co concentrations in Al-Mukalla are 6.0–27.5 µg g^{-1} (average 12.8±6.0 µg g^{-1}), 15.2–33.6 µg g^{-1} (average 24.4±5.6 µg g^{-1}) in Aden and 6.3–23.0 µg g^{-1} (average 14.4±6.4 µg g^{-1}) in Al-Hodeidah. The concentrations of Ni in Al-Mukalla range from 5.8 µg g^{-1} to 24.0 µg g^{-1} (average 12.9±4.8 µg g^{-1}), in Aden from 23.8 µg g^{-1} to 48.1 µg g^{-1} (average 36.0±7.2 µg g^{-1}) and in Al-Hodeidah range from 7.1 µg g^{-1} to 116.4 µg g^{-1} (average 57.3±31.0 µg g^{-1}). The Cu concentrations vary from 3.1 µg g^{-1} to 51.0 µg g^{-1} (average 40.4±14.2 µg g^{-1}) in Al-Mukalla, 9.5 µg g^{-1} to 22.8 µg g^{-1} (average 16.1±4.0 µg g^{-1}) in Aden, and from 36.7 µg g^{-1} to 79.9 µg g^{-1} (average 61.0±13.8 µg g^{-1}) in Al-Hodeidah. The concentrations of Zn range from 10.6 µg g^{-1} to 90.8 µg g^{-1} (average 55.6±18.7 µg g^{-1}) in Al-Mukalla, 21.9 µg g^{-1} to 263.5 µg g^{-1} (average 130.3±79.4 µg g^{-1}) and from 4.0 µg g^{-1} to 18.3 µg g^{-1} (average 11.5±4.5 µg g^{-1}) in Al-Hodeidah. The Cd concentrations range from 0.3 µg g^{-1} to 2.6 µg g^{-1} (average 1.5±0.6 µg g^{-1}), in Al-Mukalla, 6.0 µg g^{-1} to 3.3 µg g^{-1} (average 1.9±1.1 µg g^{-1}) in Aden and from 0.2 µg g^{-1} to 5.8 µg. g^{-1} (average 3.0±1.5 µg g^{-1}) in Al-Hodeidah. The concentrations of Pb differ from 5.3 µg g^{-1} to 23.0 µg g^{-1} (average 10.3±4.8 µg g^{-1}) in Al-Mukalla, 42.3 µg g^{-1} to 136.2 µg g^{-1} (average 78.0±25.7 µg g^{-1}) in Aden and from 5.0 µg g^{-1} to 6.3 µg g^{-1} (average 5.6±0.3 µg g^{-1}) in Al-Hodeidah.

The heavy metal concentrations in the coastal sediments of Yemen are generally close to the levels in other coastal regions. For example, the concentrations of Co, Zn, Cd and Pb in the sediments from Al-Mukalla, Aden and Al-Hodeidah are similar to the values reported from the Taranto Gulf in Italy (Buccolieri et al. 2006), Victoria Harbour in Hong Kong (Tanner et al. 2000), the Tupilipalem coast and the Bay of Bengal in India (Raj and Jayaprakash 2008; Ganugapenta et al. 2018), the Beibu Gulf and the East China Sea in China (Zhang et al. 2018), Sfax in Tunisia (Gargouri et al. 2011), Asaluyeh in the Arabian/Persian Gulf (Delshab et al. 2017), Mimika in Indonesia (Tanjung et al. 2019) and other parts of the Red Sea (Abu-Hilal 1987, 1993; Abu-Hilal and Badran 1990; El Nemr and El-Said 2014; Abd El-Wahab et al. 2005; Saad and Fahmy 1994; El-Moselhy and Gabal 2004; El-Metwally et al. 2017). The concentrations of Mn in the sediments of Al-Mukalla and Al-Hodeidah are much lower than the levels in the Taranto Gulf, Victoria Harbour, the Bay of Bengal and the Tupilipalem coast in India and Asaluyeh in the Arabian/Persian Gulf and similar to the concentration in Aden (Tanner et al. 2000; Buccolieri et al. 2006; Raj and Jayaprakash 2008; Ganugapenta et al. 2018). The Fe concentrations in the sediments of Al-Mukalla, Aden and Al-Hodeidah are relatively lower than the concentrations measured in the Taranto Gulf (Buccolieri et al. 2006), Victoria Harbour (Tanner et al. 2000) and the Tupilipalem coast (Raj and Jayaprakash 2008).

The contamination factors (CF) (Equation 7.1; Table 7.3a) are in the order of Cd > Zn > Pb > Co > Ni > Mn > Fe in the sediments from Al-Mukalla, Cd > Pb > Zn > Co > Ni > Mn > Cu > Fe from Aden and Cd > Cu > Ni > Pb > Co > Zn > Mn > Fe from Al-Hodeidah. The CF values of the heavy metals

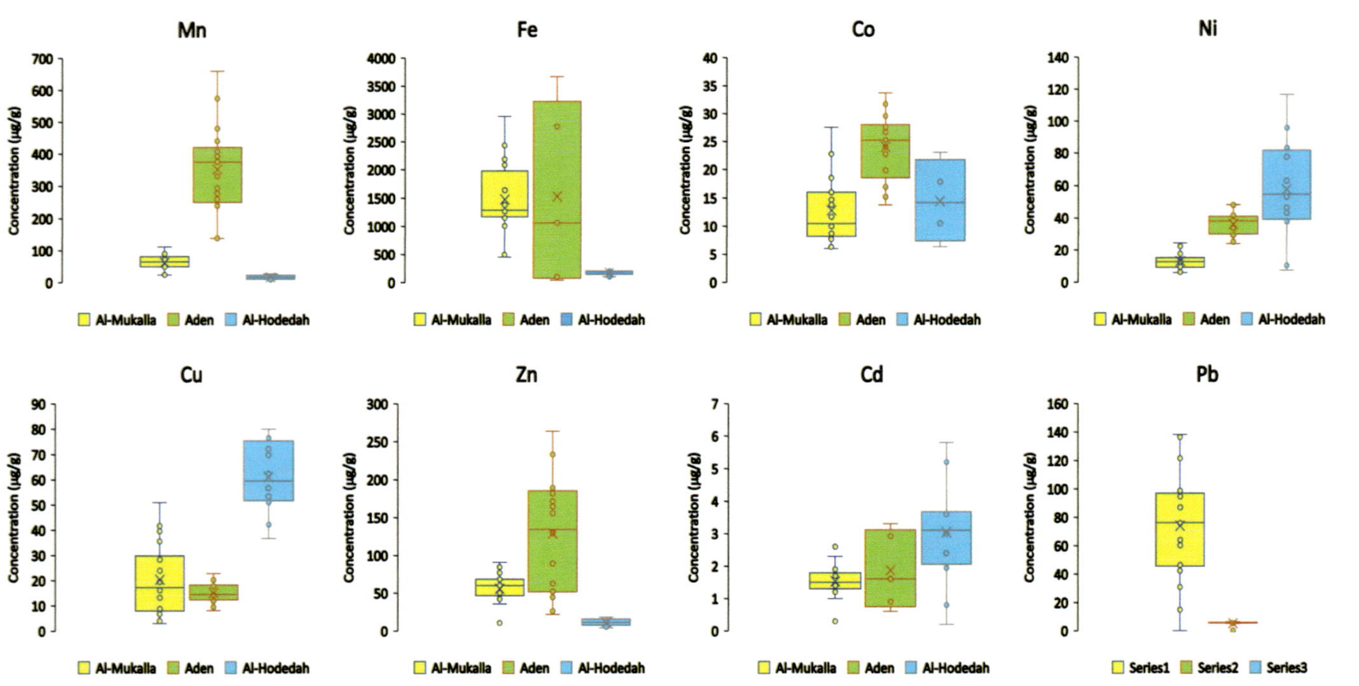

FIGURE 7.3 Plots showing the concentrations of various heavy metals in sediments from Al-Mukalla, Aden (Gulf of Aden) and Al-Hodeidah (Red Sea).

TABLE 7.3 (a) Contamination factor (CF) and contamination degree (CD); (b) geo-accumulation index (I_{geo}) and pollution load index (PLI) values of various heavy metals in sediments from Al-Mukalla, Aden (Gulf of Aden) and Al-Hodeidah (Red Sea).

(A)	*CF*							
	MN	*FE*	*CO*	*NI*	*CU*	*ZN*	*CD*	*PB*
Al-Mukalla	0.06	0.05	0.43	0.14	0.31	0.71	10.23	1.03
Aden	0.32	0.02	0.80	0.40	0.24	1.65	12.43	7.73
Al-Hodeidah	0.01	0.00	0.48	0.64	0.90	0.14	20.24	0.56
	I_{GEO}							
	MN	*FE*	*CO*	*NI*	*CU*	*ZN*	*CD*	*PB*
Al-Mukalla	−4.85	−5.68	−1.95	−3.48	−2.70	−1.21	2.64	−0.69
Aden	−2.35	−7.47	−0.95	−1.94	−2.75	−0.22	2.76	2.21
Al-Hodeidah	−6.75	−9.14	−1.81	−1.58	−0.78	−3.55	3.40	−1.42

(B)	*CD*		*PLI*	
	AVERAGE	*SD*	*AVERAGE*	*SD*
Al-Mukalla	13.0	5.1	0.38	0.20
Aden	23.6	12.8	0.74	0.32
Al-Hodeidah	22.3	11.6	0.26	0.08

for Al-Mukalla sediments indicate low contamination by Mn, Fe, Co, Ni, Cu and Zn (CF < 1); moderate contamination by Pb (1 > CF < 3) and high contamination by Cd (CF > 6). The values for Aden show that the sediment contaminations are low for Mn, Fe, Co and Ni; moderate for Zn and high for Pb and Cd. The contamination of sediments from Al-Hodeidah is low for all metals except Cd, which is high. The contamination degree (CD) (Equation 7.2; Table 7.3b) values of the heavy metals range from 13 to 24, indicating that the sediments are moderately contaminated.

The estimation of the geo-accumulation index (I_{geo}) (Equation 7.3; Müller 1969) values in Table 7.1a shows that all the sediments of Al-Mukalla are uncontaminated by all heavy metals ($I_{geo} < 0$), except for being moderately contaminated by Cd ($2 > I_{geo} < 3$). Sediments from Aden are uncontaminated by all heavy metals except for being moderately contaminated by Cd ($I_{geo} = 2.76$) and Pb ($I_{geo} = 2.21$). Al-Hodeidah sediments are uncontaminated by Mn, Fe, Co, Ni, Cu and Zn ($I_{goe} < 0$), but they are moderately to heavily contaminated by Cd ($I_{geo} = 3.40$) (Table 7.1a). The computed pollution load index (PLI) (Equation 7.4; Müller 1969) values of the heavy metals in the sediments (Table 7.3b) are 0.38±0.20 for Al-Mukalla, 0.74±0.32 for Aden and 0.26±0.08 for Al-Hodeidah, where all values are less than 1, indicating that the sediments are unpolluted.

7.3.3 Biota

7.3.3.1 *Algae*

The data on heavy metals in marine algae from Yemen are limited; the only published data were from Aden in the Gulf of Aden and Al-Hodeidah in the Red Sea. The concentrations of heavy metals in the coastal algae of Yemen differ between the Gulf of Aden and the Red Sea, as shown in Table 7.4. Their levels are in the order of Fe > Cu > Mn > Ni > Zn > Pb > Cd > Co in the algae from Aden and Fe > Zn > Mn > Cu > Ni > Co > Cd from Al-Hodeidah. The reported values in Table 7.4 show that the heavy metals in algae are higher in Aden than in Al-Hodeidah, except Zn.

TABLE 7.4 Concentrations (μg g^{-1}) of heavy metal in algal species from Aden (Gulf of Aden) and Al-Hodeidah (Red Sea).

	ADEN				*AL-HODEIDAH*			
	MIN.	*MAX.*	*AVG.*	*SD*	*MIN.*	*MAX.*	*AVG.*	*SD*
Mn	2.20	13.70	6.42	3.25	1.60	7.10	3.85	1.82
Fe	13.6	90.1	36.2	21.0	7.90	20.35	15.53	3.58
Co	0.14	1.00	0.58	0.26	0.10	0.66	0.26	0.18
Ni	0.80	13.30	4.39	3.47	0.35	2.07	1.16	0.54
Cu	2.31	20.50	10.79	5.62	0.35	3.68	1.74	0.91
Zn	2.35	9.20	4.87	2.25	1.53	9.65	5.72	2.62
Cd	0.17	2.30	0.76	0.57	0.10	0.33	0.19	0.07
Pb	0.35	6.00	2.16	1.66	0.23	3.01	1.04	0.83

Source: Al-Shwafi and Rushdi (2008)

The concentrations of the heavy metals in algae from Yemen are in the range of the values reported from other regions, such as Iran (Dadolahi-Sohrab et al. 2011; Alahverdi and Savabieasfahani 2012), Saudi Arabia (Al-Homaidan 2007, 2008; Alnashiri 2022), Qatar (Kureishy 1991) and Turkey (Topcuoğlu et al. 2003).

7.3.3.2 *Fish*

The concentrations of heavy metals in the fish species from the coastal zone of Yemen are shown in Table 7.5 and Figure 7.4. There is no data for Mn, Fe, Co, Ni, Cu or Zn from Al-Mukalla and nothing for the first four heavy metals from Aden. They range from 1.50 μg g^{-1} to 6.30 μg g^{-1} (average 3.65±3.3 μg g^{-1}) for Mn, 1.10 μg g^{-1} to 22.60 μg g^{-1} (average 8.57±7.87 μg g^{-1}) for Fe, 6.25 μg g^{-1} to 11.25 μg g^{-1} (average 8.01±1.32 μg g^{-1}) for Co and from 4.95 μg g^{-1} to 9.50 μg g^{-1} (average 7.33±1.12 μg g^{-1}) for Ni in fish from Al-Hodeidah. The Cu concentrations vary from 0.05 µg g^{-1} to 0.40 µg g^{-1} (average 0.25±0.11 µg g^{-1}) in Aden and from 0.08 µg g^{-1} to 0.48 µg g^{-1} (average 0.30±0.12 µg g^{-1}) in Hodeidah. The concentrations of Zn range from 0.50 µg g^{-1} to 3.05 µg g^{-1} (average 2.00±0.09 µg g^{-1}) in Aden and from 0.67 µg g^{-1} to 3.71 µg g^{-1} (average 2.26±0.97 µg g^{-1}) in Hodeidah. The concentrations of Cd in Al-Mukalla range from 0.006 µg g^{-1} to 0.008 µg g^{-1} (average 0.007±0.001 µg g^{-1}), in Aden range from 0.06 µg g^{-1} to 1.09 µg g^{-1} (average 0.42±0.39 µg g^{-1}) and in Hodeidah range from 0.01 µg g^{-1} to 0.17 µg g^{-1} (average 0.09±0.04 µg g^{-1}). The concentrations of Pb range from 0.03 µg g^{-1} to 0.09 µg g^{-1} (average 0.06±0.02 µg g^{-1}) in Al-Mukalla, from 0.02 µg g^{-1} to 0.35 µg g^{-1} (average 0.21±0.11 µg g^{-1}) in Aden and from 0.02 µg g^{-1} to 0.34 µg g^{-1} (average 0.17±0.10 µg g^{-1}) in Hodeidah.

The concentrations of heavy metals in fish from the coastal zones of Yemen are comparable to the values reported in other regions such as in the northeast of India (Kumar et al. 2012), the Mediterranean Sea (Turkmen et al. 2008), the Gulf of Aqaba (Abu-Hilal and Ismail 2008; Ahmed and Naim 2008), the coasts of Ghana (Gbogbo et al. 2018) and Malaysia (Bashir et al. 2013), the Red Sea coasts (El-Moselhy et al. 2014; Alnashiri 2022; Zaghloul et al. 2022) and Saudi Arabia (Younis et al. 2021). All the concentrations of heavy metals in fish from Yemen are below the permitted values given by the FAO and WHO (FAO 1983; WHO 1989; FAO/WHO 1984). Therefore, commercial fish from the coastal waters of Yemen is safe for human consumption.

TABLE 7.5 Concentrations (µg g^{-1}) of heavy metals in fish from the coastal water of Al-Mukalla, Aden (Gulf of Aden) and Al-Hodeidah (Red Sea), Yemen.

	AL-MUKALLA		ADEN		AL-HODEIDAH		
	AVG.	SD	AVG.	SD	AVG.	SD	REFERENCE
Mn	ND		ND		3.650	1.707	(1, 2)
Fe	ND		ND		8.567	7.872	(1, 2)
Co	ND		ND		8.014	1.324	(1, 2)
Ni	ND		ND		7.332	1.119	(1, 2)
Cu	ND		0.3	0.1	0.295	0.125	(1–4)
Zn	ND		2.0	0.9	2.260	0.969	(1–4)
Cd	0.007	0.001	0.415	0.386	0.090	0.043	(1–4)
Pb	0.055	0.020	0.214	0.110	0.173	0.102	(1–4)

(1) Heba et al. (2015), (2) Omar et al. (2014), (3) Al Shwafi (2002), (4) Al-Qadasy et al. (2018)

7.3.3.3 Molluscs

The heavy metal concentrations in various molluscs from the coast of Yemen are shown in Table 7.6 and Figure 7.5. The Mn concentrations in the mollusc species range from 0.82 µg g^{-1} to 2.20 µg g^{-1} (average 1.29±0.48 µg g^{-1}) in Al-Hodeidah, 2.11 µg g^{-1} to 27.2 µg g^{-1} (average 11.09±7.67 µg g^{-1}) in Aden, and from 0.30 µg g^{-1} to 4.8 µg g^{-1} (average 2.94±1.28 µg g^{-1}) in Al-Mukalla. The concentrations of Fe vary from 250 µg g^{-1} to 525 µg g^{-1} (average 384±82 µg g^{-1}) in Al-Mukalla and from 5.53 µg g^{-1} to 11.16 µg g^{-1} (average 7.73±2.03 µg g^{-1}) in Al-Hodeidah. The concentrations of Co range from 2.77 µg g^{-1} to 9.85 µg g^{-1} µg g^{-1} (average 6.36±2.1 µg g^{-1}) in Aden. The concentrations of Ni range from 0.5 µg g^{-1} to 6.0 µg g^{-1} (average 2.63±1.5 µg g^{-1}) in Al-Mukalla, from 3.90 µg g^{-1} to 30.15 µg g^{-1} (average 16.95±7.2 µg g^{-1}) in Aden and from 0.24 µg g^{-1} to 1.6 µg g^{-1} (average 0.91±0.62 µg g^{-1}) in Al-Hodeidah. The Cu concentrations vary from 3.80 µg g^{-1} to 111.00 µg g^{-1} (average 46.98±31.52 µg. g^{-1}) in Al-Mukalla, from 1.1 µg g^{-1} to 462.2 µg g^{-1} (average 117.3±131.6 µg g^{-1}) in Aden and from 3.59 µg g^{-1} to 11.15 µg g^{-1} (average 6.85±3.18 µg g^{-1}) in Al-Hodeidah. The concentrations of Zn range from 2.40 µg g^{-1} to 17.50 µg g^{-1} (average 10.01±4.59 µg g^{-1}) in Al-Mukalla, from 9.20 µg g^{-1} to 121.20 µg g^{-1} (average 67.52+36.69 µg g^{-1}) Aden and from 9.20 µg g^{-1} to 30.5 µg g^{-1} (average 20.66±7.98 µg g^{-1}) in Al-Hodeidah. The concentrations of Cd in Al-Mukalla range from 1.00 µg g^{-1} to 2.10 µg g^{-1} (average 1.29±0.57 µg g^{-1}), in Aden range from 0.20 µg g^{-1} to 10.45 µg g^{-1} (average 3.10±2.76 µg g^{-1}) and in Al-Hodeidah range from 0.55 µg g^{-1} to 3.25 µg g^{-1} (average 2.45±1.07 µg. g^{-1}). The concentrations

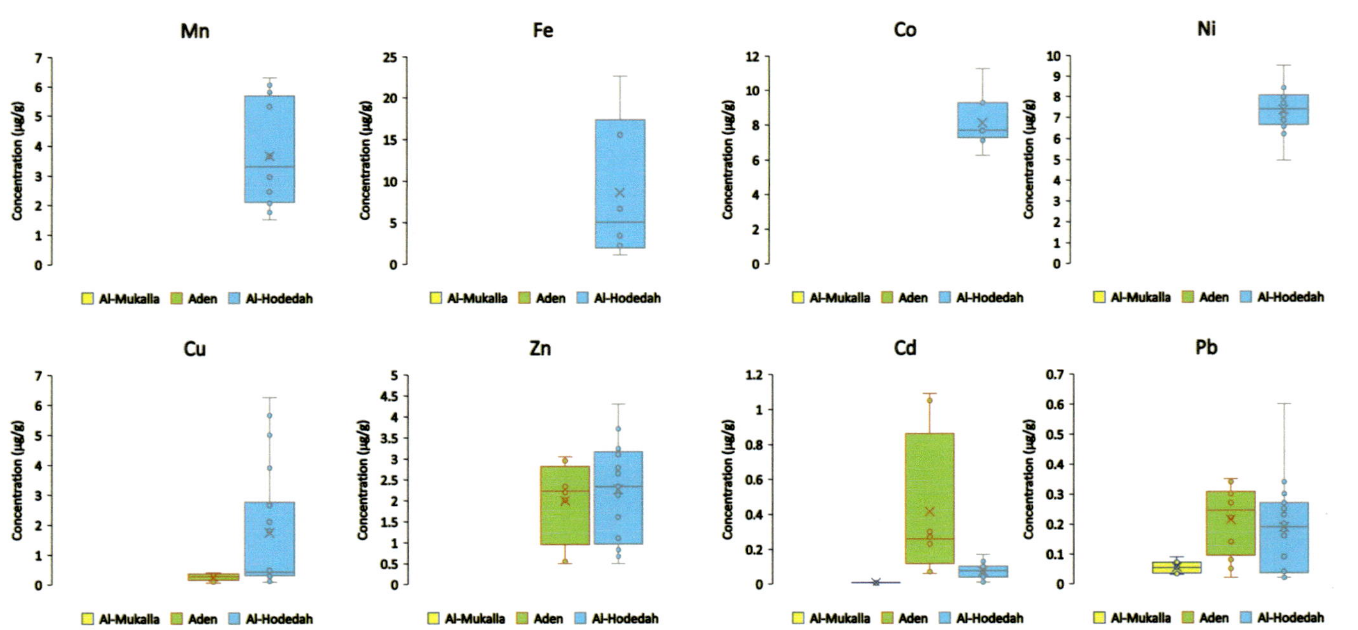

FIGURE 7.4 Plots showing the concentrations of heavy metals in marine fish from the coastal zone of Yemen: Al-Mukalla, Aden (Gulf of Aden) and Al-Hodeidah (Red Sea).

TABLE 7.6 Concentrations (µg g^{-1}) of heavy metals in mollusc species from the coastal water of Al-Mukalla, Aden (Gulf of Aden) and Al-Hodeidah (Red Sea), Yemen.

	AL-MUKALLA		*ADEN*		*AL-HODEIDAH*		
	AVG.	*SD*	*AVG.*	*SD*	*AVG.*	*SD*	*REFERENCE*
Mn	7.14	9.75	12.36	9.10	1.29	0.48	(1–4)
Fe	383.50	82.09			7.73	2.03	(1–4)
Co			7.01	3.37			(1)
Ni	2.63	1.50	16.95	7.20	0.91	0.62	(1–4)
Cu	46.98	31.52	117.34	131.60	6.85	3.18	(1–4)
Zn	11.68	7.06	67.52	36.69	20.66	7.98	(1–4)
Cd	1.63	0.93	3.33	3.04	2.45	1.07	(1–4)
Pb	8.14	5.91	14.93	5.08	2.55	0.75	(1–4)

(1) Mostafa et al. (2009), (2, 3) Heba et al. (2003, 2014), (4) Al-Edreesi et al. (2002).

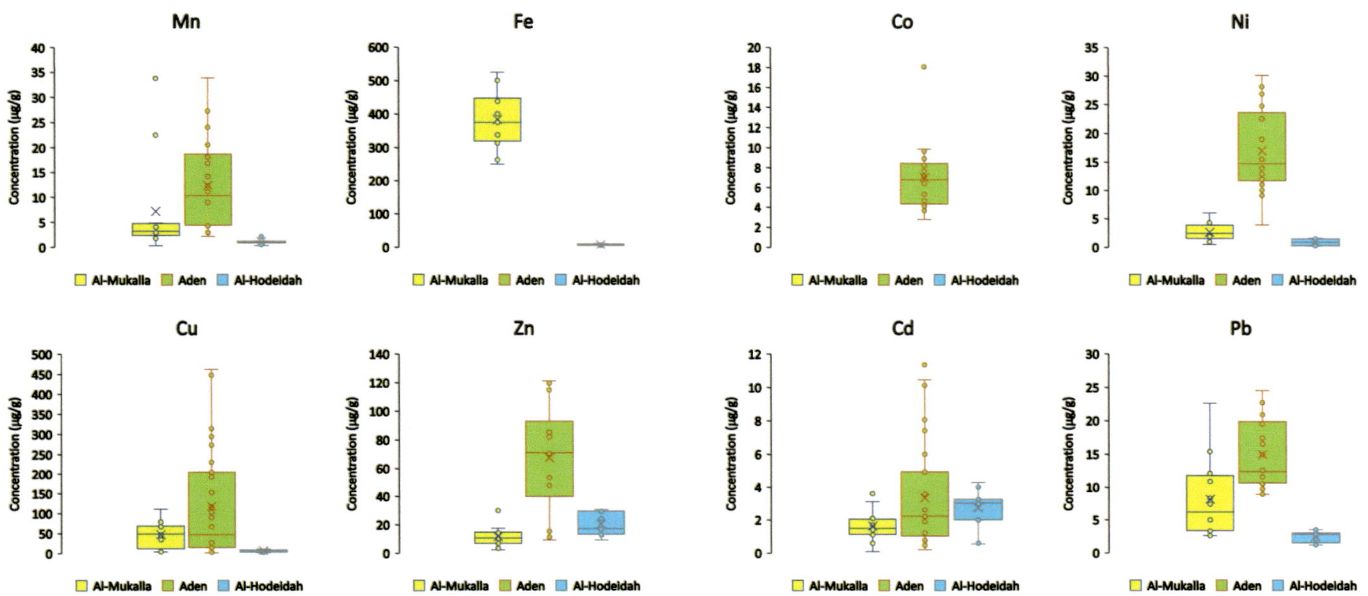

FIGURE 7.5 Plots showing the concentration of heavy metals in molluscs from the coastal zone of Yemen: Al-Mukalla, Aden (Gulf of Aden) and Al-Hodeidah (Red Sea).

of Pb range from 2.6 µg g^{-1} to 22.6 µg g^{-1} (average 8.14±5.91 µg g^{-1}) in Al-Mukalla, from 8.90 µg g^{-1} to 24.50 µg g^{-1} (average 14.93±5.08 µg g^{-1}) in Aden and from 1.49 µg g^{-1} to 3.49 µg g^{-1} (average 2.55±0.75 µg g^{-1}) in Al-Hodeidah.

The heavy metal concentrations in molluscs from the three coastal zones of Yemen are similar to the levels measured in other regions, such as Sao Paulo in Brazil (Avelar et al. 2000), the Black Sea in Turkey (Bat and Şahin 2018), Alexandria in Egypt (Abdallah 2013; Abd-Ellah et al. 2020), Dar es Salaam in Tanzania (Rumisha et al. 2012) and the Mediterranean Sea (Abdallah 2013; Zaidi et al. 2022).

Heavy metals are naturally occurring elements in the aquatic environment and important micronutrients for marine biota but can become toxic at certain levels. Anthropogenic sources from coal mining, fossil fuel emission and oil spills (e.g., shipping tankers, shore refinery operations etc.) as well as industrial manufacturers elevate their concentrations in marine environments, causing adverse effects on coastal ecosystems and human health. Therefore, assessing their different sources such as human and developmental activities as well as natural inputs from ore-bearing rock erosion, wind and dust storm transports and biogenic origins is critical.

7.4 CONCLUSION

The literature results and data on heavy metals in seawater, sediments and different biota of the coastal zones of Yemen reveal that their levels vary between the Gulf of Aden and the Red Sea. Continental shelf sediments are important additional sources of heavy metals in seawater, as shown by the high concentrations of Fe, Co, Ni, Zn and Cd in seawater relative to other heavy metals, as indicated by their high concentrations (mainly Fe, Co, Zn, Pb, Mn and Cu) in Al-Hodeidah and Aden relative to Al-Mukalla seawater.

The concentrations of the heavy metals in the sediments from the three coastal cities differ: sediments are moderately contaminated by only Cd and Pb, based on the applications of CF, I_{geo} and unpolluted based on PLI. The concentrations of heavy metals in algae, commercial fish and molluscs are in the range of the values reported in other coastal regions, where the concentrations are at the acceptable levels determined by international organizations such as the FAO and WHO. This indicates that the marine species of the Red Sea coast of Yemen are safe for human utilization and dietary consumption. However, comprehensive studies are needed to distinguish between the different sources and characterize the fates and biogeochemical cycles of these heavy metals in coastal and marine environments of the Gulf of Aden, Red Sea and the Arabian Sea and their impact on their coastal habitats.

REFERENCES

Abdallah, M.A.M. 2013. Bioaccumulation of heavy metals in Mollusca species and assessment of potential risks to human health. *Bull Environ Contam Toxicol* 90, no. 5: 552–557.

Abd-Ellah, S.M., El-Sherif, S., and R. El-Morshedy. 2020. Seasonal effects of heavy metals on the date mussel Lithophaga lithophaga (Mollusca: Bivalvia) at Eastern Harbor, Alexandria, Egypt. *Swed J Bio Sci Res* 1, no. 1: 62–77.

Abd El-Wahab, M., Dar, M.A., and T.A. Mohammad. 2005. Sediments, coral reefs and seawater interactions in some coastal lagoons, Red Sea, Egypt. *Egypt J Aquat Res* 31(Special Issue): 69–85.

Abu-Hilal, A.H. 1987. Distribution of trace elements in nearshore surface sediments from the Jordan Gulf of Aqaba (Red Sea). *Mar Pollut Bull* 18, no. 4: 190–193.

Abu-Hilal, A.H. 1993. Observations on heavy metal geochemical association in marine sediments of the Jordan Gulf of Aqaba. *Mar Pollut Bull* 26, no. 2: 85–90.

Abu-Hilal, A.H., and M.M. Badran. 1990. Effect of pollution sources on metal concentration in sediment cores from the Gulf of Aqaba (Red Sea). *Mar Pollut Bull* 21, no. 4: 190–197.

Abu-Hilal, A.H., and N.S. Ismail. 2008. Heavy metals in eleven common species of fish from the Gulf of Aqaba, Red Sea. *Jordan J Biol Sci* 1, no. 1: 13–18.

Ahmed, H.A.H., and S.I. Naim. 2008. Heavy metals in eleven common species of fish from the Gulf of Aqaba, Red Sea. *Jordan J Biol Sci* 1, no. 1: 13–18.

Alahverdi, M., and M. Savabieasfahani. 2012. Metal pollution in seaweed and related sediment of the Persian Gulf, Iran. *Bull Environ Contam Toxicol* 88: 939–945.

Al-Edreesi, M.A.M., M.A. Abdelmoneim, H.T. Al-Saad, and H.M.A. Heba, 2002. Heavy metal contents of some molluscs and crustaceans along Al-Hoodeidah Red Sea Coast of Yemen. *Bull Nat Inst Oceanogr Fish A.R.E.* 28: 319–331.

Al Hagibi, H.A., Al-Selwi, K.M., Nagi, H.M., and N.A. Al-Shwafi. 2018. Study of heavy metals contamination in mangrove sediments of the Red Sea Coast of Yemen from Al-Salif to Bab-el-Mandeb Strait. *J Ecol Nat Resour* 2, no. 1: 121.

Alharbi, T., Alfaifi, H., and A. El-Sorogy. 2017. Metal pollution in Al-Khobar seawater, Arabian Gulf, Saudi Arabia. *Mar Pollut Bull* 119, no. 1: 407–415.

Al-Homaidan, A.A. 2007. Heavy metal concentrations in three species of green algae from the Saudi coast of the Arabian Gulf. *J Food Agric Environ* 5, nos. 3/4: 354.

Al-Homaidan, A.A. 2008. Accumulation of nickel by marine macroalgae from the Saudi coast of the Arabian Gulf. *J Food Agric Environ* 6, no. 1: 148–151.

Alnashiri, H.M. 2022. A brief review on heavy metal bioaccumulation studies from Red Sea. *Adsorp Sci Technol* 2022: 1–8.

Al-Qadasy, M.K.O., Babaqi, A.S., Al-Abyadh, M.M., and A.G.A. Al-Kaf. 2017. Trace metals in surface seawaters in the Red Sea and Gulf of Aden-Yemen. *Pharm Res* 2, no. 6: 53–61.

Al-Qadasy, M.K.O., Babaqi, A.S., Al-Abyadh, M.M., and A.G.A. Al-Kaf. 2018. The effects of lead, cadmium, mercury and arsenic on fish and seawater in Red Sea and the Gulf of Aden at three different locations in Yemen. *SF J Pharm Anal Chem* 1, no. 1: 1–13.

Al Shwafi, N.A., 2002. Heavy metals concentration levels in some fish species in the Red Sea and Gulf of Aden-Yemen. *Qatar Univ Sci J* 22: 171–176.

Al-Shwafi, N.A., Al-Wosabi, M.A., Nagi, H.M., and N.S. Mol-Aldwila. 2017. Assessment of heavy metal pollution in the seawater of Hadhramout coastal area, Yemen. *Hadhramout Univ J Natl Appl Sci* 14, no. 1: 59–69.

Al-Shwafi, N.A., Al-Wosabi, M.A., Nagi, H.M., and N.S. Mol-Aldwila. 2021. Assessment of heavy metal pollution in the seawater of Hadhramout coastal area, Yemen. *Hadhramout Univ J Natl Appl Sci* 14, no. 1: 7.

Al-Shwafi, N.A., and A.I. Rushdi. 2008. Heavy metal concentrations in marine green, brown, and red seaweeds from coastal waters of Yemen, the Gulf of Aden. *Environ Geol* 55: 653–660.

Al-Taani, A.A., Batayneh, A., Nazzal, Y., Ghrefat, H., Elawadi, E., and H. Zaman. 2014. Status of trace metals in surface seawater of the Gulf of Aqaba, Saudi Arabia. *Mar Pollut Bull* 86, nos. 1–2: 582–590.

Arfaeinia, H., Dobaradaran, S., Moradi, M., Pasalari, H., Mehrizi, E.A., Taghizadeh, F., Esmaili, A., and M. Ansarizadeh. 2019. The effect of land use configurations on concentration, spatial distribution, and ecological risk of heavy metals in coastal sediments of northern part along the Persian Gulf. *Sci Total Environ* 653: 783–791.

Arikibe, J.E., and S. Prasad. 2020. Determination and comparison of selected heavy metal concentrations in seawater and sediment samples in the coastal area of Suva, Fiji. *Mar Pollut Bull* 157: 111157.

Ashbaugh, L.L., Carvacho, O.F., Brown, M.S., Chow, J.C., Watson, J.G., and K.C. Magliano. 2003. Soil sample collection and analysis for the fugitive dust characterization study. *Atmos Environ* 37: 1163–1173.

Avelar, W.E.P., Mantelatto, F.L.M., Tomazelli, A.C., Silva, D.M.L., Shuhama, T., and J.L.C. Lopes. 2000. The marine mussel Perna perna (Mollusca, Bivalvia, Mytilidae) as an indicator of contamination by heavy metals in the Ubatuba Bay, Sao Paulo, Brazil. *Water Air Soil Pollut* 118, no. 1: 65–72.

Bashir, F.H., Othman, M.S., Mazlan, A.G., Rahim, S.M., and K.D. Simon. 2013. Heavy metal concentration in fishes from the coastal waters of Kapar and Mersing, Malaysia. *Turk J Fish Aquat Sci* 13, no. 2: 375–382.

Bat, L., and F. Şahin. 2018. Non-essential heavy metals in Mollusca fauna from İğneada shores of the Black Sea. *Fish Aquat Sci* 31: 211–214.

Bazzi, A.O. 2014. Heavy metals in seawater, sediments and marine organisms in the Gulf of Chabahar, Oman Sea. *J Oceanogr Mar Sci* 5, no. 3: 20–29.

Bradl, H. (ed.). 2002. *Heavy Metals in the Environment: Origin, Interaction and Remediation*, vol. 6. London: Academic Press.

Buccolieri, A., Buccolieri, G., Cardellicchio, N., Dell'Atti, A., Di Leo, A., and A. Maci. 2006. Heavy metals in marine sediments of Taranto Gulf (Ionian Sea, southern Italy). *Mar Chem* 99, nos. 1–4: 227–235.

Chen, S., Jiang, N., Huang, J., Xu, X., Zhang, H., Zang, Z., Huang, K., Xu, X., Wei, Y., Guan, X., and X. Zhang. 2018. Quantifying contributions of natural and anthropogenic dust emission from different climatic regions. *Atmos Environ* 191: 94–104.

Clark, R.B. 1989. *Marine Pollution*, 2nd ed. New York: Oxford Science Publication, 220 p.

Dadolahi-Sohrab, A., Nikvarz, A., Nabavi, S.M.B., Safahyeh, A., Ketal-Mohseni, M., et al. 2011. Environmental monitoring of heavy metals in seaweed and associated sediment from the Strait of Hormuz, IR Iran. *World J Fish Mar Sci* 3: 576–589.

Dekker, H.F.G., and F.G. de Ceuninck van Capelle. 1994. Survey of Yemen Red Sea shells, collected by the Tibia-I Expedition. *De Kreukel* 30, nos. 7–10: 79–147.

Delshab, H., Farshchi, P., and B. Keshavarzi. 2017. Geochemical distribution, fractionation and contamination assessment of heavy metals in marine sediments of the Asaluyeh port, Persian Gulf. *Mar Pollut Bull* 115, nos. 1–2: 401–411.

Dick, B. 1987. Pollution. In: *The Red Sea*. Eds. A.J. Edwards and S.M. Head, 383–404. Oxford: Pergamon Press.

DouAbul, A.A.-Z., and H.A. Al-Shiwafi. 1998. Dissolved/dispersed hydrocarbons in Arabian region. *Mar Poll Bull* 36: 844–850.

DouAbul, A.A.-Z., Hebba, H.M.A., and K.H. Fareed. 1997. Polynuclear aromatic hydrocarbon (PAHs) in fish from the Red Sea coast of Yemen. *Hydrobiologia* 352: 251–262.

El-Metwally, M.E., Abouhend, A.S., Dar, M.A., and K.M. El-Moselhy. 2017. Effects of urbanization on the heavy metal concentrations in the Red Sea coastal sediments, Egypt. *Int J Mar Sci* 7, no. 13: 114–124.

El-Moselhy, K.M., and M.N. Gabal. 2004. Trace metals in water, sediments and marine organisms from the northern part of the Gulf of Suez, Red Sea. *J Mar Syst* 46, nos. 1–4: 39–46.

El-Moselhy, K.M., A.I. Othman, H. Abd El-Azem, and M.E.A. El-Metwally. 2014. Bioaccumulation of heavy metals in some tissues of fish in the Red Sea, Egypt. *Egypt J Basic Appl Sci* 1, no. 2: 97–105.

El Nemr, A., and G.F. El-Said. 2014. The pollution status along the Red Sea: A review. *Blue Biotechnol J* 3, no. 4: 403.

FAO. 1983. Compilation of legal limits for hazardous substances in fish and fishery products. In: *FAO Fishery Circular No. 464*, 5–100. Rome: Food and Agriculture Organisation of the United Nations.

FAO/WHO. 1984. *List of Maximum Levels Recommended for Contaminants by the Joint FAO/WHO Codex Alimentarius Commission, Second Series. CAC/FAL, Rome 3: 1–8. Food and Drugs*. Kuala Lumpur: Malaysia Law Publisher.

Fergusson, J.E. (ed.). 1990. *The Heavy Elements: Chemistry, Environmental Impact and Health Effects*. Oxford: Pergamon Press.

Ganugapenta, S., Nadimikeri, J., Chinnapolla, S.R.R.B., Ballari, L., Madiga, R., and L.P. Tella. 2018. Assessment of heavy metal pollution from the sediment of Tupilipalem Coast, southeast coast of India. *Int J Sediment Res* 33, no. 3: 294–302. http://doi.org/10.1016/J.IJSRC.2018.02.004.

Gargouri, D., Serbaji, C.M., Jedoui, Y., and M. Montacer. 2011. Heavy metal concentrations in the surface marine sediments of Sfax Coast, Tunisia. *Environ Monit Assess* 175, no. 1: 519–530.

Gbogbo, F., Arthur-Yartel, A., Bondzie, J.A., Dorleku, W.P., Dadzie, S., Kwansa-Bentum, B., Ewool, J., Billah, M.K., and A.M. Lamptey. 2018. Risk of heavy metal ingestion from the consumption of two commercially valuable species of fish from the fresh and coastal waters of Ghana. *PLoS ONE* 13, no. 3: e0194682.

Gerlach, S.A. 1976. *Marine Pollution. Diagnosis and Therapy*. Berlin, Heidelberg and New York: Springer, 218 p.

Hakanson, L. 1980. Ecological risk index for aquatic pollution control, a sedimentological approach. *Water Res* 14: 975–1001.

Haskoning. 1991. *Proposal for Developing a Coastal Management Plan in the Republic of Yemen*. Report by Haskoning (Royal Dutch consulting Engineering and Architects). Support to Secretariate of the Environmental Protection Council, Yemen, 53 p.

He, Z.L., Yang, X.E., and P.J. Stoffella. 2005. Trace elements in agroecosystems and impacts on the environment. *J Trace Elem Med Biol* 19, nos. 2–3: 125–140.

Heba, H.M.A., Abu Zeid, I., Abuzinadah, O., Farajalla, A., and Z. Al-Hasawi. 2015. Determination of some heavy metals in tissues and organs of three commercial fish species at Al-Hudaydah, Red Sea Coast of Western Yemen. *J World* 7: 198–208.

Heba, H.M.A., Abuzinadah, O., Al-Hamadi, M., Al-Nedhary, A., Zeid, I.A., Saini, K., Farajalla, F., and M.M. Ahmed. 2014. Detection of heavy metals contamination in greasy grouper (*Epinephelus tauvina*) and striped mackerel (*Rastrelliger kanagurta*) from Al Hodeidah, Red Sea coast of Yemen. *J Food Agric Environ* 12, no. 2: 845–850.

Hebba, H.M.A., Al-Edresi, M.A.M., Al-Saad, H.T., and M.A. Abdemomeim. 2004. Background levels of heavy metals in dissolved, particulate phases of water and sediment of Al-Hodeidah Red Sea coast of Yemen. *JKAU Mar Sci* 15: 53–71.

Heba, H.M.A., Saeed, M.A., Al Shawafi, N.A., and H.T. Al Saad. 2003. Petroleum hydrocarbons and trace metals in Mollusca (*T. ponderoso*) from the Gulf of Aden. *J KAU Mar Sci* 14: 77–86.

IUCN (International Union for Conservation of Natural Resources). 1987a. *Yemen Arab Republic: Marine Conservation Survey, Vol 1. Distribution of Habitats and Species Along the YAR Coastline*. IUCN Red Sea and Gulf of Aden Environmental Program/TMRU, York, UK, 100 p.

IUCN (International Union for Conservation of Natural Resources). 1987b. *Yemen Arab Republic: Marine Conservation Survey, Vol 2. Distribution of Habitats and Species Along the YAR Coastline*. IUCN Red Sea and Gulf of Aden Environmental Program/TMRU, York, UK and Gland, Switzerland.

Kalenderski, S., Stenchikov, G., and C. Zhao. 2013. Modelling a typical winter-time dust event over the Arabian Peninsula and the Red Sea. *Atmos Chem Phys* 13: 1999–2014.

Krupp, F., Apel, M., Hamoud, A., Schneider, W., and U. Zajonz. 2006. Zoological survey in the Red Sea coastal zone of Yemen. *Fauna of Arabia* 21: 11–32.

Kumar, B., Sajwan, K.S., and D.P. Mukherjee. 2012. Distribution of heavy metals in valuable coastal fishes from North East Coast of India. *Turk J Fish Aquat Sci* 12, no. 1: 81–88.

Kureishy, T.W. 1991. Heavy metals in algae around the coast of Qatar. *Mar Pollut Bull* 22: 414–416.

Law, R. 1991. Fishing in evolutionary waters. *New Sci* 1758: 35–37.

Li, L., Hughes, I., and J.A. Arnot. 2021. Addressing uncertainty in mouthing-mediated ingestion of chemicals on indoor surfaces, objects, and dust. *Environ Int* 146: 106266.

Mahboob, S., Ahmed, Z., Khan, M.F., Virik, P., Al-Mulhm, N., and A.A. Baabbad. 2022. Assessment of heavy metals pollution in seawater and sediments in the Arabian Gulf, near Dammam, Saudi Arabia. *J King Saud Univ Sci* 34, no. 1: 101677.

Mostafa, A.R., Al-Alimi, A.K.A., and A.O. Barakat. 2009. Metals in surface sediments and marine bivalves of the Hadhramout coastal area, Gulf of Aden, Yemen. *Mar Pollut Bull* 58: 290–311.

Müller, G. 1969. Index of geo-accumulation in sediments of the Rhine River. *Geo J* 2: 109–118.

Nagi, H.M.H. 2021. Delineating and calculating the length of Yemen's mainland shoreline. *Int J Altern Fuels Energy* 5, no. 1: 1–9.

Nagi, H.M.H., and M.M. Abubakr. 2013. Threats status to the mangrove ecosystem along the coastal zone of Yemen. *J King Abdulaziz Univ* 24, no. 1: 101.

Nasr, S.M., Okbah, M.A., and S.M. Kasem. 2006. Environmental assessment of heavy metal pollution in bottom sediments of Aden Port, Yemen. *Int J Oceans Oceanogr* 1, no. 1: 99–109.

Nomaan, M.H., Panaskar, D.B., and R.S. Pawar. 2021. Analysis of heavy metal pollutants in the sediments from coastal sites of Al-Hodeidah Governorates, Yemen. In: *Techno-Societal 2020*, 911–921. Cham, Switzerland: Springer.

Omar, W.A., Saleh, Y.S., and M.A.S. Marie. 2014. Integrating multiple fish biomarkers and risk assessment as indicators of metal pollution along the Red Sea coast of Hodeida, Yemen Republic. *Ecotoxicol Environ Saf* 110: 221–231.

Prakash, P.J., Stenchikov, G., Kalenderski, S., Osipov, S., and H. Bangalath. 2015. The impact of dust storms on the Arabian Peninsula and the Red Sea. *Atmos Chem Phys* 15, no. 1: 199–222.

Rahman, M.A., and H. Ishiga. 2012. Trace metal concentrations in tidal flat coastal sediments, Yamaguchi Prefecture, southwest Japan. *Environ Monit Assess* 184: 5755e5771.

Raj, S.M., and M. Jayaprakash. 2008. Distribution and enrichment of trace metals in marine sediments of Bay of Bengal, off Ennore, south-east coast of India. *Environ Geol* 56, no. 1: 207–217.

Rouphael, T., Turak, E., and J. Brodie. 1998. Seagrasses and mangroves of Yemen's Red Sea. In: *Protection of Marine Ecosystems of the Red Sea Coast of Yemen*. Eds. A. DouAbal, et al., 41–49. Hassell & Assoc., AMSAT and UNOPS.

Rumisha, C., Elskens, M., Leermakers, M., and M. Kochzius. 2012. Trace metal pollution and its influence on the community structure of soft bottom molluscs in intertidal areas of the Dar es Salaam coast, Tanzania. *Mar Pollut Bull* 64, no. 3: 521–531.

Rushdi, A.I. 2021. Lipid tracers in coastal sediments of the northwestern Arabian/Persian Gulf: Characteristics, sources, distribution and possible ecological impacts. In: *The Arabian Seas: Biodiversity, Environmental Challenges and Conservation Measures*. Ed. L.A. Jawad, 1185–1185. Cham, Switzerland: Springer.

Rushdi, A.I., Abubakr, M.M., and H.M.A. Hebba. 1994. *Marine Habitats of the Red Sea at AlUrj-Alsalif and Dhubab-Yaktul Areas: Their Ecology, Environment and Management Recommendations Preliminary Investigations*. UN – Faculty of Sciences, Sana'a University, Yemen, 117 p.

Rushdi, A.I., Ba-Issa, A.A., and A. Ba-Bagi. 1991. Preliminary investigations of oil pollution along the Red Sea coast of Yemen. In: *Proc Seminar on the Status of the Environment in the Republic of Yemen, Sana'a, January 1–5*. EPC, Sana'a, Yemen.

Rushdi, A.I., Chase, Z., Simoneit, B.R.T., and A. Paytan. 2019. Sources of organic tracers in atmospheric dust, surface seawater particulate matter and sediment of the Red Sea. In: *Oceanographic and Biological Aspects of the Red Sea*. Eds. N.M.A. Rasul and I.C.F. Stewart, 75–88. Cham, Switzerland: Springer Oceanography, Springer Nature.

Saad, M.A.H., and M.A. Fahmy. 1994. Heavy metal pollution in coastal Red Sea waters, Jeddah. *J KAU Mar Sci* 7, Spec Issue: Symp Red Sea Mar Environ, Jeddah.

Saleh, S.M., Amer, A.T., and A.H. Al-Alawi. 2018. Potential ecological risk of heavy metals in surface sediments from the Aden coast, Southern Yemen. *J Environ Sci Toxicol Food Technol* 12, no. 10: 42–55.

Smith, J.D., Butler, E.C.V., Grant, B.R., Little, G.W., Millis, N., and P.J. Milne. 1981. Distribution and significance of copper, lead, zinc and cadmium in the Corio Bay ecosystem. *Aust J Mar Freshw Res* 32: 151–164.

Tanjung, R.H.R., Hamuna, B., and M.N. Yonas. 2019. Assessing heavy metal contamination in marine sediments around the coastal waters of Mimika Regency, Indonesia. *J Ecol Eng* 20, no. 11: 35–42.

Tanner, P.A., Leong, L.S., and S.M. Pan. 2000. Contamination of heavy metals in marine sediment cores from Victoria Harbour, Hong Kong. *Mar Pollut Bull* 40, no. 9: 769–779.

Taylor, S.R. 1964. Abundance of chemical elements in the continental crust: A new table. *Geochim Cosmochim Acta* 28: 1273–1285.

Tchounwou, P.B., Yedjou, C.G., Patlolla, A.K., and D.J. Sutton. 2012. Heavy metal toxicity and the environment. *Molecular Clin Environ Toxicol* 101: 133–164.

Tomlinson, D.L., Wilson, J.G., Harris, C.R., and D.W. Jeffrey. 1980. Problem in the assessment of heavy metals levels in estuaries and the formation of a pollution index. *Helgoländer Meeresuntersuchungen* 33: 566–575.

Topcuoğlu, S., Güven, K.C., Balkıs, N., and Ç. Kırbaşoğlu. 2003. Heavy metal monitoring of marine algae from the Turkish Coast of the Black Sea, 1998–2000. *Chemosphere* 52, no. 10: 1683–1688.

Turekian, K.K., and K.H. Wedepohl. 1961. Distribution of the elements in some major units of the earth's crust. *Bull Geol Soc Am* 72, no. 2: 175–192. https://doi.org/10.1130/0016-7606(1961)72[175:DOTEIS]2.0.CO;2

Turkmen, M., Turkmen, A., and Y. Tepe. 2008. Metal contaminations in five fish species from Black, Marmara, Aegean and Mediterranean Seas, Turkey. *J Chil Chem Soc* 53, no. 1: 1424–1428.

Walker, D., and R.F.G. Ormond. 1982. Coral death from sewage and phosphate pollution at Aqaba, Red Sea. *Mar Poll Bull* 13: 21–25.

Wang, J., Liu, R.H., Yu, P., Tang, A.K., Xu, L.Q., and J.Y. Wang. 2012. Study on the pollution characteristics of heavy metals in seawater of Jinzhou Bay. *Procedia Environ Sci* 13: 1507–1516.

Wasel, S.O., and M.A. Al-Edresy. 2016. An assessment of heavy metals contamination in surface sediment. *Tehama J* 10, no. 10: 1–1.

Whitfield, M. 1979. The mean oceanic residence time (MORT) concept – a rationalization. *Mar Chem* 8: 101–123.

World Health Organization (WHO). 1989. *Heavy Metals Environmental Aspects. Environment Health Criteria. No. 85*. Geneva, Switzerland.

Xu, F., Liu, Z., Cao, Y., Qiu, L., et al. 2017. Assessment of heavy metal contamination in urban river sediments in the Jiaozhou Bay catchment, Qingdao, China. *Catena* 150: 9–16.

Younis, E.M., Abdel-Warith, A.W.A., Al-Asgah, N.A., Elthebite, S.A., and M.M. Rahman. 2021. Nutritional value and bioaccumulation of heavy metals in muscle tissues of five commercially important marine fish species from the Red Sea. *Saudi J Biol Sci* 28, no. 3: 1860–1866.

Yu, G.B., Liu, Y., Yu, S., Wu, S.C., Leung, A.O.W., Luo, X.S., Xu, B., Li, H.B., and M.H. Wong. 2011. Inconsistency and comprehensiveness of risk assessments for heavy metals in urban surface sediments. *Chemosphere* 85, no. 6: 1080–1087.

Yunus, K., Zuraidah, M.A., and A. John. 2020. A review on the accumulation of heavy metals in coastal sediment of Peninsular Malaysia. *Ecofeminism Clim Chang* 1, no. 1: 21–35.

Zaghloul, G.Y., El-Din, H.M.E., Mohamedein, L.I., and K.M. El-Moselhy. 2022. Bio-accumulation and health risk assessment of heavy metals in different edible fish species from Hurghada City, Red Sea, Egypt. *Environ Toxicol Pharmacol* 95: 103969.

Zaidi, M., Athmouni, K., Metais, I., Ayadi, H., and V. Leignel. 2022. The Mediterranean limpet Patella caerulea (Gastropoda, Mollusca) to assess marine ecotoxicological risk: A case study of Tunisian coasts contaminated by metals. *Environ Sci Pollution Res* 29, no. 19: 28339–28358.

Zhang, J., Zhou, F., Chen, C., Sun, X., Shi, Y., Zhao, H., and F. Chen. 2018. Spatial distribution and correlation characteristics of heavy metals in the seawater, suspended particulate matter and sediments in Zhanjiang Bay, China. *PLoS ONE* 13, no. 8: e0201414.

Zulkifli, S.Z., Ismail, A., Mohamat-Yusuff, F., Arai, T., and N. Miyazaki. 2010. Johor Strait as a hotspot for trace elements contamination in Peninsular Malaysia. *Bull Environ Contam Toxicol* 84, no. 5: 568–573.

A lead isotopic study of pollution distribution and sourcing from nearshore sediments of Jeddah, Saudi Arabia

8

Riyadh F. Halawani, Moahd K. Alghuson, Myra E. Wilson and Andrew N. Quicksall

8.1 INTRODUCTION

It is essential in environmental research to recognize the source of pollution and determine the transport history of pollutants. Given the broad range of uses and transport mechanisms, determining the sources of lead in the environment can be difficult. Pb can be found naturally, via rock weathering and aerosol deposition (Monastra et al. 2004; Choi et al. 2007; Monna et al. 2000). Common Pb ore minerals in nature are galena (PbS), cerussite ($PbCO_3$) and anglesite ($PbSO_4$) (Faure 1977); additionally, Pb is extracted as a co-product from ores of copper, zinc and silver (Cheng and Hu 2010). Because of its unique physical and chemical characteristics, lead has been extensively extracted from these ores and used in different industries such as construction, battery production and the manufacturing of ammunition, paint and fusible alloys (Wani et al. 2015; Al-Saleh et al. 1993; Barbosa et al. 2005; Kim and Kang 2015; Lee et al. 2019). Anthropogenic contamination of Pb is likely produced in urban areas by industrial sources, exhaust from gasoline vehicles, paint and pesticides, thus complicating the identification of lead pollution sources (Cheng and Hu 2010; Komárek et al. 2008; Chen et al. 2016b).

Pb is recognized as toxic to humans, typically enters the human body via ingestion and/or inhalation and affects most of the functions of the body (Yu et al. 2016). A variety of diseases such as anemia, central nervous system disorders and immune system disorders are directly attributed to lead exposure (Kosnett 2007; Wani et al. 2015; Needleman et al. 1990). Moreover, sleeplessness, tiredness, loss of memory, weight loss, increased blood pressure and cardiovascular diseases are linked to high Pb concentrations in human blood (Yap et al. 2016). The World Health Organization (WHO) set 10 μg/l as a limit for Pb in drinking water, and the Panel on Contaminants in the Food Chain at the European Food Safety Authority (EFSA) set 25 μg/week/kg body weight as a maximum total weekly ingestion of Pb (WHO 2011; EFSA 2010). World associations, including the United Nations Environment Program (UNEP) and the World Health Organization (WHO), supported the idea of worldwide elimination of Pb paint by 2020 because of the documented severe danger of lead to humans (O'Connor et al. 2018).

Lead is also well documented as a toxin to fauna. Fish are among the most abundant marine animals and impact humans directly through the food chain. Therefore, most of the research on lead toxicity to marine animals has been carried out in fish (Lee et al. 2019; Vinodhini and Narayanan 2008; Cheng and Hu 2010). Pb is known as one of the extremely toxic metals to fish due, in part, to its bioaccumulation in tissues, influence on physiological and biochemical functions and neurotoxicity (Taylor and Maher 2014; Lee et al. 2019). The liver, kidney and spleen are the most common tissues in fish for accumulation after long-term Pb exposure (Vinodhini and Narayanan 2008; Javed 2012; Kim and Kang 2015). Even low concentrations of lead in the marine environment have been shown to have adverse effects on the gills of fish (Sures and Siddall 1999; Hwang et al. 2016; Macirella et al. 2019). Additionally, lead exposure in marine algae negatively affects photosynthetic activities and net total chlorophyll concentrations, thereby damaging the food web of marine ecosystems (Gan et al. 2019).

Stable lead isotopes are valuable for discerning Pb sources, including when they are significantly derived from anthropogenic sources. Isotopic ratios are frequently a more effective tracking method than utilizing their concentrations alone to recognize sources of Pb pollution (Alyazichi et al. 2016; Chen et al. 2015). The four known stable lead isotopes are non-radiogenic

DOI: 10.1201/9781003321408-11

^{204}Pb as well as radiogenic ^{206}Pb, ^{207}Pb and ^{208}Pb, produced by the radioactive decay of ^{238}U, ^{235}U and ^{232} Th, respectively (Faure 1977). Various studies have evaluated the adequacy of their ratios, $^{208}Pb/^{206}Pb$ and $^{206}Pb/^{207}Pb$, as indicators of pollution sources; they include studies in Scotland (Bacon et al. 1995), France (Monna et al. 2000), East Antarctica (Townsend and Snape 2002), the Czech Republic (Komárek et al. 2008), China (Cheng and Hu 2010; Yu et al. 2016; Liu et al. 2019), Australia (Alyazichi et al. 2016) and Singapore (Chen et al. 2016a).

Some geological and mineralogical studies have analyzed Pb concentrations and isotopic ratios in deep Red Sea basins and brine pools (Dupré et al. 1988; Baumann 1994; Altherr et al. 1990; Pierret et al. 2010; Laurila et al. 2014). Pierret et al. (2010) consider Atlantis II Deep as a reference location for the most metal-rich sediments, which assembled through chemical deposition from overlying mineralized, salt-rich hydrothermal brines.

There is strong recent interest in studying lead in marine sediments from the countries bordering the Red Sea due to the rapid development and lack of prior research in the area. This recent body of literature shows lead concentrations were high in some marine sediment locations along the coastal areas of Egypt (Mansour et al. 2011; Salem et al. 2014; El-Taher et al. 2018; Badawy et al. 2018;), Sudan (Idris et al. 2007), Jordan (Al-Najjar et al. 2011) and Saudi Arabia (Ruiz-Compean et al. 2017; Karuppasamy et al. 2017; Kahal et al. 2018), including the Jeddah region specifically (Pan et al. 2011; Abu-Zied et al. 2013; Halawani et al. 2022). Despite these recent studies, there remains a serious lack of data and doubt concerning the distribution and, specifically, the sourcing of Pb. The aims of this study are 1) to set natural Pb isotopic ratio background values for the study area and 2) to evaluate the spatial distribution of Pb isotopic ratios to identify and, if possible, to quantify sources of Pb pollution.

8.2 MATERIALS AND METHODS

8.2.1 Study area and sample collection

Three locations were selected in the study area as shown in Figure 8.1 for Pb isotopic and total abundance analyses. A total of forty submarine surface sediment samples were collected. The ten samples collected from the North location of Jeddah were taken from an area remote from heavy human activities. Twenty samples were collected from the Middle location near the fish market and the main seaport of the city. The remaining ten samples were taken near Jeddah Industrial City and a refinery plant from the South location. All the samples were collected by stainless steel soil scoops, stored in clear, flat polyethylene zipper bags and held at 4°C until transported to the laboratory.

8.2.2 Sample preparation

Sediment samples were analyzed for total Pb concentration and Pb isotopic composition following an adapted US EPA method 3050B (USEPA 1996). The method was modified by utilizing approximately 0.20 g of sediment. Prior to analyses, samples were placed in 50 mL tubes, and 5 mL of 5% HNO_3 was added to remove significant carbonate. Then 20 mL additions of 70% HNO_3 were added inside a fume hood and digested at 85°C on a hot block until dry. Samples were resolubilized in a known mass of 5% HNO_3 for analysis. Analyses were run using collision cell technology with kinetic energy discrimination (CCT-KED) using an X-series 2 inductively coupled plasma-mass spectrometer (ICP-MS) made by Thermo Scientific. To establish calibration curves for Pb concentrations, a standard CLMS-2A from SPEX Certi Prep (Metuchen, NJ, USA) was used, with a range from 0.1 to 10 ppb (six points). All ICP operations and sample dilutions used a matrix of 5% HNO_3, mixed from concentrated 70% HNO_3 certified ACS plus (A200C-212) from Fisher Scientific and 18.2 MΩ nanopure water. Calibration curves with r^2 values of 0.999 or larger were applied to all samples. For quality, each sample was analyzed in triplicate, and a blank sample (5% HNO_3) was measured every 12 samples to verify sufficient washout time between runs. All ICP-MS plasticware used in the analyses was acid washed (5% HCl) for 24 hours and rinsed with 18.2 MΩ nanopure water prior to use.

8.2.3 Enrichment factors

The metal enrichment factor (EF) reflects the status of the environmental contamination by indicating whether a specific measured metal concentration is from natural weathering processes or anthropogenic sources. The EF is calculated using the following equation:

$$EF = (C_i / C_{Rb})_S / (C_i / C_{Rb})_B \tag{8.1}$$

where EF is an enrichment factor for a particular metal, $(C_i / C_{Rb})_S$ is the metal concentration to rubidium ratio (both in mg/kg) in the studied sediment sample (Grant and Middleton 1990; Alyazichi et al. 2016), and the average background shale values adopted from Turekian and Wedepohl (1961) are given by $(C_i / C_{Rb})_B$. Amin et al. (2009) differentiated contamination into categories based on enrichment factor values, with an EF value less than 1 showing no enrichment, an EF value between 1 and 3 showing minor enrichment, an EF value between 3 and 5 showing moderate enrichment, an EF value from 5 to 10 showing moderately severe enrichment, an EF value between 10 and 25 showing severe enrichment, and an EF value greater than 25 showing extremely severe enrichment.

8.2.4 Pb Sources and contributions

It is essential in most Pb studies to estimate and quantify the contributions from different sources (Kersten et al. 1997; Yu et al. 2016; Xu et al. 2019). Numerous potential anthropogenic Pb sources could discharge and reach the Jeddah coastal region. The use of total concentration data solely provides limited understanding of sourcing. Each Pb source has its preserved isotopic signature and does not change under common transport processes (Cheng and Hu 2010; Wang et al. 2014; Monna et al. 1997). It is common in environmental studies to plot two isotopic ratios, $^{208}Pb/^{206}Pb$ and $^{206}Pb/^{207}Pb$, against each other to relate samples to the isotopic compositions of different potential Pb sources (Carrasco et al. 2018;

FIGURE 8.1 Regional map of the eastern margin of the central Red Sea showing the distribution of sampling locations relative to the Jeddah coastline. The green dots show sample stations for the North location, the red dots for the Middle location and blue dots for the South location.

Yu et al. 2016; Cheng and Hu 2010; Komárek et al. 2008; Flament et al. 2002).

Further, various researchers have used mixing models to analyze such isotopic data (Monna et al. 2000; Komárek et al. 2008; Xu et al. 2019). Here, a two-member, or binary, mixing model is used for the low Pb concentration (<100 mg/kg) samples. A fractional contribution model of three components, or a ternary model, is used for high Pb concentration (>100 mg/kg) samples. Specific methods for each model and the rationale for each is discussed later in this chapter.

A linear mixing model was used in the current study to calculate the estimated contributions of Pb from natural sources and gasoline by using the $^{206}Pb/^{207}Pb$ ratio (Komárek et al. 2008; Monna et al. 1997). The model equations used were:

$$X_{\%}^{G} = \left[\left(\frac{206_{Pb}}{207_{Pb}}\right)_{S} - \left(\frac{206_{Pb}}{207_{Pb}}\right)_{Nat}\right] \Big/ \left[\left(\frac{206_{Pb}}{207_{Pb}}\right)_{Gas} - \left(\frac{206_{Pb}}{207_{Pb}}\right)_{Nat}\right] * 100 \quad (8.2)$$

$$X_{\%}^{Gas} + X_{\%}^{Nat} = 100 \quad (8.3)$$

where $X_{\%}^{Gas}$ and $X_{\%}^{Nat}$ are the contributions of gasoline and natural sources in percentage, $(^{206}Pb/^{207}Pb)_S$ is the Pb isotopic composition in the sample, $(^{206}Pb/^{207}Pb)_{Nat}$ is the Pb isotopic composition of the natural sources (1.22) (Carrasco et al. 2018; Chen et al. 2016a; Monna et al. 1995; Komárek et al. 2008), and $(^{206}Pb/^{207}Pb)_{Gas}$ is the Pb isotopic composition of the gasoline (1.10) (Bollhöfer and Rosman 2001; Carrasco et al. 2018; Monna et al. 1995).

The fractional contribution of the three-component model was used to quantify the input from three Pb sources to the sediment samples with high Pb loading (>100 mg/kg) (Gobeil et al. 1995; Chiaradia et al. 1997; Cheng and Hu 2010). The following three equations were employed in the model:

$$\left(\frac{208_{Pb}}{206_{Pb}}\right)_S = F1\left(\frac{208_{Pb}}{206_{Pb}}\right)_{US} + F2\left(\frac{208_{Pb}}{206_{Pb}}\right)_{Nat} + F3\left(\frac{208_{Pb}}{206_{Pb}}\right)_{Gas} * 100 \quad (8.4)$$

$$\left(\frac{206_{Pb}}{207_{Pb}}\right)_S = F1\left(\frac{206_{Pb}}{207_{Pb}}\right)_{US} + F2\left(\frac{206_{Pb}}{207_{Pb}}\right)_{Nat} + F3\left(\frac{206_{Pb}}{207_{Pb}}\right)_{Gas} * 100 \quad (8.5)$$

$$F1 + F2 + F3 = 100 \quad (8.6)$$

where $(^{208}Pb/^{206}Pb)_S$ and $(^{206}Pb/^{207}Pb)_S$ were the measured ratios of the samples while F1, F2 and F3 were the percentage contributions of Pb from unknown sources, natural sources and gasoline, respectively. The values of $^{208}Pb/^{206}Pb$ and $^{206}Pb/^{207}Pb$ for the unknown sources were calculated as described by Monna et al. (2000) and Gallon et al. (2011). Simply, linear regressions of $^{206}Pb/^{207}Pb$ and $^{208}Pb/^{206}Pb$ versus 1/Pb concentration should yield intercepts representing the isotopic ratio values of a contaminating source (Figures 8.2a and b). Here, the intercepts of the best fit lines ($^{206}Pb/^{207}Pb = 1.183$; $^{208}Pb/^{206}Pb = 2.068$) provided an estimate of the isotopic ratios of an unknown contamination source of Pb in the area. The values of the $^{206}Pb/^{207}Pb$ for natural and gasoline components were the same values used in equation 2. Singular values of $^{208}Pb/^{206}Pb$ for natural and gasoline sources were not available from the literature, so they were calculated from the linear regression of the plot of $^{208}Pb/^{206}Pb$ and $^{206}Pb/^{207}Pb$ in Figure 8.3, yielding $^{208}Pb/^{206}Pb$ values of 2.01 for natural sources and 2.11 for gasoline. Both values fall well within the range of reported values for such sources.

8.3 RESULTS AND DISCUSSION

The data from Table 8.1 shows that the average concentration of Pb in the North location of the study area was 4.72 mg/kg, with a range between 3.57 and 6.96 mg/kg; the average for the Middle location was 80.44 mg/kg, with a range between 1.70 and 372.71 mg/kg; and the average for the South location was 147.91 mg/kg, with a range between 6.36 and 578.47 mg/kg. The calculated EF values suggest anthropogenic sources for all locations; Pb enrichment factors range from 30.03 to 130.84, with an average of 71.50; 10.18 to 946.31, with an average of 243.99; and 123.74 to 9607.91, with an average 2397.99 for North, Middle and South locations, respectively (Table 8.1). The results indicate that more than 80% of the sediment samples were considered extremely severely enriched with Pb (EF >50). Given the ubiquitous enrichment across the entire region, it is likely that the region has been impacted by an anthropogenic source that has elevated the entire region's background. That background would be composed of two major components: naturally sourced Pb and a widespread anthropogenic source. That latter source is almost certainly gasoline-derived Pb, given the environmental history of the region and the omnipresence of enrichment. Meanwhile, there are sample stations

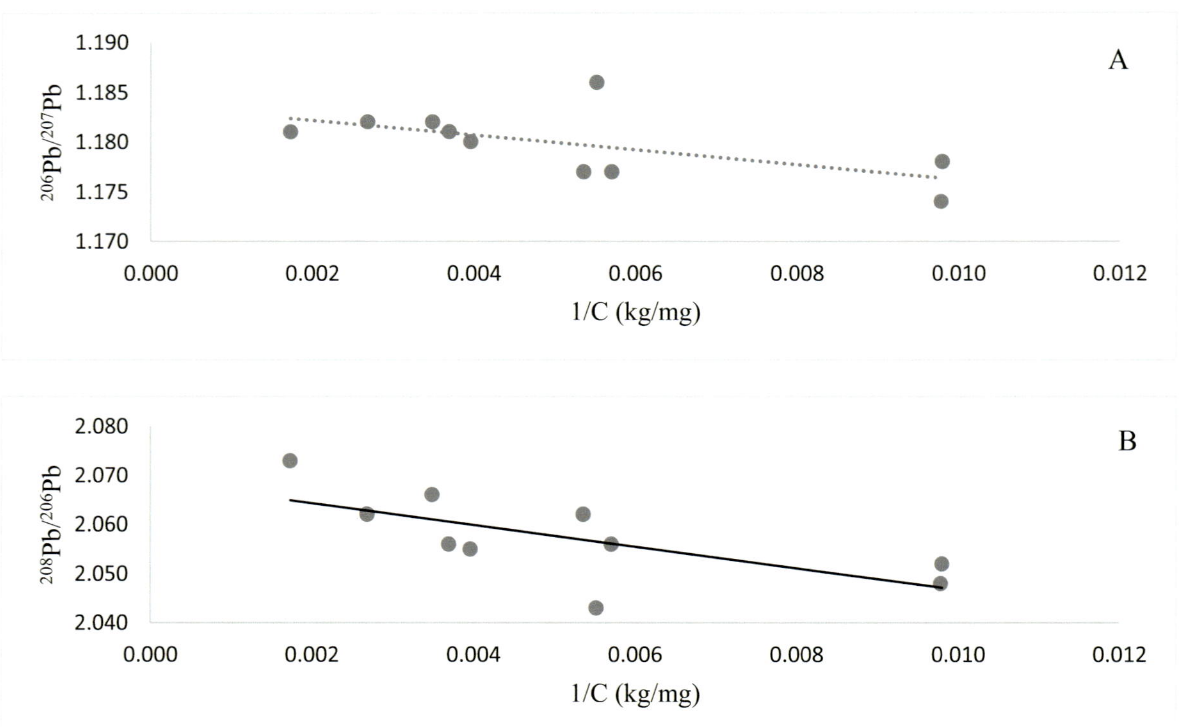

FIGURE 8.2 $^{206}Pb/^{207}Pb$ (a) and $^{208}Pb/^{206}Pb$ (b) vs 1/Pb concentration. Both panels plot solely higher concentration (> 100 mg/kg) sediment samples from the region. Solid lines in both panels show least squared regressions of best fit with intercepts of 1.184 (a) and 2.069 (b).

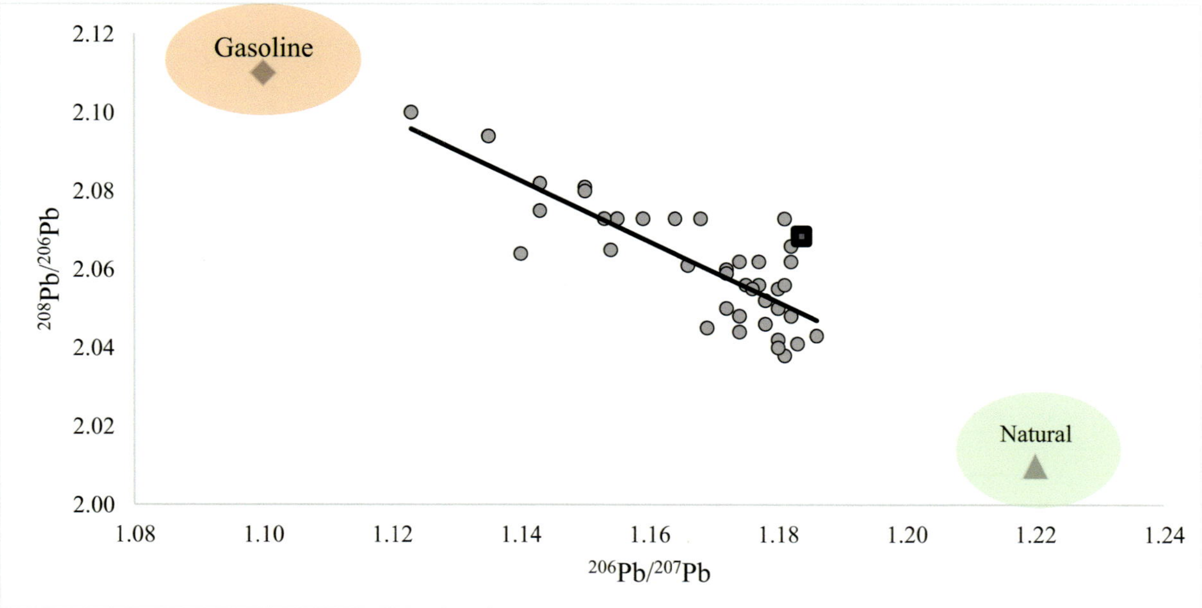

FIGURE 8.3 Distribution of Pb isotopic ratios $^{208}Pb/^{206}Pb$ vs $^{206}Pb/^{207}Pb$ of surface sediments from the Jeddah shoreline. The solid black line provides a least squared regression of best fit with $r^2 = 0.66$. The black square shows the unknown source of Pb, constrained as described in the text and graphically shown in Figure 8.2. The triangle displays the centroid of natural sources while the blue ellipse shows the approximate boundaries of the range of natural sources. Similarly, the diamond denotes the centroid of gasoline sources, and the associated red ellipse outlines a range of gasoline sources from the literature.

TABLE 8.1 Overview of Pb concentrations (mg/kg), Pb isotopic ratios and enrichment factors (EF) for Jeddah region coastal sediments organized by location.

		MG/KG	$^{208}PB/^{206}PB$	$^{206}PB/^{207}PB$	EF
North	Average	4.72	2.059	1.168	71
	Min	3.57	2.041	1.153	30
	Max	6.96	2.073	1.183	130
Middle	Average	80.44	2.066	1.163	243
	Min	1.70	2.042	1.123	10
	Max	372.71	2.100	1.186	946
South	Average	147.91	2.053	1.179	2397
	Min	6.36	2.038	1.172	123
	Max	578.47	2.073	1.182	9607

with excessively enriched Pb values. It is likely that there exist one or more other anthropogenic sources in addition to the new combined natural-gasoline background. The Pb concentrations of the Jeddah nearshore sediment samples were, therefore, split into two groups: one with lower concentration (<100 mg/kg) and the other with a higher concentration (>100 mg/kg).

8.3.1 Sourcing of background Pb in sediments

Figure 8.3 shows the distribution of Pb isotopes in the study area. All samples fall between well-established values for natural sources from the region and typical gasoline-derived Pb. Despite falling broadly between these values, a number of samples are off a linearity between the end members and trend toward an unidentified source discussed later in this chapter (Figure 8.3). This graphically supports the separation of the sample set into two populations as discussed earlier.

One method of delineating differences in the populations is to observe isotopic variance by location. The average ratio for $^{206}Pb/^{207}Pb$ was 1.17, with a range of 1.15–1.18 for the North location. The Middle location average was 1.16, with a range of 1.12–1.19. For the South location, the average was 1.179, and the range was 1.17–1.18 (Table 8.1). $^{206}Pb/^{207}Pb$ ratio values decrease from north to south for the North location (1–10) and increase from north to south for the South locations (31–40) (Figure 8.4). The Middle location samples show minimal values in that region's central area. Generally, then, the entire study area has lower $^{206}Pb/^{207}Pb$ values near central latitudes, with increasing values to the north and south (Figure 8.4).

As implied earlier, assessing concentration control on isotopic expression is required to fully interpret the data. By using the two end-member model (equations 2 and 3), we found the relative Pb contribution for gasoline and natural sources (Figure 8.5) for the low concentration samples (< 100 mg/kg). The average percentage contribution of Pb from natural sources was 53% in the range of 69–19%, and the average for gasoline was 46% with range of 80–30%. There is a significant positive correlation between $^{206}Pb/^{207}Pb$ and $^{208}Pb/^{206}Pb$ (r^2 =0.79) for the sediment samples from the study area that have the lowest Pb concentrations (<100 mg/kg) (Figure 8.6). This quantitatively shows the linear combination of the two sources creating a new Pb isotopic background spectrum of combined gasoline-derived and natural sources. These data now enable us to better interpret latitudinal variance. Samples from the far north and south are likely dominated by natural Pb (the green zone for natural Pb > 50%); alternatively, the effect of legacy gasoline-derived Pb in the modern background distribution is confirmed by the proximity of gasoline-dominated samples (the white zone for gasoline-derived Pb > 50%) to the urban core of Jeddah found in the Middle section of Figure 8.4.

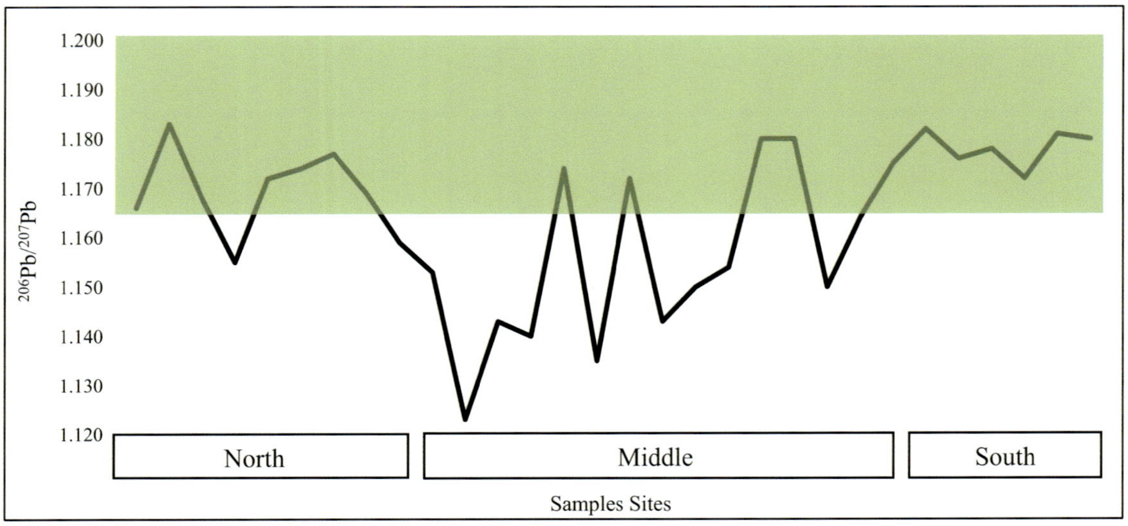

FIGURE 8.4 $^{206}Pb/^{207}Pb$ for the low Pb concentration (<100 mg/kg) nearshore sediment samples arranged by location progressing North to South. The green zone shows where natural Pb is greater than 50%, whereas the white zone shows samples dominated by a gasoline-derived source.

For the North location, the average contributions of Pb from the natural sources was 56%, and it was 43% for gasoline. The points N4, N9 and N10 (Table 8.2) showed a high contribution of Pb gasoline (54, 50 and 55%, respectively), which could be due to localized heavy traffic activities (Figure 8.5A). The Middle location showed approximately the inverse of the North location, as the average contributions of Pb was 46% for the natural sources and 53% for the gasoline. Some points, such as M1, M2, M3, M5 and M7, showed a contribution of Pb gasoline of greater than 60% (Table 8.2). The sources of Pb gasoline in this location could be the heavy traffic, the diesel fuel–powered desalination plant and the activities of Jeddah's seaport (Figure 8.5B). The average contribution of Pb from gasoline was 34% and, from the natural sources, 56% in the South location. As shown in Figure 8.5C, the contribution from the natural sources was very high – more than 60% – which indicates that natural Pb dominates the area.

The model showed that some points from the North location (N1, N2, N3, N5, N6, N7 and N8), Middle location (M4, M6, M11, M13, M19 and M20) and South location (S4, S5, S6, S8, S9 and S10) had a greater than 50% contribution from natural sources rather than gasoline derived (Figure 8.4). Additionally, all these points' samples were in the $^{206}Pb/^{207}Pb$ range of 1.16–1.18 (Table 8.2). This finding is broadly consistent with that of Zhu et al. (2010), who reported the natural $^{206}Pb/^{207}Pb$ in the sediments from the South China Sea (1.17); Carrasco et al. (2018), who explored sediments and soil from Singapore (1.21–1.18); and Xu et al. (2019), who found values in sediments from the eastern Beibu Gulf, South China Sea, of less than 1.17.

On the other hand, the model showed some points with a contribution from gasoline greater than 50% in the North location (N4, N9 and N10) and Middle location (M1, M2, M3, M5, M7, M8, M9, M17) (Figure 8.5A and B). These points were in the range between 1.16–1.12 of $^{206}Pb/^{207}Pb$ (Table 8.2). These results are consistent with data obtained by Carrasco et al. (2018), who reported the ratio of the $^{206}Pb/^{207}Pb$ for leaded gasoline around at 1.12; Chen et al. (2016b), who reported the ratio in the South China Sea at 1.137; and Chen et al. (2016a), who stated the ratio in Singapore as less than 1.14.

The two end-member model confirms that most marine sediment Pb in the region is derived from natural weathering (terrestrial sources) and sediment accumulation of legacy

TABLE 8.2 Isotopic ratio values for samples with concentrations below 100 mg/kg (n=30) and modeled lead source percentage contributions using the two end-member model.

SAMPLE NO.	MG/KG	% GASOLINE	% NATURAL	208PB/206PB	206PB/207PB
N 1	4.02	45	55	2.061	1.166
N 2	3.57	30	69	2.041	1.183
N 3	5.57	43	56	2.073	1.168
N 4	4.40	54	45	2.073	1.155
N 5	6.96	40	60	2.050	1.172
N 6	4.48	38	61	2.044	1.174
N 7	4.51	35	64	2.056	1.177
N 8	5.21	42	57	2.045	1.169
N 9	3.86	50	49	2.073	1.159
N 10	4.62	55	44	2.073	1.153
M 1	1.70	80	19	2.100	1.123
M 2	6.70	64	35	2.082	1.143
M 3	5.04	66	33	2.064	1.140
M 4	37.44	38	61	2.062	1.174
M 5	21.64	70	29	2.094	1.135
M 6	39.63	40	60	2.060	1.172
M 7	4.56	64	35	2.075	1.143
M 8	2.95	58	41	2.081	1.150
M 9	8.32	55	45	2.065	1.154
M 11	75.65	33	66	2.042	1.180
M 13	60.39	33	66	2.050	1.180
M 17	8.06	58	41	2.080	1.150
M 19	36.02	46	53	2.073	1.164
M 20	80.49	37	62	2.056	1.175
S 4	40.37	31	68	2.048	1.182
S 5	75.81	36	63	2.055	1.176
S 6	20.30	35	65	2.046	1.178
S 8	21.43	40	60	2.059	1.172
S 9	26.38	32	67	2.038	1.181
S 10	6.36	33	66	2.040	1.180
Average	20.88	46	53	2.06	1.16
Min	1.70	30	19	2.04	1.12
Max	80.49	80	69	2.10	1.18

(N) = North location samples, (M) = Middle location samples, (S) = South location samples.

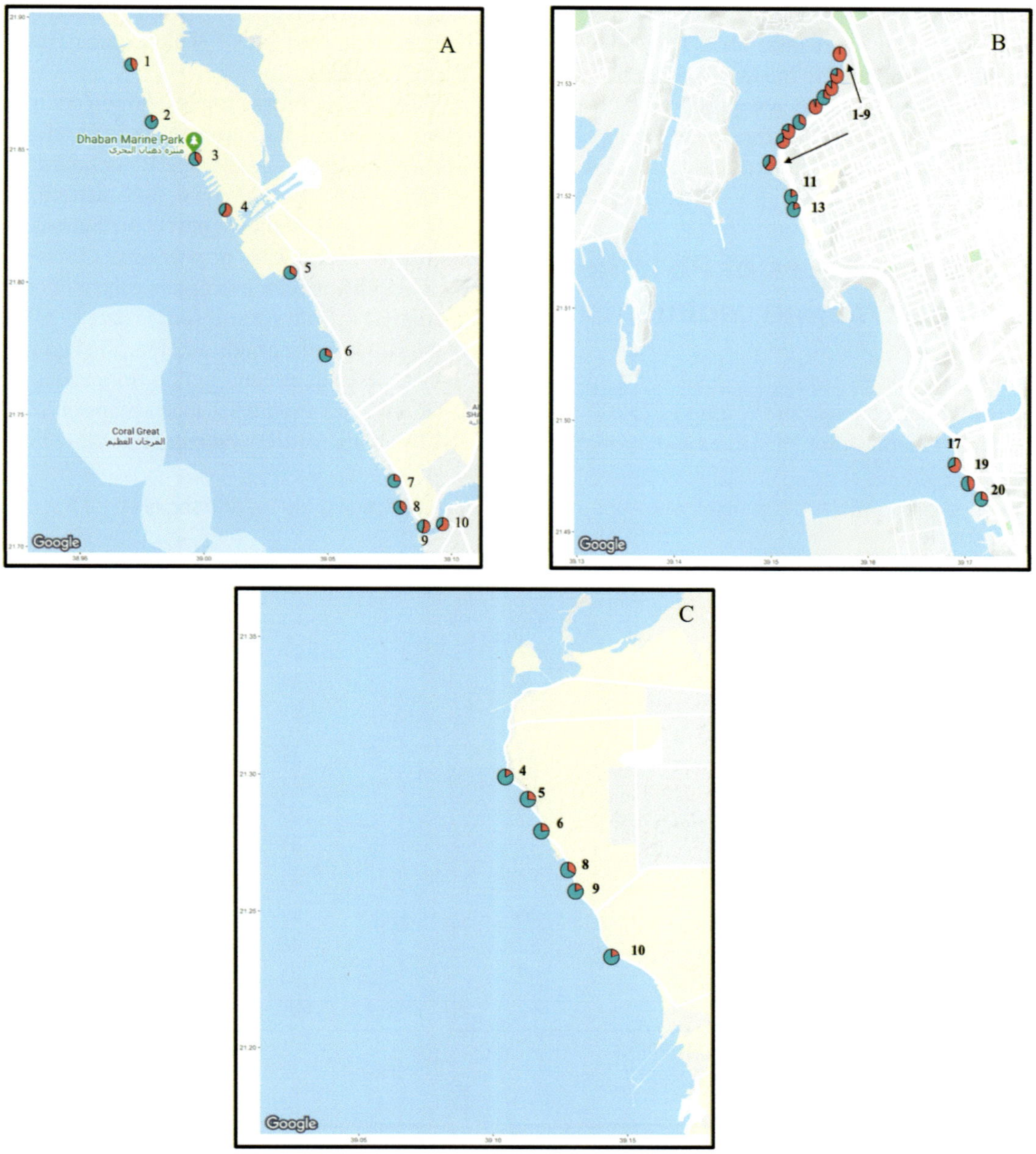

FIGURE 8.5 Spatial distribution of percentage contributions of gasoline-derived (red) and natural (green) source of Pb for North (A), Middle (B) and South (C) locations in sediment samples with Pb concentration < 100 mg/kg (n=30).

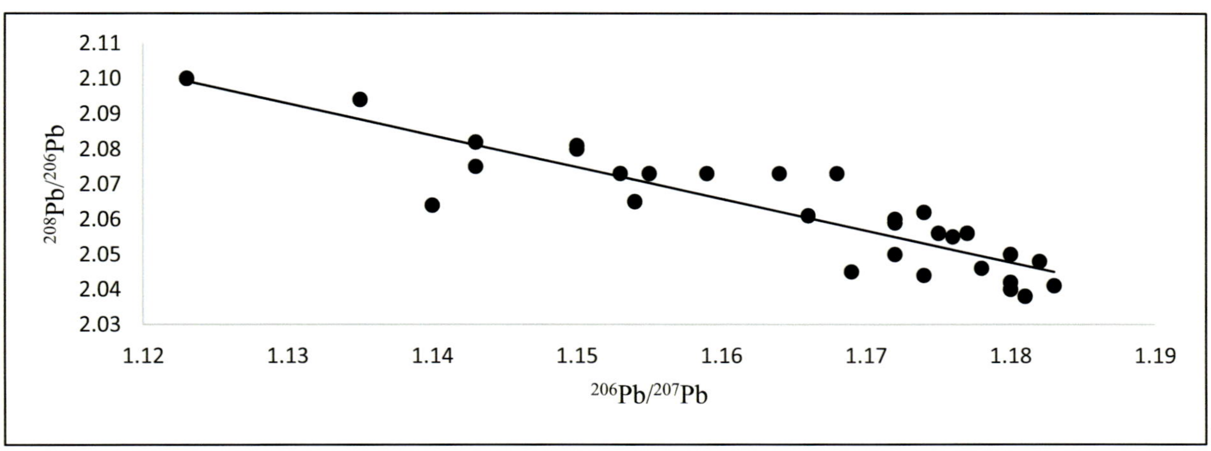

FIGURE 8.6 $^{208}Pb/^{206}Pb$ vs $^{206}Pb/^{207}Pb$ from < 100 mg/kg sediment samples from this study (n=30). The solid black line provides a least squared regression of best fit showing a significant positive correlation (r^2=0.79).

gasoline-derived sources. This interpretation best describes the majority of the samples in the region, which have relatively low total concentrations compared to the heavily enriched samples. This broad elevated background composed of both gasoline and natural sources favours the anthropogenic gasoline-derived source in the mid-latitudinal sections of the area centred on the urban core of Jeddah.

8.3.2 Possible sources of Pb in heavily enriched marine sediments

The three-member mixing model (equations 4, 5 and 6) was applied to higher concentration samples (>100 mg/kg). Table 8.3 shows the modeled Pb percentage contributions using the fractional contribution of three sources for enriched Jeddah sediment, all of which come from the Middle and South locations.

Six of the twenty samples showed contributions of Pb from gasoline, natural backgrounds and the unknown source in the Middle location. The average Pb contributions in the Middle location were 47% unknown, 34% natural, and 18% gasoline. M14 and M18 had the highest contributions from the unknown source (86% and 71%, respectively) (Table 8.3). For the South location, 48% was the average Pb contribution from the unknown source, 32% from natural source and 19% from gasoline. One of the points in the South location, S7 (Figure 8.6B), showed the highest percentage contribution from unknown sources for that location (57%). Figure 8.7 shows maps of Middle (A) and South (B) locations for the percentage contributions of the Pb from

TABLE 8.3 Isotopic ratio values for samples with concentrations above 100 mg/kg (n=10) and modeled Pb fractional contributions from the three-source model.

SAMPLE NO.	*MG/KG*	*% GASOLINE*	*% NATURAL*	*UNKNOWN SOURCES*	*^{208}PB/^{206}PB*	*^{206}PB/^{207}PB*
M 10	102	38	62	−1	2.048	1.174
M 12	102	27	47	24	2.052	1.178
M 14	286	5	8	86	2.066	1.182
M 15	175	24	39	35	2.056	1.177
M 16	181	23	60	16	2.043	1.186
M 18	372	9	18	71	2.062	1.182
S 1	578	−0.2	−7.	107	2.073	1.181
S 2	252	20	37	41	2.055	1.180
S 3	270	18	34	47	2.056	1.181
S 7	186	18	24	57	2.062	1.177
Average	250	18	32	48	2.057	1.180
Min	102	−0.2	−7	−1	2.043	1.174
Max	578	38	62	107	2.073	1.186

(M) = Middle location samples, (S) = South location samples.

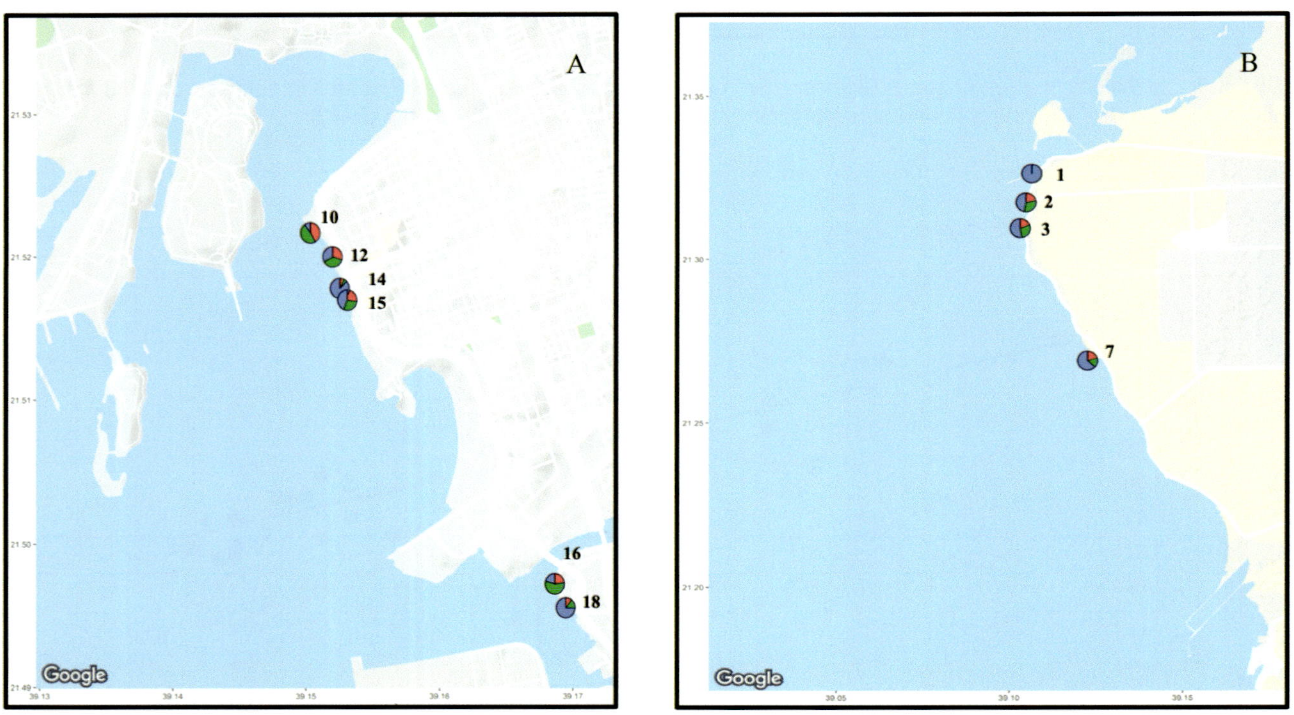

FIGURE 8.7 Spatial distribution of percentage Pb contributions of gasoline-derived (red), natural source (green), and unknown source (blue) for Middle (A) and South (B) locations in sediment samples with Pb concentrations > 100 mg/kg (n=10).

gasoline, natural sources and unknown sources in the sediment samples with Pb concentration > 100 mg/kg.

The ternary diagram for Jeddah sediment samples (Pb concentration > 100 mg/kg), including 8 out of 10 points, is displayed in Figure 8.8. The points M10 and S10 were out of the range of 0–100%. There is a clear trend from a blended natural source and gasoline background toward the unknown source (Figure 8.8). Moreover, the positive correlation (r^2=0.77) between the highest 10 Pb concentrations and the percentage contribution of the unknown source of the study area supports the strong control of this source on total loading (Figure 8.9).

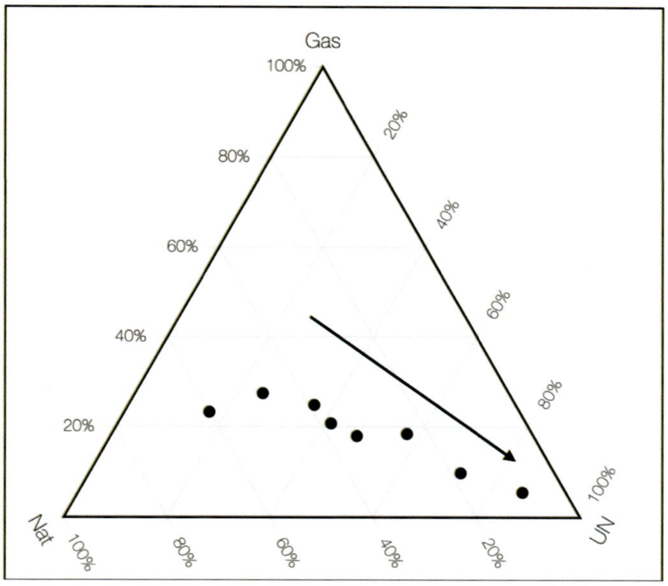

FIGURE 8.8 Ternary diagram for the fractional contribution of the three sources – gasoline (Gas), natural (Nat) and the unknown source (UN) – to the study's high-concentration sediment samples (Pb concentration >100 mg/kg). The black arrow provides a general progression from the left axis toward the lower right vertex. That progression moves away from the natural-gas background linearity toward the unknown contaminant.

The unknown source might be derived from industrial or domestic discharge from the surrounding urban area. Some of the highest lead concentrations were from the Middle location (Figure 8.7A) and were located near industrial and sewage wastewater outflow pipes (Abu-Zied and Orif 2019). Similarly, other points were from the South location (Figure 8.7B), which was near the treated and semi-treated sewage pipe outflow discharge from the Al Kumarra municipal treatment plant (Basaham et al. 2009). Also, the city of South Jeddah has an industrial waste treatment plant; the treated discharges from industrial activities are released nearby (Ramadan and Halawani 2010).

8.4 CONCLUSION

The aim of the present research was to evaluate sediment-bound Pb in the Red Sea near the Jeddah coastal area. For the first time in the region, sediment Pb isotopic data was employed to constrain Pb sourcing. The Middle location of the study area was the most Pb-impacted location, with decreasing pollution to the North and South. While the central latitudes are the most heavily Pb loaded, the entire region was found to be impacted as the EF results showed that more than 80% of the sediment samples were considered extremely severely enriched with Pb.

A two end-member model and a three-component fractional contribution model were used to identify possible Pb sources and their percentage contributions in the study area. The linear distribution of the two end members to lesser enriched sites clearly identified a baseline for the region that records a heterogeneous blending of a true natural source and gasoline-derived lead. The study further found a strong positive correlation between the highest Pb concentration and an additional unknown anthropogenic source which controls the overall pollution in the study area. While the exact source is unknown, it can be confirmed to be anthropogenic, and,

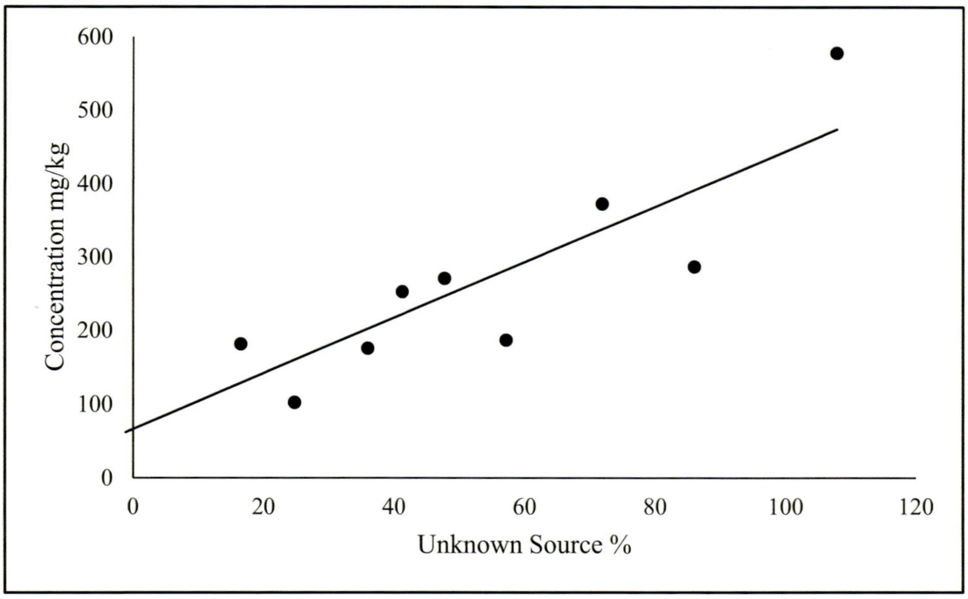

FIGURE 8.9 Regression analysis showing significant positive correlation between the highest 10 Pb concentration samples (>100 mg/kg) and the percentage of the unknown source for the study area (n=10). The solid black line provides a least squared regression of best fit with r^2 =0.77.

further, the isotopic composition is now known ($^{206}Pb/^{207}Pb$= 1.183, $^{208}Pb/^{206}Pb$=2.068). Consequently, this research serves as benchmark data and useful documentation of both legacy and modern environmental Pb pollution along the Saudi Arabian coastal areas of the Red Sea.

REFERENCES

Abu-Zied, R.H., Basaham, A.S., and M.A. El Sayed. 2013. Effect of municipal wastewaters on bottom sediment geochemistry and benthic foraminifera of two Red Sea coastal inlets, Jeddah, Saudi Arabia. *Environ Earth Sci* 68, no. 2: 451–469. http://doi.org/10.1007/s12665-012-1751-7.

Abu-Zied, R.H., and M.I. Orif. 2019. Recent environmental changes of Al-Salam Lagoon as inferred from core sediment geochemistry and benthic foraminifera, Jeddah City, Saudi Arabia. *Environ Earth Sci* 78, no. 3: 60. http://doi.org/10.1007/s12665-019-8057-y.

Al-Najjar, T., Rasheed, M., Ababneh, Z., Ababneh, A., and H. Al-Omarey. 2011. Heavy metals pollution in sediment cores from the Gulf of Aqaba, Red Sea. *Natl Sci* 3, no. 9: 775. http://doi.org/10.4236/ns.2011.39102.

Al-Saleh, I.A., Fellows, C., Delves, T., and A. Taylor. 1993. Identification of sources of lead exposure among children in Arar, Saudi Arabia. *Ann Clin Biochem* 30, no. 2: 142–145. http://doi.org/10.1177/000456329303000205.

Altherr, R., Henjes-Kunst, F., and A. Baumann. 1990. Asthenosphere versus lithosphere as possible sources for basaltic magmas erupted during formation of the Red Sea: Constraints from Sr, Pb and Nd isotopes. *Earth Planet Sci Lett* 96, nos. 3–4: 269–286.

Alyazichi, Y.M., Jones, B.G., and E. McLean. 2016. Lead isotope fingerprinting used as a tracer of lead pollution in marine sediments from Botany Bay and Port Hacking estuaries, southern Sydney, Australia. *Reg Stud Mar Sci* 7: 136–141. http://doi.org/10.1016/j.rsma.2016.06.006.

Amin, B., Ismail, A., Arshad, A., Yap, C.K., and M.S. Kamarudin. 2009. Anthropogenic impacts on heavy metal concentrations in the coastal sediments of Dumai, Indonesia. *Environ Monit Assess* 148, nos. 1–4: 291–305. http://doi.org/10.1007/s10661-008-0159-z.

Bacon, J.R., Beffow, M.L., and C.A. Shand. 1995. The use of isotopic composition in field studies of lead in upland Scottish soils (UK). *Chem Geol* 124, nos. 1–2: 125–134. http://doi.org/10.1016/0009-2541(95)00030-P.

Badawy, W.M., El-Taher, A., Frontasyeva, M.V., Madkour, H.A., and A.E. Khater. 2018. Assessment of anthropogenic and geogenic impacts on marine sediments along the coastal areas of Egyptian Red Sea. *Appl Radiat Isot* 140: 314–326. http://doi.org/10.1016/j.apradiso.2018.07.034.

Barbosa, F., Tanus-Santos, J.E., Gerlach, R.F., and P.J. Parsons. 2005. A critical review of biomarkers used for monitoring human exposure to lead: Advantages, limitations, and future needs. *Environ Health Perspect* 113, no. 12: 1669–1674. http://doi.org/10.1289/ehp.7917.

Basaham, A.S., Rifaat, A.E., El-Mamoney, M.H., and M.A. El Sayed. 2009. Re-evaluation of the impact of sewage disposal on coastal sediments of the southern Corniche, Jeddah, Saudi Arabia. *J King Abdulaziz Univ Mar Sci* 20: 109–126. www.kau.edu.sa/Files/320/Researches/54566_24891.pdf.

Baumann, A. 1994. Lead and strontium isotopes in metalliferous and calcareous pelitic sediments of the Red Sea axial trough. *Miner Deposita* 29, no. 1: 81–93. http://doi.org/10.1007/BF03326398.

Bollhöfer, A., and K.J.R. Rosman. 2001. Isotopic source signatures for atmospheric lead: The Northern Hemisphere. *Geochim Cosmochim Acta* 65, no. 11: 1727–1740. http://doi.org/10.1016/S0016-7037(00)00630-X.

Carrasco, G., Chen, M., Boyle, E.A., Tanzil, J., Zhou, K., and N.F. Goodkin. 2018. An update of the Pb isotope inventory in post leaded-petrol Singapore environments. *Environ Pollut* 233: 925–932. http://doi.org/10.1016/j.envpol.2017.09.025.

Chen, M., Boyle, E.A., Switzer, A.D., and C. Gouramanis. 2016a. A century long sedimentary record of anthropogenic lead (Pb), Pb isotopes and other trace metals in Singapore. *Environ Pollut* 213: 446–459. http://doi.org/10.1016/j.envpol.2016.02.040.

Chen, M., Goodkin, N.F., Boyle, E.A., Switzer, A.D., and A. Bolton. 2016b. Lead in the western South China Sea: Evidence of atmospheric deposition and upwelling. *Geophys Res Lett* 43, no. 9: 4490–4499. http://doi.org/10.1002/2016GL068697.

Chen, M., Lee, J.M., Nurhati, I.S., Switzer, A.D., and E.A. Boyle. 2015. Isotopic record of lead in Singapore Straits during the last 50 years: Spatial and temporal variations. *Mar Chem* 168: 49–59. http://doi.org/10.1016/j.marchem.2014.10.007.

Cheng, H., and Y. Hu. 2010. Lead (Pb) isotopic fingerprinting and its applications in lead pollution studies in China: A review. *Environ Pollut* 158, no. 5: 1134–1146. http://doi.org/10.1016/j.envpol.2009.12.028.

Chiaradia, M., Chenhall, B.E., Depers, A.M., Gulson, B.L., and B.G. Jones. 1997. Identification of historical lead sources in roof dusts and recent lake sediments from an industrialized area: Indications from lead isotopes. *Sci Total Environ* 205, nos. 2–3: 107–128. http://doi.org/10.1016/S0048-9697(97)00199-X.

Choi, M.S., Yi, H.I., Yang, S.Y., Lee, C.B., and H.J. Cha. 2007. Identification of Pb sources in Yellow Sea sediments using stable Pb isotope ratios. *Mar Chem* 107, no. 2: 255–274. http://doi.org/10.1016/j.marchem.2007.07.008.

Dupré, B., Blanc, G., Boulègue, J., and C.J. Allègre. 1988. Metal remobilization at a spreading centre studied using lead isotopes. *Nature* 333, no. 6169: 165.

El-Taher, A., Zakaly, H.M., and R. Elsaman. 2018. Environmental implications and spatial distribution of natural radionuclides and heavy metals in sediments from four harbours in the Egyptian Red Sea coast. *Appl Radiat Isot* 131: 13–22. http://doi.org/10.1016/j.apradiso.2017.09.024.

European Food Safety Authority (EFSA)m Alexander, J., Benford, D., Boobis, A., Ceccatelli, S., Cravedi, J.P., Di Domenico, A., et al. 2010. EFSA panel on contaminants in the food chain; scientific opinion on marine biotoxins in shellfish – Emerging toxins: Ciguatoxin group. *EFSA J* 1627: 1–38.

Faure, G. 1977. *Principles of Isotope Geology*, 4th ed. New York: Wiley.

Flament, P., Bertho, M.L., Deboudt, K., Véron, A., and E. Puskaric. 2002. European isotopic signatures for lead in atmospheric aerosols: A source apportionment based upon 206Pb/207Pb ratios. *Sci Total Environ* 296, nos. 1–3: 35–57. http://doi.org/10.1016/S0048-9697(02)00021-9.

Gallon, C., Ranville, M.A., Conaway, C.H., Landing, W.M., Buck, C.S., Morton, P.L., and A.R. Flegal. 2011. Asian industrial lead inputs to the North Pacific evidenced by lead concentrations and isotopic compositions in surface waters and aerosols. *Environ Sci Technol* 45, no. 23: 9874–9882. http://doi.org/10.1021/es2020428.

Gan, T., Zhao, N., Yin, G., Chen, M., Wang, X., Liu, J., and W. Liu. 2019. Optimal chlorophyll fluorescence parameter selection for rapid and sensitive detection of lead toxicity to marine microalgae Nitzschia closterium based on chlorophyll fluorescence technology. *J Photochem Photobiol B Biol* 111551. http://doi.org/10.1016/j.jphotobiol.2019.111551.

Gobeil, C., Johnson, W.K., Macdonald, R.W., and C.S. Wong. 1995. Sources and burden of lead in St. Lawrence estuary sediments: Isotopic evidence. *Environ Sci Technol* 29, no. 1: 193–201. http://doi.org/10.1021/es00001a025.

Grant, A., and R. Middleton. 1990. An assessment of metal contamination of sediments in the Humber Estuary, UK. *Estuarine Coastal Shelf Sci* 31, no. 1: 71–85. http://doi.org/10.1016/0272-7714(90)90029-Q.

Halawani, R.F., Wilson, M.E., Hamilton, K.M., Aloufi, F.A., Taleb, M.A., Al-Zubieri, A.G., and A.N. Quicksall. 2022. Spatial distribution of heavy metals in near-shore marine sediments of the Jeddah, Saudi Arabia region: Enrichment and associated risk indices. *J Mar Sci Eng* 10, no. 5: 614.

Hwang, I., Kim, K., Kim, J., and J. Kang. 2016. Toxic effects and depuration after the dietary lead(II) exposure on the bioaccumulation and hematological parameters in starry flounder (Platichthys stellatus). *Environ Toxicol Pharmacol* 45: 328–333. http://doi.org/10.1016/j.etap.2016.06.017.

Idris, A.M., Eltayeb, M.A.H., Potgieter-Vermaak, S.S., Van Grieken, R., and J.H. Potgieter. 2007. Assessment of heavy metals pollution in Sudanese harbours along the Red Sea Coast. *Microchem J* 87, no. 2: 104–112. http://doi.org/10.1016/j.microc.2007.06.004.

Javed, M. 2012. Effects of zinc and lead toxicity on the growth and their bioaccumulation in fish. *Pak Vet J* 32: 357–362. www.pvj.com.pk/. . ./357-362.pdf.

Kahal, A.Y., El-Sorogy, A.S., Alfaifi, H.J., Almadani, S., and H.A. Ghrefat. 2018. Spatial distribution and ecological risk assessment of the coastal surface sediments from the Red Sea, northwest Saudi Arabia. *Mar Pollut Bull* 137: 198–208. http://doi.org/10.1016/j.marpolbul.2018.09.053.

Karuppasamy, M.P., Qurban, M.A., Krishnakumar, P.K., Mushir, S.A., and N. Abuzaid. 2017. Evaluation of toxic elements As, Cd, Cr, Cu, Ni, Pb and Zn in the surficial sediments of the Red Sea (Saudi Arabia). *Mar Pollut Bull* 119, no. 2: 181–190. http://doi.org/10.1016/j.marpolbul.2017.04.019.

Kersten, M., Garbe-Schönberg, C.D., Thomsen, S., Anagnostou, C., and A. Sioulas. 1997. Source apportionment of Pb pollution in the coastal waters of Elefsis Bay, Greece. *Environ Sci Technol* 31, no. 5: 1295–1301. http://doi.org/10.1021/es960473z.

Kim, J.H., and J.C. Kang. 2015. The lead accumulation and hematological findings in juvenile rock fish Sebastes schlegelii exposed to the dietary lead (II) concentrations. *Ecotoxicol Environ Safety* 115: 33–39. http://doi.org/10.1016/j.ecoenv.2015.02.009.

Komárek, M., Ettler, V., Chrastný, V., and M. Mihaljevič. 2008. Lead isotopes in environmental sciences: A review. *Environ Internat* 34, no. 4: 562–577. http://doi.org/10.1016/j.envint.2007.10.005.

Kosnett, M.J. 2007. *Heavy Metal Intoxication and Chelators. Basic and Clinical Pharmacology*, 10th ed. New York: McGraw-Hill, 945–957.

Laurila, T.E., Hannington, M.D., Petersen, S., and D. Garbe-Schönberg. 2014. Trace metal distribution in the Atlantis II Deep (Red Sea) sediments. *Chem Geol* 386: 80–100. http://doi.org/10.1016/j.chemgeo.2014.08.009.

Lee, J.W., Choi, H., Hwang, U.K., Kang, J.C., Kang, Y.J., Kim, K.I., and J.H. Kim. 2019. Toxic effects of lead exposure on bioaccumulation, oxidative stress, neurotoxicity, and immune responses in fish: A review. *Environ Toxicol Pharmacol* 68: 101–108. http://doi.org/10.1016/j.etap.2019.03.010.

Liu, J., Yin, M., Luo, X., Xiao, T., Wu, Z., Li, N., et al. 2019. The mobility of thallium in sediments and source apportionment by lead isotopes. *Chemosphere* 219: 864–874. http://doi.org/10.1016/j.chemosphere.2018.12.041.

Macirella, R., Sesti, S., Bernabò, I., Tripepi, M., Godbert, N., and E. Brunelli. 2019. Lead toxicity in seawater teleosts: A morphofunctional and ultrastructural study on the gills of the Ornate wrasse (Thalassoma pavo L.). *Aquat Toxicol* 211: 193–201. http://doi.org/10.1016/j.aquatox.2019.04.009.

Mansour, A.M., Nawar, A.H., and H.A. Madkour. 2011. Metal pollution in marine sediments of selected harbours and industrial areas along the Red Sea coast of Egypt. *Ann Naturhist Mus Wien Ser A* 113: 225–244. www.jstor.org/stable/41701739.

Monastra, V., Derry, L.A., and O.A. Chadwick. 2004. Multiple sources of lead in soils from a Hawaiian chronosequence. *Chem Geol* 209, nos. 3–4: 215–231. http://doi.org/10.1016/j.chemgeo.2004.04.027.

Monna, F., Ben Othman, D., and J.M. Luck. 1995. Pb isotopes and Pb, Zn and Cd concentrations in the rivers feeding a coastal pond (Thau, southern France): Constraints on the origin(s) and flux(es) of metals. *Sci Total Environ* 166, no. 1: 19–34. http://doi.org/10.1016/0048-9697(95)04514-2.

Monna, F., Clauer, N., Toulkeridis, T., and J.R. Lancelot. 2000. Influence of anthropogenic activity on the lead isotope signature of Thau Lake sediments (southern France): Origin and temporal evolution. *Appl Geochem* 15, no. 9: 1291–1305. http://doi.org/10.1016/S0883-2927(99)00117-1.

Monna, F., Lancelot, J., Croudace, I.W., Cundy, A.B., and J.T. Lewis. 1997. Pb isotopic composition of airborne particulate material from France and the southern United Kingdom: Implications for Pb pollution sources in urban areas. *Environ Sci Technol* 31, no. 8: 2277–2286. http://doi.org/10.1021/es960870+.

Needleman, H.L., Schell, A., Bellinger, D., Leviton, A., and E.N. Allred. 1990. The long-term effects of exposure to low doses of lead in childhood: An 11-year follow-up report. *New Engl J Med* 322, no. 2: 83–88. http://doi.org/10.1056/NEJM199001113220203.

O'Connor, D., Hou, D., Ye, J., Zhang, Y., Ok, Y.S., Song, Y., et al. 2018. Lead-based paint remains a major public health concern: A critical review of global production, trade, use, exposure, health risk, and implications. *Environ Int* 121: 85–101. http://doi.org/10.1016/j.envint.2018.08.052.

Pan, K., Lee, O.O., Qian, P.Y., and W.X. Wang. 2011. Sponges and sediments as monitoring tools of metal contamination in the eastern coast of the Red Sea, Saudi Arabia. *Mar Pollut Bull* 62, no. 5: 1140–1146. http://doi.org/10.1016/j.marpolbul.2011.02.043.

Pierret, M.C., Clauer, N., Bosch, D., and G. Blanc. 2010. Formation of Thetis Deep metal-rich sediments in the absence of brines, Red Sea. *J Geochem Explor* 104, nos. 1–2: 12–26. http://doi.org/10.1016/j.gexplo.2009.10.001.

Ramadan, M.H., and R.F. Halawani. 2010. Study on wastewater in paper recycling plants case study. *J King Abdulaziz Univ Meteorol Environ Arid Land Agriculture Sci*, 21, no. 2.

Ruiz-Compean, P., Ellis, J., Cúrdia, J., Payumo, R., Langner, U., Jones, B., and S. Carvalho. 2017. Baseline evaluation of sediment contamination in the shallow coastal areas of Saudi Arabian Red Sea. *Mar Pollut Bull* 123, nos. 1–2: 205–218. http://doi.org/10.1016/j.marpolbul.2017.08.059.

Salem, D.M.A., Khaled, A., El Nemr, A., and A. El-Sikaily. 2014. Comprehensive risk assessment of heavy metals in surface sediments along the Egyptian Red Sea coast. *Egypt J Aquatic Res* 40, no. 4: 349–362. http://doi.org/10.1016/j.ejar.2014.11.004.

Sures, B., and R. Siddall. 1999. Pomphorhynchus laevis: The intestinal Acanthocephalan as a lead sink for its fish host, Chub (Leuciscus cephalus). *Exp Parasitol* 93, no. 2: 66–72. http://doi.org/10.1006/expr.1999.4437.

Taylor, A.M., and W.A. Maher. 2014. Exposure – dose – response of Tellina deltoidalis to metal contaminated estuarine sediments 2. Lead spiked sediments. *Comp Biochem Physiol C Toxicol Pharmacol* 159: 52–61. http://doi.org/10.1016/j.cbpc.2013.09.006.

Townsend, A.T., and I. Snape. 2002. The use of Pb isotope ratios determined by magnetic sector ICP-MS for tracing Pb pollution in marine sediments near Casey Station, East Antarctica. *J Anal At Spectrom* 17, no. 8: 922–928. http://doi.org/10.1039/B203449M.

Turekian, K.K., and K.H. Wedepohl. 1961. Distribution of the elements in some major units of the earth's crust. *Geol Soc Am Bull* 72, no. 2: 175–192. http://doi.org/10.1130/0016-7606(1961)72[175:DOTEIS]2.0.CO;2.

US Environmental Protection Agency (EPA). 1996. Method 3050B: Acid digestion of sediments, sludges, and soils. In: *Test Methods for Evaluating Solid Waste, Physical/Chemical Methods*. Washington, DC: US Environmental Protection Agency.

Vinodhini, R., and M. Narayanan. 2008. Bioaccumulation of heavy metals in organs of fresh water fish Cyprinus carpio (Common carp). *Int J Environ Sci Technol* 5, no. 2: 179–182. http://doi.org/10.1007/BF03326011.

Wang, D., Zhao, Z., and M. Dai. 2014. Tracing the recently increasing anthropogenic Pb inputs into the East China Sea shelf sediments using Pb isotopic analysis. *Mar Pollut Bull* 79, nos. 1–2: 333–337. http://doi.org/10.1016/j.marpolbul.2013.11.032.

Wani, A.L., Ara, A., and J.A. Usmani. 2015. Lead toxicity: A review. *Interdiscip Toxicol* 8, no. 2: 55–64. http://doi.org/10.1515/intox-2015-0009.

World Health Organization (WHO). 2011. Guidelines for drinking-water quality. *WHO Chron* 38, no. 4: 104–108.

Xu, D., Wang, R., Wang, W., Ge, Q., Zhang, W., Chen, L., and F. Chu. 2019. Tracing the source of Pb using stable Pb isotope ratios in sediments of eastern Beibu Gulf, South China Sea. *Mar Pollut Bull* 141: 127–136. http://doi.org/10.1016/j.marpolbul.2019.02.028.

Yap, C.K., Cheng, W.H., Karami, A., and A. Ismail. 2016. Health risk assessments of heavy metal exposure via consumption of marine mussels collected from anthropogenic sites. *S Total Environ* 553: 285–296. http://doi.org/10.1016/j.scitotenv.2016.02.092.

Yu, R., Hu, G., Yang, Q., He, H., and C. Lin. 2016. Identification of Pb sources using Pb isotopic compositions in the core sediments from Western Xiamen Bay, China. *Mar Pollut Bull* 113, nos. 1–2: 247–252. http://doi.org/10.1016/j.marpolbul.2016.09.027.

Zhu, L., Guo, L., Gao, Z., Yin, G., Lee, B., Wang, F., and J. Xu. 2010. Source and distribution of lead in the surface sediments from the South China Sea as derived from Pb isotopes. *Mar Pollut Bull* 60, no. 11: 2144–2153. http://doi.org/10.1016/j.marpolbul.2010.07.026.

Temporally resolved lead sourcing via isotopic analysis along the Jeddah, Saudi Arabian coast

9

Hope Rasmussen, Bandar Almur and Andrew N. Quicksall

9.1 INTRODUCTION

Lead (Pb) pollution is a persistent, critical issue worldwide. The exponential population growth and move toward urbanization over the last century have greatly influenced the spread of Pb in the environment from various sources, such as the industrial, energy and transportation sectors, along with increased atmospheric dust loads from the disturbance of natural systems. The persistence and bioaccumulation of Pb in the environment, combined with its toxic effects, make it a major hazard to human and ecological health. Clean air, water and soil are vital resources, and even low levels of Pb entering these systems can be detrimental to public health (Wani et al. 2015). Information on Pb inputs into the environment can influence public health considerations, such as decision making on environmental regulations, best practices for industries, remediation techniques and pollution control methods.

This study focuses on a region of high interest, the Red Sea off the coast of Jeddah, Saudi Arabia. The area lacks environmental Pb sourcing data despite the industrialization and urbanization of the Jeddah coast, which has increased exponentially over recent decades (Atalla et al. 2018). This study employs the analysis of marine sediment cores as recent historical and current records of Pb concentrations and isotopic ratios to characterize sources of Pb pollution on the eastern coast of Saudi Arabia near Jeddah.

9.1.1 Pb isotopes for environmental sourcing

Analyses of sediment cores from aquatic settings serve as a useful tool for studying the fate and transport of Pb in the environment over time (e.g., Orani et al. 2019; Palchan et al. 2013; Xu and Gao 2020). In addition to studying overall lead concentrations, measuring Pb isotopic signatures can give insight into sources of Pb input into sediments (Erel et al. 1997; Komárek et al. 2008). Anthropogenic pollutants, such as gasoline, industrial-used ores and fly ash, as well as natural geological sources of lead, have unique regional isotopic Pb ratios. Even after processing and use, sources such as gasoline maintain their initial signatures (Monna et al. 2006; Komárek et al. 2008). Weathered rocks have evolved Pb signatures based on the initial uranium and thorium values in the parent rock, which can be used to identify distinct natural sources (Morton-Bermea et al. 2011; Komárek et al. 2008). The non-radiogenically derived ^{204}Pb maintains its original abundance while the other radiogenic isotopes increase over time, making it very useful as a normalizing value (Monna et al. 1997; Sutherland et al. 2003).

Organic ligands such as ethylenediaminetetraacetic acid (EDTA) are often used to sequester extractable concentrations of Pb from total concentrations for comparison of data (e.g., Xu et al. 2013; Chrastný et al. 2008). The extractable fractions of sediments have been shown to be closely related to anthropogenically influenced lead while more recalcitrant portions are typically more geological in nature (Komárek et al. 2008; Teutsch et al. 2001).

9.1.2 Background on the Red Sea region

The Red Sea is a semi-enclosed basin, with research showing that there is little mixing and exchange of sediments between it and its surrounding water bodies, the Gulf of Aqaba, the Gulf of Aden and the Gulf of Suez (Chow 1968). Mid-ocean ridge basalts (MORB) in the Red Sea are constantly being erupted

DOI: 10.1201/9781003321408-12

and creating new crustal material, and numerous studies have characterized the Pb isotopic signature of the raw basalts (Volker et al. 1993; Dupré et al. 1988). However, many studies have hypothesized that the vast majority of sediment source materials in the Red Sea come from aerosol transport of dust or terrigenous runoff of the surrounding drainage area from flooding, rather than weathering and spread of MORB-derived particles (Palchan et al. 2013, 2018; Dupré et al. 1988; Chow 1968; Langodan et al. 2018). Prakash et al. (2016) estimate up to 6 Mt of mineral dust are deposited into the Red Sea off the coast of Saudi Arabia each year from major dust storms.

Beyond natural geological influences, it is likely that there are significant anthropogenic Pb contributions to Red Sea sediments. Leaded and unleaded gasoline, coal fly ash, industrial emissions and purposeful wastewater discharge often contribute substantial Pb to the surrounding environment. Gasoline signatures are well characterized in the literature for European, North American and Southeast Asian countries (e.g., Monna et al. 1997; Doe and Delevaux 1972; Yao et al. 2015) but not for the Saudi Arabian region; however, Pb inclusions in gasoline often utilize ore from other regions (e.g., Kersten et al. 1997; Graney et al. 1995). Lead isotopic ratios from highly utilized raw ores have been studied extensively and are distinct from one another in source and region (Doe and Delevaux 1972; Ostic et al. 1967).

Leaded gasoline was once believed to be the major source of lead exposure in Saudi Arabia; however, it was mostly phased out by 2001, and many other current anthropogenic sources remain (Shaik et al. 2014). Regional industries that continue to emit lead into the environment as point sources are businesses that manufacture batteries, cement and paint, as well as battery recycling companies (Shaik et al. 2014).

There is no threshold level of lead that is safe for human exposure (Wani et al. 2015). While there is some limited data on Pb sourcing from blood analyses, environmental studies characterizing potential sources in this region have been severely lacking (Ahmad et al. 1988; Shaik et al. 2014). This study presents evidence of Pb inputs over time to the Jeddah coastline from both anthropogenic, gasoline-based sources and from the wind-blown transport of uniquely signatured dust via analysis of Pb isotopic signatures in sediment cores.

9.2 MATERIALS AND METHODS

9.2.1 Sampling sites and procedures

Three 50 cm deep, 4.5 cm diameter Red Sea sediment cores were collected by SCUBA divers in January 2015 off the shore of Jeddah (Al-Mur et al. 2017). Each core was split into 2 cm–depth aliquots, resulting in 75 total subsamples. These subsamples were each homogenized and kept refrigerated at 4°C. Table 9.1 presents details on the core locations, and the sites are shown relative to one another in Figure 9.1.

9.2.2 Reagents and sample materials

All solutions were prepared using doubly deionized (DDI) water at 18.2 mΩ resistivity. Trace metal–grade acids (Fisher Scientific, Pittsburgh, Pennsylvania) were used for heavy acid digestions, as well as dilutions, calibration standard preparation and blanks. 0.05 M Na-EDTA solution was prepared by dissolving the appropriate amount of ACROS Organics disodium EDTA dihydrate salt (99+% analysis-grade purity) in DDI water. Glassware and plasticware were cleaned between uses by soaking in 10% Liquinox overnight, rinsing, soaking 24 hours in a 10% HNO_3 bath and rinsing five times with DDI water. Calibration solutions for inductively coupled plasma mass spectrometry (ICP-MS) analyses were serial diluted from Claritas PPT grade Multi-element Solution 2A purchased from Fisher Scientific.

9.2.3 Digestions and extractions

Total Pb in this study was obtained by reverse *aqua regia* (3:1 HNO_3:HCl) digestion of solids. Aliquots of approximately 0.1 g

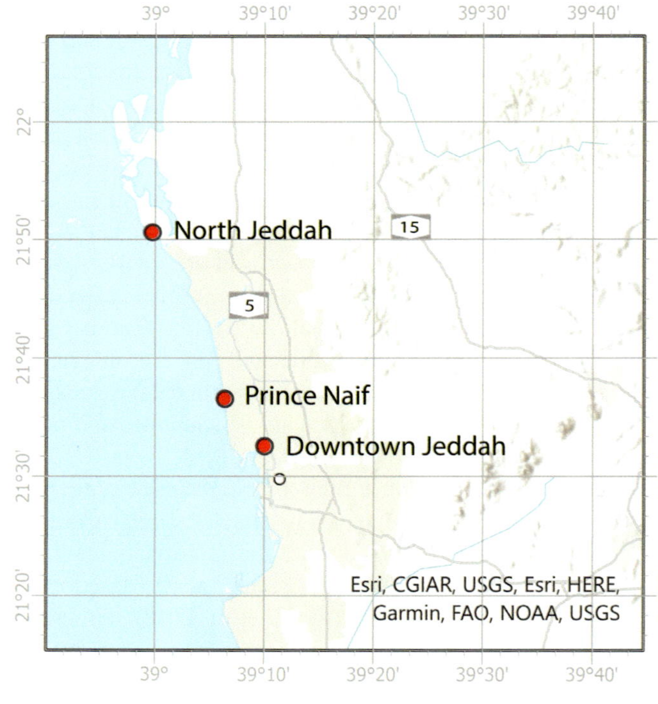

FIGURE 9.1 Core locations adjacent to Jeddah, Saudi Arabia.
Source: Al-Mur et al. (2017)

TABLE 9.1 Locations and collection data of sediment cores.

SITE	*LATITUDE*	*LONGITUDE*	*DISTANCE FROM SHORE (M)*	*WATER DEPTH (M)*
North Jeddah at Salman Gulf	21°51′2″	38°58′38″	1,000	2
Prince Naif	21°35′33″	39°6′21″	1,000	25
Downtown Jeddah	21°31′45″	39°9′26″	10	2

of sediment sample for each 2 cm core section were placed in 30 ml ultra-pure PFA vials (Savillex Corporation) and massed on a microbalance to a precision of 10^{-4} g. 8 ml of acid mixture was added to each sample and heated to dryness at 85°C in aluminum digestion blocks on a hot plate. The process was repeated twice for each sample set. Residues were redissolved into 10 ml of 5% HNO_3 and stored in a refrigerator at 4°C until ICP-MS analysis (described later in this chapter).

In this study, 0.05 M Na-EDTA was used to extract readily bioavailable fractions of Pb from aliquots of each sub-sample. A 10:1 w/v ratio of 0.05 M Na-EDTA was used by adding 5 ml of 0.05 M Na-EDTA solution to 0.5 g of solid. Mixtures were shaken on a shaker table for one hour in 50 ml centrifuge tubes and centrifuged for ten minutes at 3,800 rpm; the supernatant syringe was filtered through 0.45 μm PTFE filters to remove any excess particles and organics from the supernatant that might affect ICP-MS analysis. Supernatants were immediately diluted at least tenfold using 5% HNO_3 and refrigerated at 4°C prior to analysis.

9.2.4 Pb isotope analyses by ICP-MS

All instrumental analyses were performed at the Aqueous and Soil Chemistry Research Laboratory at Southern Methodist University in Dallas, Texas. Samples were analyzed for ^{204}Pb, ^{206}Pb, ^{207}Pb and ^{208}Pb with a Thermo X–Series 2 ICP-MS using trace metal–certified argon gas in collision cell mode with kinetic energy dispersion (CCT-KED). The instrument ran samples in triplicate, and the average counts of these triplicates were used to perform calculations. A 5% HNO_3 blank was also analyzed every ten samples for quality control. For matrix-matching purposes, 8% 0.05 M Na-EDTA was included in the standards and blanks when analyzing extraction supernatants. Ratios were determined by dividing blank-subtracted raw counts for each isotopic measurement. The uncertainty for each isotopic ratio was calculated using the standard deviation between instrumental counts of the triplicate run divided by the square root of the number of times each sample was run and propagated using Poisson uncertainty calculations. To assess heterogeneity among sample aliquots, three samples were split for each analysis process into duplicates or triplicates, which were subjected to the same digestion, extraction and dilution procedures. Calibrations used for the calculation of total Pb were accepted where the linear regression fit of standard counts versus concentrations had an $r^2 > 0.999$.

The instrument detection level (IDL) was determined as the average of the blank result plus three times the standard deviation of blank result (in μg/l) for n = 10 blanks (US EPA 2014). The limit of quantification (LOQ) was determined by multiplying the result of 10 times the average standard deviation of counts over ten runs of the 0.5 μg/l standard. The IDLs for ^{204}Pb, ^{206}Pb, ^{207}Pb and ^{208}Pb were 0.029, 0.020, 0.011 and 0.010 μg/l and LOQs were 0.570, 0.107, 0.149 and 0.084 μg/l, respectively.

9.2.5 Statistical analysis and modeling

Sets of anthropogenic and geological lead isotopic data were compiled from literature. These sets were condensed into smaller groupings using K-means clustering, performed in JMP Pro 15. The clusters were plotted as groups in bivariate plots where multiple data points existed, using ranges of maximum, minimum and average values from the literature data to determine region boundaries.

A simple two end-member mixing model was derived from prospective geological and anthropogenic sources (Figure 9.2a) to determine sediment fractionation by depth (Monna et al. 2000; Komárek et al. 2008; Brugam et al. 2012). The centroids of the two end members were derived from the weighted averages of the compilations of literature data for $^{206/204}Pb$ values. The geological component, $(^{206/204}Pb)_{natural}$, captures groups 1 and 2 of Figure 9.2a while the anthropogenic end member, $(^{206/204}Pb)_{anth}$, simulates group 3 of Figure 9.2a.

Pollution-derived concentrations of lead were determined as shown in equations 1–3.

$$Pb_{anth\text{-}fraction} = [(^{206/204}Pb)_{sample} - (^{206/204}Pb)_{natural}] / [(^{206/204}Pb)_{anth} - (^{206/204}Pb)_{natural}] \quad (9.1)$$

$$Pb_{anth\text{-}conc} = Pb_{anth\text{-}fraction} \text{ x } Pb_{sample\text{-}total} \quad (9.2)$$

$$Pb_{natural\text{-}conc} = Pb_{sample\text{-}total} - Pb_{anth\text{-}conc} \quad (9.3)$$

where $Pb_{anth\text{-}fraction}$ is the fraction of Pb concentration from the sample influenced by the anthropogenic end member; $^{206/204}Pb$ refers to the ratio of the individual sample, the natural end member or the anthropogenic end member, as denoted by subscripts sample, natural and anth, respectively; $Pb_{anth\text{-}conc}$ and $Pb_{natural\text{-}conc}$ are the concentrations of anthropogenically and naturally derived Pb, respectively, in mg/kg, from each end member; and $Pb_{sample\text{-}total}$ is the Pb concentration in each sample, in mg/kg.

9.3 RESULTS AND DISCUSSION

The Pb isotopic signatures of the three sites indicated all three are influenced by both anthropogenic and natural Pb inputs (Figure 9.2a, b, c). The Downtown site has the highest overall anthropogenic contributions from ore and gasoline along with highest overall Pb concentrations while the less industrial or urbanized area of North Jeddah shows the lowest influence from pollutant sources and serves as an example of a relatively unpolluted control site for this region.

9.3.1 Pb isotopic sourcing from data

Pb isotopic data from total digestions and EDTA extractions were superimposed over isotopic regions of potential anthropogenic and natural sources of influence compiled from an extensive review of the literature. The overview of the groups of sourced data along with their published sources are presented in Figure 9.2a. The data from this study superimposed over literature source data is presented in Figure 9.2b, referencing the groups from Figure 9.2a as a key to the plot. The Pb isotopic ratios shown include ore and gasoline signatures from sources and products worldwide, as well as natural signatures of geographical areas neighbouring the Red Sea.

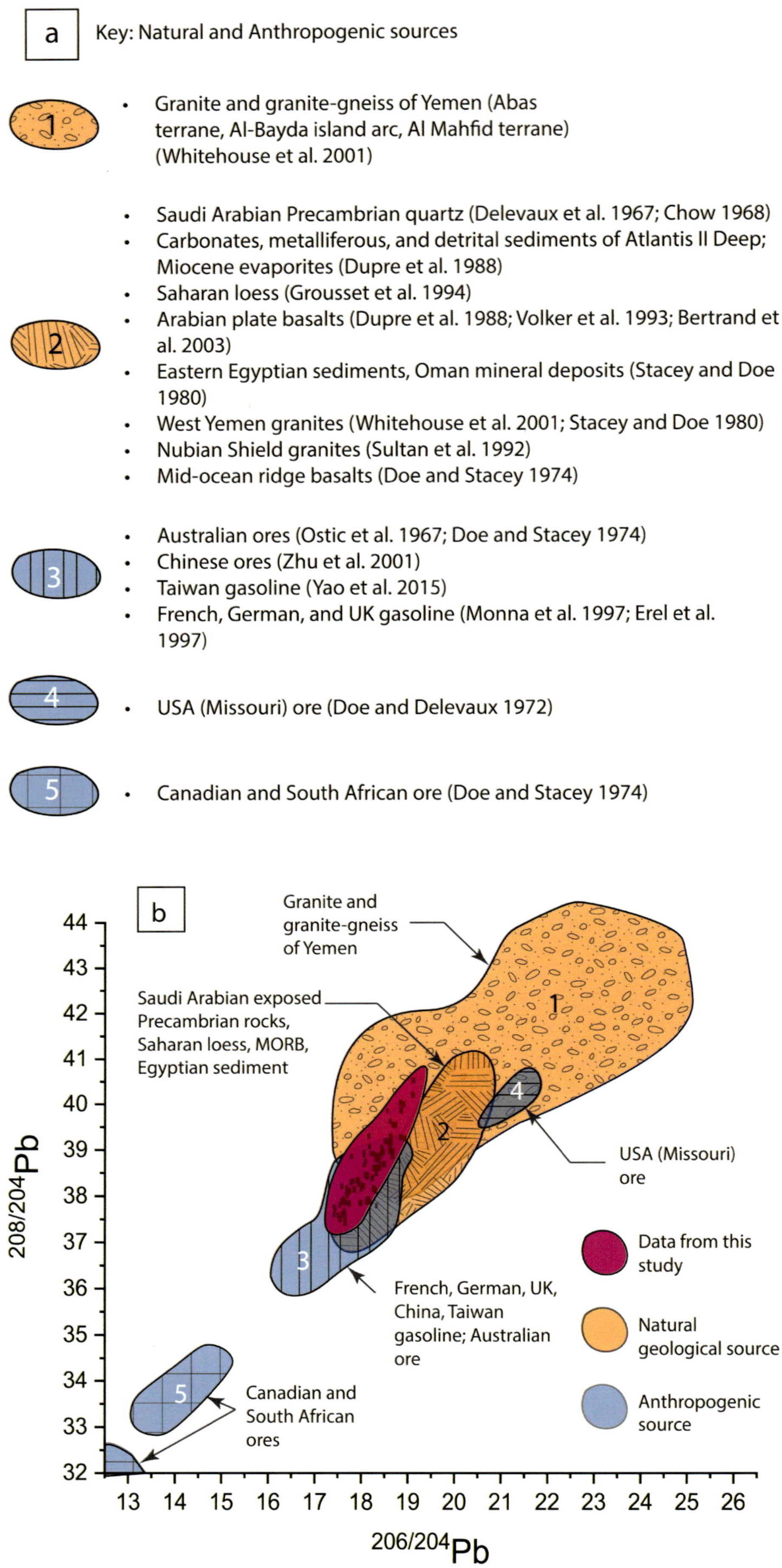

FIGURE 9.2 a) Relevant geological and anthropogenic sources from literature as they are clustered for this study; b) data from this study plotted alongside Pb isotopic ratios for relevant geological and anthropogenic sources grouped and compiled from literature; c) extractable and total digestion data for each core plotted over a visual representation of Pb isotopic signatures for major sources of Pb in the environment.

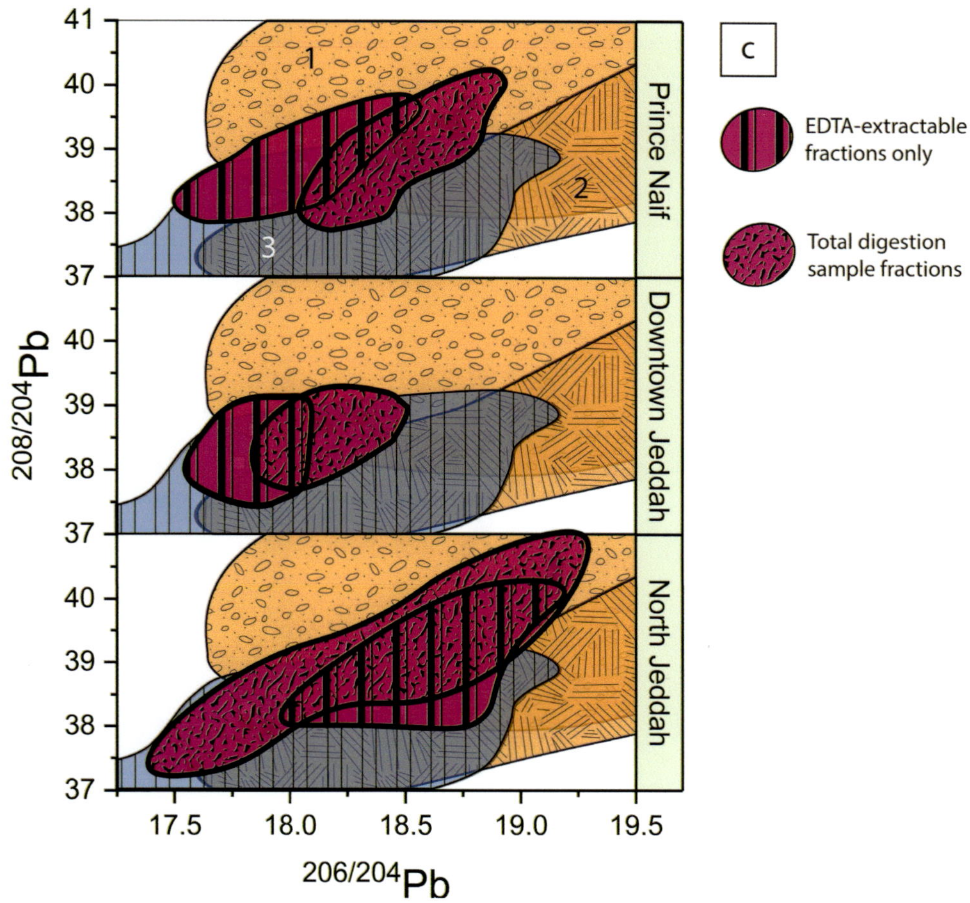

FIGURE 9.2 (Continued)

After comparing the collective sources with the data from this study, sources were abridged further to identify a narrower isotopic range of sources for each site, shown in Figure 9.2c. EDTA-extractable data was plotted separately from total digestion data as the extractable regions of Pb are typically associated strongly with pollution sources.

It is evident from the linearity of the data in Figure 9.2 that at all three sites, there are two dominant end members influencing the isotopic signatures along a mixing line. At the lower end of each ratio are anthropogenic gasoline and ore signatures while at the upper end, geological signatures appear to influence the ratios.

To understand how natural and anthropogenic particles may come together to influence signatures for these sites, wind patterns and geography were examined. The fate and transport of dust particles is based on their shape and diameter combined with the prevailing climate systems, and anthropogenic gasoline- and ore-based particles from combustion are often very fine. Figure 9.3 presents an overview of the major wind systems of the Red Sea. There is a seasonal trough of wind blowing directly from the Gulf of Aden across Yemen and down the longitudinal axis of the Red Sea. Air masses blow toward the Gulf of Aqaba in the Red Sea region from continental Europe across the Mediterranean Sea (Palchan et al. 2013; Langodan et al. 2018). During the winter and spring, air masses arrive from the Saharan Desert, Arabian Peninsula and other local sources, all of which may contain high loads of windblown fine particulate matter, including trace metals such as lead (Chien et al. 2019). Rainstorms are infrequent, but when they occur, the runoff from the high clay soils often leads to flash floods, which can deposit fine fractions of silt and clay particles onto the coastal plain (Prakash et al. 2016). These particles can then become aerosolized and form fine particulate matter as the land remains dry for the rest of the year. Yemen, a source of highly radiogenic signatures (Whitehouse et al. 2001), experiences frequent extreme dust storms that can transport dust long distances into Saudi Arabia, due to topography and typical drought conditions (Alghamdi et al. 2015).

The coastal area of Jeddah has been described as covered with thick alluvial deposits from the fringing mountains, which are composed of granites and granitic gneiss (Basaham et al. 2006); however, Pb isotopic data from Saudi Arabian mountainous granites is sparse. Floods are known to carry water through the ancient wadis of the mountains out to the Red Sea coastline (Youssef et al. 2016; Basaham et al. 2015; Sultan et al. 1992). The Arabian Peninsula is one of the major sources of dust in the Middle East and can contribute to the transport of dust particles around the world (Palchan et al. 2013; Rezazadeh et al. 2013).

After examining the potential for spread of particles by natural processes, a mixing model was used to qualitatively assign contributions from natural and/or anthropogenic particles as inputs to Pb in the sediments. The proportions of Pb concentration for total digestions and EDTA extractions as aligned primarily with pollution or natural signatures from the mixing model results are shown in Figure 9.4, broken down by site, depth and overall concentration at each point.

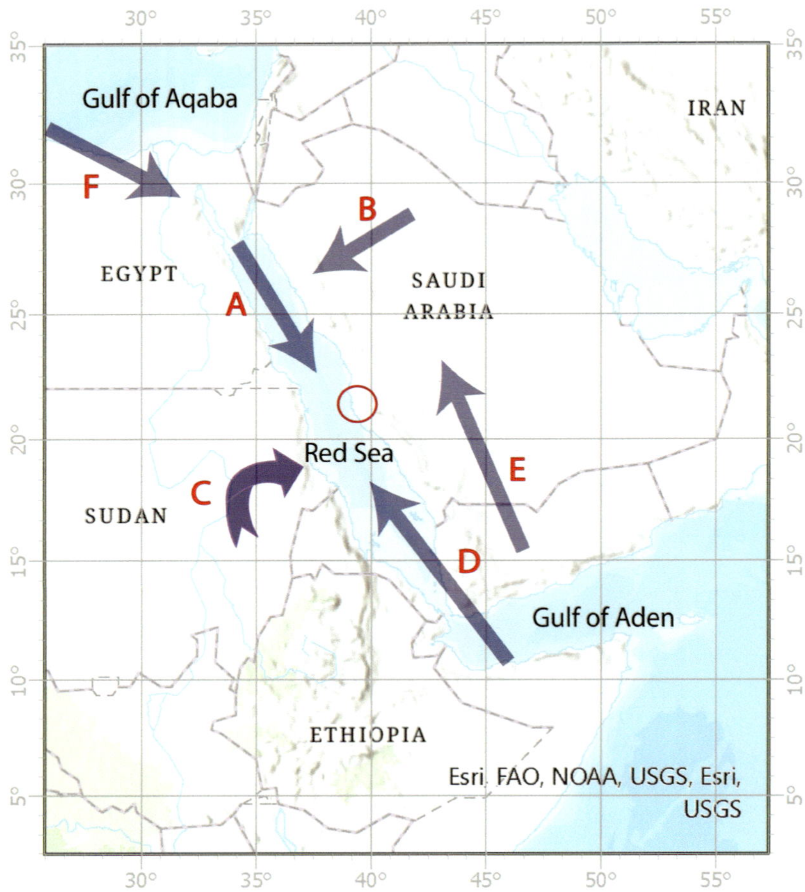

FIGURE 9.3 Diagram of dominant wind systems of the Red Sea. The study area is circled in red. Winds are identified by letters: A: Year-round wind system that stretches from the north to the south of the Red Sea; B: Year-round winds from valleys across the northern Arabian Peninsula; C: High winds generated by the Tokar Gap wind from the African coast during summer monsoon season; D: Winds associated with the Indian monsoon season (winter); E: Red Sea trough winds; F: Winds from Mediterranean cyclones.

(*Source:* A, B, C, D: Langodan et al. 2018; D, F: Palchan et al. 2013; E, F: Palchan et al. 2018)

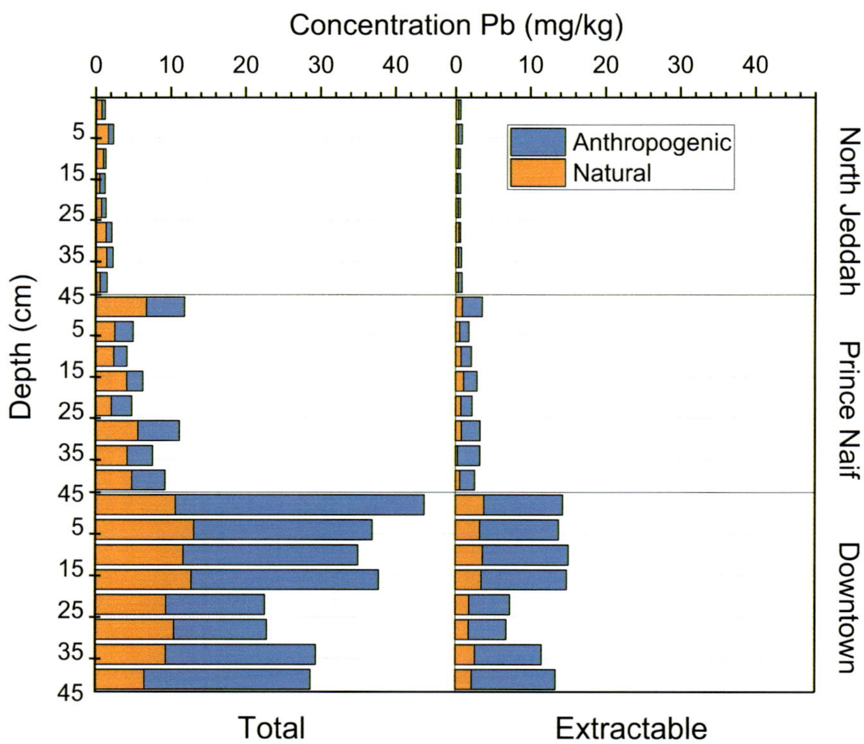

FIGURE 9.4 Mixing model results for the three sites, indicating the concentrations associated with the majority of anthropogenic or natural signatures.

Downtown Jeddah shows the highest influence from anthropogenic pollution in both concentration and composition. Despite dust storms depositing natural dust evenly across the region, the data in this study shows that Pb concentrations are not being dispersed and captured in the coastal sediments homogenously. Youssef et al. (2016) aggregated rainfall data collected between 1970 and 2011 from the Jeddah region and used it in climatic change models to determine if patterns of intense rainfall have been statistically increasing. They found a predicted 27.8% temporal increase in precipitation trends. Extreme rainfall events (> 50 mm over a few hours) are rare, but five events occurred in the 1970s, two events occurred in the 1990s, and two catastrophic flash-flooding events occurred in 2009 and 2011. Figure 9.4 indicates that the Downtown site has an overall positive trend in Pb concentrations from the lower to the upper depths, with increasing contributions of Pb from anthropogenic sources. The EDTA-extractable Pb data also shows a slight overall increasing trend at the Downtown site from the lower to the upper depths while maintaining steady inputs over the 0–20 cm depths, indicating consistent inputs of the regional source of this extractable anthropogenic Pb over recent times. Fine, Pb-containing combustion-based particles, as well as natural particles spread by dust storms, may be captured in greater quantities in the Downtown sediments because of an increased impervious cover from development and expansion, which makes those urban areas more susceptible to extreme flooding (Youssef et al. 2016; Basaham et al. 2006). The increases in precipitation over the past 50 years could explain why there is an increase in overall Pb concentrations, particularly Downtown.

Major exposure routes of this Pb can be from inhalation of the airborne component, ingestion of fish or other marine-derived food contaminated with lead, ingestion of contaminated soil, desalination of water and subsequent rerelease of concentrated brine or skin exposure and ingestion through swimming. The airborne component likely has much higher peak concentrations of Pb that are not retained in the sediment. Of particular concern to public health is the quantity of EDTA-extractable anthropogenic Pb found in the Downtown region (Figure 9.2c). This is more likely to dissolve in the bloodstream when inhaled and would be associated with greater nervous system effects than insoluble particles that would pass through the body more easily (Wani et al. 2015).

9.4 CONCLUSION

This chapter demonstrates that the Pb captured in the coastal Red Sea marine environment near Jeddah shows major contributions from anthropogenic sources such as ore and gasoline, as well as natural sediments transported from weathered rock deposited by dust storms and runoff. The Downtown site shows the highest influence in the sediment record of concentrated Pb, with the majority of Pb coming from pollutant sources which are increasing over time from the lower to the upper depths of the sediment. All sites show influence from the spread of natural, highly radiogenic particles dispersed across the Arabian Peninsula. Changes in development patterns and subsequent increases in impervious cover, exacerbated by changes in precipitation patterns over decades, have led to the steadily increasing deposition of anthropogenic and natural Pb in the Red Sea. North Jeddah represents an appropriate example of a control site, as it does not receive as much Pb input into the coastal sediment, likely due to less industrial development and impervious cover. The contribution of dust-blown particles to the overall Pb load in the region is an important discovery because, while air quality has improved worldwide through the shift to unleaded gasoline with enhanced air emission standards, airborne dust containing Pb from weathering goes largely ignored. Particulate matter overall is a significant contributor to loss of quality of life and premature death around the world, and both soluble and insoluble Pb in these particles add an additional risk (e.g., Kim et al. 2015). There are limited studies on particulate matter composition in the Jeddah region (Alghamdi et al. 2015). Geological data of Pb isotopic signatures in this region is particularly lacking. This data could provide more specific sources of the region. Further work should be done to identify the most appropriate natural and anthropogenic sources of Pb to Jeddah that are missing from scientific literature and to address signatures of airborne particulate matter, particularly during large dust storms.

REFERENCES

Ahmad, P., Kutbi, I.I., Abulfaraj, W.H., and M. Ahmed. 1988. Measurements of hair lead concentrations in children of four cities in Saudi Arabia. *Environ Internat* 14: 237–242. http://doi.org/10.1016/0160-4120(88)90144-4.

Alghamdi, M.A., Almazroui, M., Shamy, M., Redal, M.A., Alkhalaf, A.K., Hussein, M.A., and M.I. Khoder. 2015. Characterization and elemental composition of atmospheric aerosol loads during springtime dust storm in western Saudi Arabia. *Aerosol Air Qual Res* 15, no. 2: 440–453. http://doi.org/10.4209/aaqr.2014.06.0110.

Al-Mur, B.A., Quicksall, A.N., and A.M.A. Al-Ansari. 2017. Spatial and temporal distribution of heavy metals in coastal core sediments from the Red Sea, Saudi Arabia. *Oceanologia* 59, no. 3: 262–270. http://doi.org/10.1016/j.oceano.2017.03.003.

Atalla, T.N., Gasim, A.A., and L.C. Hunt. 2018. Gasoline demand, pricing policy, and social welfare in Saudi Arabia: A quantitative analysis. *Energy Policy* 114: 123–133. http://doi.org/10.1016/j.enpol.2017.11.047.

Basaham, A.S., El Sayed, M.A., Ghandour, I.M., and H. Masuda. 2015. Geochemical background for the Saudi Red Sea coastal systems and its implication for future environmental monitoring and assessment. *Environ Earth Sci* 74, no. 5: 4561–4570. http://doi.org/10.1007/s12665-015-4477-5.

Basaham, A.S., Rifaat, A.E., El-Sayed, M.A., and N. Rasul. 2006. Sharm Obhur: Environmental consequences of 20 Years of uncontrolled coastal urbanization. *J King Abdulaziz Univ Mar Sci* 17, no. 1: 129–152. http://doi.org/10.4197/mar.17-1.8.

Bertrand, H., Chazot, G., Blichert-Toft, J., and S. Thoral. 2003. Implications of widespread high-μ volcanism on the Arabian Plate for Afar mantle plume and lithosphere composition. *Chem Geol* 198, nos. 1–2: 47–61. http://doi.org/10.1016/S0009-2541(02)00418-7.

Brugam, R.B., Ketterer, M., Maines, L., Lin, Z.Q., and W.A. Retzlaff. 2012. Application of a simple binary mixing model to the reconstruction of lead pollution sources in two Mississippi River floodplain lakes. *J Paleolimnol* 47, no. 1: 101–112. http://doi.org/10.1007/s10933-011-9562-5.

Chien, C.T., Benaltabet, T., Torfstein, A., and A. Paytan. 2019. Contributions of atmospheric deposition to Pb concentration and isotopic composition in seawater and particulate matters in the Gulf of Aqaba, Red Sea. *Environ Sci Technol* 53, no. 11: 6162–6170. http://doi.org/10.1021/acs.est.9b00505.

Chow, T.J. 1968. Lead isotopes of the Red Sea region. *Earth Planet Sci Lett* 5: 143–147.

Chrastný, V., Komárek, M., Jrovcová, E., and J. Štíchová. 2008. A critical evaluation of the 0.05 M EDTA extraction of Pb from forest soils. *Int J Environ Analytical Chem* 88, no. 6: 385–396. http://doi.org/10.1080/03067310701744283.

Delevaux, M.H., Doe, B.R., and G.F. Brown. 1967. Preliminary lead isotope investigations of brine from the Red Sea, galena from the Kingdom of Saudi Arabia, and galena from United Arab Republic (Egypt). *Earth Planet Sci Lett* 3: 139–144. http://doi.org/10.1016/0012-821x(67)90025-8.

Doe, B.R., and M.H. Delevaux. 1972. Source of lead in southeast Missouri galena ores. *Econ Geol* 67, no. 4: 409–425. http://doi.org/10.2113/gsecongeo.67.4.409.

Doe, B.R., and J.S. Stacey. 1974. The application of lead isotopes to the problems of ore genesis and ore prospect evaluation: A review. *Econ Geol* 69, no. 6: 757–776. http://pubs.geoscienceworld.org/segweb/economicgeology/article-pdf/69/6/757/3484826/757.pdf.

Dupré, B., Blanc, G., Boulègue, J., and C.J. Allègre. 1988. Metal remobilization at a spreading centre studied using lead isotopes. *Nature* 333, no. 6169: 165–167. http://doi.org/10.1038/333165a0.

Erel, Y., Veron, A., and L. Halicz. 1997. Tracing the transport of anthropogenic lead in the atmosphere and in soils using isotopic ratios. *Geochim Cosmochim Acta* 61, no. 21: 4495–4505. http://doi.org/10.1016/S0016-7037(97)00353-0.

Graney, J.R., Halliday, A.N., Keeler, G.J., Nriagu, J.O., Robbins, J.A., and S.A. Norton. 1995. Isotopic record of lead pollution in lake sediments from the Northeastern United States. *Geochim Cosmochim Acta* 59, no. 9: 1715–1728. http://doi.org/10.1016/0016-7037(95)00077-D.

Grousset, F.E., Quétel, C.R., Thomas, B., Buat-Ménard, P., Donard, O.F.X., and A. Bucher. 1994. Transient Pb isotopic signatures in the western European atmosphere. *Environ Sci Technol* 28, no. 9: 1605–1608. https://pubs.acs.org/sharingguidelines.

Kersten, M., Garbe-Schönberg, C.D., Thomsen, S., Anagnostou, C., and A. Sioulas. 1997. Source apportionment of Pb pollution in the coastal waters of Elefsis Bay, Greece. *Environ Sci Technol* 31, no. 5: 1295–1301. http://doi.org/10.1021/es960473z.

Kim, K.H., Kabir, E., and S. Kabir. 2015. A review on the human health impact of airborne particulate matter. *Environ Int* 74: 136–143. http://doi.org/10.1016/j.envint.2014.10.005.

Komárek, M., Ettler, V., Chrastný, V., and M. Mihaljevič. 2008. Lead isotopes in environmental sciences: A review. *Environ Int* 34, no. 4: 562–577. http://doi.org/10.1016/j.envint.2007.10.005.

Langodan, S., Cavaleri, L., Pomaro, A., Portilla, J., Abualnaja, Y., and I. Hoteit. 2018. Unraveling climatic wind and wave trends in the Red Sea using wave spectra partitioning. *J Clim* 31, no. 5: 1881–1895. http://doi.org/10.1175/JCLI-D-17-0295.1.

Monna, F., Hamer, K., Lévêque, J., and M. Sauer. 2000. Pb isotopes as a reliable marker of early mining and smelting in the Northern Harz Province (Lower Saxony, Germany). *J Geochem Explor* 68, no. 3: 201–210. http://doi.org/10.1016/S0375-6742(00)00005-4.

Monna, F., Lancelot, J., Croudace, I.W., Cundy, A.B., and J.T. Lewis. 1997. Pb isotopic composition of airborne particulate material from France and the southern United Kingdom: Implications for Pb pollution sources in urban areas. *Environ Sci Technol* 31, no. 8: 2277–2286. http://doi.org/10.1021/es960870+

Monna, F., Poujol, M., Losno, R., Dominik, J., Annegarn, H., and H. Coetzee. 2006. Origin of atmospheric lead in Johannesburg, South Africa. *Atmos Environ* 40, no. 34: 6554–6566. http://doi.org/10.1016/j.atmosenv.2006.05.064.

Morton-Bermea, O., Rodríguez-Salazar, M.T., Hernández-Alvarez, E., García-Arreola, M.A., and R. Lozano-Santacruz. 2011. Lead isotopes as tracers of anthropogenic pollution in urban topsoils of Mexico City. *Chem Erde* 71, no. 2: 189–195. http://doi.org/10.1016/j.chemer.2011.03.003.

Orani, A.M., Vassileva, E., Renac, C., Schmidt, S., Angelidis, M.O., Rozmaric, M., and D. Louw. 2019. First assessment on trace elements in sediment cores from Namibian coast and pollution sources evaluation. *Sci Total Environ* 669: 668–682. http://doi.org/10.1016/j.scitotenv.2019.03.059.

Ostic, R.G., Russell, R.D., and R.L. Stanton. 1967. Additional measurements of the isotopic composition of lead from stratiform deposits. *Can J Earth Sci* 4, no. 2: 245–269. http://doi.org/10.1139/e67-012.

Palchan, D., Stein, M., Almogi-Labin, A., Erel, Y., and S.L. Goldstein. 2013. Dust transport and synoptic conditions over the Sahara-Arabia Deserts during the MIS6/5 and 2/1 transitions from grain-size, chemical and isotopic properties of Red Sea cores. *Earth Planet Sci Lett* 382: 125–139. http://doi.org/10.1016/j.epsl.2013.09.013.

Palchan, D., Stein, M., Goldstein, S.L., Almogi-Labin, A., Tirosh, O., and Y. Erel. 2018. Synoptic conditions of fine-particle transport to the last interglacial Red Sea-Dead Sea from Nd-Sr compositions of sediment cores. *Quatern Sci Rev* 179: 123–136.

Prakash, P.J., Stenchikov, G., Tao, W., Yapici, T., Warsama, B., and J.P. Engelbrecht. 2016. Arabian Red Sea coastal soils as potential mineral dust sources. *Atmos Chem Phys* 16: 11991–11994. http://doi.org/10.5194/acp-16-11991-2016.

Rezazadeh, M., Irannejad, P., and Y. Shao. 2013. Climatology of the Middle East dust events. *Aeolian Res* 10: 103–109. http://doi.org/10.1016/j.aeolia.2013.04.001.

Shaik, A.P., Sultana, S.A., and A.H. Alsaeed. 2014. Lead exposure: A summary of global studies and the need for new studies from Saudi Arabia. *Dis Markers* 2014: 415160. http://doi.org/10.1155/2014/415160.

Stacey, J.S., Doe, B.R., Roberts, R.J., Delevaux, M.H., and J.W. Gramlich. 1980. A lead isotope study of mineralization in the Saudi Arabian shield. *Contrib Mineral Petrol* 74, no. 2: 175–188. http://doi.org/10.1007/BF01132003.

Sultan, M., Bickford, M.E., El Kaliouby, B., and R.E. Arvidson. 1992. Common Pb systematics of Precambrian granitic rocks of the Nubian Shield (Egypt) and tectonic implications. *Geol Soc Am Bull* 104, no. 4: 456–470. http://doi.org/10.1130/0016-7606(1992)104<0456:CPSOPG>2.3.CO;2.

Sutherland, R.A., Day, J.P., and J.O. Bussen. 2003. Lead concentrations, isotope ratios, and source apportionment in road deposited sediments, Honolulu, Oahu, Hawaii. *Water Air Soil Pollut* 142, no. 1: 165–186. http://doi.org/10.1023/A:1022026612922.

Teutsch, N., Erel, Y., Halicz, L., and A. Banin. 2001. Distribution of natural and anthropogenic lead in Mediterranean soils. *Geochim Cosmochim Acta* 65, no. 17: 2853–2864. http://doi.org/10.1016/S0016-7037(01)00607-X.

US EPA. 2014. Method 6020B-1 revision 2. In: *Inductively Coupled Plasma – Mass Spectrometry*. Washington, DC.

Volker, F., McCulloch, M.T., and R. Altherr. 1993. Submarine basalts from the Red Sea: New Pb, Sr, and Nd isotopic data. *Geophys Res Lett* 20, no. 10: 927–930. http://doi.org/10.1029/93GL00050.

Wani, A.L., Ara, A., and J.A. Usmani. 2015. Lead toxicity: A review. *Interdiscip Toxicol* 8, no. 2: 55–64. http://doi.org/10.1515/intox-2015-0009.

Whitehouse, M.J., Windley, B.F., Stoeser, D.B., Al-Khirbash, S., Ba-Bttat, M.A.O., and A. Haider. 2001. Precambrian basement character of Yemen and correlations with Saudi Arabia and Somalia. *Precamb Res* 105, nos. 2–4: 357–369. http://doi.org/10.1016/S0301-9268(00)00120-0.

Xu, D., and B. Gao. 2020. Lead isotopes combined with geochemical baseline in sediments: A novel tool to trace anthropogenic Pb sources. *Int J Environ Res Public Health* 17, no. 3: 1112. http://doi.org/10.3390/ijerph17031112.

Xu, D., Zhou, P., Zhan, J., Gao, Y., Dou, C., and Q. Sun. 2013. Assessment of trace metal bioavailability in garden soils and health risks via consumption of vegetables in the vicinity of Tongling mining area, China. *Ecotoxicol Environ Saf* 90: 103–111. http://doi.org/10.1016/j.ecoenv.2012.12.018.

Yao, P.-H., Shyu, G.-S., Chang, Y.-F., Chou, Y.-C., Shen, C.-C., Chou, C.-S., and T.-K. Chang. 2015. Lead isotope characterization of petroleum fuels in Taipei, Taiwan. *Int J Environ Res Public Health* 12, no. 5: 4602–4616. http://doi.org/10.3390/ijerph120504602.

Youssef, A.M., Sefry, S.A., Pradhan, B., and E.A. Alfadail. 2016. Analysis on causes of flash flood in Jeddah City (Kingdom of Saudi Arabia) of 2009 and 2011 using multi-sensor remote sensing data and GIS. *Geomatics Natural Hazard Risk* 7, no. 3: 1018–1042. http://doi.org/10.1080/19475705.2015.1012750.

Zhu, B., Chen, Y.-W., and J.-H. Peng. 2001. Lead isotope geochemistry of the urban environment in the Pearl River delta. *Appl Geochem* 16, no. 4: 409–417. http://doi.org/10.1016/S0883-2927(00)00047-0.

Mangrove sediment lead apportionment along the Saudi Arabian Red Sea coast using isotopic methods

10

Abdullah S. Alnasser, Riyadh F. Halawani and Andrew N. Quicksall

10.1 INTRODUCTION

Mangroves cover about 60%–70% of the worldwide tropical and subtropical shorelines and nearly 135 km^2 of the Saudi Arabian Red Sea coastline (Alzubaidy et al. 2016; Almahasheer et al. 2016). Globally, the Red Sea is unusual, given the absence of other water bodies flowing into it (Shukri and Higazy 1944). This limits the diversity of both natural and human-induced sources of lead (Al-Mutairi and Yap 2021). Mangroves benefit the coastal environment in different ways. They can protect the area from storms, the rise of sea levels and tsunamis (Alzubaidy et al. 2016; Khalil 2015). They also offer food, medicine and habitat for marine organisms and act as a sink for heavy metals in the marine ecosystem as they trap sediments in their complicated root structure (Almahasheer et al. 2016; Ramos e Silva et al. 2006; Marchand et al. 2006; Kathiresan 2003). Furthermore, mangrove can capture atmospheric carbon and sequester it in associated sediments (Marchand et al. 2006). However, the mangrove forests of the Red Sea live in tough conditions where there is no freshwater input, yielding high salinity and surface water temperatures that exceed 31°C in the summer (Almahasheer et al. 2016). Hence, they are known to be fragmented forests in intertidal areas of the Red Sea coast of Saudi Arabia. Additionally, Red Sea mangrove environments are under stress due to anthropogenic inputs from mining, coastal growth, petrochemical industries and industrial waste. Further, infrastructure development for tourism and harbour activities threatens the health of mangrove systems (Almahasheer et al. 2016). Mangrove stands on the Saudi Arabian Red Sea shoreline extend from Jordan to Jizan; however, due to the desert environment, they are not continuous (Shaltout et al. 2020). Further trends in the region show greater densities in the nearshore versus distal areas and greater abundance from southern to northern latitudes. Most of the mangrove forests in the Red Sea consist of a monospecific population capable of thriving in the arid conditions (Mandura 1997).

Lead (Pb) is known to be a toxic element that puts both human health and the environment at risk (Wang et al. 2019). Indeed, chronic exposure to Pb can have deleterious effects on the human body, causing problems in the nervous, immune, skeletal and glandular systems (Lei et al. 2016). Since Pb is persistent in the environment, its concentration may gradually grow in sediments, water or biological tissues. High Pb concentrations in marine systems may negatively impact marine life (Soto-Jiménez et al. 2008). For example, Pb causes behavioural disorders and influences the growth of multiple types of fish (Jakimska et al. 2011). Pb can also adversely impact the growth of plants by causing disturbances in water balance and enzyme activities (Hadi and Aziz 2015). Lead can enter the marine environment naturally through bedrock weathering. It is also introduced via anthropogenic activities such as mining and smelting, coal and leaded gasoline combustion, industrial uses and sewage sludge applications (Hao et al. 2008; Komárek et al. 2008). Furthermore, Pb can be incorporated into or sorbed to airborne particulate matter yielding the potential for long distance transport before settling (Shetaya et al. 2018).

Pb has four stable isotopes (percent abundance): ^{204}Pb (1%), ^{206}Pb (24%), ^{207}Pb (23%) and ^{208}Pb (52%). ^{206}Pb ^{207}Pb, and ^{208}Pb are radiogenic isotopes that are generated from the radioactive decay of ^{238}U, ^{235}U and ^{232}Th, respectively, while ^{204}Pb is the only primordial isotope (Monna et al. 1997; Komárek et al. 2008). The goal of analyzing Pb isotopic ratios is to determine the route of a contaminant transport from its origin to deposition (Charalampides and Manoliadis 2002). Lead isotopes have been used by different researchers. Monna et al. (1997) obtained the Pb isotopic signatures of airborne particulate matter, incinerator ashes and gasoline samples from France, as well as airborne particulate matter and gasoline samples from the southern United Kingdom to determine Pb pollution sources in those regions. Furthermore, Yao et al. (2015) collected gasoline and diesel samples from service stations in Taiwan to determine the Pb isotopic signatures of domestic fuel and then compared

DOI: 10.1201/9781003321408-13

them with other possible sources of atmospheric Pb in Taiwan. Also, Pb isotopic signatures of fly ash from waste incinerators, sewage sludge from a treatment plant and particulates from auto exhaust from Switzerland were reported by Hansmann and Köppel (2000) and compared to Pb in Swiss soils.

The first goal of this chapter is to measure the concentration and isotopic ratios of Pb from Red Sea mangrove sediments by analyzing samples from three regions on the Saudi Arabian coast (Figure 10.1). Such values are as yet unreported and will prove a valuable academic resource as a baseline. The second goal is to identify and, if possible, quantify the sources of Red Sea mangrove sediment Pb.

10.2 MATERIALS AND METHODS

10.2.1 Study area

Three regions along the Saudi Arabian Red Sea coastline were chosen for the current study, ranging from north to south as follows: Yanbu, Jeddah and the Farasan Islands (Figure 10.1). Yanbu is one of the most industrialized cities on the Saudi Arabian Red Sea coast, having the largest oil shipping centre along the Red Sea, and is home to various petrochemical and mineral activities (Alharbi et al. 2019). The Jeddah region is an industrial city and a rapidly developing harbour along the central eastern margin of the Saudi Arabian Red Sea, with multiple oil refineries and numerous harbour activities (Mannaa et al. 2021; Mandura 1997; Bantan et al. 2020). The Farasan Islands are a marine sanctuary off the southern part of the Saudi Arabian Red Sea coastline. The nearest industrial activities are about 40 km away in Jizan City, which carries on both industrial and agricultural activities (Bantan and Abu-Zied 2014; Said et al. 2014). The main human activity in the Farasan Islands is fishing although minimal other activities, such as rain-fed agriculture, exist (Usman et al. 2013). The Farasan Islands comprise more than a hundred islands, and the largest two, Farasan and Sajid Islands, were selected for the current study. Seven surficial mangrove sediment samples were taken from Farasan Island, and three samples were chosen from Sajid Island.

10.2.2 Sampling, preparation and chemical analyses

Ten surficial sediment samples were taken from three mangrove sites in Yanbu, Jeddah and the Farasan Islands, each with a depth

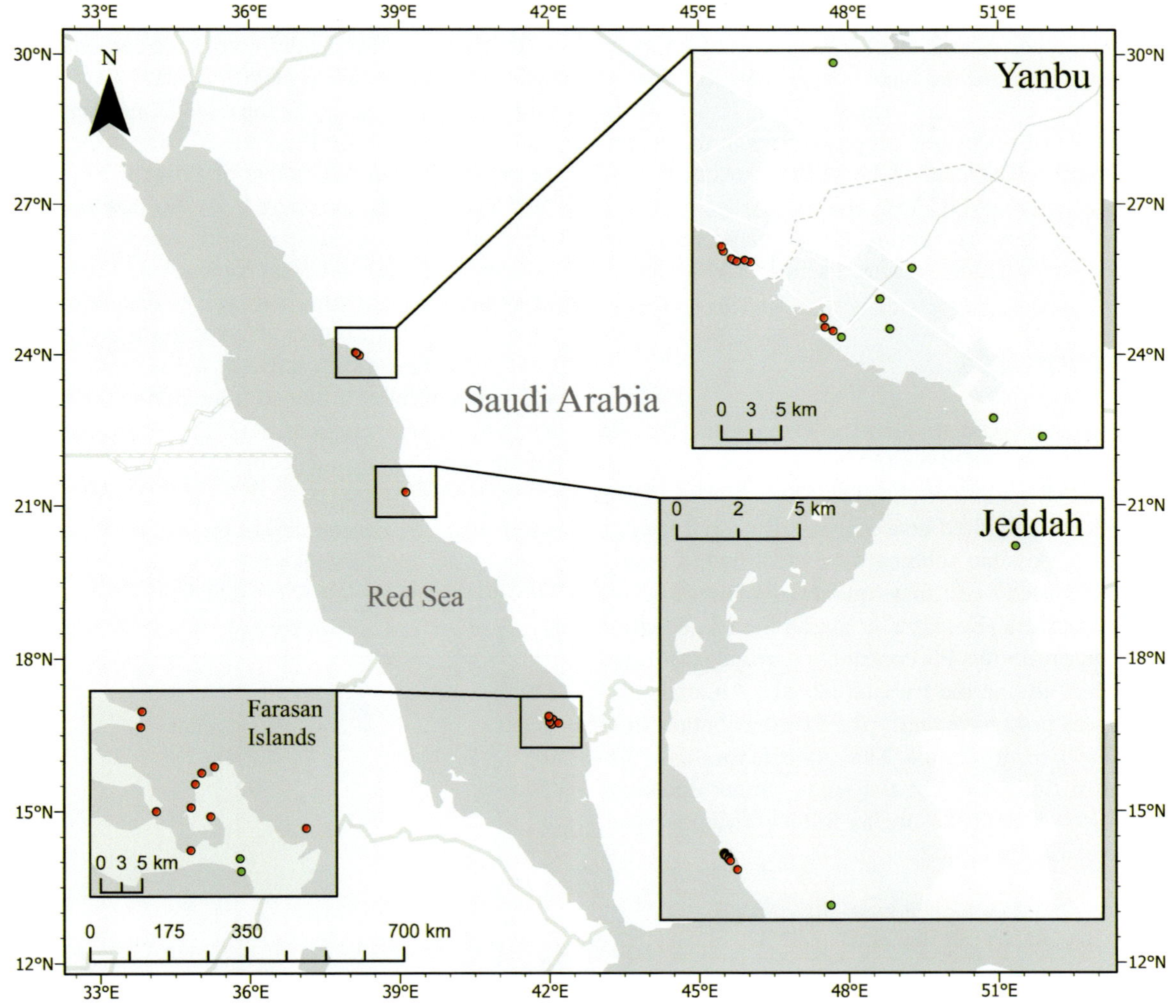

FIGURE 10.1 Sampling regions on the eastern margin of the Red Sea along the coast of Saudi Arabia. Red markers show specific sample locations for each region. Green markers represent known potential anthropogenic sources.

from 0 to 10 cm, using a stainless-steel scoop and marked by GPS (Figure 10.1). Each sample was then homogenized, resulting in 30 samples. Sediments were oven dried overnight at 105°C, disaggregated via mortar and pestle, then sieved through a 0.074 mm sieve to separate fine particulate matter from coarse debris. Then, an adapted USEPA 3050B method was followed to determine Pb concentrations and isotopic ratios (US Environmental Protection Agency 1996). Briefly, 0.15 to 0.5 g of prepared samples were placed into Teflon vessels, based on sample availability after sieving. Next, 6 mL of 5% nitric acid (HNO_3) was added to remove carbonates. Finally, 20 mL of 70% HNO_3 was added to the Teflon vessels in a fume hood and brought to and maintained at 85°C in aluminum hot blocks for digestion until the samples were fully dried.

Trace metal–grade 70% HNO_3 was diluted to 5% using 18.2 MΩ deionized water. Digested sediment sample salts were redissolved using the prepared 5% nitric acid. Aqueous digestate samples were diluted and analyzed in triplicate using a Thermo X–series II ICP-MS in collision cell mode with kinetic energy discrimination to determine total Pb concentrations and isotopic abundances (^{206}Pb, ^{207}Pb and ^{208}Pb). ICP-MS is highly sensitive for this type of analysis since it can provide a count rate of nearly 5 x 10^5 cps/ppb for lead (Monna et al. 1997). Calibration curves were built using a standard solution from Spex CertiPrep (CLMS-2AN) Claritas PPT. Curves were prepared with six calibration points from 0.5 to 20 ppb; all curves had r^2 values greater than 0.999 for each Pb isotope. For quality control purposes, instrumental and experimental blank samples were analyzed every ten samples. Instrumental blanks were analyzed tubes of 5% HNO_3 used for dilutions and resuspension. Experimental blanks were tubes that underwent all procedures in the experiment and then had 5% HNO_3 added. Using the instrumental and experimental blanks assured ICP-MS run quality against possible problems such as cross-contamination. All plasticware used was acid washed for 24 hours and rinsed with 18.2 MΩ water.

10.2.3 Pb sources

Common approaches used to identify Pb sources include plotting $^{206/207}$Pb versus $^{208/206}$Pb and Pb isotopic ratios against Pb concentrations (Komárek et al. 2008; Shetaya et al. 2018). Both techniques were applied here and are discussed later in this chapter. Once possible sources were identified, a binary mixing model was used to estimate quantitative inputs of two Pb source end members (Komárek et al. 2008). The equation was used to determine the Pb contribution from petrol and natural Pb for Jeddah and the Farasan Islands. An analogous mixing model was used to quantify the Pb contribution from ore and natural sources for Yanbu. The rationale for this is discussed later in this chapter. The two end-member equations used for both sets of analyses are as follows (Monna et al. 2000; Emmanuel and Erel 2002):

$$Pb_{anthro}\% = 100 * [(^{208/206}Pb_S - ^{208/206}Pb_N)] / [(^{208/206}Pb_{anthro} - ^{208/206}Pb_N)] \quad (10.1)$$

where $Pb_{anthro}\%$ is the contribution of the anthropogenic source (gasoline and ores). $^{208/206}Pb_S$ is the measured Pb isotopic ratio in each sample. $^{208/206}Pb_N$ is the average (2.05) of the natural Pb isotopic ratios from regional literature (Bertrand et al. 2003; Volker et al. 1993; Stuckless et al. 1984; Dupré et al. 1988; Delevaux et al. 1967). $^{208/206}Pb_{anthro}$ is the average (2.15) of isotopic ratios of gasoline-derived sources from worldwide literature since there are no gasoline Pb isotopes studies specific to the region (Erel et al. 1997; Hansmann and Köppel 2000; Monna et al. 1997; Yao et al. 2015). $^{208/206}Pb_{anthro}$ is the average (2.10) of the isotopic ratios of ore sources from local and international literature used during the analyses of Yanbu samples, described later in this chapter (Doe and Delevaux 1972; Doe and Stacey 1974; Bokhari and Kramers 1982; Brevart et al. 1982).

10.2.4 Statistical analyses

Regression analyses were done using a statistical analysis system (SAS JMP 15.1.0). Regression aided in comparing the Pb isotopic ratios from the current study to previous studies, as well as comparing regions with each other.

10.3 RESULTS AND DISCUSSION

The average concentrations of Pb (mg/kg) from the study area were highest in Yanbu as follows: 7.39 for Yanbu, 3.82 for Jeddah and 3.44 for the Farasan Islands. The average background worldwide crustal value used here is 9 mg/kg, as reported by Turekian and Wedepohl (1961) for carbonate sedimentary rocks. The results from Table 10.1 show that three mangrove sediment samples in the Yanbu region, one sample from Jeddah and two samples from the Farasan Islands exceeded the average crustal background value of Pb (Y1, Y4, Y6, J1, F9 and F10). Yanbu sediments were a few kilometres away from an industrial zone, an industrial port, and a commercial port. It was expected, therefore, that Yanbu would have more samples exceeding the background average, given its closer proximity to potential pollution sources relative to the Jeddah samples and the near absence of such activities near Farasan. The ^{208}Pb/^{206}Pb of the area of interest ranged from 2.06 to 2.08 in Yanbu sediments, with an average of 2.07. The ^{208}Pb/^{206}Pb values of the Jeddah samples varied from 2.07 to 2.12, with an average

TABLE 10.1 Pb concentrations (mg/kg) and isotopic ratios of mangrove sediment samples from the study area. Averages and standard deviations for each region are included.

SAMPLER ID	*CONCENTRATION*	*$^{206/207}$PB*	*$^{208/206}$PB*
YANBU			
Y1	10.10	1.19	2.08
Y2	6.39	1.19	2.07
Y3	5.97	1.19	2.07
Y4	14.50	1.18	2.08
Y5	8.15	1.19	2.07
Y6	9.10	1.18	2.07
Y7	2.87	1.20	2.06
Y8	5.32	1.19	2.07
Y9	8.41	1.20	2.06
Y10	3.07	1.19	2.07
Average	7.39	1.19	2.07
SD	3.48	0.00	0.01

(Continued)

TABLE 10.1 (Continued)

SAMPLER ID	*CONCENTRATION*	$^{206/207}$Pb	$^{208/206}$Pb
JEDDAH			
J1	20.74	1.15	2.12
J2	3.78	1.16	2.10
J3	2.71	1.16	2.09
J4	1.95	1.16	2.08
J5	1.24	1.18	2.08
J6	1.47	1.18	2.08
J7	1.23	1.17	2.08
J8	1.31	1.18	2.07
J9	1.44	1.20	2.07
J10	2.34	1.16	2.09
Average	3.82	1.17	2.09
SD	6.00	0.01	0.02
FARASAN ISLANDS			
F1	3.20	1.18	2.06
F2	2.05	1.18	2.06
F3	1.81	1.19	2.06
F4	1.48	1.18	2.07
F5	1.43	1.19	2.06
F6	1.29	1.19	2.06
F7	2.30	1.18	2.06
F8	1.23	1.18	2.08
F9	9.79	1.16	2.10
F10	9.78	1.15	2.10
Average	3.44	1.18	2.07
SD	3.40	0.01	0.02

of 2.09, and from 2.06 to 2.10 for the Farasan Islands, with an average of 2.07 (Table 10.1). Yanbu has a range of ^{206}Pb/^{207}Pb from 1.18 to 1.20, with an average of 1.19. Jeddah has an average of 1.17 for ^{206}Pb/^{207}Pb, with a range between 1.15 and 1.20. For the Farasan Islands, the lowest ^{206}Pb/^{207}Pb value was 1.15, and the highest was 1.19, with an average of 1.18 (Table 10.1).

Figure 10.2 shows a mixing line with two end members of gasoline and natural Pb signatures from other studies for comparison with the current study. Literature used for the comparison is categorized into two groups: natural and anthropogenic. The first represents Pb that may have been transported from regional natural sources, including Arabian plate basalts, Red Sea basalts, granites from the Arabian Shield of Saudi Arabia, Red Sea Atlantis II deep sediments and quartz from Saudi Arabia (Bertrand et al. 2003; Volker et al. 1993; Stuckless et al. 1984; Dupré et al. 1988; Delevaux et al. 1967, respectively). The second group provides isotopic Pb signatures derived from gasoline sources from various countries, including Israel, Switzerland, France and Taiwan (Erel et al. 1997; Hansmann and Köppel 2000; Monna et al. 1997; Yao et al. 2015, respectively). Given the lack of lead isotopic studies in the Red Sea region, petrol-Pb signatures were employed from the countries mentioned earlier. While both potential sources cover a range of ratios, weighted averages of all listed literature for each source were calculated to provide a centroid to be used in modeling and defining the mixing line plotted in Figure 10.2. Figure 10.2 also shows a mixing line of industrial ore sources and natural Pb sources. The industrial ore isotopic ratios employed in the current study were reported by local and international studies and were similarly used in a weighted average calculation to obtain a centroid (Doe and Delevaux 1972; Doe and Stacey 1974; Bokhari and Kramers 1982; Brevart et al. 1982). Measured Pb isotopic ratios of the current study for Yanbu, Jeddah and the Farasan Islands are also plotted in Figure 10.2 and can therefore be visually compared to the theoretical mixing lines. The

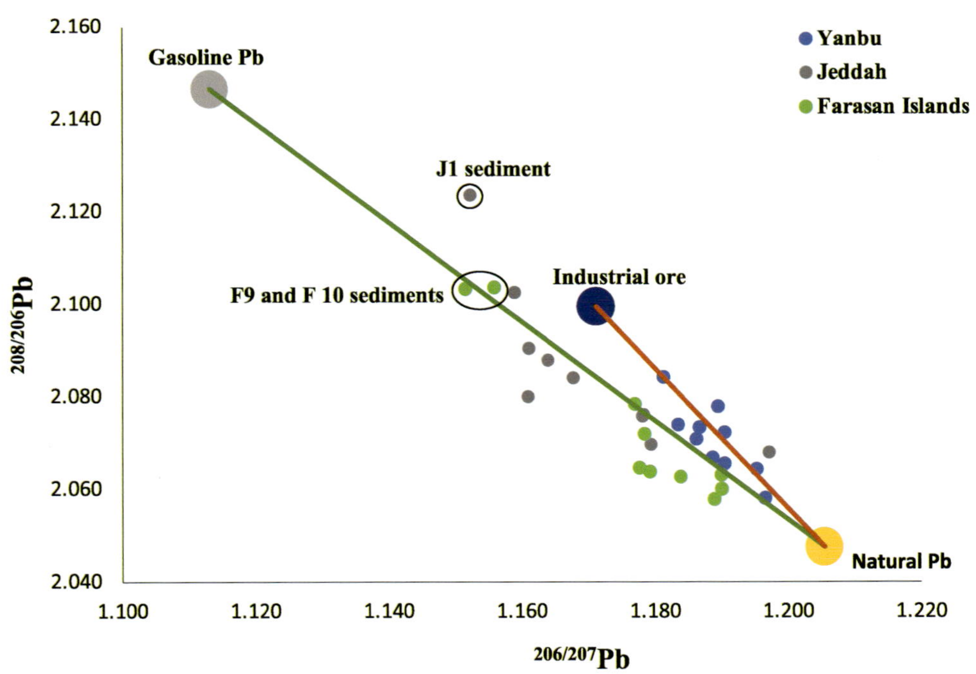

FIGURE 10.2 $^{208/206}$Pb versus $^{206/207}$Pb values of Yanbu, Jeddah and the Farasan Islands denoted in the legend. Mixing lines of gasoline and natural isotopic ratios (green line) and of industrial ore Pb and natural Pb isotopic ratios (orange line) connect the isotopic signatures of known sources. For reference values and method of exact source end-member centroid determination, see text.

linearity of the data in Figure 10.2 suggests that the Jeddah and Farasan Islands samples are a mix of gasoline and natural Pb sources. In addition, it suggests that Yanbu sediments have a mixture of gasoline and natural sources as well as industrial ore as a third Pb source. Figure 10.3 shows the gasoline-derived source and Pb concentrations of the three regions. It implies that Jeddah has the highest gasoline Pb load in its sediments of the three. It also indicates, however, that despite Yanbu sediments having the highest total Pb concentrations across the study area, they have a lower gasoline loading than Jeddah. This leads to an agreement with Figure 10.2 that Yanbu sediments are affected by an industrial ore Pb source which must be highly elevated in its Pb concentration.

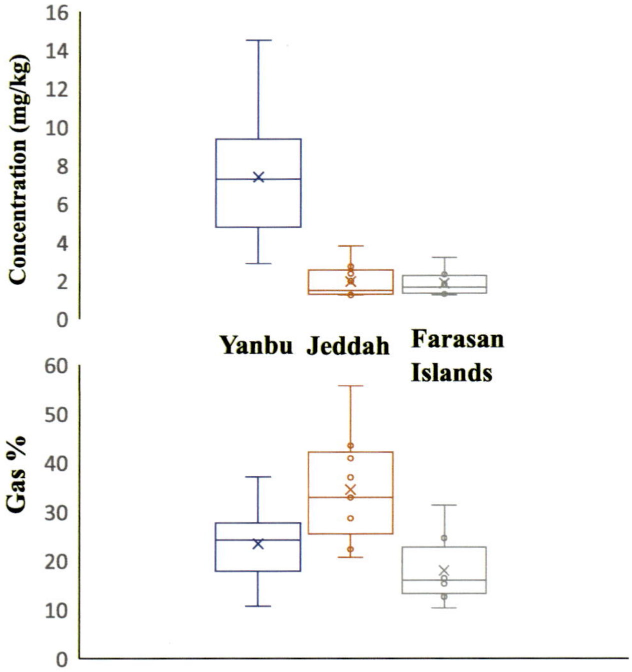

FIGURE 10.3 Box and whisker plots of concentrations (top) and percent gas contribution (bottom) for Yanbu, Jeddah and the Farasan Islands.

Further evidence is gathered by regression analyses of $^{206/207}Pb$ and $^{208/206}Pb$ versus Pb concentrations for Yanbu, Jeddah and the Farasan Islands (Figure 10.4 A) in similar fashion to Shetaya et al. (2018). Yanbu sediments have a clear correlation between isotopic signature and concentration, which is confirmed by the r^2 values reported in Table 10.2. This anthropogenic source elevates the Yanbu sediment's concentrations well above those of the other regions. Its unique signature off axis of the natural-gas mixing line (Figure 10.2) is recorded in the isotopic ratio trends with concentration (Figure 10.4). Meanwhile, the interpretation that Jeddah and the Farasan Islands sediments are solely composed of gasoline and natural Pb sources is supported

TABLE 10.2 Linear regression r^2 values of $^{208/206}Pb$ and $^{206/207}Pb$ ratios versus Pb concentrations for mangrove sediments from Yanbu, Jeddah and the Farasan Islands regions represented in Figure 10.4.

REGION	*R^2 OF FIGURE 10.4 A*	*R^2 OF FIGURE 10.4 B*
Yanbu	0.74	0.36
Jeddah	0.78	0.44
Farasan Islands	0.08	0.01

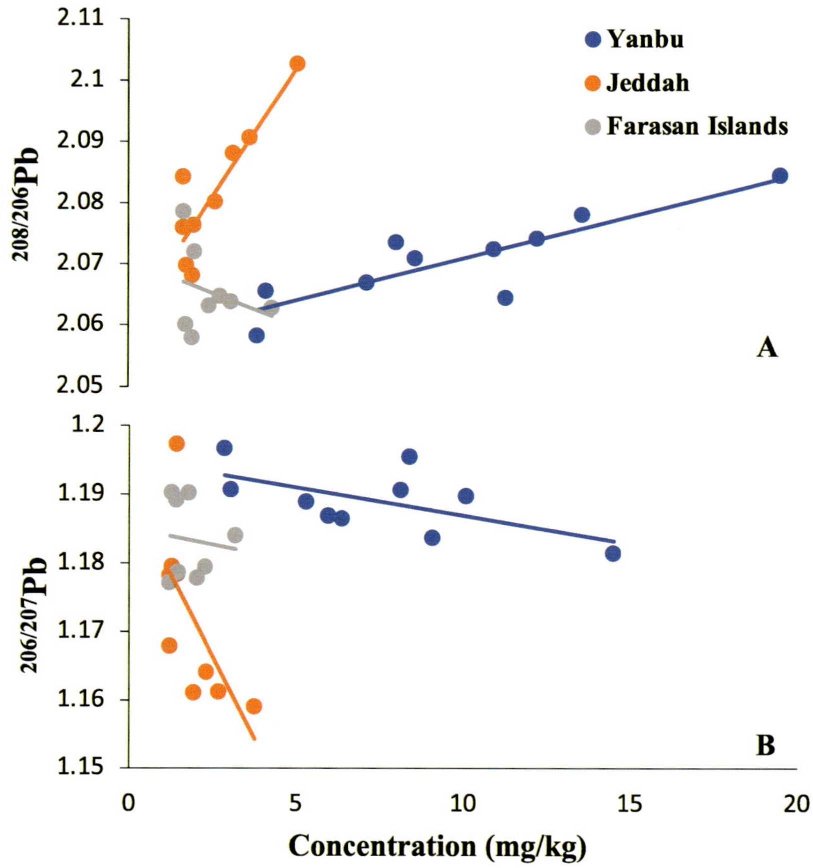

FIGURE 10.4 $^{208/206}Pb$ (A) and $^{206/207}Pb$ (B) ratios versus Pb concentrations for Yanbu, Jeddah and the Farasan Islands. Lines show the least squares regression of best fit; r^2 values for these fits are reported in Table 10.2.

by the low concentrations (Figure 10.3) and the poorer correlation between isotopic ratios and concentration (Figure 10.4). Of note, in addition to the strong correlation of Yanbu isotopic ratios with concentrations, Jeddah also has a strong correlation (Table 10.2). Given the relatively low concentrations there, this further supports a background contaminant that alters the isotopic signature with small additions but has little impact on overall loading. This is consistent with the gasoline-derived end member.

Three samples are identified in Figure 10.2 (J1, F9 and F10) from Jeddah and Farasan that are outliers. They have been dropped from the analyses and are interpreted to be recording the stochastic impact of another source; however, with so few data points of the set of 30, no further interpretation of that source is possible.

Table 10.3 shows the percentages of gasoline-derived and natural Pb sources for Jeddah and the Farasan Islands. The gasoline-derived average in the Jeddah samples is higher (34%) than in Farasan Islands sediments (18%). For Yanbu sediments, Table 10.4 reports the industrial ore and natural composition percentages with an average of 44% for the industrial ore. Tables 10.3 and 10.4 show that the entire investigated area has more natural Pb than gasoline or ore components, which affirms Figure 10.2 since it shows that most of the sediment samples, regardless of location, fall closer to the natural end member than the gasoline and the industrial ore sources. Figure 10.5 shows that the higher the concentration, the greater the industrial ore input to Yanbu sediments, confirming that source's control on total lead loading for that location. Also, it shows that higher percentages of gasoline Pb are associated with higher Pb concentrations in Jeddah sediments. As suggested earlier, the Jeddah samples record a strong history of legacy gasoline-derived Pb accumulation; however, they have not yielded significant total Pb contamination. For Jeddah and the Farasan Islands, the outlier samples are most likely from unknown anthropogenic sources since they plot off the mixing line in Figure 10.2 paired with Pb concentrations exceeding the average background crustal value.

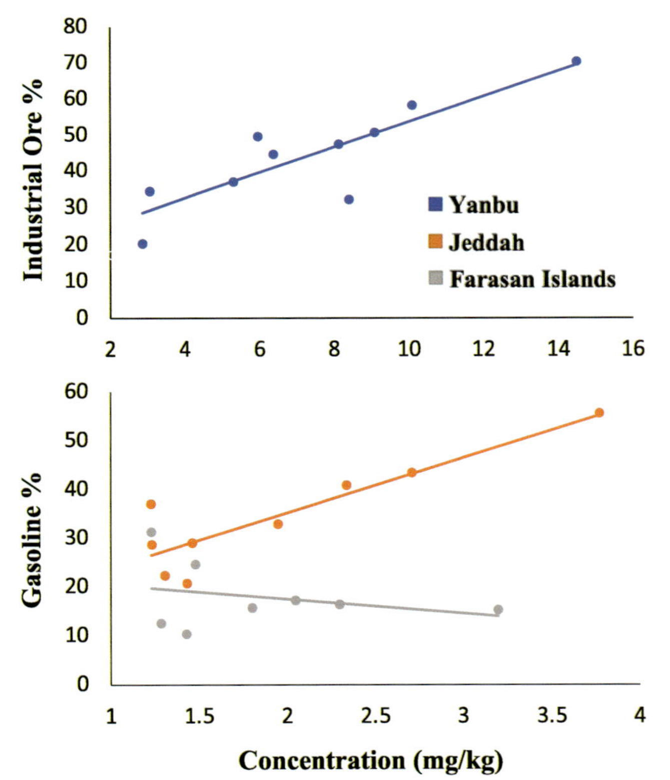

FIGURE 10.5 Percentage of gasoline-derived sources versus concentrations for Jeddah and the Farasan Islands (top). Industrial ore percentage contribution versus concentrations for Yanbu (bottom). Lines show the least squares regression of best fit.

TABLE 10.3 Percentages of gasoline and natural Pb sources of Jeddah and Farasan Islands sediments.

SAMPLE ID	*GAS %*	*NATURAL %*
J2	56	44
J3	43	57
J4	33	67
J5	29	71
J6	29	71
J7	37	63
J8	22	78
J9	21	79
J10	41	59
Average	34	66
F1	15	85
F2	17	83
F3	16	84
F4	25	75
F5	10	90
F6	13	87
F7	16	84
F8	31	69
Average	18	82

TABLE 10.4 Percentages of ore and natural contributions to total Pb loading of Yanbu samples.

SAMPLE ID	*ORE %*	*NATURAL %*
Y1	58	42
Y2	45	55
Y3	50	50
Y4	71	29
Y5	47	53
Y6	51	49
Y7	20	80
Y8	37	63
Y9	32	68
Y10	34	66
Average	44	56

10.4 CONCLUSION

Throughout the study area, Pb concentrations are relatively low when compared to the average preindustrial background, and Jeddah and the Farasan Islands have lower concentrations in general than Yanbu, as Table 10.1 indicates. The Farasan Islands and Jeddah have Pb mostly derived from petrol and natural inputs. Jeddah sediments have the highest gasoline-derived Pb loading across the study area, although the Pb concentrations are low compared to Yanbu. For Farasan Islands sediments, the dominant Pb source is natural, as is expected, given the Islands have the least urban activities in the study area as the main industry is fishing with some agricultural activities. Pb isotopic ratios suggest that Yanbu has an unknown industrial ore source of Pb that markedly elevates concentrations proportionally to altering its isotopic signatures.

Mangrove environments are known to be ecosystems that can easily sequester heavy metals when exposed. They, therefore, act as highly sensitive indicators of a region's overall susceptibility to heavy metal pollution. Here, we find relatively low lead pollution across the region; however, there is clear documentation of what is likely a legacy atmospheric fallout of gasoline-derived lead. While this signature is ubiquitous, it is centred in the urban area of Jeddah. Despite trended gasoline isotopic values with concentrations, this Pb source has little impact on total lead loading. There is a stark contrast with the Yanbu mangrove area. There, an anthropogenic, likely a modern and point source clearly impacts both the isotopic signature and the total Pb loading. While the environmental threat of gasoline-derived Pb is likely stable and only a historical record, the industrial ore sourcing in Yanbu mangrove sediments should be a clear warning. Further work identifying the exact source and the breadth of impact should be prioritized. Additionally, management strategies to mitigate this pollution should be developed and implemented prior to its exacerbation.

REFERENCES

Alharbi, O.M.L., Khattab, R.A., Ali, I., Binnaser, Y.S., and A. Aqeel. 2019. Assessment of heavy metals contamination in the sediments and mangroves (Avicennia marina) at Yanbu coast, Red Sea, Saudi Arabia. *Mar Pollut Bull* 149: 110669.

Almahasheer, H., Aljowair, A., Duarte, C.M., and X. Irigoien. 2016. Decadal stability of Red Sea mangroves. *Estuar Coast Shelf Sci* 169: 164–172.

Al-Mutairi, K.A., and C.K. Yap. 2021. A review of heavy metals in coastal surface sediments from the red sea: Health-ecological risk assessments. *Int J Environ Res Public Health* 18, no. 6: 2798.

Alzubaidy, H., Essack, M., Malas, T.B., Bokhari, A., Motwalli, O., Kamanu, F.K., Jamhor, S.A., et al. 2016. Rhizosphere microbiome metagenomics of gray mangroves (Avicennia marina) in the Red Sea. *Gene* 576, no. 2: 626–636.

Bantan, R.A., and R.H. Abu-Zied. 2014. Sediment characteristics and molluscan fossils of the Farasan Islands shorelines, southern Red Sea, Saudi Arabia. *Arab J Geosci* 7, no. 2: 773–787.

Bantan, R.A., Al-Dubai, T.A., and A.G. Al-Zubieri. 2020. Geo-environmental assessment of heavy metals in the bottom sediments of the Southern Corniche of Jeddah, Saudi Arabia. *Mar Pollut Bull* 161: 111721.

Bertrand, H., Chazot, G., Blichert-Toft, J., and S. Thoral. 2003. Implications of widespread high-μ volcanism on the Arabian Plate for Afar mantle plume and lithosphere composition. *Chem Geol* 198, nos. 1–2: 47–61.

Bokhari, F.Y., and J.D. Kramers. 1982. Lead isotope data from massive sulfide deposits in the Saudi Arabian Shield. *Econ Geol* 77, no. 7: 1766–1769.

Brevart, O., Dupré, B., and C.J. Allegre. 1982. Metallogenic provinces and the remobilization process studied by lead isotopes; lead-zinc ore deposits from the southern Massif Central, France. *Econ Geol* 77, no. 3: 564–575.

Charalampides, G., and O. Manoliadis. 2002. Sr and Pb isotopes as environmental indicators in environmental studies. *Environ Int* 28, no. 3: 147–151.

Delevaux, M.H., Doe, B.R., and G.F. Brown. 1967. Preliminary lead isotope investigations of brine from the Red Sea, galena from the Kingdom of Saudi Arabia, and galena from United Arab Republic (Egypt). *Earth Planet Sci Lett* 3: 139–144.

Doe, B.R., and M.H. Delevaux. 1972. Source of lead in southeast Missouri galena ores. *Econ Geol* 67, no. 4: 409–425.

Doe, B.R., and J.S. Stacey. 1974. The application of lead isotopes to the problems of ore genesis and ore prospect evaluation: A review. *Econ Geol* 69, no. 6: 757–776.

Dupré, B., Blanc, G., Boulègue, J., and C.J. Allègre. 1988. Metal remobilization at a spreading centre studied using lead isotopes. *Nature* 333, no. 6169: 165–167.

Emmanuel, S., and Y. Erel. 2002. Implications from concentrations and isotopic data for Pb partitioning processes in soils. *Geochim Cosmochim Acta* 66, no. 14: 2517–2527.

Erel, Y., Veron, A., and L. Halicz. 1997. Tracing the transport of anthropogenic lead in the atmosphere and in soils using isotopic ratios. *Geochim Cosmochim Acta* 61, no. 21: 4495–4505.

Hadi, F., and T. Aziz. 2015. A mini review on lead (Pb) toxicity in plants. *J Biol Life Sci* 6, no. 2: 91–101.

Hansmann, W., and V. Köppel. 2000. Lead-isotopes as tracers of pollutants in soils. *Chem Geol* 171, no. 1–2: 123–144.

Hao, Y., Guo, Z., Yang, Z., Fan, D., Fang, M., and X. Li. 2008. Tracking historical lead pollution in the coastal area adjacent to the Yangtze River estuary using lead isotopic compositions. *Environ Pollut* 156, no. 3: 1325–1331.

Jakimska, A., Konieczka, P., Skóra, K., and J. Namieśnik. 2011. Accumulation of metals in tissues and organs of marine organisms. *Effect Metals Mar Org* 20: 1117–1125.

Kathiresan, K. 2003. How do mangrove forests induce sedimentation? *Rev Biol Trop* 51, no. 2: 355–360.

Khalil, A.S. 2015. Mangroves of the Red Sea. In: *The Red Sea: The Formation, Morphology, Oceanography and Environment of a Young Ocean Basin*. Eds. N.M.A. Rasul and I.C.F. Stewart, 585–597. Berlin and Heidelberg: Springer Earth System Sciences.

Komárek, M., Ettler, V., Chrastný, V., and M. Mihaljevič. 2008. Lead isotopes in environmental sciences: A review. *Environ Int* 34, no. 4: 562–577.

Lei, K., Giubilato, E., Critto, A., Pan, H., and C. Lin. 2016. Contamination and human health risk of lead in soils around lead/zinc smelting areas in China. *Environ Sci Pollut Res* 23, no. 13: 13128–13136.

Mandura, A.S. 1997. A mangrove stand under sewage pollution stress: Red Sea. *Mangroves Salt Marshes* 1, no. 4: 255–262.

Mannaa, A.A., Khan, A.A., Haredy, R., and A.G. Al-Zubieri. 2021. Contamination evaluation of heavy metals in a sediment core from the Al-Salam Lagoon, Jeddah coast, Saudi Arabia. *J Mar Sci Eng* 9, no. 8: 899.

Marchand, C., Lallier-Vergès, E., Baltzer, F., Albéric, P., Cossa, D., and P. Baillif. 2006. Heavy metals distribution in mangrove sediments along the mobile coastline of French Guiana. *Mar Chem* 98, no. 1: 1–17.

Monna, F., Hamer, K., Lévêque, J., and M. Sauer. 2000. Pb isotopes as a reliable marker of early mining and smelting in the Northern Harz province (Lower Saxony, Germany). *J Geochem Explor* 68, no. 3: 201–210.

Monna, F., Lancelot, J., Croudace, I.W., Cundy, A.B., and J.T. Lewis. 1997. Pb isotopic composition of airborne particulate material from France and the southern United Kingdom: Implications for Pb pollution sources in urban areas. *Environ Sci Technol* 31, no. 8: 2277–2286.

Ramos e Silva, C.A., Da Silva, A.P., and S.R. De Oliveira. 2006. Concentration, stock and transport rate of heavy metals in a tropical red mangrove, Natal, Brazil. *Mar Chem* 99, no. 1–4: 2–11.

Said, T.O., Omran, A.A., Fawy, K.F., and A.M. Idris. 2014. Heavy metals in twelve edible marine fish species from Jizan fisheries, Saudi Arabia: Monitoring and assessment. *Fresenius Environ Bull* 23, no. 3: 801–809.

Shaltout, K.H., Ahmed, M.T., Alrumman, S.A., Ahmed, D.A., and E.M. Eid. 2020. Evaluation of the carbon sequestration capacity of arid mangroves along nutrient availability and salinity gradients along the Red Sea coastline of Saudi Arabia. *Oceanologia* 62, no. 1: 56–69.

Shetaya, W.H., Marzouk, E.R., Mohamed, E.F., Elkassas, M., Bailey, E.H., and S.D. Young. 2018. Lead in Egyptian soils: Origin, reactivity and bioavailability measured by stable isotope dilution. *Sci Total Environ* 618: 460–468.

Shukri, N.M., and R.A. Higazy. 1944. Mechanical analysis of some bottom deposits of the Northern Red Sea. *J Sed Res* 14, no. 2: 43–69.

Soto-Jiménez, M.F., Páez-Osuna, F., Scelfo, G., Hibdon, S., Franks, R., Aggarawl, J., and A.R. Flegal. 2008. Lead pollution in subtropical ecosystems on the SE Gulf of California Coast: A study of concentrations and isotopic composition. *Mar Environ Res* 66, no. 4: 451–458.

Stuckless, J.S., Hedge, C.E., Wenner, D.B., and I.T. Nkomo. 1984. Isotopic studies of postorogenic granites from the northeastern Arabian Shield, Kingdom of Saudi Arabia. In: *Saudi Arabian Deputy Ministry for Mineral Resources,* 1–40. Jeddah: Open-File Report USGS-OF-04-42.

Turekian, K.K., and K.H. Wedepohl. 1961. Distribution of the elements in some major units of the earth's crust. *Geol Soc Am Bull* 72, no. 2: 175–192.

US Environmental Protection Agency (EPA). 1996. Method 3050B: Acid digestion of sediments, sludges, and soils. In: *Test Methods for Evaluating Solid Waste, Physical/Chemical Methods*. Washington, DC: US Environmental Protection Agency.

Usman, A.R.A., Alkredaa, R.S., and M.I. Al-Wabel. 2013. Heavy metal contamination in sediments and mangroves from the coast of Red Sea: Avicennia marina as potential metal bioaccumulator. *Ecotoxicol Environ Saf* 97: 263–270.

Volker, F., McCulloch, M.T., and R. Altherr. 1993. Submarine basalts from the Red Sea: New Pb, Sr, and Nd isotopic data. *Geophys Res Lett* 20, no. 10: 927–930.

Wang, Z., Dwyer, G.S., Coleman, D.S., and A. Vengosh. 2019. Lead isotopes as a new tracer for detecting coal fly ash in the environment. *Environ Sci Technol Lett* 6, no. 12: 714–719.

Yao, P.-H., Shyu, G.-S., Chang, Y.-F., Chou, Y.-C., Shen, C.-C., Chou, C.-S., and T.-K. Chang. 2015. Lead isotope characterization of petroleum fuels in Taipei, Taiwan. *Int J Environ Res Public Health* 12, no. 5: 4602–4616.

Use of recent invertebrates as a tool for monitoring environmental hazards *Strombus tricornis* Humphrey 1786 – Case study

11

Hashem Madkour and Mohamed Youssef

11.1 INTRODUCTION

The presence of heavy metals in the marine environment is partly due to natural processes such as volcanic activity and erosion, but is mostly the result of human activities (Merian 1991). They mainly enter the sea with industrial wastes, both liquid and solid, and in solid particles carried by winds that are eventually deposited in the sea. Marine organisms may take in metals through food or water. Studies on some invertebrates show that about an equal amount can be taken in via both routes. As regards metal bioaccumulation in marine organisms, several authors have proven, in experiments carried out with crustaceans, molluscs and fish, their high accumulation ability, which generally depends on exposure time and metal concentration in the water (Bryan 1984; Mance 1987). Different species have different rates of bioaccumulation, and sites of major concentration may differ from one species to another. Bryan (1984) and Mance (1987) found that metal levels in marine organisms can reached much higher values than those in the seawater. Other factors affecting bioaccumulation are physiological condition, growth, salinity and temperature, age, sex, position relative to shoreline, water depth and pollutant interactions (Phillips 1980). Heavy metals come from mining, oil refineries, shipment operations in the harbours, shipyards, power plants, domestic sewage, desalination plants and landfill and dredging. They are found in the antifouling and anticorrosive paints used to protect ships and coastal structures. Moreover, in urban areas, industrial and domestic wastewater introduces various types of pollutants such as toxic metals, which cause serious damage to aquatic and other environmental systems (Madkour 2004).

Heavy metals may occur in aquatic environments from natural processes and from discharges or leachates from several anthropogenic activities (Connell et al. 1999; Franca et al. 2005). The contamination of natural waters by heavy metals negatively affects aquatic biota and poses considerable environmental risk and concern (Cajaraville et al. 2000; Ravera 2001). Sediments are an important sink for a variety of pollutants, particularly heavy metals, and may serve as an enriched source of benthic organisms (Wang and Rainbow 2005), especially in estuarine ecosystems.

The use of recent invertebrates in the marine environment to detect the extent to which those organisms are affected by concentrations of heavy metals and their relationship and susceptibility to the concentrations of those heavy metals in the associated marine sediments have received significant attention (Youssef et al. 2017; Harayashiki et al. 2020). Previous studies related to heavy element concentrations in sea shells and their relationship to associated sediments in the Red Sea Coast are few. However, previous studies have proven the ability of molluscan shells to monitor the environmental quality on the Red Sea coast (e.g., Ziko et al. 2001; Madkour 2004, 2005, 2013, 2015; Mansour et al. 2005; Madkour and Youssef 2009; El-Taher and Madkour 2011; Mekawy and Madkour 2012, 2013; Mohamed et al. 2013a, 2013b; El-Sorogy et al. 2013; Youssef et al. 2017).

In the present chapter, we document the environmental effects of heavy metals on the molluscan shells of *Strombus tricornis*. Heavy metals such as zinc, cadmium, lead, copper, cobalt, iron, manganese and nickel are naturally absorbed by this species. The collected specimens of molluscan shells of *Strombus tricornis* in the study areas were identified according to Wye (1991). *Strombus tricornis* lives near the sediment-water interface; therefore, contaminants and heavy metals in sediments have a potential influence on them.

The main objective of this chapter is to document 1) the concentrations of heavy elements in the shell organism, *Strombus*

DOI: 10.1201/9781003321408-14

tricornis; 2) the concentrations of those heavy elements by the associated sediments; and 3) the effect of the continental contribution through estuaries of wadis and places with multiple human activities. These results were compared with a control area.

11.2 MATERIALS AND METHODS

11.2.1 Study area

The concentrations of heavy metals in the shells of *Strombus tricornis* and associated marine sediments are recorded in the Red Sea coast from the Haleib area in the south to Ras El-Behar in the north. The study area extends for a distance of 750 km and includes 15 sites (Figure 11.1), some of which were affected by the continental contribution through the estuaries of the wadis and some by multiple human activities, which were the main source of differences in the concentrations of heavy elements, whether in the accompanying sediments or in the shells of *Strombus tricornis.*

A preliminary visual survey of the Red Sea coast, using snorkeling and SCUBA diving equipment, led to selection of fifteen sites on the Egyptian Red Sea coast. These sites being arranged from south to north illustrates the various human activities found in the areas. The first, second, third and fourth sites, which are the Haleib, Abu-Ramed, Shaltein and Marsa Humeira areas, respectively, are exposed to natural inputs by wadis and human activities represented by the effects of discharge operations of desalination plants on the marine environment. The fifth site represents the Wadi al-Gemal region, which is 50 km south of Marsa Alam and represents an active area by Wadi El-Gemal in terms of the continental contribution to the marine environment. The sixth site is Marsa Alam Marina, which is a port area under construction and represents the activities of fishing and tourist boats. The seventh site represents the Abu Dabab area, which is an area of natural activities by Wadi Abu Dabab. The eighth and ninth

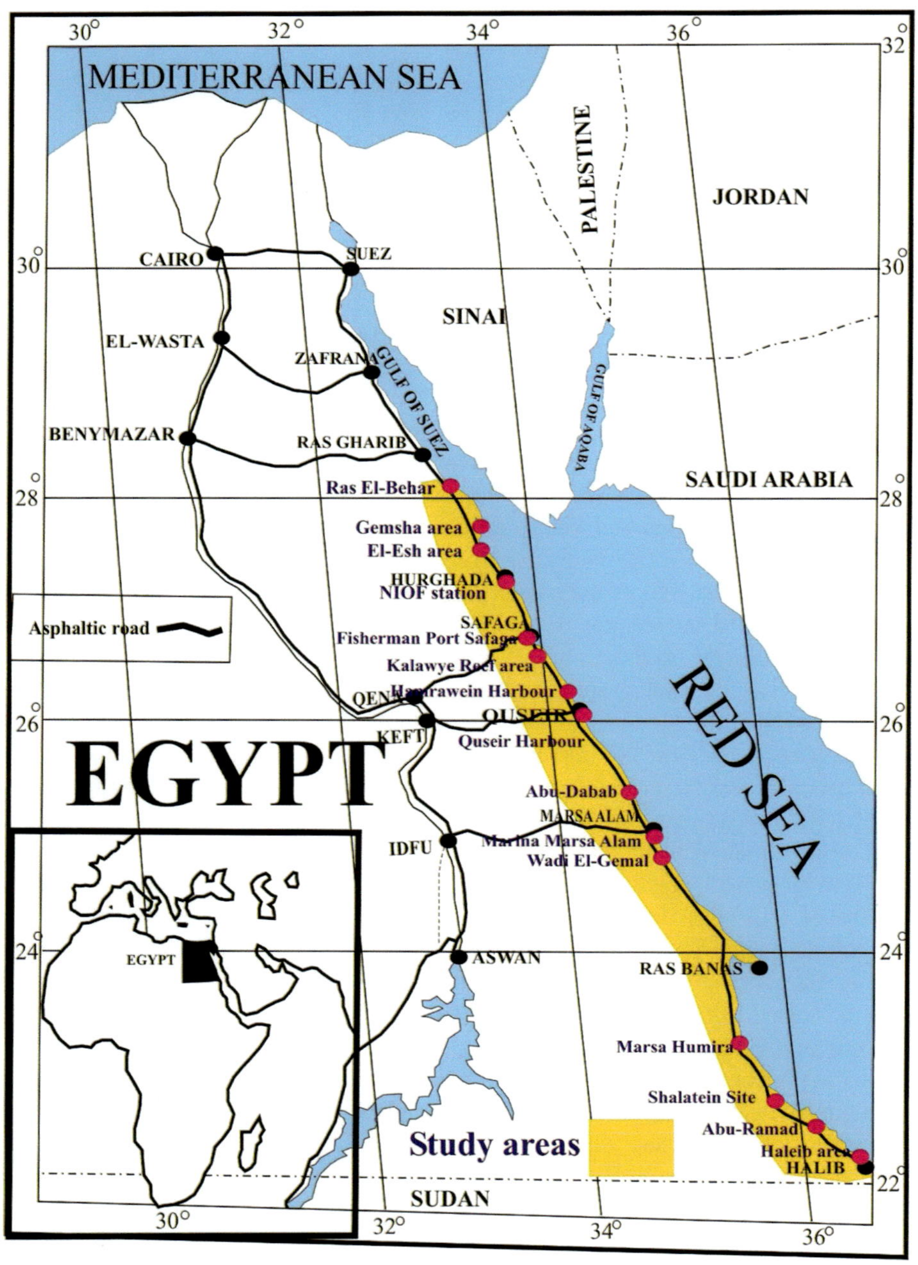

FIGURE 11.1 Location map of the study areas along the Red Sea coast, Egypt.

TABLE 11.1 Sample location, depth and the measured hydrographic parameters of water mass in the studied areas.

		POSITION						DEPTH	TEMP.	SALINITY	DO	PH	TDS	SPC	TYPES OF
		LAT.			LONG.			(M)	°C	O\OO	MG\L		PPT	MS\CM	ACTIVITIES
NO.	LOCATION	O	/	/ / N	O	/	/ / E								
1	Haleib area (H)	22	11	48.90	36	40	51.64	2.5	30.70	42.15	4.50	7.99	40.56	63.60	Natural inputs
2	Abu-Ramad area (AR)	22	24	46.5	36	25	15.41	3.0	29.90	42.00	3.20	7.95	41.01	62.85	Natural inputs
3	Shalatein area (SH)	23	0.9	18.78	35	36	06.44	4.0	32.00	43.37	5.90	7.87	41.40	64.53	Natural inputs
4	Marsa Humira area (MH)	23	28	14.20	35	29	34.41	3.0	30.60	42.87	4.50	7.90	41.47	62.23	Natural inputs
5	Wadi El-Gemal (WG)	24	40	13	35	05	30	5.0	25.60	41.10	8.70	8.20	39.50	61.25	Natural inputs
6	Marsa Alam Marina (MM)	25	4	9.27	34	54	3.07	3.0	29.70	40.90	5.98	7.91	38.97	60.95	Natural inputs with human activities
7	Abu-Dabab area (D)	25	20	14	34	44	15	2.0	29.50	42.10	6.50	8.00	40.10	63.20	Natural inputs
8	Quseir Harbor (QH)	26	05	02	34	16	58	3.0	26.30	40.80	6.70	8.50	39.01	61.20	Old phosphate shipments
9	El-Hamreawein Harbour (HH)	26	15	17	34	12	00	3.0	27.20	41.80	5.90	7.96	39.30	60.96	Phosphate shipments
10	Kalawye Reefs (KR)	26	30	35	34	04	06	6.0	26.50	40.90	6.50	8.20	39.70	61.20	Control area
11	Fishing Port Safaga (S)	26	43	57	33	56	15	3.0	25.90	41.30	5.97	8.30	39.40	61.40	Different activities
12	NIOF Station area (NB)	27	17	01.0	33	46	21.0	3.0	24.76	42.75	7.50	8.04	40.53	64.53	Natural inputs
13	El-Esh area (Eh)	27	28	30	33	37	42	5.0	23.50	42.30	6.70	8.50	40.60	62.30	Oil production
14	Gemsha area (G)	27	38	34	33	35	36	3.0	20.50	42.00	5.98	8.10	39.93	62.20	Oil production
15	Ras El-Behar area (RB)	27	43	45	33	32	58	2.0	20.50	42.50	7.50	8.30	40.10	63.00	Oil production

locations represent Quseir Harbour and Hamrawein Harbour and are exposed to mining and phosphate shipments. The tenth, the Kalawye area, is the control area. The fishing port of Safaga represents the eleventh site, and it is an area with many human activities. The Marine Science Station (NIOF Station) represents the twelfth site and is an area affected by the continental contribution through Wadi Abu Shaar Al-Qebli. The thirteenth, fourteenth and fifteenth sites represent the El-Esh, Gemsha and Ras El-Behar areas, respectively, which are exposed to oil production. The exact location and GPS coordinates, as well as the impact levels of the study sites, were determined and are provided in Table 11.1. The following is a brief description for these sites:

Haleib area is the last city in the southeastern region on the Egyptian coast of the Red Sea, 165 km south of Shalatein City. It is located between latitude 22°11′48.90″N and longitude 36°40′51.64″E. This site is affected by a desalination plant which discharges about 1,500 m^3/day of saline water into the marine environment. The desalination plant and the increase in the percentage of salts in the water discharged to the marine environment and its impact on the coastal area have led to the loss of biodiversity in the marine environments in the region near the drainage (Figure 11.2a, b).

Abu-Ramad area is located in the extreme south of the eastern desert near the border of Sudan and lies in the southern triangle (Halaib-Shalatein-Abu Ramad). It directly overlooks the Red Sea, about 125 km south of the city of Shalatein. This site is affected by a desalination plant which discharge about 1,800 m^3/day of saline water. The area near the discharge of saline water was impacted and characterized by the deterioration of the marine environment, especially coral reefs, where a type of green algae blooms in abundance, leading to the death of coral reefs (Figure 11.2c).

The Shalatein site represents a semi-closed area and is located 700 km south of Hurghada. This area is shallow and crowded with ships and boats, and the suction pump room of the main Shalatein desalination plant lies on a metal jetty in the middle of the harbour, in addition to the outlet pipeline at the middle beach (Figure 11.2d).

The Marsa Humira area is located on the shore of Marsa Humira at the southern part of the Egyptian Red Sea about 30 km north of Shalatein City. It affected by a desalination plant which discharges about 300 m^3/day of saline water into the marine environment. This site was characterized by a low diversity of coral and marine organisms. Terrigenous sediments have been transported to the marine environment by Wadi Humira, especially in the northern part of the area. The coastal zone has a gentle slope, creating a wide sabkha with accumulations of sand in the form of small dunes covered by some taller plants. Furthermore, the northern part of the coast has a high density of mangroves. Most sediment samples have dark to black colours due to the high contribution of terrigenous materials that are iron enriched and heavy elements. (Figure 11.2e, f).

Wadi El-Gemal is one of the more famous Red Sea wadis, characterized by an arid climate and dominated by hot, rainless summers and mild winters. Most of the precipitation occurs as heavy showers of short duration, resulting in flash floods during the winter season between October and February. A huge amount of rainwater can percolate and replenish the groundwater reservoir if the runoff is controlled (Mansour 2003). The source rocks of Wadi El-Gemal are composed of a complex of granites, diorites and green breccia (El-Mamoney 1995), with the sediments along this wadi having a relatively large grain size, the result of the violent water flow during heavy torrents. Two prominent items, the palm trees and the mangroves, distinguish the shore area of

FIGURE 11.2 Study areas: General view of the beach and nature of the marine environment, Haleib area (a, b); general view of the marine area of Abu-Ramad (c); general view of the Shalatein area (d); impact of materials contributed by wadis in the Marsa Humira area (e, f); beach area and collection of sediment by SCUBA diving at the Wadi El-Gemal area (g, h); Marina Marsa Alam under construction (i); beach area of Abu-Dabab (j); general view of Quseir Harbour (k); phosphate shipment at Hamrawein Harbour (l); general view and nature of marine environment in the Kalawye Reefs area (m, n); nature of marine environment of fishing Port Safaga (o); photo underwater of *Strombus tricornis* at NIOF Station (p); oil production in the El-Esh and Gemsha areas (r).

this wadi. The rocky beach drops abruptly into deep water. This area is covered by seagrasses and biogenic sand (Figure 11.2g, h).

Marina Masa Alam is characterized by a wide intertidal zone of about 500–1,000 m. The very gentle slope of this area creates a wide sabkha basin and a wider tidal flat. The beach is rocky with raised reefs in some sites, and the intertidal zone is followed by a wide tidal flat. This region has been affected by many human activities, including fishing and tourist boats, in addition to the continental contribution through Wadi El-Alam (Figure 11.2i).

Abu Dabab is a bay surrounded by raised beaches to both the north and south situated 30 km north of Marsa Alam. The Abu Dabab area lies off Wadi Abu Dabab, which extends about 17.5 km into the Eastern Desert. The marine area is relatively shallow, and the depth increases seaward, where the nearshore zone is flanked with coral reefs on the southern and northern sides while the middle has a large seagrass bed (Figure 11.2j).

Quseir Harbour is located in Quseir City and is one of the oldest harbours on the Egyptian Red Sea coast. This harbour lies in a small bay at the mouth of Wadi Ambaji. Most sediment samples have a brown colour due to phosphate shipment operations. The marine environment deterioration in the Quseir area includes the spread of algal blooms, dense seagrasses, coral bleaching and declining productivity (Figure 11.2k).

El-Hamrawein Harbour, located about 20 km north of Quseir City, is now the largest port for shipping phosphate ore on the Red Sea coast. Terrigenous sediments have been transported to the marine environment through this wadi. Vegetated coastal dunes and sabkhas occupy the lowland areas at its mouth. The supratidal area represents a sabkha area while the shoreline is occupied by fine to medium beach sands. The tidal flat, which is very narrow, extends smoothly and slopes gently seaward. Most sediment samples have a brown colour due to phosphate shipment operations (Figure 11.2l).

The Kalawye Reef area lies about 30 km to the south of Safaga City and is kidney shaped. The plentiful water drained from the wadis in the area necessitated construction of a tunnel under the asphaltic highway (Safaga-Quseir Road) so torrential water does not damage the highway. Before the shoreline, sand accumulates in the form of small dunes covered by some vegetation. The gentle slope of this area creates an extremely wide tidal flat, and the intertidal zone of this area reaches about 300 to 350 m offshore. Prominent features which characterizes the area seaward of the tidal flat include seagrass and some coral reef patches, namely fringing reefs. The coral reef in this area is healthy (Figure 11.2m, n).

Fishing Port Safaga represents a part of the Grand Port of Safaga and is located directly south of the Safaga Port for freight and passengers. The intertidal zone in the fishing port is extremely narrow. The coastal zone along the Red Sea at Safaga Town is the site of industrial development and the future establishment of dense recreational resorts. It includes Safaga Harbour, on which international trade depends. Therefore, there is an urgent need for studies in different fields to help managers identify anthropogenic impacts and better assess the need for remediation by detecting any changes in existing levels from future activities (Figure 11.2o).

NIOF Station is a marine science station area affected by the continental contribution of Wadi Abu Shaar Al-Qibli. The beach sediments are generally coarse sand, mixed with significant numbers of biogenous fragments. The intertidal zone is a rocky tidal flat covered with a thin layer of soft deposits contributed by Abu-Shaar Al-Qibili (Figure 11.2p).

The El-Esh area is about 30 km to the north of Hurghada and was subjected to some impact from drilling for oil, where an artificial tongue of sediments enters the sea for more than 500 m with pipelines for oil production. There are two potential sources of terrestrial materials entering the marine environment: anthropogenic activities and sediment input from Wadis El-Esh (Figure 11.2q).

The Gemsha area is a famous area along the Egyptian Red Sea coast located about 60 km north of Hurghada City. It used to be a mining area for the production of sulfur; now, it is an oil production area for a number of oil fields. The tidal flat area is covered by muddy sand with dead and destroyed hard coral due to the oil production, and samples of marine sediments are dark grey (Figure 11.2r).

The Ras El-Behar area, located on the Gulf of Suez about 70 km from Hurghada, is characterized by its wide tidal zone; it is an oil production area and one of the more famous areas on the Gulf of Suez. Most of the area is covered with sandy mud with hard coral reefs that are dead and destroyed due to oil production and drilling.

11.2.2 Methodology

The studied samples of marine sediments and shells of *Strombus tricornis* Humphrey, 1786 were collected from 15 sites on the Egyptian coast of the Red Sea from the Haleib area in the south to Ras el-Behar on the Gulf of Suez in the north (Table 11.1; Figure 11.1) during the period from 2012 to 2018. Using SCUBA diving and snorkeling, the shells of *Strombus tricornis* and marine sediments were collected from the study sites. The shells of *Strombus tricornis* were cleaned, and the soft tissue was removed on the beach of the sampling site. Hydrographic parameters such as temperature, salinity, dissolved oxygen (DO), pH, total dissolved salts (TDS) and specific conductivity (SPC) were measured at the surface of the water in the studied localities using the HANNA Hi 9828 instrument (PH/ORP/EC/DO; Table 11.1). The empty shells were soaked in water for 15 to 30 minutes in order to kill any clinging algae, then left to air dry. The marine sediment samples were gently washed several times with distilled water to remove soluble salts, then spread on glass sheets and left to air dry. About ten grams of shell parts and sediments were ground to a powder using an agate mortar (Retsch Mortar), passed through a 80 mesh sieve and kept in dry, clean bags waiting for analysis. 0.5 grams of the prepared ground sample were completely digested in a Teflon cup by using a mixture of concentrated nitric (HNO_3), perchloric ($HClO^4$) and hydrofluoric acids (HF) with the ratio 3:2:1, respectively, following Chester et al. (1994). Acids were slowly added to the dried samples and left overnight before heating. Samples were heated to a temperature of approximately 200°C, then left to cool and filtered to get rid of the nondigested parts. The solution was justified to a volume of 25 ml; then the concentration of elements was determined by the atomic absorption spectrophotometry (AAS) technique, using GBC-932 ver. 1.1 with detection limits 0.01 mg g^{-1} of the National Institute of Oceanography and Fisheries. Results were expressed in mg g^{-1}.

11.3 RESULTS AND DISCUSSION

11.3.1 Hydrographic parameters

11.3.1.1 Regional variations of physical parameters

Regarding temperatures, we would like to clarify that samples were collected during the period from 2012 to 2018 in different seasons of the year. The highest temperatures were recorded in the Haleib area, while Gemsha and Ras El-Behar recorded the lowest temperatures. The water salinity and TDS variations are homogeneous in the studied localities, where the highest values of salinity and TDS are 43.73 psu and 41.73 ppt, respectively, in the Shalatein area while the lowest values are 39.01 psu and 40.80 ppt in Quseir Harbour (Table 11.1). The high values of water salinity and TDS in the Shalatein area are affected by the semi-enclosed, shallow water and the presence of the desalination plant outlet. The salinity and TDS recorded low values in Quseir, Marina Marsa Alam and Kalawye areas compared to the other three studied areas, resulting from stronger mixing of water with the seawater, in addition to association with natural causes such as the shallowness, the local current system, and the arid climate. The main reason is the active valleys in these places, which have a great impact on the tidal zone. The pH fluctuated within narrow limits in the areas under study; it averaged between 7.87 and 8.5 (Table 11.1). The maximum values of pH were recorded in Quseir Harbour while low values were recorded in the Shaltein area; the differences in pH values were the result of human activities, such as sanitation work in the region and shipping of phosphate ore.

FIGURE 11.3 *Strombus tricornis* Humphrey, 1786: underwater photographs showing general or dorsal view (a) and aperture or ventral view (b).

11.3.2 Systematic description, distribution and ecology of *Strombus tricornis*

Genus: *Strombus* Linné

1758, *Strombus tricornis* Humphrey, 1786, pl. 1 figs. a, b
1979, *Strombus tricornis* Humphrey – Gert Lindner, p. 144, pl. 17
1980, *Strombus tricornis* Humphrey – Oliver, p. 70
1982, *Strombus tricornis* Humphrey – *The Macdonald Encyclopedia of Shells*, entry 288
1982, *Strombus tricornis* Humphrey – Abed, p. 273, pl. VII, figs. 1a, b
1984, *Strombus tricornis* Humphrey – Sharabati, pl. 9, figs. 4, 4a–b
1988, *Strombus tricornis* Humphrey – Ebaid-Alla, p. 109, pl. XIV, figs. 3,4
1991, *Strombus tricornis* Humphrey – Wye, *The Encyclopedia of Shells*, p. 73
2004, *Strombus tricornis* Humphrey – Madkour, p. 269, pl. 15, figs. 1, 2

TABLE 11.2 Shell measurements for *Strombus tricornis* in the study areas.

LOCATION	*LENGTH (CM)*	*WIDTH (CM)*	*HEIGHT (CM)*
Haleib area	13.56	11.5	6.89
Abu-Ramed area	11.56	9.89	6.56
Shaltein area	13.56	12.22	7.45
Marsa Humira area	10.89	9.78	5.78
Wadi El-Gemal	10.56	9.45	5.65
Marsa Alam Marina	12.56	11.35	6.45
Abu-Dabab	9.56	8.56	4.89
Quseir Harbour	10.56	9.48	5.68
El-Hamreawein Harbour	10.6	9.8	5.3
Kalawe Reefs area	11.45	9.56	6.43
Fishing Port Safaga	12.45	11.25	6.43
NIOF Station	14.56	12.58	7.45
El-Esh area	12.78	10.89	6.87
Gemsha area	10.58	8.59	5.68
Ras El-Behar area	11.85	8.5	5.5

11.3.2.1 Description

Strombus tricornis have large (4½ in), very thick, heavy, and solid shells with a fairly high spire and knobbed above the suture; the body whorl expands near the aperture and partly enfolds some of the preceding whorls; the suture is barely incised; the aperture edge is broadened and white, tending to a brightish pink at the margins; it has a strong stromboid notch and heavy columella, a parietal wall and a yellow outer lip (Figure 11.3a, b).

11.3.2.2 Distribution and ecology

This heavy and bulky species is common in very shallow inner reef flats on sandy bottoms and feeds on algae, seagrasses and detritus. It is endemic to the Red Sea and Gulf of Aden (Sharabati 1984). It occurs in rare-to-frequent numbers at the studied localities due to the collection of this species for sale, which results in destroying the ecosystem in the marine environment. Measurements were made of all the shells with respect to the length, width and height of *Strombus tricornis* in all study areas, as shown in Table 11.2.

11.3.3 Concentration of heavy metals

The main idea of this study is to determine the relationship between the concentrations of heavy metals in the shells of *Strombus tricornis* and the associated marine sediments and to see whether the effect is natural through continental contributions from wadis or through different human activities in the study areas. The results of the analysis of the shells of *Strombus tricornis* and associated marine sediment from the studied localities are shown in Tables 11.3 and 11.4.

11.3.3.1 Iron and manganese

Iron plays an important role as an essential element in all living systems from invertebrates to humans, but increasing iron

TABLE 11.3 Heavy metal concentrations in *Strombus tricornis* Humphrey, 1786, at the studied locations.

		SP.	*HEAVY METALS*						
NO.	*LOCATION*		*FE**	*MN**	*ZN**	*CU**	*PB**	*NI**	*CD**
1	Haleib area (H)	*Strombus tricornis* Humphrey, 1786	559.46	15.53	12.31	5.32	8.52	21.35	0.21
2	Abu-Ramad area (AR)		348.45	12.35	9.13	6.52	7.35	18.31	0.19
3	Shaltein area (SH)		445.68	15.39	8.75	3.93	4.31	15.29	0.14
4	Marsa Humira area (MH)		789.56	32.53	12.56	9.35	11.69	22.59	0.41
5	Wadi El-Gemal (WG)		595.31	22.89	6.45	5.32	8.31	22.51	0.35
6	Marsa Alam Marina (MM)		394.51	20.52	11.52	2.53	2.31	9.51	0.45
7	Abu-Dabab area (D)		549.56	19.91	9.51	5.32	4.51	12.51	0.32
8	Quseir Harbor (QH)		1392.81	25.53	4.67	1.95	1.85	15.67	0.79
9	El-Hamreawein Harbour (HH)		1275.31	33.51	5.91	3.29	2.58	13.98	1.28
10	Kalawye Reefs (KR)		422.52	21.52	10.51	2.95	3.52	12.45	0.12
11	Fishing Port Safaga (S)		1198.53	31.98	3.59	1.59	14.59	12.95	1.89
12	NIOF Station area (NB)		958.13	26.95	2.98	1.19	7.56	4.59	1.29
13	El-Esh area (Eh)		443.15	9.53	7.58	9.53	5.76	14.41	0.73
14	Gemsha area (G)		539.31	10.51	5.31	2.31	11.51	12.31	1.091
15	Ras El-Behar area (RB)		495.91	11.31	7.51	3.59	9.98	11.31	1.051

* = values in ppm

TABLE 11.4 Heavy metals in sediment samples at the studied locations.

		SAMPLES	*HEAVY METALS*						
NO.	*LOCATION*	*NO.*	*FE**	*MN**	*ZN**	*CU**	*PB**	*NI**	*CD**
1	Haleib area (H)	3	1185.20	353.11	65.31	22.51	31.51	22.51	0.07
2	Abu-Ramad area (AR)	3	875.60	79.50	19.51	12.15	8.79	12.11	0.04
3	Shaltein area (SH)	4	959.31	95.31	22.31	8.53	10.31	11.51	0.31
4	Marsa Humira area (MH)	3	5996.30	650.90	72.31	33.51	38.21	63.21	0.03
5	Wadi El-Gemal (WG)	5	2936.20	540.80	80.14	35.20	39.52	65.79	0.09
6	Marsa Alam Marina (MM)	3	891.51	102.31	15.91	3.53	5.31	2.98	0.52
7	Abu-Dabab area (D)	3	2759.30	339.90	75.21	25.39	45.53	55.91	0.07
8	Quseir Harbor (QH)	4	13522.80	925.10	25.91	5.31	15.39	29.59	1.09
9	El-Hamreawein Harbour (HH)	5	1885.30	503.41	75.21	19.40	40.59	44.52	2.42
10	Kalawye Reefs (KR)	3	665.10	97.31	12.65	2.15	1.53	0.98	0.79
11	Fishing Port Safaga (S)	4	12431.51	985.30	18.31	10.85	25.39	35.91	2.53
12	NIOF Station area (NB)	3	2385.20	65.97	15.37	4.59	15.34	5.93	0.19
13	El-Esh area (Eh)	3	1957.30	253.40	14.91	85.41	19.98	12.19	1.98
14	Gemsha area (G)	3	1864.52	155.30	15.31	19.51	25.38	11.51	2.54
15	Ras El-Behar area (RB)	3	1187.87	45.91	12.31	4.12	18.91	0.95	1.08

* = values in ppm

in the marine environment is reflected in bioaccumulation in marine organisms such as bivalves, gastropods, coral reefs, fish etc. The iron concentration in the shell of *Strombus tricornis* ranges from 1392.81 mg g^{-1} at Quseir Harbour to 422.52 mg g^{-1} at Kalawye Reefs. Similarly, the marine sediments samples recorded the highest values of iron content in Quseir Harbour and the lowest values in the Kalawye Reefs area (Figure 11.4). The iron content has higher values in shells of *Strombus tricornis* and marine sediments in Quseir Harbour, Fishing Port Safaga and Hamrawein Harbour than in the other studied localities (Figure 11.4). Generally, the sites with multiple human activities recorded high values for iron in marine sediment the and shells of *Strombus tricornis* compared to areas with natural activities that were affected by the continental contribution. Ramadan and Shata (1993) stated that the concentration of iron in bivalve shells depends to some extent on the iron concentration in the food supply and on the organism's growth rate. Madkour (2005) recorded iron concentration in *Tridacna maxima* varying from 216.4 mg g^{-1} at Abu-Ghusun Lagoon to 1286.4 mg g^{-1} at Safaga Harbour. The iron content in molluscan shells is between 523 mg g^{-1} in *Lambis truncata* at Quseir Harbour and 1644 mg g^{-1} in *Strombus tricornis* in the El-Esh area (Youssef et al. 2017). This variation indicates the difference in the bioaccumulation of iron in the shells of *Strombus tricotnis* in the different localities under study. Corrosion of pipelines, steel marinas and old ships are possible sources of increased Fe concentration in the studied areas. In the same manner, the Mn level in shells of *Strombus tricornis* is from 33.51 mg g^{-1} at Hamrawein Harbour to 9.53 mg g^{-1} in the El-Esh area while marine sediment recorded the highest values

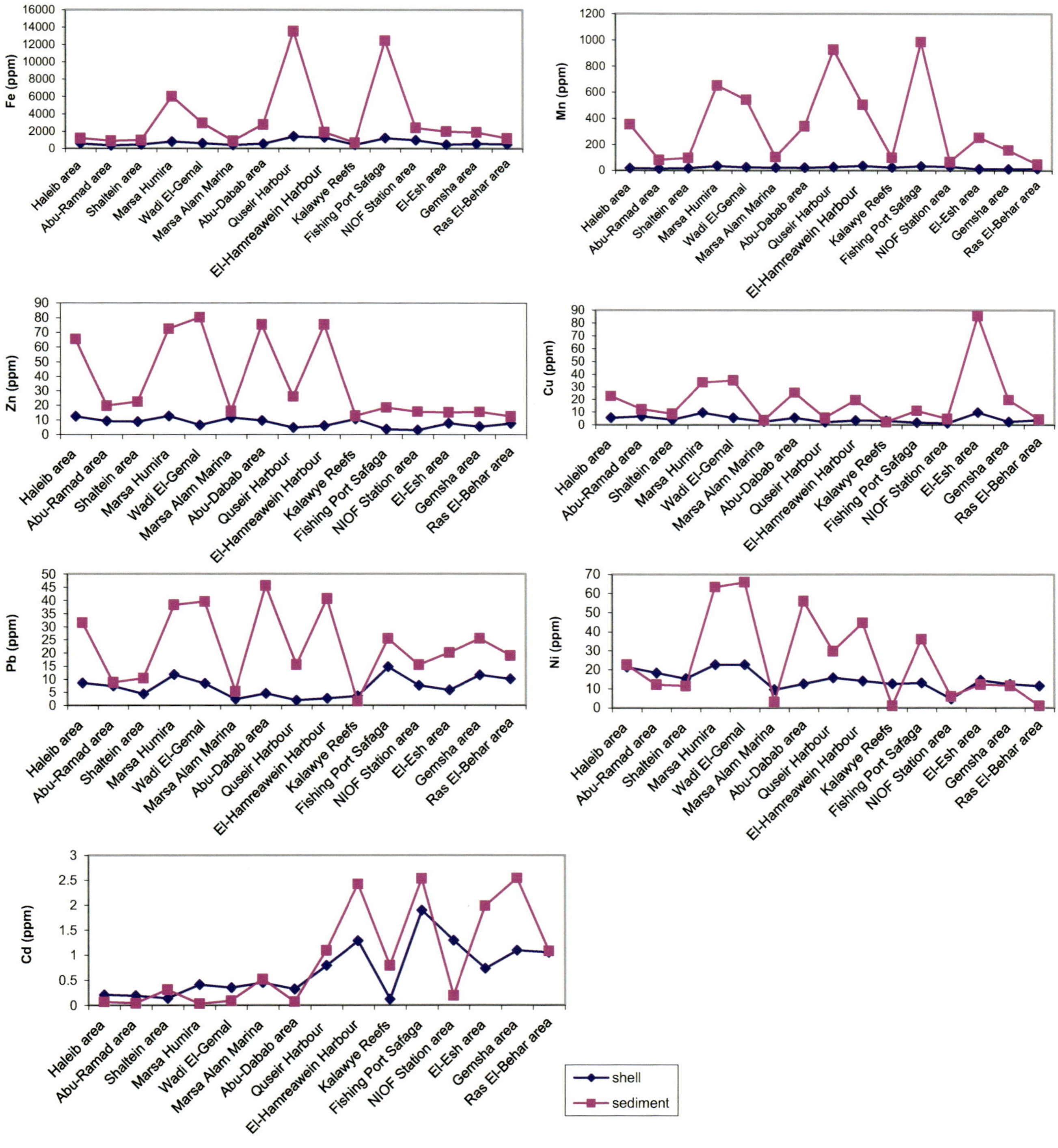

FIGURE 11.4 Relation between concentrations of heavy metals in shells of *Strombus tricornis* and in marine sediments from the studied localities.

in Fishing Port Safaga and Quseir Harbour, and the lowest values were recorded in the NIOF Station and Kalawye Reefs areas (Figure 11.4). Similar to Fe, anthropogenic sites and natural input area have higher Mn values than the control area and some natural input areas (Tables 11.3 and 11.4). Madkour (2004) attributed manganese transported to the marine environment to anthropogenic inputs from the same sources as iron. Because Mn is related to and associated with iron in the conditions of accumulation and dissolution, it seems to be associated with iron in its distribution in the studied localities. Abd El-Aal and Hassan (1988) found that the average value of Mn content in the shells collected from Al-Ghardaqa is 0.061%. Ramadan and Shata (1993) stated that the manganese level in *Anadara diluvii* varies from 12 mg g^{-1} to 50 mg g^{-1} in shells collected from the beach of Port Said. Shata and Hassan (2000) recorded that the Mn content in bivalve shells collected from the Eastern Harbour of Alexandria is 24.07, 44.38 and 40.4 mg g^{-1} in the families *Arcidae*, *Cardidae* and *Veneridae*, respectively. Ziko et al. (2001) reported that the Mn concentration in recent shells ranges from 9 mg g^{-1} in *Clanculus* (*Clanculus*) *pharaonius* at Gebel Zeit to 171 mg g^{-1} in *Turbo* (*Batillus*) *radiatus* at the same place. The Mn content in giant clam shells (*Tridacna Maxima*) ranges from 29.9 mg g^{-1} at Abu-Ghusun Lagoon to 65.2 mg g^{-1} at Safaga Harbour (Madkour 2005). Youssef et al. (2017) recorded that the Mn level in molluscan shells is from 8.2 mg g^{-1} in *Lambis truncata* at the El-Esh area to 15 mg g^{-1} in *Tridacna gigas* at Safaga Harbour. The correlation analysis observed that iron is highly positively correlated with Mn in shell samples and marine sediment samples in the studied localities (Table 11.5). Iron is negatively correlated with zinc and copper while it shows a positive correlation with cadmium (Table 11.5).

11.3.3.2 Zinc and copper

Zinc plays an important role as an essential trace element in all living systems from bacteria to humans (Merian 1991). The Zn concentration in shells of *Strombus tricornis* ranges from 12.56 mg g^{-1} at Marsa Humira to 2.98 mg g^{-1} at the NIOF Station (Figure 11.4). Shells of *Strombus tricornis* in natural input areas recorded the maximum values of Zn compared to the areas with multiple human activities (Figure 11.4) while the highest value (80.14 mg g^{-1}) of Zn content in marine sediments was recorded in the Wadi El-Gemal area and the lowest (12.31 mg/g−1) in the Ras El-Behar area. Marine sediments in the Wadi El-Gemal, El-Hamrawein, Abu-Dabab and Marsa Humira areas have very high values of Zn compared to those in the other studied localities. Thus, it is evident that the regions with continental contributions through the estuaries of the wadis recorded higher values of zinc content in the shells of *Strombus tricornis* and also samples of marine sediments than areas with human activities. Mansour et al. (2005) reported that zinc is mainly co-precipitated with calcium carbonate and substitute calcium to form isomorphous zinc carbonate. The uncontaminated sites recorded higher values of Zn than the human impact sites. This is probably due to the influence of terrigenous sediments on the marine environment. Moreover, in harbours, a large amount of chemical material (e.g., zinc sulphate) is usually used in marine painting processes, and zinc sulphate is widely distributed along the coast of Hurghada Harbour.

TABLE 11.5 Correlation coefficients between heavy metal concentrations in *Strombus tricornis* samples and associated marine sediments in the study areas.

	FE*	MN*	ZN*	CU*	PB*	NI*	CD*
Strombus tricornis samples							
Fe*	1						
Mn*	**0.76**	1					
Zn*	**−0.60**	−0.20	1				
Cu*	−0.40	−0.23	**0.55**	1			
Pb*	0.01	0.02	−0.17	0.11	1		
Ni*	−0.08	0.03	0.45	**0.63**	0.20	1	
Cd*	**0.66**	0.37	**−0.78**	−0.47	0.42	−0.48	1
Marine sediment samples							
Fe*	1						
Mn*	**0.89**	1					
Zn*	−0.01	0.38	1				
Cu*	−0.10	0.10	0.25	1			
Pb*	0.13	0.45	**0.86**	0.39	1		
Ni*	0.37	0.67	**0.87**	0.30	**0.85**	1	
Cd*	0.28	0.28	−0.29	0.17	0.06	−0.12	1

Copper is an essential and potentially toxic element (Merian 1991). The Cu content in the shells of *Strombus tricornis* ranges from 1.19 mg g^{-1} at the NIOF Station to 9.53 mg g^{-1} at El-Esh (Figure 11.4). These contents are comparable to those of the marine sediments, in which Cu ranges from 2.15 mg g^{-1} at Kalawye Reefs to 85.41 mg g^{-1} at El-Esh (Figure 11.4). The concentration of Cu in marine sediments at El-Esh is 11 times higher than in *Strombus tricornis*. Moreover, the natural sites recorded high values of Cu content in *Strombus tricornis* and marine sediments compared to the human impact sites, except in the El-Esh area. This is probably due to the influence of the terrigenous flux from wadis, derived from the nearby volcanic and ultramafic rocks. There is a positive correlation between Zn and Cu and Pb in shell samples and associated marine sediments in the study areas (Table 11.5). The observed differences between correlation coefficients refer to the change in the source of these metals in the areas under study.

11.3.3.3 Lead and nickel

Lead is the leading toxic metal in the marine environment. The highest concentration of Pb in shells of *Strombus tricornis* is 14.59 mg g^{-1} at Fishing Port Safaga while the lowest in shells of *Strombus tricornis* is 1.85 mg g^{-1} at Quseir Harbour (Figure 11.4). It is obvious that concentrations of lead content in the shells of *Strombus tricornis* show high values in the areas with human activities, except for the harbours of Quseir and El-Hamrawein and Marina Marsa Alam. High values were also recorded in the areas affected by the continental contribution through the wadis. On the other hand, in marine sediments samples, Pb ranges from 40.59 mg g^{-1} at El-Hamreawein Harbour to 1.53 mg g^{-1} at Kalawye Reefs (Figure 11.4). Quite the opposite of the low values that were recorded in *Strombus* shells in El-Hamreawein Harbour, marine sediments samples recorded the highest values compared to other areas, and this shows that the processes of extracting heavy elements for marine organisms differ from one place to another. There is a similarity in that areas with oil production recorded high values in the *Strombus* shells as well as the marine sediment samples. Milliman (1974) reported that lead tends to be concentrated relative to calcium. He mentioned that lead in carbonates and most organisms accumulates over 100 times relative to calcium. In El-Esh, Gemsha and Ras El-Behar, oil production and export activities provide pollution by hydrocarbons. On the other hand, sites affected by the continental contribution through the estuaries of the wadis also recorded high lead values in the shells of *Strombus* as well as in the samples of marine sediments.

The Ni content in the shells of *Strombus tricornis* varies from 4.59 mg g^{-1} at the NIOF Station to 22.59 mg g^{-1} at Marsa Humira while, in marine sediments, it ranges from 0.95 mg g^{-1} at Ras El-Behar to 65.79 mg g^{-1} at Wadi El-Gemal (Figure 11.4). Clearly, the nickel content values were higher in *Strombus* shells, as well as in marine sediment samples affected by the continental contribution from wadis, than in areas with human activities. This indicates that the main source of nickel is the terrestrial sediments, either naturally from wadis or by human activities in anthropogenic sites. Mansour et al. (2005) reported that the ability of invertebrates to adsorb metals is largely dependent on the physical and chemical characteristics of the metal as well as the seawater in which they live. On the other hand, De Carlo and Spencer (1995) found that Ni and Co do not display trends indicative of a large anthropogenic contribution to the sediments.

11.3.3.4 Cadmium

Cd is not an essential element for plants, animals or human beings (Merian 1991). The occurrence of cadmium in the marine environment is rare. Therefore, the impact of Cd on the environment is small. The highest concentrations of Cd content in the shells of *Strombus tricornis* and marine sediments are recorded at Fishing Port Safaga in the Safaga area at 1.89 mg g^{-1} and2.53 mg g^{-1}, respectively, while the lowest is at Kalawye Reefs (0.12 mg g^{-1}), and in the marine sediments, the Haleib and Abu-Dabab areas recorded the lowest values (0.07 mg/g^{-1}) (Figure 11.4). The anthropogenic sites recorded the highest concentrations of Cd in the shells of *Strombus tricornis* and marine sediments compared with those in the natural sites. In general, the cadmium content had high values in the *Strombus* shells and marine sediment samples compared to the areas affected by the continental contribution through the estuaries of the wadis. Cadmium sources in the anthropogenic sites come via phosphate shipment, ship traffic and coal packing, especially in the Safaga area and Quseir Harbour, and oil seepage across pipelines, in addition to drilling fluids in the oil fields at El-Esh, Gemsha and Ras El-Behar. The correlation matrix (Table 11.5) shows Cd is negatively correlated with Zn while Cd shows a positive correlation with iron, manganese and lead in the study areas. Generally, Cd is of anthropogenic origin in the study areas.

11.3.4 Cluster and component analysis

A hierarchical cluster analysis subdivided the studied heavy metals in *Strombus tricornis* and the marine sediments into two different clusters. The first contained only Fe, and the second cluster included all other recorded heavy metals (Mn, Cu, Cr, Zn, Ni, Cd and Co) (Figures 11.5 and 11.6).

A principal component analysis (PCA) was used to detect possible sources of the heavy metals in the studied sea shell and marine sediments. In the *Strombus tricornis* shell, the number of variables was divided into three principal components, covering 85.61% of the total variance (Table 11.6). The first component

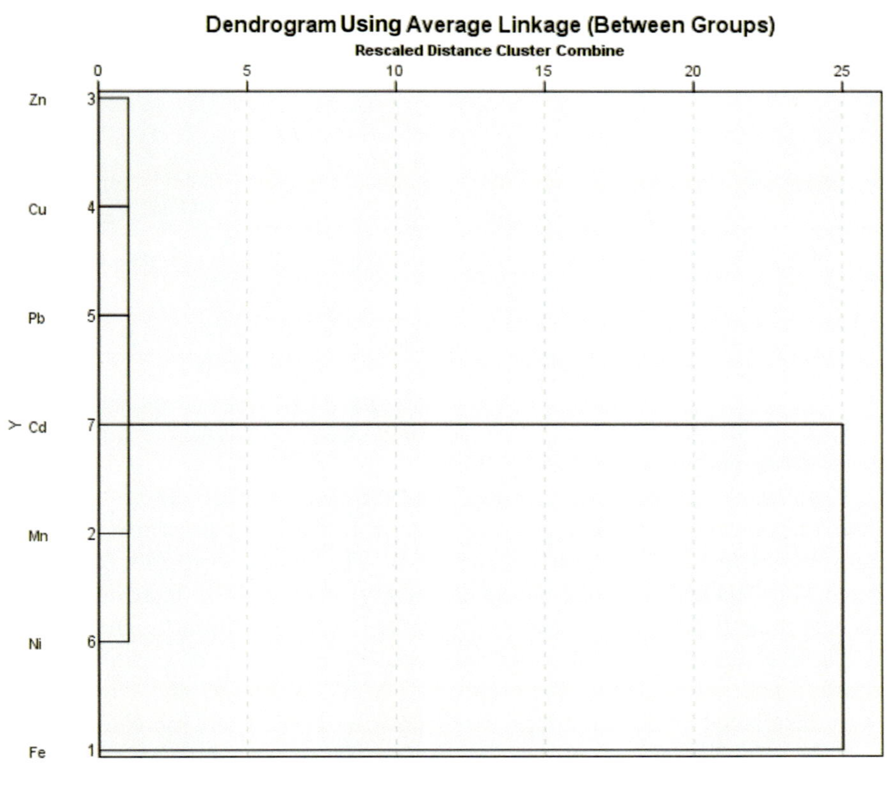

FIGURE 11.5 Dendrogram of hierarchal cluster analyses of seven metals in the shells of *Strombus tricornis*.

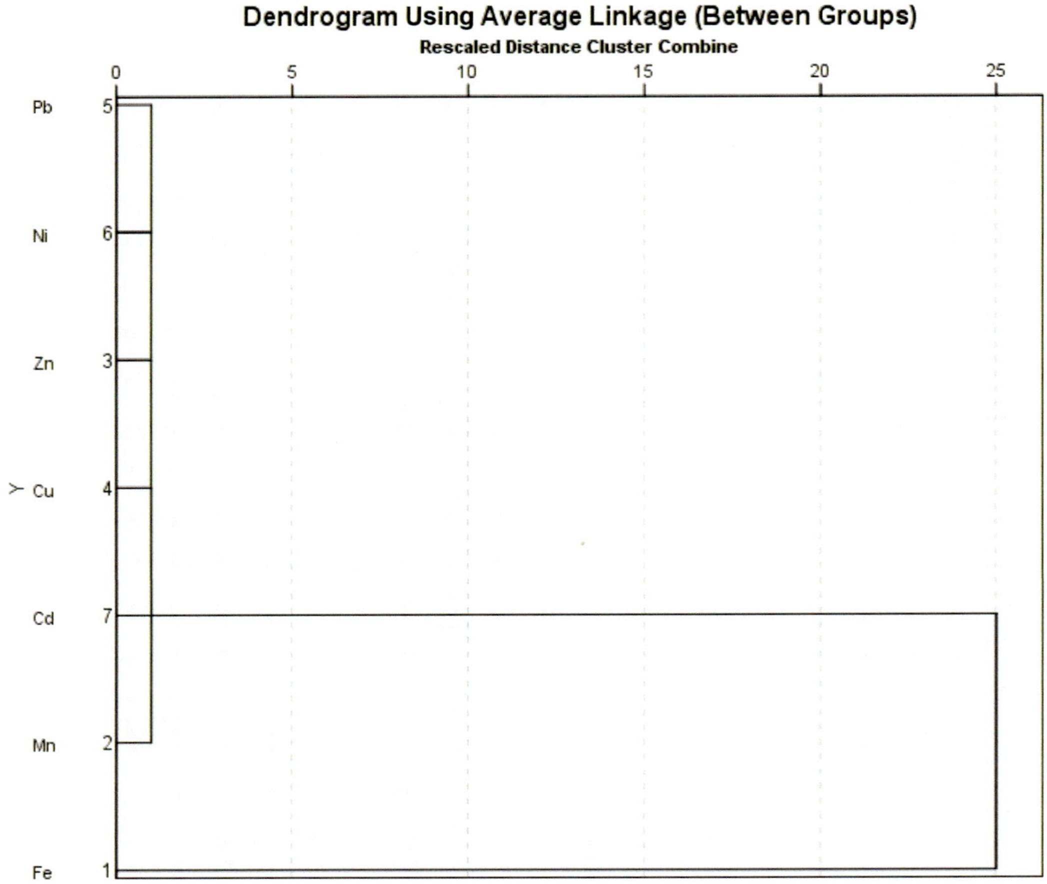

FIGURE 11.6 Dendrogram of hierarchal cluster analyses of seven metals in the sediments of the study area.

TABLE 11.6 Extraction method: Principal component analysis of *Strombus tricornis* samples. (Highly positive loadings are in bold.)

COMPONENT MATRIX[a]			
	COMPONENT		
	1	2	3
Fe	−.281	**.915**	.121
Mn	.015	**.930**	−.030
Zn	**.713**	−.350	−.382
Cu	**.822**	−.227	.106
Pb	.127	−.030	**.948**
Ni	**.900**	.128	.129
Cd	−.628	.439	**.577**
% Variance	35.490	29.798	20.322
Cumulative %	35.490	65.288	85.610

[a] 3 components extracted

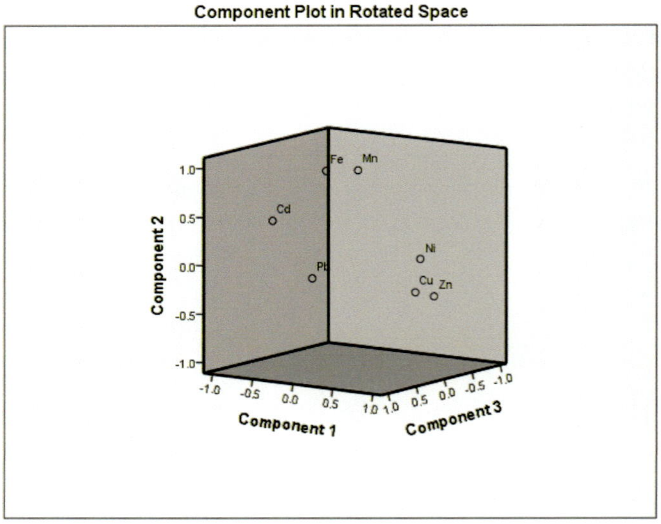

FIGURE 11.7 Three component plots using the varimax method with Kaiser normalization in the shells of *Strombus tricornis.*

covered 35.49% and presented strongly positive loadings for Zn, Cu and Ni (0.713, 0.822 and 0.900, respectively). The loading indicates that this group of elements may be homogeneous. The second component covered 29.80% and produced positive loadings for Fe (0.915) and Mn (0.930). The third component covered 20.32% and indicated strongly positive loadings for Pb (0.948) and Cd (0.577). These components can be explained using a varimax method with Kaiser normalization. The metals in the component plot were distributed into three groups, corresponding to the results of the PCA (Figure 11.7).

In the marine sediments, the number of variables was divided into three principal components, covering 85.61% of the total variance (Table 11.7). The first component covered 42.74% and presented strongly positive loadings for Mn, Zn, Pb and Ni (0.774, 0.832, 0.886 and 0.961, respectively). The loading indicates that this group of elements may be homogeneous. The second component covered 29.89% and produced positive loadings for Mn (0.744), Zn (0.832), Pb (0.886) and Ni (0.961). The third component covered 17.43% and indicated strongly positive loading for Cu (0.768) and Cd (0.650). The metals in the component plot were distributed into three groups, corresponding to the results of the PCA (Figure 11.8).

TABLE 11.7 Extraction method: Principal component analysis of marine sediments samples. (Highly positive loadings are in bold.)

COMPONENT MATRIX[A]			
	COMPONENT		
	1	2	3
Fe	.478	**.813**	−.235
Mn	**.774**	**.594**	−.134
Zn	**.832**	−.487	−.135
Cu	.371	−.232	**.768**
Pb	**.886**	−.279	.149
Ni	**.961**	−.144	−.124
Cd	.047	**.625**	**.650**
% Variance	42.736	29.890	17.432
Cumulative %	42.736	72.626	90.058

[a] 3 components extracted

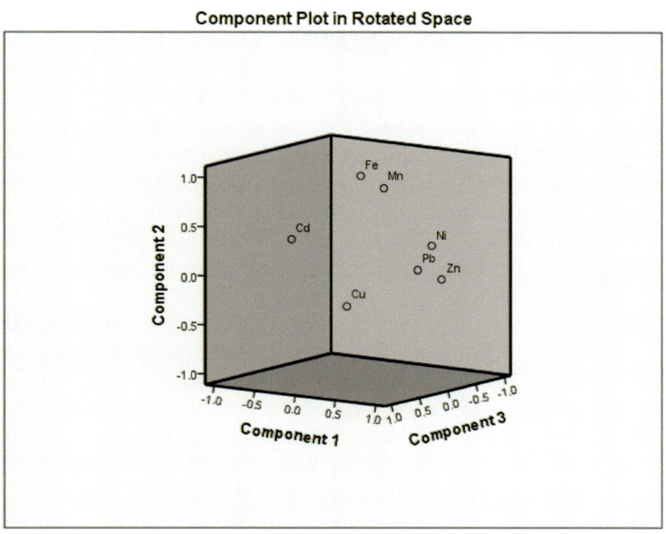

FIGURE 11.8 Three component plots using the varimax method with Kaiser normalization in the sediments of the study area.

11.3.5 Mechanisms of heavy metal accumulation in the shells of *Strombus tricornis*

Clearly, there are differences and similarities with respect to the concentrations of heavy elements in the shells of *Strombus* and marine sediments. For example, there are some heavy metals such as cadmium that recorded high values in shells and marine sediments in areas of human activity while lower values were recorded in areas affected by the continental contribution through wadi estuaries. Therefore, we can say that there is a relationship between the increase in the concentration of heavy elements in the shells of *Strombus* and marine sediments as this shell organism lives adjacent to those sediments in the marine environment. Huanxin et al. (1999) stated that in an environment affected by contaminant, both oyster tissue and shells have an opportunity to adsorb heavy metals from suspended particles. These suspended particles may be either contaminated sediment or sediment controlled by the behaviour of heavy metals themselves and the physical and chemical conditions of the environment. Therefore, there are no clear relationships between concentrations of heavy metals in the shell of *Strombus* and those in sediment.

11.4 CONCLUSIONS

Analyses of the shells of *Strombus tricornis* and marine sediment samples revealed no distinct relationship between heavy metal levels in the shells of *Strombus* and the marine sediments in which they thrive, indicating that heavy metal accumulations in *Strombus tricornis* may not be directly or solely derived from marine sediments but from other sources, such as living or dead suspended particles and dissolved metals in the water. Heavy metal concentrations in the shells of *Strombus tricornis* and marine sediment samples recorded high values in the anthropogenic areas compared to the natural input and control areas. There are no clear relationships between concentrations of heavy metals in the shells of *Strombus* and those in sediment. This chapter provides baseline information on the anthropogenic impacts of environmental pollution. Hence, it could be useful for the management and sustainable development of the studied localities. This data can serve as a guideline for future researchers and environmental managers to identify future anthropogenic impacts at the study locations with respect to the studied metals and to better assess the need for remediation by monitoring changes in the existing levels.

ACKNOWLEDGEMENT

Some of the samples were collected based on a plan of the National Institute of Oceanography and Fisheries, Red Sea Branch. The authors wish to thank the team for participating in the present work.

REFERENCES

Abd El-Aal, A.A., and A.K. Hassan. 1988. Chemical composition and wall structure of some Gastropod shells from the Mediterranean Sea, Suez-Canal and Red Sea, Egypt. *Bull Inst Oceanogr Fish, ARE* 14, no. 3: 155–166.

Bryan, G.W. 1984. Pollution due to heavy metals and their compounds. In: *Marine Ecology: A Comprehensive, Integrated Treatise on Life in Oceans and Coastal Waters*. Ed. O. Kinne. London: Wiley-Interscience.

Cajaraville, M.P., Bebianno, M.J., Blasco, J., Porte, C., Saarasquete, C., and A. Viarengo. 2000. The use of biomarkers to assess the impact of pollution in coastal environments of the Iberian Peninsula: A practical approach. *Sci Total Environ* 247: 295–311.

Chester, R.F.G., Lin, F.J., and A.S. Basaham. 1994. Trace metals solid state speciation changes associated with the down-column fluxes of oceanic particulates. *J Geol Soc London* 151: 351–360.

Connell, D., Lam, P., Richardson, B., and R. Wu. 1999. *Introduction to Ecotoxicology*. Oxford: Blackwell Science Ltd., 71 p.

De Carlo, E.H., and K.J. Spencer. 1995. Records of lead and other heavy metal inputs to sediments of the Ala Wai Canal, O'ahu, Hawai'i. *Pac Sci* 49, no. 4: 471–491.

El-Mamoney, M.H. 1995. *Evaluation of Terrestrial Contribution to the Red Sea Sediments, Egypt*. PhD Thesis, Faculty of Science, Alexandria University, 146 p.

El-Sorogy, A.S., El Kammar, A., Ziko, A., Aly, M., and H. Nour. 2013. Gastropod shells as pollution indicators, Red Sea coast, Egypt. *J Afr Earth Sci* 87: 93–99.

El-Taher, A.M., and H.A. Madkour. 2011. Distribution and environmental impacts of metals and natural radionuclides in marine sediments in front of different wadies mouth along the Egyptian red sea coast. *Appl Radiat Isot* 69: 550–558.

Franca, S., Vinagre, C., Cacador, I., and H.N. Cabral. 2005. Heavy metal concentrations in sediment, benthic invertebrates and fish in three salt marsh areas subjected to different pollution loads in the Tagus Estuary (Portugal). *Mar Pollut Bull* 50: 993–1018.

Harayashiki, C., Marquez, F., Cariou, E., and I.B. Castro. 2020. Shell alterations resulting from coastal contamination and other environmental factors. *Environ Pollut* 265: 114881. http://doi.org/10.1016/j.envpol.2020.114881.

Huanxin, W., Lejun, Z., and B.J. Presley. 1999. Bioaccumulation of heavy metals in oyster (*Crassostrea virginica*) tissue and shell. *Environ Geol* 39, no. 11: 1216–1226. http://doi.org/10.1007/s002540000110.

Madkour, H.A. 2004. *Geochemical and Environmental Studies of Recent Marine Sediments and Some Invertebrates of the Red Sea, Egypt*. PhD Thesis, South Valley University, Qena, 317 p.

Madkour, H.A. 2005. Distribution and relationships of heavy metals in the Giant Clam (*Tridacna maxima*) and associated sediments from different sites in the Egyptian Red Sea coast. *Egypt J Aquat Res* 31, no. 2: 45–59.

Madkour, H.A. 2013. Impacts of human activities and natural inputs on heavy metal contents of many coral reef environments along the Egyptian Red Sea coast. *Arab J Geosci* 6: 1739–1752.

Madkour, H.A. 2015. *Detection of Damaged Areas Due to Tourism Development Along the Egyptian Red Sea Coast Using GIS, Remote Sensing and Foraminifera*. State of the Art, Nat Instit Ocean Fish, Red Sea Branch, 148 p.

Madkour, H.A., and M. Youssef. 2009. Heavy metals in the benthic foraminifera from the coastal lagoons, Red Sea, Egypt: Indicators of anthropogenic impact on environment (case study). *Environ Geol* 58: 543–553.

Mance, G. 1987. *Pollution Threat of Heavy Metals in Aquatic Environments*. Pollution Monitoring Series. London: Elsevier.

Mansour, A.M. 2003. Pressures and impacts of coastal zone of Abu-Minqar and Giftun Islands, Hurghada, Red Sea, Egypt: A management priority. *Fifth International Conference on the Geology of the Middle East*, Ain Shams University, Cairo, Egypt, January 2003, pp. 417–430.

Mansour, A.M., Nawar, A.H., and H.A. Madkour. 2005. Metals concentration of recent invertebrates along the Red Sea Coast of Egypt: A Tool for monitoring environmental hazards. *Sediment Egypt* 13: 171–185.

Mekawy, M.S., and H.A. Madkour. 2012. Studies on the Indo-Pacific Tridacnidae (*Tridacna maxim*) from the northern Red Sea, Egypt. *Int J Geosci* 3: 1089–1095.

Mekawy, M.S., and H.A. Madkour. 2013. Oxygen and carbon stable isotope analysis on some Egyptian *Tridacna maxima* as an environmental and climatic indicator. *Egypt J Paleontol* 13: 173–183.

Merian, E. 1991. *Metals and Their Compound in the Environment*. Weinheim: Weinheim Verlag Chemie, 365 p.

Milliman, J.D. 1974. *Marine Carbonates*. Berlin, Heidelberg and New York: Springer, 375 p.

Mohamed, A.M., Madkour, H.A., and A. El-Taher. 2013a. Recent benthic foraminifera in the saline pool and its surrounding areas at Ras Shukeir and Gulf of Suez, Egypt. *Indian J Geo-Mar Sci* 42, no. 3: 293–299.

Mohammed, T.A., Mohamed, E.M., Ebrahim, Y.M., Hafez, A., and R.M. Zamzamy. 2013b. Some metal concentrations in the edible parts of Tridacna maxima, Red Sea, Egypt. *Environ Earth Sci* 71: 301–309. http://doi.org/10.1007/s12665-013-2434-8.

Phillips, D.J.H. 1980. *Quantitative Aquatic Biological Indicators*. London: Applied Science Publishers, 488 p.

Ramadan, S.E., and A. Shata. 1993. Biogeochemical studies on the mollusk bivalve *Anadara diluvii* (Lamarck, 1805) (Pteriomorpha Arcidae). *Bull Nat Inst Ocean Fish, ARE* 19: 145–157.

Ravera, O. 2001. Monitoring of the aquatic environment by species accumulator of pollutant: A review. *J Limmol* 60, Suppl 1: 63–78.

Sharabati, D. 1984. *Red Sea Shells*. London, Boston, Melbourne, and Henley: KPI, 127 p.

Shata, A., and A. Hassan. 2000. Geochemical and environmental studies on some bivalves inhabiting the intertidal zone surrounding the Eastern Harbour of Alexandria, Egypt. *J Aquat Biol Fish* 4, no. 3: 147–171.

Wang, W.X., and P.S. Rainbow. 2005. Influence of metal exposure history on trace metal uptake and accumulation by marine invertebrates. *Ecotoxicol Environ Saf* 61: 145–159.

Wye, K.R. 1991. *The Encyclopedia of Shells*. London: Headline Publishing, 288 p.

Youssef, M., Madkour, H., Mansour, A., Alharbi, W., and A. El-Taher. 2017. Invertebrate shells (mollusca, foraminifera) as pollution indicators, Red Sea Coast, Egypt. *J Afr Earth Sci* 133: 74–85.

Ziko, A., El-Sorogy, S., Aly, M.M., and H.E. Nour. 2001. Sea shells as pollution indicators Red Sea coast, Egypt. *Egypt J Paleontol* 1: 97–113.

PART 4

Marine debris

Distribution and abundance of marine litter along the central Sudanese Red Sea coast

12

Nahid A.R. Osman, Eldirdery A.A. Ibrahim, Khalid M. Elfatih, Osman M. Tiah, Haiat S. Hassan, Khadega I. Idrees and Mariam A. Eldai

12.1 INTRODUCTION

The Red Sea is a semi-enclosed, highly saline young ocean situated in the dry and hot tropics, where the temperatures are very high, and precipitation is scarce. It is a regional waterway bordered by nine countries and 22 seaports. The countries bordering the Red Sea differ in socioeconomic capabilities with regard to coastal development, waste disposal management and waste reception facilities. The latter is very important since the Red Sea is one of the major maritime shipping routes in the world, connecting Asia, Africa and Europe. As a result, numerous shipwrecks have occurred in the Red Sea since the 18th century, and many of them have now become famous recreational diving sites. Therefore, pollution with marine litter (also referred to as marine debris) is inevitable for many reasons. For instance, the southern Sudanese coastal waters and islands are heavily contaminated with various commodities due to the sinking of a container ship in October 2022 northwest of Jazan Port, Saudi Arabia (≈ 300 km southeast of Port Sudan). Most of the load of the ship, including the containers, were driven to the Sudanese coast by the currents. In another accident in June 2022, a ship sank in Suakin Harbour due to overloading of the cargo, and huge efforts were exerted to clean the harbour area and the surroundings. Moreover, the relatively long residence time of the water mass in the Red Sea (Gerges 2002) is of particular concern as contaminants may remain for a long time in the basin. Additionally, the accumulation of marine litter in the Red Sea coastal environment is aggravated by the outwash of waste dumped inland via the network of seasonal streams crossing the coastal plain to the Red Sea.

As the Red Sea is a semi-enclosed regional sea with limited connection to open oceans and high marine biodiversity, its marine litter is of paramount importance. Semi-enclosed regional seas such as the Red Sea are considered potential sinks for marine litter. Martin et al. (2019b) indicate that the circulation pattern of the Red Sea may facilitate the accumulation of floating plastic litter due to the greater likelihood of import from the Indian Ocean than export from the Red Sea.

In addition to the hydrodynamic regime, the Red Sea coasts are largely lined with fringing reefs, seagrass beds, mangrove stands and sabkha ecosystems, which are reported to act as traps of marine litter (Abu-Hilal and Al-Najjar 2009; Yin et al. 2019; Martin et al. 2019a; Li et al. 2021; John et al. 2022; De et al. 2023).

Recent incidences of ship accidents in the Red Sea indicate that the local and regional dispersion and movement of marine litter in the Red Sea are largely modulated by atmospheric and oceanographic conditions such as winds, surface currents and waves. In line with this, Veerasingam et al. (2020) reported that winds and currents are responsible for marine litter arrival in the Qatar coastal environment. In the case of the Sudanese Red Sea coast, two opposing wind streams converge in winter at the central Red Sea, around 18° to 19°N (Menezes et al. 2018), where the Sudanese coast is located. Here, the wind becomes calmer and turns westward toward the Sudanese coast to pass through the Tokar Gap, a 110-km-wide gap in the Red Sea Hills. In summer, cross-axis wind jets blow from the eastern Red Sea coast toward

DOI: 10.1201/9781003321408-16

the Sudanese coast and pass through Tokar Gap (Jiang et al. 2009), forming oceanic dipole eddies between 18°N and 20°N. This wind regime modulates local surface currents and waves along the Sudanese coast and subsequently, the movement of marine litter across coastal ecosystems. This situation raises concerns about the vulnerability of the Sudanese coast to the accumulation of marine litter originating from local and regional sources.

12.1.1 Sudanese Red Sea coast

The Sudanese Red Sea coast has three types of coastal lagoons: they are either cut deeply into the coastal plain for more than 4 km, are restricted to the fringing reef without reaching the coastline, or barely indent it (Schroeder 1982). Commonly, lagoons on the coastal plains are lined with mangroves and seagrass meadows. This type of coastal geomorphology attenuates the currents and waves and enables sedimentation and settlement of negatively buoyant objects, including marine litter.

For administrative and management purposes, the coast of Sudan is divided into three geographical areas: the northern coast extends from north of Port Sudan Harbour to the Egyptian border; the central coast is about 60 km long, located between Port Sudan and Suakin; and the southern coast lies from Suakin to the Eritrean border. Relative to the northern and southern parts, the central Sudanese coast is the most urbanized, industrialized and residential. This part contains three oil terminals, two commercial ports, salt production areas, industries, diving tourism and desalination plants, to name a few developments. Despite the considerable length of Sudan's coast, it is sparsely populated, with Port Sudan and Suakin being the major seaports and most populated cities. These two coastal cities are considered the main sources of marine litter deposited on the Sudanese coast. Port Sudan in particular is the largest and most populated port in the country and where most industries are located. The geomorphology of Port Sudan is conducive to the arrival of land-based debris in the coastal area of the Sudanese Red Sea. The city is located in a low-lying area surrounded by the Red Sea hills and seasonal streams, which flood during winter and torrential summer rains. The streams wash out all the debris generated and dumped on the land into the coastal zone. As the country's main harbour, Port Sudan receives several types of marine vessels, including cargo ships and tourist crafts, besides passing-by vessels. The destination of wastes generated by all these vessels while they are in Sudanese waters is not clearly known.

12.1.2 Marine litter research in the Red Sea

Systematic research on marine litter in the Red Sea region may date back to the work of Abu-Hilal and Al-Najjar (2004) along the Jordanian Red Sea coast. Since then, more research on marine litter from Jordan, Saudi Arabia, Israel, Yemen, Sudan, Egypt, and Eritrea has been produced. However, most of this research is from the eastern coast of the Red Sea – i.e., Saudi Arabia, Jordan and Israel – for example, by Abu-Hilal and Al-Najjar (2009), Baalkhuyur et al. (2018), Martin et al. (2019a, 2022), Ibrahim et al. (2020), Tewelde et al. (2021), Brümmer et al. (2022), and Habib and Thiemann (2022). These studies cover the composition, abundance and distribution of marine litter in the Red Sea environment's physical (beach and water column) and biological (coral, fish and mangrove) components. Most of these studies, however, researched the beached macro litter or micro plastics and did not sufficiently consider the marine litter in the sediments. The study by Martin et al. (2020) on the sequestration of plastics in mangrove sediment from Saudi Arabia revealed that small-size plastics (>0–5 mm), which do not exist in Red Sea surface water, were dominant in mangrove sediment. The study highlighted the differential size-dependent deposition of plastic litter across the coastal and marine habitats in the Red Sea. Information about marine litter in Red Sea marine sediments (i.e., buried marine litter), whether in beach, coral reef, or seafloor sediment, is scarce, but it is essential to improving our understanding of the marine litter life cycle.

Recently, it was reported that the main sources of marine litter in the Red Sea are recreational activities, fishing and shipping (Diem et al. 2022); however, data supporting a comprehensive understanding of this issue is lacking and inconsistent. It is evident on the Red Sea coast of Sudan that residential waste management, in terms of proper landfills and sewer system management, is inefficient. In many locations, accumulated residential waste is clearly apparent, and it is driven into the Sudanese coastal environment by rainwater runoffs and flash floods, suggesting that residential waste is the primary source of marine litter in Sudan.

The status of marine litter from the Sudanese Red Sea coast has started to receive noticeable attention (e.g., Ibrahim et al. 2020 and this chapter) as information on the characteristics of marine litter is crucial for national efforts to manage and maintain sustainable coastal development of the Sudanese Red Sea. Ibrahim et al. (2020) outlined the status of beach litter on the northern Sudanese Red Sea coast at the World Natural Heritage of Dungonab-Mukawar Island (DMNP). The study showed that the beaches of DMNP are clean, with a Clean-Coast Index (CCI) of 4.6, and indicated that the rate of beached litter accumulation decreased steadily over the study period (ten months) from 21.9 to 16.43 items/100 m/day. It is important to mention that information on the status of marine litter in the Sudanese coastal habitats and beach sediment is scarce, and inadequate knowledge greatly impedes our understanding of the behaviour of marine litter in the Sudanese marine ecosystem and the Red Sea in general.

Here, for the first time, an assessment of the status of marine litter on beaches, in beach sediments (buried marine litter) and in mangroves at selected sites along the central Sudanese coast is presented (Table 12.1). The purpose is to demonstrate the spatial distribution and the abundance of marine litter along the Sudanese Red Sea coast with regard to coastal development and to provide baseline information for national and regional management and monitoring. This information may help identify where, when and what management interventions should be implemented along the Sudanese Red Sea coast to combat marine litter. The selected sites are located from north to south as follows: Halout, Port Sudan (Abuhashish and Northern Berth) and Kilo Tamania. The sites differ in types and scale of uses and habitation. Both Halout and Kilo Tamania are recreational sites utilized as picnic spots by the local residents of Port Sudan as well as by fishermen. Recently, in Halout, saltpans were introduced, where workers are accommodated in small housing complexes. In Kilo Tamania, a governmental resort was established to provide local residents with recreational activities. In both sites, scrub mangroves are located between the seagrass meadows on the seaward side and halophytic vegetation on the landward side. The Abuhashish area in Port Sudan is located on the northern coast of the city. The Flamingo Navy Base and the Naval Force Housing Complex are located at the north of Abuhashish, In the middle of Abuhashish, two neighbourhoods – the abandoned northern industrial area of Port Sudan city and the famous Fish Marketplace, which features a landing site and fish restaurants – are located. Several glass-bottom boats operate in the fish restaurant area. The Green Quay occupies the southern end of the Abuhashish area.

12.2 MARINE LITTER ASSESSMENT

We strictly followed the guidelines of rapid beach assessment of PERSGA (2014) while assessing the marine litter in all the sites mentioned here. We randomly designated one or two transects in sandy beaches at all the studied sites parallel to the coastline. All transects were 100 m in length with varying widths due to differences in the coastal morphology between the sites (Table 12.1). All transects were sampled twice during this investigation. Litter items ≥ 2.5 cm were collected, sorted into the main litter classes and counted. In one case, the weight was determined in addition to the number of litter items.

12.2.1 Cleanliness of the Sudanese Red Sea coast

The cleanliness of the beaches in the central Sudanese Red Sea coast was assessed by estimating the Clean-Coast Index (CCI) according to Alkalay et al. (2007). This index was applied in only three of the countries bordering the Red Sea (Sudan, Eritrea and Israel) and has proved to be useful as a quantitative indicator of the effectiveness of preventive measures such as clean-up efforts. The following formula was applied to estimate the CCI: $CCI = D_{ML} \times K$, where D_{ML} is the density of marine litter (items/m^2) on the beach, and K is a constant, taken to be 20. The beaches were then classified based on the CCI score as very clean (0–2), clean (2–5), moderately clean (5–10), dirty (10–20) and extremely dirty (> 20).

TABLE 12.1 Transect locations at the assessment sites.

SITE	*TRANSECT LOCATION*	*TRANSECT AREA, M²*	*DISTANCE FROM PORT SUDAN (KM)*
Halout	N: 19°47′27.9″ E: 37°15′41.4″ E: 37°10′53″	100 ×15	32
Abuhashish	N: 19°386′.4″ E: 37°14′16″	100 ×10	2
Northern Berth	N: 19°37′24.8″ E: 37°13′25.4″	100 × 4	0
Kilo Tamania	N: 19°33′28.3″ E: 37°14′49.2″	100 × 20	8
	N: 19°33′0.5″ E: 37°14′55.8″	100 × 20	

12.2.2 Beach sediment litter

The frame method described by Haseler et al. (2018) was implemented to characterize buried marine litter in the sediment of the beaches. In the same transect, after collection of the beach litter, 10 squares of 1 m x 1 m were placed in two rows parallel to each other. Each row was composed of five squares placed at 20 m intervals. One row was placed close to the back of the beach, and the second row was placed just above the highest water level. Inside the squares, the surface sediment layer was scooped out with small shovels to a depth of approximately 5 cm. The sediment was sieved in situ with a 5 mm sieve. The retained litter was cleaned, dried and classified into the main litter classes (similar to the beach litter) and counted. Based on the method described by Barnes et al. (2009), the sediment plastic litter was classified into mega plastics (>100 mm), macro plastics (<100 to >20 mm), and meso plastics (<20 and >5 mm), and the shape of the litter was assigned to each piece.

12.2.3 Marine litter in the mangroves

Marine litter trapped in the mangroves was investigated in four plots (10 m x 10 m), established in the middle of the stand. The plots were placed at 40 to 60 m intervals. The guidelines of PERSGA (2014) were applied to describe the composition and abundance of the marine litter therein.

12.3 COMPOSITION AND ABUNDANCE OF BEACH LITTER ON THE CENTRAL SUDANESE RED SEA COAST

Six types of beach litter were recorded in all sites. In line with the regional and global trend, plastics constituted most of the beach litter, ranging from 93.87% in Abuhashish to 97.7% in Kilo Tamania (Table 12.2). The majority of the plastic litter consisted of beverage bottles; some of them were of local origin. The levels of other marine litter classes was extremely low as these items are collected from their dumping sites and recycled by Port Sudan–based companies. Although plastic litter is also collected and recycled, not all types of plastics are targeted, and thus, the rate of plastic litter accumulation seems to exceed collection and recycling rates. Similar patterns of beach litter composition were reported in Eritrea (Tewelde et al. 2021), Yemen (Al-Shwaf and Ahmed 2011) and Qatar (Veerasingam et al. 2020).

TABLE 12.2 Composition and abundance of marine litter in the selected sites along the central Sudanese Red Sea coast (n = 2).

	HALOUT		*ABUHASHISH*		*KILO TAMANIA*	
	ABUNDANCE					
LOCATION AND LITTER TYPE	*ITEMS/M2*	*%*	*ITEMS/M2*	*%*	*ITEMS/M2*	*%*
Plastics	1.31±1.4	95.53	1.28±1.25	93.87	1.35±2.11	97.7
Metals	0.008±0.00424	0.39	0.006±0.0053	0.44	0.0031±0.003	0.23
Glass	0.025±0.007	1.21	0.004±0.0035	0.29	0.00063±0.0008	0.045
Wood	0.031±0.035	1.5	0.044±0.042	3.20	0.004±0.002	0.28
Textile	0.0195±0.021	0.95	0.023±0.03	1.71	0.01±0.01	0.67
Paper/carton	0.009±0.013	0.44	0.007±0.006	0.49	0.02±0.015	1.08
Litter density	**1.37**		**1.35**		**1.38**	

The abundance of the plastic litter was almost comparable in all sites, ranging from 1.28±1.25 to 1.35±2.11 items/m^2. These abundance values are higher than the values previously reported (0.23 items/m^2) by Ibrahim et al. (2020) from the northern Sudanese coast at DMNP. Comparatively, beach litter density at or around urban centres on the Saudi Arabian coast on the eastern Red Sea is apparently lower than the density at the central Sudanese Red Sea coast. Along the Saudi Arabian coast, beach litter densities ranged from 0.1 to 0.5 items/m^2 (Martin et al. 2021). Similarly, the macro litter densities obtained from three locations of varied coastal activities along the Eritrean Red Sea coast (Tewelde et al. 2021) were remarkably lower (0.1 to 0.35 items/m^2) than those of the Sudanese coast. However, these values are comparable to litter densities (1 to 6 items/m^2, mean 2.8 items/m^2) from the Gulf of Aqaba (Abu-Hilal and Al-Najjar 2009). It seems that the surface currents and waves generated by the wind regime (e.g., RSCZ and westward wind jets) at the centre of the Sudanese coast keep marine litter along the coastline rather than transporting it to the open sea. As indicated in Menezes et al. (2018), the RSCZ prevails for a longer time (up to 19 days) in the central Sudanese Red Sea coast around latitude 19°N.

The high abundance of plastic litter in the investigated sites could be attributed to the proximity of these sites to Port Sudan City as the waste dumped on the outskirts of the city is washed into the coastal waters by surface runoffs and flash floods during rainy periods. As a result, the abundance of marine litter along the coast is likely to increase after such events. The suggested increase of marine litter following the winter rainy season is validated by the high abundance values reported here, as we sampled all the sites in the rainy season (March to May or November to January). This is further corroborated by the notable decrease in the abundance of marine litter in all sites in the subsequent samplings. For instance, in Kilo Tamania, the total amount of beach plastic litter decreased from 1,069 items/transect in the first sampling in November 2021 to 330 items/transect in the second sampling in January 2022 (data not shown).

The composition of marine litter in the Northern Berth (Figure 12.1) is similar to that of other sites despite the difference in the prevailing activities at each site. Again, plastic litter made up the majority of the marine litter (84%). Plastic sacks scored a higher percentage in the Northern Berth due to direct dumping from a manufacturing facility at the berth. The rest of the plastic litter seems to come from the discharges of one of the major seasonal streams running through Port Sudan City that washes one of the largest dumping areas into the berth basin. A gradual decrease in the rate of accumulation of all marine litter types was recorded over the study period (Figure 12.2). This is of significance for the efforts of marine litter management and monitoring, indicating that regular collection of land-based waste and beach clean-up would have a positive impact on reducing the abundance of marine litter.

The lower abundance of metal, glass and processed wood in the beach litter is probably because of the presence of recycling businesses for these materials, which encourages citizens to regularly recycle them. Petty trading in scrap metal is a common livelihood in Port Sudan City. The ultimate buyers of metal scrap are steel companies inside and outside Sudan. At a household level, glass is reused and is not thrown away unless broken. Local and informal markets for used wood are also common. It seems that

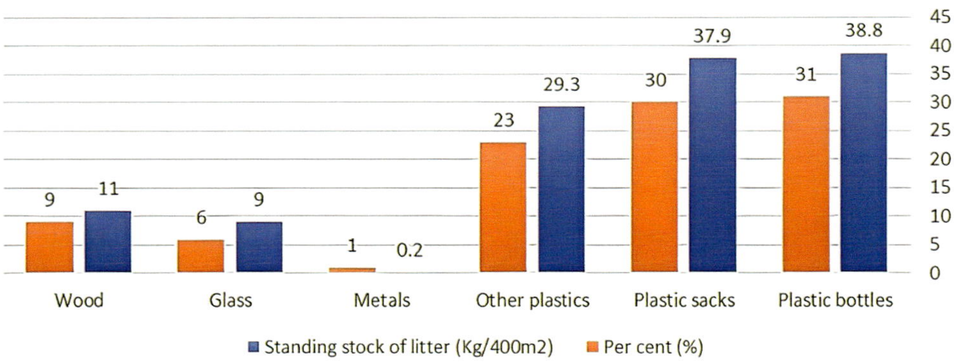

FIGURE 12.1 Composition and abundance of marine litter at Port Sudan Northern Berth (n = 4).

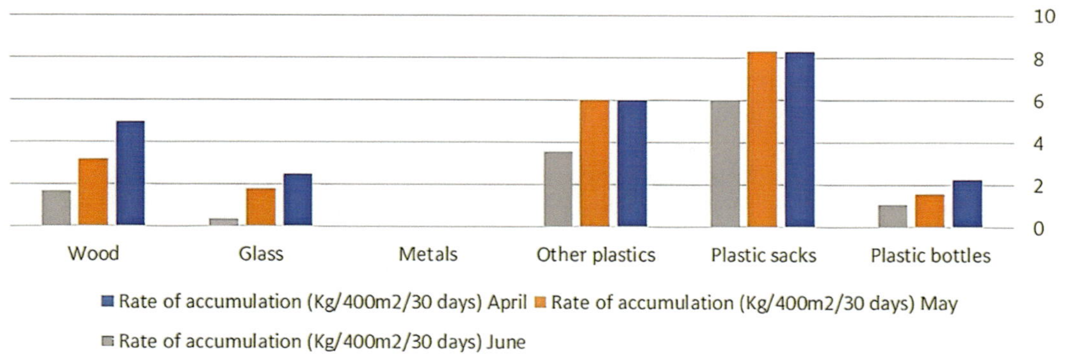

FIGURE 12.2 Rate of accumulation of marine litter in the Northern Berth of Port Sudan (April–June).

providing incentives for recycling materials that are part of the marine litter would encourage citizens to consider recycling and effectively play a role in marine litter management. Most of the plastic litter is composed of single-use, small-size plastic bottles (350 to 600 ml). Citizens tend to keep and reuse larger plastic containers, such as family-size soft drink bottles, compared to the smaller sizes. This could contribute to plastic litter management by encouraging plastic container manufacturers to consider the volumes of the containers they produce.

12.3.2 Cleanliness of the beaches on the central Sudanese Red Sea coast

Since it was proposed by Alkalay et al. (2007), the Clean-Coast Index (CCI) has found considerable application in different parts of the globe (e.g., Poluan et al. 2023; Salazar et al. 2022; Okuku et al. 2021; Perumal et al. 2021; Manullang et al. 2021; Asensio-Montesinos et al. 2021; Silva et al. 2018; Aytan et al. 2019; Portman and Brennan 2017). The CCI was applied to achieve different objectives including, but not limited to, estimation of the quality of beaches, comparison between different segments of beaches, evaluation of clean-up activities, monitoring of marine litter and awareness. Estimating CCI can provide some insights into the current status of a beach and transform measurements on beached litter abundance to a meaningful quantitative value that can be understood by a wide variety of stakeholders, including decision makers and the public.

The CCI scores of Halout (26.2), Abuhashish (25.6) and Kilo Tamania (27) indicate that these beaches are extremely dirty (Figure 12.3). These values are much higher than the CCI value from the northern Sudanese coast reported for the World Heritage Site of DMNP (Ibrahim et al. 2020), where the CCI ranged from 0.96 to 15, with an average of 4.6. In Eritrea, the CCI for three sites (0.57. 2.05 and 6.55) were lower than those of Sudan's central Red Sea coast. Regional and local variations in marine litter composition and abundance are expected and could be caused by an array of drivers, such as the scale of coastal development, the intensity and types of coastal uses at specific coastal area, transport by surface currents, waves, coastal geomorphology, the hydro-oceanographic regime, and the type of coastal habitat (e.g., seagrass and macroalgal beds).

12.3.3 Composition and abundance of marine litter in beach sediments

Marine sediment has proven to be one of the ultimate sinks of marine litter (Woodall et al. 2014; Martin et al. 2022), although no agreement has been reached on the role of sediment grain size in trapping marine litter. Some studies indicate that sediment with a higher proportion of silt/clay retains higher proportions of micro plastics (Silburn et al. 2022).

The composition and abundance of marine litter in the sediment of the beaches of the central Sudanese Red Sea coast exhibited a comparable pattern to that of beached marine litter. Plastic litter was the most abundant material in the beach sediments, ranging from a minimum of 21.13±29.95 items/m^2 in Halout to a maximum of 79.63±29.67 items/m^2 in Abuhashish (Tables 12.3, 12.4 and 12.5). Other beach litter (wood and textiles) ranked second to plastics in their abundance in beach sediment at Halout. At other sites, the abundance of metals ranked the second highest in the beach sediment, higher than wood, glass and other litter. In the Abuhashish area, the metal abundance was 32.25±46.06 items/m^2. It is noteworthy that the sediment of the upper and lower beaches differed in their marine litter abundances in all the sites tested. For instance, in sediment from the Abuhashish beach, the number of plastic fragments in the lower beach was 22 to 81 pieces/quadrat and was in the range of 83–118 pieces/quadrat in the upper beach. The number of plastic fragments in sediment from the lower beach of Halout, the least polluted site, ranged between

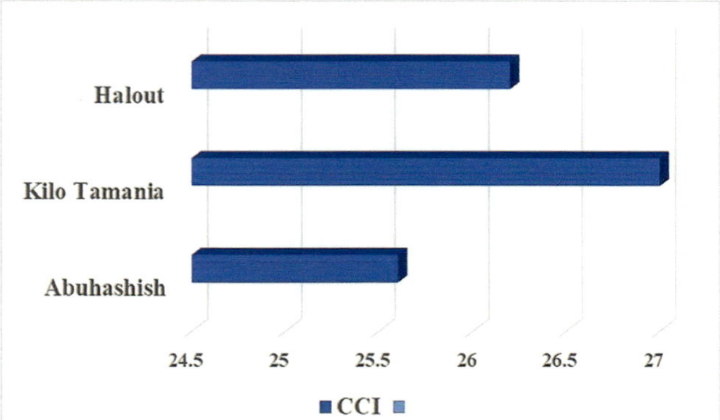

FIGURE 12.3 Clean-Coast Index (CCI) score of beaches along the central Sudanese Red Sea coast.

4 to 13 pieces/quadrat and between 2 to 81 pieces/quadrat in the sediment of the upper beach (data not shown). This difference in litter abundance might be driven by the hydrodynamic forces of the seawater, the size of plastic litter, sediment grain size and the geomorphology of the coast. Our findings are in agreement with the results of GESAMP (2019), which reported that shoreline litter exhibits a patchy distribution at fine spatial and temporal scales. A pattern of increasing abundance in beach litter from the low seawater level to the berm line is reported in some studies (e.g., Andriolo and Gonçalves 2022; Veerasingam et al. 2020; Silva et al. 2018). This pattern was attributed to the energy of the waves and tides. Interestingly, Silva et al. (2018) recorded a higher abundance of marine litter residues in dry beach sand than in wet beach sand. Waves and tidal energy not only influence the abundance of marine litter on beaches but also influence the size of marine litter across this gradient. Beach sediment litter of larger sizes was reported from the berm line and high-water mark than at low water levels (Fauziah et al. 2015).

TABLE 12.3 Marine litter in the sediment from Halout beach (items/m^2).

LITTER TYPE	LITTER ABUNDANCE (ITEMS/M^2) UPPER BEACH	LOWER BEACH	BEACH*	%**
Plastics	35±39.6	7.25±3.95	21.13±29.95	72.53
Metals	3±2.45	1.5±1.3	2.25±1.98	7.73
Glass	2.5±1.3	1±0.82	1.8±1.28	6.01
others	4.8±2.22	3.25±0.5	4±1.7	13.73

* Average abundance of marine litter type in sediment of the whole beach
** Percent of marine litter type in sediment of the whole beach

TABLE 12.4 Marine litter in the sediment from Abuhashish beach (items/m^2).

LITTER TYPES AND LOCATIONS	ABUNDANCE (ITEMS/M^2) UPPER BEACH	LOWER BEACH	BEACH*	%**
Plastics	99.8±14.3	59.5±27.74	79.63±29.67	64.02
Metals	55.25±58.61	9.25±10.14	32.25±46.06	25.93
Glass	4±3.56	2.75± 4.3	3.38±3.70	2.71
Others	7.8±7.13	10.5±4.2	9.13±5.62	7.34

* Average abundance of marine litter type in sediment of the whole beach
** Percent of marine litter type in sediment of the whole beach

TABLE 12.5 Marine litter in the sediment from Kilo Tamania beach (items/m^2).

LITTER TYPES	ABUNDANCE (ITEMS/M^2) UPPER BEACH	LOWER BEACH	BEACH*	%**
Plastics	82.5±139.94	9.75±11.61	46.13±99.81	98.4
Metals	1.25±1.9	7.25±9.91	4.5±7.34	1.067
Glass	0±0	0±0	0±0	0
Others	0.25±0.5	0.25±0.5	0.25±0.46	0.53

*Average abundance of marine litter type in sediment of the whole beach
** Percent of marine litter type in sediment of the entire beach

Overall, it has become more evident that the marine litter in sediments and its composition and abundance in the Red Sea region are far less studied than beach litter; therefore, comparing the present values to previously established regional values is not feasible. Interestingly, these values are considerably lower than the values (179–1,164 items/m^2) reported by Fauziah et al. (2015) for the abundance of buried plastic litter from recreational and fishing beaches in Malaysia, although the environmental and socioeconomic variation between the two countries might explain the observed difference.

Buried marine litter remains in the sediment and is susceptible to erosion and subsequent transportation to new locations in the coastal environment. Coastal erosion was proposed as a secondary input source of marine litter (Andriolo and Gonçalves 2022). Therefore, this shows the importance of buried litter, which is often overlooked compared to beached, benthic and floating litter. Similar to biological marine litter sinks (e.g., coral reef, fish and invertebrates), marine sediment may simultaneously act as a physical sink and a source of input due to erosion. More research is needed to investigate the life cycle of buried litter and its ultimate fate in the coastal and marine environment in order to provide better estimates of the standing stock of marine litter in the Red Sea region.

12.3.4 Size of plastic litter in beach sediments

Driven by some abiotic oceanographic variables in the marine environment, large plastic pieces fragment into smaller pieces of variable sizes in the ocean; these are the most abundant form of plastic debris in the marine environment (Kaandorp et al. 2021). Based on size, the resulting pieces are classified as mega plastics, macro plastics, meso plastics, micro plastics, and nano plastics. Each category has been assigned a size on a scale that slightly differs in the literature (c.f., Barnes et al. 2009; GESAMP 2019).

The size of marine plastics has a significant bearing on the fragmentation of plastics in the marine environment and the subsequent impacts they produce, as well as on the management and monitoring of marine plastic litter. A strong correlation between the abundance of large micro plastics and that of meso plastics was detected, suggesting that the contamination level of large micro plastics can be estimated from that of meso plastics (Lee et al. 2013, 2015; Egessa et al. 2019). This would provide an indirect yet efficient and cost-effective way to monitor the abundance of micro plastics in the coastal and marine environments. It could also be of great benefit in the monitoring of plastic fragments in remote areas, where requirements for measuring micro plastics may not be available. Furthermore, it enhances the contribution of civil society networks to marine litter monitoring and database creation.

In the sediment of Sudan's Red Sea central coast, on both the upper and lower beaches, meso plastics were, proportionally, the most abundant size category of plastic litter, followed by macro plastics and mega plastics (Figures 12.4, 12.5 and 12.6). A similar trend was reported by Manullang (2019) for meso plastics (5 items/m^2), compared to macro plastics (0.722 items/m^2) in Ambon Bay. Abuhashish scored the highest abundance of meso plastics in the sediment (55.25±28.12 items/m^2), followed by Kilo Tamania (39.4±93.93 items/m^2). While this could partly be attributed to the recreational activities in both sites and to their proximity to seasonal streams discharge points, it is also likely that they originated offshore and were transported by waves and currents to be deposited in the beach sediment. The fragmentation of plastic litter in the coastal and marine environments was attributed to photooxidation by UV radiation and abrasion, as well as to the thermal effect of temperature and other chemical and microbial drivers. Specific to plastics, the particle size and shape determine their orientation and movements under the influence of waves and inertia (Sebille et al. 2020) and, hence, their settlement location.

In a study by Brümmer et al. (2022), an investigation of the degradation of plastics on the Red Sea coast of Egypt was undertaken. Micro plastic resin pellets from a ship that was found after being grounded for 29 years were examined for their degradation status and compared to beached micro plastics. The results revealed that beached micro plastics, which are subject to UV radiation and high temperatures, degraded

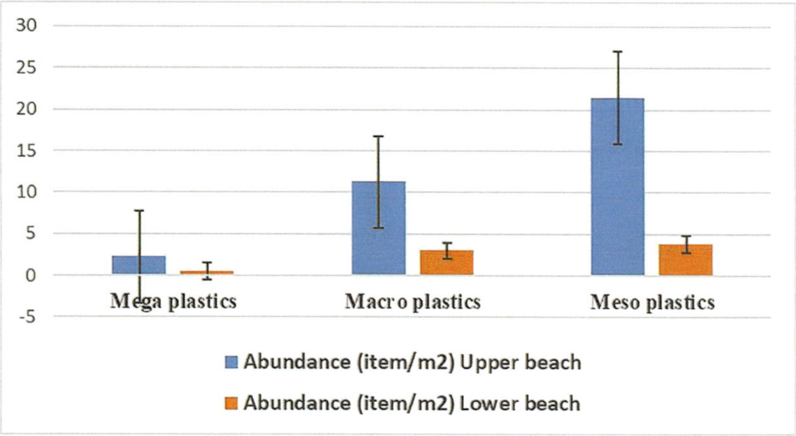

FIGURE 12.4 Size distribution of marine plastics in the sediment of Halout beach (n/m²).

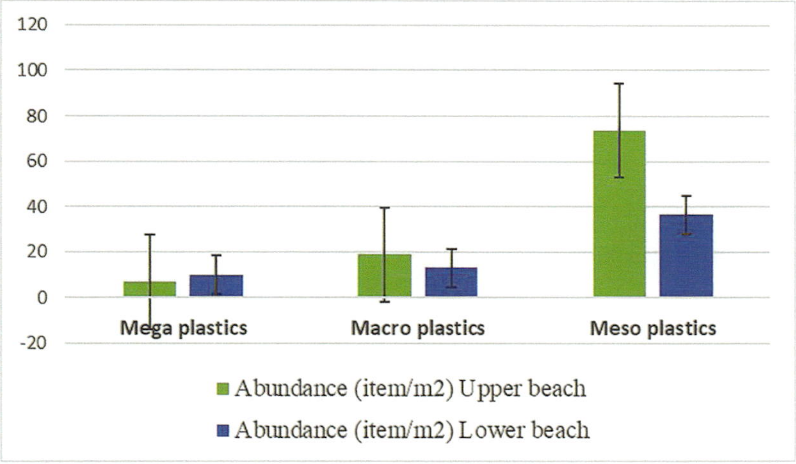

FIGURE 12.5 Size distribution of marine plastics in the sediment of Abuhashish beach (n/m²).

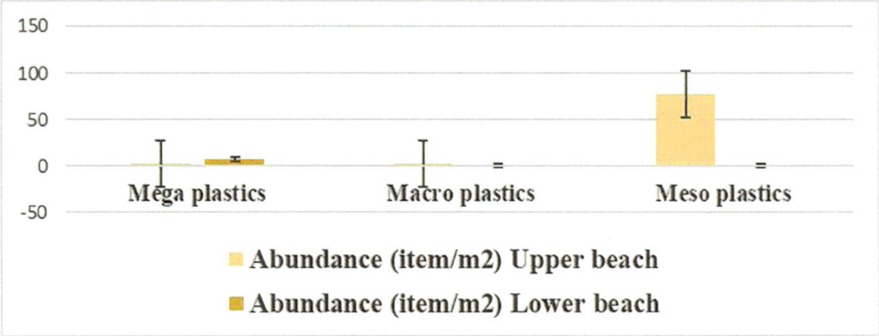

FIGURE 12.6 Size distribution of marine plastics in the sediment of Kilo Tamania beach (n/m²).

faster relative to the micro plastics from the grounded vessel. The vessel's micro plastics were found to be comparable to pristine sources of micro plastics despite the time factor, suggesting that degradation of plastics on the seafloor takes a longer time than for plastics on the beach.

12.3.5 Shapes of plastic litter in beach sediments

The shape, size and polymer composition of micro plastics, the most studied size category of marine plastics, are linked to their mobility in the marine environment (Shamskhany et al. 2021). Therefore, it is pertinent to study the shape of the plastic fragments to understand where they may end up in the marine environment and to propose management interventions. The fragments of marine plastics in the beach sediments along the Sudanese coast were found in three different shapes (Tables 12.6, 12.7 and 12.8). At Halout and Abuhashish, threads (filaments) were more abundant than spheroidal and fragmental plastics, suggesting that they originated from fishing gear. The dominance of threads in the plastic litter of fish landing sites is widely reported and is attributed to the fishing activities (Egessa et al. 2019). Additionally, this is comparable to the documented fact that fibres are the dominant shape of micro plastics on the seafloor (Kane and Clare 2019; Shamskhany et al. 2021). The abundance of spheroidal plastics in the sediment of the beaches in all sites, particularly in Abuhashish (25.25±15.11 n/m^2), is noticeable and may have a sea-based origin since the site is closer to Port Sudan harbour.

TABLE 12.6 Shape of marine plastic litter in the sediment of Halout beach (items/m^2).

PLASTIC SHAPE	ABUNDANCE (ITEMS/M^2) UPPER BEACH	LOWER BEACH	BEACH*	%**
Threads	15.25±17.5	3±2.71	9.13±13.31	43.19
Spheroids	12.25±14.29	2.75±1.26	7.5±10.68	35.50
Fragments	7.5±8.2	1.5±0.58	4.54±6.26	21.30

* Average abundance of marine litter shapes in sediment of the whole beach
** Percent of marine litter shapes in sediment of the whole beach

TABLE 12.7 Shape of marine plastic litter in the sediment of Abuhashish beach (items/m^2).

PLASTIC SHAPE	ABUNDANCE (ITEMS/M^2) UPPER BEACH	LOWER BEACH	BEACH*	%**
Threads	42.80±8.4	28.25±13.93	35.5±13.17	44.58
Spheroids	37.30±10.34	13.25±6.45	25.25±15.11	31.71
Fragments	19.80±12.81	18±16.23	18.88±13.57	23.70

* Average abundance of marine litter shapes in sediment of the whole beach
** Percent of marine litter shapes in sediment of the whole beach

TABLE 12.8 Shape of marine plastic litter in the sediment of Kilo Tamania beach (items/m^2).

PLASTIC SHAPE	ABUNDANCE (ITEMS/M^2) UPPER BEACH	LOWER BEACH	BEACH*	%**
Threads	18.5±35.67	8.25±12.68	13.4±25.4	28.99
Spheroids	41.75±64.53	1.25±1.89	21.5±47.5	46.61
Fragments	22.25±39.92	0.25±0.5	11.3±28.7	24.40

* Average abundance of marine litter shapes in sediment of the whole beach
** Percent of marine litter shapes in sediment of the whole beach

12.3.6 Composition and abundance of marine litter at Kilo Tamania mangroves

Owing to their location at the interface of land and sea and their physical structure, which slows down water mass movement, a mangrove ecosystem acts as a marine litter sink (Yin et al. 2019) as well as a filter. Larger marine litter items are retained at the front of mangrove stands while smaller objects penetrate deeper, driven in by wind and tidal forces (Debrot et al. 2013). Mangrove stands frequently line the coast of the Sudanese Red Sea, particularly at the mouths of intermittent streams as well as on offshore islands. Commonly, two zones are recognized in mangrove stands based on tree densities and sizes. The seaward zone is composed of dense and larger plants ranging in height from 3 to 7 m, with dense and longer pneumatophores (103.84 root/cm^2 and 22 to 85 cm in length, respectively). The landward zone contains smaller plants, less than 5 m in height, with sparser and shorter pneumatophores (78.1 root/cm^2 and 8 to 71 cm in length, respectively). Because of this structure and the energy of waves and wind, the sorting of marine litter in mangrove stands occurs in two dimensions: horizontally and vertically. While horizontal sorting occurs across the bottom of mangrove stands along the tide gradient, vertical sorting occurs across the different storeys of the habitat of the mangrove plants. Hence, the two zones exhibit a visual variation in the density and composition of marine litter trapped in the substrate, in their pneumatophores network and in the tree canopy. In the seaward zone, the *Avicennia marina* trees filter marine litter in two strata: upper and lower. Most of the litter items are retained in the lower strata formed by the pneumatophores system, which traps different sizes of marine litter on the bottom and between the pneumatophores. For instance, the density of plastic bottles in the lower strata of the seaward side is 1.14 item/m^2 and in the landward side is 0.042 item/m^2. In the upper strata (i.e., the lower tree branches), buoyant and larger litter items, such as foamed plastics, are trapped. Distinct differences in abundance, diversity and evenness of marine litter were recorded across mangrove stands from the lower intertidal zone to the inner stand (Chee et al. 2020). Sorting of marine litter occurs toward the coastal edge in favour of plastic items. This pattern may indicate that the main body of the mangrove efficiently retained litter with little sorting, and litter deposited closer to the mangrove edge is increasingly sorted and lost to the water body in favour of smaller plastic items.

The composition and abundance of marine litter in a stand of mangrove at Kilo Tamania on the central Sudanese coast is

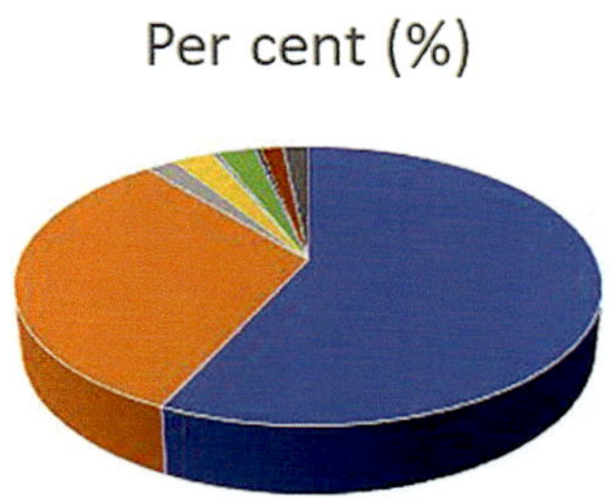

FIGURE 12.7 Abundance of marine litter in the mangrove stand of Kilo Tamania.

given in Figure 12.7. Here, plastics amounted to 96.54% of the marine litter inside the stand. The abundance of plastic bottles was the greatest at 1.71 items/m^2, followed by plastic bags at 0.64±0.16 items/m^2. The abundance of plastic litter in the mangroves of Kilo Tamania (>1.71 items/m^2) is greater than the beached plastic litter at Kilo Tamania (1.35±2.11 items/m^2). Additionally, most of the litter in the mangroves was of larger size than that on the beach. Large-size and oversize debris in the mangroves of Beilum accounted for 76.41% and 13.33% of the total, respectively (Li et al. 2021).

12.4 CONCLUSION

Based on the information presented here and the information provided previously in Ibrahim et al. (2020), it has become apparent that the marine litter on the Sudanese Red Sea coast exhibits a spatial distribution driven by natural and socioeconomic filters. The present information and the information on the marine litter available from the neighbouring countries pinpoint the need for a basin-wide investigation of marine litter, including input sources and sinks, in the Red Sea, against the prevailing climate-driven oceanographic conditions. Further, the capacity of the Red Sea coastal and marine ecosystems that represent potential sinks for marine litter should be investigated in order to estimate the litter budget in the Red Sea and to propose proper mitigation measures to preserve the biodiversity of marine life.

The findings of this investigation provide further insight into marine litter in the sediments and beaches of the central coast of Sudan and add to the existing database on marine litter contributed by neighbouring countries. The abundance of marine litter and the rate of accumulation reported here clearly indicate that the marine litter situation in the Red Sea region of Sudan is still manageable, given that integrated preventive measures will soon be put into action. In this context, the implementation of the anticipated Sudanese National Action Plan for the Sustainable Management of Marine Litter in the Republic of the Sudan (PERSGA 2023, in press), prepared by the Regional Organization for the Conservation of the Environment of the Red Sea and Gulf of Aden (PERSGA) in collaboration with the Sudanese Higher Council for Environment and Natural Resources (HCENR), would ensures that Sudan's response to management of marine litter is comprehensive and aligned with the regional efforts specified in the Regional Action Plan for the Sustainable Management of Marine Litter in the Red Sea and Gulf of Aden (PERSGA 2018). Collaboration of the countries bordering the Red Sea is crucial for better monitoring and assessment of the marine litter and for ensuring the health of the coastal ecosystems in the region.

REFERENCES

Abu-Hilal, A., and T. Al-Najjar. 2004. Litter pollution on the Jordanian shores of the Gulf of Aqaba (Red Sea). *Mar Environ Res* 58, no. 1: 39–63.

Abu-Hilal, A., and T. Al-Najjar. 2009. Marine litter in coral reef areas along the Jordan Gulf of Aqaba, Red Sea. *J Environ Manage* 90, no. 2: 1043–1049. http://doi.org/10.1016/j.jenvman.2008.03.014.

Alkalay, R., Pasternak, G., and A. Zask. 2007. Clean-coast index – a new approach for beach cleanliness assessment. *Ocean Coast Manag* 50, nos. 5–6: 352–362.

Al-Shwaf, N.A.A., and A.M. Ahmed. 2011. Litter on the beaches of the Red Sea of Yemen. *Russ J Gen Chem* 81, no. 13: 2717–2723.

Andriolo, U., and G. Gonçalves. 2022. Is coastal erosion a source of marine litter pollution? Evidence of coastal dunes being a reservoir of plastics. *Mar Pollut Bull* 174: 113307. http://doi.org/10.1016/j.marpolbul.2021.113307.

Asensio-Montesinos, F., Anfuso, G., Aguilar-Torrelo, M.T., and M.O. Ramírez. 2021. Abundance and temporal distribution of beach litter on the coast of Ceuta (North Africa, Gibraltar Strait). *Water* 13: 2739. http://doi.org/10.3390/w13192739.

Aytan, U., Sahin, F.B.E., and F. Karacan. 2019. Beach litter on Sarayköy Beach (SE Black Sea): Density, composition, possible sources and associated organisms. *Turk J Fish Aquat Sci* 20, no. 2: 137–145. http://doi.org/10.4194/1303-2712-v20_2_06.

Baalkhuyur, F.M., Bin Dohaish, E.A., Elhalwagy, M.E.A., Alikunhi, N.M., AlSuwailem, A.M., Røstad, A., Coker, D.J., Berumen, M.L., and C.M. Duarte. 2018. Microplastic in the gastrointestinal tract of fishes along the Saudi Arabian Red Sea Coast. *Mar Pollut Bull* 131: 407–415. http://doi.org/10.1016/j.marpolbul.2018.04.040.

Barnes, D.K.A., Galani, F., Thompson, R.C., and M. Barlaz. 2009. Accumulation and fragmentation of plastic debris in global environments. *Phil Trans R Soc* 364: 1985–1998. http://doi.org/10.1098/rstb.2008.0205.

Brümmer, F., Schnep, U., Resch, J., Jemmali, R., Abdi, R., Kamel, H.M., Bonten, C., and R. Müller. 2022. In situ laboratory for plastic degradation in the Red Sea. *Sci Rep* 12: 11956. http://doi.org/10.1038/s41598-022-15310-7.

Chee, S.Y., Yee, J.C., Carey, D., Yusup, Y., and J.B. Gallagher. 2020. Anthropogenic marine debris accumulation in mangroves on Penang Island, Malaysia. *J Sustain Sci Manag (Malaysia)* 15, no. 6: 36–60.

De, K., Sautya, S., Dora, G.U., Gaikwad, S., Katke, D., and A. Salvi. 2023. Mangroves in the "Plasticene": High exposure of coastal mangroves to anthropogenic litter pollution along the Central-West coast of India. *Sci Total Environ* 858: 160071. http://doi.org/10.1016/j.scitotenv.2022.160071.

Debrot, A.O., Bron, P.S., and R. de León. 2013. Marine debris in mangroves and on the seabed: Largely-neglected litter problems. *Mar Pollut Bull* 72, no. 1: 1. http://doi.org/10.1016/j.marpolbul.2013.03.023.

Diem, A., Yacob, T.T., Taly, H., Vineeta, H., and Z. Karin. 2022. Marine litter in the Red Sea: Status and policy implications. *MICRO 2022, Online Atlas Edition: Plastic Pollution from MACRO to Nano*. http://doi.org/10.5281/zenodo.7216969.

Egessa, R., Nankabirwa, A., Basooma, R., and R. Nabwire. 2019. Occurrence, distribution and size relationships of plastic debris along shores and sediment of northern Lake Victoria. *Environ Pollut* 257: 113442. http://doi.org/10.1016/j.envpol.2019.113442.

Fauziah, S.H., Liyana, I.A., and P. Agamuthu. 2015. Plastic debris in the coastal environment: The invincible threat? Abundance of buried plastic debris on Malaysian beaches. *Waste Manag Res* 33, no. 9: 812–821. http://doi.org/10.1177/0734242X15588587.

Gerges, M.A. 2002. The Red Sea and Gulf of Aden Action Plan – Facing the challenges of an ocean gateway. *Ocean Coast Manag* 45: 885–903.

GESAMP. 2019. *Guidelines or the Monitoring and Assessment of Plastic Litter and Microplastics in the Ocean*. Eds. P.J. Kershaw, A. Turra and F. Galgani. IMO/FAO/UNESCO-IOC/UNIDO/WMO/IAEA/UN/UNEP/UNDP/ISA Joint Group of Experts on the Scientific Aspects of Marine Environmental Protection. Rep Stud GESAMP no. 99, 130 p.

Habib, R.Z., and T. Thiemann. 2022. Microplastic in the marine environment of the Red Sea – A short review. *Egypt J Aquat Res* 48, no. 4: 383–388. http://doi.org/10.1016/j.ejar.2022.03.002.

Haseler, M., Schernewski, G., Balciunas, A., and V. Sabaliauskaite. 2018. Monitoring methods for large micro- and meso-litter and applications at Baltic beaches. *J Coastal Conserv* 22, no. 1: 27–50. http://doi.org/10.1007/s11852-017-0497-5.

Ibrahim, E.A.A., Osman, N.A.R., and O.A.M. Eisa. 2020. Status of the beach litter in the UNESCO World Heritage Site of Dungonab and Mukkawar Island Marine National Park in Sudan, Red Sea. *Int J Ecol* 2020, no. 2: 6904745. http://doi.org/10.1155/2020/6904745.

Jiang, H., Farrar, J.T., Beardsley, R.C., Chen, R., and C. Chen. 2009. Zonal surface wind jets across the Red Sea due to mountain gap forcing along both sides of the Red Sea. *Geophys Res Lett* 36: L19605. http://doi.org/10.1029/2009GL040008.

John, J., Nandhini, A.R., Chellam, P.V., and M. Sillanpää. 2022. Microplastics in mangroves and coral reef ecosystems: A review. *Environ Chem Lett* 20: 397–416. http://doi.org/10.1007/s10311-021-01326-4.

Kaandorp, M.L.A., Henk, A., Dijkstra, H.A., and E. Sebille. 2021. Modelling size distributions of marine plastics under the influence of continuous cascading fragmentation. *Environ Res Lett*, 16, no. 5: 054075. http://doi.org/10.1088/1748-9326/abe9ea.

Kane, I.A., and M.A. Clare. 2019. Dispersion, accumulation, and the ultimate fate of microplastics in deep-marine environments: A review and future directions. *Front Earth Sci* 7: 80. http://doi.org/10.3389/feart.2019.00080.

Lee, J., Hong, S., Song, Y.K., Hong, S.H., Jang, Y.C., Jang, M., and W.J. Shim. 2013. Relationships among the abundances of plastic debris in different size classes on beaches in South Korea. *Mar Pollut Bull* 77, nos. 1–2: 349–354. http://doi.org/10.1016/j.marpolbul.2013.08.013.

Lee, J., Lee, J.S., Jang, Y.C., Hong, S.Y., Shim, W.J., Song, Y.K., Hong, S.H., Jang, M., Han, G.M., Kang, D., and S. Hong. 2015. Distribution and size relationships of plastic marine debris on beaches in South Korea. *Arch Environ Contam Toxicol* 69, no. 3: 288–298. http://doi.org/10.1007/s00244-015-0208-x.

Li, D., Zhao, L., Guo, Z., Yang, X., Deng, W., Zhong, H., and P. Zhou. 2021. Marine debris in the Beilun estuary mangrove forest: Monitoring, assessment and implications. *Int J Environ Res Public Health*, 18: 10826. http://doi.org/10.3390/ijerph182010826.

Manullang, C.Y. 2019. The abundance of plastic marine debris on beaches in Ambon Bay. *IOP Conf Ser Earth Environ Sci* 253: 012037. http://doi.org/10.1088/1755-1315/253/1/012037.

Manullang, C.Y., Barends, W., Polnaya, D., Soamole, A., and I. Rehalat. 2021. Marine litter and grading of the coastal areas of Ambon Bay, Indonesia. *Omni-Akuatika* 17, no. 2: 70–77. http://doi.org/10.20884/1.oa.2021.17.2.903.

Martin, C., Agusti, S., and C.M. Duarte. 2019b. Seasonality of marine plastic abundance in central Red Sea pelagic waters. *Sci Total Environ* 688: 536–541. http://doi.org/10.1016/j.scitotenv.2019.06.240.

Martin, C., Almahasheer, H., and C.M. Duarte. 2019a. Mangrove forests as traps for marine litter. *Environ Pollut* 247: 499–508. http://doi.org/10.1016/j.envpol.2019.01.067.

Martin, C., Baalkhuyur, F., Valluzzi, L., Saderne, V., Cusack, M., Almahasheer, H., Krishnakumar, P.K., Rabaoui, L., Qurban, M.A., Arias-Ortiz, A., Masqué, P., and C.M. Duarte. 2020. Exponential increase of plastic burial in mangrove sediments as a major plastic sink. *Sci Adv* 6, no. 44: eaaz5593. http://doi.org/10.1126/sciadv.aaz5593.

Martin, C., Young, C.A., Valluzzi, L., and C.M. Duarte. 2022. Ocean sediments as the global sink for marine micro- and mesoplastics. *Limnol Oceanog Lett* 7, no. 3: 235–243. http://doi.org/10.1002/lol2.10257.

Martin, C., Zhang, Q., Zhai, D., Zhang, X., and C.M. Duarte. 2021. Enabling a large-scale assessment of litter along Saudi Arabian Red Sea shores by combining drones and machine learning. *Environ Pollut* 277: 116730. http://doi.org/10.1016/j.envpol.2021.1167300269-7491.

Menezes, V.V., Farrar, T.J., and A.S. Bower. 2018. Westward mountain-gap wind jets of the northern Red Sea as seen by QuikSCAT. *Remote Sens Environ* 209: 677–699. http://doi.org/10.1016/j.rse.2018.02.075.

Okuku, E.O., Kiteresi, L., Owato, G., Otieno, K., Omire, J., Kombo, M.M., et al. 2021. Temporal trends of marine litter in a tropical recreational beach: A case study of Mkomani Beach, Kenya. *Mar Pollut Bull* 167: 112273. http://doi.org/10.1016/jmarpolbul.2021.112273.

PERSGA. 2014. *Coastal Marine Litter Assessment Guidelines for the Red Sea and Gulf of Aden*. PERSGA, Jeddah, Saudi Arabia, PERSGA Guideline Number GD.0021.

PERSGA. 2018. Regional action plan for the sustainable management of marine litter in the Red Sea and Gulf of Aden. *Report Number RP.0091*. PERSGA, Jeddah, Saudi Arabia.

PERSGA. 2023. *National Action Plan for the Sustainable Management of Marine Litter in the Republic of the Sudan*. PERSGA, Jeddah, Saudi Arabia, In press.

Perumal, K., Boopathi, V., Chellaiyan, S., Muthuramalingam, S., and P. Raja. 2021. Sources, spatial distribution, and abundance of marine debris on Thondi coast, Palk Bay, southeast coast of India. *Environ Sci Eur* 33: 136. http://doi.org/10.1186/s12302-021-00576-x.

Poluan, T.A., Sangari, J.R.R., Tilaar, F.F., Lumingas, L.H.L., Pelle, W.E., and R. Lasabuda. 2023. Identification of marine debris by focusing the study of clean coast index on Karang Ria Tuminting Beach. *J Ilmiah Platax* 11, no. 1: 95–104. http://doi.org/10.35800/jip.v10i2.44018.

Portman, M.E., and E.R. Brennan. 2017. Marine litter from beach-based sources: Case study of an eastern Mediterranean coastal town. *Waste Manag* 69: 535–544. http://doi.org/10.1016/j.wasman.2017.07.040.

Salazar, J.A., Gonza, R., Navarrete, A.L., Calle, P., Alava, J.J., and G.A. Domınguez. 2022. A temporal assessment of anthropogenic marine debris on sandy beaches from Ecuador's southern coast. *Front Mar Sci* 9: 977650. http://doi.org/10.3389/fmars.2022.977650.

Schroeder, J. 1982. Aspect of coastal management of the Sudanese Red Sea. *J Fac Mar Sci* 2: 45–68.

Sebille, E.V., Aliani, S., Law, K.L., Maximenko, N., Alsina, J.M., Bagaev, A., et al. 2020. The physical oceanography of the transport of floating marine debris. *Environ Res Lett* 15: 023003. http://doi.org/10.1088/1748-9326/ab6d7d.

Shamskhany, A., Li, Z., Patel, P., and S. Karimpour. 2021. Evidence of microplastic size impact on mobility and transport in the marine environment: A review and synthesis of recent research. *Front Mar Sci* 8: 760649. http://doi.org/10.3389/fmars.2021.760649.

Silburn, B., Bakir, A., Binetti, U., Russell, J., Kohler, P., Preston-Whyte, F., Meakins, B., van Hoytema, N., Andrews, G., Carrias, A., and T. Maes. 2022. A baseline study of macro, meso and micro litter in the Belize River basin, from catchment to coast. *ICES J Mar Sci* 2022: fsab268. http://doi.org/10.1093/icesjms/fsab268.

Silva, M.L., Castro, R.C., Sales, A.S., and F.V. Araújob. 2018. Marine debris on beaches of Arraial do Cabo, RJ, Brazil: An important coastal tourist destination. *Mar Pollut Bull* 130: 153–158. http://doi.org/10.1016/j.marpolbul.2018.03.026.

Tewelde, Y., Teklemariam, Y., and K. Estifanos. 2021. Baseline of beached marine debris in three sites, Red Sea – Eritrea. *Int J Sci Eng Res* 12, no. 2: 292–308.

Veerasingam, S., Al-Khayat, J.A., Aboobacker, V.M., Hamza, S., and P. Vethamony. 2020. Sources, spatial distribution and characteristics of marine litter along the west coast of Qatar. *Mar Pollut Bull* 159: 111478. http://doi.org/10.1016/j.marpolbul.2020.111478.

Woodall, L.C., Sanchez-Vidal, A., Canals, M., Paterson, G.L.J., Coppock, R., Sleight, V., Calafat, A., Rogers, A.D., Narayanaswamy, B.E., and T. Richard. 2014. The deep sea is a major sink for microplastic debris. *Roy Soc Open Sci* 1: 140317. http://doi.org/10.1098/rsos.140317.

Yin, C.S., Chai, Y.J., Carey, D., Yusup, Y., and J.B. Gallagher. 2019. Anthropogenic marine debris and its dynamics across peri-urban and urban mangroves on Penang Island, Malaysia. *J Sustain Sci Manag* 15, no. 6: 41–67. http://doi.org/10.1101/756106.

An overview of marine litter on the Sudanese Red Sea coast

13

Awatif A. Sugga and Dirar H. Nasr

13.1 INTRODUCTION

A number of writers have contributed to the description of the Red Sea, including its physical characteristics (Morcos 1970; Sheppard et al. 1992), its key environments (PERSGA 2006; Nasr 2015; PERSGA/GEF 2001; Vine and Vine 1980; Vine 1985; Gunnar and Ormond 1981; Edwards and Head 1987) and the conservation of its natural resources (Gladstone et al. 1999; PERSGA/GEF 2001; Ormond 1980, 1987; Schroeder et al. 1980).

There is a consensus among these investigators that the Red Sea is a semi-enclosed sea and that it differs from many other tropical areas as it receives little natural fresh water drainage from adjacent lands; therefore, it has waters with relatively high salinity. Another feature is that, as the concentration of nutrients in the Red Sea is low, plankton will be scarcer, and this, in turn, leads to relatively few pelagic fish. Thus, we find that most of the various marine resources living in the Red Sea are concentrated in three natural habitats according to the nature of the bottom; these areas are mangroves in muddy areas, seagrass beds in protected sandy/muddy bottom areas and coral reefs in hard substrata up to a depth of about one hundred metres. On the other hand, the Red Sea is an important strategic route for increasing international navigation and shipping movements. Hence, there are serious concerns regarding marine environmental pollution, including oil and marine litter pollution.

Marine Litter is identified by UNEP as one of nine source categories in the Global Program of Action (GPA) for the Protection of the Marine Environment from Land-Based Activities. It was recognized as a priority issue by UNEP's Governmental Council at the Global Ministerial Environment Meeting in March 2004. The dominance of plastic marine debris and its ability to transport toxic substances are considered by UNEP (Gladstone 2013) to be primary emerging issues in the global environment. Waterborne plastics pose serious direct and indirect threats to fish, seabirds and marine mammals and reptiles, including choking, starvation, entanglement and poisoning via accumulation in marine food chains (Gladstone 2013). Plastics and other marine litter also cause a deterioration in appearance, which is of significant concern, given that coastal tourism industries are extremely important to the national economies of the region. More than 600 million people live in coastal regions that are less than 10 m above sea level, and nearly 2.5 billion people live within 100 km of the coast (UNDESA 2019). Those regions are experiencing higher rates of population growth and urbanization than inland regions (Neumann et al. 2015). Such growth has resulted in many economic benefits to coastal regions, including improved transportation and increased trade, tourism and food production, as well as social, recreational and cultural benefits (Clark and Johnston 2017).

Depending on their type, size and location, marine litter and plastics can cause lethal and sub-lethal effects on marine life. The lethal and sub-lethal effects of plastics include ingestion by whales, seals, turtles, birds and fish, potentially leading to starvation and lacerations in internal systems, and the smothering of coral reefs, causing deprivation of oxygen and light; drowning of turtles, birds and mammals due to entanglement in abandoned fishing gear and plastic packaging; and physiological stress and toxicological harm arising from the ingestion of microplastics by plankton, shellfish, fish and marine worms, all of which are critical to ecosystem functioning (Anbumani and Kakkar 2018; Duncan et al. 2018a, 2018b; Hallanger and Gabrielsen 2018; McNeish et al. 2018; Reynolds and Ryan 2018; Arias et al. 2019; Battisti et al. 2019; Donohue et al. 2019; Nelms et al. 2019; Sun et al. 2019; Landrigan et al. 2020; Vethaak and Legler 2021; Thiel et al. 2018; Bucci et al. 2019; Windsor et al. 2019; Woods et al. 2019).

The Regional Organization for the Conservation of the Environment of the Red Sea and Gulf of Aden (PERSGA) has prepared Coastal Marine Litter Assessment Guidelines for the Red Sea and Gulf of Aden; nevertheless, to date, there are insufficient studies in the region to determine the full extent of the social and economic risks emerging from litter mismanagement in coastal and marine environments. According to UNGA (2015), much of the unique marine biodiversity is facing major stresses such as overfishing, destruction and pollution on a global scale. This area includes adjacent wetlands, estuaries and the seasonal stream channels adjacent to the sea. It was recognized that new cities in many countries are established near coasts, and the populations of old coastal cities are increasing faster than in non-coastal cities, necessitating raising awareness

DOI: 10.1201/9781003321408-17

of the implications of increasing solid waste accumulation in the marine ecosystem due to human activities.

The aim of this overview is to highlight the fact that inadequate regulations or poor enforcement of regulations combined with weak public awareness in the Sudanese coastal areas allow marine litter problems to escalate. The overview covers the status of marine litter and debris in the vicinity of two main coastal cities, Port Sudan and Suakin.

13.2 STATUS OF MARINE ENVIRONMENT POLLUTION

Marine litter and plastic pollution are accumulating in the world's oceans at an unprecedented rate. The volume of plastics currently in the oceans has been estimated at between 75 million and 199 million tonnes (Jang et al. 2015; Lebreton et al. 2019; Borrelle et al. 2020; Lau et al. 2020).

Since 2000, it has been shown that 44% of the pollutants in marine waters is carried from terrestrial uplands and 10% of pollutants are dumped directly from disposal of waste in seas and oceans (El-Arosi 2000). The impacts to the marine environment from land-based sources of pollution and activities have been recognized for many years (Loya 1975; Freemantle et al. 1978; Walker and Ormond 1982). Threats from land-based sources and activities have risen rapidly in recent decades in conjunction with increasing urbanization and coastal developments. The estimated US$6–19 billion annual costs of marine plastic pollution, based on impacts including those on tourism, fisheries and aquaculture and the expense of clean-ups (Deloitte 2019), represent only a small percentage of the global plastic market, valued at more than US$579 billion in 2020 (Statista 2021).

The human health impacts of marine litter and plastic pollution arise mainly from inadequate waste handling, especially on land; ingestion of contaminated seafoods; and exposure to pathogenic bacteria and substances of concern transported into coastal waters by floating plastics (Landrigan et al. 2020). It is well known that the health of terrestrial and extremely productive marine ecosystems such as seagrass, mangroves and coral reefs are closely interrelated, and any destruction of an ecosystem will lead to the collapse of the marine community. Marine litter that accumulates in seas and oceans along coasts contaminates the water and coasts to a level that is considered pollution (Weiss et al. 2015). Mangroves retain marine litter in their extensive root systems, and this leads to adverse effects on the communities associated with mangrove ecosystems (Winarni et al. 2022).

13.3 STATUS OF MARINE LITTER IN SUDANESE COASTAL AREAS

Estuaries and coastal marine waters are persistently subjected to pollutants from sea- and land- based sources. Near human settlements, solid waste and litter are dumped on the shore and into the sea. This is likely to have significant impact on nearshore habitats and their biota as dumped waste is spread via winds and currents, often leading to pollution of offshore habitats (Gladstone 2013). The significant increase in human population and changing consumption and production patterns in the coastal zone have resulted in higher solid waste production and the consequent need for improved waste disposal. The Sudanese Red Sea coast is about 750 kilometres long, including bays and inlets. Most of the Sudanese Red Sea is still largely pristine (PERSGA/GEF 2001, 2003). Port Sudan is the largest coastal city, with a population of about 400,000. The state is well endowed with natural resources, but its inhabitants are comparatively poor, especially those living in rural and marginal urban areas (PERSGA/GEF 2001). Anthropogenic pressure is particularly high in the vicinity of the two coastal cities, Port Sudan and Suakin.

Solid waste is concentrated along much of the shoreline in the Port Sudan area along the main corniche road (Figure 13.1); in inlets near the television station at Daim Mayo, where beaches are partly used as storage areas (Figure 13.2); and at different parts of the Port Sudan shoreline. The solid waste is primarily composed of the rubber cast-offs from flip-flop stampings, various plastic bags and household garbage. This problem is due to a lack of efficient waste collection and disposal systems, aggravated by a general lack of awareness. The problem is especially

FIGURE 13.1 Solid waste accumulation along the beach at Port Sudan Corniche Road.

FIGURE 13.2 Solid waste accumulation in inlets near the television station at Daim Mayo. It seems that beaches are partly used as storage areas.

severe around Port Sudan (PERSGA/GEF 2001). The disposal of solid waste causes localized problems for habitats in coastal areas of Port Sudan and Suakin cities, where waste management systems are not well developed. Dumping of solid wastes from households and industry into coastal habitats is causing a significant problem in the coastal areas of Port Sudan City. Plastic bags appear to be the dominant form of marine debris in this area, in addition to bottles, containers, plastics, paper and fishing gear (Figure 13.3). Even in cases where pollution is not quantified, effects on marine life can be clearly visible where killed fish are observed among the dumped debris (Figure 13.4). No substantial improvements on the ground have been observed in the coastal areas of Port Sudan City.

As mentioned in the *Red Sea State Encyclopedia* (2006), marine litter generation in areas away from Suakin and Port Sudan comes from accidental loss of cargo and containers during shipping, leading to discharge of marine litter, extreme weather events, vessel collisions and improper securing of cargo, resulting in containers and cargo being lost at sea from international shipping that transits the Red Sea; fishing vessels, including those engaged in international fishing activities, can inadvertently lose or discard fishing gear, such as nets, lines and traps. Thus, coastal areas away from Suakin and Port Sudan may receive marine litter that has drifted from shipping routes.

FIGURE 13.3 Plastics are the dominant form of marine debris, in addition to bottles, containers, sheets and fishing gear at different parts of the Port Sudan coast.

FIGURE 13.4 Killed fish among the dumped debris in parts of Suakin City.

On the other hand, each neighbouring country around the Red Sea has its own unique characteristics and challenges related to marine litter. In Egypt, for instance, the tourism industry increases domestic production of plastic and marine litter. In Saudi Arabia, marine litter is a growing concern, mainly due to increasing population, urbanization and tourism. In Jordan, marine litter is a significant issue, particularly in tourist areas like Aqaba. Yemen faces numerous challenges related to marine litter, including inadequate waste management infrastructure, conflict and economic instability. However, compared to some Asian countries, the potential contribution of plastic to the ocean from PERSGA member countries is much less (Jambeck et al. 2015).

13.4 EFFORTS BY PERSGA TO MITIGATE MARINE DEBRIS AND LITTER

It should be noted that the term "marine debris" should be viewed as synonymous with the term "marine litter", used in the Jeddah Convention LBAs, and "garbage", used in MARPOL 73/78 Annex V. Much marine debris originates from land, addressed under the Jeddah Convention Protocol Concerning the Protection of the Marine Environment from Land-Based Activities in the Red Sea and Gulf of Aden (LBA Protocol) (PERSGA 2005), although a proportion originates from shipping (addressed under MARPOL 73/78 Annex V; PERSGA 2020).

With the encouragement and support of the Regional Organization for the Conservation of the Environment of the Red Sea and Gulf of Aden (PERSGA), national programmes of action (NPAs) are being developed as a prerequisite for implementing the Regional Protocol Concerning Protection of the Marine Environment from Land Based Activities signed in 2005. These plans summarize country-specific issues regarding pollutants, objectives, actions, activities, outputs, responsibility, linkages, public outreach and possible funding sources. NPAs have, to date, been completed for Sudan.

In 2008, PERSGA produced an assessment of marine litter in the PERSGA region (PERSGA/UNEP 2008), and in 2013, PERSGA produced Coastal Marine Litter Assessment Guidelines for the Red Sea and Gulf of Aden (PERSGA 2013). Surveys using these guidelines were undertaken between 2008 and 2018. A Regional Action Plan for the Sustainable Management of Marine Litter in the Red Sea and Gulf of Aden was produced in 2018 (PERSGA/UNEP 2018). Guidelines to prepare national action plans to manage marine litter in the PERSGA region were produced in 2019 (PERSGA 2020).

Based on the result of a recent regional assessment of marine litter, PERSGA has developed a Regional Action Plan for Management of Marine litter and published the plan attached to the status report in collaboration with UNEP-Regional Seas in 2009. During 2008 and 2009, PERSGA also organized several coast-cleaning campaigns under the slogan "Clean up our seas", with the distribution of educational materials in member states, including Sudan, in collaboration with

FIGURE 13.5 Beach cleaning campaign in the area of the Faculty of Marine Sciences and Fisheries as part of the activities of the national action plan (NAP) to protect the marine environment from land-based sources, implemented by PERSGA.

national focal points, local authorities and NGOs (Figure 13.5). Coastal clean-up events demonstrate an understanding of the need for and a commitment to reducing marine litter.

It is obvious that national and regional efforts to combat the marine debris (marine litter) problem, mainly by clean-up and awareness-raising campaigns, are numerous. However, there is always a need for more efforts to cure impacts that may result in degradation of the coastal and marine resources, leading to significant economic losses in tourism, fisheries and other coastal and marine activities. In 2012, a workshop aimed at assessing the status of marine litter monitoring and management in the coastal and marine environment of the Red Sea and Gulf of Aden (RSGA) was also conducted within the framework of PERSGA's efforts to establish a regional monitoring programme (Gladstone 2013).

13.5 THE ROLE OF NATIONAL NGOS IN MARINE LITTER SURVEYS AND CLEAN-UPS

A case study for identifying the role of civil society organizations in raising awareness and protecting the environment in Port Sudan City (Red Sea State) was carried out by Idris and Sugga (2020). The study showed that there are several civil societies engaged in environmental activities, including cleaning and collection of waste and plastic from the beaches and coasts of the Red Sea State. The interviewers agreed that the presence of civil society organizations is an absolute necessity (56%); the obstacles to the work of civil society organizations in the state include the lack of funding (48%), the lack of state institutions' support for environmental and health education in the state (80%), the need for environmental awareness in schools (72%) and the lack of legal and regulatory procedures for these organizations (58%). Awareness of environmental problems in general and pollution in particular, and its interaction with targeted environmental programmes among Port Sudan's citizens, is low (26%). There is a need to support organizations working in the field of the environment to work with a civil philosophy based on the support of the state and its political will.

Bashir and Abdallah (2018) found that the awareness level among Port Sudan city residents about the effects of solid waste on the surrounding environment was very high; almost all the interviewed people agreed that the solid waste had negative effects on the environment and human health, and if it reached the sea, it would adversely impact the marine life and coastal zone. On the other hand, 100% of the interviewed people agreed that the collection of solid waste and its dumping in suitable dumping sites far from residential areas is the best and most secure method for solid waste treatment. As for evaluation of the knowledge of interviewed people about the used methodologies and techniques for solid waste treatment, only 58.1% agreed that the ongoing treatment was suitable and successful in curing the problem. The awareness level about the importance of prohibiting direct dumping of solid waste in the sea or coastal zone was very high, up to 100%. A realistic view of the environmental situation is that it is not consistent with the results from the people interviewed.

Marine litter is a complex problem that spans geographical boundaries and needs to involve multiple researchers' efforts. Many researchers may not fully understand the impact of marine litter or its role in contributing to the problem. The lack of information about marine litter makes it difficult to assess its full extent along the Red Sea State coast; the main challenge facing research institutions is developing effective and efficient methods for detecting and quantifying marine litter. Collecting representative samples of marine litter and identifying and classifying the items is also difficult due to their degradation, fragmentation and transformation over time. Understanding the ecological, economic and social impacts of marine litter requires collaboration in long-term monitoring. Determining the extent to which marine litter affects marine ecosystems, wildlife, human health and coastal communities is a complex task. Harmonization of data from different sources and research studies is a challenge, and securing adequate funding and resources for marine litter research is another one.

13.6 REGULATIONS AND GOVERNANCE

13.6.1 The Ministry of Environment and Tourism

The Ministry was established in 1994 with the mandate of coordinating environmental conservation and promoting tourism. Under its umbrella, it includes the Higher Council for the Environment and Natural Resources (HCENR), which is the technical branch in charge of coordination, policy making and international cooperation; the National Tourism Corporation, which is responsible for the planning and promotion of tourism; and the General Administration for Wildlife Conservation, which is charged with the protection and management of wildlife, including protected areas.

Sudan is a signatory to a number of international conventions and agreements that generally promote the protection of the environment. National legislation in Sudan indirectly supports pollution control measures that affect the marine environment. It is suggested that the demarcation and enforcement of marine protected areas will strengthen national legislation (Pilcher and Nasr 2000).

13.6.2 International agreements

Sudan has ratified the following conventions and protocols:

- Bamako Convention on the Ban of the Import into Africa and the Control of Transboundary Movement and Management of Hazardous Wastes within Africa (1993)
- Convention Concerning the Protection of the World Cultural and Natural Heritage (1974)
- Convention on Biological Diversity (1995)
- Convention on International Trade in Endangered Species of Wild Fauna and Flora (1982)
- Protocol Concerning Regional Cooperation in Combating Pollution by Oil and Other Harmful Substances in Cases of Emergency (1984)
- Regional Convention for the Conservation of the Red Sea and Gulf of Aden Environment (1984)
- United Nations Convention on the Law of the Sea (1985)

Apart from the regional Jeddah Convention (1984) and its protocols, none of these treaties is concerned with protection of the marine environment. Even the Law of the Sea Convention does not provide either the obligations or the benefits of those international treaties which are specifically focused on the marine environment. MARPOL has not yet been ratified because of a lack of port reception facilities.

13.6.3 National legislation

The following national legislation relevant to marine litter and waste management in Sudan are primarily governed by the Environmental Protection Act (1992).

- Environmental Law of 2001 and its subsequent amendments (Establishment of National Environment Protection Council Act, 2001).
- Hazardous Waste Management Regulations (2003): These regulations provide guidelines for the management, transportation, storage and disposal of hazardous waste in Sudan.
- Solid Waste Management Regulations (2008): These regulations focus on the management of solid waste, including waste collection, transportation, recycling and disposal.
- Environmental Impact Assessment (EIA) Regulations (2016): These regulations require the assessment and evaluation of potential environmental impacts of development projects, including waste management facilities.
- Sudanese Standard and Metrology Organization (SSMO): This organization sets standards for waste management practices, including specifications for waste treatment and disposal methods, waste containers and landfill design.

The following national legislative frameworks are in place that relate to the dumping of refuse:

Sudanese Fishery Ordinances and Regulations: These date back to 1937 and was amended in 1975 and 1978. It prohibits overfishing; dumping of refuse, including oil, into the sea; and the collection of corals, shells and aquarium fish.

Environmental Health Act: This was established in 1975 and prohibits the dumping into the sea of any item that is harmful to humans or animals.

With regard to the implementation issue of international, regional or national legislation, it can be observed that there are insufficient financial resources to secure implementation procedures. In addition, some legislation did not include methods of calculating loss and compensation in environmental violations. There should be an emphasis and an enforcement of the rules stating the "polluter pays principle", which means that the person who causes the environmental pollution should pay for all the damage resulting from that pollution (PERSGA/GEF 2004).

13.6.4 Environmental Management Plan (EMP)

Specific preliminary management plans have been developed for these environmentally special areas, focusing on the environmental pressures and the responses of management to those pressures (PERSGA/ALECSO 2007). The full Environmental Management Plan (EMP) with modifications was agreed on at the end of the workshop on 29 November 1997. The EMP was devised to ensure the sustainable development of the economic free zone (EFZ) and sustainable utilization of the biological resources of the coastal stretch between Port Sudan and Suakin. A logical framework was prepared for implementation of a sustainable EMP, which had the overall goal of ensuring that the status of marine resources in the region is improved.

13.7 CONCLUSIONS

A specific preliminary management plan has been developed in Sudan to ensure sustainable utilization of the biological resources of the coastal stretch between Port Sudan and Suakin. Sudan has ratified a number of national, regional and international agreements, and PERSGA recently developed qualitative and quantitative assessments of the scale of marine litter problems in the countries of the region, including risks to key biodiversity assets, a tool for evaluating legislative and voluntary intervention strategies and its flexibility to meet local requirements and support to develop capacities which were the main pillars of the PERSGA assessment. Meanwhile, debris (litter) continues to accumulate along the shoreline of the two cities of Port Sudan and Suakin.

Several steps are essential to the solution of the marine debris (litter) problem, including integrated management of solid waste through minimizing production, increasing recycling opportunities and final safe disposal of solid waste. These steps require enhancement of human behaviour and raising awareness, together with securing financial resources for this purpose. Other treatment solutions include strengthening the environmental administration systems, monitoring and relations between concerned regulating entities.

REFERENCES

Anbumani, S., and P. Kakkar. 2018. Ecotoxicological effects of microplastics on biota: A review. *Environ Sci Pollut Res* 25: 14373–14396. http://doi.org/10.1007/s11356-018-1999-x.

Arias, A.H., Ronda, A.C., Oliva, A.L., and J.E. Marcovecchio. 2019. Evidence of microplastic ingestion by fish from the Bahía Blanca estuary in Argentina, South America. *Bull Environ Contam Toxicol* 102, no. 6: 750–756. http://doi.org/10.1007/s00128-019-02604-2.

Bashir, A., and M. Abdallah. 2018. *Management Efficiency of Shipping Waste Handling at Port Sudan Port. Study Case Port Sudan Green Port*. BSc Thesis, Faculty of Marine Sciences and Fisheries, Red Sea University.

Battisti, C., Staffieri, E., Poeta, G., Sorace, A., Luiselli, L., and G. Amori. 2019. Interactions between anthropogenic litter and birds: A global review with a 'black-list' of species. *Mar Pollut Bull* 138: 93–114. http://doi.org/10.1016/j.marpolbul.2018.11.017.

Borrelle, S.B., Ringma, J., Law, K.L., Monnahan, C.C., Lebreton, L., McGiverb, A., et al. 2020. Predicted growth in plastic waste exceeds efforts to mitigate plastic pollution. *Science* 369, no. 6510: 1515–1518. https://doi.org/10.1126/science.aba3656.

Bucci, K., Tulio, M., and C.M. Rochman. 2019. What is known and unknown about the effects of plastic pollution: A meta-analysis and systematic review. *Ecol Appl* 30, no. 2: e02044. http://doi.org/10.1002/eap.2044.

Clark, G.F., and E.L. Johnston. 2017. Australia State of the environment 2016: Coasts. *Independent Report to the Australian Government Minister for Environment and Energy*. Canberra: Australian Government Department of the Environment and Energy.

Deloitte. 2019. *The Price Tag of Plastic Pollution: An Economic Assessment of River Plastic*. www2.deloitte.com/content/dam/Deloitte/nl/Documents/strategy-analytics-and-ma/deloitte-nl-strategy-analytics-andma-the-price-tag-of-plastic-pollution.pdf. Accessed 12 February 2021.

Donohue, M.J., Masura, J., Gelatt, T., Ream, R., Baker, J.D., Faulhaber, K., et al. 2019. Evaluating exposure of northern fur seals, Callorhinus ursinus, to microplastic pollution through faecal analysis. *Mar Pollut Bull* 138: 213–221. http://doi.org/10.1016/j.marpolbul.2018.11.036.

Duncan, E.M., Arrowsmith, J., Bain, C., Broderick, A.C., Lee, J., Metcalfe, K., et al. 2018b. The true depth of the Mediterranean plastic problem: Extreme microplastic pollution on marine turtle nesting beaches in Cyprus. *Mar Pollut Bull* 136: 334–340. http://doi.org/10.1016/j.marpolbul.2018.09019.

Duncan, E.M., Broderick, A.C., Fuller, W.J., Galloway, T.S., Godfrey, M.H., Hamann, M., et al. 2018a. Microplastic ingestion ubiquitous in marine turtles. *Glob Change Biol* 25: 744–752. http://doi.org/10.1111/gcb.14519.

Edwards, A.J., and S.M. Head (eds.). 1987. *Red Sea*. Oxford: Pergamon Press, 441 p.

El-Arosi. 2000. *Environmental Pollution and Its Pollutants* (in Arabic). Alexandria: Modern Knowledge Library, 307 p.

Freemantle, M.H., Hulings, N., Mulqi, M., and E.C. Watton. 1978. Calcium and phosphate in the Jordanian Gulf of Aqaba. *Mar Pollut Bull* 9: 79–80.

Gladstone, W. 2013. *Unpublished Report Submitted to PERSGA on Marine Pollution in PERSGA Member Countries*, PERSGA, Jeddah.

Gladstone, W., Tawfiq, N., Nasr, D., Andersen, I., Cheung, C., Drammeh, H., Krupp, F., and S. Lintner. 1999. Sustainable use of renewable resources and conservation in the Red Sea and Gulf of Aden: Issues, needs and strategic actions. *Ocean Coastal Manage* 42: 671–697.

Gunnar, B., and R. Ormond. 1981. *Red Sea Coral Reefs*. London and Boston: Kegan Paul International.

Hallanger, I.G., and G.W. Gabrielsen. 2018. Plastics in the European Arctic. *Brief Report No. 045*, Norwegian Polar Institute. www.synturf.org/images/NPI_Report_-_Kortrapport45.pdf. Accessed 12 January 2021.

Idris, S.A., and A.A. Sugga. 2020. *Identifying the Role of Civil Society Organizations in Raising Awareness and Protecting the Environment in the Red Sea State (A Case Study of the City of Port Sudan): Complementary Research Submitted to the College of Graduate Studies and Scientific Research for a Master Degree in Environmental Studies*. Red Sea University, Sudan (unpublished).

Jambeck, J.R., Geyer, R., Wilcox, C., Siegler, T.R., Perryman, M., Andrady, A., Narayan, R., and K.L. Law. 2015. Plastic waste inputs from land into the ocean. *Science* 347: 768–771.

Jang, Y.C., Lee, J., Hong, S., Choi, H.W., Shim, W.J., and S.Y. Hong. 2015. Estimating the global inflow and stock of plastic marine debris using material flow analysis. *J Korean Soc Mar Environ Energy* 18: 263–273.

Landrigan, P.J., Stegeman, J., Fleming, L., Allemand, D., Anderson, D., Backer, L., et al. 2020. Human health and ocean pollution. *Ann Glob Health* 86, no. 1: 1–64. http://doi.org/10.5334/aogh.2831.

Lau, W.Y., Shiran, Y., Bailey, R.M., Cook, E., Stutchey, M.R., Koskella, J., et al. 2020. Evaluating scenarios toward zero plastic pollution. *Science* 369, no. 6510: 1455–1461. http://doi.org/10.1126/science.aba9475; or www.pewtrusts.org/en/research-and-analysis/articles/2020/10/08/plastic-pollution-rampant-worldwide-could-be-cut-by-80-percent-in-20-years.

Lebreton, L., Egger, M., and B. Slat. 2019. A global mass budget for positively buoyant macroplastic debris in the ocean. *Sci Rep* 9: 12922 [see also Lebreton, L., Egger, M., and B. Slat. 2020. Author correction: A global mass budget for positively buoyant macroplastic debris in the ocean. *Sci Rep* 10: 1841]. http://doi.org/10.1038/s41598-019-49413-5.

Loya, Y. 1975. Possible effects of water pollution on the community structure of Red Sea corals. *Mar Biol* 29: 177–185.

McNeish, R.E., Kim, L.H., Barrett, H.A., Mason, S.A, Kelly, J.J., and T.J. Hoellein. 2018. Microplastic in riverine fish is connected to species traits. *Sci Rep* 8, no. 1: 11639. http://doi.org/10.1038/s41598-018-29980-9.

Morcos, S.A. 1970. Physical and chemical oceanography of the Red Sea. *Oceanogr Mar Biol Ann Rev* 8: 73–202.

Nasr, D.H. 2015. Coral reefs of the Red Sea with special reference to the Sudanese coastal area. In: *The Red Sea: The Formation, Morphology, Oceanography and Environment of a Young Ocean Basin*. Eds. N.M.A. Najeeb and I.C.F. Stewart, 453–469. Berlin and Heidelberg: Springer Earth System Sciences.

Nelms, S.E., Barnett, J., Brownlow, A., Davison, N.J., Deaville, R., Galloway, T.S., et al. 2019. Microplastics in marine mammals stranded around the British coast: Ubiquitous but transitory? *Sci Rep* 9, no. 1: 1075. http://doi.org/10.1038/s41598-018-37428-3.

Neumann, B., et al. 2015. Future coastal population growth and exposure to sea-level rise and coastal flooding – a global assessment. *PLoS ONE* 10, no. 3: e0118571.

Ormond, R.F.G. 1980. Management and conservation of Red Sea habitats. *Proc Symp Coast Mar Environ Red Sea and Gulf of Aden and Tropical Western Indian Ocean*. University of Khartoum, Sudan, 2, pp. 137–162.

Ormond, R.F.G. 1987. Conservation and management. In: *Red Sea*. Eds. A. Edwards and S.M. Head, 405–423. Oxford: Pergamon Press.

PERSGA. 2005. *Protocol Concerning the Protection of the Marine Environment from Land-Based Activities in the Red Sea and Gulf of Aden*. PERSGA. www.persga.org/Documents/Doc_62_20090211124355.pdf.

PERSGA. 2006. State of the marine environment. *Report for the Red Sea and Gulf of Aden*. PERSGA, Jeddah.

PERSGA. 2013. Coastal marine litter assessment guidelines for the Red Sea and Gulf of Aden. *PERSGA Report No. 30*. PERSGA, Jeddah, Saudi Arabia.

PERSGA. 2020. State of the marine environment. *Report for the Red Sea and Gulf of Aden (SOMERSGA II)*. PERSGA, Jeddah. https://persga.org/Documents/Publications/QR_Downloads/English/SOMERSGA_2020.pdf.

PERSGA/ALECSO. 2007. Survey of habitats in Sudan and plans for their protection. *PERSGA Technical Series No. 13*. PERSGA, Jeddah.

PERSGA/GEF. 2001. *Strategic Action Programme for the Red Sea and Gulf of Aden, Country Reports*. PERSGA, Jeddah and the World Bank, Washington, DC, 205 p.

PERSGA/GEF. 2003. Coral reefs in the Red Sea and Gulf of Aden. Surveys 1990 to 2000 summary and recommendations. *PERSGA Technical Series No. 7*. PERSGA, Jeddah.

PERSGA/GEF. 2004. *Final Report on Marine Environmental Legislation in the Red Sea and Gulf of Aden*. PERSGA, Jeddah.

PERSGA/UNEP. 2008. *Marine Litter in the PERSGA Region*. PERSGA, Jeddah. www.persga.org//Files/Common/Flipping_Books_Downloads/Marine_Litter_in_the_PERSGA_Region.pdf.

PERSGA/UNEP. 2018. Regional action plan for the sustainable management of marine litter in the Red Sea and Gulf of Aden. *Report Number RP 0091*. PERSGA, Jeddah, Saudi Arabia.

Pilcher, N., and D. Nasr. 2000. The status of coral reefs in Sudan – 2000. Report submitted to PERSGA for the preparation of the regional status of coral reefs in the Red Sea and Gulf of Aden. In: *Status of Coral Reefs of the World 2000*. Ed. C. Wilkinson, 35–54. Townsville, QLD: Australian Institute of Marine Science.

Red Sea State Encyclopedia. 2006. *Ministry of Finance and National Economy*. Red Sea State, Sudan (in Arabic).

Reynolds, C., and P.G. Ryan. 2018. Micro-plastic ingestion by waterbirds from contaminated wetlands in South Africa. *Mar Pollut Bull* 126: 330–333. http://doi.org/10.1016/j.marpolbul.2017.11.021.

Schroeder, J.H., Nasr, D.H., and F.H. Idris. 1980. Coral reef conservation in the Sudanese Red Sea. *Proc Symposium on the Coastal and Marine Environment of the Red Sea*, Gulf of Aden and Tropical Western Indian Ocean, Khartoum, pp. 163–178.

Sheppard, C., Price, A., and C. Roberts. 1992. *Marine Ecology of the Arabian Region*. London: Academic Press.

Statista. 2021. *Global Plastic Market Size 2016–2028* (published by I. Tiseo, 24 June 2021). www.statista.com/statistics/1060583/globalmarket-value-of-plastic/. Accessed 12 September 2021.

Sun, J., Dai, X., Wang, Q., van Loosdrecht, M.C., and B.J. Ni. 2019. Microplastics in wastewater treatment plants: Detection, occurrence and removal. *Water Res* 152: 21–37. http://doi.org/10.1016/j.watres.2018.12.050.

Thiel, M., Luna-Jorquera, G., Álvarez-Varas, R., Gallardo, C., Hinojosa, I.A., Luna, N., et al. 2018. Impacts of marine plastic pollution from continental coasts to subtropical gyres – fish, seabirds, and other vertebrates in the SE Pacific. *Front Mar Sci* 5: 238. http://doi.org/10.3389/fmars.2018.00238.

UNDESA. 2019. *Percentage of Total Population Living in Coastal Areas*. United Nations, Department of Economic and Social Affairs, Population Division, New York. ST/ESA/SER.A/421. https://sedac.ciesin.columbia.edu/es/papers/Coastal_Zone_Pop_Method.pdf.

UNGA. 2015. *Transforming Our World: The 2030 Agenda for Sustainable Development*. A/Res/70/1. 25th September 2015. United Nations General Assembly. https://sustainabledevelopment.un.org/post2015/transformingourworld.

Vethaak, A.D., and J. Legler. 2021. Microplastics and human health. *Science* 371: 672–674. http://doi.org/10.1126/science.abe5041.

Vine, P.J. 1985. *The Red Sea*. London: Immel Publishing.

Vine, P.J., and M.P. Vine. 1980. Ecology of Sudanese coral reefs with particular reference to reef morphology and distribution of fishes. *Proc Symp Coast Mar Environ Red Sea, Gulf Aden Trop Western Indian Ocean* 1: 87–140.

Walker, D.I., and R.F.G. Ormond. 1982. Coral death from sewage and phosphate pollution at Aqaba, Red Sea. *Mar Pollut Bull* 13: 21–25.

Weiss, Y., McNeill, J., Pearson, D.G., Nowell, G.M., and C.J. Ottley. 2015. High saline fluids from a subducting slab as the source of fluid-rich diamond. *Nature* 524, no. 7565: 339–342. http://doi.org/10.1038/nature14857.

Winarni, I., Uju, J.S., and T. Wibowo. 2022. Bioethanol production from seaweed solid waste of agar processing biomass. IOP Conf Ser: Earth Environ Sci. IOP Publishing Ltd., IOP Conference Series: 4th Int Conf Transdisciplinary Research on Environmental Problems in Southeast Asia 16/09/2021–18/09/2021. *Earth Environ Sci* 1027: 012029. http://doi.org/10.1088/1755-1315/1027/1/012029.

Windsor, F.M., Tilley, R.M., Tyler, C.R., and S.J. Omerod. 2019. Microplastic ingestion by riverine macroinvertebrates. *Sci Total Environ* 646: 68–74. http://doi.org/10.1016/j.scitotenv.2018.07.271.

Woods, J.S., Rødder, G., and F. Verones. 2019. An effect factor approach for quantifying the entanglement impact on marine species of macroplastic debris within the life cycle impact assessment. *Ecol Indic* 99: 61–66. http://doi.org/10.1016/j.ecolind.2018.12.018.

PART 5

Hydrocarbon pollution in sediments and seawater

Oil pollution in the ROPME Sea Area with specific emphasis on the toxicity of Kuwait crude oil in the northern Arabian Gulf

14

Qusaie Karam and Abeer Al-Saleh

14.1 INTRODUCTION

Oil is one of the most essential energy commodities in the world as it supports numerous industrial, economic and daily activities of many countries around the globe. Oil exploration has been increasing in recent years to meet the energy demands of the world's countries, and as a result, oil shipping has significantly increased. Also, the demand for more oil production globally has increased in recent decades, from 500 million tonnes in the 1950s to 2,500 million tonnes in the 1990s (Çakmak and Acar 2022). In addition to energy demand, more consumption of fossil fuels has increased environmental pollution over the years (Luqman et al. 2019). Oil is shipped across the oceans by large oil tankers, and these vessels are prone to accidents while en route to deliver their cargo. Accidents may occur while transferring oil to tankers at the offshore loading piers through improper mechanical alignment of pipes connected to the vessels, resulting in a release of tonnes of oil into the marine environment. Marine pollution by accidental oil spillage has become a continuing issue, threatening aquatic life in the vicinity of spill areas. It is estimated that 706 million gallons of oil waste reached the sea in 2014 (RMSI 2020). Upon the discharge of oil into the marine environment, petroleum hydrocarbons (PH), as one of the main constituents of oil, can be responsible for causing adverse effects on the marine biota. PH is composed of highly toxic components like polycyclic aromatic hydrocarbons (PAH), mineral oil, volatile aromatic hydrocarbons (e.g., benzene, toluene, ethylbenzene and xylenes) and other metals (Kuppusamy et al. 2017). It is still debatable in the scientific community which fraction of PH has a potentially greater toxicological impact on human and marine organisms' health (Kuppusamy et al. 2020). Low-molecular-weight (LMW) components of PH, for instance, have high bioavailability and solubility in aqueous mediums; therefore, they can cause greater toxicity in marine ecosystems and are a facilitated route into aquatic organisms. In addition, long-chained hydrocarbons can exert significant adverse ecotoxicological effects on marine biota. The variations in the ecotoxicity of oil can create inconsistencies for multiple reasons, such as differences in analytical methodologies applied to selecting bioindicator species as representatives of certain marine ecosystems (Khan et al. 2018). However, these variations can enrich our understanding of spilled oil behaviour in marine waters and how PH can induce an ecotoxicological effect on the specific target organism.

The explosion of the Deepwater Horizon (DWH) oil rig in 2010 in the Gulf of Mexico is an example of how accidents in the oil exploration industry can adversely affect the marine ecosystem. The explosion resulted in the largest marine oil spill in US history, which lasted for 87 days (McNutt et al. 2012). The accident caused significant environmental impact on indigenous aquatic organisms like algae, invertebrates, birds, sea mammals and fish native to the Gulf of Mexico (Beyer et al. 2016). Also, it was estimated that two to five trillion fish, 4,803 billion oysters, 4,900 to 7,600 sea turtles and 51,000 to 84,500 birds, amongst other organisms, were lost in the Gulf of Mexico. A review study by Pasparakis et al. (2019) has demonstrated that fish in early life stages can be more susceptible to spilled oil toxicity than adults. However, adult fish can exhibit sensitivity to oil pollution even at concentrations as low as 4 μg/L (ΣPAH_{50}, total equivalent to 50 polycyclic aromatic hydrocarbon compounds).

The Arabian Gulf, which is part of the ROPME Sea Area (RSA), covers a surface area of 239,000 km^2, characterized by a shallow semi-enclosed basin and a marginal part of the Indian Ocean. Its extensions are the Shatt Al-Arab and the adjacent

DOI: 10.1201/9781003321408-19

northern waterways of the Tigris, Euphrates and Karun Rivers between Iraq and Iran. The Strait of Hormuz is the southeastern limit of this basin connecting the Arabian Gulf with the Sea of Oman. The average depth of the Gulf is 36 m, and it rarely exceeds 100 m (Purser and Seibold 1973; Reynolds 1993). The RSA is considered one of the busiest water bodies in the world, receiving almost 53,000 oil-transporting vessels every year (Al-Yamani et al. 2015). In this chapter, we present an overview of how oil pollution can adversely affect the aquatic ecosystem in the ROPME SA, with a specific emphasis on its northwestern region representing Kuwait's territorial waters.

14.2 SOURCES OF OIL POLLUTION IN THE MARINE ECOSYSTEM

In this section, we highlight the major sources of oil pollution in the marine environment. Numerous anthropogenic activities introduce oil to marine water bodies. Such sources can be summarized as follows:

- Oil tankers
- Commercial ships releasing heavy fuels
- Rupturing pipelines
- Offshore oil exploration operations
- Drilling rigs
- Spills of refined petroleum products

Based on data of Chilvers et al. (2021), the causes of worldwide oil spills from the 1970s to 2018s can be generally categorized into the following: tankers, ships, wells/platforms, pipelines and land-based facilities/tanks. Analysis of 1,702 spills was categorized in the following manner: 3% from trucks/trains and others, 6% from wells/platforms, 9% from terrestrial sources and refineries, 12% from pipelines, 23% from oil tankers and 47% from shipping, including bulk carriers, cargo and fishing vessels, cruise ships and military vessels. The study also reported that the Northern Hemisphere suffers from most of the oil spills, with 96%, while only 4% of oil spills were recorded in the Southern Hemisphere (Chilvers et al. 2021). The ROPME Sea Area has witnessed several major oil spills, resulting in detrimental environmental impact on the marine ecosystem, and these are highlighted in Table 14.1.

14.3 OIL POLLUTION IN KUWAIT TERRITORIAL WATERS

Multiple studies have investigated the effects of oil-related pollutants on marine organisms in the northern Arabian Gulf, specifically in Kuwait's territorial waters. In this section, we will highlight the major findings of those studies and their consequent impact on the aquatic ecosystem. Also, we will demonstrate that selected marine biota with biochemical and physiological capacities can be used as bioindicators in marine ecosystem health assessments.

The 1991 Gulf War devastated the Arabian Gulf waters through massive oil pollution: specifically, in the open areas of this region. An estimated 10.8 million barrels (mbbls) of oil were spilled intentionally by the Iraqi army into the Arabian Gulf waters from various oil terminals and abandoned tankers (Tawfiq and Olsen 1993). Also, 600 oil well blowouts and fires caused tremendous oil fallout from smoke plumes. These acts of vandalism by departing Iraqi troops caused one of history's largest marine environmental catastrophes (Reynolds 1993). It is estimated that the oil spills from the Gulf War have depleted approximately 2% of the State of Kuwait's oil reserves, which is equivalent to about $60 million/day in revenues ($22 billion/year), based on a price of $20 per barrel (RMSI 2020).

To assess the adverse effects of the Gulf War's resultant oil slick on the marine environment, international efforts were made to plan an assessment programme, headed and sponsored by ROPME, the United Nations Environment Programme (UNEP), the Intergovernmental Oceanographic Commission (IOC) and the US National Oceanic and Atmospheric Administration (NOAA). From February to June 1992, an elite group of 140 scientists representing 15 countries boarded the NOAA research vessel *Mt. Mitchell* to participate in this regional environmental scientific expedition. One of the major scientific outcomes of that expedition was a study by Massoud et al. (1996). They observed that sediment grain size distribution could influence the relationship between total petroleum hydrocarbon (TPH) content in a present-day and chronic

TABLE 14.1 Major historical oil spill incidents in the ROPME Sea Area.

DATE	*INCIDENT*	*REGION/COUNTRY*	*VOLUME (GALLONS)*	*CAUSE*
19/12/1972	Sea Star	Gulf of Oman		Collision
25/05/1978	Well blowout and pipeline	Ahvazin/Iran	28,000,000	Ruptured pipeline
20/08/1981	Storage tank	Shuaiba Petroleum Tank/Kuwait	31,170,000	Unknown
10/02/1983	Nowruz field	Arabian Gulf/Iran	80,000,000	Oil tanker collision with Nowruz field platform.
07/01/1983	Tanker Assimi	Ras Al-Hadd/Oman	15,800,000	Fire and explosion in tanker's engine room.
09/12/1983	Tanker Pericles GC	30 km off Doha/Qatar	14,000,000	Fire and explosion in tanker's engine room.
06/12/1985	Tanker Nova	40 km off Kharg Island/Iran	21,350,000	Unknown
19/01/1991	Gulf War oil spill disaster	Northern Arabian Gulf/Kuwait	240,000,000	War

Source: Modified from Oil Spill Intelligence Report (Integrated Coastal Area Management: Guideline for the ROPME Region, Kuwait, 2000).

pollution scenario. Muddy sediments sampled during the expedition contained the highest TPH concentrations (mean TPH 175–192 µg/g). Also, the core sediment subsamples revealed that TPH concentrations in silt/clay fractions of the cores could indicate present-day pollution levels in the study area. Those findings classified oil pollution origins as natural oil seepage, oil spillage from tankers, tanker deballasting, accidental damage to pipelines and/or the Nowruz oil slick resulting from the damaged Iranian oil well in 1983 (Table 14.1). In addition, three polluted areas were determined to be directly linked to the 1991 Kuwait oil slick. Most importantly, oil pollution guidelines were established for the bottom sediments of the Arabian Gulf, as indicated in Table 14.2.

Beg et al. (2001) have examined the toxicity and desorption by aqueous elution of metals (zinc, copper, lead, vanadium, nickel, cadmium and chromium), polycyclic aromatic hydrocarbons (PAHs) and total petroleum hydrocarbons (TPH) in marine sediments off the Shuaiba Industrial area, Kuwait, which houses three refineries, two power and desalination plants, a petrochemical company, a melamine company, a large harbour and several other industrial establishments. Solid-phase Microtox assays showed severe toxicity; however, sample elutriates were not toxic to the larval stages of brine shrimp (Beg et al. 2001). Furthermore, whole sediment suspension in seawater reduced tilapia fish (*Oreochromis spilurus*) fingerlings' survival. High metal contents and metallothionein were recorded in Yellowfin seabream (*Acanthopagrus latus*) – specifically in their liver and gills – compared to tongue sole fish (*Cynoglossus arel*). Both fish species exhibited oxidative injury in their gills; however, the injury was also recorded in *C. arel*'s liver (Beg et al. 2015a, 2015b). This study demonstrates the bioavailability of metals in marine sediments and fish samples with oxidative injury as a biomarker of metal contamination. The same authors (Beg et al. 2018) have also investigated the use of Cytochrome P450 1A as a biomarker of AHR ligand exposure to determine the effect of PAH in *A. latus* and *C. arel*. There were manifestations of seasonal PAH in marine sediments and PAH-metabolites in fish bile. In addition, a seasonal-dependent effect was observed in the ratio of petrogenic to pyrogenic hydrocarbons in fish bile that was twofold higher in winter than in summer. The study concluded that *C. arel* is a more suitable local bioindicator of oil pollution in Kuwaiti waters than *A. latus*.

In other studies (Jasperse et al. 2019; Philibert et al. 2019; Takeshita et al. 2021), it was observed that oil-related contaminant toxicity might be manifested in marine fish, which may threaten the health of the marine ecosystem. The response of fish biomarkers as a representative species of the Kuwait marine ecosystem in preserved fish samples from 2005 and 2006 through 2015 was investigated (Beg et al. 2018). PAH levels in the liver and its metabolites in the bile of *A. latus*, and cytochrome P4501A1 by assay of ethoxyresorufin-o-deethylase (EROD) were examined. Insignificant differences between EROD activity and its liver PAH content in fish sampled over the selected period were observed. Also, it was concluded that fish can metabolize PAH efficiently, thus excreting hydroxyl-metabolites in bile. Although the threat of oil pollution is constant, Kuwait's marine health is not deteriorating, and fish can be used as an effective biomarker to investigate suboptimal concentrations of oil-related contaminants (Beg et al. 2018).

TABLE 14.2 Guidelines for pollution levels in marine bottom sediments of the Arabian Gulf.

POLLUTION LEVEL	TPH (µG/G)
Unpolluted area*	10–15
Slightly polluted area	15–50
Moderately polluted area	50–200
Heavily polluted area	> 200

* Natural background levels

In recent years, Karam et al. (2014, 2019, 2021) have conducted extensive work to assess the adverse toxicological effects of Kuwait crude oil (KCO) on indigenous marine fish. The effects of dispersed and undispersed KCO on egg hatching and larval survival of the orange-spotted grouper (*Epinephelus coioides*), locally called *hamoor*, was investigated by Karam et al. (2019). Also, exposure to oil resulted in morphological/skeletal deformities of different magnitudes in *E. coioides* like lordosis-inward curvature of spine, kyphosis-hunchback, and scoliosis-lateral bending of spine (Karam et al. 2014). A life-stage dependent effect was observed, and the toxicity test chemicals were variable of *E. coioides*. Therefore, the decision to use or not to use dispersant to combat oil spills should be carefully planned, taking into consideration the subsequent toxic effects that it may exert on marine fish life stages. In another study, Karam et al. (2021) have demonstrated that test chemicals like KCO, prepared as a water-accommodated fraction (WAF) and chemically enhanced water-accommodated fraction (CEWAF) with three dispersants (Corexit 9500, Corexit 9527 and Slickgone NS), can exhibit different toxicity results when tested on larvae of the Sobaity sea bream (*Sparidentex hasta*). Therefore, dispersed oil can either be as or less toxic than undispersed oil to fish larvae, confirming the test chemical–dependent effect on marine fish. A species-dependent effect was observed for fish native to the Kuwaiti sea environment upon exposure to dispersed and undispersed KCO in an acute exposure regime.

From various studies of marine species and/or specific organ responses to oil pollution, it is evident that oil can exert an array of adverse effects on test organisms. Also, differences in toxicity exposure regimes, life stages and test chemical concentrations confirm the complex and dynamic response of marine biota to environmental stressors like oil. RMSI (2020) conducted a detailed investigation, and their findings were reported in the Development of National Hazard Profile for the State of Kuwait, and the Identification of Existing Data and Gaps for Exposure and Vulnerability Assessment Report in 2020. An assessment of oil spills in Kuwait indicated that approximately 8 onshore (areal extent 5–62 km^2) and 39 offshore (areal extent 3–745 km^2) oil spill events occurred between 2017 and 2018. Sensitive zones in Kuwait have been identified, showing that most offshore oil spill events have occurred southeast of Kuwait City (Figure 14.1). Assessed oil spills were classified into four categories: 1) 0–50 km, 2) 50–100 km, 3)100–150 km, and 4) more than 150 km from Kuwait City. Oil spill aggregation southeast of Kuwait City resulted from the effect of ships traveling in this direction, resulting in frequent oil spills. The maximum transportation of crude oil is in the direction of major seaports like Mina Abdullah, Mina Al-Shuaibah, Mina Ahmadi, Mina Suad, Khor Al-Mufattah and Al-Zour. It has been observed that the majority of offshore oil spills in Kuwait have occurred at a distance of approximately 100–150 km from Kuwait City.

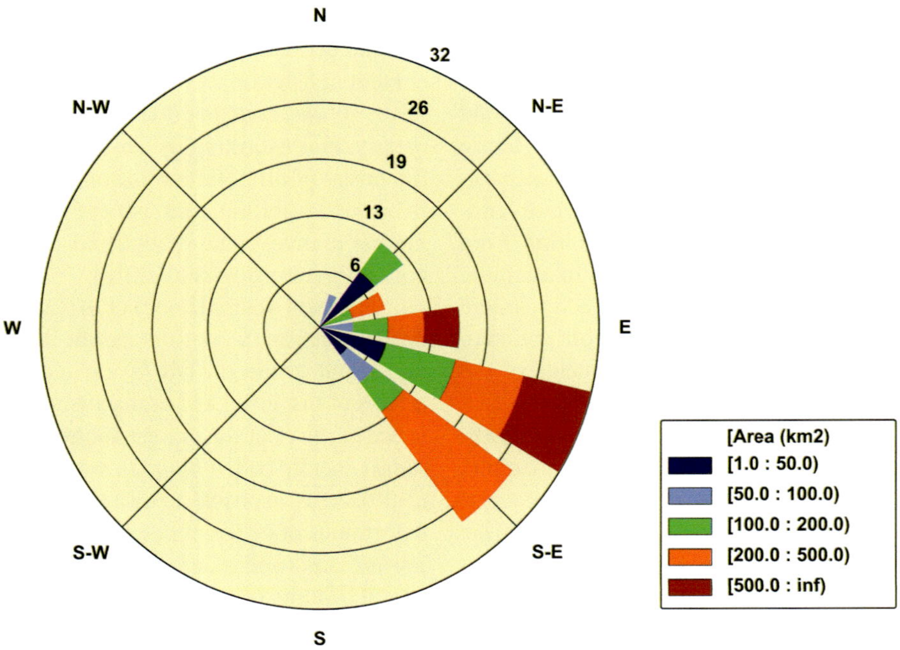

FIGURE 14.1 Rose diagram centred on Kuwait City depicting offshore oil spills.

Source: RMSI (2020)

14.4 APPLICATIONS OF SYNTHETIC APERTURE RADAR (SAR) FOR THE DETECTION OF OIL SPILLS IN KUWAIT'S TERRITORIAL WATERS

In this section, we highlight some common and recent oil spills in several geographic regions of Kuwait's territorial waters using advanced remote-sensing imagery. The techniques we have applied here to produce the images utilize radar which detects oil on the seawater surface since oil will attenuate the capillary properties on the water surface in the presence of modest winds. The acquired illuminated images result from the reflection of radar signals on the sea surface, and the oil subsequently appears as a black patch. The difference between oil and water can be classified as a maximum in the X-band, intermediate in the C-band and minimal in the L-band. The images were obtained from the USGS Sentinel satellite, the first space mission of the European Space Agency (ESA), the S1(A) mission and ground range detected (GRD) products. The approach of ESA SNAP software is constructive for efficiently detecting and processing oil spills, which requires many steps. The pre-processing image method uses image subsetting, calibration, speckle filtering and ellipsoid correction after data acquisition to address distance aberrations and bring the geometrical representation of the image as close as possible to the natural universe (Misra and Balaji 2017). Speckle noise is a characteristic of SAR imaging that affects the image quality and makes it challenging to correctly discern between image features (Misra and Balaji 2017; Gallego et al. 2018). As a result, speckle filtering is crucial for precise oil spill detection and image categorization (Fan et al. 2015; Ajadi et al. 2018). The frequency selected was 5.40 GHz, which translates to a wavelength of 5.55 cm, outfitted with C-band, IW mode, and VV polarization. The synthetic aperture radar (SAR) near real time (NRT) based oil spill detection technique maps oil spills and creates reliable and accurate results. However, the radar's ability to detect oil at sea in the presence of winds is limited and falls between 1.5 and 10 m/s (Leifer et al. 2012). All day and night and all-weather constant overflights and high spatial resolution radar sensors deployed by satellites are crucial for determining the location and size of significant oil spills and keeping track of oil discharges from offshore platforms and ships. These data are used to map out oil spills, determine sheen paths, and directly support oil spill protective measures. Figure 14.2a depicts oil spills on 24 July 2019 where the oil tanker waiting area is near the northeast of Kuwait territorial waters and south of Iraq, approximately 40 km from Failaka Island. Figure 14.2b also displays oil spills (latitude 29.847°N, longitude 48.626°E) from the route of tankers from Fao Port in Iraq over an area of 10.6 km^2. Oil discharges associated with tankers can lead to chronic water column contamination.

Between 2017 and 2022, there were four reported oil spills in the Arabian Gulf: one in front of Al Zour on 10 August 2017; another between Qaruh and Um Almaradem Islands on 7 July 2019; and a third one on 5 September 2022, which was a heavy crude oil spill in the tanker route (latitude 29.183°N, longitude 48.785°E) at a distance of 66 km from Ahmadi Port (Figures 14.3a, b and c). According to Kuwait's Environment Public Authority (KEPA), an pipeline damage caused the fourth oil spill at Al-Khiran. This spill of crude oil covered an area of 155.2 km^2 and was visible as a clear, dark patch at latitude 28.665°N, longitude 48.471°E. The volume of spilled oil was estimated as 35,000 bbl (Park et al. 2019).

These oil spills, whether accidental or natural, can exert a chronic toxicological effect on marine biota indigenous to the northern Arabian Gulf since the oil slicks can impact the ecosystem shared between three neighbouring countries: Kuwait,

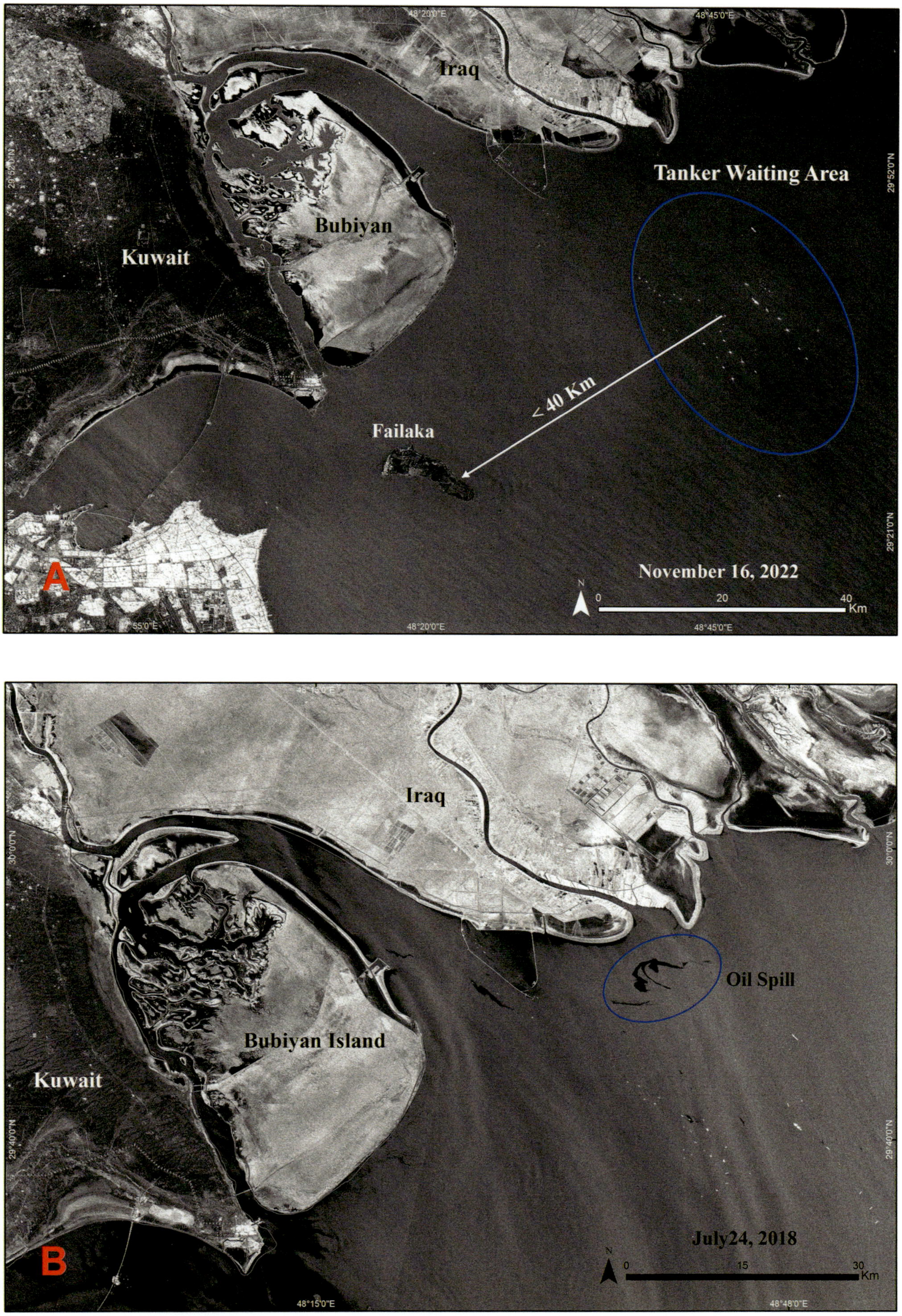

FIGURE 14.2 A) Tanker waiting area near the northeast of Kuwaiti territorial waters, south of Mina Baker, Fao area, Iraq, and B) oil spills from tankers along the route to Al-Fao Port in Iraq.

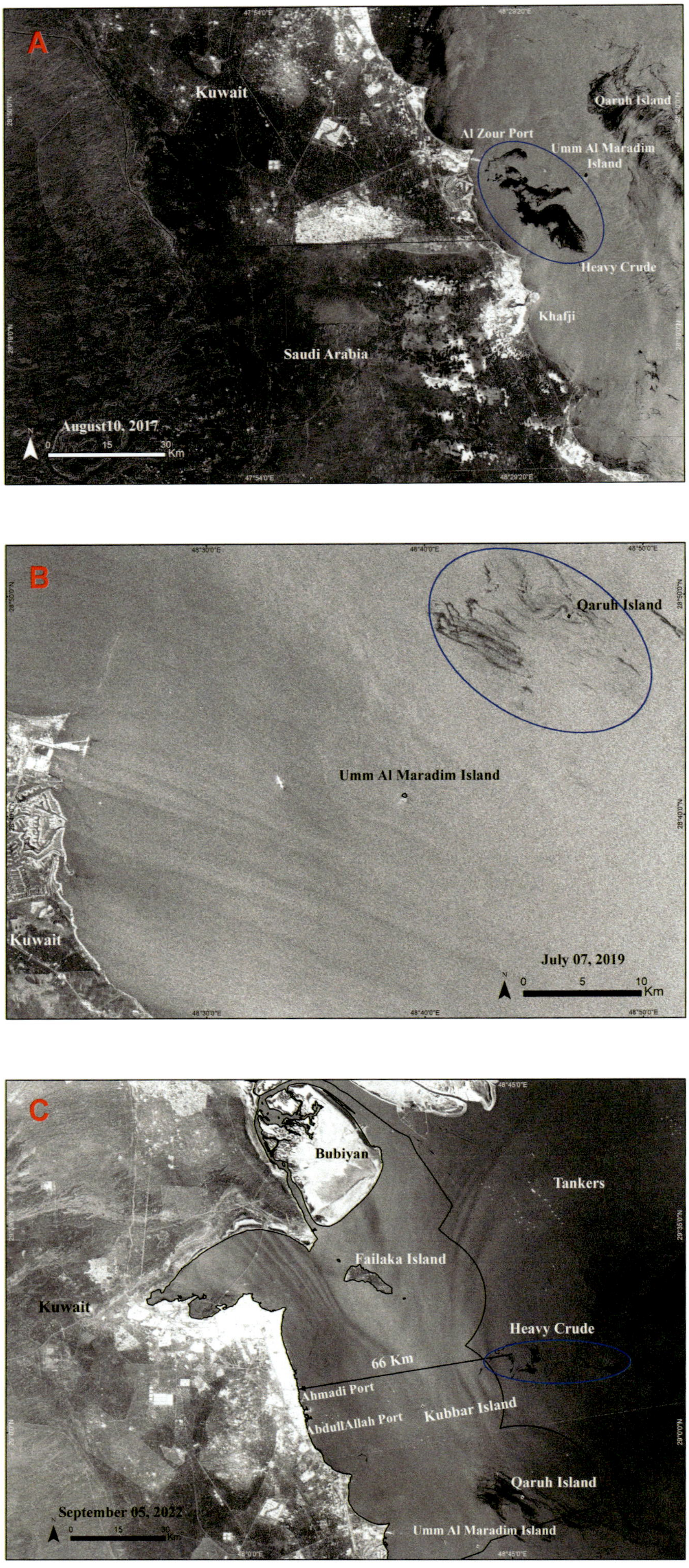

FIGURE 14.3 A) Al Zour oil spill, B) natural oil spill between Qaruh and Um Almaradem Islands, and C) spread of heavy crude oil spill on the tanker route at a distance from Ahmadi Port.

Iraq and Iran. International efforts should be dedicated to minimizing oil discharges associated with loading oil into tankers, releasing ballast water containing residual oil wastes and further investigating the environmental impacts of natural oil seepage from the seafloor.

It is important to note that to protect and preserve the national marine resources, Kuwait has developed a promising monitoring programme that has been in operation for 30 years. The programme is intended to monitor the onshore (coastal) and offshore marine environment for various pollutants discharged into marine areas via point sources from land and sea. Through its various monitoring programmes, KEPA has collected crucial monthly environmental data for nutrients, metals, physical water quality parameters and petroleum hydrocarbons (ROPME 2013; EPA 2022).

14.5 OIL POLLUTION AND PETROLEUM HYDROCARBON LEVELS IN THE ROPME SEA AREA (RSA)

The ROPME Sea Area (RSA) is characterized by natural background hydrocarbons from mixed sources including terrestrial and autochthonous. The anthropogenic input of hydrocarbons to this unique marine region stems from degraded petroleum hydrocarbons (PH). Few studies have investigated PH levels in marine sediment and seawater, specifically in the Arabian Gulf. Fowler (1988) reported one of the earliest documented studies on pollution levels and noted that PH concentrations ranged from 0.1 to 1.5 μg/g in marine sediments. However, Literathy et al. (1992) and Abdali et al. (1993) have reported much higher PH values of 10 to 20 μg/g and 5.9 to 194.8 μg/g, respectively, than those reported five years previously.

Other efforts were also directed toward estimating oil pollution levels in the marine sediments in the RSA. One of those endeavours was a 2006 winter oceanographic cruise that surveyed almost 100 selected stations in the Arabian Gulf and the Sea of Oman (also part of the RSA) (ROPME 2013). The survey revealed that TPH concentrations ranged from 7 to 80 μg/g with a mean TPH reading of 20 μg/g (ROPME 2010). Those levels were lower than levels documented in previous surveys conducted in the RSA in 1994, in which TPH concentrations averaged 33 μg/g, according to Al-Lihaibi and Ghazi (1997), and much lower than those reported by Massoud et al. (1996). Moreover, TPH levels measured after the 1991 Gulf War oil spill in the RSA also showed lower values in coastal sediments (Readman et al. 1996; ROPME/IAEA 1996, 1998, 2001, 2006). Although those readings were lower than the ones measured in the previous years, the constant threat of chronic oil pollution in marine sediments and its susceptibility to being transferred to the water column and ultimately bioaccumulated in aquatic organisms is persistent. It was concluded that the low but consistent PH levels (μg/g) in the marine sediments from all sampling stations for the 2006 ROPME expedition indicate a chronic petrogenic source for PH in the RSA, reaching the aquatic ecosystem through the discharge of oil wastes (drilling fluids), or from fuel spills (ROPME 2013). Reduced PH levels in marine sediments may also be a function of chemical and physical weathering processes (Al-Lihaibi and Ghazi 1997). The analysis demonstrated that most oil spills in the RSA region have included fuel oil discharges (including different types of oils) which have led to an increasing trend in oil pollution in the RSA region over time. However, those concentrations still do not exceed TPH background levels in surface sediments, which are 15 μg/g as chrysene equivalent unit (CEU) (ROPME 2010). Further analysis (Hong et al. 2016) revealed that marine sediments contained appreciable levels of low molecular weight (LMW) hydrocarbons compared to high molecular weight (HMW) ones, indicating a petrogenic source, possibly from fuel oil discharge incidents. Meanwhile, the presence of long-chained homologs is thought to originate from terrestrial plant higher waxes (ROPME 2006). Moreover, the trend of PH distribution in the Arabian Gulf is oriented toward the northwest direction, with fine-grained sediments containing the highest levels of contamination as PH is adsorbed onto sediment particles (Al-Lihaibi and Ghazi 1997).

14.6 MARINE POLLUTION RESPONSE STRATEGY IN THE ROPME SEA AREA

Numerous regional authorities have focused on combating oil pollution in the marine environment, and one of those efforts started with the Kuwait Regional Convention, which was held on 23 April 1978 and initiated the Regional Organization for the Protection of the Marine Environment (ROPME) to develop an agreement to protect the marine environment. The main emphasis of the agreement centred around regional cooperation in combating multiple pollution incidents by oil and other harmful substances. Therefore, in line with such circumstances, obligations were formulated with the aim of controlling oil pollution from multiple sources and applying different facets of environmental management of marine water bodies. Hence, effective response measures to combat pollution from oil or other hazardous substances discharged at sea were compiled in a marine pollution manual by the Marine Emergency Mutual Aid Centre (MEMAC), based in the Kingdom of Bahrain), an integral part of ROPME.

Articles devised by the steering committee stated various action plans that can be incorporated into the oil pollution combat protocol. For instance, among other articles, all member states of ROPME should take necessary measures to protect coastlines from pollution threats caused by oil or other hazardous chemicals that can result in marine emergencies. Also, member states shall cooperate to apply all effective means and plans to combat marine pollution, specifically by using available resources like aircraft, trained personnel, ships and equipment in the case of marine emergencies caused by pollution sources (MEMAC 2016). In addition, regional collaboration between member states is necessary if any ROPME state requires assistance in any marine emergency response. Other member states must respond to the call within their technical capacities to combat pollution, utilize facilities, transfer personnel and apply methods for the safe disposal of hazardous materials. Moreover, surveillance and monitoring of the ROPME water body are essential to ensure that every member state follows regional maritime laws and regulations within the range of their capacities. Those articles also cover special marine emergencies, in which all member states, if requested, shall mobilize their resources and render them available to other states to combat complex pollution issues caused by oil and other harmful substances that can threaten the aquatic ecosystem.

The response strategy and prioritization of the conventions encompass multiple technical work plans to combat marine pollution in the ROPME Sea Area. The response plan focuses on stopping the outflow of pollution sources causing oil spills so subsequent spillage incidents can be prevented. After that, the team should limit the spilled oil from further spreading in the vicinity of the spillage zone. Spilled oil should be recovered at sea to prevent oil slicks from reaching coastal areas and thus impacting marine biota that use coastal zones as their breeding grounds. Also, additional actions should be applied by member states to protect the main coastal facilities and installations. A detailed marine pollution combat strategy is provided in MEMAC's Marine Pollution Manual (MEMAC 2016).

14.7 CONCLUSION

The adverse environmental effects of spilled oil in the ROPME Sea Area and the northwestern Arabian Gulf are still poorly understood. Although multiple efforts have been made to assess the ecotoxicological impacts of PH on native biota, more investigative studies are still required. Regional scientific collaborations should be implemented to evaluate hazardous chemicals frequently spilled in the marine environment. Also, more toxicity tests should be conducted on dispersed and undispersed chemicals. Since new oil dispersants are always being manufactured to combat oil spills, their toxicological effects on indigenous marine organisms must be carefully assessed to establish new oil pollution standards for the ROPME Sea Area. ROPME state members, with the assistance of the Marine Emergency Mutual Aid Centre (MEMAC), should establish a regional marine ecotoxicology laboratory dedicated to the safety evaluation of hazardous chemicals that are frequently discharged into the aquatic environment.

REFERENCES

Abdali, F., Al-Najjar, H.A., Bahloul, M., Jacob, P.G., and A.N. Al-Ghadban. 1993. Pore diffusion transport of petroleum hydrocarbons within sediment column in ROPME Sea Area. *Presented at the Scientific Workshop on the Results of the Mt. Mitchell Open Sea Cruise Kuwait, Arabian Gulf, Kuwait,* 24–28 January 1993.

Ajadi, O.A., Meyer, F.J., Tello, M., and G. Ruello. 2018. Oil spill detection in synthetic aperture radar images using Lipschitz-regularity and multiscale techniques. *IEEE J Sel Top Appl Earth Obs Remote Sens* 11: 2389–2405.

Al-Lihaibi, S.S., and S.J. Ghazi. 1997. Hydrocarbon distributions in sediments of the open area of the Arabian Gulf following the 1991 Gulf War oil spill. *Mar Pollut Bull* 34, no. 11: 941–948.

Al-Yamani, F.Y., Skryabin, V., and S.R.V. Durvasula. 2015. Suspected ballast water introductions in the Arabian Gulf. *Aquat Ecosyst Health Manag* 18, no. 3: 282–289.

Beg, M.U., Al-Jandal, N., Al-Subiai, S., Karam, Q., Husain, S., Butt, S.A., Ali, A., Al-Hasan, E., Al-Dufaileej, S., and M. Al-Husaini. 2015a. Metallothionein, oxidative stress and trace metals in gills and liver of demersal and pelagic fish species from Kuwait's marine area. *Mar Pollut Bull* 100, no. 2: 662–672.

Beg, M.U., Al-Muzaini, S., Saeed, T., Jacob, P.G., Beg, K.R., Al-Bahloul, M., Al-Matrouk, K., Al-Obaid, T., and A. Kurian. 2001. Chemical contamination and toxicity of sediment from a coastal area receiving industrial effluents in Kuwait. *Environ Contam Toxicol* 41, no. 3: 289–297.

Beg, M.U., Al-Subiai, S.N., Al-Jandal, N., Butt, S.A., Beg, K.R., and M. Al-Husaini. 2015b. Seasonal effect on biomarkers of exposure to petroleum hydrocarbons in fish from Kuwait's marine area. *Mar Pollut Bull* 100, no. 2: 673–680.

Beg, M.U., Butt, S.A., Al-Dufaileej, S., Karam, Q., Al-Sharrah, T.K., and T. Saeed. 2018. Biomarkers in fish as a measure of the state of marine environment of Kuwait. *Environ Monit Assess* 190, no. 6: 1–12.

Beyer, J., Trannum, H.C., Bakke, T., Hodson, P.V., and T.K. Collier. 2016. Environmental effects of the Deepwater Horizon oil spill: A review. *Mar Pollut Bull* 110: 28–51.

Çakmak, E.E., and S. Acar. 2022. The nexus between economic growth, renewable energy and ecological footprint: An empirical evidence from most oil-producing countries. *J Clean Prod* 352: 131548.

Chilvers, B.L., Morgan, K.J., and B.J. White. 2021. Sources and reporting of oil spills and impacts on wildlife 1970–2018. *Environ Sci Pollut Res* 28, no. 1: 754–762.

EPA. 2022. *Monthly Environmental Report* (Unpublished). Environment Public Authority.

Fan, J., Zhang, F., Zhao, D., and J. Wang. 2015. Oil spill monitoring based on SAR remote sensing imagery. *Aquat Proc* 3: 112–118.

Fowler, S.W. 1988. Coastal baseline studies of pollutants in Bahrain, UAE and Oman. *Proc Symposium on Regional Marine Pollution Monitoring and Research Programmes, ROPME GC-4/2*. Regional Organization for the Protection of the Marine Environment, Kuwait, pp. 155–180.

Gallego, A.J., Gil, P., Pertusa, A., and R. Fisher. 2018. Segmentation of oil spills on sidelooking airborne radar imagery with autoencoders. *Sensors* 18: 1–16.

Hong, W.J., Jia, H., Li, Y.F., Sun, Y., Liu, X., and L. Wang. 2016. Polycyclic aromatic hydrocarbons (PAHs) and alkylated PAHs in the coastal seawater, surface sediment and oyster from Dalian, Northeast China. *Ecotoxicol Environ Saf* 128: 11–20.

Jasperse, L., Levin, M., Rogers, K., Perkins, C., Bosker, T., Griffitt, R.J., Sepúlveda, M., and S. De Guise. 2019. Hypoxia and reduced salinity exacerbate the effects of oil exposure on sheepshead minnow (Cyprinodon variegatus) reproduction. *Aquat Toxicol* 212: 175–185.

Karam, Q., Ali, M., Subrahmanyam, M.N.V., Elah, K.A.A., Bentley, M., and M.U. Beg. 2019. A comparative study on the effect of dispersed and undispersed Kuwait crude oil on egg hatching and larval survival of Epinephelus coioides. *J Environ Biol* 40, no. 2: 192–199.

Karam, Q., Annabi-Trabelsi, N., Al-Nuaimi, S., Ali, M., Al-Abdul-Elah, K., Beg, M.U., and M. Bentley. 2021. The response of Sobaity sea bream Sparidentex hasta larvae to the toxicity of dispersed and undispersed oil. *Pol J Environ Stud 30*, no. 6: 5065–5077.

Karam, Q., Beg, M.U., Al-Khabbaz, A., Al-Ballam, Z., Dakour, S., and K. Al-Abdul Elah. 2014. Effect of water accommodated fraction of Kuwait crude oil on developmental stages of orange-spotted grouper hamoor (*Epinephelus coicoides*). *Int J Adv Agric Environ Eng* 1: 105–112.

Khan, M.A.I., Biswas, B., Smith, E., Naidu, R., and M. Megharaj. 2018. Toxicity assessment of fresh and weathered petroleum hydrocarbons in contaminated soil – a review. *Chemosphere* 212: 755–767.

Kuppusamy, S., Maddela, N.R., Megharaj, M., and K. Venkateswarlu. 2020. Ecological impacts of total petroleum hydrocarbons. In:

Total Petroleum Hydrocarbons, 95–138. Cham, Switzerland: Springer Nature.

Kuppusamy, S., Thavamani, P., Venkateswarlu, K., Lee, Y.B., Naidu, R., and M. Megharaj. 2017. Remediation approaches for polycyclic aromatic hydrocarbons (PAHs) contaminated soils: Technological constraints, emerging trends and future directions. *Chemosphere* 168: 944–968.

Leifer, I., Lehr, B., Simecek-Beatty, D., Bradley, E., Clark, R., Dennison, P., Hu, Y., Matheson, S., Jones, C., Holt, B., et al. 2012. State of the art satellite and airborne oil spill remote sensing: Application to the BP Deepwater Horizon oil spill. *Remote Sens Environ* 124: 185–209.

Literathy, P., Morel, G., Zarba, M.A., Samhan, O., Bloushi, A., Hashash, H., Matrouk, K., and P.G. Jacob. 1992. Petroleum compounds in the marine environment of Kuwait. *Report No. KISR 4054*. Kuwait Institute for Scientific Research, Kuwait.

Luqman, M., Ahmad, N., and K. Bakhsh, K. 2019. Nuclear energy, renewable energy and economic growth in Pakistan: Evidence from non-linear autoregressive distributed lag model. *Renew Energy* 139: 1299–1309.

Massoud, M.S., Al-Abdali, F., Al-Ghadban, A.N., and M. Al-Sarawi. 1996. Bottom sediments of the Arabian Gulf – II. TPH and TOC contents as indicators of oil pollution and implications for the effect and fate of the Kuwait oil slick. *Environ Pollut* 93, no. 3: 271–284.

McNutt, M.K., Camilli, R., Crone, T.J., Guthrie, G.D., Hsieh, P.A., Ryerson, T.B., Savas, O., and F. Shaffer. 2012. Review of flow rate estimates of the Deepwater Horizon oil spill. *Proc Natl Acad Sci* 109: 20260–20267.

MEMAC. 2016. *MEMAC Marine Pollution Manual*. Kingdom of Bahrain: Marine Emergency Mutual Aid Centre.

Misra, A., and R. Balaji. 2017. Simple approaches to oil spill detection using Sentinel Application Platform (SNAP) – ocean application tools and texture analysis: A comparative study. *J Indian Soc Remote Sens* 45: 1065–1075.

Park, S.H., Jung, H.S., Lee, M.J., Lee, W.J., and M.J. Choi. 2019. Oil spill detection from PlanetScope satellite image: Application to oil spill accident near Ras Al Zour Area, Kuwait in August 2017. *J Coast Res* 90: 251–260.

Pasparakis, C., Esbaugh, A.J., Burggren, W., and M. Grosell. 2019. Physiological impacts of Deepwater Horizon oil on fish. *Comp Biochem Physiol Part C: Toxicol Pharmacol* 224: 108558.

Philibert, D.A., Lyons, D., Philibert, C., and K.B. Tierney. 2019. Field-collected crude oil, weathered oil and dispersants differentially affect the early life stages of freshwater and saltwater fishes. *Sci Total Environ* 647: 1148–1157.

Purser, B.H., and E. Seibold. 1973. The principal environmental factors influencing Holocene sedimentation and diagenesis in the Persian Gulf. In: *The Persian Gulf.* Ed. B.H. Purser, 1–9. Berlin and Heidelberg: Springer. http://doi.org/10.1007/978-3-642-65545-6_1.

Readman, J.W., Bartocci, J., Tolosa, I., Fowler, S.W., Oregioni, B., and M.Y. Abdulraheem. 1996. Recovery of the coastal marine environment in the Gulf following the 1991 war-related oil spills. *Mar Pollut Bull* 32, no. 6: 493–498.

Reynolds, R.M. 1993. Physical oceanography of the Gulf, Strait of Hormuz, and the Gulf of Oman: Results from the Mt. Mitchell Expedition. *Mar Pollut Bull* 27: 35–59.

RMSI. 2020. *Development of National Hazard Profile for the State of Kuwait and Identification of Existing Data and Gaps for Exposure and Vulnerability Assessment Report in 2020.* RMSI Pvt. Ltd., India.

ROPME. 1996. *ROPME 1994 Contaminant Screening Project. Mission Report by MEL/IAEA. ROPME/GC-8/001.* Regional Organization for the Protection of the Marine Environment (ROPME), Kuwait, 50 p.

ROPME. 1998. *ROPME 1998 Contaminant Screening Project. Mission Report by MEL/IAEA. ROPME/GC-8/001.* Regional Organization for the Protection of the Marine Environment (ROPME), Kuwait, 53 p.

ROPME. 2000. *Integrated Coastal Area Management: Guideline for the ROPME Region.* Regional Organization for the Protection of the Marine Environment (ROPME), Kuwait.

ROPME. 2001. *ROPME 2001–2001 Contaminant Screening Project. Mission Report by MEL/IAEA. ROPME/GC-8/001.* Regional Organization for the Protection of the Marine Environment (ROPME), Kuwait, 77 p.

ROPME. 2006. *2005 Contaminant Screening Project. Mission Report by MEL/IAEA. ROPME/GC-8/001.* Regional Organization for the Protection of the Marine Environment (ROPME), Kuwait City, Kuwait, 198 p.

ROPME. 2010. *ROPME Oceanographic Cruise – Winter 2006. Technical Report Series No. 2. Contaminants in the Marine Sediments. ROPME/GC-14/7.* Regional Organization for the Protection of the Marine Environment (ROPME), Kuwait, 139 p.

ROPME. 2013. *State of the Marine Environment Report – 2013. ROPME/GC-16/1-II.* Regional Organization for the Protection of the Marine Environment (ROPME), Kuwait, 225 p.

Takeshita, R., Bursian, S.J., Colegrove, K.M., Collier, T.K., Deak, K., Dean, K.M., et al. 2021. A review of the toxicology of oil in vertebrates: What we have learned following the Deepwater Horizon oil spill. *J Toxicol Environ Health, Part B* 24, no. 8: 355–394.

Tawfiq, N., and D.A. Olsen. 1993. Saudi Arabia's response to the 1991 Gulf oil spill. *Mar Pollut Bull* 27: 333–345.

Spatial distribution of hydrocarbons in the Shatt Al-Arab River and the Northern Coast of the Arabian/Persian Gulf

15

Characteristics, sources and environmental impacts

Ahmed I. Rushdi, Bernd R. T. Simoneit, Ali A. DouAbul, Sama Al-Maarofi, Hamid Al-Saad and Nader A. Salman

15.1 INTRODUCTION

The occurrence of organic compounds in the marine environment of the northern Arabian/Persian Gulf is mainly derived from inputs from local developmental activities such as oil-related industry, wind transport and river runoff (Rushdi et al. 2010; Harwood 2014; Dong et al. 2020). Natural oil seepage and other chemical spills from tankers are also significant sources of organic matter in the river and coastal water of the Gulf. The Shatt al-Arab River is the main river that flows into the northern part of the Gulf and deposits large amounts of suspended particles into the estuary and north coastal region (Talling 1980; DouAbul et al. 1988; Al-Yamani 2008). Ultimately, these suspended particles and other detrital matter bring diverse quantities and types of organic and inorganic substances to that marine environment (Al-Yamani 2008; Rushdi et al. 2010, 2014a, 2014b; Farid et al. 2016). Dust storms and wind-transported particulate matter are other input sources of organic matter to these marine and coastal zones (Rushdi et al. 2010). High amounts of petroleum hydrocarbons in the Arabian/Persian Gulf are introduced by naturally occurring oil seeps, oilfields (onshore and offshore), refineries and petrochemical plant discharges (Sadiq and McCain 1993; PME/UNEP 1989; Massoud et al. 1998; Rushdi et al. 2010). Oil pollution continues to be a serious problem in major areas along the coasts of the Gulf (Massoud et al. 1998; Reynolds 1993; Literathy 1993; Al-Arfaj and Alam 1993; Rushdi et al. 2014a, 2022; Al-Saad and Al-Imarah 2021; Uddin et al. 2021).

Environmental contamination by petroleum hydrocarbons, comprising high amounts of n-alkanes, biomarkers, branched and cyclic hydrocarbons (UCM) and polycyclic aromatic hydrocarbons (PAH), is a ubiquitous health issue (Corner 1979; Volkman et al. 1992; Ite et al. 2018; Cortes et al. 2012; Rushdi et al. 2014a, 2017). Iraq is one of the major oil producers in the Organization of Petroleum Exporting Countries (OPEC), where 90% of crude oil production is from oil fields in the southern part of the country (USEIA 2019). High levels of petroleum hydrocarbons have been emitted into the environment due to oil-related activities (Al-Saad et al. 2015; Aldosari 2021; Al-Saad and Al-Imarah 2021). Significant concentrations of petroleum hydrocarbons in sediments and water samples from the region have been reported lately (Rushdi et al. 2006a, 2006b, 2017, 2018; Al-Saad et al. 2011; Douabul et al. 2012; Kadhim et al. 2019; Uddin et al. 2021). The occurrence and addition of hydrocarbons to the aquatic environment eventually affect the biota of the region through adsorption, absorption and bioaccumulation processes (De Mora et al. 2005, 2010; Tolosa et al. 2005; Ashraf and Mian 2010; Al-Khion et al. 2016). Different levels and origins of natural and anthropogenic organic components will eventually impact

DOI: 10.1201/9781003321408-20

the biodiversity and biogeochemistry of the regional aquatic ecosystems.

Therefore, the goal of this chapter is mainly to report the concentrations, characteristics, spatial distributions, sources and ecological impacts of the hydrocarbons in shallow surface sediments from the southern part of the Shatt Al-Arab River, estuary, and northern Arabian/Persian Gulf, based on molecular marker analysis and biogeochemical indices.

15.2 STUDY AREAS

The study area is the southern part of the Shatt Al-Arab River from north of Al-Faw City to the estuary area and the northern part of the Arabian/Persian Gulf coast (Figure 15.1). The river deposits major amounts of silt in the estuary, causing major problems for navigation (Talling 1980; DouAbul et al. 1988; Isaev and Mikhailova 2009; Al-Dabbas et al. 2010; Al-Ramadan 2014). The drainage basin of the Shatt Al-Arab River is about 806,000 km^2, and its downstream region is influenced by semi-diurnal and diurnal tides affecting the adjoining irrigation processes on the riverbanks (Buringh 1960; Polservice 1980; CEB 2011). The width of the river varies from ~800 m at the outlet to ~232 m at the city of Basra (Isaev and Mikhailova 2009), and the depth is about 6 m to 15 m with a bed slope between 1 cm/km and 1.5 cm/km (Polservice 1980; Abdullah et al. 2015). Swamps and sedimentary muddy lands abut most of the river (Isaev and Mikhailova 2009). The coast of Iraq is narrow, consisting mainly of intertidal mudflats bordered by exposed silt flats with prevailing narrow strips of date palm stands (Maltby 1994; Partow 2001).

The main flora along the river includes shrubs, reeds and papyrus, with date palm trees all along the riverbanks. Basrah City in Iraq and Abadan City in Iran are the two major urban areas on the river, and they are the main ports for crude oil production, including storage, loading and shipping. The river is considered the main source of water in the southern part of Iraq and plays an important role in the ecosystems of the northern part of the Arabian/Persian Gulf (Al-Yamani 2008; UN-ESCWA and BGR 2013).

15.3 MATERIALS AND EXPERIMENTAL PROCEDURE

15.3.1 Sample collection

Shallow surface sediment samples from the southern part of the Shatt Al-Arab River, estuary and the coastal zone of Iraq (Figure 15.1) were collected by Van Veen grab samplers. They were collected from four zones about 17.6 kilometers north of Al-Faw City to about 18.1 kilometers off the Iraqi coast. Zone I (sites N1–N4) was north of Al-Faw City, Zone II (S1–S6) was from south of Al-Faw City to the estuary area, Zone III (E1–E3) was the estuary, and Zone IV (C1–C4) was the northern coast of the Arabian/Persian Gulf. The samples were placed in prewashed glass containers and kept in a freezer until analysis. Around 10–15 g of a sample was dried at laboratory temperature, crushed and sieved to obtain <125 μm fine particles that were taken for organic matter extraction (Rushdi et al. 2014a, 2014b, 2018).

15.3.2 Sample extraction

The extraction method was similar to the procedure described by Rushdi et al. (2014a, 2014b, 2022). About 5 g of each sample

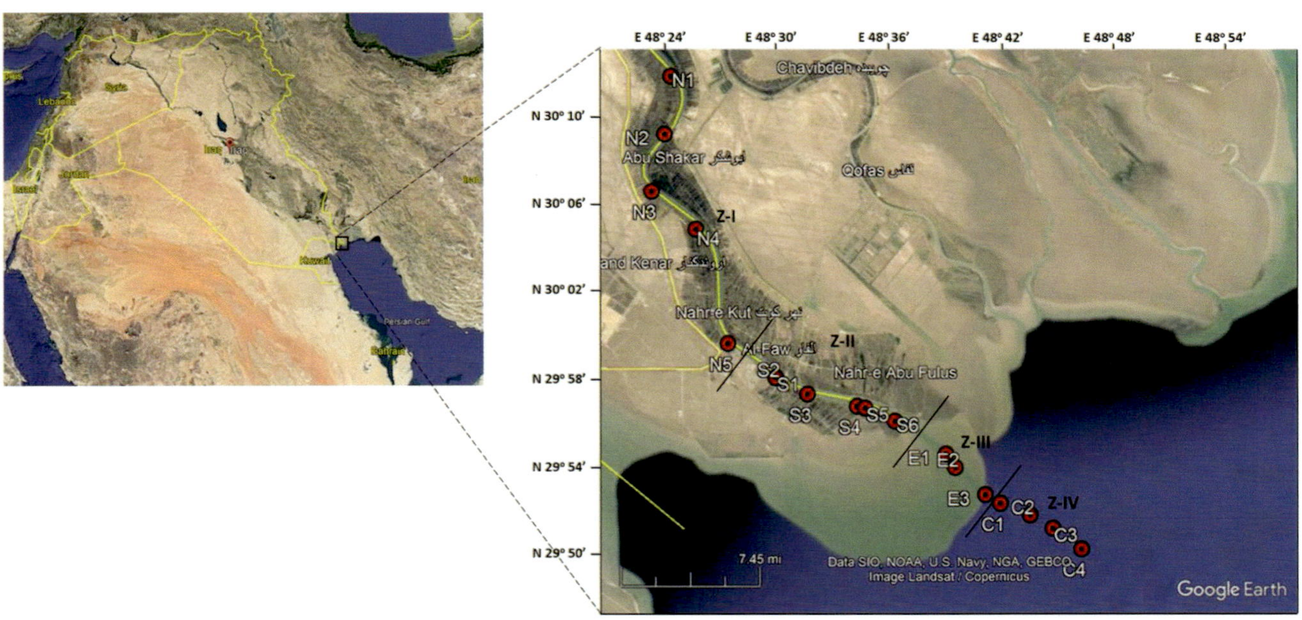

FIGURE 15.1 Location map of the sampling sites in the south of the Shatt Al-Arab River and the coastal zone of the northern part of the Arabian/Persian Gulf.

Source: Google Earth

was extracted under ultrasonication three times for 15 minutes each with 40 ml of dichloromethane (DCM) and methanol (v:v = 3:1) mixture. An organic solvent-cleaned glass fibre filter was used to separate the sediment particles by filtering the combined extract (TEOM). The filtrate was concentrated by a rotary evaporator and then concentrated to about 200 μL by blowing with nitrogen gas. The DCM and methanol mixture was used to adjust the concentrated extract to 500 μL. An aliquot sample of 100 μL was derivatized with a silylating reagent (30 μL N,O-bis(trimethylsilyl)trifluoroacetamide, BSTFA, Pierce Chemical Co.) for three hours at 70°C after drying under nitrogen flow. Before analysis by gas chromatography-mass spectrometry (GC-MS), the sample was dissolved in n-hexane after the BSTFA was evaporated under nitrogen blowdown.

15.3.3 Instrumental analysis

Instrumental analysis was carried out by gass chromoatography-mass spectrometry (GC-MS) using an Agilent 6890 gas chromatograph coupled to a 5973 mass selective detector with a DB-5MS (Agilent) fused silica capillary column (30 m length x 0.25 mm inner diameter and 0.25 μm film thickness). The carrier gas was high-grade helium where the GC was temperature programmed from 65°C (2 min initial hold) to 310°C at 6°C min^{-1} (isothermal for 20 min final time), and the MS was operated in the electron impact mode at 70 eV ion source energy. Mass spectrometric data were acquired and processed using the GC-MS ChemStation data system.

15.3.4 Identification and quantification

The hydrocarbon compounds were identified based on their GC retention time and the MS key ion pattern (i.e., fragmentograms *m/z*) of n-alkanes, pristane, phytane (*m/z* 85), hopanes (*m/z* 191) and steranes (*m/z* 217/218). The compound quantification was similar to the procedure explained by Rushdi et al. (2010), in which an authentic external standard was applied to verify the retention times of the compounds. Basically, the profile of the total ion current (TIC) relative to the external standard was utilized for quantification (Rushdi et al. 2014a, 2014b, 2010, 2017). The external standards included n-alkanes (C_8–C_{40}), and (17α(H),21β(H)-hopane, $C_{30}H_{52}$). The quantification of each compound was based on its peak area derived from the TIC trace.

15.3.5 Quality control

The technical accuracy of the method and procedure has been examined in this study. It incorporated the sampling collection, analytical chemistry, treatment and handling of the data. The background and cross-contamination due to laboratory procedures have been examined and adjusted from procedural blanks for quality and quantity control. Also, after batches of three samples, blank extracts were analyzed throughout the procedures.

15.4 RESULTS AND DISCUSSION

The most significant hydrocarbons in the total extractable organic matter (TEOM) of the sediment samples from the southern part of the Shatt Al-Arab River, the estuary and the north of the Arabian/Persian Gulf were n-alkanes; biomarkers such as pristane, phytane, hopanes and steranes; and the unresolved complex mixture of branched and cyclic hydrocarbons (UCM) (Table 15.1).

Mainly, the major sources of the total extractable organic matter (TEOM) in the river and marine sediments were natural and anthropogenic inputs (Rushdi et al. 2006a, 2006b, 2014a, 2017; Bainbridge et al. 2018). The specific inputs from point sources to sediments can be distinguished from source

TABLE 15.1 The concentrations (ng/g), distribution and geochemical parameters of the main hydrocarbons measured in the sediment samples from the Shatt Al-Arab River, the estuary and the northern part of the Arabian/Persian Gulf.

	ZONE I		*ZONE II*		*ZONE III*		*ZONE IV*	
COMPOUND	*AVG*	*SD*	*AVG*	*SD*	*AVG*	*SD*	*AVG*	*SD*
n-Alkanes								
Range	14–33		14–34		14–35		14–35	
C_{max}	19,29,29,31		29		29		29	
Total concentration (ng/g)	**13212**	**7503**	**15369**	**11399**	**9822**	**8212**	**10517**	**7165**
$CPI_{(o/e)}$[a]	1.86	0.18	2.16	0.25	2.49	0.34	2.21	0.23
$CPI_{(o/e)}$ (14–24)	0.92	0.08	0.96	0.13	1.05	0.22	0.79	0.14
$CPI_{(o/e)}$ (25–35)	3.06	0.40	3.12	0.33	3.47	0.15	3.59	0.37
T(15–24)/(25–35)	0.82	0.34	0.54	0.19	0.40	0.10	0.52	0.25
Biogenic n-Alkanes								
Total	**4271**	**1969**	**5897**	**4151**	**3843**	**2325**	**3670**	**1598**
%	32%	4%	37%	4%	42%	6%	38%	5%
Terrestrial wax n-Alkanes								
Total	**4059**	**1945**	**5419**	**3729**	**3658**	**2278**	**3558**	**1495**
%	30%	5%	34%	3%	39%	4%	37%	5%

(Continued)

TABLE 15.1 (Continued)

	ZONE I		ZONE II		ZONE III		ZONE IV	
COMPOUND	AVG	SD	AVG	SD	AVG	SD	AVG	SD
Microbial n-Alkanes								
Total	**338**	**348**	**297**	**333**	**122**	**91**	**206**	**66**
%	2%	1%	2%	1%	1%	1%	2%	1%
Aquatic algae and plankton n-Alkanes								
Total	**212**	**110**	**478**	**439**	**185**	**77**	**114**	**115**
%	2.3%	1.5%	2.6%	1.3%	2.6%	1.8%	0.8%	0.9%
Petroleum n-Alkanes								
Total	**8942**	**4822**	**9472**	**6272**	**5979**	**4394**	**6847**	**4626**
%	**68.0%**	**4.0%**	**63.3%**	**3.9%**	**58.3%**	**6.3%**	**62.5%**	**4.8%**
LMW/HMW	**0.82**	**0.34**	**0.54**	**0.19**	**0.40**	**0.10**	**0.52**	**0.25**
UCM/n-Alkanes	**13**	**2**	**25**	**16**	**14**	**2**	**14**	**4**
n-Alkanes/nC16	**255**	**293**	**65**	**17**	**53**	**53**	**232**	**210**
Pristane	**407**	**379**	**1008**	**849**	**253**	**253**	**282**	**218**
Phytane	**1334**	**927**	**1743**	**2032**	**543**	**543**	**493**	**459**
Pr/Ph	0.24	0.13	0.92	1.02	0.23	0.23	0.61	0.27
Pr/nC17	0.98	0.42	1.41	0.82	0.61	0.61	0.98	0.33
Ph/nC18	2.67	0.75	2.39	0.71	0.97	0.97	1.27	0.97
Hopanes								
Range	27–35		27–35		27–35		27–35	
C_{max}	29,30		29,30		30		29	
Total	**10648**	**7274**	**12479**	**3277**	**5161**	**2389**	**9543**	**8178**
C_{31} S/(R+S)	0.63	0.03	0.59	0.02	0.60	0.01	0.60	0.03
C_{32} S/(R+S)	0.63	0.02	0.60	0.02	0.59	0.01	0.58	0.04
Steranes								
Range	27–29		27–29		27–29		27–29	
C_{max}	29		29		29		29	
Total	**6495**	**3199**	**9762**	**4166**	**5216**	**2596**	**9911**	**4913**
C_{27} S/(R+S)	0.36	0.14	0.37	0.03	0.37	0.05	0.37	0.02
C_{29} S/(R+S)	0.47	0.05	0.47	0.03	0.45	0.02	0.43	0.03
TEOM	**285064**	**141287**	**432766**	**165901**	**244299**	**124758**	**246366**	**119200**
UCM	**177583**	**91177**	**292654**	**141044**	**129483**	**81477**	**128781**	**41701**
U:R	1.59	0.42	1.90	0.69	1.21	0.48	1.42	0.46
Terrestrial plants								
Concentration	**4059**	**2175**	**5419**	**4085**	**3658**	**2790**	**3447**	**1519**
(%)	12.5%	5.9%	12.1%	4.7%	17.5%	1.8%	12.9%	3.6%
Aquatic plants and bacteria								
Concentration	**212**	**123**	**478**	**480**	**185**	**94**	**114**	**132**
(%)	0.77%	0.43%	0.93%	0.63%	1.14%	0.91%	0.26%	0.31%
Petroleum								
Concentration	**27827**	**15400**	**34464**	**15473**	**16887**	**12365**	**27076**	**19416**
(%)	80.8%	5.2%	81.4%	7.7%	79.9%	3.5%	84.3%	2.6%

correlations of organic matter (Boot et al. 2006; Jaffé et al. 2006; Rushdi and Simoneit 2002a, 2002b; Rushdi et al. 2010, 2014a, 2014b, 2022; Stiehl et al. 2005). The concentrations of the hydrocarbon compounds in Table 15.1 were used to identify the sources of hydrocarbons in sedimentary TEOM (Rushdi and Simoneit 2002a, 2002b; Rushdi et al. 2006a, 2006b, 2018, 2019a, 2019b; Birgel et al. 2004; Gogou and Stephanou 2004; Hopmans et al. 2004; Boot et al. 2006; Medeiros and Simoneit 2008).

15.4.1 n-Alkanes

The n-alkanes ranged mostly from C_{14} to C_{35} for these sediments (Table 15.1, Figure 15.2). The maximum concentration (C_{max}) was predominantly at C_{29} in all samples, except those from Z-I, where C_{19}, C_{31} and C_{20} were major (Table 15.1). The total concentrations of n-alkanes varied from 13.2±7.5 µg/g in

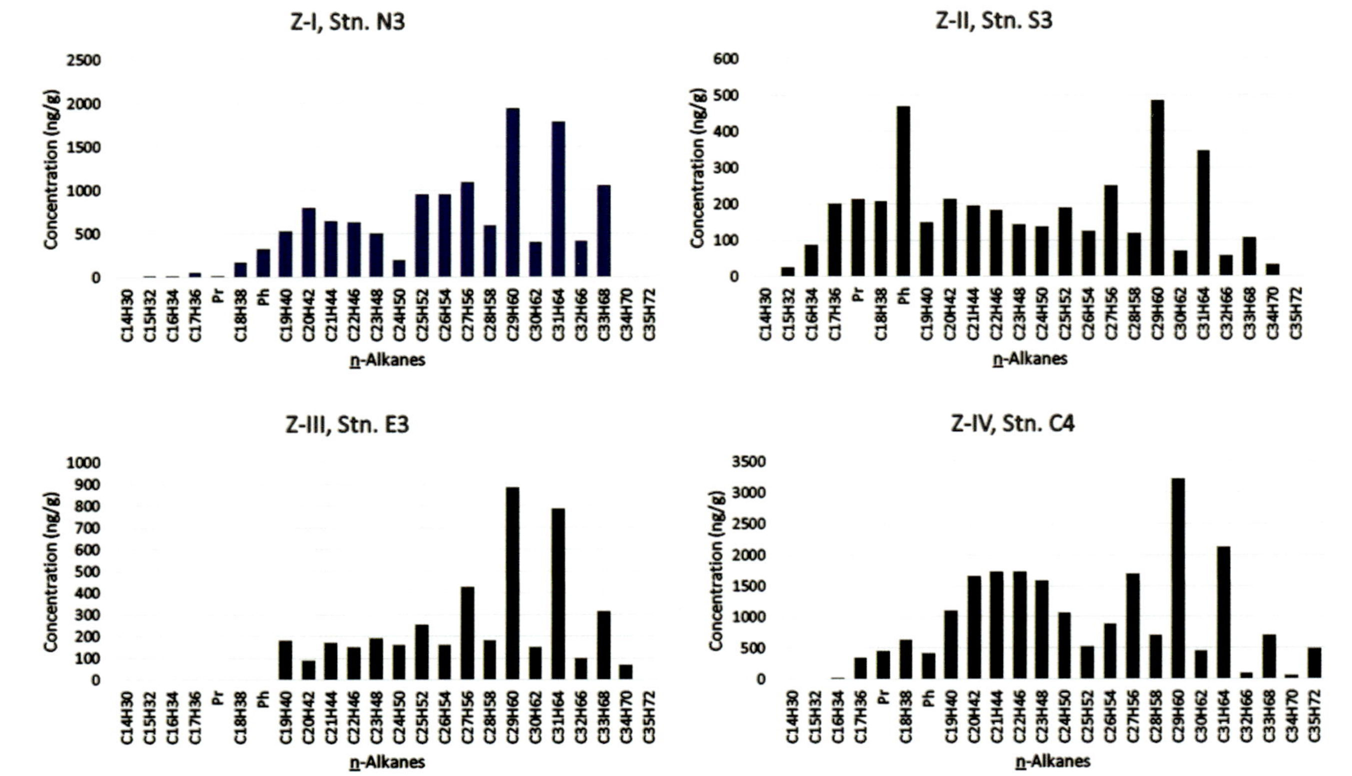

FIGURE 15.2 Distributions and concentrations of n-alkanes in selected sediment samples from the southern part of the Shatt Al-Arab River and the coastal zone in the north of the Arabian/Persian Gulf.

the sediment samples from Z-I, 15.4±11.4 µg/g for Z-II, 9.8±8.2 µg/g for Z-III and 10.5±7.2 µg/g for Z-IV (Table 15.1, Figure 15.3). The odd-numbered n-alkanes predominated over the entire range with carbon preference indices ($CPI_{o/e}$, Mazurek and Simoneit 1984) ranging from 1.86±0.18, 2.16±0.24, 2.49±0.34 and 2.21±0.23 for samples from Zones I, II, III and IV, respectively (Table 15.1).

n-Alkanes occur in all ecosystems and originate from both biogenic and anthropogenic sources. They can be utilized to evaluate the origins of organic matter (OM) in the environment according to their presence and distribution patterns (e.g., Rushdi et al. 2006a, 2014a, 2017, 2022). The fundamental parameters linked to their sources and characteristic behaviour in the environment include the n-alkane C_{max} in the homologous series, the CPI and the U:R values (Simoneit 1977, 1978, 2002; Mazurek and Simoneit 1984). These sediments generally had a C_{max} at 29, signifying that wax n-alkanes of aquatic and higher terrestrial plants were dominant (Eglinton and Hamilton 1967; Simoneit 1978; Ficken et al. 2000; Rushdi et al. 2019a, 2022). The same n-alkane distribution of higher molecular weight has also been reported in the wetland marshes of Iraq (Rushdi et al. 2006a). The contributions of biogenic and anthropogenic n-alkanes in these samples can be defined by the CPI (Simoneit 1989). The CPI values for the C_{14}–C_{35} n-alkanes ranged from 1.86 to 2.49, indicative of vascular higher plants as the dominant inputs compared with petrogenic sources in the present samples.

The ratios of LMW/HMW (i.e., total concentrations of n-C_{14} to n-C_{24} relative to the total concentrations of n-C_{25} to n-C_{35}) were estimated to be 0.82±0.34 for Z-I, 0.54±0.19 for Z-II, 0.40±0.1 for Z-III and 0.52±0.25 for Z-IV (Table 15.1). These values also implied that the major amounts of n-alkanes were from terrestrial higher plants (Commendatore et al. 2000; Gao et al. 2007).

The isoprenoids pristane and phytane were measured in all sediment samples. The concentrations of pristane were 0.40±0.38 µg/g for Z-I, 1.00±0.85 µg/g for Z-II, 0.25 µg/g for Z-III and 0.28±0.22 µg/g for Z-IV (Table 15.1, Figure 15.3). The concentrations of phytane were 1.33±0.93 µg/g, 1.74±2.03 µg/g, 0.54 µg/g and 0.49±0.46 µg/g for I, II, III and IV, respectively (Table 15.1, Figure 15.3). The ratios of pristane to phytane (Pr/Ph) were 0.24±0.13 for Z-I, 0.92±1.02 for Z-II, 0.23±0.23 for Z-III and 0.61±0.27 for Z-IV (Table 15.1). These values, which were mainly < 0.1, indicated that these compounds were altered and preserved under anoxic/evaporitic conditions (Didyk et al. 1978; ten Haven et al. 1987; Barbe et al. 1990; Peters and Moldowan 1993) with some petroleum inputs (Commendatore and Esteves 2004).

15.4.2 Biomarkers

Different organic compounds in the environment are related to specific biological sources in terms of chemical and/or basic carbon skeleton structure (Treibs 1934, 1936; Eglinton 1969; Simoneit 1977, 1986; Brassell 1992). These compounds are known as biomarkers and can be effective tracers of organic matter derived from natural sources (Brassell et al. 1983; Peters and Moldowan 1993). For example, hopanes inherit their chemical structures from the triterpenoid compounds in

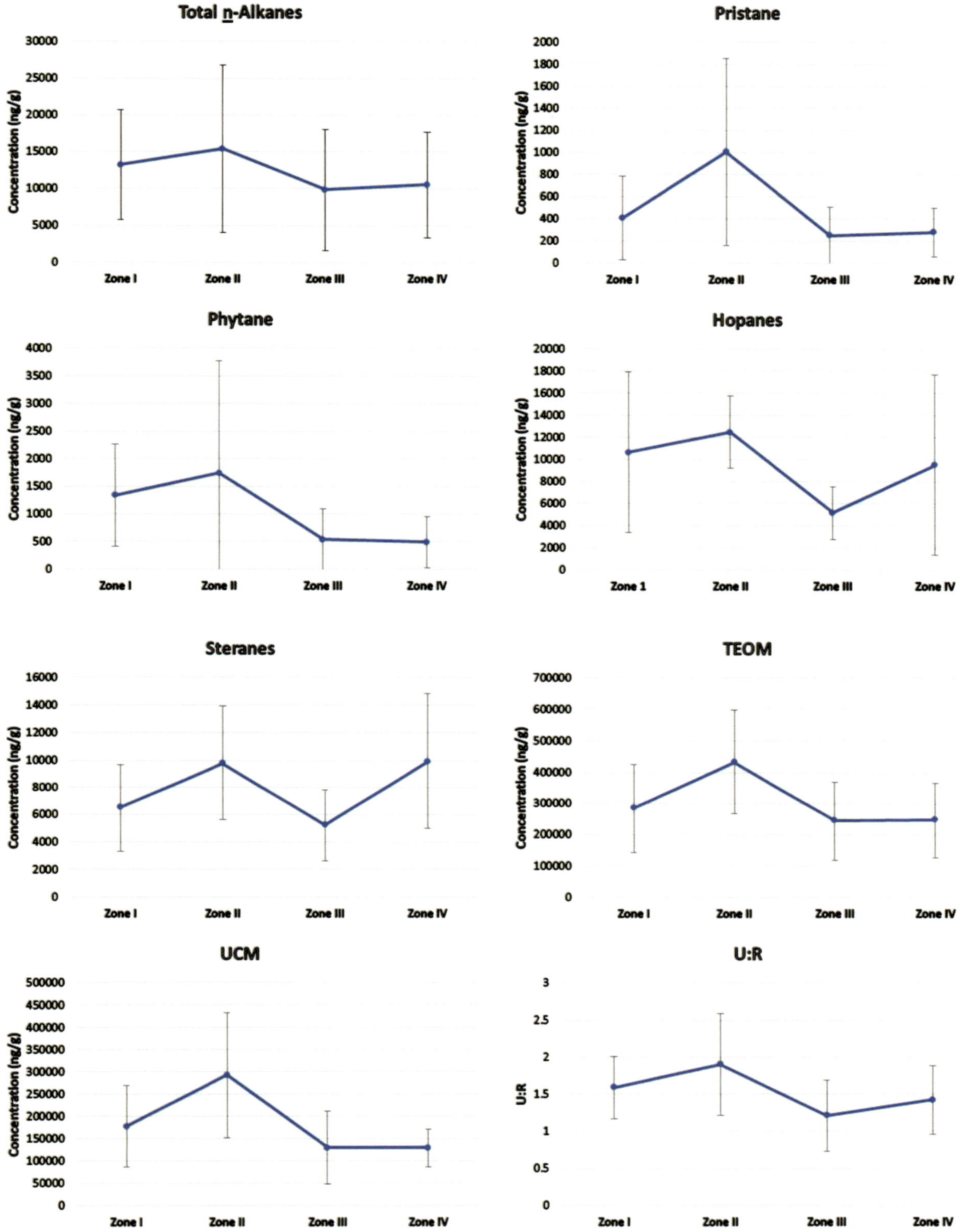

FIGURE 15.3 Plots showing the concentrations of total n-alkanes, pristanes, phytanes, hopanes, steranes, TEOM and UCM and U:R ratios in the different zones of the study area.

bacterial and some terrestrial plant resins and steranes from animal and plant sterols (Mills and Werner 1955; Ourisson et al. 1979; Mackenzie et al. 1982; Moldowan et al. 1992).

Hopanes are stable in the environment; thus, they have been utilized to indicate fossil fuel contamination (Bouloubassi et al. 2001; Peters and Moldowan 1993; Rushdi et al. 2010, 2017; Jafarabadi et al. 2019). Substantial levels of petroleum hopane biomarkers were present in the samples from the Shatt Al-Arab River, the estuary, and the coastal zone of Iraq. They were mainly 17α(H),21β(H)-hopanes, ranging from C_{27} to C_{35}, with a C_{max} at 29 or 30, with total concentrations of 10.6±7.3 µg/g for Z-I, 12.5±3.3 µg/g for Z-II, 5.1±2.4 µg/g for Z-III and 9.5±8.2 µg/g for Z-IV (Table 15.1, Figure 15.3). The presence of this isomeric configuration indicates a derivation from the diagenetic and geogenic transformation of the bacterial 17β(H),21β(H)-hopane precursors, which has been found in crude oils and mature sedimentary rocks (Peters and Moldowan 1993). The dominant distributions have shown the α,β-hopane biomarkers with the extended homologs >C_{30} as the mature C-22 *R*/*S* pairs (Simoneit 1984; Simoneit et al. 1990) confirmed that they were derived from petroleum and vehicle engine exhaust (Simoneit 1984, 1985). The C-22 $S/(S+R)$ ratios of the C_{31} and C_{32} hopanes were respectively 0.63±0.03 and 0.63±0.02 for Z-I, 0.59±0.02 and 0.60±0.02 for Z-II, 0.60±0.01 and 0.59±0.01 for Z-III and 0.60±0.03 and 0.58±0.04 for Z-IV (Table 15.1). These ratios were typical of the values found in mature crude oil (Kvenvolden et al. 1990; Rushdi and Simoneit 2002a, 2002b), confirming that these hopanes were derived from petroleum hydrocarbons.

Sterane biomarkers are rarely found in gasoline or diesel fuels, and they are generally introduced into the environment by petroleum sources and/or greases in vehicular engine emissions (Abas and Simoneit 1996). Therefore, steranes are also useful tracers for oil-product pollution in the environment (e.g., Albaiges and Cuberes 1980; Aboul-Kassim and Simoneit 1996; Rushdi et al. 2018). Steranes were detected in the sediment samples and ranged from C_{27} to C_{29}, with a C_{max} at 29 and the fully mature isomer composition – namely, the dominant 5α,14α,17β configuration as the 20*S* and 20*R* epimers. The concentrations ranged from 6.5±3.2 µg/g for Z-I, 9.8±4.2 µg/g for Z-II, 5.2±2.6 µg/g for Z-III and 9.9±4.3 µg/g for Z-IV (Table 15.1, Figure 15.3). The sterane epimerization ratios for C_{27} and C_{29} at C-20, $S/(S+R)$ varied respectively from 0.36±0.14 to 0.37±0.05 and from 0.43±0.3 to 0.47±0.5 for the sediments from the four zones Z-I, Z-II, Z-III and Z-IV (Table 15.1). These values also indicated that these steranes originated from petroleum and crude oil–related product sources.

15.4.3 Total extractable organic matter and unresolved complex mixture

The total extractable organic matter (TEOM) and the unresolved branched and cyclic compounds (UCM, as estimated by subtracting the integrated peaks of the resolved compounds from the integration of the entire chromatogram total ion traces) in these samples ranged from 244.3±124.8 µg/g (Z-III) to 432.8±165.9 µg/g (Z-II) and from 128.8±41.7 µg/g (Z-IV) to 292.7±141.0 µg/g (Z-II), respectively (Table 15.1). These concentrations of TEOM and UCM are summarized for the sediments from Z-I to Z-IV in Figure 15.3.

The occurrence of high levels of UCM in the environment indicates anthropogenic contamination by petroleum and chronic degradation of a mixture of organic matter (Volkman et al. 1992; Bouloubassi et al. 2001; Silva et al. 2013; Mille et al. 2007). The ratios of UCM to resolved compounds (U:R) (Mazurek and Simoneit 1984) can differentiate between degraded petroleum (U:R > 4) and fresh hydrocarbons (U:R < 4) (Simoneit 1982; Mazurek and Simoneit 1984; Tolosa et al. 2004; Gao et al. 2021). The ratio values of these sediments ranged from 0.98 to 2.28 (1.59±0.42) for the Z-I sediments, 1.06 to 2.68 (1.90±0.69) for Z-II, 0.54 to 1.65 (1.21±0.48) for Z-III and from 0.62 to 1.77 (1.42±0.42) for Z-IV (Table 15.1, Figure 15.3), indicating that they came from a mixture of natural and anthropogenic sources (Tolosa et al. 2004; Harji et al. 2008) The U:nCB ratios (UCM to biogenic n-alkanes) have been also applied to distinguish degraded petroleum (U:nCB > 10) from natural (U:nB < 10) sources (Peters and Moldowan 1993; Tolosa et al. 2004; Harji et al. 2008; Rushdi et al. 2014b, 2022; Gao et al. 2021). The ratio values were 13±2 for sediments from Z-I, 25±16 for Z-II, 14±2 for Z-III and 14±4 for Z-IV (Table 15.1). These estimated ratio values were > 10, suggesting that all sites were contaminated by biodegraded petroleum residues (Peters and Moldowan 1993; Tolosa et al. 2004; Harji et al. 2008; Rushdi et al. 2014b, 2022).

15.4.4 Major sources of hydrocarbons

The contribution of the biogenic n-alkanes was determined following the method, explained by Rushdi et al. (2022). We used the following equation to estimate the contribution from natural (NA_x) sources:

$$NA_x = nC_i - \frac{nC_{i-1} + nC_{i+1}}{2} \tag{15.1}$$

where C_i represents the odd carbon numbers of the entire n-alkane range and C_{i-1} and C_{i+1} are the carbon numbers before and after the C_i. The total concentrations of natural biogenic n-alkanes (i.e., aquatic, and terrestrial plants) were 4.3±2.0 µg/g (32±4%) for Z-I, 5.9±4.2 µg/g (37±4%) for Z-II, 3.8±2.3 µg/g (42±6%) for Z-III and 3.7±1.6 µg/g (38±5%) for Z-IV (Table 15.1). The C_{max} of the vascular plant wax n-alkanes commonly ranges between 25 and 31 and depends on the plant species, the time of year and the region (Eglinton and Hamilton 1967; Simoneit 1977; Simoneit and Mazurek 1982; Mazurek et al. 1991; Rogge et al. 1993; Stephanou and Stratigakis 1993; Abas and Simoneit 1998). The contribution of the terrestrial higher plant wax n-alkanes was determined following the method explained by Simoneit et al. (1991), where equation 15.1 was used for the n-alkanes ranging from nC_{25} to nC_{33}. These concentrations were 4.1±2.0 µg/g (30±5%) for Zone I, 5.4±3.7 µg/g (34±3%) for Z-II, 3.7±2.3 µg/g (39±4%) for Z-III and 3.6±1.5 µg/g (37±5%) for Z-IV (Table 15.1, Figure 15.4). The total concentrations of n-alkanes from aquatic algae, which was calculated using the same Equation (15.1) for n-alkane, range from nC_{14} to nC_{24}, and bacteria estimated using the same equation for the same range where C_i represents n-alkane of even carbon numbers were 0.34±0.35 µg/g (2±1%) in sediments from Z-I, 0.30±0.33 µg/g (2±1%) for Z -II, 0.12±0.09 µg/g (1±1%) for Z-III and 0.21±0.06 µg/g (2±1%) for Z-IV (Table 15.1, Figure 15.4). The n-alkanes from petroleum contributions were calculated by difference and are the major sources. Their concentrations were 8.9±4.8 µg/g (68±4%) for sediments from Z-I, 9.5±6.3 µg/g (63±4%) for Z-II, 6.0±4.4 µg/g (58±6%) for Z-III and 6.8±4.6 µg/g (63±5%) for Z-IV (Table 15.1, Figure 15.4).

The presence of an unresolved complex mixture (UCM) of branched and cyclic hydrocarbons is widespread in aquatic sediments and is related largely to fossil fuel product utilization and oil spills (Simoneit 1985). The levels of total oil-related sources (including petroleum n-alkanes, biomarkers and UCM) were 27.8±15.4 µg/g (81±5%) for Z-I, 34.5±15.5 µg/g (81±8%) for Z-II, 16.9±12.4 µg/g (80±4%) for Z-III and 27.1±19.4 µg/g (84±3%) for Z-IV (Table 15.1, Figure 15.5). The results indicated that anthropogenic inputs from petroleum-related products were dominant in the Shatt al-Arab river and the northern coast of the Arabian/Persian Gulf, confirming that crude oil

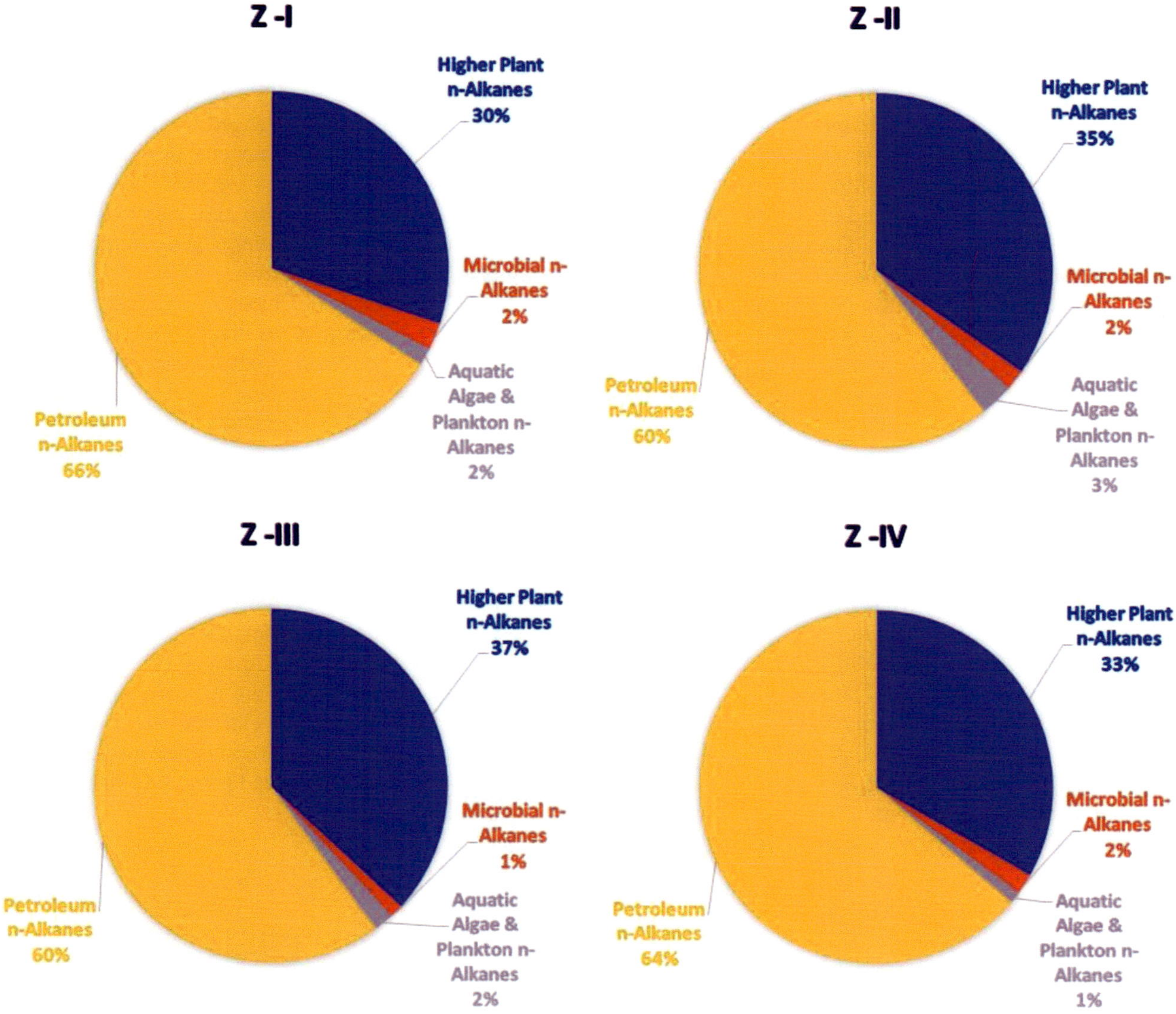

FIGURE 15.4 Plots showing the relative concentrations (%) of various sources of n-alkanes in the different zones of the study area.

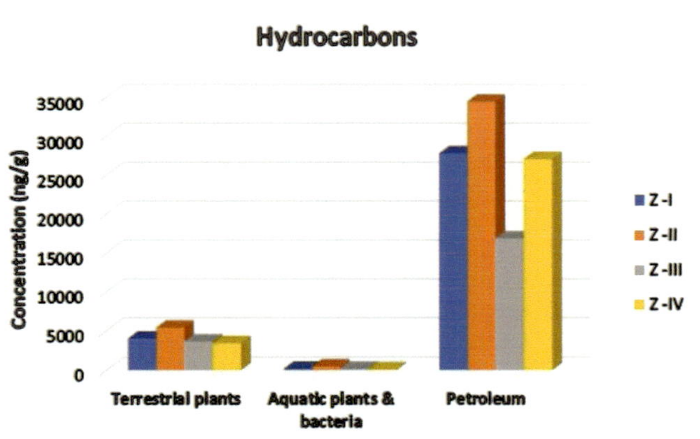

FIGURE 15.5 Plot showing the concentrations of total hydrocarbons from the three main inputs – terrestrial plants, aquatic plants and bacteria, and petroleum – to the different zones of the study area.

refineries, loading and unloading terminals, oil tanker traffic and human-related activities are the main anthropogenic organic matter inputs into the river (Alfalahi and Alfaras 2021; Al-Saad and Salman 2012; DouAbul and Al-Saad 1985; Mahdi 2015). The low levels of natural sources of hydrocarbons in the river are likely due to the recent drought in this arid region (Al-Asadi and Muttashar 2022; Al-Mahmood et al. 2015; Al-Yamani 2008; Kadhim et al. 2021).

15.5 ECOLOGICAL IMPACTS

Critical aquatic habitats are very important key areas because they provide breeding, feeding, nursing and nesting places for aquatic organisms. They include rocky shores where fauna and flora flourish, sandy beaches used by turtles as nesting sites and mangrove stands used as nursery grounds for crustaceans, shrimp and fish (Vaughan et al. 2019; Sheppard et al. 1992; Burt 2014). The ecosystems of the Arabian/Persian Gulf include benthic macroalgae, mangrove stands, coral reefs and seagrass, which are important nutrient sources in the region (Ogden and Gladfelter 1983; Crossland et al. 1987). Therefore, the contamination of sediments and water of aquatic ecosystems by petroleum hydrocarbons is a significant environmental concern. The hydrocarbons from gasoline, diesel, lubricating oils and related petroleum products are highly toxic in nature

(Ahmed and Fakhruddin 2018). Short-term exposure to oil pollution can cause narcosis (a lethal effect) in some delicate organisms, and long-term accumulation of petroleum hydrocarbons harms aquatic ecosystems (Millemann et al. 1984; Seuront and Leterme 2007). Petroleum hydrocarbons are common components of sediments and are considered to be the major causes of toxicity in benthic organisms such as the amphipod *Corophium volutator* and the polychaete *Arenicola marina* (Grant and Briggs 2002). The toxicity of petroleum hydrocarbons to organisms in an aquatic environment is more related to their bioavailability (Kuppusamy et al. 2020). Non-bioavailable hydrocarbons can also be adsorbed to particulate matter and sediments and ultimately affect benthic invertebrates, bottom-water fish, and freshly laid fish eggs (Al-Khion et al. 2021; Kuppusamy et al. 2020).

The petroleum hydrocarbon concentrations in these sediments (7.17–15.40 μg/g) were comparable to the levels measured in other regions of the world. They were lower than the concentrations reported in sediments from the Orashi River, River State, and Niger River in Nigeria (16.66–256.1 μg/g) (Edori and Edori 2021a, 2021b; Idodo-Umeh and Oronsaye 2003); rivers and canals in Tianin, the Beiluohe River and Pearl River Delta, China (72–3000 μg/g) (Ye et al. 2007; 12.1–3762 μg/g (Zhang et al. 2007); 66.6–1445 μg/g (Zhao et al. 2015); the Danube River and its main tributaries, Serbia (170–842 μg/g) (Maljević and Balać 2007); and the Karun River, Iran (13.8–234.2 μg/g) (Dehvari et al. 2023). Still, they were higher than the concentration levels measured in sediments from the St. Clair River in Canada (> 20 μg/g) (Yang et al. 2021).

Because of the low solubility of petroleum hydrocarbons in water, some of its fractions accelerate the clustering of natural and non-natural organic matter and obstruct gas exchange, including oxygen. Heavier clusters will eventually sink and accumulate in the bottom sediments (Suaria et al. 2018; Morales-McDevitt et al. 2020).

Petroleum hydrocarbons have been reported in the flesh component of fish from fresh and marine waters of Iraq (Al-Khion et al. 2021; Al-Ali et al. 2022). They were detected in economic freshwater fish from the Shatt Al-Arab River with the highest levels in the filter feeder *Palaniza abu* and lowest amounts in the omnivorous *Cyprinus carpio* (Al-Ali et al. 2022). About 80% of the n-alkanes in the tissues of Iraqi marine fish were identified as being of petroleum origin (Al-Khion et al. 2021). This indicates that contamination of aquatic species – especially fish – by crude oil hydrocarbons is a major health concern. Still, little information is available about the health risk caused by petroleum hydrocarbon contamination and its ecological effects in both freshwater and marine environments of the region, where comprehensive environmental impact assessments are needed.

15.6 CONCLUSION

Significant amounts of hydrocarbons, including n-alkanes, hopane and sterane biomarkers, and unresolved complex mixture (UCM) were characterized in sediments from the Shatt Al-Arab River, the estuary, and the northern part of the Arabian/Persian Gulf. The total n-alkanes in the sediments of the three zones varied from 9822±8212 ng/g to 15369±11399 ng/g. The hopane and sterane biomarkers ranged from 5161±2389 to 12479±3277 ng/g and from 5216±2596 ng/g to 9762±4166 ng/g, respectively. The dominant UCMs ranged from 128781±41701 ng/g to 177583±91177 ng/g. The biogenic sources of these hydrocarbons ranged between 12.1±4.7% and 17.5±1.8% from terrestrial plants and between 0.26±0.31% and 1.14%±0.91% from aquatic biota, whereas the anthropogenic sources from petroleum-related inputs ranged between 97.9±3.5% and 84.3±2.6% of TEOM. The presence of biomarkers, n-alkanes with CPI ~ 1 and UCM indicated that the main sources of anthropogenic hydrocarbons are petroleum product utilization and oil resource development activities in southern Iraq. The substantial amounts of anthropogenic hydrocarbons suggest serious threats to the aquatic habitat and biodiversity of the region. The continuous inputs of oil-related products to the environment of the Shatt Al-Arab River, estuary, and coastal region will ultimately affect the local aquatic ecosystems. Since the toxicity of petroleum hydrocarbons to aquatic organisms varies and can be due to short- or long-term exposure, it is very important to investigate the toxicity of petroleum hydrocarbons to these organisms to determine the severity of the impact on the aquatic biodiversity of the region. Further studies are required to evaluate the levels, sources, fates and physicochemical alterations of different organic matter introduced by various processes such as human and developmental activities and particulate matter transported by the wind in the region.

REFERENCES

Abas, M.R.B., and B.R.T. Simoneit. 1996. Composition of extractable organic matter of air particles from Malaysia: Initial studies. *Atmos Environ* 15: 2779–2793.

Abas, M.R.B., and B.R.T. Simoneit. 1998. Wax lipids from leaf surfaces of some common plants of Malaysia. *Pertanik J Sci Technol* 6: 171–182.

Abdullah, A.D., Masih, I., van der Zaag, P., Karim, U.F., Popescu, I., and Q. Al Suhail. 2015. Shatt al Arab River system under escalating pressure: A preliminary exploration of the issues and options for mitigation. *Int J River Basin Manag* 13, no. 2: 215–227.

Aboul-Kassim, T.A.T., and B.R.T. Simoneit. 1996. Lipid geochemistry of surficial sediments from coastal environment of Egypt I. Aliphatic hydrocarbons – characterization and sources. *Mar Chem* 54: 135–158.

Ahmed, F., and A.N.M. Fakhruddin. 2018. A review on environmental contamination of petroleum hydrocarbons and its biodegradation. *Int J Environ Sci Nal Resour* 11, no. 3: 1–7.

Al-Ali, B.S., Alkhion, D.D., Al-Saad, H.T., and A.I. Rushdi. 2022. Sources and apportionment of aliphatic hydrocarbons in freshwater fish species from Shatt Al-Arab River, Iraq. *Ecol Environ Conserv* 28: S10–S22.

Al-Arfaj, A.A., and I.A. Alam. 1993. Chemical characterization of sediments from the Gulf Area after the 1991 oil spill. *Mar Poll Bull* 27: 97–101.

Al-Asadi, S.A., and W.R. Muttashar. 2022. Impact of the environmental degradation of rivers on the reappraisal of international agreements related to the transboundary watercourse, Shatt Al-Arab River (Southern Iraq): A case study. *Sustain Water Resour Manag* 8, no. 3: 84.

Albaiges, J., and M.R. Cuberes. 1980. On the degradation of petroleum residues in the marine environment. *Chemosphere* 9: 539–545.

Al-Dabbas, M., Al-Jassim, J., and S. Al-Jumaily. 2010. Depositional environments and porosity distribution in regressive limestone reservoirs of the Mishrif Formation, Southern Iraq. *Arab J Geosci* 3: 67–78.

Aldosari, K.R. 2021. The applicability of the international and regional efforts to prevent oil pollution: Comparative analysis between the Arabian Gulf region and the North Sea. In: *Sustainability in the Maritime Domain*. Eds. A. Carpenter, T.M. Johansson and J.A. Skinner, 199–221. Cham, Switzerland: Springer.

Alfalahi, M.A., and M.S. Alfaras. 2021. Application of remote sensing technique to examine the oil pollution of the Shatt Al-Arab water. *J Appl Sci Eng* 24, no. 5: 713–718.

Al-Khion, D.D., Al-Ali, B.S., Al-Nagar, G., Al-Saad, H.T., Al-Anber, L.J., Khwadem, A.A., Saleh, S.M., and A.M. Zeidan. 2016. Polycyclic aromatic hydrocarbons in some fishes from the Iraqi marine waters. *J Natl Sci Res* 6, no. 10: 92–99.

Al-Khion, D.D., Al-Ali, B.S., Al-Saad, H.T., and A.I. Rushdi. 2021. Levels and source of aliphatic hydrocarbons in marine fishes from coast of Iraq based on biomarkers and biogeochemical indices. *Indian J Ecol* 48, no. 2: 536–544.

Al-Mahmood, H., Hassan, W., Al-Hello, A., Hammood, O., and N. Muhson. 2015. Impact of low discharge and drought on the water quality of the Shatt al-Arab and Shatt al-Basra Rivers (south of Iraq). *J Int Acad Res Multidiscip* 3, no. 1: 285–296.

Al-Ramadan, K. 2014. Diagenesis of Holocene beachrocks: A comparative study between the Arabian Gulf and the Gulf of Aqaba, Saudi Arabia. *Arab J Geosci* 7, no. 11: 4933–4942.

Al-Saad, H.T., and F.J. Al-Imarah. 2021. Hydrocarbons pollution in the north-west Arabian Gulf. In: *The Arabian Seas: Biodiversity, Environmental Challenges and Conservation Measures*. Ed. L.A. Jawad, 1187–1197. Cham, Switzerland: Springer.

Al-Saad, H.T., Farid, W.A., and A.Y. Al-Adhub. 2011. Distribution and seasonal variations of n-alkanes in some species of molluscs from Shatt Al-Arab River. *Mesop J Mar Sci* 26, no. 2: 182–196.

Al-Saad, H.T., Farid, W.A., Ateek, A.A., Sultan, A.W.A., Ghani, A.A., and S. Mahdi. 2015. n-Alkanes in surficial soils of Basrah city, southern Iraq. *Int J Mar Sci* 5, no. 52: 1–8.

Al-Saad, H.T., and N.A. Salman. 2012. Status of oil pollution in the Arabian Gulf and Shatt Al-Arab estuary: A review. In: *Coastal Environments: Focus on Asian Regions*. Ed. V. Subramanian, 36–46. Dordrecht: Springer.

Al-Yamani, F. 2008. Importance of the freshwater influx from the Shatt-Al-Arab River on the Gulf marine environment. In: *Protecting the Gulf's Marine Ecosystems from Pollution*. Eds. A.H. Abuzinada, H.-J. Barth, F. Krupp, B. Böer and T.Z. Abdessalaam, 207–222. Basel: Birkhäuser.

Ashraf, W., and A. Mian. 2010. Total petroleum hydrocarbon burden in fish tissues from the Arabian Gulf. *Toxicol Environ Chem* 92, no. 1: 61–66.

Bainbridge, Z., Lewis, S., Bartley, R., Fabricius, K., Collier, C., Waterhouse, J., Garzon-Garcia, A., Robson, B., Burton, J., Wenger, A., and J. Brodie. 2018. Fine sediment and particulate organic matter: A review and case study on ridge-to-reef transport, transformations, fates, and impacts on marine ecosystems. *Mar Pollut Bull* 135: 1205–1220.

Barbe, A., Grimalt, J.O., Pueyo, J.J., and J. Albaiges. 1990. Characterization of model evaporitic environments through the study of lipid components. *Org Geochem* 16, nos. 4–6: 815–828.

Birgel, D., Stein, R., and J. Hefter. 2004. Aliphatic lipids in recent sediments of the Farm Strait/Yermak plateau (Arctic Ocean): Composition, sources and transport processes. *Mar Chem* 88: 127–160.

Boot, C.S., Ettwein, V.J., and M.A.A. Maslin. 2006. 35,000 year record of terrigenous and marine lipid in Amazon fan sediments. *Org Geochem* 37: 208–219.

Bouloubassi, I., Fillaux, J., and A. Saliot. 2001. Hydrocarbons in surface sediments from the Changjiang (Yangtze River) Estuary, East China Sea. *Mar Pollut Bull* 42: 1335–1346.

Brassell, S.C. 1992. Biomarkers in sediments, sedimentary rocks and petroleums: Biological origins, geological fate and application. In: *Geochemistry of Organic Matter in Sediments and Sedimentary Rocks: SEPM Short Course No. 27.* Eds. L.M. Pratt, J.B. Comer and S.C. Brassell, 29–72. Tulsa: Society for Sedimentary Geology.

Brassell, S.C., Eglinton, G., and J.R. Maxwell. 1983. The geochemistry of terpenoids and steroids. *Biochem Soc Trans* 11: 575–586.

Buringh, P. 1960. *Soil and Conditions of Iraq*. Directorate General, Agricultural Research and Project, Ministry of Agriculture, Baghdad.

Burt, J.A. 2014. The environmental costs of coastal urbanization in the Arabian Gulf. *City* 18: 760–770.

CEB (Consultant Engineering Bureau). 2011. *Shatt Al-Arab, Final Report (in Arabic)*. College of Engineering, University of Baghdad, 101 (Unpublished).

Commendatore, M.G., and J.L. Esteves. 2004. Natural and anthropogenic hydrocarbons in sediments from the Chubut River (Patagonia, Argentina). *Mar Pollut Bull* 48, nos. 9–10: 910–918.

Commendatore, M.G., Esteves, J.L., and J.C. Colombo. 2000. Hydrocarbons in coastal sediments of Patagonia, Argentina: Levels and probable sources. *Mar Pollut Bull* 40: 989–998.

Corner, E.D.S. 1979. Pollution studies with marine plankton: Part I. petroleum hydrocarbons and related compounds. *Adv Mar Biol* 15: 289–380.

Cortes, J.E., Suspes, A., Roa, S., González, C., and H.E. Castro. 2012. Total petroleum hydrocarbons by gas chromatography in Colombian waters and soils. *Am J Environl Sci* 8, no. 4: 396–402.

Crossland, C.J., Dawson, S.A., Smith, S., and J.I. Marshall. 1987. Saudi Arabia: An analysis of coastal and marine habitats of the Red Sea. Saudi Arabia Marine Conservation Programme. *Synoptic Report*. National Union for Conservation of Nature, Geneva.

Dehvari, M., Takdastan, A., Jorfi, S., Ahmadi, M., Tahmasebi Birgani, Y., and A.A. Babaei. 2023. Distribution of total petroleum hydrocarbons in superficial sediments of Karun River Basin, southwest of Iran: Spatial and seasonal variations, source identification, and ecological risk. *Int J Environ Health Res* 33, no. 1: 71–82.

De Mora, S., Fowler, S.W., Tolosa, I., Villeneuve, J.P., and C. Cattini. 2005. Chlorinated hydrocarbons in marine biota and coastal sediments from the Gulf and Gulf of Oman. *Mar Pollut Bull* 50, no. 8: 835–849.

De Mora, S., Tolosa, I., Fowler, S.W., Villeneuve, J.P., Cassi, R., and C. Cattini. 2010. Distribution of petroleum hydrocarbons and organochlorinated contaminants in marine biota and coastal sediments from the ROPME Sea Area during 2005. *Mar Pollut Bull* 60, no. 12: 2323–2349.

Didyk, B.M., Simoneit, B.R.T., Brassell, S.T., and G. Eglinton. 1978. Organic geochemical indicators of palaeoenvironmental conditions of sedimentation. *Nature* 272, no. 5650: 216–222.

Dong, Y., Li, Y., Kong, F., Zhang, J., and M. Xi. 2020. Source, structural characteristics and ecological indication of dissolved organic matter extracted from sediments in the primary tributaries of the Dagu River. *Ecol Indic* 109: 105776.

Douabul, A.A., and H.T. Al-Saad. 1985. Seasonal variations of oil residues in water of Shatt Al-Arab River, Iraq. *Water Air Soil Pollut* 24, no. 3: 237–246.

Douabul, A.A., Al-Saad, H.T., Al-Timari, A., and H. al-Rakabi. 1988. Tigris-Euphrates Delta: A major source of pesticides to the Shatt al-Arab River (Iraq). *Arch Environ Con Tox* 17: 405–418.

Douabul, A.A., Farid, W.A., Al-Saad, H.T., and S.S. AlMaarofi. 2012. Hydrocarbons in soil from Basrah oil-rich governorate. *Am J Environ Sci* 8, no. 5: 563–568.

Edori, O., and E. Edori, 2021a. Profile of total petroleum hydrocarbons in the water and sediment columns of the Orashi river, Engenni, Rivers State, Niger delta, Nigeria. *J Glob Ecol Environ* 11, no. 2: 12–21.

Edori, O., and E. Edori, 2021b. Temporal (Monthly) distribution and variation of total petroleum hydrocarbons content in the water and sediments from Orashi River, Engenni, Rivers State, Nigeria. *Acad J Chem* 6, no. 2: 28–36.

Eglinton, G. 1969. Organic chemistry: The organic chemist's approach. In: *Organic Geochemistry*. Eds. G. Eglinton and M.T.J. Murphy, 20–71. New York: Springer-Verlag.

Eglinton, G., and R.J. Hamilton. 1967. Leaf epicuticular waxes. *Science* 156: 1322–1335.

Farid, W.A., Al-Salman, A.N., Ali, W.A., Al-Saad, H.T., Mahdi, S., and A.A. Al-Hello. 2016. Polycyclic aromatic hydrocarbons (PAHs) in the surface sediments of Shatt Al-Arab River, Basrah City, southern Iraq. *J Natl Sci Res* 6, no. 8: 46–55.

Ficken, K.J., Li, B., Swain, D.E., and G. Eglinton. 2000. An n-alkanes proxy for the sedimentary input of submerged/floating fresh water aquatic macrophytes. *Org Geochem* 31: 745–759.

Gao, X., Chen, S., Xie, X., Long, A., and F. Ma. 2007. Non-aromatic hydrocarbons in surface sediments near the Pearl River estuary in the South China Sea. *Environ Pollut* 148, no. 1: 40–47.

Gao, Y., Han, Y., Xia, J., Tan, J., Wang, Y.P., and S. Wang. 2021. Composition and distribution of aliphatic hydrocarbon compounds and biomarkers in seafloor sediments from offshore of the Leizhou Peninsula (South China). *ACS Omega* 6, no. 50: 34286–34293.

Gogou, A., and E.G. Stephanou. 2004. Marine organic geochemistry of the Eastern Mediterranean 2. Polar biomarkers in Cretan Sea surficial sediments. *Mar Chem* 85: 1–25.

Grant, A., and A.D. Briggs. 2002. Toxicity of sediments from around a North Sea oil platform: Are metals or hydrocarbons responsible for ecological impacts? *Mar Environ Res* 53, no. 1: 95–116.

Harji, R.R., Yvenat, A., and N.B. Bhosle. 2008. Sources of hydrocarbons in sediments of the Mandovi estuary and the Marmugoa harbour, west coast of India. *Environ Int* 34, no. 7: 959–965.

Harwood, JJ. 2014. Molecular markers for identifying municipal, domestic and agricultural sources of organic matter in natural waters. *Chemosphere* 95: 3–8.

Hopmans, E.C., Weijers, J.W.H., Schefuss, E., Herfort, L., Damste, S., Jaap, S., and S.A. Schouten. 2004. Novel proxy for terrestrial organic matter in sediments based on branched and isopernoid tetraether lipids. *Earth Planet Sci Lett* 224: 107–116.

Idodo-Umeh, G., and J.A.O. Oronsaye. 2003. Petroleum hydrocarbons in the sediments of Areba River, Olomoro, Niger Delta, Nigeria. *Tropical Freshw Biol* 12: 85–92.

Isaev, V.A., and M.V. Mikhailova. 2009. The hydrology, evolution and hydrological regime of the mouth area of the Shatt al-Arab River. *Water Resour* 36: 380–395.

Ite, A.E., Harry, T.A., Obadimu, C.O., Asuaiko, E.R., and I.J. Inim. 2018. Petroleum hydrocarbons contamination of surface water and groundwater in the Niger Delta region of Nigeria. *J Environ Pollut Human Health* 6, no. 2: 51–61.

Jafarabadi, A.R., Dashtbozorg, M., Bakhtiari, A.R., Maisano, M., and T. Cappello. 2019. Geochemical imprints of occurrence, vertical distribution and sources of aliphatic hydrocarbons, aliphatic ketones, hopanes and steranes in sediment cores from ten Iranian coral islands, Persian Gulf. *Mar Pollut Bull* 144: 287–298.

Jaffé, R., Rushdi, A.I., Medeiros, P.M., and B.R.T. Simoneit. 2006. Natural product biomarkers as indicator of sources and transport of sedimentary organic matter in subtropical rivers. *Chemosphere* 64: 1870–1884.

Kadhim, A.A., Shortridge, A., and A.K. Al-Nasrawi. 2021. Causes and consequences of environmental degradation along the Shatt Al – Arab River: A coupled human and natural systems (CHANS) perspective. *GeoJournal* 86: 2709–2722.

Kadhim, H.A., Al-Hjaje, M.M., and H.T. Al-Saad. 2019. n-Alkanes in soil of west Qurna-1 oil field in Basra City, Southern Iraq. *J Bio Innov* 8, no. 3: 273–288.

Kuppusamy, S., Maddela, N.R., Megharaj, M., and K. Venkateswarlu. 2020. Ecological impacts of total petroleum hydrocarbons. In: *Total Petroleum Hydrocarbons*, 95–138. Cham, Switzerland: Springer Nature.

Kvenvolden, K.A., Rapp, J.B., and F.D. Hostettler. 1990. Hydrocarbon geochemistry of hydrothermal generated petroleum from Escanaba Trough, offshore California, USA. *Appl Geochem* 5: 83–91.

Literathy, P. 1993. Considerations for the assessment of environmental consequences of the 1991 Gulf War. *Mar Poll Bull*. 27: 349–356.

Mackenzie, A.S., Brassell, S.C., Eglinton, G., and J.R. Maxwell. 1982. Chemical fossils: The geological fate of steroids. *Science* 217: 491–505.

Mahdi, B.A. 2015. Environmental pollution in Shatt al-Arab estuary. *J Int Acad Res Multidiscip* 3, no. 6: 32–42.

Maljević, E., and M. Balać. 2007. Determining of mineral oil – petroleum hydrocarbons in river sediments. *Desalination* 213, nos. 1–3: 135–140.

Maltby, E. (ed.). 1994. *An Environmental and Ecological Study of the Marshlands of Mesopotamia*. Draft Consultive Bulletin, AMAR Appeals Trust, London, 224 p.

Massoud, M.S., Al-Abdali, F., and A.N. Al-Ghadban. 1998. The status of oil pollution in the Arabian Gulf by the end of 1993. *Environ Int* 24: 11–22.

Mazurek, M.A., Cass, G.R., and B.R.T. Simoneit. 1991. Biological input to visibility reducing aerosol particles in the remote arid southwestern United States. *Environ Sci Technol* 25: 684–694.

Mazurek, M.A., and B.R.T. Simoneit. 1984. Characterization of biogenic and petroleum derived organic matter in aerosols over remote, rural and urban areas. In: *Identification and Analysis of Organic Pollutants in Air. American Chemical Society Symposium*. Ed. L.H. Keith, 353–370. Woburn, MA: Ann Arbor Science/Butterworth Publishers.

Medeiros, P.M., and B.R.T. Simoneit. 2008. Multi-biomarker characterization of sedimentary organic carbon in small rivers draining the northwestern United States. *Org Geochem* 39: 52–74.

Mille, G., Asia, L., Guiliano, M., Malleret, L., and P. Doumenq. 2007. Hydrocarbons in coastal sediments from the Mediterranean Sea (Gulf of Fos area, France). *Mar Pollut Bull* 54, no. 5: 566–575.

Millemann, R.E., Birge, W.J., Black, J.A., Cushman, R.M., Daniels, K.L., Franco, P.J., Giddings, J.M., McCarthy, J.F., and A.J. Stewart. 1984. Comparative acute toxicity to aquatic organisms of components of coal-derived synthetic fuels. *Trans Am Fish Soc* 113, no. 1: 74–85.

Mills, J.S., and A.E.A. Werner. 1955. The chemistry of Dammar resin. *J Chem Soc*: 3132–3140.

Moldowan, J.M., Lee, C.Y., Sundararaman, P., Salvatori, T., Alajbeg, A., Gjukić, B., Demaison, G.J., Slougui, N.-E., and D.S. Watt. 1992. Source correlation and maturity assessment of select oils and rocks from the Central Adriatic Basin (Italy and Yugoslavia). In: *Biological Markers in Sediments and Petroleum*. Eds. J.M. Moldowan, P. Albrecht and R.P. Philp, 370–471. Englewood Cliffs, NJ: Prentice-Hall.

Morales-McDevitt, M.E., Shi, D., Knap, A.H., Quigg, A., Sweet, S.T., Sericano, J.L., and T.L. Wade. 2020. Mesocosm experiments to better understand hydrocarbon half-lives for oil and oil dispersant mixtures. *PLoS ONE* 15, no. 1: e0228554.

Ogden, J.C., and E.H. Gladfelter. 1983. Coral reefs, seagrass beds and mangroves: Their interaction in the coastal zones of the Caribbean. *UNESCO* 23: 133 p.

Ourisson, G., Albrecht, P., and M. Rohmer. 1979. The hopanoids: Palaeochemistry and biochemistry of a group of natural products. *Pure Appl Chem* 51: 709–729.

Partow, H. 2001. *Demise of an Ecosystem: The Disappearance of the Mesopotamian Marshlands*. United Nations Environment Program (UNEP), Publication UNEP/DEWA/TR. 01–3, Nairobi, Kenya.
Peters, K.E., and J.M. Moldowan. 1993. *The Biomarker Guide: Interpreting Molecular Fossils in Petroleum and Ancient Sediments*. Englewood Cliffs, NJ: Prentice-Hall.
PME/UNEP. 1989. Environmental Impact Assessment- Basic Processes. *Presidency of Meteorology and Environment (PME) and United Nations Environmental Programme (UNEP)*, PME Press, Jeddah, 45 p.
Polservice, W. 1980. *Shatt Al-Arab Project – Feasibility Report, VII*. Ministry of Irrigation, Iraq.
Reynolds, R.N. 1993. Physical oceanography of the Gulf, Strait of Hormuz, and the Gulf of Oman: Results of the Mt. Mitchell expedition. *Mar Poll Bull* 27: 35–59.
Rogge, W.F., Mazurek, M.A., Hildemann, L.M., Cass, G.R., and B.R.T. Simoneit. 1993. Quantification of urban organic aerosols at a molecular level: Identification, abundance and seasonal variation. *Atmos Environ* 27A: 1309–1330.
Rushdi, A.I., Al-Mutlaq, K.F., Simoneit, B.R.T., Al-Azri, A., DouAbul, A.A.Z., Al-Zarban, S., and F. Al-Yamani. 2010. Characteristics of lipid tracers to the Arabian Gulf by run-off from rivers and atmospheric dust transport. *Arab J Geosci* 3: 113–131.
Rushdi, A.I., Al-Shaikh, I., El-Mubarak, A.H., Alnaimi, H.A., Al-Shamary, N., Hassan, H.M., and M.A. Assali. 2017. Characteristics and sources of anthropogenic and biogenic hydrocarbons in sediments from the coast of Qatar. *Mar Pollut Bull* 124, no. 1: 56–66.
Rushdi, A.I., Chase, Z., Simoneit, B.R.T., and A. Paytan. 2019a. Sources of organic tracers in atmospheric dust, surface seawater particulate matter and sediment of the Red Sea. In: *Oceanographic and Biological Aspects of the Red Sea*. Eds. N.M.A. Rasul and I.C.F. Stewart, 75–88. Cham, Switzerland: Springer Oceanography, Springer Nature. http://doi.org/10.1007/978-3-319-99417-8_4.
Rushdi, A.I., DouAboul, A.A.Z., Al-Maarofi, S.S., and B.R.T. Simoneit. 2018. Impacts of Mesopotamian wetland re-flooding on the lipid biomarker distributions in sediments. *J Hydrol* 558: 20–28.
Rushdi, A.I., DouAboul, A.A.Z, Mohammed, S.S., and B.R.T. Simoneit. 2006a. Distribution and sources of extractable organic matter in the Mesopotamian wetland marsh sediments of Iraq: I – aliphatic lipids. *Environ Geol* 50: 857–866.
Rushdi, A.I., DouAboul, A.A.Z, Mohammed, S.S., and B.R.T. Simoneit. 2006b. Compositions and sources of extractable organic matter in Mesopotamian marshland surface sediments of Iraq: II. Polar compounds. *Environ Geol* 50: 1171–1181.
Rushdi, A.I., DouAbul, A.A.Z., Simoneit, B.R.T., El-Mubarak, A.H., Al-Mutlaq, K.F., Qurban, M.A., and M.A. Goni. 2014a. Nonpolar lipid tracers in sediments from the Shatt al-Arab River of Iraq and the northwestern Arabian Gulf. *Arab J Geosci* 7, no. 12: 5495–5508.
Rushdi, A.I., El-Mubarak, A.H., Simoneit, B.R.T., Goni, M.A., Qurban, M.A., Bazeyad, A.Y., Al-Otaibi, M.T., Muhammed, Z.N., and K.F. Al-Mutlaq. 2022. Characteristics, distribution and sources of extractable organic matter in sediments from the coastal zone of the Arabian Gulf in Saudi Arabia. *Arab J Geosci* 15, no. 17: 1–17.
Rushdi, A.I., Rasul, N., Bazyad, A., and R. Dumenden. 2019b. Distribution and sources of hydrocarbon compounds in sediments from Obhur Lagoon: Red Sea coast of Saudi Arabia. In: *Oceanographic and Biological Aspects of the Red Sea*. Eds. N.M.A. Rasul and I.C.F. Stewart, 133–146. Cham, Switzerland: Springer Nature. http://doi.org/10.1007/978-3-319-99417-8_8.
Rushdi, A.I., and B.R.T. Simoneit. 2002a. Hydrothermal alteration of organic matter in sediments of the northeastern Pacific Ocean: Part 1. Middle Valley, Juan de Fuca Ridge. *Appl Geochem* 17: 1401–1428.
Rushdi, A.I., and B.R.T. Simoneit. 2002b. Hydrothermal alteration of organic matter in sediments of the northeastern Pacific Ocean: Part 2. Gorda Ridge, Escanaba Trough. *Appl Geochem* 17: 1467–1494.
Rushdi, A.I., Simoneit, B.R.T., DouAbul, A.A.Z., Al-Mutlaq, K.F., El-Mubarak, A.H., Qurban, M., and M.A. Goni. 2014b. Occurrence, distribution and sources of polar lipid tracers in sediments from the Shatt al-Arab River, Iraq – Polar compounds. *Sci Total Environ* 470–471: 180–192.
Sadiq, M., and J. McCain (eds.). 1993. *A Marine Wildlife Sanctuary for the Arabian Gulf. Environmental Research Following the 1991 Gulf War Oil Spill*. NCWCD, Riyadh and Senckenberg Research Institute, Frankfurt, Germany.
Seuront, L., and S.C. Leterme. 2007. Increased zooplankton behavioral stress in response to short-term exposure to hydrocarbon contamination. *Open Oceanogr J* 1, no. 1: 1–7.
Sheppard, C.R.C., Price, A.R.G., and C.M. Roberts. 1992. *Marine Ecology of the Arabian Region: Patterns and Processes in Extreme Tropical Environments*. London: Academic Press, 257 p.
Silva, T.R., Lopes, S.R., Spörl, G., Knoppers, B.A., and D.A. Azevedo. 2013. Evaluation of anthropogenic inputs of hydrocarbons in sediment cores from a tropical Brazilian estuarine system. *Microchem J* 109: 178–188.
Simoneit, B.R.T. 1977. Organic matter in eolian dusts over the Atlantic Ocean. *Mar Chem* 5: 443–464.
Simoneit, B.R.T. 1978. The organic chemistry of marine sediments. In: *Chemical Oceanography*, 2nd ed. Eds. J.P. Riley and R. Chester, 233–311. New York: Academic Press.
Simoneit, B.R.T. 1982. Some applications of computerized GC-MS to the determination of biogenic and anthropogenic organic matter in the environment. *Int J Environ Anal Chem* 12, nos. 3–4: 177–193.
Simoneit, B.R.T. 1984. Organic matter of the troposphere – III. Characterization and sources of petroleum and pyrogenic residues in aerosols over the western United States. *Atmos Environ* 18: 51–67.
Simoneit, B.R.T. 1985. Application of molecular marker analysis to vehicular exhaust for source reconciliation. *Int J Environ Anal Chem* 22: 203–233.
Simoneit, B.R.T. 1986. Cyclic terpenoids of the geosphere. In: *Biological Markers in the Sedimentary Record*. Ed. R.B. Johns, 43–99. New York: Elsevier.
Simoneit, B.R.T. 1989. Organic matter of the troposphere-V: Application of molecular marker analysis to biogenic emissions into the troposphere for source reconciliations. *J Atmos Chem* 8, no. 3: 251–275.
Simoneit, B.R.T. 2002. Biomass burning – a review of organic tracers from smoke from incomplete combustion. *Appl Geochem* 17: 129–162.
Simoneit, B.R.T., Cardoso, J.N., and N. Robinson. 1990. An assessment of the origin and composition of higher molecular weight organic matter in aerosols over Amazonia. *Chemosphere* 21: 1285–1301.
Simoneit, B.R.T., Crisp, P.T., Mazurek, M.A., and L.J. Standley. 1991. Composition of extractable organic matter of aerosols. from the Blue Mountains and southeast coast of Australia. *Environ Int* 17: 405–419.
Simoneit, B.R.T., and M.A. Mazurek. 1982. Organic matter of the troposphere – II. Natural background of biogenic lipid matter in aerosols over the rural western United States. *Atmos Environ* 16: 2139–2159.
Stephanou, E.G., and N.E. Stratigakis. 1993. Determination of anthropogenic and biogenic organic compounds on airborne particles: Flash chromatographic fractionation and capillary gas chromatographic analysis. *J Chromatogr* 644: 141–151.
Stiehl, T., Rullkoetter, J., and A. Nissenbaum. 2005. Molecular and isotopic characterization of lipids in cultured halophilic microorganisms from the Dead Sea and comparison with the

sediment record of this hypersaline lake. *Org Geochem* 36: 1242–1251.

Suaria, G., Aliani, S., Merlino, S., and M. Abbate. 2018. The occurrence of paraffin and other petroleum waxes in the marine environment: A review of the current legislative framework and shipping operational practices. *Front Mar Sci* 5: 94. http://doi.org/10.3389/fmars.2018.00094.

Talling, J.F. 1980. Water characteristics. In: *Euphrates and Tigris, the Mesopotamian Ecology and Destiny*. Ed. J. Rzóska, 63–79. The Hague: W. Junk Publishers.

ten Haven, H.L., De Leeuw, J.W., Rullkötter, J., and J.S. Damsté. 1987. Restricted utility of the pristane/phytane ratio as a palaeoenvironmental indicator. *Nature* 330, no. 6149: 641–643.

Tolosa, I., de Mora, S.J., Fowler, S.W., Villeneuve, J.P., Bartocci, J., and C. Cattini, 2005. Aliphatic and aromatic hydrocarbons in marine biota and coastal sediments from the Gulf and the Gulf of Oman. *Mar Pollut Bull* 50, no. 12: 1619–1633.

Tolosa, I., de Mora, S.J., Sheikholeslami, M.R., Villeneuve, J.P., Bartocci, J., and C. Cattini. 2004. Aliphatic and aromatic hydrocarbons in coastal Caspian Sea sediments. *Mar Pollut Bull* 48, nos. 1–2: 44–60.

Treibs, A. 1934. Chlorophyll und Häminderivate in bituminösen, Gesteinen, Erdölen, Erdwachen und Asphalten. *Ann Chem* 510: 42–62.

Treibs, A. 1936. Chlorophyll and hemin derivatives mineral substance. *Angew Chem* 49: 682–686.

Uddin, S., Fowler, S.W., Saeed, T., Jupp, B., and M. Faizuddin. 2021. Petroleum hydrocarbon pollution in sediments from the Gulf and Omani waters: Status and review. *Mar Pollut Bull* 173: 112913.

UN-ESCWA and BGR. 2013. Inventory of Shared Water Resources in Western Asia. *United Nations Economic and Social Commission for Western Asia; Bundesanstaltfür Geowissenschaften und Rohstoffe*. Beirut, Lebanon: Salim Dabbous Printing Co., 626 p.

USEIA (U.S. Energy Information Administration). 2019. *Independent Statistics and Analysis. New Mexico – State Energy Profile Overview*. U.S. Energy Information Administration (EIA).

Vaughan, G.O., Al-Mansoori, N., and J.A. Burt. 2019. *World Seas: An Environmental Evaluation*, 2nd ed. Chapter 1 in vol. II. The Indian Ocean to the Pacific, 1–23. Cambridge: Academic Press.

Volkman, J.K., Holdsworth, D.G., Neill, G.P., and H.J. Bavor. 1992. Identification of natural, anthropogenic and petroleum hydrocarbons in aquatic sediments. *Sci Total Environ* 112, nos. 2–3: 203–219.

Yang, Z., Shah, K., Laforest, S., Courtemanche, C., Durand, W., Lambert, P., Hollebone, B.P., Brown, C.E., Goldthorp, M., Watson, K., Yang, C., Dey, D., and V. Beaulac. 2021. Occurrence, characterization, source, and risk assessment of petroleum-related hydrocarbons in sediments along St. Clair River, Ontario, Canada. *Mar Pollut Bull* 173: 113037.

Ye, B., Zhang, Z., and T. Mao. 2007. Petroleum hydrocarbon in surficial sediment from rivers and canals in Tianjin, China. *Chemosphere* 68, no. 1: 140–149.

Zhang, L., Shi, K., and L. Yue. 2007. Chemical characteristics and pollution sources of petroleum hydrocarbons and PAHs in sediments from the Beiluohe River, Northern China. *Environ Geol* 53: 307–315.

Zhao, J., Yin, P., Zhao, L., Yu, Q., and G. Lu. 2015. Spatial and seasonal variations of total petroleum hydrocarbon in surface water and sediment in Pearl River Delta. *Bull Environ Contam Toxicol* 95: 298–303.

Four-dimensional radar hologram interferometry for monitoring oil and gas exploration near Kharg Island

16

Maged Marghany

16.1 INTRODUCTION

Oil and gas exploration might well boost the national economy, but it also increases the risk of land subsidence. Despite the fact that oil and gas exploration mechanisms are widely understood, many nations do not believe that disasters adhere to exploration techniques. For instance, millimetre-scale surface displacements are brought on by the subsurface injection and extraction of oil and gas. A thorough grasp of the oil and gas formation beneath the subsurface layers of the ground is necessary for forecasting and mapping. Forecasting the subsurface compaction that would occur in hydrocarbon reservoirs is a crucial topic in the framework of general reservoir management. Casing deformation and wellbore failures, however, are significant risks that should be avoided to ensure a high level of safety in drilling operations and reduce significant financial risks (Sayers and Schutjens 2007; Zoback and Kohli 2019; Miah 2020).

A primary concern is the forecasting of land subsidence brought on by the compaction of rock formations; anthropogenic settlements might contribute to platforms sinking. Additionally, extreme deformations of pipelines can have significant environmental impact, particularly in coastal locations, and can be a major obstruction to oil and gas exploration. This is because pipeline bends, valves and other obstructions create turbulence in the fluid being transported, which can lead to increased stress on the infrastructure and damage marine life if the pipelines are located near aquatic environments. In other words, pipeline deformation can be caused by several factors, including ground movement, corrosion and external pressure. If left unchecked, these deformations can cause severe structural damage, making it difficult to accurately transport oil and gas products without risking potential contamination. The deep-sea blowout and subsequent explosion that resulted in the Deepwater Horizon oil spill in 2010 killed 11 workers and released nearly five million barrels of crude oil into the Gulf of Mexico. This devastating event highlighted the need for more stringent regulations surrounding pipeline deformation.

Utilizing subterranean resources improperly can cause changes in the geological structure, which shows different rates of surface deformation in the oil production zone (Wenrui et al. 2013; Zoback and Kohli 2019). In this sense, oil exploitation will reduce reserves, resulting in a corresponding drop in reservoir pressure. A decrease in pore pressure causes an increase in operational stress, which leads to a subsequent decrease in reservoir size and ground subsidence (Xuejun et al. 2015; Miah 2020; Xunyu et al. 2020).

Following CO_2 or liquid injection, stress changes propagate from the deep portion of the reservoir to the surface. The frequency of surface movement or deformation is also influenced by geological conditions, mining equipment and other factors. The protection of equipment used in oil and gas production and the sustainability of oil and gas infrastructure are severely impacted by asymmetrical ground deformation brought on by human actions (Chiaramonte et al. 2008; Verdon et al. 2013). Hence, land subsidence due to oil and gas exploration requires a standard and accurate technology to be applied.

Microwave remote-sensing technology has demonstrated promising uses in the field of oil and gas exploration for many years. However, the potential synthetic aperture radar (SAR) application of the surface information has not been fully utilized throughout the previous 40 years of oil and gas operations and exploration (Wu et al. 2009; Zhang et al. 2020; Sabzevari et al. 2020). In contrast, subsurface analyses are more commonplace now that we fully comprehend what occurs below the earth's surface. Consistent with this perspective, the ability to accurately simulate oil or gas flows in reservoirs leads to more informed decisions and the capacity to capture extra energy more economically. Gravimetric data-gathering technologies hold great promise in helping us comprehend the dynamic fluctuations of fluid flow in oil and gas reservoirs (Sayers and Schutjens 2007; Zhang et al. 2020; Cheng et al.

DOI: 10.1201/9781003321408-21

2021). Conventional approaches are based on magnetic principles. These techniques are, nevertheless, time and money consuming. Furthermore, due to challenging weather and topographical conditions, many parts of the earth are difficult to reach. In other words, it has taken several decades for the development of inquiry roadmaps and modeling tools for the oil and gas industry to be standardized and appealing to the oil and gas industry sectors.

What the most effective remote-sensing technology that provides correct information for oil and gas exploration is is now a crucial question. The technology known as interferometric synthetic aperture radar (InSAR) has several applications in examining surface distortion and is capable of producing distinct precision, large-scale and surface-based distortion mapping. Utilizing InSAR technology to monitor deformation during oil and gas exploration would significantly aid in the understanding of oil and gas field performance. In other words, what significance does microwave remote-sensing technology play in resolving the aforementioned problems? Surface and above-surface remote-sensing results have had flaws in practice. This chapter is devoted to presenting a brand-new method for oil and gas exploration based on quantum interferometric synthetic aperture radar. At present, high revisit rates of satellite constellations and new SAR sensors like Sentinel-1 provide data free or at a very low cost and could be the cornerstone of oil and gas operations.

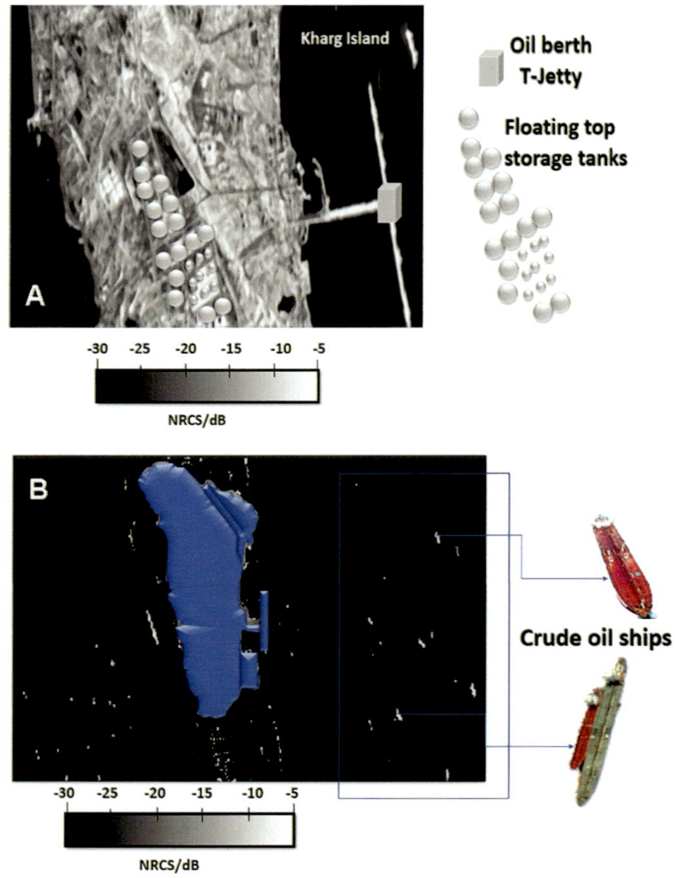

FIGURE 16.1 TerraSAR-X stripmap for oil and gas exploration: A storage tanks and B crude oil ships.

16.2 WHAT ROLE DO SATELLITES PLAY IN OIL EXPLORATION?

The potential of satellites enables oil companies to control their operations remotely and securely, tracking everything from the direction of the drill bit piercing the ocean floor to the volume of oil flowing from offshore wells through a large network of pipelines. The satellite also uses gravitational forces to create a map that can be used to pinpoint the location of oil deposits (Velotto et al. 2011; Bryksin et al. 2012; Shi et al. 2019; Marghany 2019a). Because oil is frequently surrounded by lithologies such as limestone and clay, which are light and have less gravitational force than surrounding materials, these forces can be used effectively (Zoback and Kohli 2019). In this approach, the key question is: Are satellites emerging further along the stage in the oil and gas industry? In certain circumstances, this appears to be the case: for example, when investors and purchasers gain a more immediate understanding of how inventories and capacities are acting to propel the market. With this understanding, the oil sector uses SAR data to monitor the oil storage on Kharg Island, Iran, which has reached 81% capacity (Firouzbakht et al. 2017).

In this view, Figure 16.1a indicates the TerraSAR-X stripmap mode with HH-polarization and a maximum backscatter of 5dB across floating top storage tanks and several crude oil ships (Figure 16.1b). Accordingly, the usefulness of SAR satellite-based services is being expanded to provide security for the oil and gas industries. Therefore, the stripmap data has a resolution of up to 3 m and a scene size of 30 km (width) by 50 km (length). This means that each image contains 30,000 by 50,000 pixels of resolution. As a result, each pixel of the stripmap data has a resolution of 1 cm. Stripmap data has an extremely high resolution, allowing small buildings and cars to be distinguished in the imagery.

16.3 HOW DOES INSAR WORK?

The essential principle of InSAR radar technology is the interference of two or more coherence pulses. The backscattering of the phase difference of these pulses as a consequence is applied to quantify millimetre-scale deformation variations across time scales of days to years. Let's assume that, given a real variable such as a time t, the phase of the periodic function F is an angle-like quantity corresponding to the number of periods spanned by that variable, which is indicated by $\varphi(t)$, to explain the phase mechanism:

$$\varphi(t) = 2\pi \left[\frac{t - t_0}{f^{-1}} \right]. \tag{16.1}$$

The phase difference, according to equation 1, is a function of time changes per inverse of pulse frequency f. The phase deviation (t) grows linearly with the argument t when the frequencies are different. According to this perspective, the irregularities in confrontation and reinforcement lead to the phenomenon known as beating (Ferretti et al. 2007). A beat is an interference pattern between two signals that are slightly different in frequency, and it is perceived as a periodic change in volume whose rate is determined by the difference between the two frequencies. Radar interferometry is just a simple extension of Young's interferometry theory. In this view, the two radar (SAR) antennas function as coherent point sources. Given the length of the radar wavelengths, it is simple to digitize and process the signal coherently, immediately measuring the phase information (Ferretti et al. 2007; Marghany 2013; Yu et al. 2019). When imaging a surface, the phase fronts from the two sources interact, and the recorded phase differences yield topographic data because the interference pattern is sliced by the surface topography.

Based on the baseline vector's geometric configuration, the two fundamental categories of interferometric radar can be distinguished from one another. In other words, the baseline is the vector that separates the interferometric pair's antenna positions. In this context, there are two types of interferometers for radar: cross-track interferometers and along-track interferometers (Ferretti et al. 2007).

Antennas are nominally spaced apart in the cross-track direction when using cross-track interferometers with a single antenna and repeat passes to monitor topography and surface deformation. On the other hand, along-track interferometers, in which the antennas are separated in the along-track direction, are used to detect radial velocity. Dual antenna single-pass interferometers are used in both sorts of radar interferometry (Ferretti et al. 2007; Marghany 2014a; Woodhouse 2017).

Phase information from a single image cannot be used directly, as was already shown. However, phase changes between two (or more) disparate acquisitions matter; this is the main problem with InSAR. It is possible to create a DEM by computing the phase differences between two collections from two different sites. It is likely to detect the ground deformation while calculating phase fluctuations between two acquisitions at two changing times (Marghany 2014b; Yu et al. 2019; Yu and Hu 2021).

According to this theory, understanding phase differences between two acquisitions made at two different periods can reveal millimeter-scale ground distortion. The technique's level of proficiency over large, difficult-to-access zones, day or night, and in every kind of weather, is its primary reward, in addition to its superb precision (Pinel et al. 2011; Marghany 2014b; Yu and Hu 2021). However, the limitations that must always be kept in mind when examining InSAR results. The first difficulty is linked to shifts that affect the line-of-sight (LOS) direction, which depends on the satellite's flight position and the radar beam's incidence angle. In this regard, the accuracy of the sensitivity to vertical dislocations increases with increasing incidence angle steepness (Yu et al. 2019). As a result, when gazing westward, for example, a descending orbit largely gives non-distorted coverage in the west-facing slope, whereas an ascending orbit primarily covers the east-facing slopes (Ferretti et al. 2007; Marghany 2014b; Woodhouse 2017). When the directly exposed dislocation vector is almost vertical to the LOS, the sensitivity in this perspective is also weak. Because of the direction of the orbit, which is north-south, the sensitivity to surface movement in this plane is almost negligible (Ferretti et al. 2007; Pinel et al. 2011; Woodhouse 2017).

Due to the wave's cyclic pattern, the second challenge involves phase ambiguity. There is a phase difference of more than half an effective wavelength between the two acquisitions. According to this understanding, a moving area with a velocity higher than 2.8 cm in 24 days will cause a phase ambiguity, and InSAR cannot deliver useful information about ground or object distortion. A sensor with a wavelength of 5.55 cm and a revisit time of 24 days, such as RADARSAT-2, would then experience these effects. Finally, due to the influence of the atmosphere, a signal delay is the third difficulty in the InSAR process (Pinel et al. 2011).

Advanced algorithms and techniques are essential for improving the accuracy of InSAR technology. Marghany pioneered the application of holographic interferometry theory (HIT) to enhance InSAR techniques. HIT offers several advantages over traditional InSAR algorithms. It enables faster conveyance of 4-D information and provides significantly higher resolution and improved temporal accuracy when quantum computing techniques are employed. HIT excels in resolving sub-pixel features, making it highly effective for monitoring land changes with precision. Compared to InSAR, HIT delivers more precise measurements, offers better resolution, covers a larger area and captures a wider range of surface deformations (Marghany, 2019a, 2019b, 2021a, 2021b, 2022).

16.4 HYPOTHESES AND OBJECTIVE

Prior to their applications in 4-D oil and gas storage reconstruction from sequences of the same frequency bands of SAR data, speculation is necessary for a perfect understanding of four-dimensional effects. A hologram is a promising method for converting two-dimensional objects into three dimensions. In actuality, a two-dimensional item has three dimensions embedded in it. The main query that emerges in this context is how 4-D is encoded in 3-D. In addition, how may holographic interferometry be generated from oil and gas storage coherence pixels in SAR data? In reality, interferometric SAR measurements show high coherence near the surface that covers oil and gas storage. However, because phase information is missing, it is impossible to recover the hologram fringes from the low coherence zone. Thus, by taking into account the techniques of coherence holograms, it may be possible to apply holographic interferometry to oil and gas storage covers.

Delivering a novel method for automatic oil and gas storage identification in an SAR image based on 4-D holographic interferometry is the focus of this chapter. In this understanding, how 4-D can be reconstructed from 3-D is the main question. Inferring 4-D from 3-D phase unwrapping of SAR holographic interferometry is the hypothesis described in this chapter. In reality, SAR interferometry is a potent method for detecting changes in the stability of electromagnetic wavelength spectra. Common interferometric techniques have a significant flaw in that they need specular reflectors. With the aid of holography, this restriction can be overcome by being able to detect even the smallest movements of randomly reflecting objects.

Deriving a new formula for 4-D holographic interferometry phase unwrapping using a Marghany 4-D quantized hologram interferometry algorithm is the main innovation of this chapter. The major objective is to optimize 4-D holographic interferometry to reconstruct the 4-D of oil and gas exploration using SAR satellite data.

16.5 SPECULATION FOR QUANTIZATION OF HOLOGRAM INTERFEROMETRY

Professor Susskind is one of the most prominent figures in the world of holograms, having pioneered this field with his research and theories. Professor Susskind's influence in the field of holograms cannot be overstated; he has played an integral role in the advancement of holograms, pioneering research on the subject and producing groundbreaking results.

The quantum algorithm is a complex and interesting system, allowing computers to solve problems exponentially faster than traditional algorithms. The potential implications of this are immense, allowing for vast amounts of data to be analyzed quickly and efficiently. The quantum algorithm has opened up a world of possibilities in fields ranging from physics and chemistry to artificial intelligence and cryptography. By using the principles of quantum mechanics, the quantum algorithm takes advantage of the behaviour of particles on a microscopic level to create solutions that are exponentially faster than traditional algorithms, especially in solving the uncertainty of InSAR procedures. With the quantum algorithm, complex problems can now be solved much more quickly, opening up new possibilities for research and development in a variety of fields.

Marghany (2021a) has innovated speculation quantization of hologram interferometry. In this scenario, Young's interferometry theory is a straightforward expansion of the classical version of wave interference known as InSAR (Marghany 2019a). The dual SAR antenna wave function is obtained as a specific solution of the Schrödinger wave equation as the superposition of waves produced from dual SAR antennas from the point of view of quantum mechanics. The NOON state, which is widely acknowledged as the state of a radar interferometer, is represented by the following wave function in this understanding (Marghany 2021a):

$$\left|\psi^{\phi}_{NOON}\right\rangle = \frac{1}{\sqrt{2N!}}\left(\left|N\right\rangle_1\left|0\right\rangle_2 + e^{iN\phi}\left|0\right\rangle_1\left|N\right\rangle_2\right). \tag{16.2}$$

The superposition of N photon particles in mode 1 $\left|0\right\rangle_2$ with zero particles in mode 2 $\left|0\right\rangle_2$ is represented by equation 2. Therefore, the phase measurements ϕ changes from 0 to $\frac{\pi}{N}$. In this view, NOON states are essential in quantum sensing since they enable precise phase measurements ϕ when applied to interferometry techniques.

On the other hand, the NOON state without attenuation can be obtained by:

$$\left|\psi_a\right\rangle = \frac{1}{\sqrt{2^N}}\left(\left|10\right\rangle + \left|01\right\rangle\right)^{\otimes N}. \tag{16.3}$$

Even in the absence of decoherence caused by oil-covered water, equation 3 demonstrates that separable states cannot be used to implement the conventional quantum constraint. In terms of phase change, equation 3 can be written as:

$$\lim_{\alpha_1\to 1}\lim_{\alpha 2\to 1} \partial\phi = \frac{1}{\sqrt{N}}. \tag{16.4}$$

The Heisenberg limit or beating of the shot-noise limit is now referred to as "super-sensitivity" and is based on equations 3 and 4. The best performance is anticipated for the lowest uncertainty condition when the photon number uncertainty is the greatest. However, in this case, the quantitative uncertainty is spread out over the dual-photon paths A and B (Ou 2007; Marghany 2019b; Marghany 2021b). If all N photons are in mode A (and none in B) or if all N photons are in mode B (and none in A), where the periods in both A and B would be terminated, then the situation in NOON states is fully unclear. In this view, the NOON state does not become much more confusing than that!

In this understanding, the mathematical description of the two complex SAR hologram interferometry image $H(x,y,z,t)$ is formulated as:

$$\begin{aligned} H(x,y,z,t) &= \varepsilon_0 c\left|S^2\right| = \varepsilon_0 c\left|\left(S_1 + \boldsymbol{S}_2\right)^2\right| \\ &= \varepsilon_0 c\left|S_1^{\,2} + \boldsymbol{S}_2^{\,2} + \boldsymbol{2S}_1 \bullet \boldsymbol{S}_2\right| \\ &= \varepsilon_0 c\left(S_1^{\,2} + S_2^{\,2} + 2S_1S_2 cos\vartheta_{12}\right) \\ &= \varepsilon_0 c\begin{bmatrix} S_{01}^{\,2} cos^2(kz - \omega_1 t) + \dfrac{S_{02}^{\,2}}{r^2} cos^2(kr - \omega_2 t) + \\ \dfrac{2S_{01}S_{02}}{r} cos(kz - \omega_1 t)cos(kr - \omega_2 t)cos(\vartheta_{12}) \end{bmatrix} \end{aligned} \tag{16.5}$$

Equation 5 says that the two complex radar images S_1 and S_2 are implemented to construct a 3-D hologram radar image. This hologram radar image involves the two signal wavenumber k, the baseline between both acquired SAR images r over the time of acquisition t and the radial frequency ω. Additionally, equation 5 has a portion that only uses the radar complex signals and a term for interferometry that uses the cosine of the angle of 12 between the two vector amplitudes (which may be a function of position x,y,z). Then wrapped phase $\phi_i(x,y,z)$ is sought, and it is provided by:

$$\phi_i(x,y,z) = \arctan\left(\frac{H_0(x,y,z) - H_2(x,y,z)}{H_1(x,y,z) - H_3(x,y,z)}\right), \quad i = 1,2,3. \tag{16.6}$$

The estimation of phase maps of various holographic radar images can therefore be accomplished as follows:

$$\phi_{12}(x,y,z) = \arctan\left(\frac{\sin(\phi_2(x,y,z))\times\cos(\phi_1(x,y,z))}{\cos(\phi_2(x,y,z))\times\cos(\phi_1(x,y,z))}\;\frac{-\cos(\phi_2(x,y,z))\times\sin(\phi_1(x,y,z))}{+\sin(\phi_2(x,y,z))\times\sin(\phi_1(x,y,z))}\right). \tag{16.7}$$

$$\phi_{23}(x,y,z) = \arctan\left(\frac{\sin(\phi_3(x,y,z))\times\cos(\phi_2(x,y,z))}{\cos(\phi_3(x,y,z))\times\cos(\phi_2(x,y,z))}\;\frac{-\cos(\phi_3(x,y,z))\times\sin(\phi_2(x,y,z))}{+\sin(\phi_3(x,y,z))\times\sin(\phi_2(x,y,z))}\right). \tag{16.8}$$

$$\phi_{123}(x,y,z) = \arctan\left(\frac{\sin(\phi_{12}(x,y,z))\times\cos(\phi_{23}(x,y,z))}{\cos(\phi_{12}(x,y,z))\times\cos(\phi_{23}(x,y,z))} \frac{-\cos(\phi_{12}(x,y,z))\times\sin(\phi_{23}(x,y,z))}{+\sin(\phi_{12}(x,y,z))\times\sin(\phi_{23}(x,y,z))}\right). \quad (16.9)$$

In this view, formulas 7 to 9 say that $\phi_{123}(x,y,z)$ varies from 0 to 2π without 2π cutoffs, and then the image of the wrapped phase can be created. Consequently, reconstructing the 3-D surface precisely using $\phi_{123}(x,y)$ may not yield significant results because $\phi_{123}(x,y)$ does not exhibit 2π discontinuities. This is due to SAR long-wavelength often experiencing low precision in its coherence signal. In other words, using a specific method $\phi_{123}(x,y)$ for reconstructing the 3-D surface may not produce accurate results because this method does not handle discontinuities well, especially in SAR long-wavelength signals, which tend to have lower precision.

In other words, the phase $\phi_1(x,y)$ has the wavelength ratio of three complex SAR images; there are $\frac{S_{12}(\lambda)}{S_1(\lambda)}$ discontinuities with an identical layout $S_{12}(\lambda)$. In this circumstance, inside every $S_{12}(\lambda)$, there are discontinuities of $\frac{S_{123}(\lambda)}{S_{12}(\lambda)}$ on $\phi_{12}(x,y,z)$. In other words, the conjugate of three complex SAR images $\frac{S_{123}(\lambda)}{S_{12}(\lambda)}\otimes\frac{S_{12}(\lambda)}{S_1(\lambda)}$ forms discontinuities $\phi_1(x,y,z)$. Overall, the hologram interferometry image can be mathematically expressed as:

$$H_k(x,y,z) = \alpha(x,y,z) + \beta(x,y,z)\sin(\phi(x,y,z) + 0.5(\pi k)) + G(x,y,z). \quad (16.10)$$

Equation 10 demonstrates the occurrence of white Gaussian noise $G(x,y,z)$ since equations 3 to 5 contain degradation in the computing of the absolute phases (Ou 2007; Marghany 2021a).

16.6 MARGHANY 4-D QUANTIZED HOLOGRAM INTERFEROMETRY ALGORITHM

Marghany (2021b) has delivered a novel algorithm for 4-D phase unwrapping for 4-D object reconstruction. Following Marghany (2021b), the phase unwrapping technique is needed to estimate the 3-D object reconstruction in the complex SAR data, and the absolute phase cannot be used for this. For this purpose, the 4-D phase unwrapping can be stated as:

$$\Psi_3(x,y,z,t) = \phi(x,y,z,t) + 2\pi\left(\begin{matrix}\text{int}\left(\frac{\phi_{12}(x,y,z,t)}{2\pi}\otimes\left(S_1(\lambda)\times S_2(\lambda)^{-1}\right) \otimes\left(S_2(\lambda)\times S_3(\lambda)^{-1}\right)\right) + \\ \text{int}\left(\frac{\phi_{23}(x,y,z,t)}{2\pi}\right)\otimes\left(S_2(\lambda)\times S_3(\lambda)^{-1}\right) + G(\sigma)\end{matrix}\right). \quad (16.11)$$

The innovation in equation 11 is based on the exchanging of the absolute phase estimates from the three complex SAR images, each with a unique acquisition time. In other words, the white Gaussian noise as a function of standard deviation σ is also taken into account in the innovative equation of the 4-D phase unwrapping (Marghany 2021b). In addition to reconstructing phase unwrapping in 4-D, equation 11 also seeks to eliminate a discontinuity. Let the wrapped phase ϕ_w be expressed in this case as the Laplacian of the real phase, which is derived by:

$$\nabla^2\phi = \cos\Psi_w\nabla^2(\sin\Psi_w) - \sin\Psi_w\nabla^2(\cos\Psi_w). \quad (16.12)$$

where ∇^2 is the Laplace operator, and by using an inverse of ∇^{-2}, the real phase ϕ'can be approximated in 4-D by assembling quantum Fourier-domain forward and ∇^{-2} as follows (Marghany 2022):

$$\phi'(i,j,k,t)_Q = QFT^{-1}\left[\frac{QFT[\cos\phi_w(QFT^{-1}[\left|p^2+q^2+r^2+s^2\right\rangle QFT(\sin\phi_w)])}{p^2+q^2+r^2+s^2}\right] - QFT^{-1}\left[\frac{QFT[\sin\phi_w(QFT^{-1}[(p^2+q^2+r^2+s^2)_Q QFT(\cos\phi_w)])}{\left|p^2+q^2+r^2+s^2\right\rangle}\right] \quad (16.13)$$

Equation 13 reveals that QFT^{-1} and QFT are quantum Fourier inverse and forward cosine or sine transforms, respectively, in 4-D of time-domain coordinates(i, j, k, t) and the quantum Fourier-domain coordinates $\left|p^2+q^2+r^2+s^2\right\rangle$. Equation 13 comprises three sorts of operations: quantum forward and inverse cosine transforms, trigonometric operations and the masking expression $\left|p^2+q^2+r^2+s^2\right\rangle$. Contrary to the conventional discrete Fourier transform, which requires $O(n^2)$ gates, the discrete Fourier transform on 2^n amplitudes can be implemented as a quantum circuit using only $O(n^2)$ Hadamard gates and controlled phase shift gates, where n is the number of qubits. Furthermore, the exact QFT algorithms only need O *(nlogn)* gates to perform a calculation (Marghany 2021b, 2022).

The simulation of 4-D oil spill decoherence is computed based on the standard deviation of its geometry shape displacement over the three SAR image acquisitions enclosed in the 4-D phase unwrapping Ψ as $g_{123}(\sigma)\subset\Psi(x,y,z,t)$. This insight allows us to represent the 4-D variations into holographic interferometry as:

$$\begin{pmatrix}\Delta\Phi_1\\ \Delta\Phi_2\\ \Delta\Phi_3\\ \Delta\Phi_4\end{pmatrix} = \begin{pmatrix}\vec{g}_{1i} & \vec{g}_{1j} & \vec{g}_{1_k} & \vec{g}_{1t}\\ \vec{g}_{2i} & \vec{g}_{2j} & \vec{g}_{2_k} & \vec{g}_{2t}\\ \vec{g}_{3i} & \vec{g}_{3j} & \vec{g}_{3_k} & \vec{g}_{3t}\\ \vec{g}_{4i} & \vec{g}_{4j} & \vec{g}_{4_k} & \vec{g}_{4t}\end{pmatrix}\otimes\frac{2\pi}{\lambda} \begin{pmatrix}\vec{\Psi}_{1i} & \vec{\Psi}_{1j} & \vec{\Psi}_{1k} & \vec{\Psi}_{1t}\\ \vec{\Psi}_{2i} & \vec{\Psi}_{2j} & \vec{\Psi}_{2k} & \vec{\Psi}_{2t}\\ \vec{\Psi}_{3i} & \vec{\Psi}_{3j} & \vec{\Psi}_{3k} & \vec{\Psi}_{3t}\\ \vec{\Psi}_{4i} & \vec{\Psi}_{4j} & \vec{\Psi}_{4k} & \vec{\Psi}_{4t}\end{pmatrix}\otimes\begin{pmatrix}W^x_{i,j,k,t}\\ W^y_{i,j,k,t}\\ W^z_{i,j,k,t}\\ W^t_{i,j,k,t}\end{pmatrix}. \quad (16.14)$$

In this context, W stands for the user-defined weights and Cartesian coordinates *I, j, k*, and *t* for the 4-D vector representation in the hyperspace holographic interferometry. Consistent with Wnuk et al. (2019), the spatial variance of surface topography deformation fluctuates significantly over time in this example of a 3-D volume of a hologram image that is to be stored in 4-D. Equation 14 shows how the entanglement of the wave function can be used to determine the areas with the highest and lowest entanglement levels based on the wave function of non-unwrapped areas. In other words, the wave function exhibits two unwrapping phase states, lowest-entanglement |0> and highest-entanglement |1>, and is a superb estimator of the quality of phase unwrapping images generated from InSAR data (Marghany 2019b).

This novel formula, called Marghany 4-D quantized hologram interferometry, was developed by Marghany and exploited in recent publications (Marghany 2019a, 2021a, 2022).

16.7 4-D QUANTIZED HOLOGRAM INTERFEROMETRY FOR OIL AND GAS STORAGE MONITORING: CASE STUDY KHARG ISLAND

Kharg Island is situated in the northern Persian Gulf, some 35 miles off the Iranian coast, and has emerged from the water as a coral island with the shape of a turtle shell. It is 57 kilometres northwest of Bushehr and is regarded as one of the province's coastal suburbs. The location is about 29°15′N latitude and 50°20′E longitude (Figure 16.2). The island has a humid climate with an average elevation of three metres above sea level. Oil and gas are produced there (Sani et al. 2016), and the largest export terminal in Iran is this 12.3 square mile island from which almost 90% of Iranian crude shipments leave. Crude inventory levels should change on Kharg Island partly based on export flows. The Kharg Petrochemical Complex on Kharg Island can be seen in Figure 16.1. Additionally, oil is transported to Kharg Island by pipelines from fields used for on- and offshore production and stored there. A crude tanker arrives at one of two jetties to be loaded. On the east side of the island, the larger of the two, called T-jetty, is situated.

The topographic elevation and deformation of the earth's surface are often determined through interferograms, which are produced by combining at least two separate images captured at various times (Marghany 2013). A mixed image pair with a four-day temporal separation, obtained by TerraSARX stripmap scenes on 15 August, 28 September and 5 November 2016, respectively, were used to construct this flattened cross-sensor interferogram, which includes the area of Kharg Island. With HH polarization, the first TerraSAR image is ascending while the other two images are descending (Figure 16.3).

The stripmap mode allows for images with a resolution of 2.5–2.6 m in a 30 km swath. As a consequence, there is no azimuth steering in the stripmap modes because the SAR antenna beam is always orthogonal to the direction of flight. When compared to ERS and Envisat, TerraSAR-X offers various novel views of SAR interferometry (Bhattacharya and Mukherjee 2017). The stripmap has a high resolution of 3 m or better with azimuth and range coverages of 50 x 30 km^2. A 15-km wide swath has an incidence range angle that ranges from 20° to 45° (Hennings and Herbers 2007). The electronically steerable antenna and the relatively brief orbit repeat cycle of 11 days enable quick and frequent imaging of a particular site. An interferometric pair can be acquired in just 12 days at normal latitudes. Furthermore, it is theoretically possible to obtain 12 interferometric pairs with various observation geometries in 22 days (Werninghaus 2004). Note that the direction of the nominal antenna view is to the right. Although the left-looking mode is feasible, it has several operational flaws that limit its application to high-priority circumstances (Marghany 2019a).

The Kharg Petrochemical Complex and other infrastructure, including a T-jetty, have the greatest coherence rating of one, according to the coherence map (Figure 16.4). The direct DInSAR phase unwrapping pattern consequently reveals minor deformation around the Kharg petrochemical complex and monitors significant variations along the T-jetty, possibly as a result of ship activity and water dynamic oscillations. Marghany's 4-D quantized hologram interferometry, on the other hand, recognizes completed phase unwrapping cycles (Figure 16.5). In contrast to the 2-D phase unwrapping technique, Marghany's 4-D quantum phase unwrapping algorithm can precisely determine the continuous path of every phase unwrapping cycle, allowing it to establish a clear boundary of fringes (Marghany 2021a).

The full cycle of InSAR fringe patterns is particularly well identified by 4-D quantum phase unwrapping. Additionally,

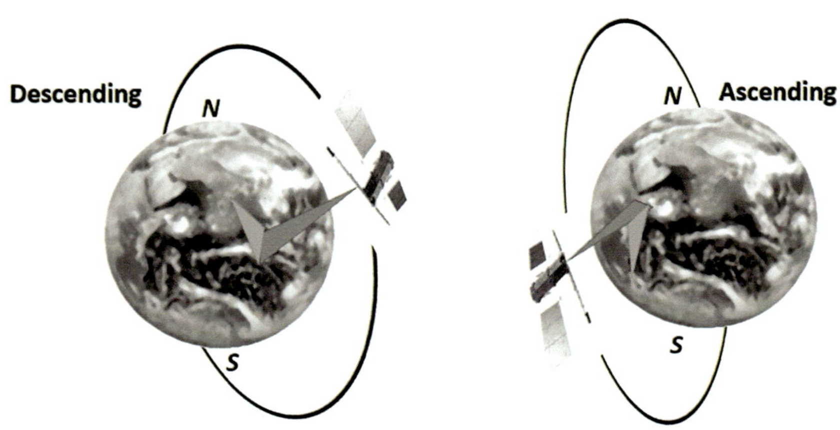

FIGURE 16.2 Descending and ascending orbits.

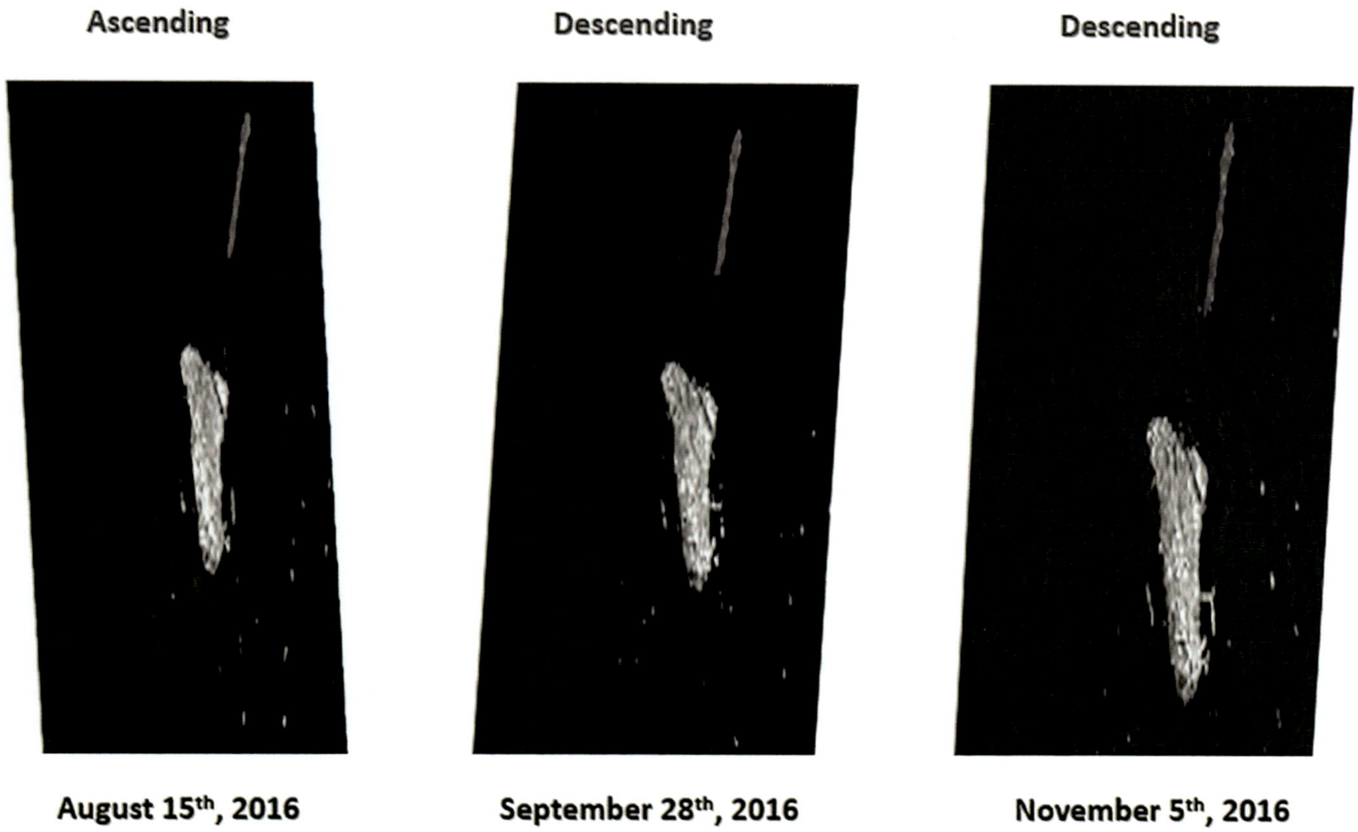

FIGURE 16.3 TerraSAR-X stripmap scenes of different acquisition dates.

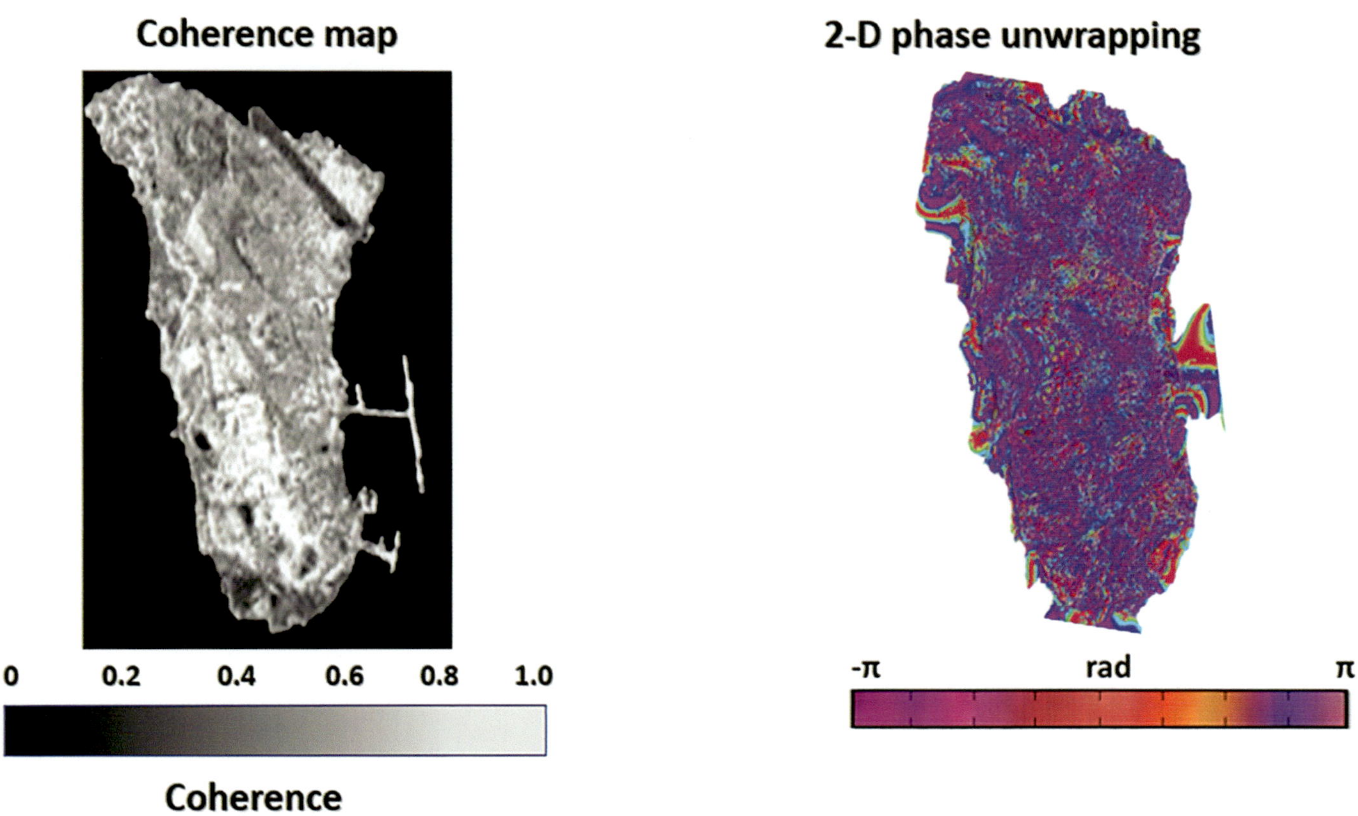

FIGURE 16.4 2-D phase unwrapping compared to coherence map.

the suggested technique has reduced the error in the interferogram cycle because it uses fringe patterns as an optimization toolkit for noise reduction (Marghany 2019a).

Marghany's 4-D quantum image distortion is shown in Figure 16.4. This view utilizes 4-D to perfectly detect morphological features from a variety of viewpoints. This includes a 4 mm/year deformation of the land's geomorphology along the Kharg Petrochemical Complex. Furthermore, the superiority of the substructures as depicted by land deformation is seen in the 4-D image. With a rotation angle of 180°

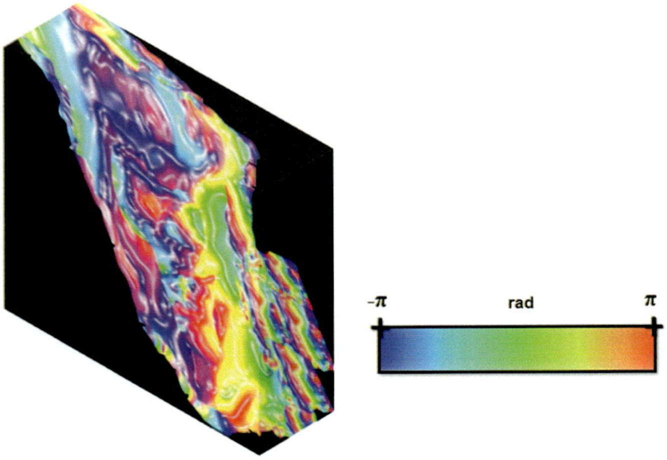

FIGURE 16.5 4 D quantum phase unwrapping.

to 360°, the geomorphology of land displacement around the Kharg Petrochemical Complex will be more visible. As a result, with different view angles ranging from 0° to 360° (Figure 16.6), the geomorphology of land subsidence in the direction of LOS can be easily identified. Therefore, the fourth dimension, which stands for the time fluctuation of data acquisitions, is indicated by the bright colour around the object's edges.

The geometrical structure of the Kharg Petrochemical Complex can be easily discerned from the perspective of the hypercube (Hartnoll et al. 2018; Marghany 2022), as is widely acknowledged (Figure 16.7). This is supported by several views in Figure 16.6, as each side of the hypercube reveals various modifications along the Kharg Petrochemical Complex (Figure 16.7). More specifically, the architecture of oil storage tanks is stretched in the fourth dimension, which is represented by tesseracts up and down, right and left. Finally, they create a cubic mass of identical trivial tesseracts as a result of which they can execute on the surfaces enclosing the right and left dimensions as well as the upward dimension when the tesseract is prearranged in the space world (Zucchini 2021).

Therefore, Figure 16.8 investigates the levels of oil and gas inside the circular tanks. This suggests that both oil and gas are filling the oil storage tanks because they have the same level. It is worth mentioning that the Marghany 4-D quantized holographic interferometry can assess the level of oil and gas and determine whether a tank is full or empty. This is demonstrated by the silver running colour, which forms equal-level alliances with all tanks. In this understanding, the rate of filled volume ranged from 11,870,370.01 m^3 to 14,253.5517 m^3. This means the capacity of filled tanks ranged from 80.1% to 96.2%.

In other words, Kharg Island's crude storage had filled to 96.2% of its capacity, leading to the conclusion that Iranian petroleum exports had been declining more quickly than the rate of production. In this view, the ROC curves of Marghany 4-D quantized hologram interferometry are presented with a higher true positive rate (TPR). In terms of the ROC area, oil and gas levels can be clustered by 99% among the 2-D phase unwrapping and classical 3D hologram interferometry.

Oil storage facilities and the local geological environment pattern appear to extend on one side of the hypercube and topological objects on the other from the perspective of duality. According to this perspective, the architecture of oil storage units is constructed as the unlinking of two connections (Moghaddam et al. 2012). Additionally, the propagator across the n-dimensional vertices of the hypercube provides the basis for how well a quantum random walk performs. A quantum random walk search self-loop, which establishes the edges of hypercube vertices as shown in Figures 16.7 and 16.8, serves as the foundation for the 4-D construction of land deformation caused by the impact of oil storage units. In this case, the 2-D Terra-SAR-X data is converted into 4-D images using Marghany 4-D quantized hologram interferometry.

In essence, there are technical inaccuracies in conventional InSAR technology. The phase unwrapping process, which exposes ambiguities in the measured surface deformation, is one of the predominant errors. Furthermore, using the 4-D images produced by Marghany 4-D quantized hologram

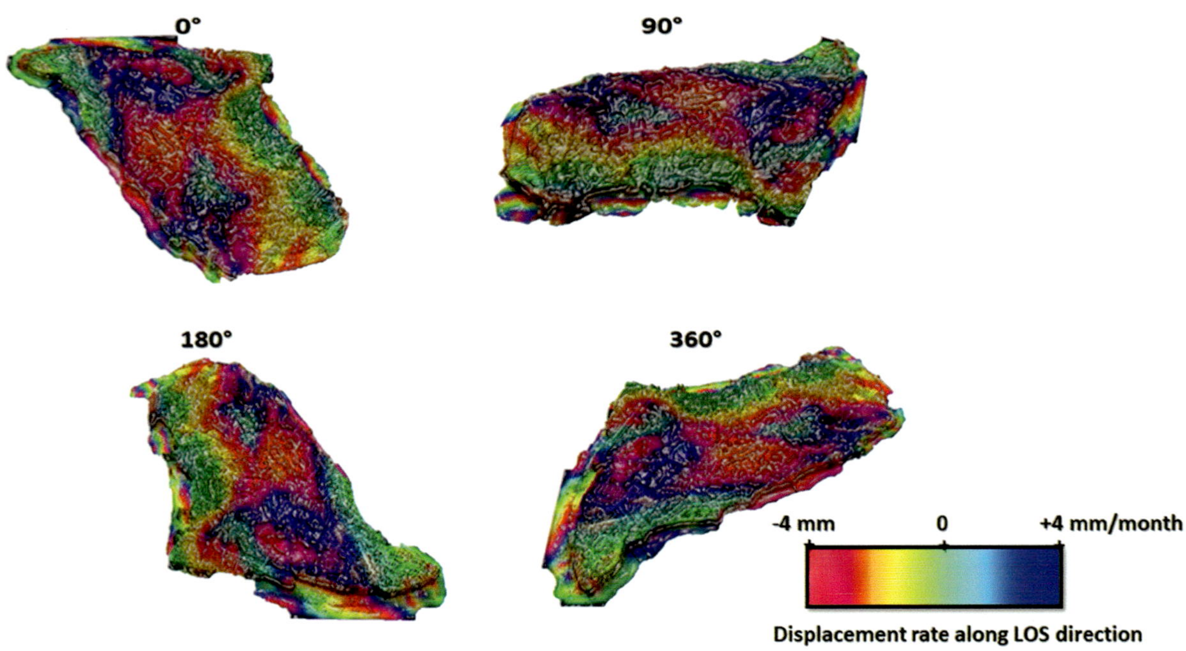

FIGURE 16.6 Different view angles of land deformation along the Kharg Petrochemical Complex.

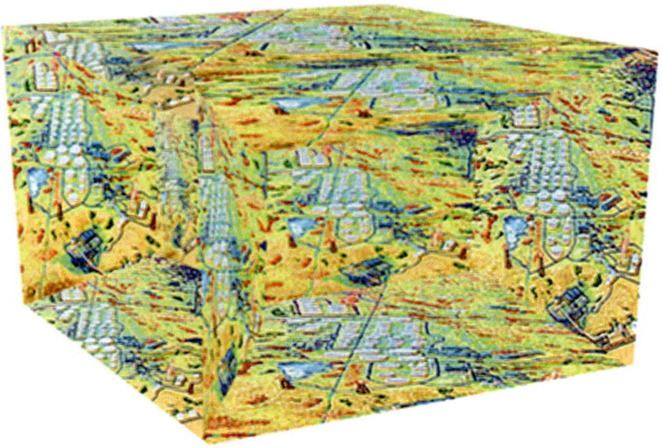

FIGURE 16.7 Hypercube of oil storage units and land deformation in TerraSAR-X data.

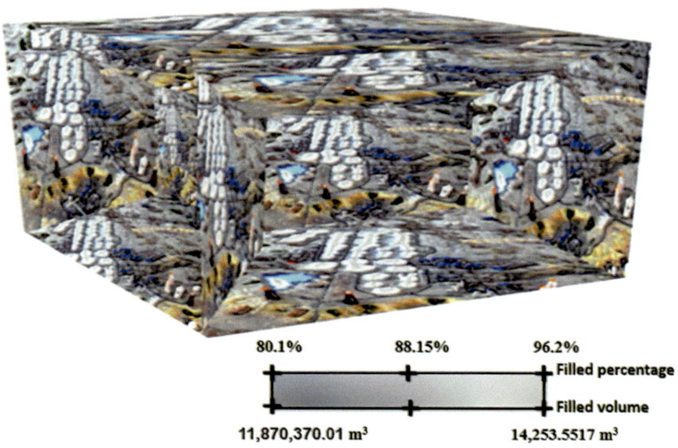

FIGURE 16.8 4-D hologram interferometry for detection of oil and gas levels at storage tanks.

interferometry, discrepancies caused by temporal and environmental noise can be sieved out. In order to provide highly accurate surface deformation while minimizing the impact of such ambiguities, an advanced InSAR system has been developed that combines Marghany 4-D quantized hologram interferometry with a quantum random walk search self-loop. The Marghany 4-D quantized hologram interferometry, it suffices to say, offers a more accurate assessment of the land deformation with better control of these errors. This cutting-edge system has a high degree of accuracy and can measure land deformation over a longer time period.

Moreover, RADARSAT-2 or other low-resolution SAR data sources, such as Terra-SAR, can be used with the Marghany 4-D quantized holographic interferometry algorithm. Because of its capability to work with a variety of low-resolution SAR data, the Marghany 4-D quantized holographic interferometry algorithm is particularly useful in Earth observation applications. As the resolution required for this algorithm is significantly lower than that of other methods, making it a more effective and economical choice, this capability is extremely advantageous for the rapid assessment of changes in geophysical parameters. Multi-temporal datasets can also be utilized with the Marghany 4-D quantized holographic interferometry algorithm. This provides for the analysis of dynamic phenomena, such as vegetation growth or urban development, in a more efficient and cost-effective manner.

16.8 CONCLUSION

This chapter explains how Marghany 4-D quantized hologram interferometry shows the interference between land deformation and oil storage facilities. One of the proofs for quantum entanglement theory is the 4-D oil storage unit geometry. The results show that the 4-D oil storage units' influence on ground deformation extended to the morphological features of nearby infrastructures. It can be deduced that Iranian petroleum exports have been dropping faster than the rate of production since Kharg

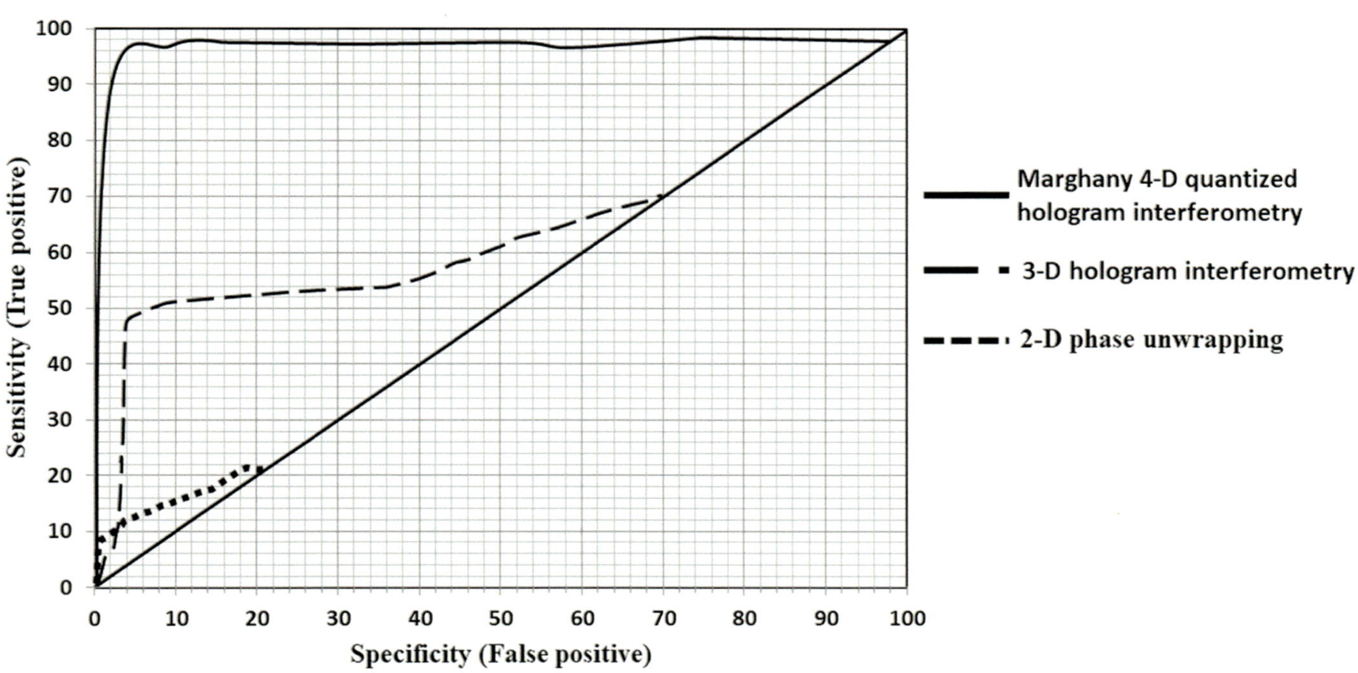

FIGURE 16.9 ROC for 4-D quantized hologram interferometry.

Island's crude storage has reached 100% of its maximum capacity. The true positive rate of the ROC curves (Figure 16.9) for Marghany 4-D quantized holographic interferometry is higher in this view (TPR). Oil and gas levels can be 99% aggregated in the ROC area between 2-D phase unwrapping and traditional 3-D holographic interferometry. Incorporating quantum theories with the 4-D phase unwrapping of TerraSAR-X satellite data led to Marghany's 4-D quantized phase unwrapping algorithm, which is a high-quality, promising technique for 4-D reconstruction of oil and gas exploration activities.

REFERENCES

Bhattacharya, A., and K. Mukherjee. 2017. Review on InSAR based displacement monitoring of Indian Himalayas: Issues, challenges and possible advanced alternatives. *Geocarto Int* 32, no. 3: 298–321.

Bryksin, V.M., Filatov, A.V., and A.V. Yevtyushkin. 2012. Using of SAR data and DInSAR-PSInSAR technique for monitoring Western Siberia and Arctic. *Журнал радиоэлектроники* 6: 1–53.

Cheng, W.C., Li, D.W., Xiao, W.Z., Chen, G.Y., Cheng, C.H., and J.S. Sun. 2021. *Design of 24 GHz Antenna Array for Powder Materials Flow Measurement Applications*. IEEE, 2021 International Symposium on Antennas and Propagation (ISAP), pp. 1–2.

Chiaramonte, L., Zoback, M.D., Friedmann, J., and V. Stamp. 2008. *Geomechanical Characterization and Reservoir Simulation of a CO2-EOR and Sequestration Project in a Mature Oil Field*. Teapot Dome, WY. AGU Fall Meeting Abstracts, 2008: H23D-1002.

Ferretti, A., Monti-Guarnieri, A.V., Prati, C.M., Rocca, F., and D. Massonnet. 2007. *INSAR Principles: Guidelines for SAR Interferometry Processing and Interpretation*. Noordwijk, The Netherlands: ESA Publications.

Firouzbakht, S., Gitipour, S., and A.F.H. Rooz. 2017. Cement-based and AC-based solidification/stabilization processes of aliphatic hydrocarbons contaminated soils in Kharg Island oil terminal. *Int J Environ Res* 11, no. 4: 439–448.

Hartnoll, S.A., Lucas, A., and S. Sachdev. 2018. *Holographic Quantum Matter*. Cambridge, MA: MIT Press.

Hennings, I., and D. Herbers. 2007. The expected potential of TerraSAR-X high resolution spotlight mode data for shallow sea bottom topography imaging: A preview. *EARSeL eProc* 6, no. 2: 67–81.

Marghany, M. 2013. DInSAR technique for three-dimensional coastal spit simulation from Radarsat-1 fine mode data. *Acta Geophysica* 61, no. 2: 478–493.

Marghany, M. 2014a. A Three-dimensional of coastline deformation using the sorting reliability algorithm of ENVISAT interferometric synthetic aperture radar. In: *Advanced Geosci Remote Sensing*, 105–121. IntechOpen. http://doi.org/10.5772/58571.

Marghany, M. 2014b. Hybrid genetic algorithm of interferometric synthetic aperture radar for three-dimensional coastal deformation. *SoMeT* 29: 116–131. http://doi.org/10.3233/978-1-61499-434-3-116.

Marghany, M. 2019a. *Synthetic Aperture Radar Imaging Mechanism for Oil Spills*. Cambridge, MA: Gulf Professional Publishing.

Marghany, M. 2019b. *Automatic Detection Algorithms of Oil Spill in Radar Images*. Boca Raton, FL: CRC Press.

Marghany, M. 2021a. *Advanced Algorithms for Mineral and Hydrocarbon Exploration Using Synthetic Aperture Radar*. Amsterdam: Elsevier.

Marghany, M. 2021b. Four-dimensional Covid-19 simulation in slums using hologram interferometry of Sentinel-1A satellite. In: *Geospatial Technology and Smart Cities*, 167–185. Cham, Switzerland: Springer.

Marghany, M. 2022. *Remote Sensing and Image Processing in Mineralogy*. Boca Raton, FL: CRC Press.

Miah, M.I. 2020. Predictive models and feature ranking in reservoir geomechanics: A critical review and research guidelines. *J Natl Gas Sci Eng* 82: 103493.

Moghaddam, F.F., Moghaddam, R.F., and M. Cheriet. 2012. Curved space optimization: A random search based on general relativity theory. *arXiv preprint arXiv:1208.2214*.

Ou, Z.Y.J. 2007. *Multi-Photon Quantum Interference*, vol. 43. Berlin: Springer.

Pinel, V., Hooper, A., De la Cruz-Reyna, S., Reyes-Davila, G., Doin, M.P., and P. Bascou. 2011. The challenging retrieval of the displacement field from InSAR data for andesitic stratovolcanoes: Case study of Popocatepetl and Colima Volcano, Mexico. *J Volcanol Geotherm Res* 200, nos. 1–2: 49–61.

Sabzevari, F.M., Winter, R.S., Oloumi, D., and K. Rambabu. 2020. A microwave sensing and imaging method for multiphase flow metering of crude oil pipes. *IEEE J Sel Top Appl Earth Obs Remote Sens* 13: 1286–1297.

Sani, F.M., Afshar, A., and M. Mohammadi. 2016. Evaluation of the simultaneous effects of sulfate reducing bacteria, soil type and moisture content on corrosion behavior of buried carbon steel API 5L X65. *Int J Electrochem Sci* 11, no. 5: 3887–3907.

Sayers, C.M., and P.M. Schutjens. 2007. An introduction to reservoir geomechanics. *Lead Edge* 26, no. 5: 597–601.

Shi, J., Yang, H., Peng, J., Wu, L., Xu, B., Liu, Y., and B. Zhao. 2019. InSAR monitoring and analysis of ground deformation due to fluid or gas injection in Fengcheng Oil Field, Xinjiang, China. *J Indian Soc Remote Sens* 47, no. 3: 455–466.

Velotto, D., Migliaccio, M., Nunziata, F., and S. Lehner. 2011. Dual-polarized TerraSAR-X data for oil-spill observation. *IEEE Trans Geosci Remote Sens* 49, no. 12: 4751–4762.

Verdon, J.P., Kendall, J.M., Stork, A.L., Chadwick, R.A., White, D.J., and R.C. Bissell. 2013. Comparison of geomechanical deformation induced by megatonne-scale CO2 storage at Sleipner, Weyburn, and in Salah. *Proc Natl Acad Sci* 110, no. 30: E2762–E2771.

Wenrui, H., Jingwei, B., and H. Bin. 2013. Trend and progress in global oil and gas exploration. *Pet Explor Dev* 40, no. 4: 439–443.

Werninghaus, R. 2004. TerraSAR-X mission. In: *SAR Image Analysis, Modeling, and Techniques VI*, vol. 5236, 9–16. Bellingham, Washington, DC: SPIE.

Wnuk, K., Walton, G., and W. Zhou. 2019. Four-dimensional filtering of InSAR persistent scatterers elucidates subsidence induced by tunnel excavation in the Sri Lankan highlands. *J Appl Remote Sens* 13, no. 3: 034508.

Woodhouse, I.H. 2017. *Introduction to Microwave Remote Sensing*. Boca Raton, FL: CRC Press.

Wu, Z., McCann, H., Davis, L.E., Hu, J., Fontes, A., and C.G. Xie. 2009. Microwave-tomographic system for oil-and gas-multiphase-flow imaging. *Meas Sci Technol* 20, no. 10: 104026.

Xuejun, W., Baoli, Y., Xiaohui, Z., Xiling, C., Hong, L., Yong, F., and L. Yulian. 2015. Development and application of "2W1H" technique in oil and gas exploration. *China Pet Explor* 20, no. 5: 41.

Xunyu, C., Jinlian, L., Peirong, Z., Chaoying, L., and C. Zhe. 2020. Oil and gas exploration progress and upstream development strategy of Sinopec. *China Pet Explor* 25, no. 1:11.

Yu, H., and X. Hu. 2021. Knowledge-aided InSAR phase unwrapping approach. *IEEE Trans Geosci Remote Sens* 60: 1–8.

Yu, H., Lan, Y., Yuan, Z., Xu, J., and H. Lee. 2019. Phase unwrapping in InSAR: A review. *IEEE Geosci Remote Sens Mag* 7, no. 1: 40–58.

Zhang, Y., Cui, Y., Liu, S., Fan, L., Zhou, N., Peng, P., Wang, Y., Guo, F., Min, M., Cheng, Y., and Y. Liu. 2020. Fast microwave-assisted pyrolysis of wastes for biofuels production – A review. *Bioresour Technol* 297: 122480.

Zoback, M.D., and A.H. Kohli. 2019. *Unconventional Reservoir Geomechanics*. Cambridge: Cambridge University Press.

Zucchini, R. 2021. 4-d Chern-Simons theory: Higher gauge symmetry and holographic aspects. *J High Energy Phys* 2021, no. 6: 1–83.

PART 6

Conservation and the marine ecosystem

Sanganeb Atoll *From lighthouse to inscription on the World Heritage List – A case study from the Sudanese Red Sea* 17

Dirar H. Nasr and El Hag A. El Hag

17.1 INTRODUCTION

Sanganeb is a Bejawian name, from the Sanganibi tree that grows in the plains of Jibait and Sinkat Cities; the tree resembles the shape of an island. The original name given by the Beja tribe was Sanganibi; it was later changed to Sanganeb (Eisaib Eisa, Dungonab Mayor, pers. comm.). Sanganeb Atoll is located in the central Red Sea at 19°42′N, 37°26′E, approximately 30 km northeast of Port Sudan (Figure 17.1). It is located among some of the diverse offshore reefs in the Red Sea, well away from most human impact and close to the Red Sea's centre of biodiversity. The atoll is entirely submerged except during the lowest tides, the only part evident above the sea's surface being the magnificent Sanganeb lighthouse and the few buildings that surround it (PERSGA/GEF 2004). Sanganeb Atoll rises from a seafloor of more than 800 m depth; the atoll is characterized by very steep slopes on all sides, with terraces in their upper parts and occasional spurs and pillars (Sheppard and Wells 1988). The dominant coral reef ecosystem harbours significant unimpacted populations of fauna and flora and iconic and endangered species such as sharks, bumphead parrotfish (*Bolbometopon muricatus*) and grouper (Nasr 2015a).

It is true that biological research on reefs in the 1960s and early 1970s led to the view that reefs were fragile ecosystems, particularly vulnerable to human activities and slow to recover if damaged (Johannes 1975). A contrasting view was that reef communities are dynamic and unstable and that self-replacement and recovery from natural disturbance are normal and contribute to the maintenance of high diversity on the reef (Connell 1978). These views have been reviewed by Pearson (1981), Brown and Howard (1985) and Brown (1987), with the consensus being that reefs are perhaps not as fragile as was previously thought.

Various ecological studies of biota and habitats in both the littoral and shallow sub-zone have been carried out in the Sudanese Red Sea, with an emphasis on Sanganeb Atoll (Krupp et al. 1994; Head 1980; Gladstone 2000; Gladstone et al. 1999; Ergmer and Schumacher 1985; Edwards and Rosewell 1981; PERSGA/GEF 2002, 2003a, 2003b; PERSGA 2006, 2016; PERSGA/ALECSO 2007; Pilcher and Nasr 2000; Reinicke et al. 2003; Sheppard and Sheppard 1991; Cousteau Society 2013; Vine and Schmid 1987; Wilkinson 2002).

Sanganeb Atoll has been the focus of various conservation efforts since the early 70s since its reefs, with their steep walls, were recognised as among the most beautiful of the world. An ad-hoc committee was formed initially under the leadership of Dr. Rupert Ormond of the Cambridge Coral Reef Research Group (CCRRG) and Commander Noureldeen Farah (Sudanese Navy), Captain Abdulhalim (private tourism sector), Dr. Dirar H. Nasr (Institute of Oceanography) and Dr. Peter J. Vine (CCRRG). The committee proposed, among other conservation measures, that Sanganeb Atoll should be established as a marine national park, and it secured funding from the WWF/IUCN for further studies and initial management work (Davis 1982). Such efforts led to its declaration in 1990 by the government of Sudan as the Sanganeb Marine National Park (SMNP), the country's first marine national park. Subsequently, during a Convention on Biological Diversity (CBD) regional workshop held in Dubai in 2015 to facilitate the description of ecologically or biologically significant marine areas (EBSAS) in the northwest Indian Ocean and adjacent gulf areas, SMNP was accepted as one of the EBSAS.

The Sudanese reefs have attracted several local and foreign scientists who carried out surveys and research work. The fringing reef off Port Sudan was assessed in the context of coral bleaching (Nasr and Al-Sheikh 2000); the overall bleached corals were estimated to cover 14% of the substrate.

DOI: 10.1201/9781003321408-23

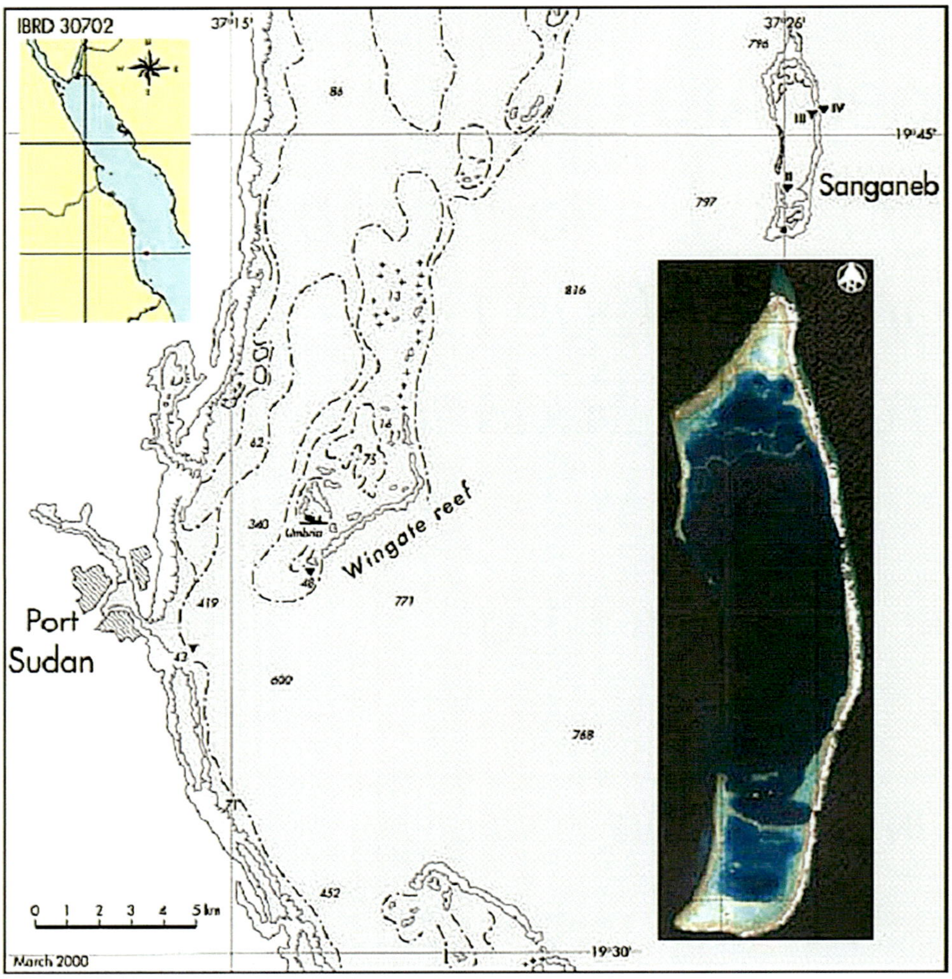

FIGURE 17.1 Location of Sanganeb Atoll in relation to adjacent reef areas. (Inset: a satellite image of the atoll).
Source: PERSGA/GEF (2004)

The morphology, sedimentology and zonation of the fringing reefs of Port Sudan were studied by Schroeder and Nasr (1983). Ormond and his colleagues, for instance, did many surveys for *Acanthaster planci* around Wingate reefs and on other reefs from Dungonab to the Southern Suakin Archipelago and Sanganeb reef, but their detailed work was all done on Towartit reef (Ormond and Campbell 1974; Roads and Ormond 1971).

Since at least the early 1980s, Sanganeb Atoll was also identified as a potential World Heritage Site, and more recently, this idea was suggested again (Pilcher 2002). The outstanding universal value of Sanganeb (i.e., it contains the most important and significant natural habitats for in-situ conservation of biological diversity, including those containing threatened species of outstanding universal value from the point of view of science or conservation) was recognised to be its location, 20 km offshore in the central Red Sea, well away at the time from any significant human impact, except, since 1998, limited coral bleaching caused by elevated ocean temperatures resulting from anthropogenic climate change (R. Ormond, pers. comm.). Bleaching of the very shallowest corals can also occur when they are exposed during the very lowest tides in summer, but otherwise, the reef is a fully submerged atoll, as opposed to emergent atolls such as Aldabra in the southwest Seychelles. The atoll is dominated by a coral reef ecosystem which harbours significant populations of a wide diversity of marine fauna and flora, including many Red Sea endemic and globally endangered species (PERSGA/GEF 2004). A detailed study concluded that Sanganeb met most of the critical World Heritage List (WHL) criteria (for instance, Criteria vii, viii, ix and x), leading to approval for Sanganeb to be inscribed on the WHL.

17.2 THE LIGHTHOUSE AND THE SURROUNDING SUBMERGED REEF

Sanganeb is a small atoll by global standards; its maximum length along the north-south axis is 6.5 km, and its maximum width is 1.6 km, making it comparable to some of the smaller atolls in the Pacific Ocean. The area of reef flat and shallow fore reef is approximately 2 km², and the area of enclosed lagoon is approximately 4.6 km² (PERSGA/GEF 2004). The rim of the atoll reaches the surface on all but the western side, where it is submerged (Figure 17.1).

Darwin's theory of atoll formation has long been shown to be correct (Veron 2000). This theory for the development of barrier reefs and atolls generated considerable debate, and even when real evidence began to appear, much of it was contradictory (Bemert and Ormond 1981). Sanganeb Atoll has been described as the only atoll in the Red Sea (Guilcher 1988). However, calling this reef an atoll is largely a question of semantics because numerous similar structures occur off the central Saudi Arabian coast too but are not usually described as atolls (Sheppard et al. 1992). Why Sanganeb is the only atoll in the Red Sea raises interesting questions about the formation of atolls in general and about the geological structure of Sanganeb Atoll in particular (PERSGA/GEF 2004).

Construction of the modern Sanganeb lighthouse was started in 1950, but there was an older structure present at this site from at least 1907. The adjacent buildings originally provided accommodation for the lighthouse keepers, and although the lights were automated, the accommodation remained. Its navigational role apart, the lighthouse also offers visitors a unique opportunity to view the whole atoll from a spectacular vantage point and, in addition, gives park rangers an excellent point from which to monitor activities within the park's boundaries (PERSGA/GEF 2004) (Figure 17.2).

The water around the atoll is typically extremely clear, and visibility of up to 46 m has been recorded (Schroeder et al. 1980 and records of the Institute of Oceanography). It is the only true atoll in the Red Sea, although there are well-developed ring reefs in both the Suakin Archipelago, farther south in Sudan, and on the Farasan Bank on the opposite shore of the Red Sea. Consequently, the atoll is among the unique structures in the Red Sea, providing a complex array of coral reef–associated habitats, including not only the wall-like reef outer faces but also shallow reef-top flats and the broad central lagoon. In addition to the occasional spurs and pillars (Sheppard and Wells 1988), deeper grooves are observed on the wave-exposed western side, and here, large aggregations of commercially significant fishes such as snapper (*Lutjanidae*) and black-spotted grunts (*Plectorhynchus gaterinus*) shelter by day (Figure 17.3).

FIGURE 17.3 Spurs and grooves with aggregation of black-spotted grunts.

Source: Hans Sjoholmn and Dirar H. Nasr

FIGURE 17.2 The lighthouse with visitors.

Source: Hans Sjoholmn and Dirar H. Nasr

The reef community displays a high diversity of fish, corals and presumably other invertebrates, including Red Sea endemic species, such as the brown-faced butterfly fish, (*Chaetodon larvatus*) and endangered species, including large sharks, such as the oceanic white-tip (*Carcharinus longimanus*) and large grouper, such as *Epinephelus tauvina*; dense schools of scale-fin Anthias (*Anthias sqammipinnis*) are a conspicuous feature of the reef edge zone (Nasr 2015a) (Figure 17.4), and schools of bumphead parrotfish (*Bolbometopon muricatus*) now scarce outside the Red Sea, are present. Importantly, outer sections of the reef serve as spawning grounds for snapper and grouper and, notably, for hump-headed wrasse (*Cheilinus undulatus*), now also threatened throughout much of their range (Figure 17.5). The atoll also supports several resident populations of dolphins, marine turtles and sharks, and large schools of scalloped hammerhead sharks (*Sphyrna lewini*) aggregate on a daily basis at the northern plateau (Nasr 2015a) (Figure 17.5).

The atoll has highly diverse coral fauna (Figure 17.6) occupying 13 different bio-physiographic reef zones, each characterised by its own typical coral assemblages (Sheppard and Wells 1988). These communities have been described by Mergner and Schumacher (1985). A total of 86 coral species in 35 genera have been recorded (Sheppard and Wells 1988). To date, including these Scleractinia corals, a total of 124 cnidarian species have been recorded (Mergner and Schumacher 1985). The structure and zonation of the reef have been reasonably well studied and are considered typical for offshore reefs of the central Red Sea (Vine and Vine 1980; Mergner and Schumacher 1985). The majority of the reef flat is in shallow water where coral cover is moderate (10–30%) and dominated by massive and encrusting growth forms, particularly colonies of *Porites*, *Goniastrea* and *Montipora*, but there is a greater diversity of coral species than in the back reef (PERSGA/GEF 2004).

FIGURE 17.4 Rich biological diversity with large numbers of *Anthias squamipinnis*.

Source: Hans Sjoholmn and Dirar H. Nasr

FIGURE 17.5 Juvenile fish, indicating SMNP serves as a spawning and nursery area. The disc-like coral on the right is *Podabacia crustacea*.

Source: Hans Sjoholmn and Dirar H. Nasr

FIGURE 17.6 Highly diverse coral fauna.

Source: Hans Sjoholmn and Dirar H. Nasr

FIGURE 17.7 The reef slope with diversity of fish life and coral.

Source: Hans Sjoholmn and Dirar H. Nasr

Around most of the outer rim of Sanganeb, the reef drops vertically by 5–10 m to a slope that continues to fall away at a steep angle to a shelf at 20–30 m. In some areas, these drop-offs continue down to depths of at least 50 m. The drop-off from the reef flat to the reef slope is spectacular because of the diversity of fish life and the variety of coral (Figure 17.7). From a diver's perspective, these drop-offs represent some of the best dive sites because of the rich marine life in very clear water (Krupp et al. 1994; PERSGA/GEF 2004). The coral reef fish fauna of Sanganeb have been surveyed by Krupp et al. (1994) and have been shown to be highly diverse, with over 251 species so far identified and an estimate of the actual number of species put at over 300. The most important reef fish to be found in the atoll are bumphead parrotfish (*Bolbometopon muricatum*), Napoleon wrasse (*Cheilinus undulates*) and grouper (*Serranidae*), all of which are subject to serious overfishing elsewhere in the world and which are becoming increasingly rare on a global scale. The Napoleon wrasse and bumphead parrotfish are classified as vulnerable on the IUCN Red List. So far, a total of nine species of grouper have been recorded in SMNP, of which the spotted coral grouper (*Plectropomus maculatus*), known locally as *najil*, is the most important commercially and from a conservation perspective. *Genicanthus melanospilus* (*Pomacanthidae*), which has a patchy distribution in the Red Sea, occurs on the deep-water reefs (Ormond 1980a). In addition to reef-associated species, a large number of pelagic fish can be observed in the open waters, including species of tuna, barracuda, sailfish, manta rays and sharks (Figures 17.8 and 17.9). The SMNP is also home to large numbers of cetaceans and, especially, a resident population of dolphins (Figure 17.10). A habitat map derived from satellite imagery is presented in Figure 17.11 (PERSGA/GEF 2004).

Vine and Vine (1980) also describe a pinnacle rising to within 20 m of the sea surface, immediately to the northeast of the atoll, that attracts large numbers of schooling pelagic fish. The open water surrounding the atoll is also an important component of SMNP. Average water depths away from the reef are 500–800 m, so pelagic species, including cetaceans, turtles and commercially important fish, predominate (PERSGA/GEF 2004).

FIGURE 17.8 School of barracuda in the open waters of Sanganeb.

Source: Hans Sjoholmn and Dirar H. Nasr

FIGURE 17.9 Grey reef shark (*Carcharhinus amblyrhynchos*) in the open waters surrounding Sanganeb.

Source: Hans Sjoholmn and Dirar H. Nasr

17.3 THE MARINE NATIONAL PARK

Beside the work of CCRRG on *Acanthaster planci*, as stated earlier, this group played an important role in stimulating coral reef conservation in Sudan (Ormond 1976, 1980b, 1987).

Following the activities of CCRRG, the driving force in reef conservation, at least during the 70s and 80s, was the Sudan Marine Conservation Committee (SMCC). This is composed mainly of representatives of government departments that have jurisdiction or authority over some aspect of the marine environment or maritime affairs, such as the Sea Ports Corporation, Fisheries Department, Fisheries Research Centre, government of the Red Sea Province, navy, Tourist Corporation, and Suakin Marine Biological Laboratory (University of Khartoum), under the leadership of the former Institute of Oceanography (presently the Institute of Marine Science and Fisheries, Red Sea University). The role of this committee was advisory, educational and coordinating, having conservation activities covering a broad spectrum ranging from legal protection to public education and from marine recreation to environmental monitoring and scientific research (Schroeder et al. 1980; Schroeder and Scheer 1981; Nasr 1985; PERSGA/GEF 2004). The activities of the SMCC were carried out with substantial support from the WWF/IUCN through Project No. 1259, "Establishment and development of a proposed Sanganeb Marine National Park", with the aim of protecting and conserving the marine habitat (mainly coral reefs), promoting public awareness, scientific research and offering recreational facilities. HRH the Duke of Edinburgh paid a visit to Sanganeb in March 1982 in the framework of this WWF/IUCN project.

The donations from IUCN/WWF in the framework of Projects No. 1259 and 1163 included slides, films and a mobile unit reflecting the beauty of reefs in general and SMNP in particular and the need for their conservation (IUCN 1977). A slide projector and a collection of high-quality slides were used to illustrate talks on coral reef ecology and SMNP. The target sector of the community included schoolchildren, fishermen, government officials and navy

FIGURE 17.10 Resident population of spinner dolphins and famous Napoleon wrasse (*Cheilinus undulatus*).

Source: Sjoholmn Hans and Dirar H. Nasr

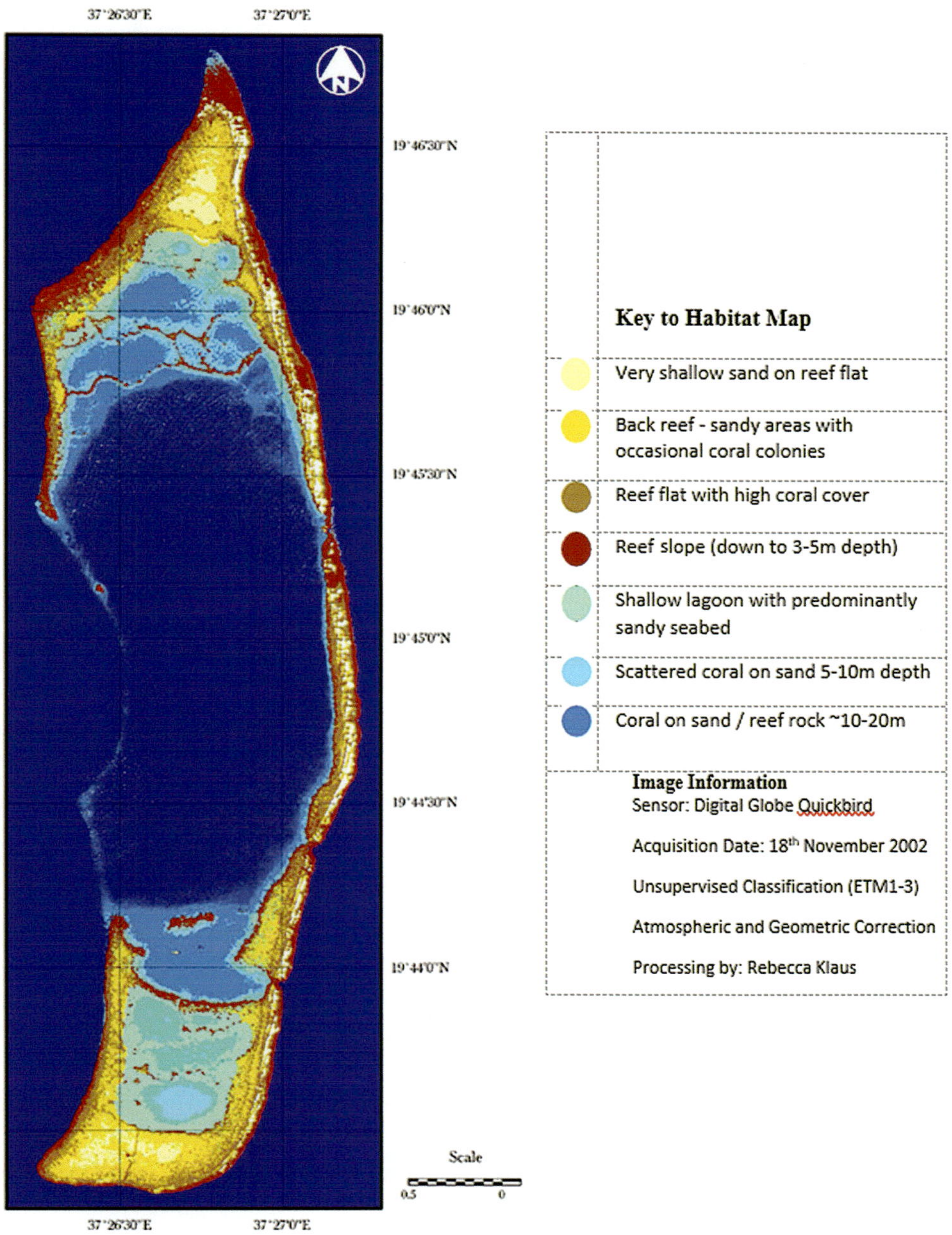

FIGURE 17.11 Habitat mapping of Sanganeb Atoll.

Source: PERSGA/GEF (2004)

and custom officers. Presentations to these community sectors were given in simple, non-scientific language, supplemented by the production of leaflets, posters and sign boards. The leaflets contained a description of the significance of SMNP and its unique features. A combination of posters with colourful images of marine life and bilingual (Arabic and English) text in addition to specimens of marine organisms (e.g., coral skeletons and shells) were presented on boards and housed in the Red Sea University's facility adjacent to the lighthouse as educational tools for the park's visitors. Similar presentations were given by a Sudanese NGO, Sudanese Development Initiative (SUDIA), in support of SMNP. Such combined efforts led to the declaration of Sanganeb by the government of Sudan as the country's first marine national park in 1990.

17.4 AN ECOLOGICALLY OR BIOLOGICALLY SIGNIFICANT MARINE AREA (EBSA)

Sudan signed the Convention on Biological Diversity (CBD) on 9 June 1992 and became a party on 30 October 1995. Since then, the Higher Council for Environment and Natural Resources (HCENR) has developed a national biodiversity strategy and action plan as part of its commitment to the Secretariat of the CBD. During the CBD Regional Workshop to Facilitate the Description of Ecologically or Biologically Significant Marine Areas (EBSAS) in the Northwest Indian

Ocean and adjacent Gulf areas, held in Dubai (19–25 April 2015), SMNP was proposed by Sudan as an EBSA. The presentation of SMNP as a candidate for EBSA followed the template prepared by the workshop secretariat for submission. It described the uniqueness of its geological structure, the unusual nature of its reefs in that they are in excellent condition, their high species richness with a large number of flagship species (e.g., large fish and sharks) and that they are not immediately threatened by human activity as there is no resident population, and they are not an important traditional fishing ground.

17.4.1 Zoning plan

PERSGA advised establishing a zoning plan for SMNP (PERSGA/GEF 2004) with the following objectives in mind:

- Conservation of Sanganeb Atoll and its natural resources
- Regulation of the use of SMNP to protect the atoll while allowing reasonable multiple uses of SMNP
- Regulation of activities that exploit the resources of SMNP to minimise the effect on the park
- Reservation of some areas of SMNP for the appreciation and enjoyment of the public
- Preservation of some areas of Sanganeb Atoll in their undisturbed natural state except for the purposes of scientific research

The plan marks nature sanctuaries, dive sites, monitoring sites, mooring buoys, an anchoring area and swimming and snorkeling areas, as shown in Figure 17.12. Having been shown to be a good candidate for recognition as an EBSA, the park was accepted by the workshop and endorsed by the CBD.

17.4.2 Becoming a member of the World Heritage List (WHL)

All documents relevant to the nomination of SMNP for inclusion on the WHL were prepared by the first and second authors while coordination with UNESCO was conducted by the National Commission for UNESCO. These documents included a statement of the outstanding universal value of SMNP; information on its area and geographical coordinates; and a description of its habitats, including fauna and flora, in addition to its history and development, supplemented by maps, figures and photos.

Most of the information provided by the authors met the relevant criteria of the WHL as follows:

1. SMNP is a submerged and overhanging predator-dominated coral reef ecosystem which harbours significant populations of fauna and flora in a stable equilibrium with numerous endemic and endangered species. It lies in a marine area that resulted from a striking geological process of continuing slow plate tectonic movement, and the atoll is the only geological structure of its type in the Red Sea (**Criterion viii**).

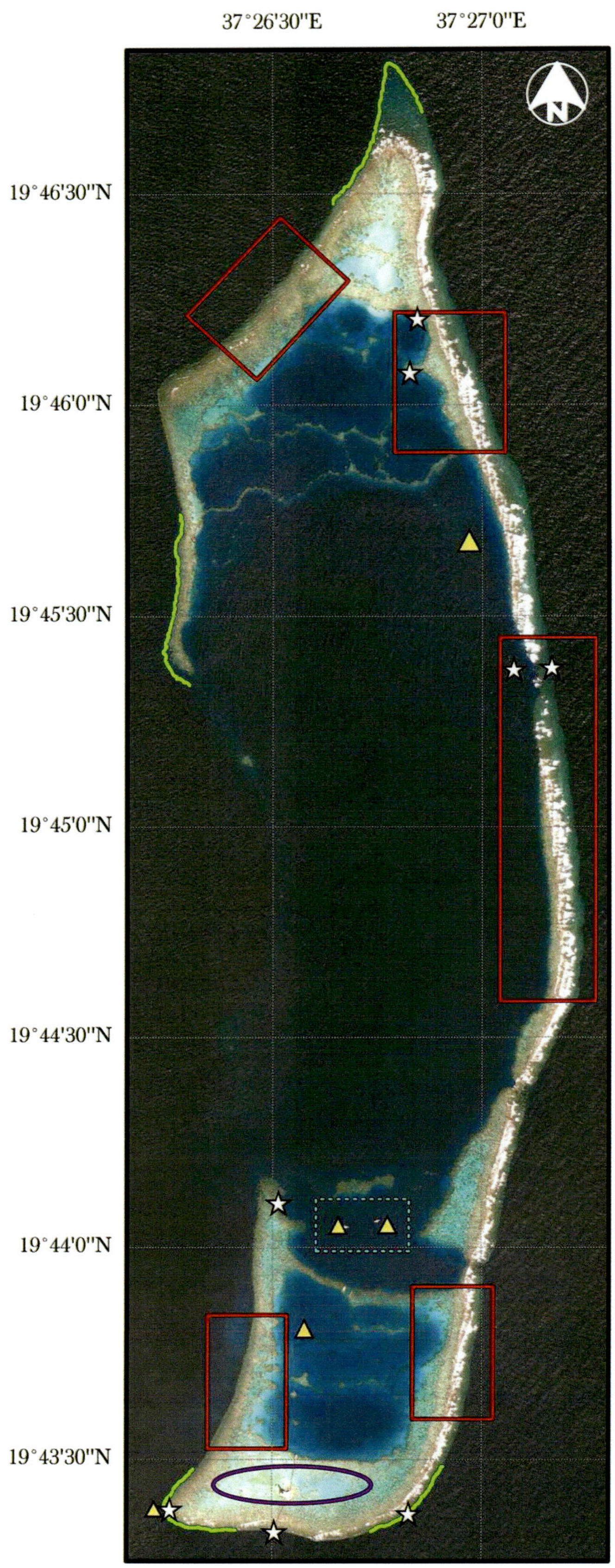

FIGURE 17.12 Zoning plan for SMNP's core zone and nature sanctuaries. KEY: Red rectangles: nature sanctuaries, green lines: dive sites, white stars: monitoring sites, yellow triangles: mooring buoys, blue dashed square: anchoring area, purple oval: swimming and snorkelling area (Sensor: Digital Globe Quickbird, 18 November 2002; marine area: true colour (ETM1–3) (1m); atmospheric and geometric correction processing by R. Klaus).

Source: PERSGA/GEF 2004

2 The combination of optimal temperature, visibility and salinity conditions creates an environment of exceptional coral growth and reef development within the nominated sites, and recovery from the 1998 coral bleaching impacts is evident. Furthermore, SMNP serves as an important larvae export area, acting as a source of recruits for all species of plants and animals present in and around the reef, including invertebrates and fish species; it also hosts important spawning sites for the key fishery species (**Criterion ix**).
3 The rise of 800 m from the seafloor to form the only atoll enclosing a large central lagoon in the Red Sea is exceptional, creating a vertical slope that exhibits a spectacular diversity of fish, coral and other invertebrate species. This topographical feature, the rich biodiversity and the crystal-clear Red Sea water render the area among the best dive sites worldwide (**Criterion vii**).
4 Sanganeb Atoll serves as a biodiversity "hotspot", lying at or close to the centre of marine biodiversity in the Red Sea and boasting hard and soft coral fauna that is amongst the richest in the region. Thriving on this rich reef ecosystem, over 300 fish species (including pelagic) inhabit SMNP. The park also hosts significant populations of globally important and endangered species. including sharks, grouper, Napoleon wrasse (a vulnerable species according to the 1996 IUCN Red List of species), marine turtles and cetaceans. SMNP also acts as an important spawning ground for key fishery species as well as a larvae export area (**Criterion x**).

In SMNP, the lack of terrestrial fresh water runoff, combined with relatively low productivity, gives rise to low levels of turbidity, and visibility can reach 70 m (PERSGA 2001). The semi-enclosed nature of the Red Sea has buffered the atoll from the more severe effects that could have impact on the area: for example, from high waves, strong currents, tsunami and other natural phenomena such as earthquakes and volcanoes. Sanganeb Atoll is rather remote from land-based activities, and the traditional artisanal fishing around it is under the control of the Fisheries Administration (Nasr 2015b). Illegal fishing by visiting fishing fleets is uncommon because of the cost of travel to the site and because the Wildlife Administration and the Sudanese Navy patrol the area regularly. These factors support the integrity of Sanganeb Atoll.

17.4.3 The legislative framework

Sudan is a party to the following regional and international conventions:

- The Regional Convention for the Conservation of the Red Sea and Gulf of Aden Environment (Jeddah Convention), 1982, signed and ratified on 5 June 1984.
- The Protocol Concerning Regional Cooperation in Combating Pollution by Oil and Other Harmful Substance in Cases of Emergency, 1982. Sudan became a party to the Protocol on 5 June 1984.
- Protocol Concerning the Conservation of Biological Diversity and the Establishment of a Network of Protected Areas in the Red Sea and Gulf of Aden, 2007.
- Protocol Concerning Technical Cooperation in Transboundary Transfer of Equipment, Experts and Technicians in Cases of Emergencies, 2009.
- Convention for Biological Diversity (CBD), 1992: Sudan became a party to the Convention on 30 October 1995.
- Convention on International Trade in Endangered Species of Wild Fauna and Flora (CITES), 1973: Sudan became a party to the Convention on 26 October 1982.
- United Nations Convention on the Law of the Sea (UNCLOS), 1982: Sudan became a party to the Convention on 23 January 1985.
- African Convention on the Conservation of Nature and Natural Resources, 1968: Sudan became a party to the Convention on 21 October 1973.

Sudan has already started the procedure for accession to the International Convention for the Prevention of Pollution from Ships (MARPOL), as well as related conventions such as the International Convention on Oil Pollution Preparedness, Response and Cooperation (1990); the International Convention Relating to Intervention on the High Seas in Cases of Oil Pollution Casualties (1969); the Civil Liability Convention (1969); and the International Oil Pollution Compensation Fund.

17.5 NATIONAL LEGISLATION

Several national legislative frameworks are in place that relate to coral reefs:

- Sudanese Fishery Ordinances and Regulations: Dates back to 1937 and was amended in 1975 and 1978. It prohibits overfishing; dumping of refuse, including oil, into the sea; and the collection of coral, shells and aquarium fish.
- Environmental Health Act: Established in 1975, it prohibits the dumping into the sea of any item that is harmful to humans or animals.
- Marine Fisheries Ordinance: Gives police, customs officers and local authorities the right to board and search a vessel and detain any craft accused of violating these regulations.
- Maritime Law: Drafted by the Maritime Administration and approved in 2011.
- Comprehensive National Strategy: Through this, Sudan has committed to the pursuit of sustainable development and environmentally sound resource management.

The Red Sea State Environment Protection Act (2006) addresses the environment, pollution, fisheries, marine parks, the coastal zone and the coastal environment. Its objectives are as follows:

> [P]rotection of the purity and natural equilibrium and conservation of the environment and the social and cultural systems thereof, to achieve safety and sustainable development for the benefit of generations; promotion of the environment for sustainable development; linking development and the environment and spreading awareness; and ascertaining and activating the role of the competent authority and its delegates and preventing laxity and short comings of performance.

17.6 REGIONAL AND NATIONAL ACTION PLANS

PERSGA has taken several steps to conserve coral reefs, mangroves, seabirds and marine turtles in the Red Sea and Gulf of Aden. Surveys were conducted to determine the status of habitats and species in the region, resulting in the development of country status reports. This step was followed by the development of regional action plans (RAPs) for conservation of those key habitats and species. However, it was necessary to develop national action plans (NAPs) for the conservation of those habitats and species by the PERSGA countries, in the light of the RAPs, as a fundamental and harmonizing step. The NAPs also facilitate implementation of actions at the national level. Accordingly, Sudan has formulated NAPs for the conservation of its key habitats and species. The implementation of these NAPs will occur through integrated networks of national and local working groups, government departments, agencies and personnel, non-governmental organisations and other stakeholders.

Sudan has also prepared the following national action plans:

- National Oil Spill Contingency Plan for Sudan
- National Integrated Coastal Zone Management
- National Plan of Action for the Protection of the Marine Environment from Land-Based Activities

17.6.1 Monitoring

A biodiversity index is the primary ecological indicator used to measure the state of conservation at the site. Biodiversity measurements are obtained through regular surveys and the use of standardised monitoring protocols.

17.6.2 Management plan

PERSGA prepared a site-specific master plan with management guidelines for SMNP. This document was prepared by Dr. Simon Wilson under contract to PERSGA with the cooperation of the Wildlife Conservation Administration (Sudan). This plan is not yet in place; however, the administrative arrangements for monitoring SMNP are the collective responsibility of the following:

- The Wildlife Conservation General Administration
- The Fisheries Administration
- The State Ministry of Environment and Tourism

17.6.3 Reporting exercise

Previous reporting exercises have included routine measuring of the physical properties of the waters surrounding SMNP by the Institute of Marine Science and Fisheries (Red Sea University), joint projects with regional and international organizations such as PERSGA and IUCN/WWF and joint projects with overseas institutions such as the Senckenberg Research Institute and Natural History Museum, in addition to university post-graduate students' studies.

The submission of the aforementioned information, maps, photographs and audiovisual images to UNESCO through the National Commission for UNESCO triggered the next stage of the process: an IUCN technical evaluation of SMNP. This evaluation process involved the following:

External review: The nomination was sent to independent experts knowledgeable about Sanganeb Atoll and its natural values.

Field mission: An independent technical adviser from UNESCO and IUCN expert, Dr. Ameer Abdulla, visited the candidate site to discuss the nomination with the relevant national and local authorities, local communities, NGOs and other stakeholders.

IUCN World Heritage Panel review: The panel intensively reviewed the SMNP nomination, field mission report and comments from external reviewers and provided its technical advice to IUCN on recommendations for the nomination. A final report was then prepared and forwarded to the World Heritage Centre for distribution to the World Heritage Committee.

Final recommendations: IUCN presented the results and recommendations of its evaluation process to the World Heritage Committee at its annual session. The World Heritage Committee made the final decision on whether or not to add the nominated property to the World Heritage List.

ACKNOWLEDGEMENT

We would like to thank PERSGA for its continuous support during all stages of SMNP development, the IUCN/WWF for providing public awareness materials and equipment in addition to their support and encouragement, UNESCO for its cooperation during the stages of nominating the SMNP for the WHL and all the national and overseas scientists who contributed to the knowledge base of SMNP.

REFERENCES

Bemert, G., and R. Ormond. 1981. *Red Sea Coral Reefs*. London: Kegan Paul International Ltd.

Brown, B.E. 1987. Worldwide death of corals: Natural cyclical events or man-made pollution? *Mar Poll Bull* 18, no. 1: 9–13.

Brown, B.E., and S. Howard. 1985. Responses of coelenterates to trace metals: A field and laboratory evaluation. *Proc 5th Int Coral Reef Congress, Tahiti* 6: 465–470.

Connell, J. 1978. Diversity in tropical rainforests and coral reefs. *Science* 199: 1302–1310.

Cousteau Society. 2013. *Toward a Sustainable Future for the Red Sea Coast of Sudan. Part 1: Coastal and Marine Habitats Survey.* Eds. T. Chekchak and R. Klaus. New York, NY: The Cousteau Society, Custodians of the sea.

Davis, G.E. 1982. *Feasibility of a Marine National Park at Sanganeb Atoll, Red Sea Province, Sudan.* Unpubl Rept, WWF/IUCN Project No. 1259. IUCN/WWF, Gland.

Edwards, A., and J. Rosewell. 1981. Vertical zonation of coral reef fishes in the Sudanese Red Sea. *Hydrobiologia* 79: 21–31.

Ergmer, U., and H. Schuhmacher. 1985. Quantitave analysis of coral communities on Sanganeb Atoll (central Red Sea): Comparison with Aqaba Reefs (northern Red Sea). *Proc 5th Int Coral Reef Congress*, Tahiti.

Gladstone, W. 2000. Ecological and social basis for management of a Red Sea marine protected area. *Ocean Coast Manag* 43: 1015–1032.

Gladstone, W., Tawfiq, N., Nasr, D., Andersen, I., Cheung, C., Drammeh, H., Krupp, F., and S. Lintner. 1999. Sustainable use of the renewable resources and conservation in the Red Sea and Gulf of Aden: Issues, needs and strategic actions. *Ocean Coast Manag* 42: 671–697.

Guilcher, A. 1988. A heretofore neglected type of coral reef: The ridge reef. Morphology and origin. *Proc 6th Int Coral Reef Symp, Australia* 3: 399–402.

Head, S.M. 1980. *The Ecology of Corals in the Sudanese Red Sea.* PhD Thesis, University of Cambridge, UK.

IUCN. 1977. *Conservation of Coral Reefs of the Sudanese Red Sea.* Unpubl Rept, IUCN/WWF Project No. 1163. IUCN/WWF, Gland.

Johannes, R.E. 1975. Pollution and degradation of coral reef communities. In: *Tropical Marine Pollution.* Eds. E.J. Ferguson Wood and R.E. Johannes, 13–50. Amsterdam: Elsevier Scientific Publishing.

Krupp, F., Türkay, M., El Hag, A.G.D., and D. Nasr (eds.). 1994. *Comparative Ecological Analysis of Biota and Habitats in Littoral and Shallow Sublittoral Waters of the Sudanese Red Sea.* Forschungsinstitut Senckenberg, Frankfurt and Faculty of Marine Science and Fisheries, Port Sudan.

Mergner, H., and H. Schumacher. 1985. Quantitative analysis of coral communities on Sanganeb Atoll (central Red Sea). Comparison with Aqaba reefs (Northern Red Sea). *Proc Fifth Int Coral Reef Congress, Tahiti* 6: 243–248.

Nasr, D.H. 1985. Coral reef conservation in Sudan. *Proc Fifth Int Coral Reef Congress, Tahiti* 4: 243–246.

Nasr, D.H. 2015a. Sanganeb Marine Protected Area and Sha'ab Rumi atoll. *Presented in the CBD Regional Workshop to Facilitate the Description of Ecologically or Biologically Significant Marine Areas (EBSAs) in the North-West Indian Ocean and Adjacent Gulf Areas and Training Session on EBSAs, 19–25 April 2015*, Dubai, UAE.

Nasr, D.H. 2015b. Coral reefs of the Red Sea with special reference to the Sudanese coastal area. In: *The Red Sea: The Formation, Morphology, Oceanography and Environment of a Young Ocean Basin.* Eds. N.M.A. Najeeb and I.C.F. Stewart, 453–469. Berlin and Heidelberg: Springer Earth System Sciences.

Nasr, D.H., and K. Al-Sheikh. 2000. Assessment of coral reefs in the Sudanese Red Sea in the context of coral bleaching. In: *Proc International Workshop on the Extent and Impact of Coral Bleaching in the Arabian Region.* Ed. H. Tatwany. National Commission for Wildlife Conservation and Development, Riyadh.

Ormond, R.F.G. 1976. The Red Sea. In: *Promotion of the Establishment of Marine Parks and Reserves in the Northern Indian Ocean including the Red Sea and Persian Gulf.* Papers and Proceedings of the Regional Meeting held at Tehran, Iran, 6–10 March 1975. IUCN Publications (New Series) 35, pp. 115–123.

Ormond, R.F.G. 1980a. Aggressive mimicry and other interspecific feeding associations among Red Sea coral reef predators. *J Zool Lond* 192: 323–350.

Ormond, R.F.G. 1980b. Management and conservation of Red Sea habitats. In: *Proc Symp Coast Marine Environ Red Sea and Gulf of Aden and Tropical Western Indian Ocean.* University of Khartoum, Sudan, 2, pp. 137–162.

Ormond, R.F.G. 1987. Conservation and management. In: *Red Sea.* Eds. A. Edwards and S.M. Head, 405–423. Oxford: Pergamon Press.

Ormond, R.F.G., and A.C. Campbell. 1974. Formation and breakdown of Acanthaster planci aggregations in the Red Sea. *Proc 2nd Int Coral Reef Symp* 1: 596–619.

Pearson, R. 1981. Recovery and recolonisation of coral reefs. *Mar Ecol Prog Ser* 4: 105–122.

PERSGA. 2006. *The State of the Marine Environment Report for the Red Sea and Gulf of Aden.* PERSGA, Jeddah, 241 p.

PERSGA. 2016. *Guidelines for the Management of Marine Protected Areas. Strategic Ecosystem Management (SEM) of the Red Sea and Gulf of Aden, Component 1.* PERSGA, Jeddah.

PERSGA/ALECSO 2007. Survey of Habitats in Sudan and plans for their protection. *PERSGA Technical Series No. 13.* PERSGA, Jeddah.

PERSGA/GEF. 2001. *Strategic Action Programme for the Red Sea and Gulf of Aden, Country Reports.* PERSGA, Jeddah and the World Bank, Washington, DC, 205 p.

PERSGA/GEF. 2002. *The Red Sea and Gulf of Aden Regional Network of Marine Protected Areas. Regional Master Plan. PERSGA Technical Series No. 1.* PERSGA, Jeddah.

PERSGA/GEF. 2003a. Regional action plan for the conservation of coral reefs in the Red Sea and Gulf of Aden. *Technical Report Series No. 3.* PERSGA Jeddah.

PERSGA/GEF. 2003b. Coral reefs in the Red Sea and Gulf of Aden. Surveys 1990 to 2000 summary and recommendations. *PERSGA Technical Series No. 7.* PERSGA, Jeddah.

PERSGA/GEF. 2004. *Sanganeb Marine National Park. Draft Site-Specific Master Plan with Management Guidelines.* PERSGA, Jeddah.

Pilcher, N. 2002. *Potential Tropical Coastal, Marine and Small Island World Heritage Sites in the Middle East Region. World Heritage Biodiversity: Filling Critical Gaps and Promoting Multi-Site Approaches to New Nominations of Tropical Coastal, Marine and Small Island Ecosystems.* UNESCO World Heritage Convention. Workshop, Hanoi, Vietnam.

Pilcher, N., and D. Nasr. 2000. The status of coral reefs in Sudan. In: *Regional Status of Coral Reefs in the Red Sea and the Gulf of Aden. Status of Coral Reefs of the World.* Eds. N. Pilcher and A. Alsuhaibany, 35–54. Townsville: Australian Institute of Marine Science.

Reinicke, G., Kroll, D., and H. Schuhmacher. 2003. Patterns and changes of reef-coral communities at the Sanganeb-Atoll (Sudan, central Red Sea): 1980 to 1991. *Facies* 49, no. 1: 271–297.

Roads, C.H., and R.F.G. Ormond. 1971. *New Studies on the Coral-Predating Crown-of-Thorns Starfish.* Cambridge: Coral Starfish Research Group, 124 p.

Schroeder, J.H., and D.H. Nasr. 1983. The fringing reefs of Port Sudan, Sudan. 1. Morphology, Sedimentology, Zonation. *Essener Geographische Arbeiten* 6: 29–44.

Schroeder, J.H., Nasr, D.H., and F.H. Idris. 1980. Coral reef conservation in the Sudanese Red Sea. In: *Proc Symposium on the Coastal and Marine Environment of the Red Sea*, Gulf of Aden and Tropical Western Indian Ocean, Khartoum, pp. 163–178.

Schroeder, J.H., and G. Scheer. 1981. *Corals of Sanganeb Reef, Collected by Schroeder J.H., Identified by Scheer G.* Institute of Oceanography, Port Sudan, 6 p.

Sheppard, C.R.C., Price, A.R.G., and C.M. Roberts. 1992. *Marine Ecology of the Arabian Region.* London and San Diego: Academic Press, 359 p.

Sheppard, C.R.C., and A.L.S. Sheppard. 1991. Corals and coral communities of Arabia. *Fauna Saudi Arabia* 12: 3–170.

Sheppard, C.R.C., and S.M. Wells (eds.). 1988. *Coral Reefs of the World. Volume II: Indian Ocean, Red Sea and Gulf.* UNEP Regional Seas Directories and Bibliographies, no. 27WCMC. IUCN, Gland, Switzerland, and UNEP, Nairobi, 389 p.

Veron, J.E.N. 2000. *Corals of the World*, vol. 1. Townsville: Australian Institute of Marine Science.

Vine, P.J., and H. Schmid. 1987. *Red Sea Explorers*. London: Immel Publishing.

Vine, P.J., and M.P. Vine. 1980. Ecology of Sudanese coral reefs with particular reference to reef morphology and distribution of fishes. *Proc Symp Coastal Mar Environ Red Sea, Gulf Aden Trop Western Indian Ocean* 1: 87–140.

Wilkinson, C. 2002. *Status of Coral Reefs of the World, 2002*. Townsville, Queensland: Australian Institute of Marine Science.

Trends in marine fisheries of the Arabian/Persian Gulf

18

Abdulrahman Ben-Hasan

18.1 INTRODUCTION

The Arabian/Persian Gulf ("Gulf") is a semi-enclosed basin surrounded by eight countries: Iran, Iraq, Kuwait, the Kingdom of Saudi Arabia (KSA), Bahrain, Qatar, the United Arab Emirates (UAE) and Oman's Musandam Peninsula (Figure 18.1). Each of these countries has its exclusive economic zone (EEZ), with Iran's EEZ extending along the entire eastern part of the Gulf. The Gulf is a relatively shallow body of water (average depth ranging between 35 and 60 m) located in an arid region characterized by dry, hot weather (Vaughan et al. 2019). The marine system of the Gulf experiences extreme oceanographic conditions: In some locations, the difference between the mean maximum and minimum temperatures is ~20°C annually; this difference far exceeds what other water bodies experience, which is a difference of 2.3–7.4°C (Hume et al. 2013). Because of the low precipitation and high evaporation levels, the average salinity surpasses 40 ppt in the Gulf (Vaughan et al. 2019; Hume et al. 2013; Hereher 2020). Despite these extreme environmental conditions, the region comprises a suite of productive habitats, such as mudflats, seagrass, coral reefs and mangroves, as well as exploitable natural resources (Vaughan et al. 2019; Grandcourt 2012; Bayani 2016).

Fishery resources in particular are important renewable resources in the Gulf (Sale et al. 2011; Grandcourt 2012). Fisheries bear cultural significance in the region; for example, before the discovery and exploitation of fossil fuels, coastal communities relied on the pearl oyster fishery to maintain their livelihoods (Grandcourt 2012). Archeological records suggest that fishing has been a crucial practice for indigenous people since the sixth millennium BC (Beech 2002). More recently, Gulf fisheries have applied different fishing gears, such as traps, trawling and gillnets, to catch a variety of species, from bottom-dwelling crustaceans to highly migratory fish (Grandcourt 2012; Al-Husaini et al. 2015). The fisheries' supply of seafood has expanded dramatically in recent years. The catch of the artisanal sector, for example, amounted to around half a million tonnes in 2018, a 167% increase over the 1950 production level (Ben-Hasan and Daliri 2022). In addition to strengthening food security in the region, fisheries employ tens of thousands of people; in one of Iran's coastal provinces, ~22,500 fishers depend on the sea for food and income (Daliri et al. 2016).

Despite the cultural, economic and nutritional importance of fisheries in the Gulf, examination of the total Gulf fisheries' catch and management is limited. Here, I first examine the amount of aquatic food produced by wild capture fisheries (or simply fisheries) and aquaculture. I identify the most important fishing gear and targeted groups, as well as the amount of discarded catch. Second, I provide an overview of some primary management regulations applied to conserve fish stock and sustain fisheries, along with discussing core issues in fisheries' sustainability. I conclude with some considerations for sustainable fisheries in the Gulf.

18.2 AQUATIC PRODUCTION

Next, I examine the trend in aquatic food production obtained from fisheries and aquaculture. This, to my knowledge, is the first attempt to determine the magnitude of aquatic food production in the Gulf by combining and comparing trends in fisheries and aquaculture production.

I used the freely accessible Sea Around Us (SAU) database (www.seaaroundus.org) to obtain the reconstructed catch time series for the Gulf fisheries as a marine ecoregion from 1950 to 2018. Generally, the catch reconstructions by SAU follow two principles: 1) unavailable data (i.e., "NA" or "no data") are not turned into zero; 2) instead, an approximate and conservative estimate of catch, based on, for example, the number of fishing vessels and average catch rates, is inserted in all such cases (Pauly and Zeller 2015, 2016). A full account of SAU catch reconstructions is provided by Pauly and Zeller (2015). To gain more insights into the most important fishing countries, fishing gear and exploited taxa, the total catch time series were categorized based on country, applied gear and functional group. The trend of discarded catch was also examined since it can reflect how has the amount of "wasted" catch changed over time.

DOI: 10.1201/9781003321408-24

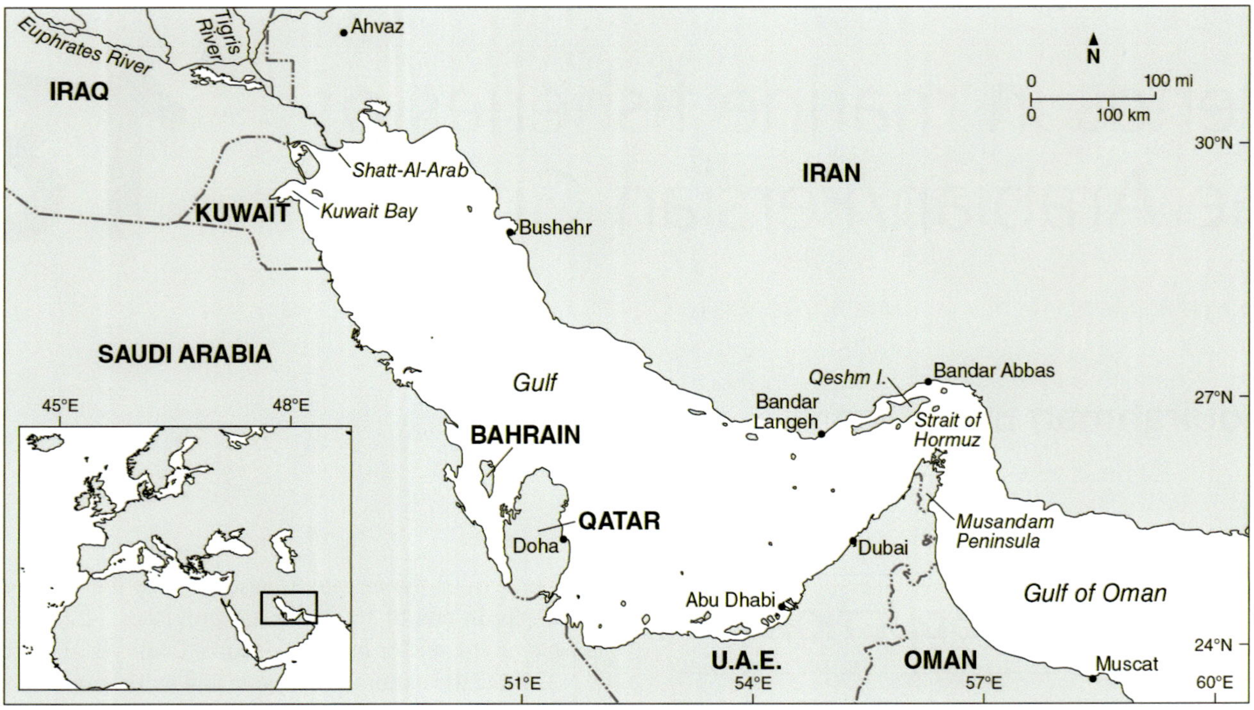

FIGURE 18.1 Map of the Arabian/Persian Gulf.

The aquaculture production for each country was obtained from Ritchie and Roser (2021), which was compiled from the Food and Agriculture Organization of the United Nations through the World Bank.

18.2.1 Fisheries and aquaculture production

The total aquatic production trend from fisheries and aquaculture in the Gulf increased dramatically between 2003 and 2018 and currently amounts to more than a million tonnes (Figure 18.2). Between the 1950s and the mid-2010s, fisheries' catch has consistently been higher than aquaculture production. However, the contribution of fisheries and aquaculture to the total aquatic production in the Gulf is now much more comparable. Indeed, the total aquaculture production in particular underwent remarkable growth between 2000 and 2018, from around 49,000 to 541,000 tonnes in about two decades (Figure 18.2). These growing trends in fisheries and aquaculture production are largely influenced by the production of one country, Iran, which accounted for 66% of total fisheries' catch and 81% of total aquaculture production in 2018 (Figures 18.3A and B). However, fisheries' and/or aquaculture productions are also experiencing a general upward trend in some of the other countries: e.g., Iraq, Oman (Musandam only), Qatar, KSA and the UAE (Figure 18.3A and B).

Generally, fisheries remain the dominant sector in most countries and higher than aquaculture (Figure 18.3B). It seems that aquaculture, at least from a production standpoint, is still developing in the region; however, there are plans by some of the Gulf countries to accelerate aquaculture operations and production (Shams 2007).

18.2.1.1 Fishing gear and exploited taxa

The Gulf fishing fleets fish predominantly with gillnets, traps, lines (e.g., hand lines, long lines), shrimp trawls and encircling nets (Figure 18.4). Collectively, these five types of fishing gear provided 74% of the total fisheries' catch in 2018. Gillnets are the most important fishing gear, contributing about a quarter (26%) of the 2018 total catch. In general, gillnets are widely applied in coastal waters of many countries due to their simplicity, efficiency and relatively low operating costs (Northridge et al. 2017; Valdemarsen 2001). The contribution of some fishing gear has experienced an overall decline (Figure 18.4). For example, the production from trawling increased sharply after the mid-1950s, yielding the largest catch of about 125,000 tonnes in 1988, after which it decreased by half in 2018, likely due to banning or the imposition of various restrictions (Figure 18.4).

The top five exploited functional groups amounted to 387,000 tonnes or 63% of the total production in 2018 (Figure 18.5). Most of the production consisted of demersal fish, followed by large pelagic and reef-associated fish. The total production was mostly composed of medium and large fish; in 2018, more than half (57%) of the fish caught were categorized as medium (30–89 cm) and large (≥ 90 cm). For instance, medium fish, both demersal and reef-associated, as well as large pelagic fish, comprise the largest fraction with growing trends (Figure 18.5). In general, larger fish have a higher market value, and therefore, they are some of the most targeted groups. They are also more vulnerable to a range of fishing gear (Dulvy and Reynolds 2002; Sadovy de Mitcheson et al. 2013; Pauly et al. 2005). It is well recognized that larger fish are the first to be depleted; however, fisheries can maintain or enhance their production by expanding to previously unfished areas (Bhathal and Pauly 2008; Pauly et al. 1998).

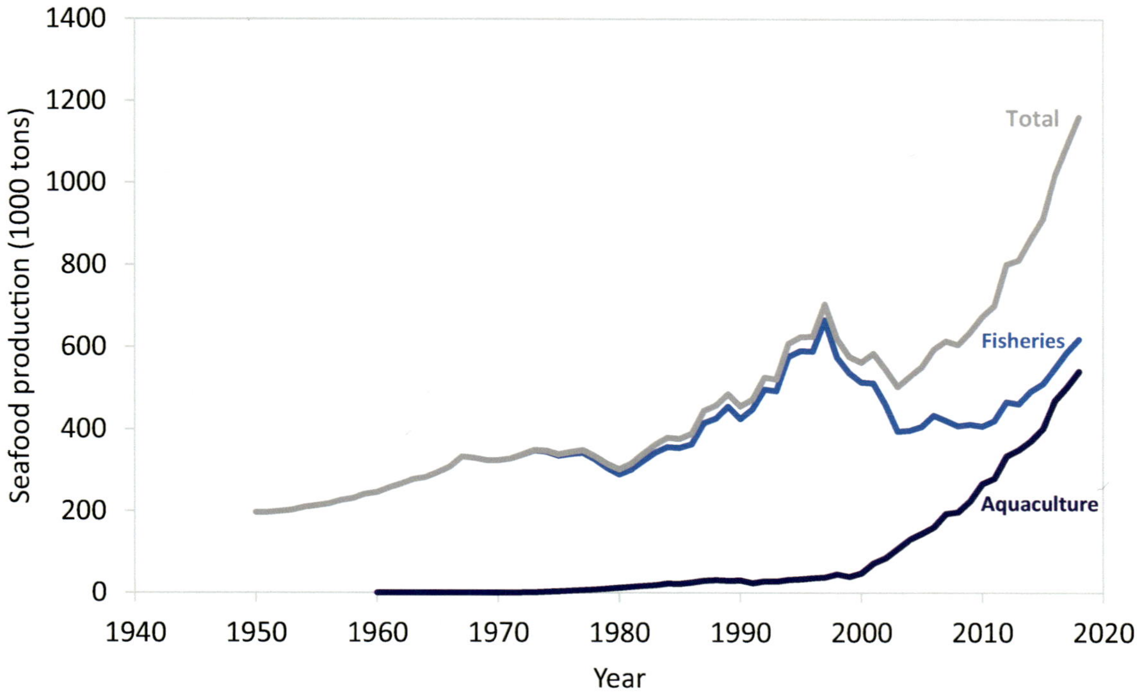

FIGURE 18.2 The total fisheries' and aquaculture production in the Arabian/Persian Gulf.

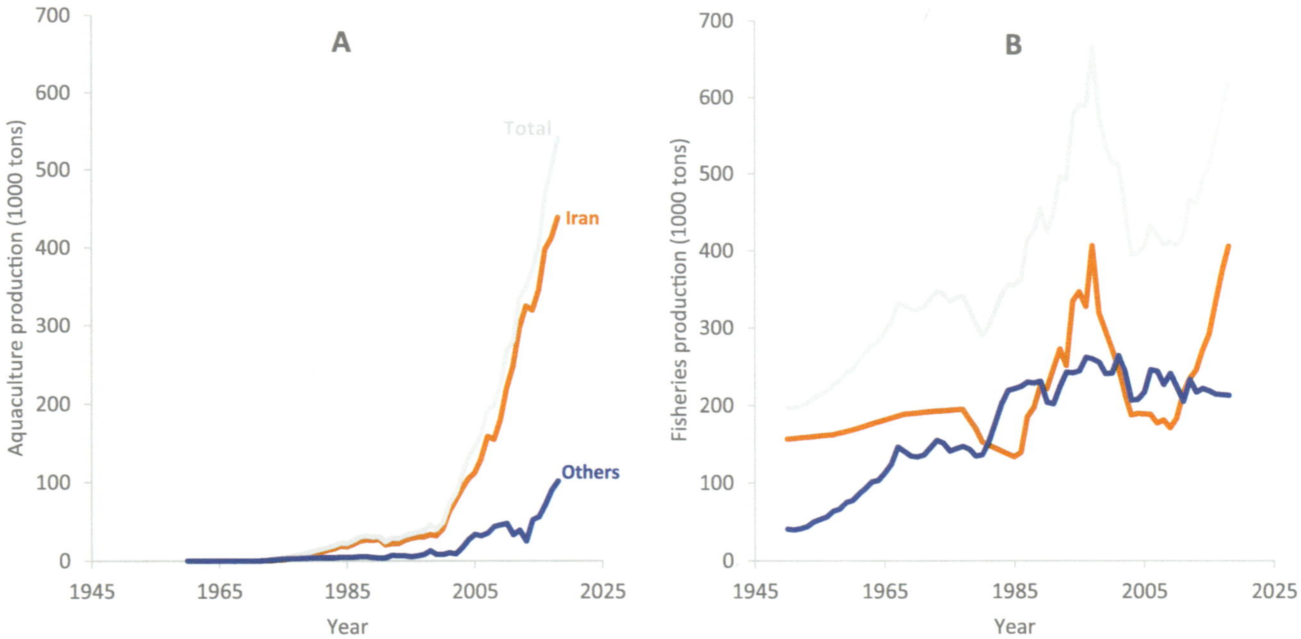

FIGURE 18.3 Fisheries' and aquaculture production in the Arabian/Persian Gulf. (A) Aquaculture production and (B) fisheries production. (The "others" include Bahrain, Iraq, Kuwait, Oman (Musandam only), Qatar, Saudi Arabia and the United Arab Emirates.)

18.2.1.2 Discarded catch

One of the most wasteful human marine practices is the discard of some or all of the production. In the Gulf, the discarding trend has generally declined from 1989 onward (Figure 18.6A). Much of the discard was driven by trawling (shrimp and bottom), which constituted, on average, around 57% the total discards in the last decade (Figure 18.6B). Over recent years, however, the bottom trawl contribution to discards has declined significantly, while the contribution of shrimp trawling has stabilized over the 2007–2018 period after a long-term decline. The decline in discards could be in part due to the ban on trawling in some countries and/or the implementation of some restrictions in countries that did not ban it (GCC 2017). Another reason for the reduced discards over time could be increased market demand for the fish that used to be discarded.

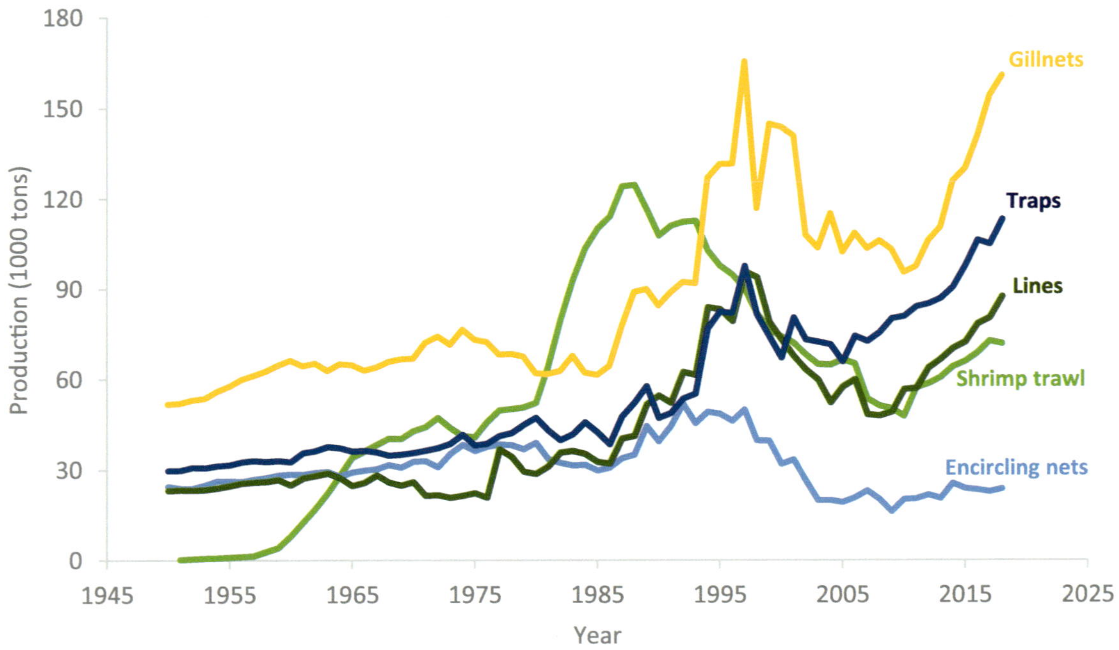

FIGURE 18.4 Fisheries' production by the top five types of fishing gear in the Arabian/Persian Gulf.

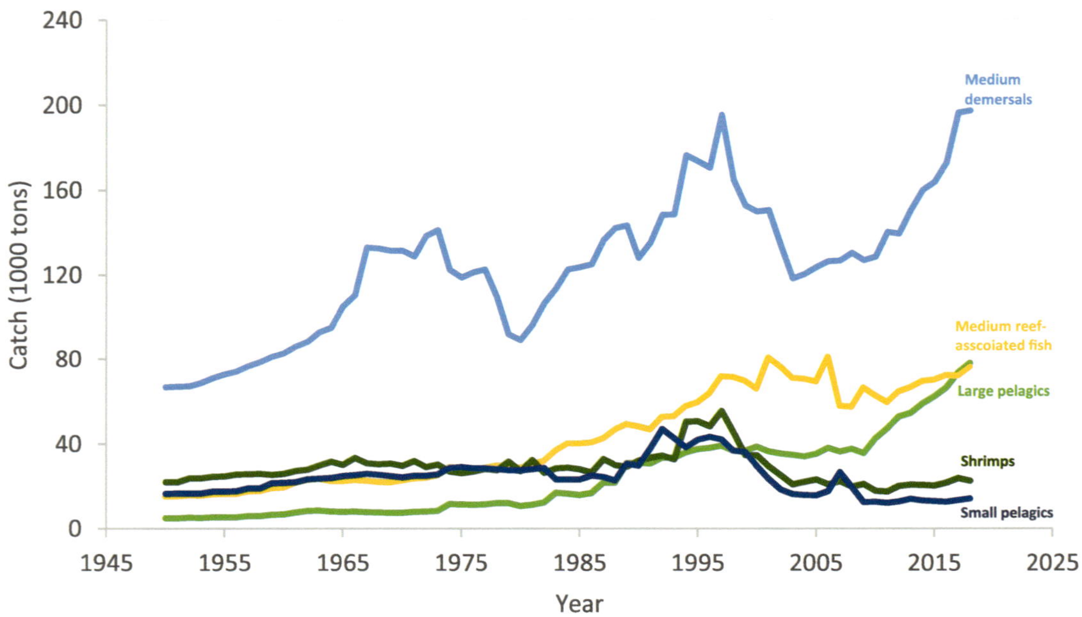

FIGURE 18.5 Fisheries' production by the top five exploited functional groups in the Arabian/Persian Gulf. The size ranges of "Large", "Medium" and "Small" correspond to ≥90 cm, 30–89 cm and <30 cm, respectively.

18.3 COMMON MANAGEMENT REGULATIONS

Some of the most common or increasingly implemented fisheries regulations in the Gulf are based on 1) managing the size selectivity of fishing, 2) marine protected areas (MPAs) and 3) temporal closures (Van Lavieren and Klaus 2013; Siddeek et al. 1999; Al-Husaini et al. 2015; Ben-Hasan and Christensen 2019; GCC 2017).

Size selectivity is commonly managed in the Gulf, including via gear regulations (e.g., gear bans or gear modification through imposing minimum codend and gillnet mesh sizes) and minimum size of capture (i.e., only fish above the minimum size can be retained) (Siddeek et al. 1999; Ben-Hasan et al. 2021). Gear-based regulations in particular are widespread in the Gulf but differ from one country to another (Table 18.1). The use of set gillnets is regulated in Iran, KSA and Kuwait while banned in Qatar and the UAE. In addition, drift gillnets are regulated by specifying mesh sizes in Kuwait while in Bahrain, specific fishing areas are designated for the use of this fishing gear (GCC 2017).

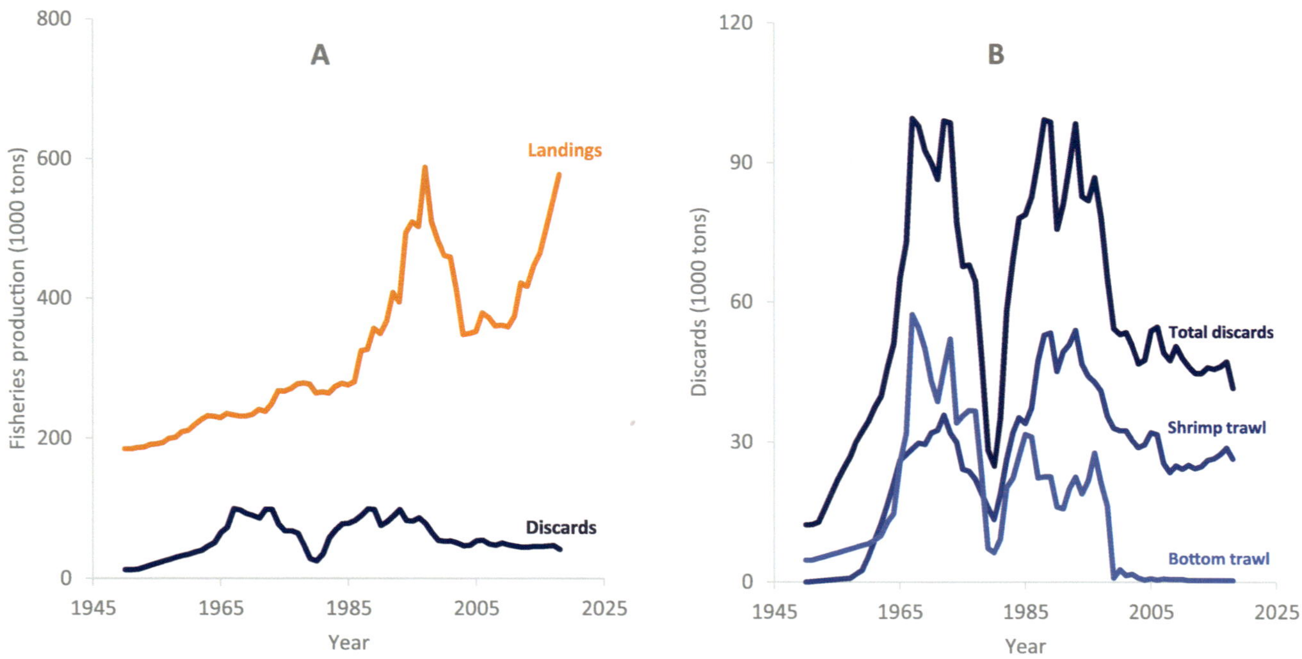

FIGURE 18.6 Fisheries' production in the Arabian/Persian Gulf. (A) Fisheries' production by landings and discards and (B) discards by major fishing gears.

TABLE 18.1 Some common fishing gear and management regulations in the Gulf countries. Green indicates that the fishing gear status is "regulated", grey indicates "allowed", red indicates "banned", white indicates that the fishing gear is not used and dark grey indicates "unregulated". "Regulated" describes fishing gear that is mentioned in the legislation and has laws regulating its use; "unregulated" describes fishing gear that is not mentioned in the legislation; "banned" describes fishing gear that is prohibited from being used by law; and "allowed" implies that fishing gear is mentioned in the legislation, but there are no laws that regulate its use. These descriptions were obtained from GCC (2017) after translation, except for the term "allowed", which did not have a clear description in the reference. More details are found in Ben-Hasan and Daliri (2022). KSA and UAE are Saudi Arabia and the United Arab Emirates, respectively.

COUNTRY	*SURROUNDING NET*	*TRAWL NET*	*DRIFT GILLNET*	*SET GILLNET*	*TRAP*	*STAKE NET*	*TRAMMEL NET*	*LONGLINE*	*BEACH SEINE*
Bahrain	Green	Green	Green	Grey	Green	Green	Green	Green	Red
Iran	Green	Green	Green	Green	Green	Green	White	Green	Green
KSA	White	Green	Red	Green	Green	Green	Red	Grey	Red
Kuwait	Red	Green	Green	Green	Green	Green	Red	Red	Red
Oman*	Green	Red	Green	Dark grey	Green	White	White	Grey	Green
Qatar	Red	Red	Green	Red	Green	Green	Red	Green	Red
UAE	White	Red	Red	Red	Green	Green	Red	Red	Green

*The major fisheries of Oman are in the Sea of Oman and the Arabian Sea

MPAs are implemented in most Gulf countries; however, the fraction of MPAs relative to the territorial waters differs across countries (Figures 18.7A and B). The UAE has expanded its MPA area significantly, from around 4% in 2016 to 11.5% in 2021 (Figure 18.7A); currently, it has the highest fraction of MPA in the region (Figure 18.7B). In contrast, the other countries have much smaller MPAs, which still occupy roughly the same amount of the territorial waters as in 2016, except for Qatar, where the MPA area has expanded from almost zero to 2.3% in a five-year period (Figure 18.7A).

Temporal closures, including closed seasons, are imposed to safeguard the spawning stock of commercially important shrimp and fish species in the Gulf. For example, shrimping is allowed in Kuwait from August/September to January/February but banned outside this period (Al-Husaini et al. 2015). Temporal closure is one of the few measures in the Gulf that is being implemented to manage some internationally shared fish stocks. For instance, all the Gulf countries have recently agreed to ban catching the narrow-barred Spanish mackerel (*Scomberomorus commerson*) between mid-August and mid-October (RECOFI 2021).

18.4 PROBLEMS UNDERMINING SUSTAINABILITY

Ensuring the sustainability of fisheries is crucial for the continued provision of food and employment. Although fish

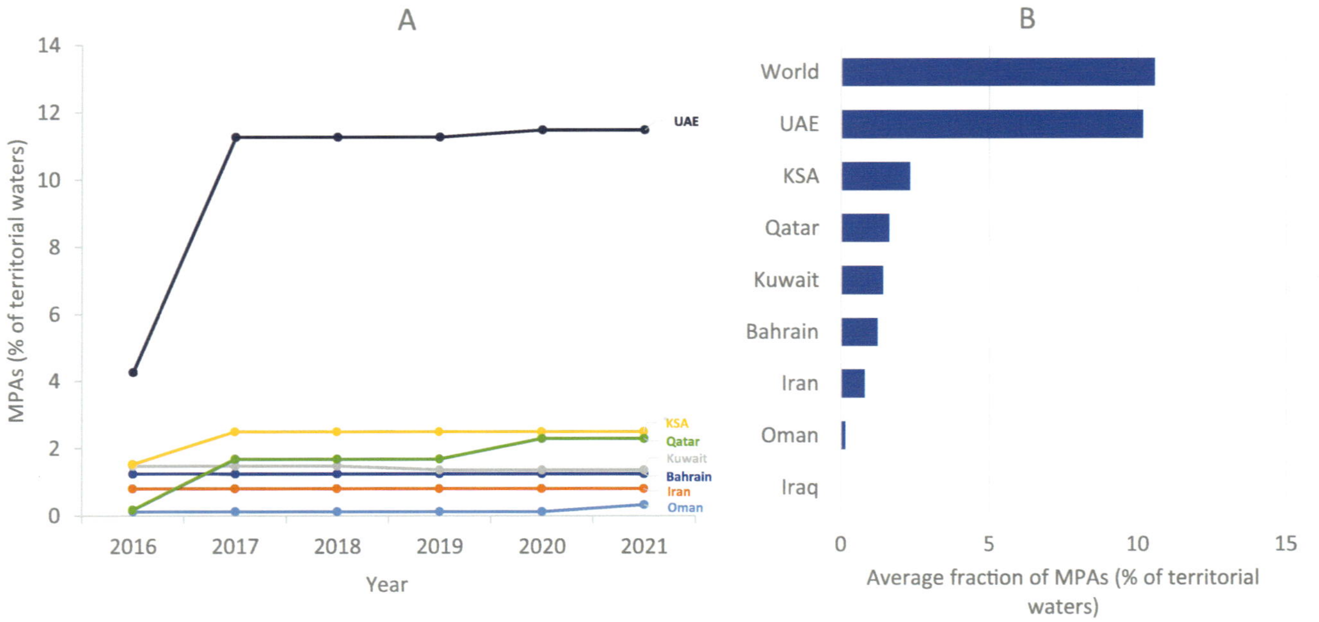

FIGURE 18.7 The fraction of marine protected areas (MPAs) relative to the territorial waters in the Arabian/Persian Gulf. (A) The fraction of MPAs over time, and (B) the average fraction of MPAs between 2016 and 2021. (KSA and UAE are Saudi Arabia and the United Arab Emirates, respectively. The world average fraction of MPAs was calculated for the available period 2016–2018.)

Source: Ritchie and Roser (2021)

production in the Gulf is generally increasing, the trend does not reflect fisheries' sustainability – exploitation rates are also likely climbing, and therefore biomass is expected to decline over time (Pauly et al. 2013; Hilborn et al. 2015; Worm and Branch 2012). Next, I highlight how the sustainability of many fisheries in the Gulf is threatened by overexploitation due to limited or absent fundamental practices underpinning sustainable fisheries (Figure 18.8). Besides the fishing impact on fish stocks, I also discuss the environmental problems affecting fish and fisheries in the Gulf, notably habitat degradation due to coastal development.

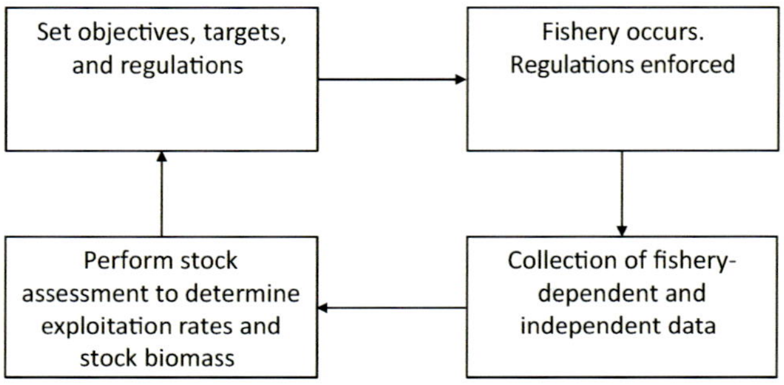

FIGURE 18.8 Outline of the cycle of fisheries management.

Source: Modified from: Hilborn and Hilborn (2019)

18.4.1 Limited fisheries management

There are two basic problems in fisheries management in the Gulf: a lack of routine data collection, which is important for stock assessment procedures, and poor enforcement of regulations (Ben-Hasan et al. 2020; Ben-Hasan and Christensen 2019; Al-Abdulrazzak et al. 2015; Siddeek et al. 1999; Morgan 2004; Jabado and Spaet 2017). The collection of various datasets and conducting of stock assessment, as well as enforcing fisheries regulations, are core elements in the typical cycle of effective fisheries management (Hilborn and Hilborn 2019; Melnychuk et al. 2021) (Figure 18.8).

First, many of the fisheries in the Gulf are poorly assessed or unassessed. If fish stocks are not assessed (e.g., stock biomass and status are unknown), it would be challenging to implement robust management actions like catch limits and rebuilding plans (typically applied as emergency measures that entail fishing bans or reductions in catch limits for depleted stocks) (Melnychuk et al. 2021). Imposing a science-based limit on the total annual catch requires long-term data (e.g., historical information on the total removal of fish and indices of relative abundance) and requires stock assessment models that use such data to estimate a sustainable harvest level. Similarly, the lack of data collection programmes and stock assessment complicates designing and setting rebuilding plans, which, along with catch limits, are essential instruments in sustaining fisheries (Worm et al. 2009; Melnychuk et al. 2012, 2021). Assessments have indicated that some of the Gulf's fish stocks are generally in a precarious condition. For example, studies assessing important finfish stocks in the Gulf, like the Malabar blood snapper, orange-spotted grouper and silver pomfret, suggest that excessive fishing pressure is one of the main drivers of declining catches and abundance.

Second, poor enforcement of fisheries' regulations, which is apparently commonplace, is seriously undermining their sustainability in the Gulf (Siddeek et al. 1999; Morgan 2004; Al-Abdulrazzak et al. 2015; Al-Abdulrazzak and Pauly 2014; Ben-Hasan et al. 2021; Jabado et al. 2018; Jabado and Spaet 2017; Al-Abdulrazzak 2013). For example, illegal fishing occurs during seasonal closures in Bahrain and Kuwait, and undersize fish are likely captured frequently, despite the existence of legal minimum size limits (Siddeek et al. 1999; Ben-Hasan et al. 2021). Additionally, even though several types of fishing gears are banned, some major fisheries continue to use them (Siddeek et al. 1999; Al-Husaini et al. 2015; Jabado and Spaet 2017; Al-Abdulrazzak and Pauly 2014; Richer 2009). For example, although stake nets have been banned in Qatar since 1994, satellite imagery in 2014 revealed that 17 stake nets continued to operate illegally (Al-Abdulrazzak and Pauly 2014). It should be noted that the currently applied management actions can be effective in sustaining food production from fisheries, but the lack of strict enforcement is potentially undermining their benefits (Worm et al. 2009; Prince and Hordyk 2019; Cabral et al. 2019). Size composition data for the Malabar blood snapper in Kuwait shows that fish are caught at much smaller sizes than the legal minimum (Ben-Hasan et al. 2021). Catching small and immature fish in conjunction with high pressure led to growth and recruitment overfishing (Ben-Hasan et al. 2021). In contrast, if high minimum size limits were rigorously enforced, future projections indicate a considerable rebuilding of the stock biomass and catch (Ben-Hasan et al. 2021).

18.4.2 Environmental degradation

The coastal area of the Gulf comprises a range of productive habitats – for example, rocky shores, mudflats, mangrove forests and coral reefs – and some of these habitats are considered important nursery and rearing areas for various commercial fish and invertebrate stocks (Al-Yamani et al. 2021; Vaughan et al. 2019; Edmonds et al. 2021; Lin et al. 2021; Al-Husaini et al. 2015).

Dam construction in the basins of the Tigris and Euphrates Rivers and extensive dredging and land reclamation activities are significant drivers of coastal habitat degradation and modification (Van Lavieren et al. 2011; Ben-Hasan et al. 2017, 2018; Burt 2014; Sheppard et al. 2010; Al-Yamani et al. 2021). In the northern Gulf, the discharge of the Shatt Al-Arab River, formed by the confluence of the Tigris and Euphrates Rivers, creates the largest and most important estuarine system (Sheppard et al. 2010). This system provides critical nursery grounds for several fish and invertebrate species (Al-Husaini et al. 2015). Because of extensive dam construction, however, the flow of the Shatt Al-Arab River has weakened drastically, with salinity levels apparently rising over the last three decades (Sheppard et al. 2010; Lawler 2016; Al-Yamani et al. 2017, 2021). Besides overfishing, the deteriorating discharge of the Shatt Al-Arab, and hence the alteration of the estuarine system, is suspected to be a chief cause driving catch declines of important fish stocks in the northern Gulf (Al-Husaini et al. 2015; Ben-Hasan et al. 2017, 2018).

In addition, the coastal regions of most Gulf countries have undergone extensive changes due to urbanization and industrialization (Burt 2014). A significant part of coastal development projects is dredging and land reclamation activities. In Dubai alone, around 559 million m^3 of marine sediments were dredged for material in the construction of Palm Jumeirah, Palm Jebel Ali and the World projects (Sale et al. 2011). To put this in context, around 55 million m^3 was dredged to build Busan Port (South Korea), the fourth-largest container terminal globally (Van Lavieren et al. 2011). Because coastal developments have often involved little environmental consideration, there has been a fast and widespread degradation of vital coastal habitats (Burt 2014). Dredging and reclamation activities can be environmentally deleterious since they generally involve removing and burying sensitive habitats (Van Lavieren et al. 2011). Such coastal developments have caused severe impacts on many crucial habitats in the Gulf: for example, the fragmentation and degradation of thousands of hectares of seagrass beds, the burial of coral reef habitats and a substantial decline in mangrove forest areas in some countries (Burt 2014; Erftemeijer and Shuail 2012; Burt et al. 2013; Van Lavieren et al. 2011; Spalding 2010).

Overall, the proportion of threatened fish species (8.2%) in the Gulf is twice that reported in other regions, with threats related to coastal development and habitat loss coming second to fisheries (Buchanan et al. 2019).

18.5 MOVING TOWARD SUSTAINABILITY

Aquatic food production is rising in the Gulf; it has recently surpassed a million tonnes a year. Fisheries represent a major sector of seafood production in the Gulf region, and the trend of fish catch is rising. The continued supply of wild-captured fish and the maintenance of numerous livelihoods hinge on the sustainability of fisheries. Yet fisheries management in the Gulf is largely inadequate for fisheries' sustainability; essential data and routine assessments are limited, and enforcement of current regulatory actions is generally weak. Besides poor fisheries management, some critical coastal areas, comprising primary nursery grounds and other sensitive habitats, have experienced substantial modification or degradation.

Typically, the basic elements of fisheries management include 1) establishment of specific objectives and targets for exploitation rates and fish biomass; 2) monitoring of exploitation rates and fish biomass; 3) application of fisheries stock assessment to assess if the targets are being fulfilled; 4) a feedback system to adjust regulations according to the outcomes from the assessment models: for example, reducing exploitation rates if they are excessive; and 5) enforcement of regulations (Hilborn et al. 2015) (Figure 18.8). It worth noting that elements 1 to 4 largely depend on fisheries' stock assessment. While most fisheries in the Gulf are data limited, which complicates the evaluation of the impacts of fishing using traditional stock assessment, there are alternative approaches designed to assess such fisheries. Current data-limited modeling approaches enable fish stock assessment using readily available data (Costello et al. 2012; Martell and Froese 2013; Froese et al. 2017; Thorson et al. 2012; Carruthers and Hordyk 2018). One simple form of such an approach is catch-based models (CBMs), which infer fishing impact on fish stocks by relying on historical catch data and simple life-history information (Martell and Froese 2013; Froese et al. 2017), although management guidance of CBMs for individual stocks may need to be interpreted cautiously (Ovando et al. 2022). Other, more comprehensive data-limited approaches include MSEtool, which takes advantage of the power of the risk-based management strategy evaluation (MSE) to examine different management tactics and underscore the most effective one (Hordyk et al. 2022; Carruthers and Hordyk 2018). In short, a suite of approaches has been developed to assess data-limited fisheries, which are showing increasing promise as cost-effective assessment approaches.

The existing management actions in the Gulf can be adequate to sustain fisheries and protect fish stocks against overexploitation. For example, managing the selectivity of fisheries and implementation of marine protected areas have been identified as essential tool in the management of data- and resource-limited fisheries (Prince and Hordyk 2019; Cabral et al. 2019). Importantly, managing size selectivity and MPAs do not require fundamental changes in the existing management regime, although a sound design of these regulations and strong enforcement are generally necessary. To strengthen enforcement, one effective approach is to involve stakeholders in collaborative management arrangements that consider fishers' goals and enable their participation in developing rules about extractive activities (Worm and Branch 2012; Cinner et al. 2012; Gutiérrez et al. 2011). Studies have highlighted that such arrangements contribute to compliance by fishers (McClanahan et al. 2006, 2011; Gutiérrez et al. 2011).

The rebuilding of fisheries in the UAE is a good case study that can provide critical insights into sound fisheries management in the Gulf region. For around two decades, the UAE has been pursuing best practices in fisheries management, implementing a series of steps to recover overexploited fishery resources through 1) systematic collection of biological and fishery data; (2) routine assessment of fish stocks using, e.g., CMSY (Froese et al. 2017), a data-limited assessment approach, along with the inclusion of traditional fishing knowledge; 3) the launch of a long-term strategy aimed at mitigating overexploitation and attaining sustainable fish stocks; and 4) strict implementation of the strategy through a series of fisheries regulations: e.g., banning or regulating the most productive fishing gear to reduce the catch of overexploited stocks (EAD 2021). In recent years, the average biomass of assessed fish stocks in Abu Dhabi waters has been steadily recovering (EAD 2021). Assessments conducted in 2020 and in previous years indicated that the majority of assessed species (22 out of 32) are sustainably exploited (EAD 2021). The case of fisheries management in the UAE, therefore, underscores the efficacy of strict enforcement of regulations, the feasibility of stock assessment using data-limited approaches and the importance of engaging stakeholders in rebuilding fisheries (EAD 2021).

Habitat modification and deterioration, which are unrelated to fishing practices, are among the biggest threats to fish species in the Gulf (Buchanan et al. 2019). Even if there is good control of fishing pressures, the destruction of essential fish habitats can potentially affect stock productivity and, ultimately, fisheries' production (Brown et al. 2019; Ben-Hasan et al. 2017). Restoration and protection of essential fish habitats are critical, at least for the continued supply of recruitment to fisheries. Restoration activities are costly to conduct, even for fairly accessible habitats like seagrass and coral reefs, and a complete failure of restoration projects may occur (Bayraktarov et al. 2016). However, the obstacle of cost can be partially overcome if restoration efforts target areas with the largest benefits to fisheries' catch (e.g., Gilby et al. 2018). A better understanding of the life-history trajectory of exploited species in the Gulf and evaluating the impacts of habitat alteration on fisheries can be valuable initial steps in the effective design of protection and restoration projects.

REFERENCES

Al-Abdulrazzak, D. 2013. Reconstructing Kuwait's marine fishery catches: 1950–2010. In: *From Dhows to Trawlers: A Recent History of Fisheries in the Gulf Countries, 1950–2010*. Eds. D. Al-Abdulrazzak and D. Pauly, 23–29. Vancouver: Fisheries Centre, University of British Columbia.

Al-Abdulrazzak, D., and D. Pauly. 2014. Managing fisheries from space: Google Earth improves estimates of distant fish catches. *ICES J Mar Sci* 71, no. 3: 450–454.

Al-Abdulrazzak, D., Zeller, D., Belhabib, D., Tesfamichael, D., and D. Pauly. 2015. Total marine fisheries catches in the Persian/Arabian Gulf from 1950 to 2010. *Reg Stud Mar Sci* 2: 28–34.

Al-Husaini, M., Bishop, J.M., Al-Foudari, H.M., and A.F. Al-Baz. 2015. A review of the status and development of Kuwait's fisheries. *Mar Pollut Bull* 100, no. 2: 597–606.

Al-Yamani, F., Polikarpov, I., and M. Saburova. 2021. Northern Gulf marine biodiversity in relevance to the river discharge. In: *Southern Iraq's Marshes: Their Environment and Conservation*. Ed. L.A. Jawad, 379–437. Cham, Switzerland: Springer.

Al-Yamani, F., Yamamoto, T., Al-Said, T., and A. Alghunaim. 2017. Dynamic hydrographic variations in northwestern Arabian Gulf over the past three decades: Temporal shifts and trends derived from long-term monitoring data. *Mar Pollut Bull* 122, nos. 1–2: 488–499.

Bayani, N. 2016. Ecology and environmental challenges of the Persian Gulf. *Iran Stud Bull Soc Iran Cult Soc Stud* 49, no. 6: 1047–1063.

Bayraktarov, E., Saunders, M.I., Abdullah, S., Mills, M., Beher, J., Possingham, H.P., Mumby, P.J., and C.E. Lovelock. 2016. The cost and feasibility of marine coastal restoration. *Ecol Appl* 26, no. 4: 1055–1074.

Beech, M. 2002. Fishing in the 'Ubaid: A review of fish-bone assemblages from Early Prehistoric coastal settlements in the Arabian Gulf. *J Oman Stud* 12: 25–40.

Ben-Hasan, A., Al-Husaini, M., and C. Walters. 2017. Adaptive management of declining fisheries: When is it worth trying to rebuild stocks through fishery regulation? *Mar Policy* 85: 107–113.

Ben-Hasan, A., and V. Christensen. 2019. Vulnerability of the marine ecosystem to climate change impacts in the Arabian Gulf – an urgent need for more research. *Glob Ecol Conservat* 17: e00556.

Ben-Hasan, A., and M. Daliri. 2022. Arabian/Persian Gulf artisanal fisheries: Magnitude, threats, and opportunities. *Rev Fish Biol Fish* November: 1–19. http://doi.org/10.1007/s11160-022-09737-4.

Ben-Hasan, A., Walters, C., Christensen, V., Al-Husaini, M., and H. Al-Foudari. 2018. Is reduced freshwater flow in Tigris-Euphrates Rivers driving fish recruitment changes in the northwestern Arabian Gulf? *Mar Pollut Bull* 129, no. 1: 1–7.

Ben-Hasan, A., Walters, C., Christensen, V., Munro, G., Sumaila, U.R., and A. Al-Baz. 2020. Age-structured bioeconomic model for strategic interaction: An application to Pomfret stock in the Arabian/Persian Gulf. *ICES J Mar Sci* 77, no. 5: 1787–1795. http://doi.org/10.1093/icesjms/fsaa049.

Ben-Hasan, A., Walters, C., Hordyk, A., Christensen, V., and M. Al-Husaini. 2021. Alleviating growth and recruitment overfishing through simple management changes: Insights from an overexploited long-lived fish. *Mar Coast Fish Dynam Manag Ecosyst Sci* 13, no. 2: 87–98.

Bhathal, B., and D. Pauly. 2008. Fishing down marine food webs and spatial expansion of coastal fisheries in India, 1950–2000. *Fish Res* 91, no. 1: 26–34.

Brown, C.J., Broadley, A., Adame, M.F., Branch, T.A., Turschwell, M.P., and R.M. Connolly. 2019. The assessment of fishery status depends on fish habitats. *Fish Fish* 20, no. 1: 1–14.

Buchanan, J.R., Ralph, G.M., Krupp, F., Harwell, H., Abdallah, M., Abdulqader, E., Al-Husaini, M., et al. 2019. Regional extinction risks for marine bony fishes occurring in the Persian/ Arabian Gulf. *Biol Conserv* 230: 10–19.

Burt, J.A. 2014. The environmental costs of coastal urbanization in the Arabian Gulf. *Cityscape* 18, no. 6: 760–770.

Burt, J.A., Feary, D.A., Cavalcante, G., Bauman, A.G., and P. Usseglio. 2013. Urban breakwaters as reef fish habitat in the Persian Gulf. *Mar Pollut Bull* 72, no. 2: 342–350.

Cabral, R.B., Halpern, B.S., Lester, S.E., White, C., Gaines, S.D., and C. Costello. 2019. Designing MPAs for food security in open-access fisheries. *Sci Rep* 9, no. 1: 8033.

Carruthers, T.R., and A.R. Hordyk. 2018. The Data-Limited Methods Toolkit (DLM Tool): An R package for informing management of data-limited populations. Ed: Samantha Price. *Methods Ecol Evol* 9, no. 12: 2388–2395.

Cinner, J.E., McClanahan, T.R., MacNeil, M.A., Graham, N.A.J., Daw, T.M., Mukminin, A., Feary, D.A., et al. 2012. Comanagement of coral reef social-ecological systems. *Proc Natl Acad Sci US* 109, no. 14: 5219–5222.

Costello, C., Ovando, D., Hilborn, R., Gaines, S.D., Deschenes, O., and S.E. Lester. 2012. Status and solutions for the World's unassessed fisheries. *Science* 338, no. 6106: 517–520.

Daliri, M., Kamrani, E., Jentoft, S., and S.Y. Paighambari. 2016. Why is illegal fishing occurring in the Persian Gulf? A case study from the Hormozgan Province of Iran. *Ocean Coast Manag* 120, Supplement C: 127–134.

Dulvy, N.K., and J.D. Reynolds. 2002. Predicting extinction vulnerability in skates. *Conserv Biol* 16, no. 2: 440–450.

EAD. 2021. *Fisheries & Aquaculture Bulletin 2020*. Environment Agency, Abu Dhabi. https://ead.gov.ae/-/media/Project/EAD/EAD/Documents/Resources/Fisheries-and-Aquaculture-Bulletin-2020.pdf.

Edmonds, N.J., Al-Zaidan, A.S., Al-Sabah, A.A., Le Quesne, W.J.F., Devlin, M.J., Davison, P.I., and B.P. Lyons. 2021. Kuwait's marine biodiversity: Qualitative assessment of indicator habitats and species. *Mar Pollut Bull* 163: 111915.

Erftemeijer, P.L.A., and D.A. Shuail. 2012. Seagrass habitats in the Arabian Gulf: Distribution, tolerance thresholds and threats. *Aquat Ecosyst Health Manag* 15, Suppl. 1: 73–83.

Froese, R., Demirel, N., Coro, G., Kleisner, K.M., and H. Winker. 2017. Estimating fisheries reference points from catch and resilience. *Fish Fish* 18, no. 3: 506–526.

GCC. 2017. *Fishing Approaches and Gears in the Cooperation Council for the Arab States of the Gulf (In Arabic)*. Cooperation Council for the Arab States of the Gulf.

Gilby, B.L., Olds, A.D., Peterson, C.H., Connolly, R.M., Voss, C.M., Bishop, M.J., Elliott, M., Grabowski, J.H., Ortodossi, N.L., and T.A. Schlacher. 2018. Maximizing the benefits of oyster reef restoration for finfish and their fisheries. *Fish Fish* 19, no. 5: 931–947.

Grandcourt, E.M. 2012. Reef fish and fisheries in the Gulf. In: *Coral Reefs of the Gulf: Adaptations to Climatic Extremes*. Eds. B.M. Riegl and R.E. Dodge, 127–161. Berlin: Springer.

Gutiérrez, N.L., Hilborn, R., and O. Defeo. 2011. Leadership, social capital and incentives promote successful fisheries. *Nature* 470, no. 7334: 386–389.

Hereher, M.E. 2020. Assessment of climate change impacts on sea surface temperatures and sea level rise – the Arabian Gulf. *Climate* 8, no. 4: 50.

Hilborn, R., Fulton, E.A., Green, B.S., Hartmann, K., Tracey, S.R., and R.A. Watson. 2015. When is a fishery sustainable? *Can J Fish Aquat Sci* 72: 1433–1441.

Hilborn, R., and U. Hilborn. 2019. *Ocean Recovery*. London: Oxford University Press.

Hordyk, A., Huynh, Q., and T. Carruthers. 2022. MSEtool: Management Strategy Evaluation Toolkit. R Package version 3.6.1. *MSEtool*. https://msetool.openmse.com/.

Hume, B., D'Angelo, C., Burt, J., Baker, A.C., Riegl, B., and J. Wiedenmann. 2013. Corals from the Persian/Arabian Gulf as models for thermotolerant reef-builders: Prevalence of Clade C3 Symbiodinium, host fluorescence and ex situ temperature tolerance. *Mar Pollut Bull* 72, no. 2: 313–322.

Jabado, R.W., Kyne, P.M., Pollom, R.A., Ebert, D.A., Simpfendorfer, C.A., Ralph, G.M., S.S. Al Dhaheri, et al. 2018. Troubled waters: Threats and extinction risk of the sharks, rays and chimaeras of the Arabian Sea and adjacent waters. *Fish Fish* 67: 1043–1062.

Jabado, R.W., and J.L.Y. Spaet. 2017. Elasmobranch fisheries in the Arabian Seas region: Characteristics, trade and management. *Fish Fish* 18, no. 6: 1096–1118.

Lawler, A. 2016. Iraq confronts a new enemy – the Persian Gulf. *Science* 351, no. 6269: 111–112.

Lin, Y.-J., Roa-Ureta, R.H., Pulikkoden, A.R.K., Premlal, P., Nazeer, Z., Qurban, M.A., and L. Rabaoui. 2021. Essential fish habitats of demersal fish in the western Arabian Gulf. *Mar Pollut Bull* 173, Pt A: 113013.

Martell, S., and R. Froese. 2013. A simple method for estimating MSY from catch and resilience. *Fish Fish* 14, no. 4: 504–514.

McClanahan, T.R., Graham, N.A.J., MacNeil, M.A., Muthiga, N.A., Cinner, J.E., Bruggemann, J.H., and S.K. Wilson. 2011. Critical thresholds and tangible targets for ecosystem-based management of coral reef fisheries. *Proc Natl Acad Sci US* 108, no. 41: 17230–17233.

McClanahan, T.R., Marnane, M.J., Cinner, J.E., and W.E. Kiene. 2006. A comparison of Marine protected areas and alternative approaches to coral-reef management. *Curr Biol* 16, no. 14: 1408–1413.

Melnychuk, M.C., Essington, T.E., Branch, T.A., Heppell, S.S., Jensen, O.P., Link, J.S., Martell, S.J.D., Parma, A.M., Pope, J.G., and A.D.M. Smith. 2012. Can catch share fisheries better track management targets? *Fish Fish* 13: 267–290.

Melnychuk, M.C., Kurota, H., Mace, P.M., Pons, M., Minto, C., Osio, G.C., Jensen, O.P., et al. 2021. Identifying management actions that promote sustainable fisheries. *Nat Sustain* 4: 440–449. http://doi.org/10.1038/s41893-020-00668-1.

Morgan, G. 2004. Country review: Iran (Islamic Republic Of). In: *Review of the State of World Marine Capture Fisheries*. FAO Fisheries Technical Paper 488. https://books.google.com/books?hl=en&lr=&id=_7JD1V3PijUC&oi=fnd&pg=PA237&dq=enforce+regulations+iran+fisheries&ots=ADcbym7HMd&sig=5VjdSQJBOKERcVpO_7SkDG54zLo.

Northridge, S., Coram, A., Kingston, A., and R. Crawford. 2017. Disentangling the causes of protected-species bycatch in gillnet fisheries. *Conserv Biol* 31, no. 3: 686–695.

Ovando, D., Free, C.M., Jensen, O.P., and R. Hilborn. 2022. A history and evaluation of catch-only stock assessment models. *Fish Fish* 23, no. 3: 616–630.

Pauly, D., Christensen, V., Dalsgaard, J., Froese, R., and F. Torres. 1998. Fishing down marine food webs. *Science* 279, no. 5352: 860–863.

Pauly, D., Hilborn, R., and T.A. Branch. 2013. Fisheries: Does catch reflect abundance? *Nature* 494, no. 7437: 303–306.

Pauly, D., Watson, R., and J. Alder. 2005. Global trends in world fisheries: Impacts on marine ecosystems and food security. *Phil Trans Roy Soc London. Series B, Biol Sci* 360, no. 1453: 5–12.

Pauly, D., and D. Zeller. 2015. Catch reconstruction: Concepts, methods and data sources. *Sea Around Us*. www.seaaroundus.org.

Pauly, D., and D. Zeller. 2016. Catch reconstructions reveal that global marine fisheries catches are higher than reported and declining. *Nat Comm* 7: 10244.

Prince, J., and A. Hordyk. 2019. What to do when you have almost nothing: A simple quantitative prescription for managing extremely data-poor fisheries. *Fish Fish* 20, no. 2: 224–238.

RECOFI. 2021. *Working Group on Fisheries Management Twelfth Meeting*. FAO. www.fao.org/fileadmin/user_upload/faoweb/FI_Meetings/RECOFI/2021/RECOFI_WGFM12_2_OCTOBER_2021.pdf.

Richer, R. 2009. *Conservation in Qatar: Impacts of Increasing Industrialization*. Centre for International and Regional Studies, Georgetown University. https://repository.library.georgetown.

edu/bitstream/handle/10822/558296/CIRSOccasionalPaper2 ReneeRicher2009.pdf?sequence=5&isAllowed=y.

Ritchie, H., and M. Roser. 2021. Biodiversity. *Our World in Data*, April. https://ourworldindata.org/biodiversity.

Sadovy de Mitcheson, Y., Craig, M.T., Bertoncini, A.A., Carpenter, K.E., Cheung, W.W.L., Choat, J.H., Cornish, A.S., et al. 2013. Fishing groupers towards extinction: A global assessment of threats and extinction risks in a billion dollar fishery. *Fish Fish* 14, no. 2: 119–136.

Sale, P.F., Feary, D.A., Burt, J.A., Bauman, A.G., Cavalcante, G.H., Drouillard, K.G., Kjerfve, B., et al. 2011. The growing need for sustainable ecological management of marine communities of the Persian Gulf. *Ambio* 40, no. 1: 4–17.

Shams, A.J. 2007. *Status of Aquaculture in Gulf Cooperation Council Countries*. www.raisaquaculture.net/uploads/media/Status%20of%20Aquaculture%20in%20Gulf%20Cooperation%20Council%20Countries-April07.pdf.

Sheppard, C., Al-Husiani, M., Al-Jamali, F., Al-Yamani, F., Baldwin, R., Bishop, J., Benzoni, F., et al. 2010. The Gulf: A young sea in decline. *Mar Pollut Bull* 60, no. 1: 13–38.

Siddeek, M.S.M., Fouda, M.M., and G.V. Hermosa. 1999. Demersal fisheries of the Arabian Sea, the Gulf of Oman and the Arabian Gulf. *Estuar Coast Shelf Sci* 49: 87–97.

Spalding, M. 2010. *World Atlas of Mangroves*. London, England: Routledge.

Thorson, J.T., Branch, T.A., and O.P. Jensen. 2012. Using model-based inference to evaluate global fisheries status from landings, location, and life history data. *Can J Fish Aquat Sci* 69, no. 4: 645–655.

Valdemarsen, J.W. 2001. Technological trends in capture fisheries. *Ocean Coast Manag* 44, no. 9: 635–651.

Van Lavieren, H., Burt, J., Feary, D.A., Cavalcante, G., Marquis, E., Benedetti, L., Trick, C., Kjerfve, B., and P.F. Sale. 2011. Managing the growing impacts of development on fragile coastal and marine ecosystems: Lessons from the Gulf. *United Nations University Institute for Water, Environment, and Health*. https://collections.unu.edu/eserv/UNU:2919/policyreport_lessonsfromthegulf.pdf.

Van Lavieren, H., and R. Klaus. 2013. An effective regional Marine Protected Area network for the ROPME Sea area: Unrealistic vision or realistic possibility? *Mar Pollut Bull* 72, no. 2: 389–405.

Vaughan, G.O., Al-Mansoori, N., and J.A. Burt. 2019. The Arabian Gulf. In: *World Seas: An Environmental Evaluation*. Ed. C. Sheppard, 1–23. Cambridge: Academic Press.

Worm, B., and T.A. Branch. 2012. The future of fish. *Trends Ecol Evol* 27, no. 11: 594–599.

Worm, B., Hilborn, R., Baum, J.K., Branch, T.A., Collie, J.S., Costello, C., Fogarty, M.J., et al. 2009. Rebuilding global fisheries. *Science* 325, no. 5940: 578–585.

PART 7

Mangroves and the coastal ecosystem

Coastal dredging and reclamation in the Arabian Gulf

Impacts and management

19

Humood A. Naser and Khalil H. Abdulla

19.1 INTRODUCTION

Owing to the growing population and increasing economic development, coastal urbanization is rapidly expanding worldwide (Wenger et al. 2018). Coastal urbanization is globally considered one of the main driving forces for environmental change (UN 2021). Coastlines have been extensively modified by activities like port development, seabed mining and land reclamation (Marin-Anton et al. 2016). Dredging and reclamation activities are typically associated with most of the infrastructure developments along coastlines (Wenger et al. 2017). Dredging involves the excavation and relocation of sediments from the seabed and other water bodies. The primary purposes for dredging include creating and maintaining waterways, utilizing dredged material for beach nourishment and coastal reclamation, mining of sand and gravel and removing contaminants from sediment during restoration projects (Salomon and Markus 2018). Reclamation is the process of creating new land from the sea by filling the area with large amounts of material, including heavy rocks and sediment (Stauber et al. 2016). The reclaimed area can be an expansion of the coastline or an artificial island (Chee et al. 2017). Reclamation is conducted globally to accommodate infrastructure, industrial, recreation and tourism projects (Sengupta et al. 2018).

Coastal habitats are ecologically and economically important ecosystems (Seitz et al. 2014). Dredging and reclamation activities can have an adverse impact on coastal and marine environments. Direct effects of dredging and reclamation projects may emerge as physical removal of substratum and associated marine organisms from the seabed and burial due to deposition of dredged material (Erftemeijer and Lewis 2006). Additionally, these activities may increase turbidity and sedimentation, reduce light availability, smother benthic communities, alter bathymetry and release contaminants and nutrients (Fraser et al. 2017; Trevisan et al. 2020). Dredging and reclamation activities can affect several ecosystems and commercially important species, including seagrasses (Erftemeijer and Lewis 2006), mangroves (Aljenaid et al. 2022a), coral reefs (Erftemeijer et al. 2012), fish and fisheries (Wenger et al. 2017), waterbirds (Ma et al. 2019) and marine mammals (Todd et al. 2015).

The Arabian Gulf is a recognized region with extensive coastal development (Nasser et al. 2014; Marin-Anton et al. 2016; AlQahtany et al. 2022). Rapid economic, industrial and social developments coupled with rapid population growth in the Arabian Gulf have intensified dredging and reclamation activities along the coastlines (Burt and Bartholomew 2019). Large-scale coastal developments are created by dredging and reclamation along the coastlines of the Arabian Gulf for residential, economic and recreational purposes. Examples of these mixed-use developments include Tarut Bay and Half Moon Bay (Saudi Arabia), the Pearl (Qatar), the Palm Islands and the World developments (Dubai, UAE), Amwaj Islands and Durrat Al Bahrain (Bahrain) and Sabah Al Ahmed City (Kuwait) (Burt 2014; Vaughan et al. 2019; Afzal et al. 2022). Reclamation is typically conducted by extraction of sand and mud from designated borrow areas, which are then dumped into shallow subtidal areas along the coastline or at offshore sites to create artificial islands (Naser 2014).

The Arabian Gulf supports a variety of habitats that provide invaluable ecosystem goods and services, including mangrove forests, sabkha and salt marshes, seagrass meadows, coral reefs and oyster beds (Vaughan et al. 2019). Dredging and reclamation activities have been reported as key causes of degradation in coastal and marine habitats in the Arabian Gulf (Al-Madany et al. 1991; Abahussain et al. 2005; Burt 2014).

DOI: 10.1201/9781003321408-26

Achieving a balance between conservation of coastal and marine ecosystems and desired coastal development is a challenge that must be addressed through evidence-based decision making (Wenger et al. 2018; Trevisan et al. 2020). Therefore, development of management procedures to prevent or mitigate environmental degradation associated with dredging and reclamation activities requires measures to holistically characterize the impact of these activities. Hence, the aim of this chapter is to construct a holistic view of dredging and reclamation in the Arabian Gulf countries. It investigates the drivers and impacts of dredging and reclamation, as well as their management in the region. The objectives of the present chapter are 1) to examine the different reasons behind undertaking dredging and reclamation, 2) to investigate the reported impacts of dredging and reclamation and the linkages among them; and 3) to investigate the measures that have been taken to address the impacts of dredging and reclamation.

19.2 STUDY APPROACH

In order to achieve the objectives, relevant peer-reviewed journal articles, edited chapters and conference papers were reviewed. A search on SCOPUS was carried out in February 2022 using relevant keywords including, but not limited to, names of the Gulf Cooperation Council (GCC) countries, "reclamation", "dredging", "infill" and "artificial island". The search resulted in 412 research items. Abstracts of these research items were carefully read to screen out irrelevant items. Items that fulfilled at least one of the following criteria were included in this review: 1) the manuscript deals with environmental or socio-economic aspects of coastal reclamation or dredging in the GCC countries, 2) the manuscript presents empirical evidence on the adverse impacts of coastal reclamation or dredging in the GCC countries or 3) the manuscript describes actions/measures to mitigate or prevent the adverse impacts of coastal reclamation or dredging in the GCC countries.

After the screening process, 63 items published between 1980 and 2021 were found to be eligible for a full-text review and represented the basis of this review. To ensure that this review is evidence based, a great deal of attention was paid to studies supported by empirical evidence on the impacts or effectiveness of management actions. It is worth mentioning that these kinds of studies are limited and do not cover all the aspects of dredging and reclamation in the GCC countries. It is also important to note that this review might not be fully comprehensive because some impacts or management actions might not be published.

19.3 THE NEED FOR DREDGING AND RECLAMATION IN THE ARABIAN GULF

The main drivers for dredging and reclamation are summarized in Figure 19.1. Generally, dredging is practised in the Arabian Gulf for navigational purposes or for sand extraction or mining. Dredging is needed to deepen or widen waterways to facilitate navigation and to maintain channels, ports and jetties (Neelamani and Al-Hulail 2011; Al Hosani 2014). It is also an important means of extracting sand needed for reclamation projects and other construction purposes (Zainal et al. 2012). For instance, 186.5 million m^3 of sand were needed for reclamation works for the Palm Jumeirah in the UAE (Peduzzi 2014). Other purposes of dredging include improving water circulation (Shanableh et al. 2003), trenching for submarine pipelines and cables (Lo 1987) and constructing underwater intakes and outfalls (Neelamani and Al-Hulail 2011).

Population growth, rapid urbanization and economic development are the main drivers for reclamation in the Arabian Gulf (Figure 19.1). Reclamation may seem inevitable in small island developing states (SIDS) such as Bahrain, which suffers from severe land scarcity, with a very small land area (less than 800 km^2) and a fast-growing population of over 1.5 million (Abdulla and Naser 2021). The severe land scarcity has led Bahrain to build multiple cities on reclaimed land to provide residential units for the public alongside the many artificial islands built by the private sector that offer a range of real-estate products. Some of Bahrain's major industrial facilities, ports, commercial buildings, luxurious recreational areas, malls, highways and causeways were constructed over reclaimed land (Al-Madany et al. 1991; Zainal et al. 2012). With the same pace of development, it was expected that

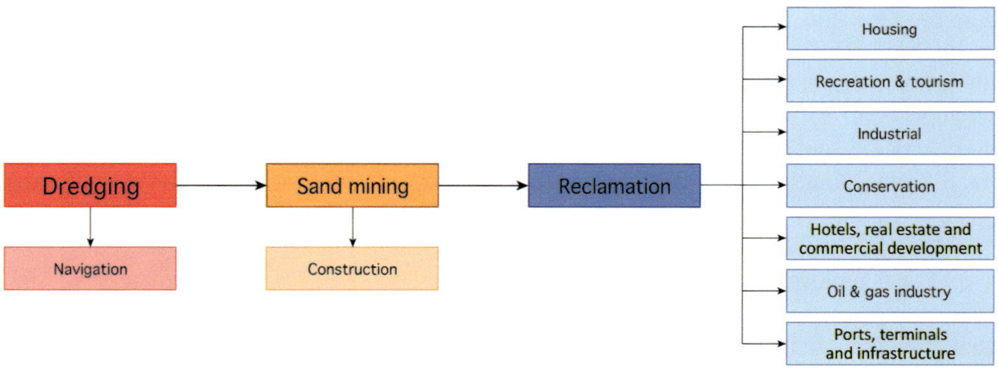

FIGURE 19.1 The main drivers of dredging and reclamation in the Arabian Gulf.

intensive reclamation of coastal and marine areas would continue in Bahrain (Naser 2015a).

The case in larger, less populated countries is different. These countries may choose to reclaim coastal areas to accommodate projects that are difficult or impossible to build inland, such as seawater desalination plants, ports and terminals (Yagoub and Kolan 2006; Al Hosani 2014). In addition, some countries use artificial islands for oil and gas production instead of marine platforms because such islands offer more flexibility and pose lower risks (Modavi et al. 2008). Interestingly, artificial islands may also be utilized for conserving wildlife. For instance, an artificial island in Al Wathba Wetland Reserve in the UAE is a dedicated breeding site for some marine birds (Khan et al. 2017).

19.4 ADVERSE IMPACTS OF DREDGING AND RECLAMATION IN THE ARABIAN GULF

Adverse impacts of dredging and reclamation can be classified as primary, secondary or long term. They include environmental and socioeconomic impacts (Figure 19.2). Other environmental pressures such as climate change, overfishing and release of pollutants may increase the severity of these impacts.

19.4.1 Environmental impact

19.4.1.1 Direct habitat loss

Dredging and reclamation activities have been considered a prime cause for the loss and degradation of coastal and marine habitats in the Arabian Gulf, including reducing mangrove swamps (Abido et al. 2011; Aljenaid et al. 2022a), fragmenting seagrass beds (Erftemeijer and Shuail 2012), smothering coral reefs (Burt et al. 2013a), shrinking sabkha flats (Lokier 2013) and altering productive mudflats (Al-Sayed et al. 2008; Van Lavieren et al. 2011).

Mangrove habitats in the Arabian Gulf occupy around 165 km^2 (Almahasheer 2018). Coastal reclamation has resulted in altering several mangrove areas along the Arabian Gulf (Friis and Burt 2020). Several studies have utilized remote-sensing techniques to monitor the spatial and temporal changes in mangrove habitats in the Arabian Gulf. Although some studies reported an increase in the spatial coverage of mangroves due to conservation and afforestation measures (Almahasheer 2018; Li et al. 2019; Elmahdy et al. 2020), substantial reductions in mangrove habitats have been detected along the coastlines of the Arabian Gulf (Aljenaid et al. 2022a). Khan and Kumar (2009) compared the coverage of mangroves in selected sites along the coastlines of the Arabian Gulf between 1971 and 2001 and revealed substantial losses of mangroves in 60% of the sites that were attributed primarily to coastal development, in comparison with 40% of undisturbed sites that showed a considerable increase in coverage. In Bahrain,

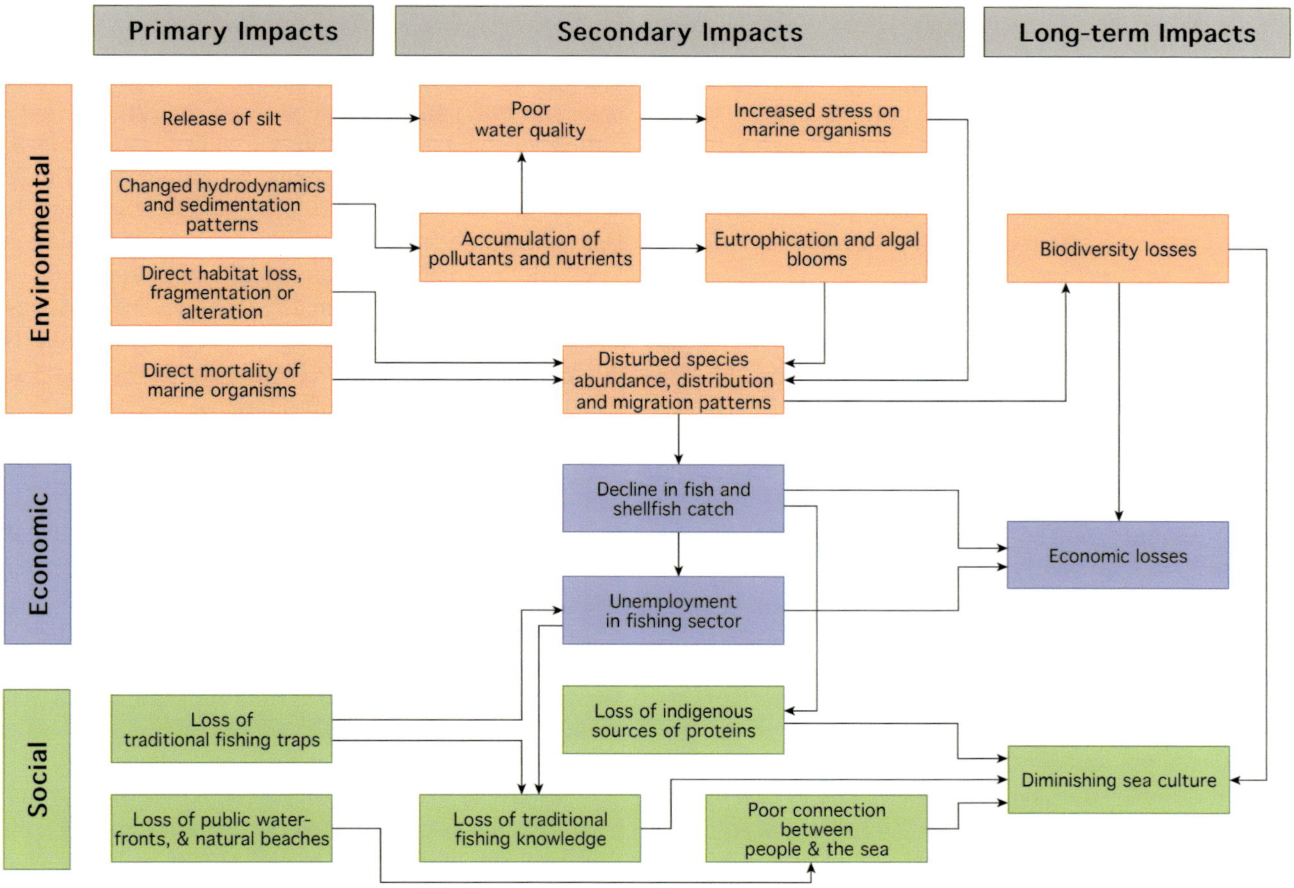

FIGURE 19.2 Impacts of dredging and reclamation on the environment and society in the Arabian Gulf.

mangroves occur naturally in Tubli Bay, which is a sheltered and shallow bay located in the northeast of the country (Abou Seedo et al. 2017). The marine area of Tubli Bay has been substantially reduced from 25 to 12 km^2 over the last 50 years due to intensive coastal reclamation, which resulted in the reduction of the spatial distribution of mangrove stands along the bay (Abahussain and Al-Sabbagh 2010; Naser 2016). Aljenaid et al. (2022a) investigated the spatial changes of mangroves in Bahrain during the period 1967–2020 using remote-sensing techniques and showed a significant decline in the coverage of mangroves in Tubli Bay; around 85% of the natural mangroves were lost since 1967 (declining from 328 to 48 ha) due to coastal reclamation.

Seagrass habitats in the Arabian Gulf support the region's productive fisheries. Losses of seagrass habitats have been reported in different areas of the Arabian Gulf. For instance, an estimated loss of 1,020 ha of seagrass habitat off the eastern coastline of Bahrain was reported by Zainal et al. (1993) during the period 1985–1992. Additionally, an estimated total cumulative loss of major habitats due to coastal reclamation in Bahrain, including seagrass and algal beds, was around 15,358 ha (Zainal et al. 2012). In Tarut Bay, Saudi Arabia, a loss of 6,446 ha of coastal habitats, including seagrass areas, was detected during the period 1973–1990 (Al-Thukair et al. 1995). Additionally, dredging of shipping channels and construction of causeways have resulted in a considerable fragmentation in nearshore seagrass beds (Erftemeijer and Shuail 2012).

Coral reefs have been buried due to coastal reclamation and the construction of artificial islands in the Arabian Gulf (Erftemeijer et al. 2012; Vaughan et al. 2019). For instance, remote-sensing imagery detected a loss of 22 ha of coral and a degradation of an additional 8 ha in the largest reef in Bahrain (Fasht Al-Adhm) due to increased coastal development in the period 1985–1992 (Zainal et al. 1993). Additionally, a loss of 2,327 ha of nearshore coral habitats due to reclamation activities was reported in the period 2004–2014 in Bahrain. These habitats were classified as mixed sand and coral, rock and sand with coral, and geological-aged coral reefs (fashts) (Naser 2015b). Sheppard et al. (2010) reported a loss of large stretches of coral reefs in the Arabian Gulf during the period 1990–2008 that was attributed to large-scale projects. Similarly, ROPME (2020) reported losses of coral cover in fringing reefs due to the construction of infrastructure projects and associated dredging and reclamation activities in the Arabian Gulf. The proximity of populations to coral reefs can cause detrimental changes to water quality by coastal modification, sewage and industrial effluents, leading to loss or degradation of coral reefs. According to a global assessment for people living by coral reefs, Bahrain ranked the highest in 2020, followed by the United Arab Emirates, when considering population densities within 5 km of a coral reef (Wong et al. 2022).

Coastal habitats harbour a great variety of flora and fauna that are significantly affected by coastal modification through direct physical removal or burial by dredging and reclamation activities (Godson et al. 2022). Nearly two-thirds of the sabkha habitats in the Arabian Gulf were lost as a result of direct physical destruction and indirect changes due to channelization caused by tidal flushing (Lokier 2013). Coastal reclamation was attributed to the loss of sabkha and salt marsh areas in Saudi Arabia during the period 1991–2000, with a decrease of 2,143.2 ha and 257.8 ha, respectively (Al-Otaibi et al. 2006). Similarly, Aljenaid et al. (2022b) reported an annual decreasing trend of 2.22% in wetland/sabkha areas in Bahrain during the period 1986–2020 due to coastal reclamation. Additionally, the biodiversity and abundance of macrobenthos are severely affected by mortality and smothering associated with mud burial in the reclaimed areas in the Arabian Gulf. For instance, the response of three common macrobenthic organisms to mud burial using marine sediment collected from a designated borrow area revealed significant impact, with the highest mortality occurring among gastropods, followed by bivalves and polychaetes (Naser 2011).

Habitat destruction and modification pose the greatest threat to coastal and marine diversity (Luypaert et al. 2020). Therefore, habitat loss due to dredging and reclamation activities is considered a primary cause of the loss of biological diversity, including species, genetic variation and ecosystem services in the Arabian Gulf.

19.4.1.2 Gradual habitat degradation

Dredging and reclamation activities may increase turbidity and sedimentation, reduce light availability, alter bathymetry and hydrodynamics and release contaminants and nutrients (Fraser et al. 2017). These physical and chemical changes can cause a gradual degradation in coastal and marine ecosystems (Trevisan et al. 2020). The Arabian Gulf has witnessed intensive dredging activities that are associated with elevated levels of sedimentation and turbidity. For instance, the construction of Jebel Ali Port in Dubai, UAE, required the disposal of 110 million m^3 of sand and mud material (Summers and Fleming 1980), and the construction of the recreational waterfront in the city of Kuwait resulted in adding more than 300,000 m^3 of sand to the beach zone (Kana et al. 1986). Similarly, construction of a pipeline in the Shuaiba area of Kuwait required dredging the sea bottom and dumping 350,000 m^3 in offshore areas (Lo 1987). In Saudi Arabia, more than 200 million m^3 of sediments were removed during the construction of Jubail City (IUCN/MEPA 1987), and around 60 million m^3 of mud and sand were dredged during the construction of the King Fahad Causeway between Saudi Arabia and Bahrain (Al-Ghadban and Price 2002). Additionally, around 327 million m^3 of dredged material was used to reclaim 62 km^2 along the coastlines and subtidal areas of Bahrain during the period 2004–2014 (Naser 2015a). Further, more than 6.2 million m^3 of dredged material was used as a landfill for Hamad International Airport in Qatar (Al-Naimi et al. 2018). Such activities have contributed to the degradation of coastal and marine ecosystems in the Arabian Gulf (Vaughan et al. 2019). Coral reefs have been degraded due to increased levels of sedimentation and turbidity (ROPME 2020). For instance, Burt et al. (2013a) surveyed the main coral reefs in Bahrain and reported a substantial degradation that was attributed to frequent bleaching events and sedimentation associated with coastal development. AlQahtany et al. (2022) analyzed reclamation projects in Dammam, Saudi Arabia, during the period 2000–2020 and reported a degradation in mangrove swamps and nearshore seagrass beds.

Suspended sediments from dredging activity can smother benthic habitats, leading to physical and chemical alterations that can cause a reduction in biodiversity and abundance in the associated macrobenthos. Oyster beds located near coastal reclamation activities showed a significantly lower number of fouling species associated with pearl oysters in Qatar (Smyth et al. 2016). Similarly, elevated suspended sediments were identified as a main factor that severely degraded pearl oysters in areas associated with dredging and reclamation activities in Kuwait (Al-Yamani et al. 2020). Additionally, lower levels of biodiversity of macrobenthic invertebrates in reclaimed coastal areas and on an artificial island have been detected in comparison with those of a natural mudflat in Bahrain (Naser 2022).

Reclamation changes hydrodynamics around newly developed areas, which can create favourable conditions for sedimentation and pollutant accumulation due to alteration of the tidal cycle in some parts of the development (Cavalcante et al. 2012). Additionally, dredging and reclamation can increase nutrients' mobility. Higher concentrations of silicate were reported in reclaimed areas (average 3.73 mg l^{-1}) than in a natural mudflat (average 0.79 mg l^{-1}) in Bahrain (Naser 2022).

Dredging and coastal reclamation activities accompanied by other stressors, including climate change, sewage, industrial pollution and overexploitation of marine resources, have contributed to the degradation of important coastal ecosystems such as sabkhas, mudflats and mangroves (Vaughan et al. 2019) and coral reefs (Lincoln et al. 2021) and the depletion of many fish populations (Grandcourt 2012). This degradation can affect marine biodiversity in the Arabian Gulf.

19.4.1.3 Implications of habitat loss and degradation

Habitat loss and ecosystem degradation associated with dredging and reclamation activities have disturbed the breeding, feeding and migration patterns of endangered species, including dugongs and sea turtles, and many commercial crustaceans, finfish and shellfish, as well as wader birds. Seagrass habitats in the Arabian Gulf serve as a primary food source for the world's second-largest population of dugongs and the endangered green turtles. Degradation and fragmentation of seagrass beds due to the construction of bridges and dredging waterways can affect dugong and turtle populations and reduce their foraging habitats (Erftemeijer and Shuail 2012; Pilcher et al. 2015). Additionally, coastal modification can reduce the beach area for turtle nesting. A global analysis of anthropogenic development of turtle nesting beaches revealed that Kuwait, Saudi Arabia and Qatar host nesting beaches that are highly developed, with development percentages of 41%, 27% and 18%, respectively (Biddiscombe et al. 2020). Coastal reclamation has resulted in the loss of many productive mudflats, which are considered important grounds for the feeding and roosting of wader populations (Mohamed 1998). The disappearance of these habitats has resulted in a decline of wintering and migratory waders in the Arabian Gulf (Al-Obaid et al. 2017).

Coastal development has a profound impact on commercially important fish. According to a regional assessment for marine bony fish in the Arabian Gulf (Buchanan et al. 2019), coastal development and habitat loss were the second most reported threat, affecting nearly 32% of the evaluated fish. It was found that 152 fish species were threatened by direct and indirect disturbances due to habitat degradation and loss (Buchanan et al. 2019). Additionally, coastal reclamation can lead to the loss of spawning grounds, nurseries and feeding areas of fish and shrimp (Zainal et al. 2012), which contributes to the decline in fisheries in the Arabian Gulf (Alqattan and Gray 2021). Therefore, habitat loss and ecosystem degradation can lead to significant impacts on the survival of endangered species, biodiversity and fisheries productivity. Additionally, coastal marshes, mangrove forests and seagrass beds in the Arabian Gulf are major blue carbon sinks that play an important role in absorbing and storing CO_2 (Schile et al. 2017). Large-scale decline or degradation of these habitats can affect a climate mitigation service and potentially release additional carbon into the atmosphere (Lincoln et al. 2021).

19.4.2 Socioeconomic impacts

Although dredging and reclamation generate considerable economic gains, they may also lead to undesired economic consequences such as a decline in fish and shellfish landings and unemployment in the fishing sector (Al-Madany et al. 1991; Zainal et al. 2012). It is suggested that the deterioration of fisheries in the Arabian Gulf may be attributed to some extent to dredging and reclamation, which caused the destruction of the nursery grounds of fish (Sheppard et al. 2010). Other pressures such as unsustainable fishing practices, marine pollution and climate change intensify this impact and make it hard to reverse. The cumulative economic consequences of deteriorated fisheries may be significant in the long run.

In addition, dredging and reclamation may have direct and indirect impacts on society. Many natural waterfronts have disappeared or been altered due to reclamation. Most coastal areas have been modified or replaced by artificial waterfronts (Burt et al. 2017). Additionally, reclamation has been observed near some historical and cultural monuments (El Safadi et al. 2022). Arguably, these modifications in the coastal and marine environments can change the human-sea relationship. They change how people interact with the sea and how they construct its meaning in their culture (El-Ghonaimy and Javed 2018).

Some argue that sea-related traditional knowledge may gradually fade away as some traditional fishing methods such as barriers (known locally as *Hadra*) have been affected or removed due to dredging and reclamation. It is also suggested that a severe decline in some fish species at the local level may lead to the gradual disappearance of some traditional seafood dishes (Al-Madany et al. 1991). The sea has always been present in the heritage of the Arabian Gulf countries, especially for communities inhabiting the coastal areas and islands. Ongoing coastal development is continuously reshaping this relationship, and it is hard to judge whether this change is desirable or not.

19.5 MANAGEMENT OF DREDGING AND RECLAMATION IN THE ARABIAN GULF

19.5.1 Specific measures

The specific measures to address the impacts of dredging and reclamation in the Arabian Gulf can be classified into four

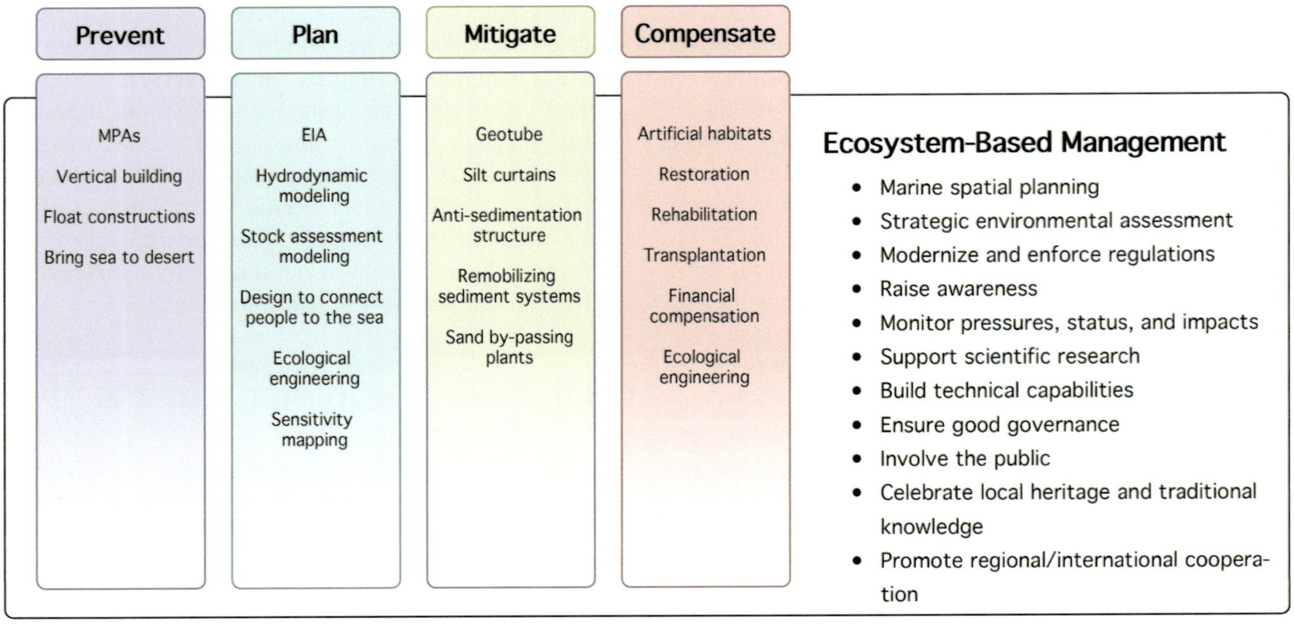

FIGURE 19.3 The main strategies for managing dredging and reclamation in the Arabian Gulf.

main categories: prevention, planning, mitigation and compensation (Figure 19.3). The first category involves preventing dredging and reclamation or finding feasible alternatives to minimize these activities. Marine protected areas (MPAs) are considered an important approach to protecting sensitive coastal and marine ecosystems. MPAs and associated buffer zones can contribute to the protection of sensitive habitats such as coral reefs, seagrass beds and mangroves from the impacts of dredging and reclamation activities (Lamine et al. 2020). The Arabian Gulf countries are increasingly designating MPAs to protect major habitats, ecosystems and species communities. However, there is a recognized need for comprehensive management plans to increase the effectiveness of MPAs in the Arabian Gulf (Van Lavieren and Klaus 2013).

Constructing cities on land by excavating desert areas of low productivity to create marine waterways is considered an alternative approach to minimizing the impacts of dredging and coastal reclamation on marine ecosystems (Jones et al. 2012). Major projects that involved bringing the sea to the desert include the Al-Khiran development in Kuwait (Ealey and Holmes 2001), the Al Khaleej Village development in Saudi Arabia (Ealey et al. 2001) and Doha West Bay in Qatar (Al-Jamali et al. 2005). Several man-made habitats, including intertidal mud and sand flats, salt marshes, mangroves and subtidal hard and soft benthic areas, have been created in the inland artificial waterways. Ecological surveys through the phases of the Al-Khiran development in Kuwait indicated that the diversity and abundance of macrobenthos in the artificial sandy and rocky areas were close to or exceeded those of similar open-sea natural habitats in Kuwait (Jones et al. 2007, 2012; Jones and Nithyanandan 2013). Likewise, the artificial lagoons of the Doha West Bay development project in Qatar supported diverse and abundant macrobenthos that were comparable to open-sea habitats (Al-Jamali et al. 2005). However, this approach might be costly or constrained by the limited availability of inland coastal zones in some countries.

Due to land scarcity in Bahrain, vertical building has been proposed as an option to accommodate the increasing population with less land area. It is argued that vertical cities are a sustainable option to meet the growing demand for housing and to reduce the need for coastal reclamation associated with environmental degradation (El-Ghonaimy and Javed 2018; Yusuf and Elghonaimoy 2020). However, vertical building might not meet the social norms and cultural aspects of the society in comparison with traditional horizontal housing.

Floating offshore structures are considered an alternative for coastal reclamation in the Arabian Gulf as well. Floating architecture is used to construct buildings for living and/or working space on floatation systems. Such constructions have been implemented in developmental projects in Dubai and Bahrain. Nonetheless, reclamation has a longer design life than floating solutions (Ghafooripour et al. 2014).

The second category of measures involves planning coa stal developments based on an environmental impact assessment (EIA), adopting appropriate modeling techniques, enforcing human-sea relationships, utilizing principles of ecological engineering and sensitivity mapping. An EIA includes identifying and mitigating effects of development proposals before major decisions are taken (Morrison-Saunders 2018). The purpose of an EIA is to inform planning decisions for individual projects, including dredging and reclamation activities (Shaw et al. 2003). The EIA has been adopted by all countries in the Arabian Gulf (Al-Azri et al. 2014). Reviews of selected EIA reports related to the dredging and reclamation activities in the Arabian Gulf revealed considerable impacts on coastal and marine habitats (Zainal et al. 2012; Naser 2015a). Given the scale and frequency of dredging and reclamation activities, the effectiveness of project-level EIAs might be limited in addressing the cumulative impacts of these activities (Naser et al. 2008). Therefore, there is a need for a holistic management tool that addresses dredging and reclamation projects on strategic levels such as plans and programmes. Strategic environmental assessment (SEA) as a tool to integrate environmental considerations into new or existing plans and programmes can provide the insight required for sustainable management of dredging and coastal reclamation (Tamis et al. 2016).

The human-sea relationship is an integral part in the identity and history of people in the Arabian Gulf. Therefore, there is a need to integrate the culture and heritage of local society into the design and construction of coastal developments. Designers, architects and engineers can incorporate nature and integrate cultural and social contexts into coastal projects. This approach involves designing artificial islands in a way to connect people with the sea and their history (El-Ghonaimy and Javed 2018).

Ecological engineering can help in mitigating some impacts and potentially create greater ecological benefits for reclamation projects in the Arabian Gulf (Burt and Bartholomew 2019). Hard artificial infrastructure such as breakwaters and seawalls surrounding coastal developments can support diverse and abundant marine organisms (Burt et al. 2011, 2013b). Feary et al. (2011) indicated that breakwaters in Abu Dhabi and Dubai (UAE) showed diverse and abundant benthic communities dominated by several species of algae, sponges, ascidians, bryozoans and urchins that were representatives of many other areas in the Arabian Gulf. Breakwaters associated with reclamation projects also showed a rapid growth of coral within five years of construction and supported abundant and diverse fish communities that were similar to natural reef systems (Burt et al. 2011; Vaughan et al. 2019).

Hydrodynamic and sediment transport modeling is considered an important tool to plan and forecast the effects of coastal modifications in the Arabian Gulf. Hydraulic modeling was applied to produce an adequate flushing regime for artificial lagoons and waterways in Qatar (Al-Jamali et al. 2005). Similarly, a hydrodynamic model was applied to determine future water circulation and renewal time due to the actions of wind and tides in the Durrat Alkhobar development in Saudi Arabia (Stamou et al. 2007). A hydrodynamic model also was used to predict the impact of the reclamation activities on the flow pattern in the vicinity of a desalination plant in Abu Dhabi, UAE (Mohamed 2007). Sedimentation modeling was applied to detect the potential impact of altering sediment behaviour associated with dredging and reclamation on coastal and marine ecosystems along the northeastern Qatar nearshore (Yousif et al. 2018). Additionally, groundwater flow modeling was used to assess the impact of port dredging on coastal hydrogeology, revealing potential perturbation in the groundwater level and flow (Lachaal and Gana 2016). Stock assessment modeling can be used to determine the impact of fishing and other anthropogenic activities on fish stocks and to support the sustainable exploitation of marine resources (Al-Qishawe et al. 2014). Modeling can provide scientific approaches to inform stakeholders of potential impacts and mitigation measures. However, the confidence in modeling outputs relies on the availability and accuracy of data.

Sensitivity mapping of coastal and marine habitats is critical for effective management of dredging and coastal reclamation in the Arabian Gulf (Butler et al. 2020, 2021; Mateos-Molina et al. 2021). These maps can be used to identify areas of vulnerable species and sensitive ecosystems, which helps in avoiding these areas in the planning stages of the development. Additionally, these maps can allocate the potential "borrow areas" for sand dredging.

The third category of measures involves minimizing impacts associated with dredging and reclamation during the implementation phases of projects. Engineering solutions can be applied to reduce the impacts of dredging and reclamation on coastal and marine ecosystems. Geotubes are dredging instruments that can dewater and consolidate the sediment and limit the siltation associated with dredging activities. These tubes were used to limit the adverse impact of dredging during the construction of the Amwaj Islands in Bahrain (Fowler et al. 2002). Innovations in dredging technology are focused on reducing and controlling suspended solids and associated contaminants in the water column during dredging and disposal operations (Salomon and Markus 2018). Silt trapping and silt curtains are frequently used in dredging and reclamation projects to restrict the movement of suspended sediment (Zainal et al. 2020). Alternative engineering technologies can be applied to reduce sedimentation in the surrounding environment, including anti-sedimentation structures, remobilizing sediment systems and sand by-passing plants (Bianchini et al. 2019).

The fourth category of measures involves utilizing the principles of ecological restoration to enhance degraded ecosystems and compensate for lost or degraded habitats. Such restoration programmes are increasingly conducted in the Arabian Gulf. Mangrove afforestation programmes have been initiated in most of the countries, resulting in local growth of mangroves along the coastlines of the Arabian Gulf (Almahasheer 2018; Li et al. 2019). Mangrove plantations have recently been used as a mitigation strategy to enhance carbon sequestration (Alsabbagh and Alnaser 2022) and to promote biodiversity and habitat complexity on artificial islands (Loughland et al. 2020). Mangroves (*Avicennia marina*) have been successfully established on the man-made islands of Sabah Al-Ahmad Sea City along the southern coast of Kuwait, with an overall germination rate of 71.8% (Loughland et al. 2020). Mangroves have been planted along 6.7 km of dredged material shoreline in Abu Dhabi, UAE, creating a habitat for a variety of birds and macrobenthic organisms. Potential ecological benefits of planting mangroves include enhancing biodiversity and protecting the shoreline from erosion (Erftemeijer et al. 2020). The restoration of the brown algae *Sargassum* is proposed as an important mitigation measure in the future coastal development in the Arabian Gulf. The beds of *Sargassum* serve as an important nursery habitat for many species, including commercially important shrimp and crabs (Alghunaim et al. 2020). The seagrass *Halodule uninervis* was transplanted into a lagoon in Qatar and found to facilitate the rapid colonization of macrobenthic species (Al-Jamali et al. 2005). Artificial reef deployment has increased in the Arabian Gulf as a measure for habitat restoration and enhancement. Artificial reefs can protect biodiversity and enhance ecological services. However, they are also associated with potential detrimental effects, including intensifying regional overfishing in the long term. Therefore, achieving the beneficial objectives of artificial reefs requires appropriate design, regulation and management (Burt et al. 2021). Coral transplantation can help in maintaining coral species populations. Coral translocation from impacted sites to offshore reef habitats has been conducted in the Arabian Gulf (Sheppard et al. 2010). Coastal habitats in the Arabian Gulf are considered important grounds for traditional fishing by barrier traps (*Hadra*) and metal-wire fishing traps (*Gargoor*). Financial compensation measures for fishermen who might be affected by the loss and degradation of coastal habitats should be explored (Zainal et al. 2012).

19.5.2 The need for holistic approaches

A considerable part of the reviewed literature criticized the sectorial nature of coastal management in the Arabian Gulf and emphasize the need for adopting holistic management approaches to address all aspects of dredging and reclamation and their interactions with other environmental pressures. Burt et al. (2017) argue that despite improvements in marine environment management across the region, current sectorial, project-level management approaches are not sufficient or effective, and there is a need to adopt the ecosystem-based management (EBM) approach. This approach requires proper marine spatial planning, which involves zoning marine areas and controlling human activities in each area (Abdulla and Naser 2021). This can ensure sustainable use of marine resources and help in protecting ecologically important marine areas against dredging and reclamation.

In addition, Naser et al. (2008) pointed out that a project-level EIA usually does not capture long-term cumulative impacts. Hence, SEA should be undertaken especially at the strategic level when formulating policies or preparing master plans. This may help in addressing or eliminating possible impacts of dredging and reclamation at early stages of the decision-making process.

An EBM approach is an evidence-based approach that necessitates continuous monitoring and assessment of pressures and the state of the environment. This should be complemented by sound scientific research to address knowledge gaps regarding the impact and management of dredging and reclamation in the Gulf, a unique, very stressed marine environment. Some examples of scientific research endeavours include an experimental study to investigate the impact of reclamation on benthic fauna (Naser 2011), hydrodynamic and sediment transport modeling studies (Stamou et al. 2007; Cavalcante et al. 2012; Yousif et al. 2018) and species recolonization and transplantation follow-up studies (Bale and Holmes 1999). Although these studies are insufficient, they help build scientific knowledge regarding the impacts of dredging and reclamation and represent a useful tool to prevent or mitigate these impacts. Burt et al. (2017) argue that empowering and building local expertise in the field of dredging and reclamation are of prime importance. Lastly, an EBM approach also advocates for effective stakeholder participation, including the public. One way to empower stakeholders and the public is raising awareness of the importance of the ecosystem services delivered by the Arabian Gulf marine environment and how dredging and reclamation are threatening this fragile environment. Stakeholder and public participation in the EIA process may help in addressing socioeconomic impacts in particular.

The GCC countries have been collaborating at many levels since the establishment of the council in 1981. In addition, the six countries, including Iraq and Iran, are members of the Regional Organization for the Protection of the Marine Environment (ROPME). Nevertheless, some point out that the collaboration on environmental management is currently inadequate, and there are great opportunities to boost this collaboration much further (Sheppard et al. 2010; Burt et al. 2017). Such collaboration can foster knowledge transfer and may help in developing a region-wide strategy to manage coastal and marine areas.

All the GCC countries are striving toward achieving the Global Agenda 2030 and its sustainable development goals (SDGs) (Azar and Raouf 2017). Sustainable reclamation practices that consider the forecasted potential increase in the sea level can provide opportunities for climate change adaptation, thus protecting vulnerable ecosystems and coastal communities (i.e., SDG 13: Climate action, SDG 14: Life below water, SDG 15: Life on land).

19.6 CONCLUSION

Dredging and reclamation activities are considered among the main pressures causing loss or degradation of coastal and marine habitats in the Arabian Gulf. These activities have contributed to the loss of mudflats and mangrove swamps, fragmentation of seagrass beds, alteration of sabkha and salt marshes and degradation of coral reefs. Dredging and reclamation might be inevitable in the Arabian Gulf; however, these activities must be managed properly to avoid or minimize associated impacts on the coastal and marine environment. Interlinks between impacts, synergy with other pressures and cumulative impacts must be understood through conducting effective EIAs for dredging and reclamation projects, assisted by technological innovation in the fields of statistical and mathematical modeling and ecological engineering. Approaches to the design and implementation of dredging and reclamation projects can include alignment between the natural environment and engineering processes to achieve a harmony between environmental and social benefits.

Management should use different tools to address the impacts of dredging and reclamation, which should be underpinned by overarching ecosystem-based approaches. An EBM can provide an important platform to holistically integrate the sectorial efforts of managing dredging and reclamation activities. An EBM encompasses a legal framework associated with strategic assessment and marine spatial planning to ensure the protection of fragile marine ecosystems in the Arabian Gulf.

Scientific research can provide evidence-based information for the decision-making process. Knowledge gaps include studies related to the recovery and success of organisms in newly reclaimed coastal areas and threshold values for adverse impacts of sedimentation on sensitive marine organisms and ecosystems, including fish, shrimp, and coral. Additionally, there is a need for studies investigating the impacts of reclamation on climate change and carbon sequestration and ecological restoration of lost and degraded ecosystems.

ACKNOWLEDGEMENT

Support provided by the Department of Biology, University of Bahrain, and the Supreme Council for Environment is greatly appreciated. Thanks are extended to Ahmed Khamis for his technical assistance.

REFERENCES

Abahussain, A., Abdu, A., Mussa, N., and A. Al-Haisabi. 2005. The environmental consequences associated with dredging marine sediment in the Kingdom of Bahrain. *J Gulf Arab Peninsula Stud* 117: 85–110.

Abahussain, A., and M. Al-Sabbagh. 2010. Integrated environmental assessment of Tubli Bay, Bahrain: Policy analysis and future scenarios. *J Gulf Arab Peninsula Stud* 36, no. 136: 245–286.

Abdulla, K.H., and H.A. Naser. 2021. Protection of marine environmental quality in the Kingdom of Bahrain. *Ocean Coast Manag* 203: 105520.

Abido, M., Abahussain, A., and H. Abdel Munsif. 2011. Status and composition of mangrove plant community in Tubli Bay of Bahrain during the years 2005 and 2010. *Arab Gulf J Sci Res* 29: 100–111.

Abou Seedo, K., Abido, M., Salih, A., and A. Abahussain. 2017. Structure and composition of mangrove associations in Tubli Bay of Bahrain as affected by municipal wastewater discharge and anthropogenic sedimentation. *Int J Biodivers*. 2017, no. 1: 1–9. http://doi.org/10.1155/2017/2084256.

Afzal, M., Tahir, F., and S. Al-Ghamdi. 2022. Recommendations and strategies to mitigate environmental implications of artificial island developments in the Gulf. *Sustainability* 14: 5027.

Al-Azri, N., Al-Busaidi, R., Sulaiman, H., and A. AlAzri. 2014. Comparative evaluation of EIA systems in the Gulf Cooperation Council States. *Impact Assess Proj Apprais* 32: 136–149.

Al-Ghadban, A., and A. Price. 2002. Dredging and infilling. In: *The Gulf Ecosystems: Health and Sustainability*. Eds. N. Khan, M. Munawar and A. Price, 207–217. Leiden, The Netherlands: Backhuys Publishers.

Alghunaim, A., Yamamoto, T., Madhusoodhanan, R., Polikarpov, I., Chen, W., Al-Said, T., Al-Zekri, W., Alenezi, Y., and F. Al-Yamani. 2020. Evaluating the potential area and environmental key factors for Sargassum bed restoration in highly turbid waters of the northwestern Arabian Gulf. *Mar Pollut Bull* 153: 110949.

Al Hosani, J. 2014. Combining development with environmental friendly solution (GASCO Artificial Islands). In: *Abu Dhabi International Petroleum Exhibition and Conference*. OnePetro, Abu Dhabi, UAE, November.

Al-Jamali, F., Bishop, J., Osment, J., Jones, D., and L. LeVay. 2005. A review of the impacts of aquaculture and artificial waterways upon coastal ecosystems in the Gulf (Arabian/Persian) including a case study demonstrating how future management may resolve these impacts. *Aquat Ecosyst Health Manag* 8: 81–94.

Aljenaid, S., Abido, M., Redha, G., AlKhuzaei, M., Marsan, Y., Khamis, A., Naser, N., Al-Rumaiad, M., and M. Alsabagh. 2022a. Assessing the spatiotemporal changes, associated carbon stock, and potential emissions of mangroves in Bahrain using GIS and remote sensing data. *Reg Stud Mar Sci* 52: 102282.

Aljenaid, S., Khadhem, G., AlKhuzaei, M., and J. Alam. 2022b. Detecting and assessing the spatio-temporal land use land cover changes of Bahrain Island during 1986–2020 using remote sensing and GIS. *Earth Syst Environ* 6: 787–802.

Al-Madany, I., Abdalla, A., and A. Abdu. 1991. Coastal zone management in Bahrain: An analysis of social, economic and environmental impacts of dredging and reclamation. *J Environ Manag* 32: 335–348.

Almahasheer, H. 2018. Spatial coverage of mangrove communities in the Arabian Gulf. *Environ Monit Assess* 190: 85. http://doi.org/10.1007/s10661-018-6472-2.

Al-Naimi, A., Karani, G., and J. Littlewood. 2018. Stakeholder views on land reclamation and marine environment in Doha, Qatar. *J Agric Environ Sci* 7: 32–39.

Al-Obaid, S., Samraoui, B., Thomas, J., El-Serehy, H.A., Alfarhan, A.H., Schneider, W., and M. O'Connell. 2017. An overview of wetlands of Saudi Arabia: Values, threats, and perspectives. *Ambio* 46, no. 1: 98–108.

Al-Otaibi, Y., Ait Belaid, M., and A. Abdu. 2006. Impact assessment of human activities on coastal zones of eastern Saudi Arabia using remote sensing and geographic information systems techniques. In: *Proc ISPRS Commission VII Symposium 'Remote Sensing: From Pixels to Processes'*, vol. XXXVI, Part 7, 8–11 May 2006, Enschede: International Society for Photogrammetry and Remote Sensing (ISPRS).

AlQahtany, A., Dano, U., Abdalla, E., Mohammed, W., Abubakar, I., Al-Gehlani, W., Akbar, N., and M. Alshammari. 2022. Land reclamation in a coastal metropolis of Saudi Arabia: Environmental sustainability implications. *Water* 14: 2546.

Alqattan, M., and T. Gray. 2021. Marine pollution in Kuwait and its impacts on fish-stock decline in Kuwaiti waters: Reviewing the Kuwaiti Government's policies and practices. *Front Sustain* 2: 1–17.

Al-Qishawe, M., Ali, T., and A. Abahussain. 2014. Stock assessment of white spotted rabbitfish (*Siganus canaliculatus* Park, 1797) in Jubail marine wildlife sanctuary, Saudi Arabia. *Int J Fish Aquat Stud* 1: 48–54.

Alsabbagh, M., and W. Alnaser. 2022. Transitioning to carbon neutrality in Bahrain: A policy brief. *Arab Gulf J Sci Res* 40: 25–33.

Al-Sayed, H., Naser, H., and K. Al-Wedaei. 2008. Observations on macrobenthic invertebrates and wader bird assemblages in a protected marine mudflat in Bahrain. *Aquat Ecosyst Health Manag* 11, no. 4: 450–456.

Al-Thukair, A., Khan, M., and K. Al-Hinai. 1995. Monitoring of coastline and habitat changes in Tarut Bay, Saudi Arabia using satellite images. In: *Proc ASCE-ASA Second Regional Conference and Exhibition: Save the Environment, 16–18 November 1995*, Beirut, Lebanon.

Al-Yamani, F.Y., Polikarpov, I., and M. Saburova. 2020. Marine life mortalities and harmful algal blooms in the northern Arabian Gulf. *Aquat Ecosyst Health Manag* 23, no. 2: 196–209.

Azar, E., and M. Raouf (eds.). 2017. *Sustainability in the Gulf-Challenges and Opportunities*. Routledge Explorations in Environmental Studies Series. Oxfordshire, UK: Taylor & Francis.

Bale, T.A., and P. Holmes. 1999. Lagoon residential and recreational developments case study 2: Al Khiran Pearl City, Kuwait. *WIT Trans Built Environ* 43.

Bianchini, A., Cento, F., Guzzini, A., Pellegrini, M., and C. Saccani. 2019. Sediment management in coastal infrastructures: Techno-economic and environmental impact assessment of alternative technologies to dredging. *J Environ Manag* 248: 109332.

Biddiscombe, S.J., Smith, E.A., and L.A. Hawkes. 2020. A global analysis of anthropogenic development of marine turtle nesting beaches. *Remote Sens* 12, no. 9: 1–10.

Buchanan, J., Ralph, G., Krupp, F., Harwell, H., Abdallah, M., Abdulqader, E., Al-Husaini, M., Bishop, J., Burt, J., Choat, J., Collette, B., Feary, D., Hartmann, S., Iwatsuki, Y., Kaymaram, F., Larson, H., Matsuura, K., Motomura, H., Munroe, T., Russel, B., Smith-Vaniz, W., Williams, J., and K. Carpenter. 2019. Regional extinction risks for marine bony fishes occurring in the Persian/Arabian Gulf. *Biol Conserv* 230: 10–19.

Burt, J. 2014. The environmental costs of coastal urbanization in the Arabian Gulf. *City* 18: 760–770.

Burt, J., Al-Khalifa, K., Khalaf, E., Alshuwaikh, B., and A. Abdulwahab. 2013a. The continuing decline of coral reefs in Bahrain. *Mar Pollut Bull* 72: 357–363.

Burt, J., and A. Bartholomew. 2019. Towards more sustainable coastal development in the Arabian Gulf: Opportunities for ecological engineering in an urbanized seascape. *Mar Pollut Bull* 142: 93–102.

Burt, J., Bartholomew, A., and L. Frith. 2021. Policy and management considerations for artificial reefs in the Arabian Gulf. Policy Paper 51. *Sheikh Saud bin Saqr Al Qasimi Foundation for Policy Research*, Ras Al Khaimah, United Arab Emirates.

Burt, J., Bartholomew, A., and P. Sale. 2011. Benthic development on large-scale artificial reefs: A comparison of communities among breakwaters of different age and natural reefs. *Ecol Eng* 37: 191–198.

Burt, J., Ben-Hamadou, R., Abdel-Moati, M.A.R., Fanning, L., Kaitibie, S., Al-Jamali, F., Range, P., Saeed, S., and C.S. Warren. 2017. Improving management of future coastal development in Qatar through ecosystem-based management approaches. *Ocean Coast Manag* 148: 171–181.

Burt, J., Feary, D., Cavalcante, G., Bauman, A., and P. Usseglio. 2013b. Urban breakwaters as reef fish habitat in the Persian Gulf. *Mar Pollut Bull* 72: 342–350.

Butler, J., Purkis, L., Purkis, S., Yousif, R., and I. Al-Shaikh. 2021. A benthic habitat sensitivity analysis of Qatar's coastal zone. *Mar Pollut Bull* 167: 112333.

Butler, J., Purkis, S., Yousif, R., Al-Shaikh, I., and C. Warren. 2020. A high-resolution remotely sensed benthic habitat map of the Qatari coastal zone. *Mar Pollut Bull* 160: 111634.

Cavalcante, G., Kjerfve, B., and D. Feary. 2012. Examination of residence time and its relevance to water quality within a coastal mega-structure: The Palm Jumeirah Lagoon. *J Hydrol* 468–469: 111–119.

Chee, S., Othman, A., Sim, Y., Adam, A., and L. Firth. 2017. Land reclamation and artificial islands: Walking the tightrope between development and conservation. *Glob Ecol Conserv* 12: 80–95.

Ealey, T.A., and P. Holmes. 2001. Lagoon residential and recreational development case study 2: Al Khiran Pearl City, Kuwait. In: *Coastal Engineering and Marina Development*. Eds. C.A. Brebbia and P. Anagnostopoulos, 635–646. Southampton: Wessex Inst Tech Press.

Ealey, T.A., Holmes, P., Abu Sitta, H., Kelly, P., and I. Williams. 2001. Lagoon residential and recreational development case study 1: Al Khaleej Village, Saudi Arabia. In: *Coastal Engineering and Marina Development*. Eds. C.A. Brebbia and P. Anagnostopoulos, 615–633. Southampton: Wessex Inst Tech Press.

El-Ghonaimy, I., and N. Javed. 2018. Concomitant of global warming and land reclamation: Designing a new interface between human kind and sea. *Proc Int Conf Ind Eng Oper Manag* 2018: 2175–2186.

Elmahdy, S., Ali, T., Mohamed, M., Howari, F., Abouleish, M., and D. Simoner. 2020. Spatiotemporal mapping and monitoring of mangrove forests changes from 1990 to 2019 in the northern Emirates, UAE using random forest, kernel logistic regression and naive bayes tree models. *Front Environ Sci* 8: 102.

El Safadi, C., Ray, N., Ortiz-Vazquez, R., Nikolaus, J., Blue, L., Westley, K., Andreou, G., and C. Breen. 2022. Maritime cultural heritage and urbanization in the Middle East and North Africa. *Hist Environ Policy Pract* 13, no. 3: 318–340.

Erftemeijer, P., Agastian, T., Yamamoto, H., Cambridge, M., Hoekstra, R., Toms, G., and S. Ito. 2020. Mangrove planting on dredged material: Three decades of nature-based coastal defence along a causeway in the Arabian Gulf. *Mar Freshw Res* 71: 1062–1072.

Erftemeijer, P., and R. Lewis. 2006. Environmental impacts of dredging on seagrasses: A review. *Mar Pollut Bull* 52: 1553–1572.

Erftemeijer, P., Riegel, B., Hoeksema, B., and P. Todd. 2012. Environmental impacts of dredging and other sediment disturbances on corals: A review. *Mar Pollut Bull* 64: 1737–1765.

Erftemeijer, P., and D. Shuail. 2012. Seagrass habitats in the Arabian Gulf: Distribution, tolerance thresholds and threats. *Aquat Ecosyst Health Manag* 15: 73–83.

Feary, D., Burt, J., and A. Bartholomew. 2011. Artificial marine habitats in the Arabian Gulf: Review of current use, benefits and management implications. *Ocean Coast Manag* 54: 742–749.

Fowler, J., Stephens, T., Santiago, M., and P. De Bruin. 2002. Amwaj Islands constructed with Geotubes, Bahrain. In: *Proc CEDA Conference*, Denver, USA, pp. 1–14.

Fraser, M., Jessie, S., Kendrick, G., Mclean, D., Keesing, J., Byrne, M., Calery, M., Clarke, D., Davis, A., Erftemeijer, P., Field, S., Gustin-Craig, S., Huisman, J., Keough, M., Lavery, P., Masini, R., McMahon, K., Mengersen, K., Rasheed, M., Statton, J., Stoddart, J., and P. Wu. 2017. Effects of dredging on critical ecological processes for marine invertebrates, seagrasses and macroalgae, and the potential for management with environmental windows using Western Australia as a case study. *Ecol Indic* 78: 229–242.

Friis, G., and J. Burt. 2020. Evolution of mangrove research in an extreme environment: Historical trends and future opportunities in Arabia. *Ocean Coast Manag* 195: 105288.

Ghafooripour, A., Nezamifar, E., Ogwuda, O., and M. Sun. 2014. The effectiveness of sustainable floating structures in the Gulf region (Middle East). *WIT Trans Built Environ* 137: 653–664.

Godson, P., Vincent, S., and S. Krishnakumar. 2022. *Ecology and Biodiversity of Benthos*. Amsterdam, The Netherlands: Elsevier.

Grandcourt, E. 2012. Reef fish and fisheries in the Gulf. In: *Coral Reefs of the Gulf: Adaptation to Climatic Extremes*. Eds. B.M. Riegl and S. Purkis, 127–161. Dordrecht, The Netherlands: Springer.

IUCN/MEPA. 1987. Arabian Gulf. Saudi Arabia: An assessment of biotopes and coastal zone management requirements for the Arabian Gulf. MEPA coastal and marine management series, *Report No. 5*. Gland, Switzerland: IUCN.

Jones, D., Ealey, T., Baca, B., Livesey, S., and F. Al-Jamali. 2007. Gulf desert developments encompassing a marine environment, a compensatory solution to the loss of coastal habitats by infill and reclamation: The case of the Pearl City Al-Khiran, Kuwait. *Aquat Ecosyst Health Manag* 10: 268–276.

Jones, D., and M. Nithyanandan. 2013. Recruitment of marine biota onto hard and soft artificially created subtidal habitats in Sabah Al-Ahmad Sea City, Kuwait. *Mar Pollut Bull* 72: 351–356.

Jones, D., Nithyanandan, M., and I. Williams. 2012. Sabah Al-Ahmad Sea City Kuwait: Development of a sustainable man-made coastal ecosystem in a saline desert. *Aquat Ecosyst Health Manag* 15: 84–2012.

Kana, T., Al-Sarawi, M., and M. Holland. 1986. Design and performance of artificial beaches for the Kuwait waterfront project. *Coastal Eng Proc* 1, no. 20: 187.

Khan, M., and A. Kumar. 2009. Impacts of urban development on mangrove forest along the west coast of the Arabian Gulf. *Earth Sci India* 2: 159–173.

Khan, S.B., Javed, S., Ahmed, S., Shah, J.N., Al Hammadi, A.A., and E.A. Al Hammadi. 2017. Greater Flamingo (Phoenicopterus roseus): Important wintering sites and breeding records in the United Arab Emirates. *Zool Middle East* 63, no. 3: 194–201.

Lachaal, F., and S. Gana. 2016. Groundwater flow modeling for impact assessment of port dredging works on coastal hydrogeology in the area of Al-Wakrah (Qatar). *Model Earth Syst Environ* 2: 1–15.

Lamine, E., Mateos-Molina, D., Antonopoulou, M., Burt, J.A., Das, H.S., Javed, S., Muzaffar, S., and S. Giakoumi. 2020. Identifying coastal and marine priority areas for conservation in the United Arab Emirates. *Biodivers Conserv* 29: 2967–2983.

Li, W., El-Askary, H., Qurban, M., Li, J., Manikandan, K., and T. Piechota. 2019. Using multi-indices approach to quantify mangrove changes over the Western Arabian Gulf along Saudi Arabia coast. *Ecol Indic* 102: 734–745.

Lincoln, S., Buckley, P., Howes, E., Maltby, K., Pinnegar, J., Ali, T., Alosairi, Y., Al-Ragum, A., Baglee, A., Balmes, C., Hamadou, R., Burt, J., Claereboudt, M., Glavan, J., Mamiit, R., Naser, H., Sedighi, O., Shokri, M., Shuhaibar, B., Wabnitz, C., and W. Le Quesne. 2021. A regional review of marine and coastal impacts of climate change on the ROPME Sea area. *Sustainability* 13, no. 24: 13810.

Lo, J. 1987. Assessment of turbidity at Shuaiba offshore seawater intakes due to dredging new pipeline trench. In: *Proc Second International Conf on Coastal and Port Engineering in Developing Countries*. Beijing, China, 7–11 September, 1987, pp. 1768–1780.

Lokier, S. 2013. Coastal sabkha preservation in the Arabian Gulf. *Geoheritage* 5: 11–22.

Loughland, R., Butt, S., and M. Nithyanandan. 2020. Establishment of mangrove ecosystems on man-made islands in Kuwait: Sustainable outcomes in a challenging and changing environment. *Aquat Bot* 167: 103273.

Luypaert, T., Hagan, J., McCarthy, M., and M. Poti. 2020. Status of marine biodiversity in the Anthropocene. In: *YOUMARES 9 – The Oceans: Our Research, Our Future*. Eds. S. Jungblut, V. Liebich and M. Bode-Dalby, 57–82. Cham, Switzerland: Springer.

Ma, T., Li, X., Bai, J., and B. Cui. 2019. Habitat modification in relation to coastal reclamation and its impacts on waterbirds along China's coast. *Glob Ecol Conserv* 17: e00585.

Marin-Anton, M., Negro, V., del Campo, J., Lopez-Gutierrez, J., and M. Esteban. 2016. Review of coastal land reclamation situation in the World. *J Coastal Res* 75: 667–671.

Mateos-Molina, D., Lamine, E., Antonopoulou, M., Burt, J., Das, H., Haved, S., Judas, J., Khan, S., Muzaffar, E., Pilcher, N., Rodriguez-Sarate, C., Taylor, O., and S. Giakoumi. 2021. Synthesis and evaluation of coastal and marine biodiversity spatial information in the United Arab Emirates for ecosystem-based management. *Mar Pollut Bull* 167: 112319.

Modavi, C., Martin, W.W., Muflehi, A., Hakeem, A., Walters, J.V., and G. Ismail. 2008. Field development expansion of a giant oil field in Abu Dhabi using artificial islands as drilling and production centers. In: *Abu Dhabi International Petroleum Exhibition and Conference*. Abu Dhabi: OnePetro.

Mohamed, K.A. 2007. Model investigation on the impact of Raha Beach development on Umm Al Nar power and desalination plant. *Desalination* 212, nos. 1–3: 357–366.

Mohamed, S.A. 1998. Density and distribution of migratory waders along the shores of Bahrain islands. *Arab Gulf J Sci Res* 16: 145–157.

Morrison-Saunders, A. 2018. *Advanced Introduction to Environmental Impact Assessment*. Cheltenham: Edward Elgar.

Naser, H. 2011. Effects of reclamation on macrobenthic assemblages in the coastline of the Arabian Gulf: A microcosm experimental approach. *Mar Pollut Bull* 62: 520–524.

Naser, H. 2014. Marine ecosystem diversity in the Arabian Gulf: Threats and conservation. In: *Ecosystem Biodiversity – The Dynamic Balance of the Planet*. Eds. G. Oscar, 297–328. Croatia, Rijeka: INTECH Publishing.

Naser, H. 2015a. The role of environmental impact assessment in protecting coastal and marine environments in rapidly developing islands: The case of Bahrain, Arabian Gulf. *Ocean Coast Manag* 104: 159–169.

Naser, H. 2015b. Effects of coastal developments on corals in Bahrain. *Conference of Coral Reefs of Arabia: Exploring New Insights and Opportunities on Our Regional Reefs. New York University – Abu Dhabi, 15–17 February 2015*. Abu Dhabi, UAE.

Naser, H. 2016. Management of marine protected zones – case study of Bahrain, Arabian Gulf. In: *Applied Studies of Coastal and Marine Environments*. Ed. M. Marghany, 322–350. Croatia, Rijeka: INTECH Publishing.

Naser, H. 2022. Community structures of benthic macrofauna in reclaimed and natural intertidal areas in Bahrain, Arabian Gulf. *J Mar Sci Eng* 10: 945.

Naser, H., Bythell, J., and J. Thomason. 2008. Ecological assessment: An initial evaluation of the ecological input in environmental impact assessment reports in Bahrain. *Impact Assess Proj Apprais* 26, no. 3: 201–208.

Nasser, A., Blackburn, G., and D. Whyatt. 2014. Developing the desert: The peace and process of urban growth in Dubai. *Systems* 45: 50–62.

Neelamani, S., and F. Al-Hulail. 2011. Technical feasibility analysis of submarine pipelines for intake of sea water with high total suspended sediment loads. *ISH J Hydraul Eng* 17, no. 2: 99–109.

Peduzzi, P. 2014. Sand, rarer than one thinks. *Environ Dev* 11: 208–218.

Pilcher, N., Al-Maslamani, I., Williams, J., Gasang, R., and A. Chikhi. 2015. Population structure of marine turtles in coastal waters of Qatar. *Endanger Species Res* 28: 163–174.

ROPME. 2020. *Status and Trends of Coral Reefs in the ROPME Sea Area: Past, Present and Future*. Regional Organization for the Protection of the Marine Environment, Kuwait.

Salomon, M., and T. Markus. 2018. *Handbook on Marine Environmental Protection: Science, Impacts and Sustainable Management*. Berlin and Heidelberg, Germany: Springer.

Schile, L., Kauffman, J., Crooks, S., Fourqurean, J., Glavan, J., and P. Megonigal. 2017. Limits on carbon sequestration in arid blue carbon ecosystems. *Ecol Appl* 27: 859–874.

Seitz, R., Wennhage, H., Bergstrom, U., Lipicus, R., and T. Ysebaert. 2014. Ecological value of coastal habitats for commercially and ecologically important species. *ICES J Mar Sci* 71: 648–665.

Sengupta, D., Chen, R., and M. Meadows. 2018. Building beyond land: An overview of coastal and reclamation in 16 global megacities. *Appl Geogr* 90: 229–238.

Shanableh, A., Omar, M., Al-Midfa, A.A., and A. Dolan. 2003. Baseline assessment of water quality and circulation patterns in the Khalid and Al-Khan adjacent coastal lagoons in Sharjah, UAE. In: *Thirteenth International Offshore and Polar Engineering Conference*. OnePetro.

Shaw, A., Hendry, M., Allen, B., Genina, W., and A. Mujadaba. 2003. Dredging, spoil disposal and environmental mitigation: Marine environmental assessment in Abu Dhabi. In: *Breakwaters, Coastal Structures and Coastlines*. Ed. N. Allsop, 161–174. London: Thomas Telford Publishing.

Sheppard, C., Al-Husiani, M., Al-Jamali, F., Al-Yamani, F., Baldwin, R., Bishop, J., Benzoni, F., Dutrieux, E., Dulvy, N.K., Durvasula, S.R.V., Jones, D.A., Loughland, R., Medio, D., Nithyanandan, M., Pilling, G.M., Polikarpov, I., Price, A.R.G., Purkis, S., Riegl, B., Saburova, M., Namin, K.S., Taylor, O., Wilson, S., and K. Zainal. 2010. The Gulf: A young sea in decline. *Mar Pollut Bull* 60: 13–38.

Smyth, D., Al-Maslamani, I., Giraldes, B., Chatting, M., Al-Ansari, E., and L. Le Vay. 2016. Anthropogenic related variations in the epibiotic biodiversity and age structure of the 'Pearl Oyster' Pinctada radiata within the eulittoral zone of Qatar. *Reg Stud Mar Sci* 5: 87–96.

Stamou, A.I., Memos, C., and K. Spanoudaki. 2007. Estimating water renewal time in semi-enclosed coastal areas of complicated geometry using a hydrodynamic model. *J Coastal Res* 50: 282–286.

Stauber, J., Chariton, A., and S. Apte. 2016. Global change. In: *Marine Ecotoxicology: Current Knowledge and Future Issues*. Eds. J. Nlasco, P. Chapman, O. Campana and M. Hampel, 273–313. London: Elsevier.

Summers, L., and C. Fleming. 1980. For disposal: 10 M.cu.m. of calcium carbonate slurry. *Coastal Eng Proc* 1, no. 17: 170. http://doi.org/10.9753/icce.v17.170.

Tamis, J., de Vries, P., Jongbloed, R., Lagerveld, S., Jak, R., Karman, C., Van der Wal, J., Slijkerman, D., and C. Klok. 2016. Toward a harmonized approach for environmental assessment of human activities in the marine environment. *Integr Environ Assess Manag* 12, no. 4: 632–642.

Todd, V., Todd, I., Gardiner, J., Morrin, E., MacPherson, N., DiMarzio, N., and F. Thomsen. 2015. A review of impacts of marine dredging activities on marine mammals. *ICES J Mar Sci* 72: 328–240.

Trevisan, C., Vicentle, M., Rocha, B., and J. Wasserman. 2020. Development of a dredging sensitivity index, applied to an industrialized coastal environment in Brazil. *Sci Total Environ* 748: 141294.

UN. 2021. *The Second World Ocean Assessment*. New York: United Nations.

Van Lavieren, H., Burt, J., Feary, D., Cavalcante, G., Marquis, E., Benedetti, L., Trick, C., Kjerfve, B., and P. Sale. 2011. *Managing the Growing Impacts of Development on Fragile Coastal and Marine Ecosystems: Lessons from the Gulf.* Hamilton, ON, Canada: UNU-INWEH.

Van Lavieren, H., and R. Klaus. 2013. An effective regional marine protected area network for the ROPME Sea area: Unrealistic vision or realistic possibility? *Mar Pollut Bull* 72: 389–405.

Vaughan, G., Al-Mansoori, N., and J. Burt. 2019. The Arabian Gulf. In: *World Seas: An Environmental Evaluation*, 2nd ed. Ed. C. Sheppard, 1–23. Amsterdam, The Netherlands: Elsevier Science.

Wenger, A., Harvey, E., Wilson, S., Rawson, C., Newman, S., Clarke, D., Saunders, B., Browne, N., Travers, M., McIlwain, J., Erftemeijer, P., Hobbs, J., Mclean, D., Depczynski, M., and R. Evans. 2017. A critical analysis of the direct effects of dredging on fish. *Fish Fish* 18: 967–985.

Wenger, A., Rawson, C., Wilson, S., Newman, S., Travers, M., Atkinson, S., Browne, N., Clarke, D., Depczynski, M., Erftemeijer, P., Evans, R., Hobbs, J., McIlwain, J., McLean, D., Saunders, B., and E. Harvey. 2018. Management strategies to minimize the dredging impacts of coastal development on fish and fisheries. *Conserv Lett* 11: e12572.

Wong, A., Vrontos, S., and M. Taylor. 2022. An assessment of people living by coral reefs over space and time. *Glob Change Biol* 28, no. 23: 7139–7153. http://doi.org/10.1111/gcb.16391.

Yagoub, M.M., and G.R. Kolan. 2006. Monitoring coastal zone land use and land cover changes of Abu Dhabi using remote sensing. *J Indian Soc Remote Sens* 34, no. 1: 57–68.

Yousif, R., Warren, C., Ben-Hamadou, R., and S. Husrevoglu. 2018. Modeling sediment transport in Qatar: Application for coastal development planning. *Int Environ Assess Manag* 14: 240–251.

Yusuf, M.B., and I. Elghonaimoy. 2020. Housing dilemma and vertical dimensions. *Civil Eng Arch* 8, no. 5: 1107–1118.

Zainal, A., Dalby, D., and I. Robinson. 1993. Monitoring marine ecological changes on the east coast of Bahrain with Landsat TM. *Photogram Eng Remote Sens* 59: 415–421.

Zainal, K., Al-Madany, I., Al-Sayed, H., Khamis, A., Al-Shuhaby, S., Al-Hisaby, A., Elhoussiny, W., and E. Khalaf. 2012. The cumulative impacts of reclamation and dredging on the marine ecology and land-use in the Kingdom of Bahrain. *Mar Pollut Bull* 64: 1452–1458.

Zainal, K., Isa, A., and Q. Mandeel. 2020. Spatial and temporal variation patterns of total suspended solids around the coastal areas of Bahrain, a water quality guideline. *Aquat Ecosyst Health Manag* 23: 136–144.

Mapping the mangroves in the Arabian Gulf, Saudi Arabia

20

Karuppasamy Manikandan, Mohammed Qurban, Jinoy Gopalan, Reynaldo Lindo, Thadickal V. Joydas and Mohamed Asharaf

20.1 INTRODUCTION

Mangroves are a unique type of forest found in coastal regions and estuaries, consisting of trees and shrubs that are adapted to live in saline water. These forests are vital ecosystems that provide numerous benefits to both the environment and the communities that depend on them. One of the most significant benefits of mangroves is their role in protecting coastal communities from natural disasters. They act as a natural buffer, reducing the impact of storm surges and tsunamis on the coast. They also stabilize shorelines, preventing erosion and protecting coastal infrastructure such as harbours and seawalls (Dahdouh-Guebas et al. 2005). Mangroves are also important habitats for a variety of wildlife, including migratory birds, fish and crustaceans. The roots of mangrove trees provide a nursery ground for fish while the leaves and branches serve as a food source for many species (Alongi 2002). Mangroves are also significant carbon sinks, with the ability to sequester up to four times more carbon than other tropical forests (Donato et al. 2011). Since the destruction of mangrove forests can cause significant carbon emissions into the atmosphere, they are essential for reducing climate change. Despite their importance, mangrove forests are under threat from a variety of factors, including deforestation, pollution and climate change. Therefore, it is crucial to estimate the extent of mangroves and monitor their changes over time to better understand their current status and plan for effective conservation strategies.

Satellite imagery has proven to be an efficient and reliable tool for mapping and monitoring mangroves at a large scale. Remote-sensing techniques have the advantage of providing accurate and objective measurements of mangrove extent, which can be difficult to obtain through ground-based surveys. Additionally, satellite data can be used to detect changes in mangrove cover over time, which is essential for assessing the impact of human activities and climate change on these ecosystems.

Several studies have used satellite imagery to estimate the extent of mangroves on local, regional and global scales. For instance, Giri et al. (2011) used Landsat data to map mangroves at the global scale and estimated their coverage to be around 137,760 km^2. Similarly, Alongi and Murdiyarso (2015) used remote-sensing techniques to map mangroves in the Asia-Pacific region and found that the total area covered by mangroves in this region was 92,335 km^2. In addition to providing estimates of mangrove extent, satellite imagery can also be used to identify changes in mangrove cover over time. Friess et al. (2016) used satellite imagery to assess mangrove cover change in Southeast Asia in the last two decades and found that the region had experienced a net loss of approximately 40,000 hectares per year. Apart from mapping and monitoring mangroves, satellite data can also provide insights into the drivers of mangrove loss and degradation. For instance, Richards and Friess (2016) used satellite imagery to investigate the factors driving mangrove loss in Southeast Asia and found that the primary drivers were aquaculture and agriculture expansion, as well as urbanization and infrastructure development. Similarly, Nagendra et al. (2013) used satellite data to study the impact of shrimp farming on mangroves in India and found that shrimp farms had led to significant mangrove loss in the region. In 2022, an updated view of the distribution of mangrove forests was released in the Global Mangrove Watch (GMW) Version 3.0. This map includes a detailed historical census and shows some 147,000 km^2 of mangroves in 2020 (Bunting et al. 2022).

There have been several studies conducted to estimate the area of mangroves in the Middle East using remote sensing. Habshi et al. (2007) and Howari et al. (2009) estimated mangrove areas in selected sites in the UAE, Etemadi et al. (2021) studied the mangrove ecosystems from Nayband and Mond Protected Area in the southwest region of Iran, Aljenaid et al. (2022) assessed the spatio-temporal changes in the mangrove habitat around Tubli Bay, Kingdom of Bahrain, while Khan and Kumar (2009) and Almahasheer et al. (2013) estimated mangrove areas in selected sites in Saudi Arabia. In addition, Abido et al. (2011) reported countrywide mangrove areas in

DOI: 10.1201/9781003321408-27

Bahrain, and Elmahdy et al. (2020) monitored changes in mangrove forests from 1990 to 2019 in the Northern Emirates, UAE, using random forest, Kernel logistic regression and Naive Bayes Tree models.

The Arabian Gulf coast of Saudi Arabia is home to several mangrove forests that are concentrated in a few areas along the coastline. The most extensive mangrove forests in the country are found in the Eastern Province, particularly in the Jubail Marine Wildlife Sanctuary, which is the largest protected area in Saudi Arabia. Other significant mangrove areas include Al-Ha'ir, Al-Qatif and Al-Zour. Satellite imagery has been used to map the extent of mangroves in Saudi Arabia since the early 1990s. One of the most comprehensive studies of mangrove extent in Saudi Arabia was conducted by Almahasheer (2018) using Landsat 8 imagery. The study used 25 satellite images obtained from Landsat 8 data in 2017 and examined the distribution and spatial coverage of mangroves along the Gulf coastlines. The results showed that there were approximately 165 km^2 of fragmented and scattered mangroves, which were predominantly concentrated in the United Arab Emirates. Li et al. (2019) estimated the extent of mangroves in Saudi Arabia using supported vector machine (SVM) and decision tree (DT), referred to as classification and regression trees (CART) and random forest (RF) models. In Saudi Arabia, several other studies have used remote-sensing techniques to map and monitor mangroves, including studies using Landsat ETM imagery (Kumar 2011) and synthetic aperture radar (SAR) data (Almahasheer et al. 2016). These investigations have given us important information about the dynamics and spatial distribution of mangrove ecosystems throughout the Kingdom.

20.2 MATERIALS AND METHODS

20.2.1 Study area

Our study area covers eight regions in the Eastern Province of Saudi Arabia: Abu Ali, Furay, Gurmah, Rahima, Ras Tanura, Safwa, Saihat and Tarout (Figure 20.1). These regions were chosen based on the results of a survey conducted by KFUPM. Google Earth Engine (GEE), a cloud-based platform that provides access to a vast amount of satellite imagery and geospatial data, allows users to process, analyze and visualize large datasets quickly and efficiently, making it an important tool for various applications, including mangrove mapping. In the context of mangrove mapping, GEE enables researchers to access Sentinel-2 multispectral data and apply various remote-sensing techniques to detect and map mangroves. These techniques include spectral analysis, classification and change detection, which can help identify changes in mangrove extent over time. The use of GEE in mangrove mapping has several advantages, including its ability to process large datasets quickly and efficiently, its user-friendly interface and its integration with other

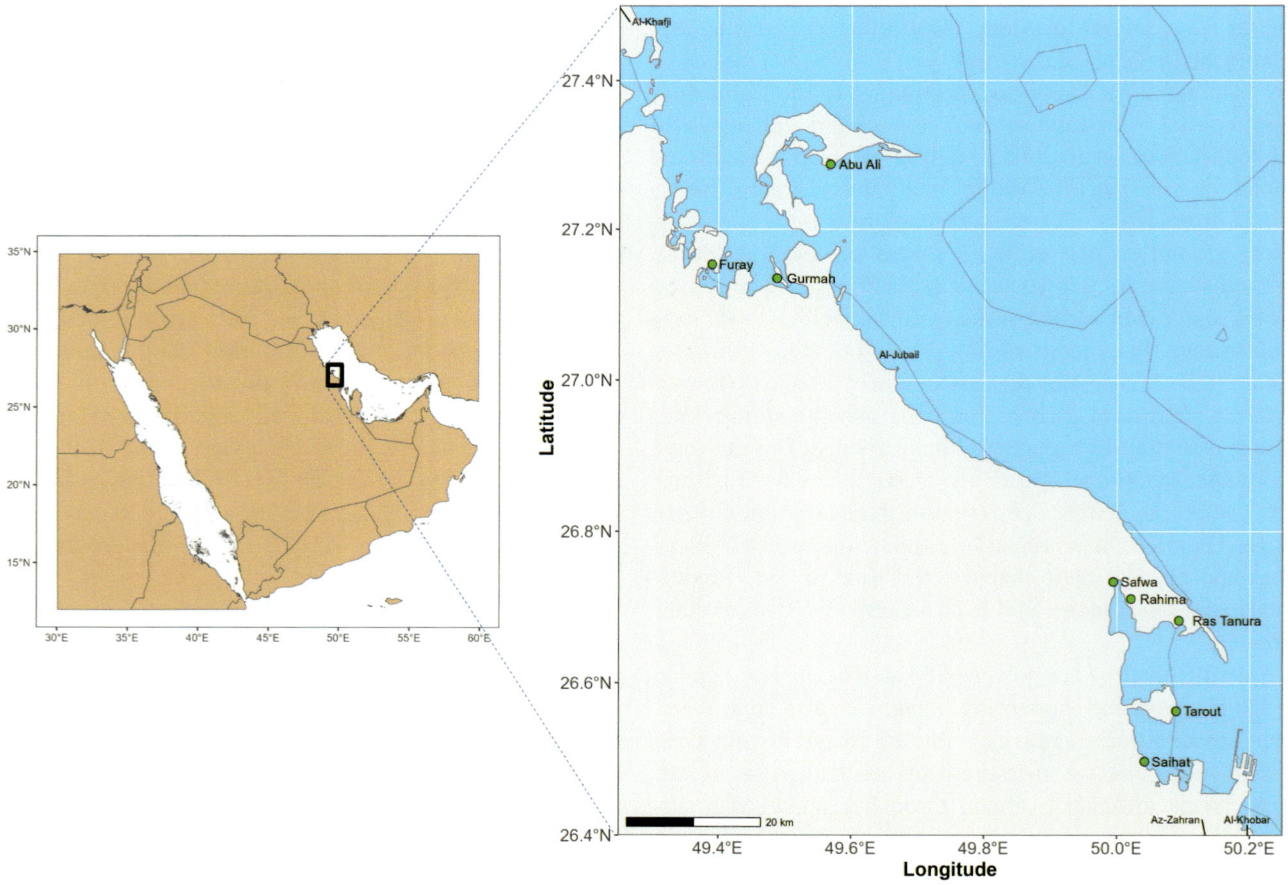

FIGURE 20.1 Study area.

geospatial tools such as Google Earth and Google Maps. This makes it a valuable platform for researchers and practitioners working on mangrove conservation and management.

Mangrove indices are tools used in remote sensing to measure the extent and health of mangrove forests. These indices are derived from satellite imagery and are based on the unique spectral properties of mangrove vegetation. Supervised classification is a technique used to categorize pixels in an image into different land cover classes based on their spectral properties. This technique involves training a classifier using a set of samples, in which each sample represents a different land cover class. The classifier then uses these samples to classify pixels in the image into the different land cover classes. Supervised classification can be used to map the extent and distribution of mangrove forests, as well as to monitor changes in mangrove cover over time. It is a powerful tool for environmental management and conservation as it allows for the identification of areas that need protection and management.

20.2.2 Satellite datasets

The Sentinel-2 multi-spectral instrument (MSI) can collect multispectral data with 13 bands in the visible/near infrared (VNIR) and shortwave infrared (SWIR) regions of the spectrum at a frequency of every five days or less. This contrasts with traditional multispectral satellite sensors like Landsat as the MSI includes three vegetation red edge bands. The spatial resolution for the blue (490 nm), green (560 nm), red (665 nm) and NIR (842 nm) bands is 10 m while the three vegetation red edge bands (705 nm, 740 nm and 783 nm), narrow NIR band (865 nm) and two shortwave infrared bands (1610 nm and 2190 nm) have a resolution of 20 m. This study utilized Sentinel-2 top of atmosphere (TOA) reflectance data from the COPERNICUS/S2 image collection in GEE. The data was filtered to remove scenes with a cloud cover percentage higher than 10%, and then the best image with lowest cloud percentage was used for classification tasks.

The Combined Mangrove Recognition Index (CMRI) developed by Gupta et al. (2018) and Modular Mangrove Recognition Index (MMRI) developed by Diniz et al. (2019) are two commonly used mangrove-specific indices that are derived from satellite imagery. These indices are useful for mapping and monitoring mangrove forests as they enable the identification and quantification of mangroves from other types of vegetation.

The Combined Mangrove Recognition Index (CMRI) is a spectral index (Gupta et al. 2018) for mapping mangrove forests using satellite imagery. It combines several spectral bands from optical sensors, such as Sentinel-2 and Landsat, to highlight the unique spectral characteristics of mangrove forests. The index is based on the fact that mangrove forests reflect more near-infrared (NIR) and red-edge (RE) radiation and absorb more blue and green radiation than non-mangrove vegetation or water. The CMRI formula uses four spectral bands: blue (B), green (G), red-edge (RE) and near-infrared (NIR). The formula for CMRI is:

$$\text{CMRI} = (\text{NDVI} - \text{NDWI})$$

where NDVI and NDWI are calculated by:

$$\text{NDVI} = (\text{Red} - \text{NIR})/(\text{Red} + \text{NIR})$$

$$\text{NDWI} = (\text{Green} - \text{NIR})/(\text{Green} + \text{NIR})$$

CMRI values greater than 0.47 were considered to be mangrove forests, CMRI values between −0.25 and 0.47 represented non-mangrove forests, and CMRI values less than or equal to −0.25 were classified as water-covered areas (Pham Duy et al. 2022).

The Modular Mangrove Recognition Index (MMRI) is a remote sensing–based index designed to identify and map mangrove forests with high accuracy. It was developed by Diniz et al. (2019) as an improvement over previous mangrove indices by using a modular approach. The MMRI combines a vegetation index and a water index, which increases the contrast of mangrove cover. It can be defined as:

$$\text{MMRI} = \text{MNDWI} - |\text{NDVI}|/\text{MNDWI} + |\text{NDVI}|$$

where MNDWI is calculated as:

$$\text{MNDWI} = (\text{Green} - \text{SWIR})/(\text{Green} + \text{SWIR})$$

Thresholds for the distinction of mangroves were based on Sahadevan et al. (2021). In order to calculate CMRI and MMRI, Sentinel-2 datasets were filtered for each year from 2018 to 2022 and were further filtered to remove cloudy pixels using Google Earth Engine. The first image with the lowest cloud percentage was selected, and all bands were resampled to 10 metres resolution. The equations for CMRI and MMRI were then applied to the Sentinel-2 bands to generate mangrove-specific indices. These indices were then used to estimate the area of mangroves for each location and year.

Machine learning tools such as classification and regression trees (CART) and random forest (RF) have been widely used in GEE for the classification of mangroves. In recent years, several research studies have applied these tools to analyze satellite images and identify changes in mangrove ecosystems. The CART model has been implemented in GEE for mapping and monitoring mangroves in various regions. The CART model divides the data into smaller groups recursively, based on the best attributes that can discriminate between the various classes. It generates a decision tree that can be used to classify new data. On the other hand, random forest is an ensemble learning technique that blends various decision trees to increase classification accuracy. A study by Zhao et al. (2014) used the CART algorithm in GEE to map the extent of mangroves in the in Hainan Dongzhai Harbour, China. Similarly, Phan and Stive (2022) used the CART algorithm in GEE to map the extent of mangroves in the Mekong Delta region of Vietnam. Ghorbanian et al. (2021) used CART and RF algorithms in GEE to map the extent and distribution of mangroves in the Sundarbans delta region. Kolli et al. (2022) estimated the change in the extent of mangrove ecosystems in the Andaman Islands using different spectral indices in Google Earth Engine based on a random forest model. Based on the best Sentinel-2 cloud-free imagery, the samples were divided into training samples (70%) and testing samples (30%) in order to assess the model's accuracy. The total number of samples utilized for training and testing in the study ranged from 50 to 60 for each class. The final classification was produced by combining indices including NDVI, NDWI, CMRI, MNDWI and MMRI. A noise filter was applied to the categorized image to lower the number of unconnected pixels, and a mask function was added to separate the non-mangrove pixels from the mangrove pixels.

20.3 RESULTS

The results in Table 20.1 may be recast based on overall gain or loss of mangrove coverage as %/year in different locations by CMRI and MMRI. The present result description is confined to 2021 and 2022, and there is no insight into the change of cover over the last five years.

TABLE 20.1 Temporal variations in the mangrove cover from 2018 to 2022 based on CMRI and MMRI indices.

		MANGROVE COVERAGE (SQUARE KM)	
LOCATION	*YEAR*	*MMRI*	*CMRI*
Abuali	2018	0.5889	0.5076
	2019	0.4163	0.3629
	2020	0.4994	0.6946
	2021	0.4134	0.3278
	2022	0.708	0.3983
Furay	2018	0.3101	0.1464
	2019	0.1191	–
	2020	0.2433	0.0218
	2021	0.0468	–
	2022	0.1687	–
Gurmah	2018	0.9679	0.4179
	2019 to 2022	–	–
Rahimah	2018	3.0938	2.008
	2019	2.725	1.2198
	2020	3.2064	2.4029
	2021	2.4408	2.2508
	2022	2.3464	0.8649
RT	2018	0.8284	0.446
	2019	0.3251	0.1343
	2020	0.266	0.1925
	2021	0.2271	0.2276
	2022	0.3	0.1302
Safwah	2018	1.6394	1.3874
	2019	1.6558	1.8426
	2020	1.8048	2.4602
	2021	1.251	2.3788
	2022	1.2659	1.3898
Saihat	2018	0.1104	0.1198
	2019	0.0746	0.1111
	2020	0.0829	0.1319
	2021	0.0786	0.1656
	2022	0.1003	0.1237
Tarout	2018	2.923	4.7637
	2019	1.8504	4.4462
	2020	1.9	6.4627
	2021	2.4255	6.0006
	2022	2.0143	3.246

Table 20.1 shows the mangrove extent in square kilometres for the eight studied locations using the Combined Mangrove Recognition Index (CMRI) and Modular Mangrove Recognition Index (MMRI) over a five-year period from 2018 to 2022. Based on the CMRI and MMRI, Abu Ali experienced an increase in the area estimated by CMRI from 0.3278 km^2 in 2021 to 0.3983 km^2 in 2022 and an increase in the area estimated by MMRI from 0.4134 km^2 in 2021 to 0.708 km^2 in 2022. Furay exhibited an increase in the estimated area by both CMRI and MMRI in 2022 compared to 2021. CMRI showed an increase from 0 to 0.1687 km^2, and MMRI showed an increase from 0.0468 km^2 to 0.1687 km^2. Gurmah had no mangroves in the area from 2018 to 2022 as shown by both the CMRI and MMRI. In Rahimah, the area estimated by CMRI decreased from 2.2508 km^2 in 2021 to 0.8649 km^2 in 2022 while the area estimated by MMRI decreased from 2.4408 km^2 in 2021 to 2.3464 km^2 in 2022. In Ras Tanura (RT), the area estimated by CMRI increased from 0.2276 km^2 in 2021 to 0.1302 km^2 in 2022, while the area estimated by MMRI increased from 0.2271 km^2 in 2021 to 0.3 km^2 in 2022. Safwah experienced a decrease in the area estimated by CMRI from 2.3788 km^2 in 2021 to 1.3898 km^2 in 2022 while the area estimated by MMRI increased from 1.251 km^2 in 2021 to 1.2659 km^2 in 2022. In Saihat, the area estimated by CMRI increased from 0.1656 km^2 in 2021 to 0.1237 km^2 in 2022 while the area estimated by MMRI increased from 0.0786 km^2 in 2021 to 0.1003 km^2 in 2022. Finally, in Tarout, the area estimated by CMRI decreased from 6.0006 km^2 in 2021 to 3.246 km^2 in 2022 while the area estimated by MMRI decreased from 2.4255 km^2 in 2021 to 2.0143 km^2 in 2022.

Table 20.2 presents the area of mangroves in square kilometres in different locations, classified using two techniques, CART and RF, for the years 2018 and 2022. Figures 20.2 to 20.9 show the classification based on random forest and CART for 2018 and 2022. This study employed 50 to 60 sample observation points for each of the three identified categories, which include non-mangrove vegetation (NMV), mangrove vegetation (MV), land and water bodies. The accuracy of values ranged from > 95% for CART and > 90% for others, applied on Sentinel-2 data.

Overall, the results show that the area of mangroves in most locations increased from 2018 to 2022. The highest increase was observed in Gurmah, where the area of mangroves increased by 53% from 0.39 km^2 in 2018 to 0.99 km^2 in 2022. The area of mangroves in Safwah remained unchanged between 2018 and 2022, indicating that conservation efforts have been successful in maintaining the mangroves in this location.

In Abu Ali, the area of mangroves increased by 73% from 0.33 km^2 in 2018 to 0.57 km^2 in 2022 when classified using the CART technique and by 39% from 0.36 km^2 to 0.51 km^2 when classified using the RF technique. Similarly, in Furay, the area of mangroves increased by 19% from 0.47 km^2 in 2018 to 0.56 km^2 in 2022 when classified using the CART technique while it remained constant at 0.48 km^2 when classified using the RF technique.

In Tarout, the area of mangroves decreased by 3% from 1.31 km^2 in 2018 to 1.27 km^2 in 2022 when classified using the

TABLE 20.2 Temporal variations in the mangrove cover between 2018 and 2022 based on CART and random forest models.

LOCATION	*CLASSIFICATION TYPE*	*YEAR*	*AREA OF MANGROVES (SQ KM)*
AbuAli	CART	2018	0.33
		2022	0.57
	RF	2018	0.36
		2022	0.51
Furay	CART	2018	0.47
		2022	0.56
	RF	2018	0.48
		2022	0.48
Gurmah	CART	2018	0.32
		2022	0.79
	RF	2018	0.39
		2022	0.99
Rahimah	CART	2018	0.98
		2022	0.82
	RF	2018	0.92
		2022	1.23
RT	CART	2018	0.43
		2022	0.65
	RF	2018	0.42
		2022	0.52
Safwah	CART	2018	2.09
		2022	2.09
	RF	2018	2.01
		2022	1.98
Saihat	CART	2018	0.11
		2022	0.11
	RF	2018	0.10
		2022	0.10
Tarout	CART	2018	1.31
		2022	1.27
	RF	2018	1.17
		2022	0.79

CART technique while it decreased by 32% from 1.17 km^2 to 0.79 km^2 when classified using the RF technique. In Saihat, the area of mangroves remained constant between 2018 and 2022 at 0.11 km^2 for both classification techniques.

Furthermore, comparing the two classification techniques, the results show that the RF technique generally yielded lower estimates of the area of mangroves than the CART technique. This was particularly evident in Tarout, where the area of mangroves estimated using the RF technique was significantly lower than that estimated using the CART technique.

20.4 DISCUSSION

From the results of the present study, the total coverage of mangroves in the Arabian Gulf in Saudi Arabia ranged from 6.6 km^2 (based on random forest classification) to 6.86 km^2 (based on CART classification). Almahasheer (2018) estimated the total coverage of fragmented mangroves in the Arabian Gulf was about 165 km^2, with the highest coverage of 78.76 km^2 in the UAE and the lowest coverage of 0.58 km^2 in Kuwait. Saudi Arabia ranked third in terms of mangrove coverage, with an estimated area of 10.36 km^2. Li et al. (2019) used Landsat 7 and 8 data to estimate mangrove cover in the Eastern Province of Saudi Arabia while the current study used Sentinel-2 data. In terms of the specific locations where the studies overlap, there are some differences in the estimates of mangrove area. In Gurmah, Li et al. (2019) estimated mangrove cover to be between 0.348 and 0.389 km^2, while the current study reports a wider range of 0.32–0.99 km^2. Similarly, for Safwah, Li et al. (2019) estimated mangrove cover to be between 1.516 and 1.806 km^2 while the current study reports a narrower range of 1.98–2.09 km^2. For Rahimah, Li et al. (2019) estimated mangrove cover to be between 1.902 and 2.687 km^2 while the current study reports a narrower range of 0.82–1.23 km^2. Finally, for Abu Ali, Li et al. (2019) estimated mangrove cover to be 0.1928 km^2 while the current study reports a wider range of 0.33–0.57 km^2. The differences in estimates could be due to a number of factors. One possible explanation is differences in the datasets used to estimate mangrove cover. These differences in methodology could lead to differences in the estimates of mangrove cover.

Khan and Kumar (2009) estimated the total coverage of mangroves in Tarout was 3.9 km^2 in 1999. Subsequently, Almahasheer et al. (2013) estimated the coverage in Tarout was 5.4 km^2. However, in the current study, the estimated mangrove coverage was 0.79 km^2 based on random forest classification and 1.27 km^2 based on CART classification. The decrease of 77% in total coverage could be due to urban development and land reclamation.

However, despite the ecological importance of mangroves, their sustainability in the Eastern Province of Saudi Arabia has been threatened by various factors, including human activities and climate change. To address these threats and conserve the mangroves in the Eastern Province, various conservation efforts have been undertaken. These include the establishment of protected areas, the implementation of laws and regulations to control human activities in mangrove areas and the restoration of degraded mangrove ecosystems. One of the key initiatives is the Jubail Mangrove Park, which was established in 2014 and covers an area of 12.4 km^2. Another recent initiative was from Saudi Aramco, a company which planted around 100 mangrove seedlings in 1993, followed by massive planting campaigns in a huge area of Abu Ali. The company also established a dedicated Mangrove Ecopark in Ras Tanura, which has a visitor centre and research laboratory, allowing the local community and educational institutions to actively increase their knowledge, awareness and appreciation of mangrove ecosystems.

20.5 CONCLUSION

The findings of this study offer crucial information on the condition and distribution of mangrove forests in Saudi Arabia's Eastern Province. The findings are consistent with previous studies conducted in the region, which have reported the

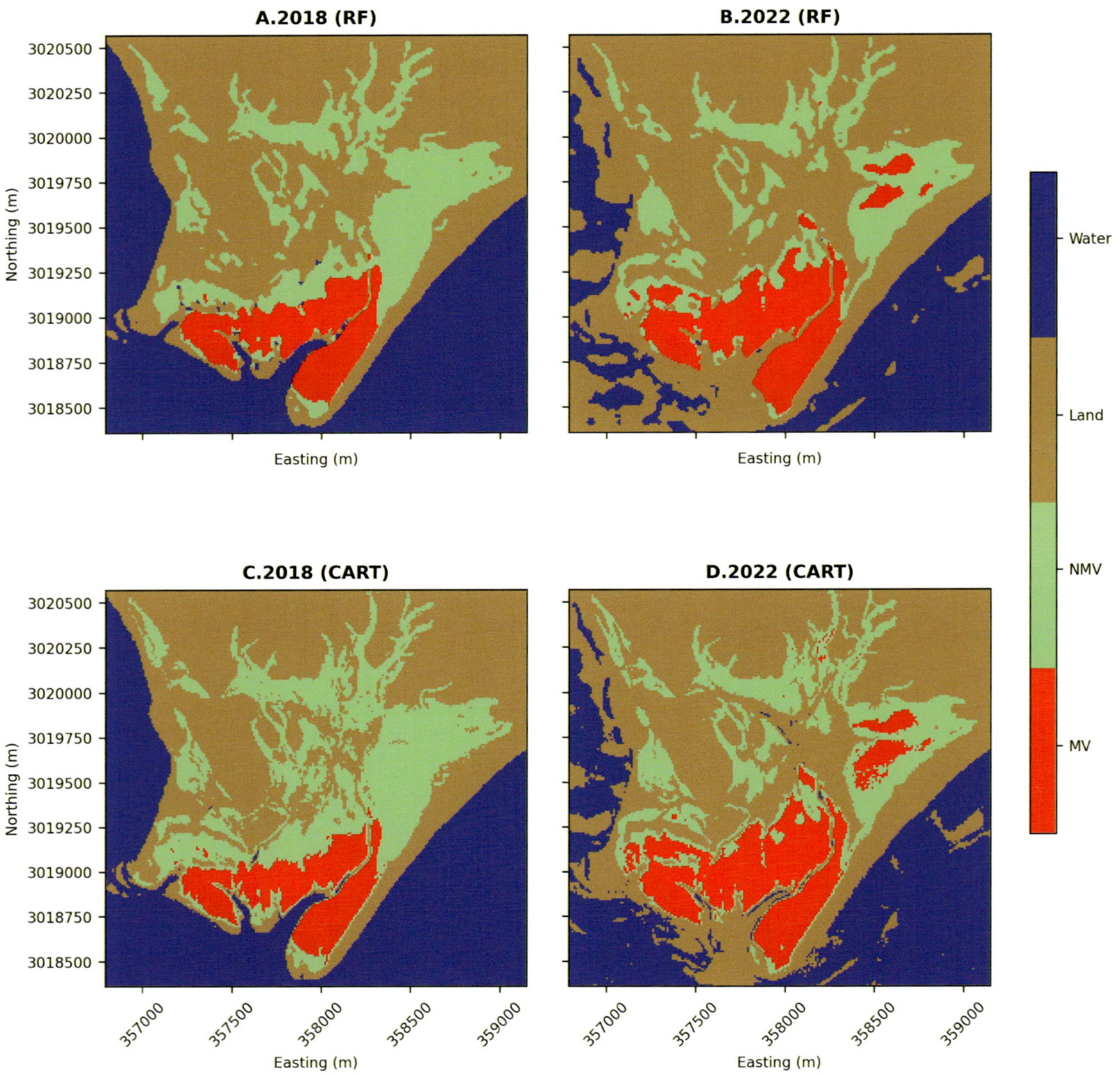

FIGURE 20.2 Land-use classification map of Abu Ali based on CART and random forest classifications (NMV: non-mangrove vegetation; MV: mangrove vegetation; coordinates (metres) for Figures 20.2 to 20.9 are in UTM zone 39).

presence of mangrove forests along the coastline of the Eastern Province. Overall, the study highlights the need for conservation efforts to protect these important ecosystems in the region. It is important to continue monitoring and mapping the extent of mangroves in this region using high-resolution satellite imagery and accurate classification algorithms as mangroves play a crucial role in the protection and preservation of the coastline and its biodiversity.

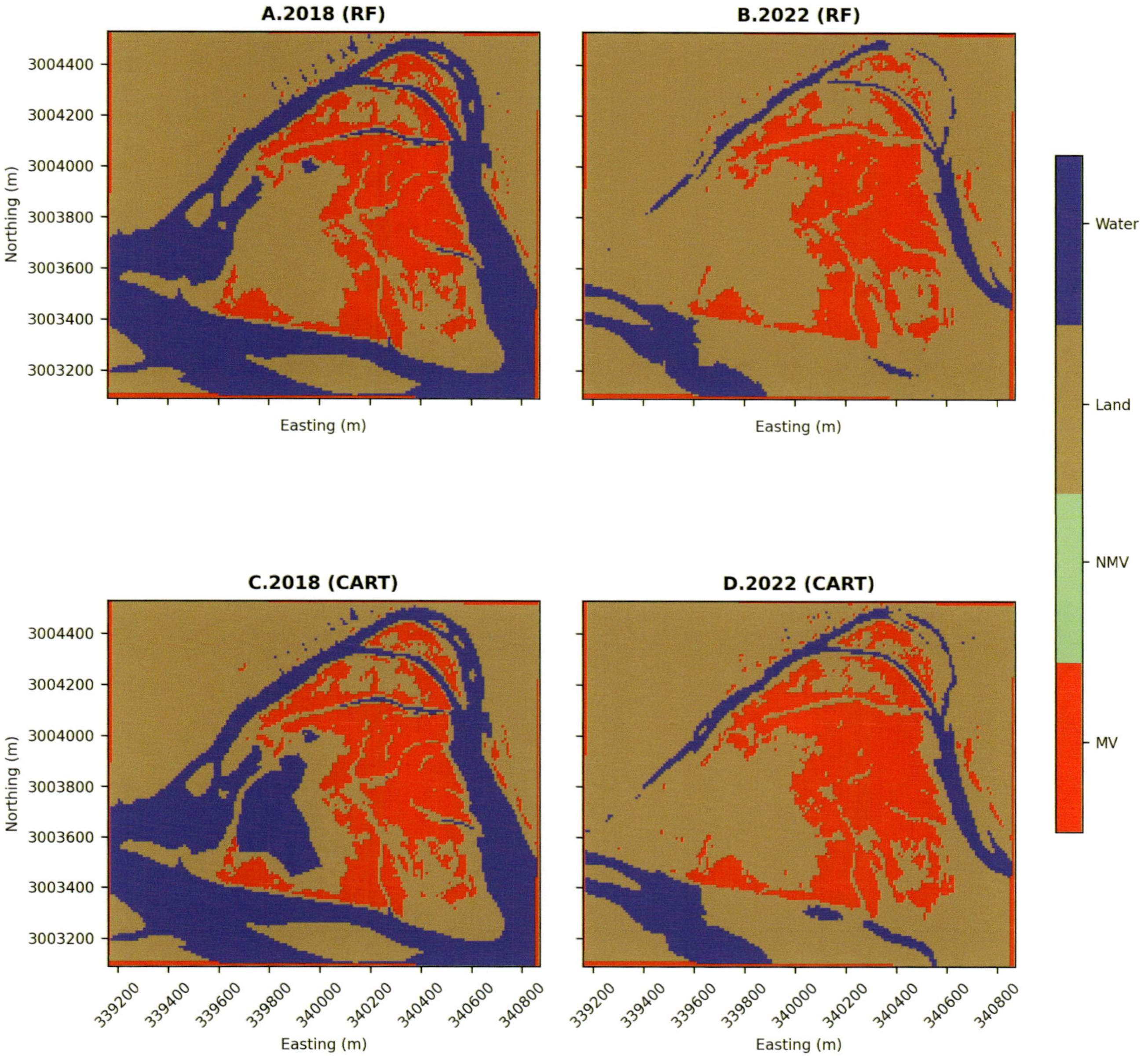

FIGURE 20.3 Land-use classification map of Furay based on CART and random forest classifications.

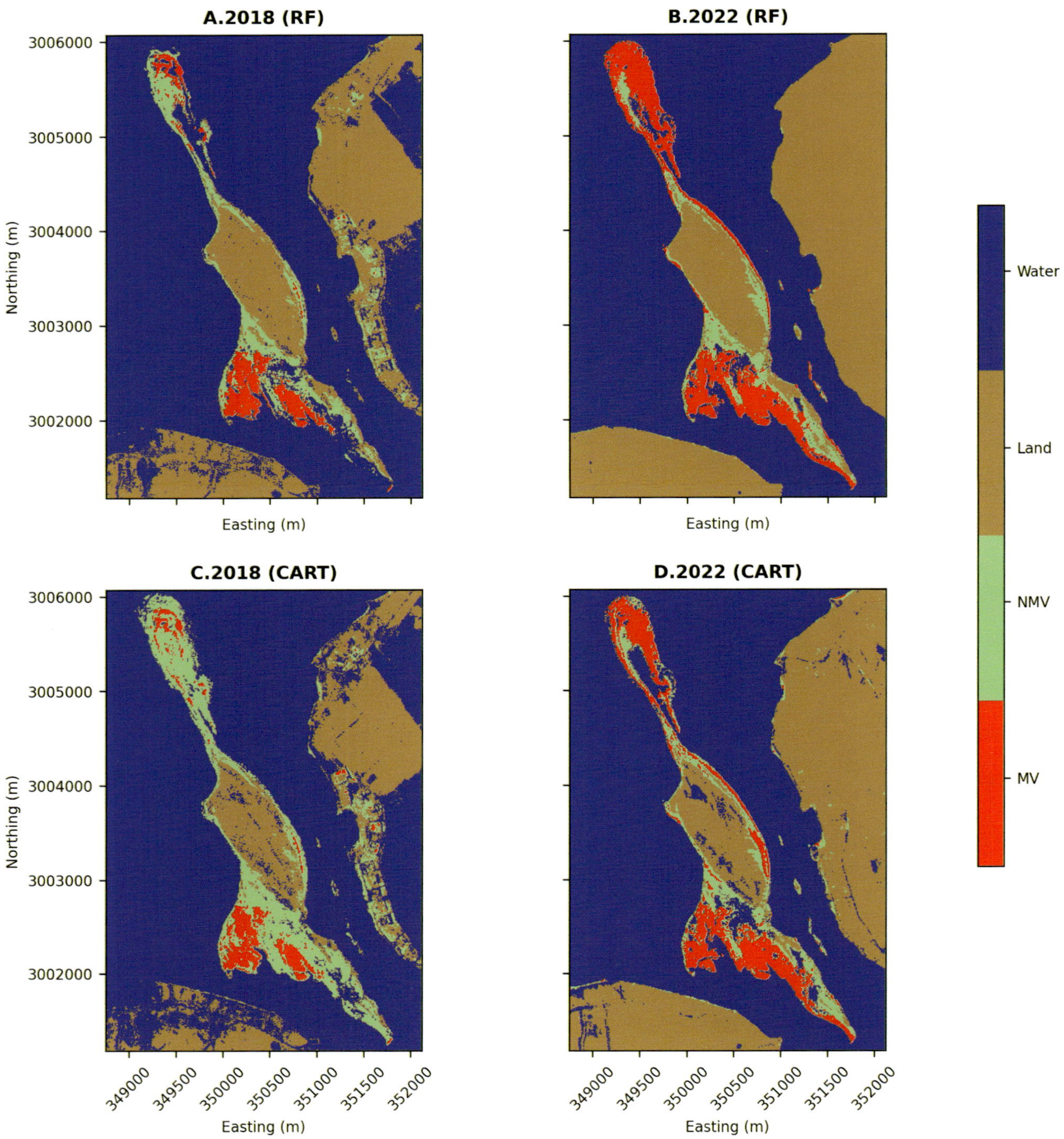

FIGURE 20.4 Land-use classification map of Gurmah based on CART and random forest classifications.

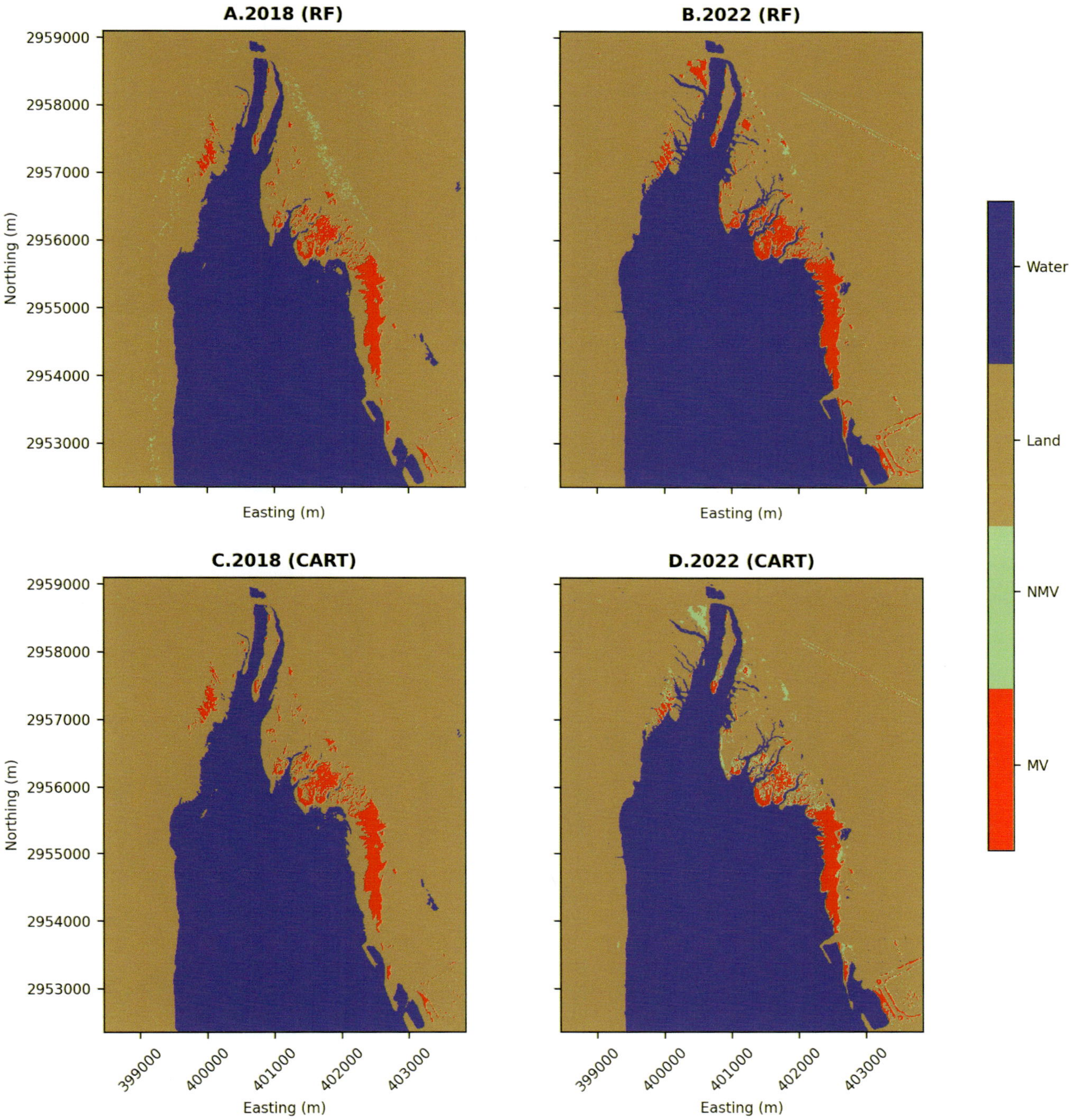

FIGURE 20.5 Land-use classification map of Rahimah based on CART and random forest classifications.

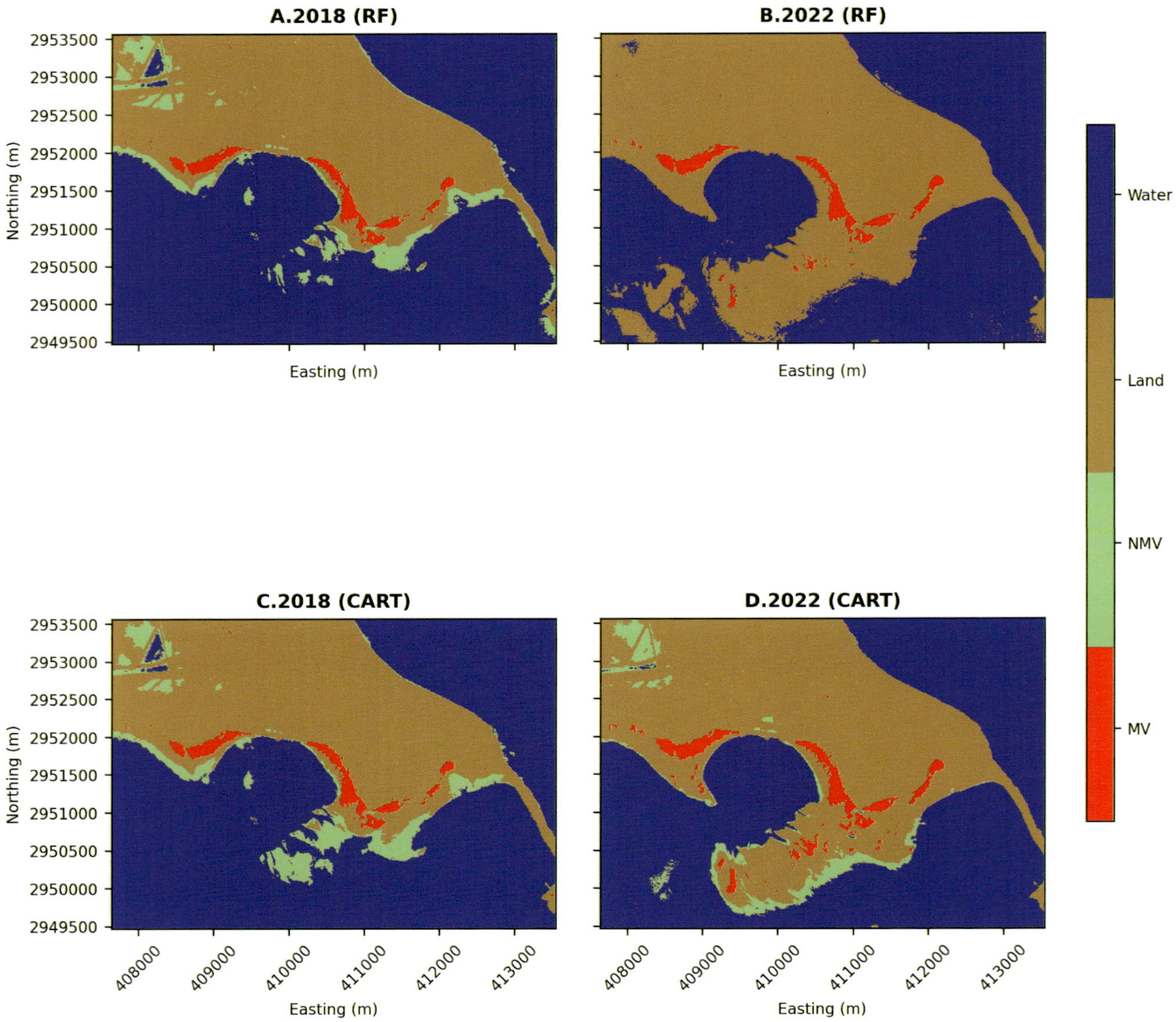

FIGURE 20.6 Land-use classification map of Ras Tanura based on CART and random forest classifications.

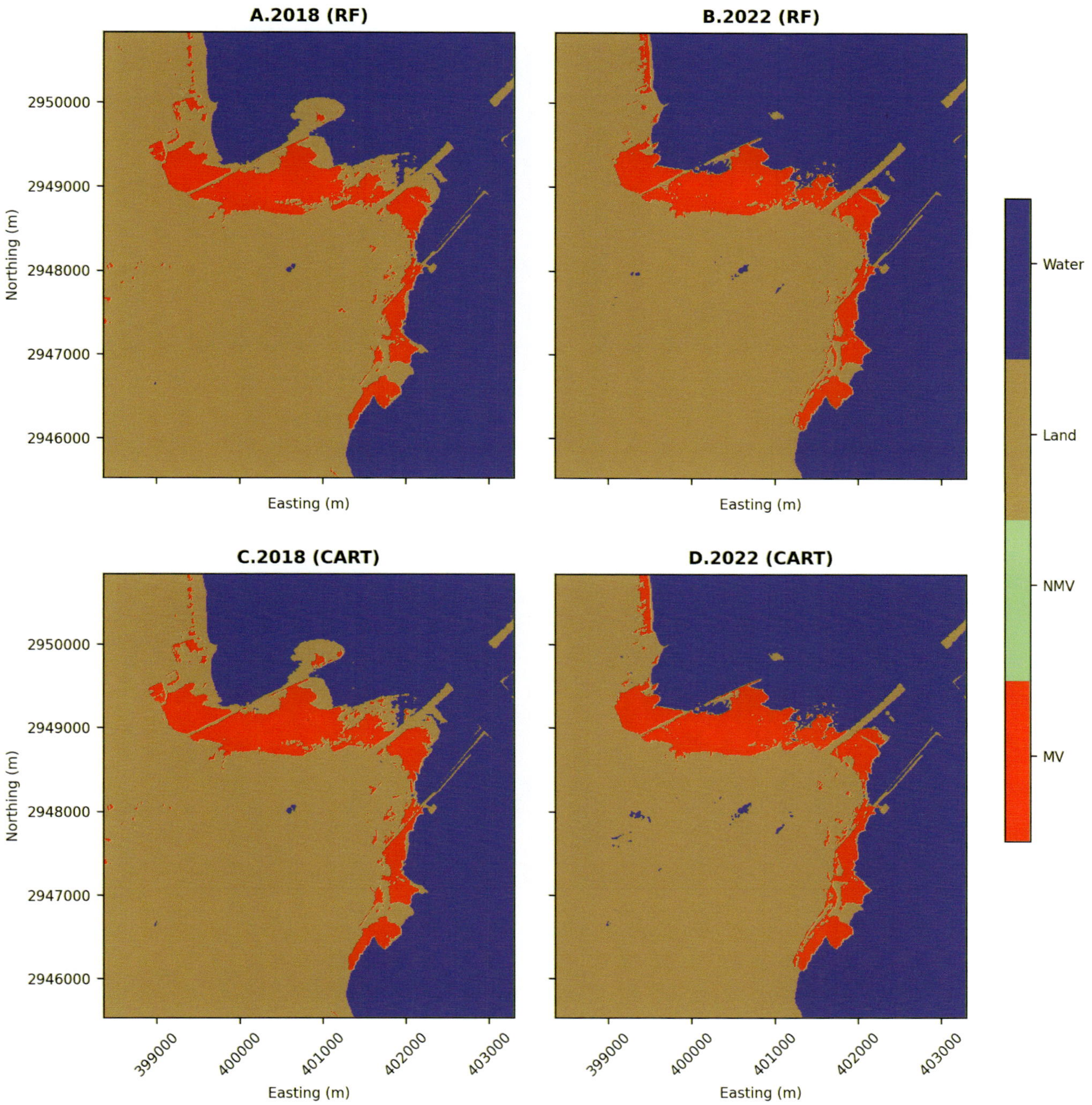

FIGURE 20.7 Land-use classification map of Safwah based on CART and random forest classifications.

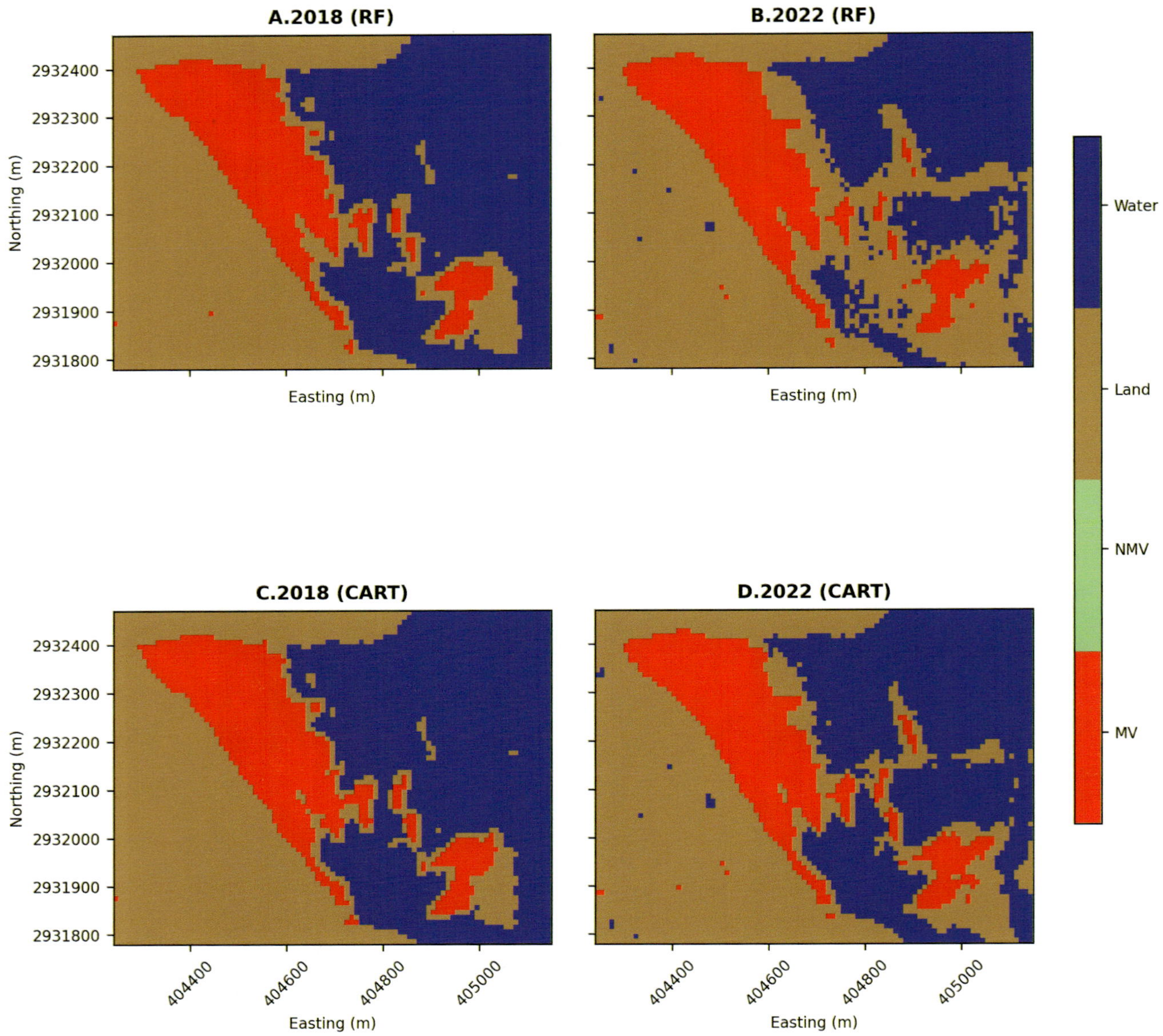

FIGURE 20.8 Land-use classification map of Saihat based on CART and random forest classifications.

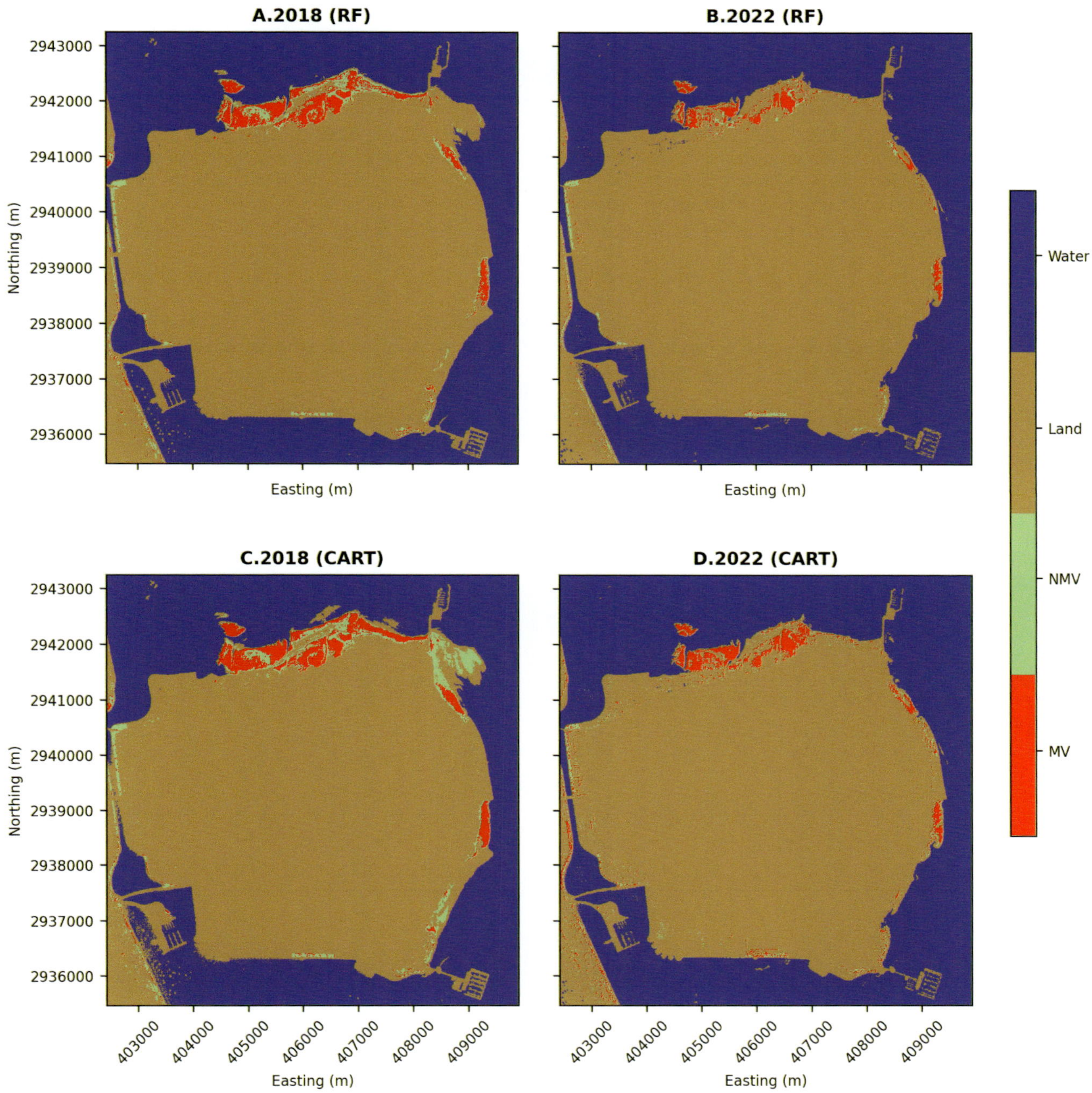

FIGURE 20.9 Land-use classification map of Tarout based on CART and random forest classifications.

REFERENCES

Abido, M.S., Abahussain, A.A., Abdel, M., et al. 2011. Status and composition of mangrove plant community in Tubli Bay of Bahrain during the years 2005 and 2010. *Arab Gulf J Sci Res* 29: 100–111.

Aljenaid, S., et al. 2022. Assessing the spatiotemporal changes, associated carbon stock, and potential emissions of mangroves in Bahrain using GIS and remote sensing data. *Reg Stud Mar Sci* 52: 102282.

Almahasheer, H. 2018. Spatial coverage of mangrove communities in the Arabian Gulf. *Environ Monit Assess* 190: 85.

Almahasheer, H., Aljowair, A., Duarte, C.M., and X. Irigoien. 2016. Decadal stability of Red Sea mangroves. *Estuar Coast Shelf Sci* 169: 164–172.

Almahasheer, H., Al-Taisan, W., and M.K. Mohamed. 2013. Mangrove deterioration in Tarut Bay on the eastern province of the Kingdom of Saudi Arabia. *Pakhtunkhwa J Life Sci* 1: 49–59.

Alongi, D.M. 2002. Present state and future of the world's mangrove forests. *Environ Conserv* 29: 331–349.

Alongi, D.M., and D. Murdiyarso. 2015. Carbon storage and fluxes in mangrove forests. *Wiley Interdiscip Rev Water* 2: 211–224.

Bunting, P., Rosenqvist, A., Hilarides, L., Lucas, R.M., Thomas, N., Tadono, T., Worthington, T.A., Spalding, M., Murray, N.J., and L.-M. Rebelo. 2022. Global mangrove extent change 1996–2020: Global Mangrove watch version 3.0. *Remote Sens* 14: 3657.

Dahdouh-Guebas, F., et al. 2005. Mangrove vegetation dynamics and coastal change in Kwazulu-Natal, South Africa (1997–2001) using aerial photography and change detection. *J Coastal Res* 39: 1347–1352.

Diniz, C., et al. 2019. Brazilian mangrove status: Three decades of satellite data analysis. *Remote Sens* 11: 808.

Donato, D.C., et al. 2011. Mangroves among the most carbon-rich forests in the tropics. *Nat Geosci* 4: 293–297.

Elmahdy, S.I., et al. 2020. Spatiotemporal mapping and monitoring of mangrove forests changes from 1990 to 2019 in the Northern Emirates, UAE using random forest, Kernel logistic regression and Naive Bayes Tree models. *Front Environ Sci* 8: 102.

Etemadi, H., Smoak, J.M., and E. Abbasi. 2021. Spatiotemporal pattern of degradation in arid mangrove forests of the Northern Persian Gulf. *Oceanologia* 63: 99–114.

Friess, D.A., et al. 2016. Mangroves are one of the most carbon-rich forests in the tropics. *Nat Clim Change* 6: 1038–1043.

Ghorbanian, A., et al. 2021. Mangrove ecosystem mapping using Sentinel-1 and Sentinel-2 satellite images and random forest algorithm in Google Earth Engine. *Remote Sens* 13: 2565.

Giri, C., et al. 2011. Status and distribution of mangrove forests of the world using earth observation satellite data. *Glob Ecol Biogeogr* 20: 154–159.

Gupta, K., et al. 2018. An index for discrimination of mangroves from non-mangroves using LANDSAT 8 OLI imagery. *MethodsX* 5: 1129–1139.

Habshi, A.A., Youssef, T., Aizpuru, M., and F. Blasco. 2007. New mangrove ecosystem data along the UAE coast using remote sensing. *Aquat Ecosyst Health Manag* 10: 309–319.

Howari, F.M., Jordan, B.R., Bouhouche, N., and S. Wyllie-Echeverria. 2009. Field and remote-sensing assessment of mangrove forests and seagrass beds in the northwestern part of the United Arab Emirates. *J Coastal Res* 25: 48–56.

Khan, M.A., and A. Kumar. 2009. Impact of urban development on mangrove forests along the west coast of the Arabian Gulf. *Earth Sci India* 2: 159–173.

Kolli, M.K., et al. 2022. Assessment of change in the extent of mangrove ecosystems using different spectral indices in Google Earth Engine based on random forest model. *Arab J Geosci* 15: 889.

Kumar, A. 2011. *Temporal Changes in Mangrove Cover Between 1972 and 2001 Along the South Coast of the Arabian Gulf.* Dhahran: King Fahd University of Petroleum and Minerals.

Li, W., et al. 2019. Using multi-indices approach to quantify mangrove changes over the Western Arabian Gulf along Saudi Arabia coast. *Ecol Indic* 102: 734–745.

Nagendra, H., Sudhira, H.S., Katti, M., and M. Schewenius. 2013. Sub-regional assessment of India: Effects of urbanization on land use, biodiversity and ecosystem services. In: *Urbanization, Biodiversity and Ecosystem Services: Challenges and Opportunities: A Global Assessment*. Eds. Thomas Elmqvist, et al., 65–74. Dordrecht: Springer.

Pham Duy, Q., Sounthala, P., and H. Nguyen Hai. 2022. Using Sentinel-2a derived CMRI for mangrove cover mapping over 7 years (2016–2022) in Quang Yen, Quang Ninh province. *J Forest Sci Technol* 14: 141–152.

Phan, M.H., and M.J.F. Stive. 2022. Managing mangroves and coastal land cover in the Mekong Delta. *Ocean Coast Manag* 219: 106013.

Richards, D.R., and D.A. Friess. 2016. Rates and drivers of mangrove deforestation in Southeast Asia, 2000–2012. *Proc Natl Acad Sci* 113: 344–349.

Sahadevan, A.S., Joseph, C., Gopinath, G., Ramakrishnan, R., and P. Gupta. 2021. Monitoring the rapid changes in mangrove vegetation of coastal urban environment using polynomial trend analysis of temporal satellite data. *Reg Stud Mar Sci* 46: 101871.

Zhao, S., Liu, Y., Jiang, J., Cheng, W., Zhou, M., Li, M., and R. Ruan. 2014. Extraction of mangrove in Hainan Dongzhai Harbor based on CART decision tree. In: *2014 22nd International Conference on Geoinformatics*, IEEE, pp. 1–6.

The Red Sea mangroves and the mangrove microbiome

21

Hanin Alzubaidy, Azad Baazeem and Alexandre S. Rosado

21.1 RED SEA CHARACTERISTICS

The Red Sea (Al-Bahr Al-Ahmar in Arabic), with a surface area of 4.5 × 1011 km^2, lies between Africa and Asia as a seawater inlet of the Indian Ocean, to which it connects through the Bab el Mandab Strait in the south. The northern area of the Red Sea is split into two narrow bays flanking the Sinai Peninsula; these are the Gulf of Suez, which is connected to the Mediterranean Sea by the Suez Canal, and the Gulf of Aqaba (Elsawy and Lakhouit 2020).

The Red Sea is surrounded by six countries. On the eastern shores, Saudi Arabia covers most of the central and northeastern shores of the Red Sea while Yemen is on the southeastern side. The western shores of the Red Sea are bordered by Egypt in the northwest; Sudan in the central west; and, in the southwest, Eritrea and Djibouti. The Red Sea also contains more than a thousand islands of various sizes, with the largest two groups being the Farasan Islands of Saudi Arabia in the southeast, and the Dahlak Archipelago of Eritrea in the west (Khalil 2015).

The Red Sea has a unique position in the arid climate belt of the Northern Hemisphere, where the major characteristics of the region are its low precipitation; high evaporation rates from water surfaces and high sea surface temperatures of 22°C and <26°C in winter and 29–30°C and >31°C in summer in the northern and southern Red Sea, respectively (Berumen et al. 2019; Eladawy et al. 2017); and the absence of major rivers that drain into the sea. These characteristics of the Red Sea contribute to its high salinity, reaching 36% or 36 practical salinity units (PSU) in the south and 41% in the northern region, compared to the average seawater salinity of 35% (Loáiciga 2006; Blanco-Sacristán et al. 2022).

Another unique characteristic of the Red Sea is its oligotrophic waters due to the lack of river input and major connection to the ocean waters. However, the connection of the Red Sea to the Indian Ocean is the primary source of nutrients for this semi-enclosed water body (Raitsos et al. 2015). In general, the Red Sea region has a humid tropical to subtropical climate, with the surface water temperatures ranging from 22°C to >31°C, and the fastest warming rate in the world of 0.40–0.45°C/decade, compared to the average global rate of 0.17±0.07°C/decade (Berumen et al. 2019; Shaltout 2019).

21.2 RED SEA MANGROVE ECOSYSTEM

The Red Sea coast has unique coastal wetlands such as seagrass beds, saltpans, salt marshes, coral reefs, and the focus of this chapter, mangroves. The term "mangrove" is used to describe shrubs and trees that are located in intertidal as well as shallow zones of tropical and subtropical tidal marshes. Their ability to thrive in brackish and salty water wetlands eliminates competition from plants that are only adapted to freshwater wetlands (Nyanga 2020). To adapt to the high salinity conditions of the Red Sea and salt wetlands in general, mangroves have salt coping mechanisms that give them the ability to filter out more than 90% of the salt in seawater. Another form of mangrove adaptation is their water-hoarding mechanism, which facilitates their thick, succulent leaves. Mangroves also breathe through their unique root structures and snorkel-like parts, depending on the species (Eid and Shaltout 2016; Nyanga 2020). Mangroves on the Red Sea coasts grow in nutrient-poor conditions, which gives them a more dwarfed morphology than mangroves thriving in nutrient-rich ecosystems, reflecting the nutrient limitation in the Red Sea, along with the scarce rainfall and semi-enclosed conditions (Eid and Shaltout 2016; Almahasheer et al. 2016a). Mangrove forests on the African shores of the Red Sea cover an area of approximately 69 km^2, and on the Arabian Peninsula shores have an area of 51 km^2 (Almahasheer et al. 2016b).

21.3 RED SEA MANGROVE PHENOTYPE, ABUNDANCE AND DISTRIBUTION

The mangrove ecosystem creates a habitat and nursery for various species of fish, crabs, shrimp, and molluscs. The mangrove species that inhabit the coasts of the Red Sea are *Avicennia marina* and *Rhizophora mucronata* (Figure 21.1). However, the mangrove species *Ceriops tagal* has been found in Eritrea

DOI: 10.1201/9781003321408-28

FIGURE 21.1 The phenotype of Red Sea mangrove ecosystem (*Rhizophora mucronata* and *Avicennia marina*).

and Djibouti, and *Bruguiera gymnorhiza* has been found only in Djibouti (Khalil 2015).

A. marina is the dominant mangrove species in the Red Sea, while *R. mucronata* is the most abundant in the south of the Red Sea (Khalil 2015). *A. marina* is characterized as the mangrove species most tolerant to high salinity and aridity due to their great leaf water storage capability, which explains their abundance in the Red Sea (Nguyen et al. 2017).

Their distribution in the Red Sea is mostly dependent on the soil characteristics such as texture, bulk density and sand and clay content (Blanco-Sacristán et al. 2022). In the northern Red Sea, there are stony corals which make it difficult for the mangrove to thrive; on the other hand, the southern Red Sea has finer sediment, a high clay content, rainwater drainage and nutrient input from the Indian Ocean, facilitating the growth of the mangroves in that region (Anton et al. 2020). It is speculated, using satellite data and metadata with a maximum entropy modeling approach, that most of the islands in the Red Sea have a high potential for hosting a greater mangrove distribution (Blanco-Sacristán et al. 2022). The current mangrove area in the Red Sea is 175.3 km^2, and the estimated area suitable for mangrove expansion is 410.9 km^2 (Blanco-Sacristán et al. 2022).

21.4 MANGROVE PHENOTYPE IN THE CENTRAL, NORTHERN AND SOUTHERN RED SEA

21.4.1 *Avicennia marina*

Even though dwarf *A. marina* and *R. mucronata* inhabit the Red Sea as mentioned earlier, they tend to have different heights and trunk girths along the coasts. It was reported by Almahasheer et al. (2016b) that the average height of *A. marina* trees in the central Red Sea was 1.847 m. Another report by Anton et al. (2020) emphasized that the shortest *A. marina* trees were observed in the central Red Sea, with an average height of 2 m.

In the northern Red Sea, our ongoing studies in Al Wajh Lagoon (25.654°N, 36.449°E) measured 479 trees with an average height of 1.70 m. However, we noticed that the tree height and trunk girth differed on each island of the lagoon as well as in the lagoon coastal area, as illustrated in Figure 21.2. The results show that Ghawar Island (25.721°N, 36.554°E) has the tallest *A. marina* trees, with a mean (±SE) of 2.79 ±13.4 m; in contrast, Abu Lahiq Island (25.522°N, 36.922°E) has the shortest *A. marina* trees with a mean of 0.88 ±2.5 m. The phenotype difference of the mangrove trees is highly correlated to the soil characteristics and topography of each island (Blanco-Sacristán et al. 2022).

In the southern Red Sea, *A. marina* was found in the height range of 0.3 m to 5 m in Sudan, Yemen and Saudi Arabia (Khalil 2004; Anton et al. 2020). Moreover, in Saudi Arabia, it was reported that some trees were above 5 m high (Anton et al. 2020). However, the tallest *A. marina* was reported in Yemen between the Al-Ruays and Yakhtul inner zone with a height of 6 m to 7 m (Khalil 2004).

21.4.2 *Rhizophora mucronata*

R. mucronata has different heights when it stands alone and when it is co-occurring with *A. marina*. In the central Red Sea, precisely 37 km south of Shalateen, Egypt, El-Khouly reported the average height of *R. mucronata* alone was 2.73±14.91 m, and they tend to be shorter when they co-occur with *A. marina*, with an average height of 1.02±42.14 m (Ahmed and Abdel-Hamid 2007).

In the northern Red Sea, at Al Wajh Lagoon, we encountered *R. mucronata* in the Um Rumah and Sheybarah Islands, standing alone or co-occurring with *A. marina* in Um Rumah Island and only co-occurring with *A. marina* in Sheybarah Island. The height of *R. mucronata* in Al Wajh Lagoon was about 1.6 m.

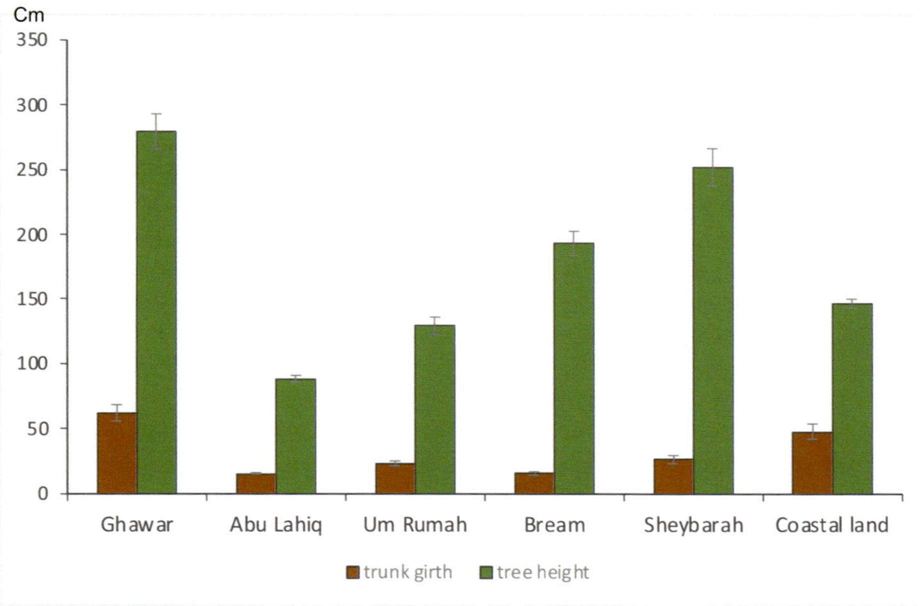

FIGURE 21.2 *Avicennia marina* phenotype in Al Wajh Lagoon.

FIGURE 21.3 Camel-grazing effect on Al Wajh Lagoon mangrove micro-forest.

In the southern Red Sea, *R. mucronata* is more abundant and well developed. In Yemen, the *R. mucronata* height on Kamaran Island was reported to be an average of 7 to 9 m. In Gandal Islet, one of Al-Hudaydah's islets, *R. mucronata* height was reported to be an average of 3 m to 6 m (Khalil 2004). Strikingly, the most well-developed *Rhizophora mucronata* trees in the southern Red Sea were observed in two locations in Djibouti – in Godoria and on Moucha Island – where *R. mucronata* height was reported to be an average of 9 m to 13 m (Khalil 2004).

21.4.3 Phenotypic variation factors

Other factors besides soil characteristics that drive the differences in the phenotypes of the mangrove trees, especially in the Red Sea, are camel grazing (Figure 21.3) and anthropogenic factors such as organic pollution by sewage effluence, browsing and/or cutting trees and oil pollution (Khalil 2004). An important factor that contributes significantly to the soil health and properties in intertidal systems is animal and microbial biocoenosis, which is the main driver of biogeochemical cycles such as sulfur cycling, nitrogen cycling and carbon cycling. In sulfur cycling, the microbes play a role in reducing sulfate to sulfide by sulfate-reducing bacteria, then oxidizing sulfide to sulfate by sulfur-oxidizing bacteria (van Vliet et al. 2021). In nitrogen cycling, microbes play a role in fixing nitrogen into ammonia by nitrogen fixers such as cyanobacteria, nitrogen-fixing bacteria and archaea, then oxidizing ammonia to hydroxylamine and further to nitrite by aerobic ammonia-oxidizing bacteria and archaea. Interestingly, some members of the *nitrospira* bacterial genus can oxidize ammonia to nitrite under ammonia-limited conditions. After that, nitrite is oxidized to nitrate by aerobic nitrite-oxidizing bacteria, anoxygenic phototrophs and anaerobic ammonium-oxidizing bacteria. As it is a cycle, nitrite is reduced to ammonium and nitric oxide by the majority of bacteria, then nitric oxide is reduced to nitrogen again (Kuypers et al. 2018). In carbon cycling, the microbes play role in sequestering carbon by photosynthetic eukaryotes, cyanobacteria and mixotrophs that consume other microbes. Then particulate organic matter from these microbes and dissolved organic matter respire carbon by heterotrophic bacteria, archaea, saprotrophs, mixotrophs and photosynthetic taxa in the absence of light (Worden et al. 2015). As mentioned earlier, these biogeochemical cycles and their produced compounds contribute greatly to the health of soil and water and, therefore, the health and phenotype of the plant.

Bioturbation by fiddler crabs in the Red Sea mangrove sediment has shown better aeration of the soil, enhanced nutrient assimilation and recycling and enhanced microbial diversity that contributes to the thriving of the central Red Sea mangrove trees (Fusi et al. 2022). Hereafter, we will focus on the role of the microbiome in plants: specifically, in the mangrove ecosystem.

21.5 THE MANGROVE MICROBIOME

Various plants have developed different strategies to exploit their beneficial microbes to fight pathogens and, simultaneously, to gain tolerance to abiotic stresses; the molecular mechanism behind this has been gaining more interest from microbiologists and plant scientists over the years (de Zélicourt et al. 2018). Recently, marine biologists and microbiologists became interested in understanding mangrove plants and their microbiome (Allard et al. 2020). We believe that exploring the existing microbiome associated with each micro-environment in different mangrove areas around the world would help us understand this symbiosis in order to protect mangrove forests and sustain their health. We anticipate that, in extreme environments of high salinity, high surface temperature and oligotrophic conditions, such as the central Red Sea, the microbiome is tailored to confer tolerance to the host. Understanding such interactions would unlock the great potential of the extremophilic microbiome to serve as a tool in sustaining healthy shores of mangrove forests and healthier water bodies for marine life.

21.6 PLANTS AND MICROBIOME

Generally, a plant's health in a natural environment depends on complex and dynamic interactions with macro- and microorganisms (Sasse et al. 2018). The complex interactions of microbes in plants and the plant ecosystem sustain a relationship that both the plants and the microbes need for nutrition and ecological balance (Muwawa et al. 2021). Given that plants grow in soils rich with residing microbes, they tend to have high microbial diversity (Edwards et al. 2015). The microbiome is defined as a group of dynamic and complex microbial communities that include bacteria, archaea, fungi, viruses, eukaryotes and their genomic material. There are many factors present in each plant's microbiome that influence the composition of the microbial communities. These factors include the biogeographical properties of the plants, as well as the ecological factors affecting the soil surrounding them, such as the nutrients present in the soil, the presence of organic or inorganic compounds (e.g., pollution from sewage drainage and oil spills) and the food web composition of the plant's ecosystem (Andreote et al. 2012).

The pathogenic roles some microbes play are well understood, but their benefit to the host's health and stress tolerance remains an active area of research (Simmons et al. 2018). Both plants and microbes have evolutionarily adapted to mutually benefit from their close relationship (Edwards et al. 2015). For example, microbes play a critical role in biogeochemical cycles by supplying nutrients that degrade complex compounds into more usable forms for plant uptake. In return, bacteria in the rhizosphere receive carbon metabolites secreted from the plant through root exudates (Edwards et al. 2015; Andreote et al. 2012). The positive influence of the plant-microbe interaction has also been observed in plant growth (Bais et al. 2006). Microbes contribute to plant growth by various mechanisms, such as atmospheric nitrogen fixation (Moulin et al. 2001) and the direct and indirect effects of plant-growth-promoting rhizobacteria (Gray and Smith 2005). An additional example of the positive relationship between plants and their microbiome is the presence of fungal endophytes in grasses. These fungal endophytes were shown to play a role in increasing biotic and abiotic stress tolerance in grass (Schardl et al. 2004). Moreover, some bacteria can form biofilms for plant protection while others produce antibiotics that defend plants against pathogens (Bais et al. 2004). Finally, some microbes have the ability to degrade compounds (many of which are allelopathic or toxic) produced by the plant or other microbes present in the soil (Bais et al. 2006).

In the following sections, a deeper view of mangrove microbiome functions will be discussed. An important question to address concerning the microbiome is whether or not the plant contributes to the way its microbiome is shaped, and if so, what mechanisms underlie the recruitment of these organisms from the plant's rhizosphere. Although the answers remain elusive, the proposed mechanisms will be further discussed in the following sections. Since the prokaryotic communities play key roles in the biogeochemical transformation and nutrient cycling of the mangrove ecosystem (Muwawa et al. 2021), it is important to understand the interaction between the host and its microbiome to get a broader view of the host's health, fitness and stress tolerance (Simmons et al. 2018).

21.7 HOW THE MICROBIOME HELPS MANGROVES

Although the mangrove trees' physiology has allowed them to adapt to tough environmental conditions, a key player in the mangrove ecosystem is the microbiome surrounding it (Alzubaidy et al. 2016). These conditions (e.g., high salt, low oxygen in its choking mud, and high sulfur) create an attractive hotspot for microbial communities, making this complex ecosystem a pool of unique microbial diversities (Andreote et al. 2012; Muwawa et al. 2021). In addition, the sediment in a mangrove ecosystem is rich in organic compounds and is anaerobic, with a thin aerobic layer around the roots. These properties of the sediment further attract various types of microbes (Reef et al. 2010), especially sulfate-reducing bacteria and methanogens (Zhuang et al. 2020). As these microbes gain an excellent source of nutrients to support their proliferation, they assist in maintaining a healthy mangrove ecosystem (Thatoi et al. 2013).

Many studies have demonstrated the importance of the mutually beneficial relationship between mangroves and their associated microbiome in the mangroves' resistance, productivity, recovery and conservation (Andreote et al. 2012; Holguin et al. 2001). An example of the positive relationship between mangrove trees and their microbiome is in the decomposition and mineralization of mangrove leaf, bark and wood (Reef et al. 2010). The leaf litter of mangroves contains mainly lignocellulose and is degradable by microorganisms. In terrestrial systems and mangrove forests, fungi are the main decomposers of lignocellulose. They decompose it using cellulases, hemicelluloses, ligninases, carbohydrate esterases, lytic polysaccharide monooxygenase and xylanase (Arfi et al. 2013; Paixão et al. 2021). After the leaves fall, their degradation process starts immediately when they are colonized by the bacteria and fungi present in the mangrove sediment: e.g., bacteria from the genera *Microbulbifer*, *Teredinibacter* and *Saccharophagus* and the mangrove fungus *Pestalotiopsis sp.* (Gontikaki et al. 2015; Arfi et al. 2013). The degradation process could last for two to six months or longer for wood (Holguin et al. 2001). Studies have reported an increase in the protein, nitrogen and phosphorus content in dead red and grey mangrove leaves, accompanied by a decrease in their weight after 27 to 105 days (Holguin et al. 2001).

Nitrogen and phosphorus were implicated as the growth-limiting factors in mangroves (Reef et al. 2010). The nitrogen present in the sediment is derived from the degradation of nitrogenous compounds from the organic litter present in the sediment. This nitrogen is converted by bacteria to ammonium to conserve the nitrogen in the ecosystem for later absorption by the plant. Bacteria could also compete with plants for the nitrogen present in the ecosystem. This highlights the complex relationship plants have with the surrounding microbiome (Holguin et al. 2001). In mangroves, the top layer of the sediment and the thin aerobic layer around the roots create a favourable environment for nitrifying bacteria that convert ammonium to nitrate for the mangrove uptake. These ammonia-oxidizing bacteria and archaea and nitrifying bacteria account for 40–60% of the nitrogen required by the mangrove trees. In addition, many intertidal mangrove habitats contain benthic microbial mats that could contribute to nitrogen fixation and the mangrove nitrogen cycle, especially when they are dominated by nitrogen-fixing cyanobacteria (Reef et al. 2010).

Another important nutrient for plant growth is phosphorus. However, phosphorus is immobile and unavailable for direct plant uptake in nutrient-limited environments such as the mangrove ecosystem. Thus, microbes that solubilize phosphate are essential for facilitating plant growth in mangroves (Reef et al. 2010). Plants residing in low salinity soils are reported to have a symbiotic relationship between the roots and arbuscular mycorrhizal (AM) fungi. AM fungi are widespread in nearly all soil types and are known for their role in the uptake of immobile nutrients – most importantly, phosphate (Bradley et al. 1982). Very few studies have addressed the occurrence of AM fungi in mangrove sediments or observed the occurrence of AM fungi in low salinity mangrove sediments (Kothamasi et al. 2006). The absence of AM fungi in high salinity soils would limit the plant's ability to absorb many nutrients (such as zinc, copper, iron and phosphorus), thereby increasing the plant's susceptibility to toxic metals (Bradley et al. 1982). Similarly, some mangroves have been associated with phosphate-solubilizing bacteria in the oxygenated area around the roots (Kothamasi et al. 2006). Moreover, the sulfur bacteria from the genera of *Beggiatoa* and *Thiomargarita* are shown to play a vital role in increasing phosphorus availability in the sediment via releasing phosphate from stored polyphosphate and phosphorite formations (Brock and Schulz-Vogt 2011) and nitrogen availability via storing nitrate when oxidizing sulfide to sulfate (Schulz and Jørgensen 2001).

21.8 HOW MANGROVES HELP MICROBES

The mangrove ecosystem also serves as a shelter for microorganisms and creates a nutrient-rich ecosystem where microbes can thrive. This section will focus on these two services that the mangrove ecosystem provides.

Even though mangroves depend on microbes for nutrient recycling, microbes benefit as well from their association with mangroves by exploiting carbon metabolites and other nutrients secreted by mangrove roots as root exudates (Zhuang et al. 2020). Bacteria colonizing the root surface induce root exudates, causing the stimulation of microbial activity in both the soils surrounding the plant and the rhizosphere (Yin and Yan 2020; Luo et al. 2018). The presence of high bacterial activity coincides with the presence of mangroves, as opposed to sediments in areas not occupied by mangroves (Yin and Yan 2020). Moreover, sulfate-reducing bacteria were also reported in greater abundance where mangrove roots were present. The mangrove tree's aerial roots in some species can also supply oxygen to the otherwise anaerobic sediment.

In addition to providing nutrients for the microbial community, mangroves can alter the physiochemical properties of the sediment, such as altering the pH and salinity caused by rainfall. During a rainy season, the sediment in an Indian mangrove ecosystem was less acidic than the sediment that was not occupied by mangroves. The reported decrease in bacterial population in the deforested areas was proposed to be caused by the increase in sediment salinity and loss of carbon input caused by the absence of mangrove trees (Holguin et al. 2001).

21.9 SHAPING THE MICROBIOME

The plant microbiome is determined by environmental factors such as climate and nutrient availability, in addition to host-related factors like the genotype, organ, species and plant health

(Ofek-Lalzar et al. 2014). Plant species and genotypes have been shown to play a significant role in shaping the rhizosphere and a plant's microbiome. Depending on these factors, different plant organs are also associated with specific microbial communities. Many studies have applied this knowledge to produce specific inoculum for protecting and improving the health of crop plants and model plants such as *Arabidopsis thaliana* (Andrés-Barrao et al. 2021; Türksoy et al. 2022; Vogel et al. 2021). However, there is limited information regarding tree species in mangrove forests, with only a few studies that consider multiple organs such as the roots, leaves, stem and fruits (Escobar Prieto 2021; Purahong et al. 2019). Therefore, the microbial community distribution within each organ remains elusive. To explore this further, recent studies have focused on several different aspects, such as the different fungal and bacterial communities in each organ of the mangrove tree (Lee et al. 2020; Purahong et al. 2019) while other studies focused on a specific organ, mainly roots and leaves, and investigated the microbial communities grouped by their presence in the endosphere or episphere of the specific organ (Yao et al. 2019; Zhuang et al. 2020). Some studies have also considered the association of more than one organ with the same bacterial or fungal community with differences in microbial abundance (Purahong et al. 2019). Through studying the diversity, assembly and function of the mangrove tree's root-associated microbiome, Zhuang et al. (2020) demonstrated the selective role of the soil-root interface in shaping the microbial community structure in the mangrove tree. This process was proposed to be done through root exudates since the bacterial operational taxonomic units (OTUs) enriched in the rhizosphere were also enriched in either the rhizoplane or the endosphere of the roots (Edwards et al. 2015; Bais et al. 2006). Plant roots secrete a large range of low molecular weight compounds into the rhizosphere, including ions, free oxygen, water, enzymes, hormones, mucilage and a diverse range of primary and secondary metabolites containing carbon (Bais et al. 2006). In this way, the roots' exudation provides a valuable source of carbon and other nutrients for microbes while the microbes colonizing the rhizosphere help the plants acquire phosphorus, potassium and nitrogen, or help the plant cope with stress (Alzubaidy et al. 2016). This mutually beneficial exchange of services between mangroves and microbes creates a symbiotic relationship that was first described by Hiltner in 1904 as the "rhizosphere effect" (Bais et al. 2006). The root microbiome assembly was proposed to follow a multistep model in which the initial step is the secretion of waste material with an unknown function, followed by the exudation of compounds with known functions. The response of the microbes to these compounds could be beneficial or pathogenic, which creates competition in the vicinity of the roots between soil microbes for nutrients (Bais et al. 2006). Other studies have also highlighted the role of the host epigenetics in fine-tuning their root-associated microbiome to adapt to their environment (Wilkinson and Ton 2020).

21.10 OMICS APPROACHES TO UNDERSTANDING THE MANGROVE MICROBIOME

The resistance of mangroves depends on the mangrove physiology and interaction with its associated microbiome. Thus, it is important to understand the mangrove ecosystem and work toward improving the mangrove's health. To do that, it is essential to understand the diversity, assembly and function of the microbiome in relation to the mangrove's location, age and organs. There are still many gaps in our current understanding of the mangrove ecosystem and its microbiome. The techniques used in previous studies were culture-dependent techniques, clone libraries and fingerprinting methods (Cleary et al. 2012; Zhang et al. 2009; Liang et al. 2007; Varon-Lopez et al. 2014; Baskaran and Prabavathy 2022; Samanta and Bhadury 2014). These techniques were considered biased and did not represent a complete view of the microbial community due to sampling and methodological limitations (Helfer and Hassenrück 2021). The major limitation in culture-dependent techniques is that 1% or <1% of environmental microorganisms are culturable, which only allows studying this small selection (Hahn et al. 2019). However, it is worth mentioning that this percentage varies based on the studied ecosystem.

The "omics" techniques are currently emerging as a tool to gain a more holistic view of various environmental systems and have gained popularity by applying high-throughput sequencing in studying microbial communities (Muwawa et al. 2021). Several studies have utilized high throughput sequencing of the 16S rRNA and ITS genes using the Illumina platforms, but each study had a different approach in the subsequent analysis steps. Some studies divided the tree's organs (Muwawa et al. 2021; Purahong et al. 2019; Yao et al. 2019; Zhuang et al. 2020) while others focused on a single organ, such as the roots (Zhuang et al. 2020) or leaves (Yao et al. 2019), then divided it into additional sub-levels. The diversity of sampling methods, sequencing techniques and analysis pipelines has generated a large amount of data. While genomics has allowed for a more holistic view of the microbial genome present, there are sequences with unassigned taxonomy that are yet to be identified (Lee et al. 2020). In addition, there is a lack of metatranscriptomic, metaproteomics and metabolomics data to further elucidate the intricacies and complexities of the microbial activity in the mangrove ecosystem (Alzubaidy et al. 2016).

REFERENCES

Ahmed, E.-K.A., and A. Abdel-Hamid. 2007. Zonation pattern of Avicennia Marina and Rhizophora Mucronata along the Red Sea coast, Egypt. *World Appl Sci J* 2, no. 4: 283–288.

Allard, S.M., Costa, M.T., Bulseco, A.N., Helfer, V., Wilkins, L.G.E., Hassenrück, C., Zengler, K., et al. 2020. Introducing the mangrove microbiome initiative: Identifying microbial research priorities and approaches to better understand, protect, and rehabilitate mangrove ecosystems. *MSystems* 5, no. 5: e00658-20. http://doi.org/10.1128/mSystems.00658-20.

Almahasheer, H., Duarte, C.M., and X. Irigoien. 2016a. Nutrient limitation in central Red Sea mangroves. *Front Mar Sci* 3: 271.

Almahasheer, H., Duarte, C.M., and X. Irigoien. 2016b. Phenology and growth dynamics of Avicennia marina in the central Red Sea. *Sci Rep* 6, no. 1: 37785. http://doi.org/10.1038/srep37785.

Alzubaidy, H., Essack, M., Malas, T.B., Bokhari, A., Motwalli, O., Kamanu, F.K., Jamhor, S.A., et al. 2016. Rhizosphere microbiome metagenomics of gray mangroves (Avicennia Marina) in the Red Sea. *Mar Genom* 576, no. 2, part 1: 626–636. http://doi.org/10.1016/j.gene.2015.10.032.

Andreote, F.D., Jiménez, D.J., Chaves, D., Dias, A.C.F., Luvizotto, D.M., Dini-Andreote, F., Fasanella, C.C., et al. 2012. The microbiome of Brazilian mangrove sediments as revealed by metagenomics. *PLoS ONE* 7, no. 6: e38600. http://doi.org/10.1371/journal.pone.0038600.

Andrés-Barrao, C., Alzubaidy, H., Jalal, R., Mariappan, K.G., de Zélicourt, A., Bokhari, A., Artyukh, O., et al. 2021. Coordinated bacterial and plant sulfur metabolism in enterobacter Sp. SA187 – Induced plant salt stress tolerance. *Proc Natl Acad Sci* 118, no. 46: e2107417118. http://doi.org/10.1073/pnas.2107417118.

Anton, A., Almahasheer, H., Delgado, A., Garcias-Bonet, N., Carrillo-de-Albornoz, P., Marbà, N., Hendriks, I.E., et al. 2020. Stunted mangrove trees in the oligotrophic central Red Sea relate to nitrogen limitation. *Front Mar Sci* 7. www.frontiersin.org/articles/10.3389/fmars.2020.00597.

Arfi, Y., Chevret, D., Henrissat, B., Berrin, J.-G., Levasseur, A., and E. Record. 2013. Characterization of salt-adapted secreted lignocellulolytic enzymes from the mangrove Fungus Pestalotiopsis Sp. *Nat Comm* 4, no. 1: 1810. http://doi.org/10.1038/ncomms2850.

Bais, H.P., Fall, R., and J.M. Vivanco. 2004. Biocontrol of Bacillus Subtilis against infection of Arabidopsis roots by Pseudomonas Syringae is facilitated by biofilm formation and surfactin production. *Plant Physiol* 134, no. 1: 307–319.

Bais, H.P., Weir, T.L., Perry, L.G., Gilroy, S., and J.M. Vivanco. 2006. The role of root exudates in rhizosphere interactions with plants and other organisms. *Ann Rev Plant Biol* 57, no. 1: 233–266. http://doi.org/10.1146/annurev.arplant.57.032905.105159.

Baskaran, V., and V.R. Prabavathy. 2022. Diverse key nitrogen cycling genes NifH, NirS and NosZ associated with Pichavaram Mangrove rhizospheres as revealed by culture-dependent and culture-independent analyses. *Arch Microbiol* 204, no. 1: 109. http://doi.org/10.1007/s00203-021-02661-4.

Berumen, M.L., Voolstra, C.R., Daffonchio, D., Agusti, S., Aranda, M., Irigoien, X., Jones, B.H., Morán, X.A.G., and C.M. Duarte. 2019. The Red Sea: Environmental gradients shape a natural laboratory in a nascent ocean. In: *Coral Reefs of the Red Sea*. Eds. C.R. Voolstra and M.L. Berumen, 1–10. Cham: Springer International Publishing. http://doi.org/10.1007/978-3-030-05802-9_1.

Blanco-Sacristán, J., Johansen, K., Duarte, C.M., Daffonchio, D., Hoteit, D.I., and M.F. McCabe. 2022. Mangrove distribution and afforestation potential in the Red Sea. *Sci Total Environ* 843: 157098. http://doi.org/10.1016/j.scitotenv.2022.157098.

Bradley, R., Burt, A.J., and D.J. Read. 1982. The biology of Mycorrhiza in the Ericaceae: VIII. The role of Mycorrhizal infection in heavy metal resistance. *New Phytologist* 91, no. 2: 197–209.

Brock, J., and H.N. Schulz-Vogt. 2011. Sulfide induces phosphate release from polyphosphate in cultures of a marine Beggiatoa strain. *ISME J* 5, no. 3: 497–506. http://doi.org/10.1038/ismej.2010.135.

Cleary, D.F.R., Smalla, K., Mendonça-Hagler, L.C.S., and N.C.M. Gomes. 2012. Assessment of variation in bacterial composition among microhabitats in a mangrove environment using DGGE fingerprints and barcoded pyrosequencing. *PLoS ONE* 7, no. 1: 1–8. http://doi.org/10.1371/journal.pone.0029380.

Edwards, J., Johnson, C., Santos-Medellín, C., Lurie, E., Podishetty, N.K., Bhatnagar, S., Eisen, J.A., and V. Sundaresan. 2015. Structure, variation, and assembly of the root-associated microbiomes of rice. *Proc Natl Acad Sci* 112, no. 8: E911–920. http://doi.org/10.1073/pnas.1414592112.

Eid, E.M., and K.H. Shaltout. 2016. Distribution of soil organic carbon in the mangrove Avicennia Marina (Forssk.) Vierh. along the Egyptian Red Sea Coast. *Reg Stud Mar Sci* 3: 76–82. http://doi.org/10.1016/j.rsma.2015.05.006.

Eladawy, A., Nadaoka, K., Negm, A., Abdel-Fattah, S., Hanafy, M., and M. Shaltout. 2017. Characterization of the northern Red Sea's oceanic features with remote sensing data and outputs from a global circulation model. *Oceanologia* 59, no. 3: 213–237. http://doi.org/10.1016/j.oceano.2017.01.002.

Elsawy, M.B.D., and A. Lakhouit. 2020. A review on the impact of salinity on foundation soil of coastal infrastructures and its implications to north of Red Sea coastal constructions. *Arab J Geosci* 13, no. 13: 555. http://doi.org/10.1007/s12517-020-05601-6.

Escobar Prieto, J.D. 2021. *Environmental Origin and Compartmentalization of Bacterial Communities Associated with Avicennia Marina Mangroves on the Red Sea Coast.* M.Sc. Thesis, KAUST, Thuwal, Saudi Arabia. http://doi.org/10.25781/KAUST-91856.

Fusi, M., Booth, J.M., Marasco, R., Merlino, G., Garcias-Bonet, N., Barozzi, A., Garuglieri, E., et al. 2022. Bioturbation intensity modifies the sediment microbiome and biochemistry and supports plant growth in an arid mangrove system. *Microbiol Spectrum* 10, no. 3: e01117–01122. http://doi.org/10.1128/spectrum.01117-22.

Gontikaki, E., Thornton, B., Cornulier, T., and U. Witte. 2015. Occurrence of priming in the degradation of lignocellulose in marine sediments. *PLoS ONE* 10, no. 12: e0143917. http://doi.org/10.1371/journal.pone.0143917.

Gray, E.J., and D.L. Smith. 2005. Intracellular and extracellular PGPR: Commonalities and distinctions in the plant – bacterium signaling processes. *Soil Biol Biochem* 37, no. 3: 395–412.

Hahn, M.W., Koll, U., and J. Schmidt. 2019. Isolation and cultivation of bacteria. In: *The Structure and Function of Aquatic Microbial Communities.* Ed. C.J. Hurst, 313–351. Cham: Springer International Publishing. http://doi.org/10.1007/978-3-030-16775-2_10.

Helfer, V., and C. Hassenrück. 2021. Microbial communities in mangrove sediments. In: *Dynamic Sedimentary Environments of Mangrove Coasts.* Eds. F. Sidik and D.A. Friess, 141–175. Elsevier. http://doi.org/10.1016/B978-0-12-816437-2.00003-3.

Holguin, G., Vazquez, P., and Y. Bashan. 2001. The role of sediment microorganisms in the productivity, conservation, and rehabilitation of mangrove ecosystems: An overview. *Biol Fertil Soils* 33, no. 4: 265–278. http://doi.org/10.1007/s003740000319.

Khalil, A.S.M. 2004. Status of mangroves in the Red Sea and Gulf of Aden. *PERSGA Technical Ser* 11.

Khalil, A.S.M. 2015. Mangroves of the Red Sea. In: *The Red Sea: The Formation, Morphology, Oceanography and Environment of a Young Ocean Basin.* Eds. N.M.A. Najeeb and I.C.F. Stewart, 585–597. Berlin and Heidelberg: Springer Earth System Sciences. http://doi.org/10.1007/978-3-662-45201-1_33.

Kothamasi, D., Kothamasi, S., Bhattacharyya, A., Kuhad, R.C., and C.R. Babu. 2006. Arbuscular Mycorrhizae and phosphate solubilising bacteria of the rhizosphere of the mangrove ecosystem of Great Nicobar Island, India. *Biol Fertil Soils* 42, no. 4: 358–361. http://doi.org/10.1007/s00374-005-0035-8.

Kuypers, M.M.M., Marchant, H.K., and B. Kartal. 2018. The microbial nitrogen-cycling network. *Nat Rev Microbiol* 16, no. 5: 263–276. http://doi.org/10.1038/nrmicro.2018.9.

Lee, N.L.Y., Huang, D., Quek, Z.B.R., Lee, J.N., and B.J. Wainwright. 2020. Distinct fungal communities associated with different organs of the mangrove Sonneratia Alba in the Malay Peninsula. *IMA Fungus* 11, no. 1: 17. http://doi.org/10.1186/s43008-020-00042-y.

Liang, J.-B., Chen, Y.-Q., Lan, C.-Y., Tam, N.F.Y., Zan, Q.-J., and L.-N. Huang. 2007. Recovery of novel bacterial diversity from mangrove sediment. *Mar Biol* 150, no. 5: 739–747. http://doi.org/10.1007/s00227-006-0377-2.

Loáiciga, H.A. 2006. Modern-age buildup of CO2 and its effects on seawater acidity and salinity. *Geophys Res Lett* 33, no. 10: L10605. http://doi.org/10.1029/2006GL026305.

Luo, L., Wu, R., Gu, J.-D., Zhang, J., Deng, S., Zhang, Y., Wang, L., and Y. He. 2018. Influence of mangrove roots on microbial abundance and ecoenzyme activity in sediments of a subtropical coastal mangrove ecosystem. *Int Biodeterioration Biodegradation* 132: 10–17. http://doi.org/10.1016/j.ibiod.2018.05.002.

Moulin, L., Munive, A., Dreyfus, B., and C. Boivin-Masson. 2001. Nodulation of legumes by members of the β-Subclass of Proteobacteria. *Nature* 411, no. 6840: 948–950. http://doi.org/10.1038/35082070.

Muwawa, E.M., Obieze, C.C., Makonde, H.M., Jefwa, J.M., Kahindi, J.H.P., and D.P. Khasa. 2021. 16S RRNA gene amplicon-based metagenomic analysis of bacterial communities in the rhizospheres of selected mangrove species from Mida Creek and Gazi Bay, Kenya. *PLoS ONE* 16, no. 3: e0248485. http://doi.org/10.1371/journal.pone.0248485.

Nguyen, H.T., Meir, P., Sack, L., Evans, J.R., Oliveira, R.S., and M.C. Ball. 2017. Leaf water storage increases with salinity and aridity in the mangrove Avicennia Marina: Integration of leaf structure, osmotic adjustment and access to multiple water sources. *Plant Cell Environ* 40, no. 8: 1576–1591. http://doi.org/10.1111/pce.12962.

Nyanga, C. 2020. The role of mangroves forests in decarbonizing the atmosphere. In: *Carbon-Based Material for Environmental Protection and Remediation*. IntechOpen. http://doi.org/10.5772/intechopen.92249.

Ofek-Lalzar, M., Sela, N., Goldman-Voronov, M., Green, S.J., Hadar, Y., and D. Minz. 2014. Niche and host-associated functional signatures of the root surface microbiome. *Nat Comm* 5, no. 1: 4950. http://doi.org/10.1038/ncomms5950.

Paixão, D.A.A., Tomazetto, G., Sodré, V.R., Gonçalves, T.A., Uchima, C.A., Büchli, F., Alvarez, T.M., et al. 2021. Microbial enrichment and meta-omics analysis identify CAZymes from mangrove sediments with unique properties. *Enzyme Microbial Technol* 148: 109820. http://doi.org/10.1016/j.enzmictec.2021.109820.

Purahong, W., Sadubsarn, D., Tanunchai, B., Wahdan, S.F., Sansupa, C., Noll, M., Wu, Y.-T., and F. Buscot. 2019. First insights into the microbiome of a mangrove tree reveal significant differences in taxonomic and functional composition among plant and soil compartments. *Microorganisms* 7, no. 12: 585. http://doi.org/10.3390/microorganisms7120585.

Raitsos, D.E., Yi, X., Platt, T., Racault, M.-F., Brewin, R.J.W., Pradhan, Y., Papadopoulos, V.P., Sathyendranath, S., and I. Hoteit. 2015. Monsoon oscillations regulate fertility of the Red Sea. *Geophys Res Lett* 42, no. 3: 855–862. http://doi.org/10.1002/2014GL062882.

Reef, R., Feller, I.C., and C.E. Lovelock. 2010. Nutrition of mangroves. *Tree Physiol* 30, no. 9: 1148–1160. http://doi.org/10.1093/treephys/tpq048.

Samanta, B., and P. Bhadury. 2014. Analysis of diversity of chromophytic phytoplankton in a mangrove ecosystem using RbcL gene sequencing. *J Phycol* 50, no. 2: 328–340. http://doi.org/10.1111/jpy.12163.

Sasse, J., Martinoia, E., and T. Northen. 2018. Feed your friends: Do plant exudates shape the root microbiome? *Trends Plant Sci* 23, no. 1: 25–41. http://doi.org/10.1016/j.tplants.2017.09.003.

Schardl, C.L., Leuchtmann, A., and M.J. Spiering. 2004. Symbioses of grasses with seedborne fungal endophytes. *Ann Rev Plant Biol* 55, no. 1: 315–340.

Schulz, H.N., and B.B. Jørgensen. 2001. Big bacteria. *Ann Rev Microbiol* 55, no. 1: 105–137. http://doi.org/10.1146/annurev.micro.55.1.105.

Shaltout, M. 2019. Recent sea surface temperature trends and future scenarios for the Red Sea. *Oceanologia* 61, no. 4: 484–504.

Simmons, T., Caddell, D.F., Deng, S., and D. Coleman-Derr. 2018. Exploring the root microbiome: Extracting bacterial community data from the soil, rhizosphere, and root endosphere. *JoVE (J Visualized Experiments)*, no. 135: e57561.

Thatoi, H., Behera, B.C., Mishra, R.R., and S.K. Dutta. 2013. Biodiversity and biotechnological potential of microorganisms from mangrove ecosystems: A review. *Ann Microbiol* 63, no. 1: 1–19.

Türksoy, G.M., Carron, R., Koprivova, A., Kopriva, S., Wippel, K., and T.G. Andersen. 2022. Modulation of Arabidopsis growth by volatile organic compounds from a root-derived bacterial community. *bioRxiv*. http://doi.org/10.1101/2022.04.12.488003.

van Vliet, D.M., von Meijenfeldt, F.A.B., Dutilh, B.E., Villanueva, L., Damsté, J.S.S., Stams, A.J.M., and I. Sánchez-Andrea. 2021. The bacterial sulfur cycle in expanding dysoxic and euxinic marine waters. *Environ Microbiol* 23, no. 6: 2834–2857. http://doi.org/10.1111/1462-2920.15265.

Varon-Lopez, M., Dias, A.C.F., Fasanella, C.C., Durrer, A., Melo, I.S., Kuramae, E.E., and F.D. Andreote. 2014. Sulphur-oxidizing and sulphate-reducing communities in Brazilian mangrove sediments. *Environ Microbiol* 16, no. 3: 845–855. http://doi.org/10.1111/1462-2920.12237.

Vogel, C.M., Potthoff, D.B., Schäfer, M., Barandun, N., and J.A. Vorholt. 2021. Protective role of the Arabidopsis leaf microbiota against a bacterial pathogen. *Nat Microbiol* 6, no. 12: 1537–1548. http://doi.org/10.1038/s41564-021-00997-7.

Wilkinson, S.W., and J. Ton. 2020. Methylation moulds microbiomes. *Nat Plants* 6, no. 8: 910–911.

Worden, A.Z., Follows, M.J., Giovannoni, S.J., Wilken, S., Zimmerman, A.E., and P.J. Keeling. 2015. Rethinking the marine carbon cycle: Factoring in the multifarious lifestyles of microbes. *Science* 347, no. 6223: 1257594. http://doi.org/10.1126/science.1257594.

Yao, H., Sun, X., He, C., Maitra, P., Li, X.-C., and L.-D. Guo. 2019. Phyllosphere epiphytic and endophytic fungal community and network structures differ in a tropical mangrove ecosystem. *Microbiome* 7, no. 1: 1–15.

Yin, Y., and Z. Yan. 2020. Variations of soil bacterial diversity and metabolic function with tidal flat elevation gradient in an artificial mangrove wetland. *Sci Total Environ* 718: 137385. http://doi.org/10.1016/j.scitotenv.2020.137385.

Zélicourt, A. de, Synek, S., Saad, M.M., Alzubaidy, H., Jalal, R., Xie, Y., Andrés-Barrao, C., et al. 2018. Ethylene induced plant stress tolerance by Enterobacter Sp. SA187 is mediated by 2-keto-4-methylthiobutyric acid production. *PLOS Genet* 14, no. 3: e1007273. http://doi.org/10.1371/journal.pgen.1007273.

Zhang, Y., Dong, J., Yang, B., Ling, J., Wang, Y., and S. Zhang. 2009. Bacterial community structure of mangrove sediments in relation to environmental variables accessed by 16S RRNA gene-denaturing gradient gel electrophoresis fingerprinting. *Sci Mar* 73, no. 3: 487–498. http://doi.org/10.3989/scimar.2009.73n3487.

Zhuang, W., Yu, X., Hu, R., Luo, Z., Liu, X., Zheng, X., Xiao, F., Peng, Y., He, Q., Tian, Y., et al. 2020. Diversity, function and assembly of mangrove root-associated microbial communities at a continuous fine-scale. *NPJ Biofilms Microbiomes* 6, no. 1: 1–10.

Mangroves in therapeutic and experimental medicine

22

Reshma Sahito, Hussain Raza, Qamar Abbas and Song J. Kim

22.1 INTRODUCTION

The Red Sea was initiated around 24 million years ago when the African and Arabian plates split apart (El-Hossary et al. 2020). Since then, the Red Sea has been distinguished by unique qualities, including a relatively high average temperature. The Suez Canal links the Red Sea to the Mediterranean Sea in the north. The Bab El-Mandeb Strait in the south connects the Red Sea to the Gulf of Aden and the Indian Ocean. In all spheres, the Red Sea presents a vast diversity of life (Raitsos et al. 2013). The Red Sea is a significant natural source of bioactive substances. The Red Sea has unique characteristics, such as significant seasonal temperature changes in the air and water as well as a high level of marine biodiversity, that have drawn natural products chemists to study marine creatures there. In fact, of the 180 species of soft coral known to exist worldwide, the Red Sea is home to over 40% of them (Abou El-Ezz et al. 2017).

As the source of the oldest descriptions of mangroves by ancient academics, the Red Sea mangroves have great historical relevance. Theophrastus and Pliny both wrote about the mangrove trees that grew along the Red Sea coast in their works *Historia Plantarum* (350 BC) and *Historia Naturalis* (AD 77), respectively. The Arabian botanist Abual-Abbas al-Nabaty described the *Rhizophora* mangrove and gave it the name "*kendela*" around 1230 AD. He also gave the name "*quorm*" to the *Avicennia* mangrove. Both names are still used today in Arabic and Swahili (Khalil 2015). The Red Sea mangroves are of important ecological and biogeographic relevance since they are the most northern part of their Indo-Pacific distribution range (PERSGA/GEF 2004). Given the prevailing aridity and lack of permanent woody vegetation in its coastal areas, mangroves may be a beneficial resource for them (Abu Ghararah and Khalil 2012).

The fundamental characteristics of mangroves are their extensive aerial roots. Since mangrove trees are typically located around beaches, their aerial roots give the impression that the trees are perched on top of the ocean. A specific root system has been established by several genera, including *Rhizophora* and *Avicennia*, to assist respiration and stability. Prop roots and pneumatophores slow down current flow, causing more suspended particles to settle out. Mangroves are particularly vital for preventing coastal erosion because they may migrate seaward to occupy new coastal habitats (Kathiresan 2003). They are known for their ability to act as a tidal shield, and sediments can be protected from erosion by mangrove plants. As the driving forces for metabolic pathway adaptations, which can direct the production of useful metabolites but can also cause diminishment of the mangrove ecosystem, mangrove ecosystems must endure frequent shifts in tides, strong winds, high temperatures, anaerobic soils and salinity gradients of the mangrove habitat (Duke and Schmitt 2015). Mangroves have developed morphological and physiological adaptations that help them overcome these challenging circumstances (Genilar et al. 2021). The ability of mangroves to keep fresh water at a constant level in their root systems is only one of their many exceptional qualities. Mangroves must maintain their freshwater level because of the unusual environments in which they are frequently located (Ball 1996). The three categories of main species, minor species and mangrove allies are commonly used to categorize mangroves; nonetheless, mangroves have proven to be challenging to categorize (Tomlinson 2016).

The three main categories of mangroves are:

Red mangroves: Red mangroves (*Rhizophora* species) are prevalent in tropical and subtropical intertidal zones. They are distinguished by their root systems, which create a thick network around the tree's stem, giving rigidity and assisting in nutrient intake. Red mangroves have dark brown or black bark, and their branches are tall and thin with a leathery appearance (Cerón-Souza et al. 2010).

Black mangroves: Although they typically grow at higher elevations, black mangroves (*Avicennia* species) can be prevalent in the same geographic areas as red mangroves. Black mangroves, in contrast to red mangroves, have a network of roots that extends deeper into the sediment, giving them the ability to penetrate soil that is rich in oxygenation. Pneumatophores – tiny, finger-like protrusions from the earth surrounding the tree that allow for gas exchange – are what distinguish them (Peterson and Bell 2012).

White mangroves: Around similar areas as red and black mangroves, white mangroves (*Laguncularia*

DOI: 10.1201/9781003321408-29

racemosa) can be observed, usually at greater heights. Their oval-shaped leaves and smooth, grey bark make them simple to recognize. White mangroves, unlike red and black mangroves, lack aerial roots and pneumatophores in favour of a network of tiny horizontal roots that assist them in absorbing minerals via the soil (Tomlinson 2016).

Avicennia marina is the predominant species in mangrove stands in the Red Sea and are primarily monospecific. This species is better suited to the challenging circumstances of the high salinity and aridity of the Red Sea coastal regions and makes up the great bulk of the monospecific mangrove stands in the area. In several regions, notably in Djibouti, Eritrea, Yemen and Saudi Arabia, significant *Rhizophora mucronata* stands coexist.

A specific root system has been established by several genera, including *Rhizophora* and *Avicennia*, to assist respiration and stability. Prop roots and pneumatophores slow down current flow, causing more suspended particles to settle out. Mangroves are particularly vital for preventing coastal erosion because they may migrate seaward to occupy new coastal habitats (Kathiresan 2003).

Rhizophora mucronata occurs in a number of mangrove habitats in Saudi Arabia, including the Wajh Bank, Umm al Qandal Island, Gharaniq Island, Zifaf Island, Shi'b Abu Al-Liqa and the Farasan Islands. *A. marina* also coexists in each of these locations; however, *R. mucronata* only grows in specific areas near the seaward location of the woods. Al-Wetaid and Khalil (2003) and other more recent studies expanded this list to include 11 regions across the nation where *R. mucronata* exists. The Farasan Islands, Al-Wajh Bank and the southern portions of the coast all have substantial *R. mucronata* stands (PERSGA/GEF 2004). *R. mucronata* was discovered in Yemen at two locations – Kamaran Island and a small islet close to Hodeida – out of the 29 mangrove sites that were studied throughout the nation. In Yemen, the former is home to the biggest *R. mucronate* stands (PERSGA/GEF 2004; Nagi et al. 2012).

The most prevalent mangrove species in Djibouti has been identified as *Avicennia marina*. Three locations – Godoria, Khor Angar and Moucha Island – have the greatest largest forests in the area (Saenger 2002), where *A. marina* cohabits with quite dense and well-grown *Rhizophora mucronata* stands. Previous studies reported the presence of *Bruguiera gymnorrhiza* and *Ceriops tagal*. A few *R. mucronate* trees were spotted in Ras Siyyan, the only other mangrove forest that is monospecific and solely made up of *Avicennia marina*. The mangroves are often found in thin or somewhat dense forests along the beach, on nearby and offshore islands and along the edges of tidal inlets. They expand landward over depressed terrain, generating shallow canals of varying sizes that are known locally as *khors* (Ibrahim et al. 2022).

In Egypt, the *Avicennia marina* stands are usually narrow, most of them between 15 and 300 metres wide. They develop along the shoreline, on small islands close to the beach and on creeks or inlets that stretch inland along depressions of different sizes known as mersas. The stands are generally tiny and rarely extend more than 1–2 km. Toward the southern shore, which supports muddier soils and gets greater freshwater intake from surface runoff, the density and size of the stands increase (Shaltout et al. 2005).

The improved oxygenation, sandier substrate and significant subsurface freshwater seepage, however, may favour *Avicennia* tree development to a greater height and girth at breast height (GBH) in particular places in the northern sections of Sudan (e.g., Arakiyai). The coastal area may be separated into three sections based on biological characteristics, including the size and distribution of mangrove stands (Kathiresan et al. 2013).

The traditional medicinal uses of mangroves are listed in Table 22.1. The locals from numerous mangrove locations responded positively when asked "Can you use this mangrove plant as medicine?" and went on to discuss their different applications. However, when a local facility is accessible, the answer to the question "What do you do when you are sick?" was frequently "We go to the hospital or to the clinic!" Some mangrove residents experience this. The majority of mangrove regions in the developing countries are inhabited by rural populations at the lowest socioeconomic level, with a high prevalence of tropical illnesses and issues with biting or noxious insects. Some of the rural communities, particularly those that are in and depend on mangroves, have increased their usage of medical facilities as a result of the popularity of Western medicine, help from wealthy nations and the implementation of a public health system (Bandaranayake 1998). This does not and should not, however,

TABLE 22.1 Traditional uses of mangrove plants.

NAME OF MANGROVE	PLANT PART	MEDICINAL USES	REFERENCES
Avicennia marina	Bark	Contraceptive agent, astringent	IMPPAT Database, Namazi et al. (2013)
	Fruit	Abscesses, furunculosis	
	Leaf	Anti-bacterial agent, antiviral agent, insecticide	
	Root	Aphrodisiac	
	Seed	Abscesses, furunculosis, wounds, injuries	
	Whole plant	Anti-inflammatory agent, diuretic, smallpox lesions	
Rhizophora mucronata	Bark	Asthma, astringent, diabetes mellitus, fever, hematuria, hemorrhage, leprosy, syphilis	IMPPAT Database, Aljaghthmi et al. (2017), Suganthy and Pandima Devi (2016)
	Root	Fractures, bones, hematemesis, hematuria, hemoptysis	

(continued)

TABLE 22.1 (Continued)

NAME OF MANGROVE	*PLANT PART*	*MEDICINAL USES*	*REFERENCES*
Acanthus ilicifolius	Aerial part	Anti-inflammatory agent, diuretic	IMPPAT Database, Firdaus et al. (2013), Singh et al. (2009)
	Flower	Antiviral agent	
	Leaf	Antirheumatic agent, antiviral agent, biliary tract diseases, headache, leishmaniasis, neuralgia, skin diseases, snake bites	
	Root	Antiviral agent, asthenia, asthma, leukorrhea, toothache	
	Whole plant	Analgesic, dyspepsia, asthma	
Aegiceras corniculatum	Leaf	Anti-infective agent, pesticide, wounds, injuries	IMPPAT Database, Roome et al. (2008)
Bruguiera gymnorhiza	Aerial part	Hypotension	IMPPAT Database, Bibi Sadeer et al. (2022)
	Bark	Astringent, diarrhea, malaria	
	Fruit	Astringent	
	Whole plant excluding root	Cardiovascular system	
Bruguiera parviflora	Leaf	Constipation	IMPPAT Database Chumkaew et al. (2005)
Acanthus ebracteatus	Fruit	Antiseptic, blood purifier, boils, volds	Bandaranayake (1998)
	Bark	Rheumatism, snake bites	
	Leaves	Skin allergies, snake bites	
Avicennia africana	Bark	Cancer, thrush, gangrenous wounds, lice, mange, ringworm, skin parasites, tumors, ulcers	Bandaranayake (1998)
Avicennia germinans	Bark	Incontinence, rheumatism, throat pain, ulcers of the mouth	IMPPAT Database, Bandaranayake (1998)
	Root	Aphrodisiac	
	Seed	Abscesses, furunculosis	
Ceriops decandra	Aerial part	Anti-inflammatory agent, diuretic, hypotension	IMPPAT Database, Aritra and Amit (2012)
	Bark	Hemorrhage, ulcer	
	Leaf	Dyspepsia	
	Whole plant excluding root	Antiviral agent, cardiovascular system, diuretic	
Concocarpus erecta	Root	Catarrh	Bandaranayake (1998); Khalil et al. (2020)
	Leaves	Febrifuge	
	Bark	Gonorrhoea, malaria, bleeding	
Derris uliginosa	Bark	Anthelmintic, antirheumatic agent, dysmenorrhea, insect repellent, antispasmodic, stimulant	IMPPAT Database, Bandaranayake (1998)
	Leaf	Dysentery, dyspepsia	
	Root	Abdominal pain, insecticide, neoplasms	
	Stem	Diabetes mellitus	
Excoecaria agallocha	Bark	Arthritis, cathartic, leprosy	IMPPAT Database, Mondal et al. (2016)
	Fruit	Antiviral agent	
	Leaf	Antirheumatic agent, antiviral agents, epilepsy, ulcers	
	Root	Edema, inflammation	
	Seed	Wounds, injuries	
	Stem	Tooth diseases	
Heritiera littoralis	Bark	Hemorrhoids	Tewtrakul et al. (2010); IMPPAT Database
	Fruit	Appetite stimulant, general tonic for rejuvenation	
	Leaf	Female urogenital diseases, headache, pain, urologic diseases	
	Root	Insecticide	
	Seed	Diarrhea, dysentery	
	Shoot	Toothbrushing, anti-bacterial agent	
Kandelia rheedii	Bark	Diabetes, antihypertensive agent	Agrawal and Mourya (2021)
Nypa fruiticans	Leaves	Asthma, diabetes, leprosy, rheumatism, snake bite, antioxidant	Hossain and Islam (2015); Prasad et al. (2013)
Sonneratia alba	Bark	Antioxidant, antimicrobial	Saad et al. (2012)

diminish the potential of mangroves as a source of medicine. In the developed world, the mangrove zone is a well-known vacation destination with high-end public health services and expensive methods to combat these diseases (Melana et al. 2000).

22.2 BIOMEDICAL IMPORTANCE

Mangroves are important habitats that serve humans as well as the environment in a number of ways. The Red Sea mangroves' potential for use in medical research has been acknowledged. Alkaloids, flavonoids and terpenoids, which have distinctive chemical compositions and physiological roles, are only a few of the bioactive substances that mangroves are known to generate. These substances have been found to display a wide variety of pharmacological characteristics, such as antiviral, anti-inflammatory, antibacterial and anticancer effects (Berumen et al. 2019)

22.3 CHEMICAL COMPOSITION

Mangroves are intricate ecosystems with a variety of plants and accompanying microbes. The presence of different plant species, the salinity of the water and the availability of nutrients may all affect the chemical makeup of Red Sea mangroves. However, research indicates that Red Sea mangroves frequently contain elevated concentrations of tannins, a kind of polyphenolic ingredient with a variety of biological qualities, such as antioxidant and antibacterial effects (Hawas et al. 2022)

Mangroves in the Red Sea are also rich in lipids, proteins, carbohydrates and micronutrients, including calcium, magnesium and potassium. According to several studies, the mangroves contain significant amounts of fatty and amino acids, which are vital for both human and animal nutrition. Alkaloids, flavonoids and terpenoids, among other secondary metabolites with putative therapeutic functions, have also been discovered in Red Sea mangroves. In conclusion, the chemical makeup of Red Sea mangroves is diversified and includes minerals and sugars. Depending on the particular mangrove species and the environmental factors where they grow, the precise composition can change (Hawas et al. 2022).

22.4 PHARMACOLOGICAL AND BIOMEDICAL SIGNIFICANCE

22.4.1 Mangroves as an anticancer agent

Mangroves contain secondary chemicals that are effective against cancerous cells. Even though numerous studies show that mangroves can be a good source of natural medicines, the quest for bioactive chemicals from coastal mangroves in the pharmacological field is still being disregarded (Das et al. 2015). Antioxidants, which can be found in the secondary metabolites, are used as drugs and are helpful for keeping the immune system's homeostasis in check. Antioxidants have some disadvantages, but it is obvious that they are useful and should not be disregarded (Thyagarajan and Sahu 2018).

22.4.2 Mangroves as an antitumor agent

Seven substances that have the potency to inhibit cancer are seen in mangrove extracts. In *Rhizophora apiculata* mangrove extracts, there are high amounts of 4-pyrrolidinyl, ketone and pyrazole derivatives, which can be used as an anticancer and anti-inflammatory drug (Yang et al. 2015).

22.4.3 Inflammation-relieving effect

Mangrove extracts have turned out to be successful and important in traditional indigenous medical treatments. Studies demonstrate that the actions discovered may result in an anti-inflammatory substance. The methanolic extracts of *Rhizophora apiculata* have been shown to have anti-inflammatory properties, which may produce a high concentration of ketone derivatives, thiazolidinediones and 4-pyrrolidinyl. The flavonoid makeup of the mangrove extract is crucial to its usage as an anti-inflammatory medication (Prabhu and Guruvayoorappan 2012). Flavonoids are a group of secondary metabolites that are frequently present in plants and typically include 15 carbon atoms. Numerous flavonoids have a direct impact on the enzymes involved in inflammatory processes. Some flavonoids can prevent inflammation by preventing the induction of adhesion molecules such blood neutrophils (Middleton et al. 2000). The relationship between the structures of the chemicals produced a favourable outcome, suggesting that the extracts may be used as anti-inflammatories, although there are still adjustments that need to be made (Cui et al. 2018).

22.4.4 Antimicrobial activity

It has been discovered that the leaves and bark of *Rhizophora apiculata* and *Rhizophora mucronata* contain an active acid polysaccharide that can stop the human immunodeficiency virus (HIV) from attaching to cells (Premanathan et al. 1999). Acid polysaccharides (sulphate polysaccharides) use the electrostatic interaction between the negative charges to inhibit the binding process (Battulga et al. 2019).

22.4.5 Mangroves as an antifungal and antibacterial agent

The actinomycetes in mangroves have significant potential to create a novel antibiotic for diseases that are difficult to control with the antibiotics now on the market (Ramasubburayan et al. 2015). It takes unique antibiotics to create a new class of drugs that can treat actinomycetes and resistant strains, and this can only be done by conducting research in uncharted or untapped territory. The mangrove is therefore likely to have more antioxidants (Rahman 2018).

22.4.6 Antioxidative stress

Antioxidants commonly come from medicinal plants, vegetables and fruits. Antioxidants have been considered effective agents in preventing and treating diseases linked to reactive oxygen species, such as cancer, cardiovascular disease, atherosclerosis, hypertension, ischemia/reperfusion injury, diabetes mellitus, neurodegenerative diseases, rheumatoid arthritis and aging. In comparison to the other mangroves in the Andaman and Nicobar Islands, endophytic colonization of Trichoderma was shown to be more prevalent in the leaves of *Aegiceras corniculatum* mangroves (Thatoi et al. 2014).

22.4.7 Organ-protective effect

The mangrove's connection to the complication of oxidative stress makes it possible to use it as an antidiabetic medication. Those phytochemicals from mangroves have antidiabetic effects that include insulin-mimetic activity, decreased intestinal glucose absorption, decreased advanced glycation end-products (AGEs) and an antioxidative effect, reducing oxidative-stress-associated diabetic problems (Das et al. 2015).

22.5 CONCLUSION

It is undeniable that mangroves are beneficial in the management of various diseases. These plants can inhibit ailments in a variety of ways. The fact that they have multiple modes of action allows them to manage conventional diseases. These plants have the potential to be useful therapeutic agents or adjuncts to the current therapies for the control of various human diseases such as cancer, tumors, inflammation, fungal diseases, bacteria and viruses and can treat diabetes. Even though evidence-based therapeutic applications of many plants are limited, the plants discussed in this chapter are those that have been traditionally associated with the effects of human disease and that have been studied either experimentally or clinically.

REFERENCES

Abou El-Ezz, R., Ibrahim, A., Habib, E., Kamel, H., Afifi, M., Hassanean, H., and S. Ahmed. 2017. Review of natural products from marine organisms in the Red Sea. *Int J Pharmaceutical Sci Res* 8, no. 3: 940–974.

Abu Ghararah, Z.H., and A.S.M. Khalil. 2012. *Ecosystem Based Adaptations in the Red Sea and Gulf of Aden*. Climate Action Publication Edit. 2012, Climate Action Program, UK UNEP.

Agrawal, O.P., and N. Mourya. 2021. Identification of bioactive compounds extracted from barks of Kandelia Rheediia and evaluation of in vitro antidiabetic potential. *J Adv Sci Res* 12, Suppl 1: 85–89.

Aljaghthmi, O.H., Heba, H.M., and I.A. Zeid. 2017. Antihyperglycemic properties of mangrove plants (Rhizophora mucronata and Avicennia marina): An overview. *Adv Biol Res* 11, no. 4: 161–170.

Al-Wetaid, A., and A. Khalil. 2003. *Status of Red Sea Mangrove in Kingdom of Saudi Arabia*. PERSGA, Jeddah. Draft Report.

Aritra, S., and R. Amit. 2012. Analysis of and correlation between phytochemical and antimicrobial constituents of Ceriops decandra, a medicinal mangrove plant, from Indian Sundarban estuary. *J Med Plants Res* 6, no. 32: 4755–4765.

Ball, M.C. 1996. Comparative ecophysiology of mangrove forest and tropical lowland moist rainforest. In: *Tropical Forest Plant Ecophysiology*. Eds. S.S. Mulkey, R.L. Chazdon and A.P. Smith, 461–496. Boston: Springer.

Bandaranayake, W.M. 1998. Traditional and medicinal uses of mangroves. *Mangroves Salt Marshes* 2, no. 3: 133–148.

Battulga, T., Tumurbaatar, O., Ganzorig, O., Ishimura, T., Kanamoto, T., Nakashima, H., et al. 2019. Analysis of interaction between sulfated polysaccharides and HIV oligopeptides by surface plasmon resonance. *Int J Biol Macromol* 125: 909–914.

Berumen, M.L., Voolstra, C.R., Daffonchio, D., Agusti, S., Aranda, M., Irigoien, X., et al. 2019. The Red Sea: Environmental gradients shape a natural laboratory in a nascent ocean. In: *Coral Reefs of the Red Sea*. Eds. C. Voolstra and M. Berumen, 1–10. Cham, Switzerland: Springer.

Bibi Sadeer, N., Sinan, K.I., Cziáky, Z., Jekő, J., Zengin, G., Jeewon, R., Abdallah, H.H., AlDhaheri, Y., Eid, A.H., and M.F. Mahomoodally. 2022. Towards the pharmacological validation and phytochemical profiling of the decoction and maceration of Bruguiera gymnorhiza (L.) Lam. – A traditionally used medicinal halophyte. *Molecules* 27, no. 6: 2000.

Cerón-Souza, I., Rivera-Ocasio, E., Medina, E., Jiménez, J.A., McMillan, W.O., and E. Bermingham. 2010. Hybridization and introgression in New World red mangroves, Rhizophora (Rhizophoraceae). *Am J Bot* 97, no. 6: 945–957.

Chumkaew, P., Kato, S., and K. Chantrapromma. 2005. A new triterpenoid ester from the fruits of Bruguiera parviflora. *Chem Pharm Bull* 53, no. 1: 95–96.

Cui, H., Liu, Y., Li, J., Huang, X., Yan, T., Cao, W., et al. 2018. Diaporindenes A–D: Four unusual 2, 3-Dihydro-1 H-indene analogues with anti-inflammatory activities from the mangrove Endophytic Fungus *Diaporthe sp.* SYSU-HQ3. *J Org Chem* 83, no. 19: 11804–11813.

Das, G., Gouda, S., Mohanta, Y.K., and J.K. Patra JK. 2015. Mangrove plants: A potential source for anticancer drugs. *Indian J Geo Mar Sci* 44, no. 5: 666–672.

Duke, N.C., and K. Schmitt. 2015. Mangroves: Unusual forests at the seas edge. Tropical forestry handbook. In: *Tropical Forestry Handbook*. Eds. M. Köhl and L. Pancel, 1–24. Berlin and Heidelberg: Springer. http://doi.org/10.1007/978-3-642-41554-8_129-1.

El-Hossary, E.M., Abdel-Halim, M., Ibrahim, E.S., Pimentel-Elardo, S.M., Nodwell, J.R., Handoussa, H., Abdelwahab, M.F., Holzgrabe, U., and U.R. Abdelmohsen. 2020. Natural products repertoire of the Red Sea. *Mar Drugs* 18, no. 9: 457.

Firdaus, M., Prihanto, A.A., and R. Nurdiani. 2013. Antioxidant and cytotoxic activity of Acanthus ilicifolius flower. *Asian Pac J Trop Biomed* 3, no. 1: 17–21.

Genilar, L.A., Kurniawaty, E., Mokhtar, R.A., and K.A. Audah. 2021. Mangroves and their medicinal benefit: A mini review. *Ann Romanian Soc Cell Biol* 7: 695–709.

Hawas, U.W., El-Kassem, L.T.A., Shaher, F.M., Al-Farawati, R., and M. Ghandourah. 2022. Phytochemical compositions of some Red Sea halophyte plants with antioxidant and anticancer potentials. *Molecules* 27, no. 11: 3415.

Hossain, M.F., and M.A. Islam. 2015. Utilization of mangrove forest plant: Nipa palm (Nypa fruticans Wurmb.). *Am J Agric For* 3, no. 4: 156–160.

Ibrahim, H.A., Abdel-Latif, H.H., and E.H. Zaghloul. 2022. Phytochemical composition of Avicennia marina leaf extract, its antioxidant, antimicrobial potentials and inhibitory properties on Pseudomonas fluorescens biofilm. *Egypt J Aquat Res* 48, no. 1: 29–35.

Kathiresan, K. 2003. How do mangrove forests induce sedimentation? *Rev Biol Trop* 51, no. 2: 355–360.

Kathiresan, K., Anburaj, R., Gomathi, V., and K. Saravanakumar. 2013. Carbon sequestration potential of Rhizophora mucronata and Avicennia marina as influenced by age, season, growth and sediment characteristics in southeast coast of India. *J Coastal Conserv* 17, no. 3: 397–408.

Khalil, A.S. 2015. Mangroves of the Red Sea. In: *The Red Sea: The Formation, Morphology, Oceanography and Environment of a Young Ocean Basin*. Ed. N.M.A. Rasul and I.C.F. Stewart, 585–597. Berlin and Heidelberg: Springer Earth System Sciences.

Khalil, R., Ali, Q., Hafeez, M.M., and A. Malik. 2020. Phytochemical activities of Conocarpus erectus: An overview. *Biol Clin Sci Res J* 2020: e008. http://doi.org/10.54112/bcsrj.v2020i1.8.

Melana, D.M., Atchue III, J., Yao, C.E., Edwards, R., Melana, E.E., and H.I. Gonzales. 2000. *Mangrove Management Handbook*. Department of Environment and Natural Resources, Manila, Philippines through the Coastal Resource Management Project, Cebu City, Philippines, 55 p.

Middleton, E., Kandaswami, C., and T.C. Theoharides. 2000. The effects of plant flavonoids on mammalian cells: Implications for inflammation, heart disease, and cancer. *Pharmacol Rev* 52, no. 4: 673–751.

Mondal, S., Ghosh, D., and K. Ramakrishna. 2016. A complete profile on blind-your-eye mangrove Excoecaria agallocha L. (Euphorbiaceae): Ethnobotany, phytochemistry, and pharmacological aspects. *Pharmacogn Rev* 10, no. 20: 123–138.

Nagi, H.M., Khanbari, K.M., and A. Al-Sameh. 2012. Estimating total area of mangrove habitats in the Republic of Yemen using remote sensing and GIS. *Faculty Sci Sana'a Uni Yemen Bull* 24: 75–84.

Namazi, R., Zabihollahi, R., Behbahani, M., and A. Rezaei. 2013. Inhibitory activity of Avicennia marina, a medicinal plant in Persian folk medicine, against HIV and HSV. *Iranian J Pharm Res* 12, no. 2: 435–443.

PERSGA/GEF. 2004. Status of mangroves in the Red Sea and Gulf of Aden. *Technical Series Number 11*. PERSGA, Jeddah, 67 p. (Prepared by A.S.M. Khalil).

Peterson, J.M., and S.S. Bell. 2012. Tidal events and salt-marsh structure influence black mangrove (Avicennia germinans) recruitment across an ecotone. *Ecology* 93, no. 7: 1648–1658.

Prabhu, V.V., and C. Guruvayoorappan. 2012. Anti-inflammatory and anti-tumor activity of the marine mangrove Rhizophora apiculata. *J Immunotoxicol* 9, no. 4: 341–352.

Prasad, N., Yang, B., Kong, K.W., Khoo, H.E., Sun, J., Azlan, A., Ismail, A., and Z.B. Romli. 2013. Phytochemicals and antioxidant capacity from Nypa fruticans Wurmb. fruit. *Evid Based Complement Alternat Med* 2013: 154606. http://doi.org/10.1155/2013/154606.

Premanathan, M., Kathiresan, K., and H. Nakashima. 1999. Mangrove halophytes: A source of antiviral substances. *South Pacific Study* 19, nos. 1–2: 49–57.

Rahman, M. 2018. *Antifungal and Antibacterial Activity of Actinomycetes Isolated from the Sundarbans Mangrove Forest Soil and Water from Bay of Bengal*. Thesis, BRAC University, Dhaka, Bangladesh.

Raitsos, D.E., Pradhan, Y., Brewin, R.J., Stenchikov, G., and I. Hoteit. 2013. Remote sensing the phytoplankton seasonal succession of the Red Sea. *PLoS ONE* 8, no. 6: e64909.

Ramasubburayan, R., Prakash, S., Iyapparaj, P., Sumathi, S., Thaddaeus, B.J., Palavesam, A., et al. 2015. Investigation on antibacterial, antifungal and cytotoxic properties of chosen mangroves. *Ind J Geo Mar Sci* 44: 1769–1777.

Roome, T., Dar, A., Ali, S., Naqvi, S., and M.I. Choudhary. 2008. A study on antioxidant, free radical scavenging, anti-inflammatory and hepatoprotective actions of Aegiceras corniculatum (stem) extracts. *J Ethnopharmacol* 118, no. 3: 514–521.

Saad, S., Taher, M., Susanti, D., Qaralleh, H., and A.F. Awang. 2012. In vitro antimicrobial activity of mangrove plant Sonneratia alba. *Asian Pac J Trop Biomed* 2, no. 6: 427–429.

Saenger, P. 2002. *Mangrove Ecology, Silviculture and Conservation*. Dordrecht: Springer. http://doi.org/10.1007/978-94-015-9962-7.

Shaltout, K.H., Khalaf-Allah, A., and M. El-Bana. 2005. *Environmental Characteristics of the Mangrove Sites Along the Egyptian Red Sea Coast*. Assessment and Management of Mangrove Forests in Egypt for Sustainable Utilization and Development: A Project Funded by ITTO (Japan) and Supervised by MALR/MSEA – EEAA, Cairo.

Singh, A., Duggal, S., and A. Suttee. 2009. Acanthus ilicifolius linn. – lesser known medicinal plants with significant pharmacological activities. *Int J Phytomed* 13, no. 1: 431–436.

Suganthy, N., and K. Pandima Devi. 2016. In vitro antioxidant and anti-cholinesterase activities of Rhizophora mucronata. *Pharmaceutical Biol* 54, no. 1: 118–129.

Tewtrakul, S., Tansakul, P., Daengrot, C., Ponglimanont, C., and C. Karalai. 2010. Anti-inflammatory principles from Heritiera littoralis bark. *Phytomed* 17, no. 11: 851–855.

Thatoi, H.N., Patra, J.K., and S.K. Das. 2014. Free radical scavenging and antioxidant potential of mangrove plants: A review. *Acta Physiol Plant* 36, no. 3: 561–579.

Thyagarajan, A., and R.P. Sahu. 2018. Potential contributions of antioxidants to cancer therapy: Immunomodulation and radiosensitization. *Integr Cancer Ther* 17, no. 2: 210–216.

Tomlinson, P.B. 2016. *The Botany of Mangroves*. Cambridge: Cambridge University Press. http://doi.org/10.1017/CBO9781139946575.

Yang, Y., Zhang, Y., Liu, D., Li-Weber, M., Shao, B., and W. Lin. 2015. Dolabrane-type diterpenes from the mangrove plant *Ceriops tagal* with antitumor activities. *Fitoterapia* 103: 277–282.

Mangroves of the Sudanese Red Sea coastal area

23

Awatif A. Sugga, Afkar A. Rahama and Dirar H. Nasr

23.1 INTRODUCTION

Globally, there has been an increasing interest in marine coastal ecosystems, particularly those existing in risky environments such as islands and lowlands subject to the impact of sea-level rise due to projected climate changes. These active ecosystems play a vital role in hosting flora and fauna forming productive coastal communities and are additional biological contributors to the biodiversity on the earth's surface. It is well recognized that many marine habitats in tropical and subtropical zones are occupied by mangroves: flowering plants that tolerate occasionally high salinity, up to 40%. In Sudan, mangroves exist only in the Red Sea State, which has a total area of 212,410 square kilometres. The Red Sea state is characterized by high temperatures that range from 15°C up to 47°C in summer, while the average annual rainfall is about 25 to 40 mm, mainly from December to February (Al-Hassan 1999).

Mangrove forests globally include approximately 70 species within 12 families, in which some species dominate others in the form of trees and shrubs. Under favourable conditions, the trees reach an average height ranging between 5 and 7 m, while shrubs range between 2.5 and 4.0 m. Upon maturation, they begin to bloom at the age of three years. The highest growth of seedlings observed was at seawater salinity levels of up to 50%. Mangrove vegetation exhibits different forms of physiological and phenological behaviour in different regions and latitudes. This is reflected in flowering periods, fruiting and seed maturity. Most mature mangrove seeds store humidity, with water content of up to 70–90%; the seeds are characterized by rapid cellular division and fast vegetative growth. They exhibit the vivipary phenomena as seeds germinate while still attached to the mother tree, and the propagule falls into the water to continue its growth under favourable conditions (Zahran 2007).

The Sudanese Red Sea coast is about 750 km long and semi-enclosed; this leads to a concentration of salts in the waters which increases from north to south, reaching up to 37% in the southern parts and 39% in the Port Sudan area. The Red Sea is also characterized by a distinctive environmental and biological diversity where many habitats exist, especially in estuaries and deltas, which receive sediments transported to the sea from upland areas (Akondo and Tomas 2013).

Halophytic vegetation, mangroves, salt marshes, algal beds, sandy beaches, mudflats, rocky shores, sabkhas and hypersaline pools, seagrass beds and coral reefs are present along the Sudanese coast. Among the halophytes, mangroves represent a very important component that closely interlinks all these ecosystems.

Historically, *Avicennia marina*, *Rhizophora mucronata* and *Brugiera gymonoriyzhia* were the three mangrove species reported along the Red Sea coast. The report of the survey carried out by PERSGA in 2004 (PERSGA/GEF 2004) mentioned that *Avicennia marina* was the only mangrove species found in Sudan during this survey. Two of mangrove species still exist in Sudan: namely, *Avicennia marina*, which is widespread and known to be highly tolerant of high temperature and salinity fluctuations, and *Rhizophora sp.* (possibly *R. mucronata*), which was reported recently at the far southern coast of Sudan at the Aqiq locality in the Ras Kasar area (Figure 23.1) (Sugga et al. 2017). It had not before been reported since 1957. Our observations of mangroves at Kilo Tammania in the Port Sudan vicinity showed continuous deterioration during the 1980s and 1990s, but they have been flourishing again since 2019, as can be seen in Figure 23.2. This suggests that mangroves are affected by favourable and unfavourable environmental conditions.

Bashir (1998) mentioned that there are 19 identified and sizeable mangrove stands along the Red Sea coast, some of which are luxuriant forests. Local topography at each site affects the forest structure and productivity. Considerable damage to mangrove swamps has been caused and is still caused by natural and biological stresses. Comparing satellite images from 1958, 1966, 1978 and 1996 shows that the average loss from 1958 to 1998 ranged between 33.5% and 60.9% in three sites representing well-established, moderate and degraded stands.

DOI: 10.1201/9781003321408-30

FIGURE 23.1 *Rhizophora sp.* fruits, at Ras Kassar, Sudan (2017).

FIGURE 23.2 Flourishing mangroves at Kilo Tammania in vicinity of Port Sudan.

23.2 MANGROVE AREAS IN SUDAN

When comparing satellite images from 1958 to 1996, the total area of mangrove in Sudan was variable (Bashir 1998). Sabeel (2015) reported that the total Sudanese mangrove area ranged between 329 and 721 ha. This indicates that information about the total area occupied by mangroves is not yet definite in Sudan; it needs further satellite and field surveys.

The mangrove spatial extent increased from 1984 to 1990 and decreased sharply during the period 1995–2000, possibly due to severe droughts caused by El Niño-Southern Oscillation events in 1997 and 1998. Between 2000 and 2010, a large increase in the mangrove area was observed, which could be attributed to increased rainfall conditions after the droughts. While the detected changes in the non-impacted mangroves followed the pattern of rainfall, the mangrove areal extent close to the shrimp farm showed a continuous increase until 2010, irrespective of the meteorological conditions; this may suggest a significant positive effect from the farm through a decrease in soil salinity and an increase in nutrients promoting the growth of mangroves during that period. However, since 2010, the area of this impacted mangrove forest has been decreasing.

23.3 CLASSIFICATION OF MANGROVE STANDS ON THE SUDANESE COAST

Sugga et al. (2017) confirmed the classification reported by Mohamed (1984) and Bashir (1998), who classified mangrove forests as 1) offshore islets forests and 2) shoreline forests. These consist of four types: 1) well-established mangroves at ten sites: Inkeifal, Maghersum, Khorelnasrani, Klanaeib, Hosheri, Guwab, Haidoub, Ashstat, KhorSrtrab and Agig; 2) moderate aggregations of mangroves represented by four sites: Maroub, Salak, Kilo Tammania and Sheik Saad; 3) small aggregations at Haliab, Shinab and Kauawi; and 4) relict sporadic populations in one site, Skiehk Ibraheim.

In Sudan, mangroves usually grow in clusters in the intertidal zone of the Red Sea; associated flora are usually non-true mangroves and halophytes that are adapted to saline soils. Halophytes in Sudan spread in the fertile transition zone between the seashore and the fertile soils toward the Red Sea hills. Mangroves grow as scattered stands, either on the coast or on islands (Mohamed 1984; Bashir 1998; Rhama 2016).

23.4 MANGROVE REPRODUCTIVE BEHAVIOUR

Clear variations in the timing of mangrove reproductive behaviour were reported to be linked with the environmental characteristics of the site. It has been proven that the beginning of mangrove flowering is associated with temperature variation at different locations, while the light period length and intensity also had a significant impact on the activation of the plant's reproductive cycle. The phenomenological stages and the reproduction process has a direct relation with temperature, where the maximum growth rate has been reported around 28°C coupled with a high rate of flowering. The number of mature flowers increases with the rise of the average daily temperature where the flowering rate reached about 70% at temperatures higher than 25°C; this rate decreased to zero with the average daily temperature falling below 0°C (Saenger and Moverly 1985).

23.5 MANGROVE HABITATS

Mangroves were reported to grow successfully on coasts with ventilated loose sediments, rich in organic matter mixed with clay granules. Climatic conditions may slow the growth of mangrove trees, evidenced by a decreased number of branches, increased drying and falling of leaves; production of very small seeds; increased exposure to injuries with insects and pest attacks; and an inability to germinate and reach maturity.

Vegetative growth of the lower branches of mangroves was also observed in some species (Jaratli 2011). Mangroves are often tall trees in the protected zone on the seaward side with many flowers and large stem diameters while mangroves in the landward direction are usually dwarfed (Bashir 1998).

Marine currents add nutrients that form rich environments which contribute to the stability, prosperity and continuity of mangrove forests, and their formation often occurred over 15 to 25 years (Abdel-Razik 1994). Mangroves usually provide large quantities of seedlings in a short time when grown in nurseries, but their nurseries need protection from herds and livestock. Insecticides and fertilizers are not often used in these nurseries (Jaratli 2011). The biological adaptation of mangroves to saline habitats is reflected in the characteristics of the green leaves, which are very similar to other plants in terrestrial saline soils or highly stressed environments, including the presence of a thick epidermis and a waxy layer covering the leathery leaves and outer corners above the surface of the leaf (Abdel-Razik 1994).

23.6 ENVIRONMENTAL ROLE OF MANGROVES

Mangroves act as protectors from wave action for the coastal zone through their dense roots, which decrease soil erosion and purify water runoff from adjacent upland. Mangroves also play an important role in absorbing carbon dioxide (CO_2), which helps reduce the impact of global warming. On the other hand, by hosting many marine organisms as fodder sources and protective secure habitats, they offer sheltered nurseries for juvenile fish and benthos. Leaves of some mangroves are used after drying to prepare traditional drinks, the fruits are often eaten, and the flowers are also used in bee breeding (Khafagi et al. 2003) In addition, *Avicennia marina* leaf extracts have pronounced effects on a number of pathogenic bacteria and fungi. *Rhiyzophora sp.* extracts (one of the globally widespread mangrove species) is used in the production of active metabolics, which have a reported activity against dengue fever (Robert 2012).

Avicennia marina wood products are used as house-building materials, for fishing vessels and as a fuel source. In the Arabian Gulf region, since ancient times, dried mangrove branches have been a source of fuel and perfume. Many researches have shown the benefits of its wood for a variety of purposes, including fabrics from wood fibres and extracting coating materials from natural components (Zahran 2007; Bharathi 2008) Active ingredients from mangrove gum, leaves and seedlings were extracted for use against some pathogenic bacteria and fungi (Abdkadir 2013; Huang et al. 2014).

23.7 SEAWATER AND SEDIMENT PROPERTIES IN MANGROVE SWAMPS IN SUDAN

At mangrove stands, water salinity of the Sudanese Red Sea ranged between 35% and 45%. The lowest values usually were detected during February, and the highest values were observed during August. The seawater pH ranged between 8.40 and 8.73; the lowest was detected in February and the highest in August. The sediment quality along the Sudanese Red Sea coasts is composed mainly of silt (39%), sand (34%) and clay (27%) (Bashir 1998); the presence of a high proportion of silt might be attributed to the quality of the fine sediments carried by the runoff during rainy seasons.

23.8 VEGETATION CHARACTERISTICS OF MANGROVE FORESTS

The botanical structure of the mangroves on the Sudanese Red Sea coast and islands includes the species *Avicennia marina*, in the form of trees with an average height ranging between 5.5 and 7.5 m, with a dominance of shrubs up to 2.5 to 3.5 m. The lanceolate leaves are usually greenish on the upper side and white on the lower side, with an average length of 5 cm. The dense respiratory roots (pneumatophores) are often up to 75 cm in height. In most mangrove stands, two true halophytes are usually associated with mangroves. *Sueda monica* (Figure 23.3) is a medium- and short-sized shrub up to 3 m; the branches are hairy, and the leaves are 10 to 25 mm in length and flat on both sides. The other species is *Arthrothnemum glaucum* (Figure 23.4), with an average height of up to 2.0 m. The majority of the living plants in the mangrove frontier zone are the species *Sueda monica* (Rhama 2016).

FIGURE 23.3 *Sueda monica* on the land side adjacent to mangrove land site.

FIGURE 23.4 *Arthrocnemum sp.*

FIGURE 23.5 *Avicennia sp.*, multi-stemmed shrubs (MSS).

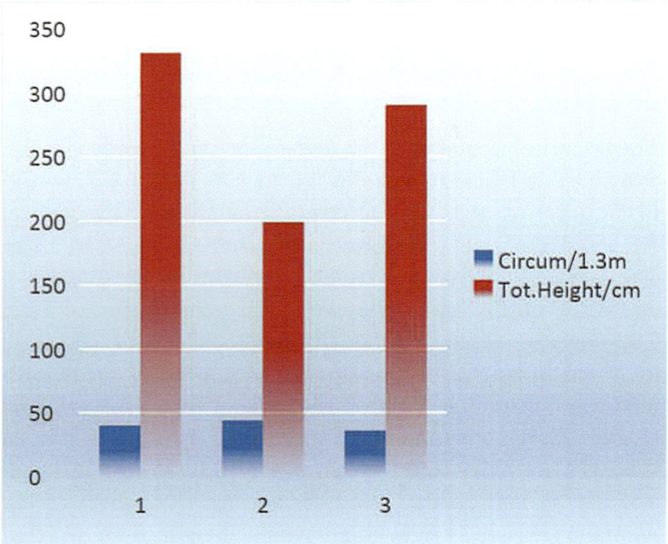

FIGURE 23.6 Comparison of the height and the stem circumference of *Avicennia marina* multi-stemmed shrubs (MSS).

Avicennia marina along the Sudanese Red Sea coast spreads to many islands and along the coastal zone in sporadic patches, represented by two apparently different forms: the multi-stemmed shrubs (MSS) (Figure 23.5) and the single-stemmed trees (SST). *Avicennia marina* exists in four different phenotypes in terms of biological characteristics that are genetically similar (Said and Ehsan 2010). A study conducted by Rhama (2016) at Mersa Durour Island, north of Port Sudan City, showed that the circumference of *Avicennia* MSS ranged between 14 and 44 cm. As for the SST, the average circumference range was 44 to 52 cm. By comparing the height of the stem and its circumference for mangrove stands, an inverse relationship was detected for samples of multi-stemmed shrubs, meaning that the greater the height of the stem, the less the circumference (Figure 23.6). When comparing the total length and the circumference of *Avicennia marina* single-stemmed trees, it was found that the correlation was weak; the reduction of the stem circumference might be attributable to the high salinity of the Red Sea water (Bashir 1998; Parida et al. 2004).

As previously mentioned, *Avicennia marina* grows naturally in Sudan in two apparently different forms: the multi-stemmed shrubs (MSS) and the single-stemmed trees (SST). The MSS originate from both vegetative and sexual reproduction and prevail over the SST. Information about the biological structure of the mangrove vegetation, especially on islands, is essential. The average leaf area of the MSS was 2.9 cm², and the SST leaf area reached an average of 10.0 cm². As for seedling growth and survival of naturally occurring *Avicennia marina* in mangrove stands, usually the total height, number of leaves and continuity across the season showed that the percentage of live seedlings reached up to 100%. The lack of tidal movement in the northern half of the Red Sea between latitudes 25°N and 28°N limits the growth of seedlings; the seedlings breathe through the stomata on the leaves and take their oxygen during the reflux process (Mohamed 1984; Bashir 1998). The leaf specific area (LSA) is one of the measures of productivity in mangrove plants where temperature, salinity and abundance of nutrients directly affect the specific area of the leaf (Huang et al. 2014).

The specific leaf area of MSS indicates an increase in weight and a greater ability for water storage as *Avicennia marina* leaves store water to increase the effectiveness of the vital processes in the plant. The leaf has a larger surface area in MSS as well, which leads to the prosperity of this species and the increase in its spread.

In Sudan, it was documented that there is a variation in height of *Avicennia marina* seedlings over the year due to variations of the rate of seedling growth during the first months of life; with the increase in the crown and number of leaves, growth is regulated. A high degree of illumination positively affects the growth of seedlings at low salinity while the effect of light is weak if the salinity rises due to the low rates of photosynthesis (Huang et al. 2014). By comparing the height of seedlings and the number of their leaves, it became clear that there is a direct correlation that is highly significant between the two variables (Figure 23.7).

23.9 BLACKNESS OF *AVICENNIA MARINA* FLOWERS

In many mangroves stands in Sudan, many of the flowers turn black, especially in the MSS. No studies have been carried out so far regarding the blackness of *Avicennia marina* flowers; this phenomenon needs more investigation. Trees with branched

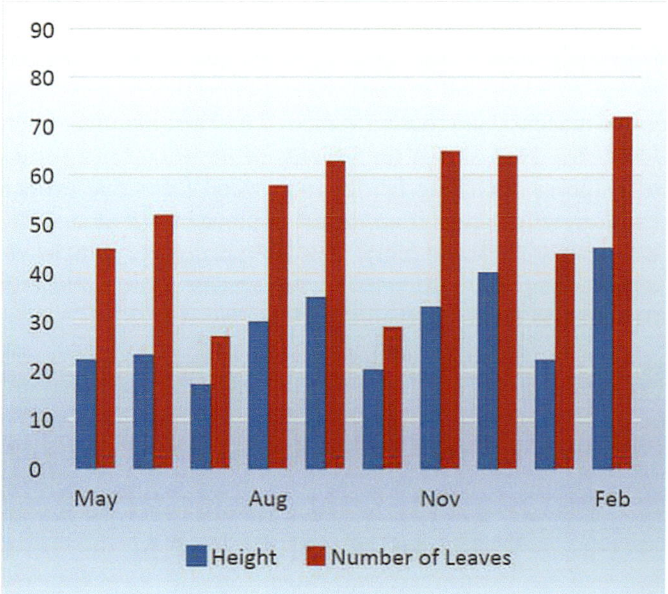

FIGURE 23.7 Comparison of the height of seedlings with the number of leaves.

stems have a very high percentage of flowers while trees with one main stem have a low percentage of flowers (Figure 23.8).

FIGURE 23.8 Blackened *Avicennia marina* flowers.

23.10 MANGROVE-ASSOCIATED FLORA AND FAUNA

Many species are associated with *Avicennia sp.*, either in the swamp or next to the mangrove stand landward, where the halophyte species *S. monica* is widely spread. The associated macro-flora in the muddy swamp is rich with many algae and seagrass taxa; this could be attributable to the availability of nutrients from adjacent lands carrying sediment through seasonal streams and protection from tidal waves in estuaries occupied by mangroves. The most dominant species associated with mangroves include *Chaetomorpha sp.*, *Cystoseira sp.*, *Cladophora sp.*, *Enteramorpha*, *Lyngbya sp.*, *Bryopsis sp. and Spirdina sp.* (Bashir 1998). Many fish, crustaceans and molluscs are associated with mangroves in Sudan. As for meiofauna, mangroves in Sudan show elevated meiofaunal and nematode abundances. The meiofauna taxa *Acari* and *Copepoda* and the nematode genera *Onchium*, *Terschellingia*, *Haliplectus*, *Syringolaimus*, *Spirinia* and *Sphaerolaimus* were absent in the sediment of the cleared mangrove, whereas *Daptonema*, *Theristus* and *Ethmolaimus* also occurred. The structural and functional diversity estimates were lower in the cleared mangrove than in the intact mangrove. Differences in community composition between the cleared and intact mangrove are at least partly explained by differences in sediment characteristics. These results highlighted that those variations in the characteristics of the meiobenthic community between the intact mangrove and the other habitats, when assessed by various taxonomic and functional attributes concomitant with changes in sediment properties, can be helpful in understanding the ecological value of this habitat (Sabeel 2015).

23.11 THREATS TO MANGROVES

Halophytes and mangroves in Sudan are threatened with extinction due to natural and human stresses, such as changes in climatic factors that induce drought cycles and desertification. On the other hand, additional non-natural factors threaten halophytes and mangrove vegetation, including grazing, illegal wood cutting and oil pollution, in addition to competition imposed by invasive species, such as *Prospis sp.* (Muskeet trees).

Threats to mangrove forests, including natural and man-made factors, are encountered globally, contributing to deterioration of these fragile ecosystems. The scarcity of fresh water deposited during rainy seasons increases the salinity of the estuaries, coupled with the loss of silt and organic matter that drifts from the land to the coast and has an adverse impact on mangrove growth (PERSGA 2001; Jaratli 2011). A decrease in temperature and variations in water temperature between seasons hinder the ideal growth of mangroves. The structural composition of the sediment also affects mangrove growth, and significant effects of sediment composition and land topography on the mangrove distribution were reported in Sudan (Bashir 1998).

In spite of the effect of muddy sediments at the coast occupied by mangroves, mangrove seedlings are subject to burial by desert encroachment and wind-borne sand. Natural and human stresses had severe effects on the coasts of the Red Sea in Sudan. Some mangrove forests were seriously damaged. The average damage to mangrove forests was found to range from 33% in a previously well-established stand to 60% in a small stand from 1957 to 1997 (Bashir 1998).

A plan is needed for the establishment of a tissue-propagation programme for the MSS that reproduce vegetatively on the Sudanese coast and islands and for development of cell tissue culture laboratories for production of seedlings in sufficient quantities to enhance the cultivation of *Avicennia marina* in its natural sites. On the other hand, training at fish and crustacean farms to increase the benefits from mangrove swamps, especially on islands, will enhance the livelihoods of coastal communities. The endorsement of the environmental and coastal protection laws and legislation, coupled with the development of scientific manuals to clarify the indicators used for monitoring the environmental status of mangroves are the main milestones for sustainability of this unique ecosystem.

REFERENCES

Abdel-Razik, M.S. 1994. *The Garum Plants (Avicennia Marina) and Its Propagation Experiences in Qatar.* Center for Scientific and Applied Research in the State of Qatar (In Arabic).

Abdkadir, W.J. 2013. Anti-inflammatory and anti-tumor activity of the marine mangrove *Rhizophora Apiculata*. *J Immune Toxicol* 9, no. 2: 341–352.

Akondo, J., and J. Tomas. 2013. The chemical investigations of the mangrove plant *Avicennia marina* and its entophytes. *Open Natl Prod J* 2: 24–32.

Al-Hassan, B.M. 1999. Biodiversity in rangeland plants, pasture and forage management, Red Sea State. *Reports of the Ministry of Agriculture*, Sudan (In Arabic).

Bashir, A.A. 1998. *On the Ecology of Sudanese Red Sea Coastal Vegetation with Special Emphasis on Mangrove Ecosystem.* Ph.D. Thesis, University of Khartoum, Sudan.

Bharathi, V., 2008. Optimization of extraction of phenolic compounds from *Avicennia marina* (Forssk.) Vierh. using response surface. *Aust J Plant Physiol* 9, no. 5: 519–528.

Huang, J., Lu, X., Zhang, W., Huang, R., Chen, S., and Y. Zheng. 2014. Transcriptome sequencing and analysis of leaf tissue of *Avicennia marina* using the Illumina platform. *PLoS ONE* 9, no. 9: e108785. http://doi.org/10.1371/journal.pone.0108785.

Jaratli, M. 2011. The importance and benefits of producing mangrove seedlings in tissue cell culture laboratories. *J Agric Stud Res*, no. 31 (In Arabic).

Khafagi, I., Gab-Alla, A., Salama, W., and M. Fouda. 2003. Biological activities and phytochemical constituents of the gray mangrove *Avicennia marina* (Forssk.) Vierh. *Egypt J Biol* 5: 62–69.

Mohamed, B.F. 1984. Ecological observations on mangroves of the Red Sea shores of the Sudan. *Hydrobiologia* 110: 109–111.

Parida, A.K, Das, A.B., and B. Mittra. 2004. Effects of salt on growth, ion accumulation, photosynthesis and leaf anatomy of the mangrove, *Bruguiera parviflora*. *Trees* 18: 167–174.

PERSGA. 2001. Strategic action program for the Red Sea and Gulf of Aden. *Country Report*. Jeddah, Saudi Arabia, 205 p.

PERSGA/GEF. 2004. Status of mangroves in the Red Sea and Gulf of Aden. *PERSGA Technical Series No. 11*. PERSGA, Jeddah.

Rhama, A.A. 2016. *Characteristics of Mangrove Vegetation at Darour Island*. M.Sc. Thesis. Red Sea University, Faculty of Higher Education, Sudan (Unpublished).

Robert, R. 2012. Phytochemical characterization and antimicrobial efficiency of mangrove plants *Avicennia marina* and *Avicennia officinalis*. *Int J Pharm Biol Arch* 3, no. 2: 348–351.

Sabeel, R.A.O. 2015. *Variation in Distribution of Sudanese Mangroves and Their Ecological Significance for Benthic Fauna*. Thesis submitted for PhD fulfillment, Universiteit Gent, Faculty of Science.

Saenger, P., and J. Moverly. 1985. Vegetative phenology of mangrove along the Queensland coastline. *Proc Ecological Soc Aust* 13: 257–265.

Said, W.M., and N.O.M. Ehsan. 2010. Morphological and molecular evidences among four hetero-forms of *Avicennia marina* (Forssk) Vierh. *J Am Sci* 6, no. 11: 843–856.

Sugga, A.B., Rabhaa, K., and M. Abu-Aldardaa. 2017. Characteristics of mangrove vegetation at Ras-Kassr coasts south Red Sea shore line in Sudan. *Unpublished Report*, Red Sea University, Institute of Marine Research (IMR).

Zahran, M.A. 2007. Status of the mangrove forests of the coastal belts of the Arabian Peninsula. *Unpublished Report*, Taif University, Saudi Arabia.

The Red Sea sabkhas, seaweeds and seagrasses

24

Nahid A.R. Osman

24.1 INTRODUCTION

The Red Sea and Gulf of Aden (RSGA) basin, including the coastal plain, is located in a tough bioclimatic zone with extreme temperatures, low precipitation and high evapotranspiration (Le Houerou 2003). The coastal environment exhibits remarkably high seawater salinity, temperature and transparency. The coastal plain of the RSGA is, therefore, an extensive saline pan and sabkha land with a dendritic drainage system formed by many seasonal freshwater streams that end up in the sea. The exceptions are the Tokar and Arbaat deltas on the Sudanese Red Sea coast, formed by the Baraka and Arbaat streams, respectively. For some organisms (e.g., mangroves, dugongs and coral reefs), the RSGA basin may represents the end of their ecological range. The coastal vegetation is xero-halophytic in nature and is found in a narrow longitudinal strip along the seashore and in the water courses of the coastal streams. These factors make the RSGA basin a unique ecosystem of significant scientific and socioeconomic interest.

24.2 COASTAL SABKHAS

Coastal sabkhas, the dominant ecosystems along the Red Sea coasts, appear as whitish, wet and flat lowlands occupying vast areas along the coast. Sabkha lands were estimated to cover over 50% of the coast of Yemen (PERSGA 2006). Sabkhas developed due to the coexistence of certain environmental filters that characterize semi-arid and arid climates, including high temperatures (Yechieli and Wood 2002) and drought. It has been proposed that, under these conditions, soil capillary forces bring solutes and water to the top soil layer, where the water evaporates, and evaporites precipitate either within the sediment or at the surface (Engel et al. 2021).

Generally, sabkhas are distinguished as coastal or continental. Most coastal sabkhas along the Red Sea coast extend from the supratidal zone landward. Therefore, they are periodically flooded by seawater during high tides and/or fresh water during the short rainy period in winter and by the torrential rains in summer. When coastal sabkhas dry out during the summer season, the high temperature and the evapotranspiration rate lead to accumulation of salts and evaporites in their sediment. Coastal sabkhas are located immediately behind the mangrove stands at the mouths of seasonal streams (locally known as *wadis/khors*), and some are located behind inlets, embayments and lagoons. These sabkhas are also subjected to seawater seepage and seawater spray (Zahran 1982).

Broadly, coastal sabkha soil is described as a heterogenous, salt-bearing and waterlogged formation (Mohamedzein and Al-Rawas 2011). The soil of Red Sea coastal sabkhas is composed of variable sediment types and constituents. For instance, the soil of the Towaratit sabkha in the Sudanese Red Sea coastal plain is composed of a mixture of three facies – sandy mud, muddy sand and sandy gravel – with organic matter and aeolian dust (Omer 2001). Similarly, Basyoni and Aref (2016) described the soil texture of Jizan on the western Red Sea coast of Saudi Arabia as composed of sand, silt and evaporites. In line with this, high contents of fine sand, mud, organic matter and salts were reported from sabkhas along the Suez Canal (El-Omla and Abou-Lela 2012). The sediment constituents of coastal sabkhas allow for longer retention of water than other surrounding environments, to the extent that the water table is located a few centimetres to one or two metres below the ground surface.

The sources of water in coastal sabkhas are sea flooding, capillary rise of groundwater by evaporation and rainfall (Sabtan et al. 1995). With the exception of the pH, most variables of a coastal sabkha's groundwater are comparatively higher than those of seawater. These variables tend to exhibit an increasing gradient in the landward direction. The salinity of the groundwater of coastal sabkhas along the Red Sea is two- to fourfold higher than the salinity of seawater (Banat et al. 2005; Saeed et al. 2020), with a remarkable difference between minimum and maximum values. For instance, the water salinity in coastal sabkhas along the eastern Red Sea varied from a minimum value of 3.2 ppm to a maximum value of 201.21 ppm (Banat et al. 2005). According to Sabtan et al. (1995), salinity increases the ability of soil to retain water. Therefore, the higher the salinity in coastal sabkha soil, the higher the moisture content in the soil. In addition, some of the salts precipitated on the sabkha soil surface, such as sodium and calcium chlorides, are hygroscopic salts (Moon et al. 2019), absorbing moisture from the air when the relative humidity is high.

Despite the seemingly inhospitable conditions, coastal sabkhas are productive ecosystems accommodating a

DOI: 10.1201/9781003321408-31

FIGURE 24.1 Shore birds in a coastal sabkha along the Sudanese Red Sea coast.

Source: The author

considerable diversity of specialized plants and animals which utilize smaller ecological niches within the extensive area of sabkha. Both higher vascular and microscopic plants are the major components of the coastal sabkha plant community. For some animals, coastal sabkhas are temporary refuge and nurture areas. Resident and migratory shore birds (Figure 24.1), mosquitoes, domestic dogs and camels were observed in coastal sabkhas during different seasons along the Sudanese Red Sea coast.

24.2.1 Vascular plants of coastal sabkhas

Vegetated coastal sabkhas are regarded as highly saline wetland ecosystems comparable to salt marsh ecosystems (El-Magsodi and Haddoud 2010; Kearney and Fagherazzi 2016). The vegetation of coastal sabkhas in the Red Sea region is visually dominated by xerohalophytes. According to Gonzalez (2019), xerohalophytes are those terrestrial halophytes that grow in saline environments under dry conditions. However, taking into account the environmental setting of the Red Sea region, xerohalophyte plants in coastal sabkhas are both xerophytes and terrestrial halophytes that adapt to grow in saline, heterogenous and periodically flooded soil under arid conditions, typically like those in the coastal sabkhas of the Red Sea region. In this region, both the arid climate and the saline soils are major components of the natural environment where xerophytes and halophytes coexist. Both plant groups need to adapt to one similar environmental stressor, which is the remarkable water shortage in their surroundings. While xerophytes need to adapt to physical drought due to the scarcity of water, halophytes need to adapt to physiological drought due to the high soil salinity. These plants have developed comparable strategies to cope with water shortages in their environment. These are, for instance, osmotic adjustment, reduction of water loss by transpiration and water storage. The mechanisms developed to achieve these strategies vary on interspecific as well as intraspecific levels in the two groups. Xerohalophytes are, therefore, a heterogenous group of plants highly tolerant of drought and salinity and are able to overcome water shortages (Le Houerou 1993) by developing specific adaptations at morphoanatomical, morphophysiological, physiological, biochemical and genetic levels (Priyashree et al. Pattanayak 2010; Gonzalez 2019; Hafeez et al. 2021). Examples of these adaptations are presented in Table 24.1. In addition, xerohalophytes contain specific functional genes such as ion transporter genes and phytohormones biosynthesis pathway regulatory genes (Chai et al. 2019) that enable them to grow successfully under arid and saline conditions.

TABLE 24.1 Examples of adaptations of xerohalophytes to water shortage.

LEVEL OF ADAPTATION	*EXAMPLES OF XEROHALOPHYTES ADAPTATIONS*
Morphoanatomical	Formation of salt glands/bladders, epidermal and cuticle thickening, succulence and sunken stomata
Morphophysiological	Heteromorphic and dimorphic seeds
Physiological	Accumulation of inorganic ions (Na^+ and Cl^-) and organic solutes of low molecular weight such as proline and glycine betaine, accumulation of Na^+ to increase osmosis, latex production to reduce transpiration and mucilage production to improve water retention
Biochemical	Synthesis of antioxidant compounds (e.g., flavonoids, glutathione and polyphenols)

The plant composition of coastal sabkhas of the RSGA was investigated by Dar et al. (2022), Ghazanfar et al. (2019), El Shaer (2011), Hegazy et al. (2007), and Zahran (1982). Most of these studies were undertaken in Saudi Arabia and Egypt while studies from Sudan, Eretria, Somalia, Djibouti, Yemen, Jordon and Israel are very scarce or unavailable. This may reflect the paucity of information on vegetation of the coastal sabkha of the RSGA and reveals the need for dedicating special attention to sabkhas as a unique and common ecosystem of the Red Sea coastal environment. In almost all these studies, similar plant community compositions, at least on a family level, were reported for Red Sea coastal sabkhas. Common plant families reported include Asteraceae, Poaceae, Amaranthaceae, Juncaceae, Cyperaceae, Chenopodiaceae, Zygophyllaceae, and Plumbaginaceae.

24.2.2 Vegetation of coastal sabkhas along the Sudanese Red Sea coast

The plant composition of two coastal sabkhas along the Sudanese Red Sea coast was investigated. In each sabkha, three belt transects were surveyed, and the species encountered were either identified in situ or taken to the laboratory to be identified according to Farag and Kordofani (2012). The identification of the plant species and their scientific names were confirmed by consulting the following websites: African Plants Database (http://africanplantdatabase.ch), Global Biodiversity Information (www.gbif.org), World Flora Online (www.worldfloraonline.org) and The Plant List (www.theplantlist.org). Overall, 11 species were recorded (Table 24.2), belonging to the following growth forms: shrubs, herbs, rushes, sedges and grasses. The species *Odyssea*

TABLE 24.2 Species composition of coastal sabkhas along the Sudanese Red Sea coast.

NO.	SPECIES NAME	FAMILY
1	*Aeluropus lagopoides*	Poaceae
2	*Arthrocnemum macrostachyum*	Chenopodiaceae
3	*Cyperus leavigatus*	Cyperaceae
4	*Hyphaene thebiaca*	Palmae
5	*Juncus rigidus*	Juncaceae
6	*Odyssea mucronata*	Poaceae
7	*Phoenix dactylifera*	Palmae
8	*Pluchea dioscoridis*	Asteraceae
9	*Prosopis juliflora*	Leguminosae
10	*Sueada vermiculata*	Chenopodiaceae
11	*Portulaca quadrifida*	Portulaceae

mucronata is a new record added by this study to the flora of the Red Sea state, Sudan, from its coastal sabkha.

As the list indicated, true halophytic species (e.g., *A. macrostachyum*) and xerophytics species (e.g., *P. juliflora*) coexist in the investigated coastal sabkhas along the Sudanese Red Sea (Figure 24.2). This is corroborated by Guvensen et al. (2006), who reported the presence of xerophytes in sabkha from Turkey. The species *O. mucronata* tends to form continuous and monospecific mounds on the ground surface of the sabkha (Figure 24.3). Where it occurs, the growth habit of this plant

FIGURE 24.2 *Hyphaene thebiaca* and *Prosopis juliflora* in a coastal sabkha at the Sudanese Red Sea Coast.

Source: The author

FIGURE 24.3 Mounds of *Odyssea mucronata* from a coastal sabkha along the Sudanese Red Sea coast.

Source: The author

influences the movement of water and shapes the morphology of the water drainage system of coastal sabkhas.

24.2.3 Microbial mats of coastal sabkha

The microscopic forms of plants growing in the benthic environment in coastal sabkhas form natural aggregates conspicuous to the naked eye. These halophilic aggregates are present either at the surface of the sediment or in ephemeral and perennial pools. Depending on the complexity of these aggregates, growth of these microscopic plants is referred to as microbial mat or biofilm (Carreira et al. 2020).

Similar to their xerohalophytic counterpart, microbes of sabkha are well adapted to living at the lowest possible water levels by developing survival strategies such as producing extracellular polymeric compounds to act as reservoirs of water and means of protection from irradiation as well as from total desiccation (Krumbein et al. 2004). Microbial mats exhibit a vertically stratified growth of microorganisms in which the upper layer is occupied by oxygenic photosynthetic microorganisms represented by microalgae, nematodes and cyanobacteria (Al-Thani et al. 2014; Manaa and Aref 2022), and the lower layers are occupied by anoxygenic sulfur bacteria, heterotrophs and sulfate-reducing bacteria (Taher 2014). In addition, fungi and viruses were also reported on microbial mats from a hypersaline environment (Carreira et al. 2020). A detailed account on the biodiversity of a microbial mat from Jazan, Saudi Arabia, was provided in Albokari et al. (2017). The two bacterial families Salinibacteraceae and Rubidibacteraceae dominated the bacterial community, while the archaeal community was dominated by Halobacteriaceae.

Microbial mats play a significant role in the decay of organic matter, nitrogen fixation, sediment biostabilization and mineralization processes. According to Aref and Taj (2018) and Manaa and Aref (2022), sabkha microbial mats develop different surface morphologies known as microbially induced sedimentary structures (MISS). These MISS may include reticulate, tuft, pinnacle, wrinkle, gas dome and shrinkage crack patterns. Specifically, the organic matter produced by some of these microorganisms may contribute to the formation, development and erosion of some of these patterns (Lokier et al. 2017). The formation of dolomite in coastal sabkha is thought to be mediated by the action of the microbes in the mat (Bontognali et al. 2010; Al Disi et al. 2017; Diloreto et al. 2021).

The colour of the water of the ephemeral and perennial pools in coastal sabkha was observed to turn a pink-red colour in the summer season when the salinity of these pools reaches higher levels due to excessive evaporation. This colouration is attributed to the growth of microorganisms such the dinoflagellate *Dunaliella sp.* and halophylic archaea (Oren 2009).

24.2.4 Impact on coastal sabkha

Coastal sabkhas may act as sinks for litter from the land and sea. Litter is carried to coastal sabkhas by the water inflow from the hinterland during the rainy season and gets trapped in

FIGURE 24.4 Litter trapped by vegetation on a coastal sabkha along the Sudanese Red Sea coast.

Source: The author

the sabkha vegetation (Figure 24.4). Similarly, marine litter is carried to sabkhas during high tides and remains trapped after the ebb. Most of the litter is plastics, mainly bottles. In addition, dredging and filling due to construction of various infrastructure projects impede the hydrology of coastal sabkhas and impact their biodiversity and geochemistry.

24.2.5 Coastal sabkhas – conclusion

Taking into account the climatic and edaphic characteristics of the Red Sea region together with the definition of coastal sabkha and the different types of coastal habitats to which it applies, it could be concluded that the whole coast of the Red Sea region is a sabkha ecosystem with some environmental variations that lead to the creation of ecological niches. Considerable literature on sabkhas from Saudi Arabia, Egypt and Jordan is available, compared to Sudan, Eritrea, Djibouti, Somalia, Yemen and Israel, where more research is needed.

The perception of the coastal sabkhas of the Red Sea region as a wasteland with no direct and tangible benefits to humans has contributed to their underestimation as productive ecosystems. This is manifested in the amount of research undertaken on this prominent habitat with extensive spatial coverage compared to other coastal habitats such as mangroves, which have received much attention.

24.3 SEAWEED OF THE RSGA

Red Sea seaweed has contributed significantly to the knowledge of seaweed since the earlier stages of collection and identification of species (see Ateweberhan and Van Reine 2005). Species of seaweed were first collected from the Red Sea by travelers who supplied European scientists with specimens to describe, as was the practice at that time. Then, collection of seaweed from the Red Sea continued by scientific expeditions (e.g., S.M.S POLA, 1859–1896 and 1897–1898, and the International Indian Ocean Expedition 1959–1965) undertaken at different times (Sonnewald and El-Sherbiny 2017). Some of these specimens form the type specimens for the first description of the species (e.g., *Portieria sp.*).

The Red Sea has a high biodiversity, and many of its biota have shown a remarkable endemism level (DiBattista et al. 2016), which is about 9% for seaweed (Papenfuss 1968; PERSGA 2003). Previous records mentioned in Walker (1987) indicated that the Red Sea contains about 485 macroalgal species that generally belong to the circumtropical and subtropical marine flora of the Indo-Pacific, the Mediterranean and the Caribbean. In a recent review of Red Sea macroalgae, a total of 576 species were reported (Einav et al. 2021).

The Red Sea is generally nutrient deficient (oligotrophic) and depends on inflow from the Indian Ocean, via Bab al Mandab, for most of its nitrogen and phosphorus (Rasul and Stewart 2019). Therefore, its macroalgal communities are characterized by patchy distribution and strong seasonality in species biomass and cover. The composition, distribution and diversity of the Red Sea macroalgae seem to follow the natural latitudinal gradient in salinity, temperature and nutrient richness (Kürten et al. 2014). Also, geological events such as tectonic or paleoclimatic events were claimed as possible causes of the disjunct distribution of Red Sea seaweed (Ateweberhan and Van Reine 2005). The diversity of macroalgal species is greater in the northern and southern parts of the Red Sea than in the central part (Walker 1987; Sheppard et al. 1992). However, this assumption was based on previous records of collections undertaken in the Gulfs of Suez and Aqaba in the north and in the Dahlak archipelago in the south. Following the latitudinal gradient in the oceanographic conditions, the Red Sea is traditionally divided into the following parts:

PART	*DESCRIPTION*
The northern Red Sea	Includes the Gulf of Aqaba, the Suez Canal and the Egyptian coast
The central Red Sea	Includes the Sudanese and Saudi Arabian coasts
The southern Red Sea	Includes the coasts of Eritrea, Djibouti, Yemen and Somalia

According to Walker (1987) the distribution and composition of the seaweed of the RSGA is influenced by temperature and salinity as follows:

PART	SEAWEED COMPOSITION
The northern Red Sea	Contains tropical species with some taxa from slightly cooler regions, mostly green (34.6%) with red (30.2%) and brown (17.6%) species.
The central Red Sea	Contains a mix of the species of the northern and southern Red Sea macroalgal flora. The green macroalgae account for 5.6%, the browns 4.8%, and the red are 3.6%.
The southern Red Sea	Contains tropical species of warmer and saline tropics, mostly brown (42%), with red (18.3%) and green (8.9%) species.

This distribution pattern was based on old investigations undertaken mostly in the northern and southern RSGA (Walker 1987). Wider investigation of the overall geographical distribution of seaweed in the RSGA is lacking and is critically needed for conservational and management purposes.

The geographical distribution of RSGA macroalgae has been investigated on a local level. Spatial and temporal variations in seaweed communities along the Egyptian coast of the Red Sea were attributed to variations in seawater surface temperature (Mohamed et al. 2022). Similarly, spatio-temporal variation in seaweeds was also recorded in the Gulf of Aqaba, Jordan (Al-Zibdah and Damhoureyeh 2006). The study of Issa et al. (2014) on the Red Sea coast of Egypt reported a distinct spatial pattern with significant differences between sites located from the north to the south, although this was attributed to chemical contamination at some of the study sites.

Knowledge of the seaweed from other coastal ecosystems, such as mangroves and coral reefs of the Red Sea, is scarce. However, El-Sharouny et al. (2001) reported 23 seaweed species over a period of four seasons from the mangroves of the Red Sea, Egypt, compared to 7 species from an adjoining mangrove site (Mohamed et al. 2022). The investigation of the floristic composition and zonation of a northern Red Sea coral reef island revealed the diversity of seaweed functional forms and the influence of biotic and abiotic factors on species composition and distribution (El-Manawy 2001). Generally, frondose and leathery seaweed such as *Sargassum spp.* and *Turbinaria spp.* survive well and flourish in the surf areas of the reefs (El-Manawy and Rashedy 2022), together with calcareous seaweed such as *Halimeda spp.* The presence of a healthy population of these taxa is an indication of the integrity of the reef ecosystem. Additionally, almost all coralline and crustose red seaweed (e.g., *Corallina spp.* and *Porolithon spp.*) are associated with reef ecosystems forming noticeable algal ridges in the seaward front of the reef ecosystem of the Red Sea. The dominance of turf and fleshy seaweed in coral ecosystems is a sign of degradation. Some of the coral reef ecosystem in Eilat, Israel, is covered with up to 72% of turf seaweed and was considered degraded when compared with adjoining reefs with only 6% cover of turf seaweed (Bahartan et al. 2010). Dead coral stones at the reef crest are mostly colonized with turf seaweed, such as the reds *Gelidiella spp.* and *Ceramium spp.* and the filamentous brown *Sphacelaria spp.*

24.3.1 Seaweed zonation in the RSGA

Localized growth and distribution of seaweed is mainly influenced by the synergetic effect of abiotic (e.g., temperature and type of substrate) and biotic (e.g., pigmentation and grazing) filters. An example of the zonation of a seaweed community on the Sudanese Red Sea coast is given in Table 24.3. The subtidal zone was occupied by green seaweed (nine species) followed by the browns (three species) and the reds (one species). The filamentous greens (*Boodlea sp.*, *Cladophota sp.* and *Enteromorpha spp.*) were found in tufts in the upper shallow subtidal zone attached to solid substrata at the water edge or forming mats on the seafloor at greater depths. The frondose green macroalgae such as *Halimeda* species were frequently encountered in the lower subtidal and back reef. In the lower subtidal zone, brown macroalgae such as *Padina*, *Polycladia* and *Sargassum* species flourished, attaining remarkable frond size and abundance (Figure 24.5), both of which gradually decrease toward the reef back where the thalli tend to diminish in size.

TABLE 24.3 Example of zonation of seaweed community along the Sudanese Red Sea coast.

	SEAWEED				
ZONE	GREEN	BROWN	RED	ASSOCIATED FLORA	ASSOCIATED FAUNA
Subtidal	*Acetabularia sp., Boodlea composita, Enteromorpha intestinalis, E. compressa, Cladophora prolifera., Halimeda macroloba, H. opuntia, Caulerpa racemosa, Avrainvilea sp.*	*Padina pavonica, Polycladia myrica, Sargassum wightii*	*Palisada perforata*	*Cymodacea rotundata, Thalassia hemprichii, Halophila ovalis, H. stipulacea, Halodule uninervis*	*Holothuria atra*, red sponge, black sponge
Back reef	*Caulerpa serrulata, C. racemosa, Halimeda tuna, H. opuntia, H. macroloba*	*Turbinaria ornata, P. myrica, Dictyota dichotoma, P. pavonica, Sargassum sp.*	*Digenia simplex, Tricleocarpa fragilis, Gracilaria crassa, Hypnea valentia, Jania rubens*	*C. rotundata, T. hemprichii, H. uninervis, H. stipulacea*	*H. atra*, elephant ear sponge, black sponge

FIGURE 24.5 Seaweed community in the lower subtidal area of the Sudanese Red Sea coast with visual dominance of brown seaweed.

Source: The author

However, some of the browns in the back reef exhibit visual dominance due to their larger sizes. In particular, the canopies of *T. ornata* and *P. myrica* could be seen floating on the water surface. Notably, the brown macroalga *P. pavonica* was able to establish itself and grow in all water depths, starting from very shallow depths of few centimetres at the upper subtidal to a depth of more than one metre in the back reef. Most of the red macroalgae were found from the back reef to the reef crest, except for *P. perforata*, which was encountered in the upper subtidal at a shallow depth, just enough to cover the dwarf prostrate thalli. The two red species *H. valentia* and *J. rubens* exhibited abundant growth in the back reef. All the other reds were found around the reef crest attached to coral stones or growing in crevices of coral rocks, where the wave action is stronger. Specifically, the two red species *T. fragilis* and *D. simplex* seem to be restricted to the reef crest – an indication of tolerance to wave action. Here, seagrass was confined to depressions with a thin sediment layer.

Little information is available on the biodiversity of Red Sea seaweed, and some valuable old literature is inaccessible at both regional and national levels. Table 24.4 shows the total number of the seaweed species for each country bordering the RSGA as recorded in Guiry and Guiry (2022). Records of Saudi Arabia, Yemen and Egypt may contain records of other marine habitats such as the Arabian Gulf, the Arabian Sea and the Mediterranean Sea, as well as records on microalgae. Generally, there is a paucity of information on seaweed from the southern RSGA region compared to the northern parts.

24.3.2 Utilization and threats to seaweed in the RSGA

To date, macroalgae are not utilized for any kind of human or animal consumption by the nations of the RSGA. However, the Kingdom of Saudi Arabia is developing a project on algae and seaweed biotechnology with a prime goal of boosting the fisheries sector and also to harness other benefits of marine algae through sustainable farming (www.arabnews.com/node/2020451/business-economy). Therefore, there is no targeted or overharvested macroalgal species to be reported at this time. Nevertheless, the seaweed in the RSGA may be threatened by the following human impacts:

1 **Negligence:** The value of the macroalgal resources is underestimated in the region despite its importance to other coastal ecosystems and the global demand for seaweed for various industries. Consequently, no efforts are being made to utilize or to protect this resource for the benefit of the nations.
2 **Dredging and filling:** Dredging and filling are identified as key environmental issues for the region of the RSGA (UNEP 2015), as well as for some other countries such as Egypt (Mansour et al. 2007) and Saudi Arabia (Husain and Khalil 2013). Both operations destroy macroalgal lawns and eliminate their services from the marine environment. The writer has witnessed the complete disappearance of *Ulva lactuca* and *Pryopsis sp.* from a stretch of coast in Sudan since 2006, in the aftermath of a dredging operation in the harbour. Besides the complications of water turbidity caused by dredging and filling, they also change the water depth, sediment compaction and sediment granulometric properties required for macroalgae to reestablish later on. Dredging and filling may be considered the key issues threatening macroalgal lawns in the RSGA.
3 **Coastal discharges:** Generally, the impact of coastal effluents on the RSGA environment is very localized, confined to the vicinity of big coastal cities like Jeddah (Mohorjy and Khan 2006; Pena-Garcia et

TABLE 24.4 Total number of seaweed species in the countries bordering the Red Sea and the Gulf of Aden.

RSGA PART	*COUNTRY*	*NO. OF SEAWEED SPECIES*	*TOTAL BY PART*
Northern Red Sea	Egypt	1,034	**1,749**
	Jordan	67	
	Israel	648	
Central Red Sea	Sudan	144 (Osman and Mohammed 2016)	**376**
	Saudi Arabia	232	
Southern Red Sea	Eritrea	179	**679**
	Djibouti	58	
	Somalia	209	
	Yemen	233	

Source: Guiry and Guiry (2022)

al. 2014) or in proximity to areas of major shipping activities (Fahmy et al. 2016). Local eutrophication caused by these effluents reduces macroalgal diversity and produces blooms of a single algal species.

4 **Climate change:** Krokos et al. (2019) provided strong evidence that over the next decades, the sea surface temperature (SST) trend in the Red Sea purely related to global warming is expected to be counteracted by the cooling Atlantic multidecadal oscillation phase. Regardless of the current positive trends, projections incorporating long-term natural oscillations suggest a possible decreasing effect on SST in the near future. Even within the current prevailing prediction of the fast warming of the Red Sea, research has predicted that some of the Red Sea organisms (e.g., coral reefs) may naturally adapt to an increase in sea temperature (Cantin et al. 2010). Taking into consideration the apparent seasonality of Red Sea macroalgae in response to the natural change in sea surface temperature, it could be anticipated that their present latitudinal distribution may experience a shift in its boundaries. The macroalgae of the southern Red Sea may extend further into the central and possibly the northern Red Sea. Red Sea warming may also influence the abundance of some species as has been widely documented for macroalgal assemblages (Piñeiro-Corbeira et al. 2018) in response to seawater warming.

5 **Marine-introduced species:** As one of the major and busiest shipping routes, the RSGA region is susceptible to receiving alien species introduced by ballast water and ship hulls. Fortunately, no records on marine-introduced species were found for the region. Instead, species of the Red Sea have been reported to migrate to the Mediterranean Sea (Galil et al. 2019) through the Suez Canal.

Macroalgae are good indicators of human land-based activities and marine ecosystem health. For instances, blooming of the filamentous green alga *Enteromorpha* species occurs at sites receiving coastal effluents of an organic nature. Also, the presence of healthy populations of calcareous green (*Halimeda* species) and brown (*Turbinaria* and *Sargassum* species) macroalgae at the reef flat or back reef indicates the integrity of the Red Sea coral reef ecosystems.

24.3.3 Conclusion on seaweed of the RSGA

The present knowledge of Red Sea macroalgae contains some gaps. Importantly, knowledge of species systematics, biogeography and genetic diversity is incomplete. Checklists, field guides, and catalogues are lacking or geographically restricted, mainly due to a lack of phycologists, proper research infrastructure and scientific collaboration between countries. On another level, little research has been done on macroalgae as a source of natural products with industrial, environmental, pharmaceutical and medicinal potential. This kind of research helps unravel the benefits of macroalgae to human well-being and thus supports management of the resource. Also, knowledge of the response of Red Sea macroalgae to the impact of stressors – for instance, climate change, dredging and filling, and nutrients overload – is critically needed.

Taking into consideration the advancement in the field of molecular biology that has strongly influenced the systematics of seaweed, Red Sea seaweed is, at present, among the least studied organisms in the Red Sea. The recent achievement in determining the taxonomic status of some species in dispute, the establishment of evolutionary relationships which led to resurrection of old names, the creation of new combinations and the erection of taxonomic ranks need to be taken into consideration when revising or examining the Red Sea seaweed flora.

24.4 SEAGRASSES OF THE RSGA

Seagrass meadows cover about 6,963 km^2 of the African coast of the RSGA and have an economic value of US$21,752 million per year (Tregarot et al. 2020). The RSGA contains 12 species of seagrass, compared to 72 species present globally and 12 to 14 species present in the western Indo-Pacific region (Short et al. 2016). These 12 species, which belong to the two families of Cymodoceaceae and Hydrocharitaceae (El Shaffai 2016), are *Cymodocea rotundata*, *Cymodocea serrulata*, *Halodule uninervis*, *Halodule pinifolia*, *Syringodium isoetifolium*, *Thalassodendron ciliatum*, *Enhalus acoroides*, *Halophila ovalis*, *Halophila decipiens*, *Halophila minor*, *Halophila stipulacea* and *Thalassia hemprichii.*

Seagrass beds in the RSGA extend from the subtidal zone to a depth of 70 m (PERSGA 2006). Commonly, Red Sea seagrass meadows, composed of one or two species, extend parallel to the coastline in soft bottoms. The patterns of distribution of seagrass in the RSGA are not fully resolved yet. Nevertheless, existing knowledge indicates that the central Red Sea (between 25° and 18°N) has a higher diversity of seagrass (10 to 12 seagrass species) than the northern and the southern Red Sea. The latter has more seagrass species (9 to 10 species) than the former, which hosts 5 to 8 species. The Gulf of Suez has 5 seagrass species, the Gulf of Aden has 6 species (Bawazir 2009), and Gulf of Aqaba has 7 seagrass species (Price et al. 1988; Al-Rousan et al. 2011). Of the 7 species in the Gulf of Aqaba, only *H. stipulacea*, *H. ovalis* and *Halodule uninervis* penetrate to the northern end of the Gulf (Khalaf et al. 2012).

The biogeographic distribution of individual seagrass species in the RSGA is yet unclear. However, existing data may help construe a preliminary pattern. The native species *H. stipulacea* that occurs in all the RSGA is reported to dominate in the northern Red Sea (Al-Rousan et al. 2011), forming discontinuous monospecific meadows in Eilat. This species has been reported to be of wide ecological occurrence and has recently been recorded as an invasive species from the Caribbean and South American seas (Winters et al. 2020). Similarly, the species *H. uninervis* is widely spread all over the RSGA. The two species *C. rotundata* and *C. serrulata* are common in the central and southern Red Sea (El-Shaffai 2016). In addition, the abundance of some species increases northward (e.g., *T. ciliatum and S. isoetifolium*) while that of some others increase southward (e.g., *C. rotundata* and *C. serrulata*).

The species composition of a seagrass meadow in a specific location reflects the succession phase of that meadow, which is the function of the synergetic effect of abiotic and biotic factors, including anthropogenic pressures (Greve and Binzer 2004; Turschwell et al. 2021; Moreira-Saporiti et al. 2021). In this regard, a seagrass species could either be a pioneer species, an opportunistic species or a climax species. Pioneer species of seagrass are generally small in size and fast growing (e.g., *H. ovalis* and *H. uninervis*). Climax species (e.g., *T. hemprichii* and *E. acoroides*) are slow growing, long lived and highly resistant to pressures. Seagrass species designated as opportunistic (e.g., *H. stipulacea* and *C. rotundata*) have mixed traits of both pioneer and climax species. Climax species are established in stable substrates, whereas pioneer and opportunistic species predominate in dynamic substrates (Di Carlo and Tombolahy 2011). This information is of significance for the monitoring, restoration and management of seagrass ecosystems of the RSGA and other coastal ecosystems. Additionally, it is relevant to environmental assessments of ecosystems' health/status as these species could be used as bioindicators or biological quality elements.

An example of the zonation of seagrasses in a site along the Sudanese Red Sea coast is given in Table 24.3. In this site, *C. rotundata* tend to colonize in the soft bottom at the subtidal zone, forming a semi-continuous monospecific belt parallel to the coastline dotted with remarkable clusters of the calcareous green alga *H. opuntia*. This association is a common feature in the subtidal zone of some localities along the Sudanese Red Sea. In some places, *T. hemprichii* may occupy the same position as *C. rotundata*, forming a thick, monospecific belt (Figure 24.6). The rest of the seagrasses in this zone tend to exhibit a dispersed and patchy growth in the empty spaces left by the two species. A sparsely populated meadow of the opportunistic species *H. stipulacea* was observed growing in a well-oxygenated sandy sediment in the aftermath of dredging and filling operations in Suakin Harbour, Sudan. Most of the associated fauna were found in the lower subtidal region.

Table 24.5 illustrates the current knowledge of seagrass species composition in the RSGA countries, as well as where data are not available or accessible (e.g., Djibouti). It is important to take into account that systematic studies of the biogeography and taxonomy of Red Sea seagrass are yet to be undertaken. This is because vast coastal areas along the Red Sea remain to be examined due to their remoteness from major urban centres, such as the southern coast of Sudan.

24.4.1 Status of seagrasses of the RSGA

As mentioned earlier, the deficiency of data on Red Sea seagrass habitats limits the provision of a comprehensive account of their current status or the threats they actually face. Therefore, information on threatened species is deduced from existing literature and the status of species associated with seagrass, such as turtles and dugongs, due to the perceived link between them. Fortunately, the status of the seagrass in the vast Indo-Pacific region (i.e., Seagrass Bioregion 5, which includes the Red Sea and Gulf of Aden) was designated as of least concern when the Red List criteria of the IUCN were applied for the 72 species of seagrass (Short et al. 2011). However, some species may be subject to anthropogenic impacts as described next.

FIGURE 24.6 Meadow of the climax seagrass species *Thalassia hemprichii* in the subtidal area of the Sudanese Red Sea coast.

Source: The author

TABLE 24.5 Seagrass species composition in RSGA countries.

COUNTRY	SPECIES NO.	SPECIES NAMES	REFERENCE(S)
Egypt	11	*Halophila stipulacea, Halophila ovalis, Halophila decipiens, Thalassia hemprichii, Enhalus acoroides, Halodule uninervis, Halodule pinifolia, Cymodocea rotundata, Cymodocea serrulata, Syringodium isoetifolium, Thalassodendron ciliatum*	El Shaffai et al. (2014)
Jordan	3	*Halophila stipulacea, Halodule uninervis, Halophila ovalis*	Al-Rousan et al. (2011)
Israel	1	*Halophila stipulacea*	Winters et al. (2015)
Sudan	10	*Thalassia hemprichii, Halophila stipulacea, Halophila ovalis, Halophila minor, Enhalus acoroides, Halodule uninervis, Cymodocea rotundata, Cymodocea serrulata, Syringodium isoetifolium, Thalassodendron ciliatum*	GaibAlla (2005)
Saudi Arabia	10	*Thalassodendron ciliatum, Syringodium isoetifolium, Enhalus acoroides, Halophila ovalis, Halophila stipulacea, Halophila decipiens, Thalassia hemprichii, Cymodocea rotundata, Cymodocea serrulata, Halodule uninervis*	Qurban et al. (2019)
Eritrea	8	*Halophila ovalis, Halophila stipulacea, Thalassia hemprichii, Halodule uninervis, Syringodium isoetifolium, Thalassodendron ciliatum, Enhalus acoroides, Cymodocea rotundata*	Ghebretinsae et al. (2019)
Djibouti		**No data available**	
Somalia	7	*Halophila ovalis, Halophila stipulacea, Cymodocea rotundata, Cymodocea serrulata, Thalassia hemprichii, Halodule uninervis, Syringodium isoetifolium*	Lugendo (2016)
Yemen	9	*Halophila stipulacea, Halophila ovalis, Halodule uninervis, Thalassodendron ciliatum, Thalassia hemprichii, Cymodocea serrulata, Cymodocea rotundata, Enhalus acoroides, Syringodium isoetifolium*	Bawazir (2009), Barratt et al. (1987)

***Enhalus acoroides*:** This unique Indo-Pacific and climax species has a very limited distribution in the RSGA (El-Shaffai 2016). It was observed at one single location in Sudan. Unfortunately, the species disappeared after the site was dredged and has never been seen again in any other locations. Accordingly, and based on criterion B (restricted area of occupancy or extent of occurrence) for the evaluation of species inclusion on the IUCN Red List, this species is strongly recommended to be assessed and included on the list.

***Cymodocea* species, *Thalassia hemprichii* and *Halodule uninervis*:** On a regional scale, seagrass meadows in the RSGA are subjected to the physical impacts of land-based activities, in particular removal by dredging and burial by filling operations during coastal development. Meadows of the species belonging to the genera *Cymodocea*, *Thalassia* and *Halodule* that grow in shallow subtidal water are threatened by these activities.

24.4.2 *Dugong dugon*

The RSGA is home to significant dugong populations (Shawky 2019; Nasr et al. 2019) where their distribution matches the availability of seagrass meadows. Given that the Red Sea and the Persian Gulf host the last remaining healthy populations of this species on the eastern coast of Africa (PERSGA 2006), this can be taken as an indicator of the good health of the seagrass in the RSGA.

REFERENCES AND FURTHER READING

African Plant Database (version 4.0.0). *Conservatoire et Jardin botaniques de la Ville de Genève and South African National Biodiversity Institute*, Pretoria. http://africanplantdatabase.ch. Accessed July 2022.

Ahmad, F., Azman, S., Said, M., and L. Baloo. 2015. Tropical seagrass as a bioindicator of metal accumulation. *Sains Malays* 44: 203–210. http://doi.org/10.17576/jsm-2015-4402-06.

Albokari, M.A., Çinar, S., and M.B. Mutlu. 2017. Microbial characterization of Jazan sabkha in Saudi Arabia. *Appl Ecol Environ Res* 15, no. 3: 1069–1077. http://doi.org/10.15666/aeer/1503_10691077.

Al Disi, Z.A., Jaoua, S., Bontognali, T.R.R., Attia, E.S.M., Al-Kuwari, H.A.A.S., and N. Zouari. 2017. Evidence of a role for aerobic bacteria in high magnesium carbonate formation in the evaporitic environment of Dohat Faishakh Sabkha in Qatar. *Front Environ Sci* 5: 1. http://doi.org/10.3389/fenvs.2017.00001.

Al-Rousan, S., Al-Horani, F., Eid, E., and M. Khalaf. 2011. Assessment of seagrass communities along the Jordanian coast of the Gulf of Aqaba, Red Sea. *Mar Biol Res* 7: 93–99. http://doi.org/10.1080/17451001003660319.

Al-Shamsi, N., Hussain, M.I., and A. El-Keblawy. 2020. Physiological responses of xerohalophytic Suaeda vermiculata to salinity in its hyper-arid environment. *Flora* 273: 151705.

Al-Thani, R., Al-Najjar, M.A.A., Al-Raei, A.M., Ferdelman, T., Thang, N.M., et al. 2014. Community structure and activity of a highly dynamic and nutrient-limited hypersaline microbial mat in Um Alhool Sabkha, Qatar. *PLoS ONE* 9, no. 3: e92405. http://doi.org/10.1371/journal.pone.0092405.

Al-Zibdah, M., and S. Damhoureyeh. 2006. Spatial and temporal distribution of macroalgae along the Jordanian coast of the Gulf of Aqaba, Red Sea. *Dirasat Pure Sci* 33, no 1.

Aref, M.A., and R.J. Taj. 2018. Recent evaporite deposition associated with microbial mats, Al-Kharrar supratidal – intertidal sabkha, Rabigh area, Red Sea coastal plain of Saudi Arabia. *Facies* 64: 28. http://doi.org/10.1007/s10347-018-0539-y.

Ateweberhan, M., and P.W.F.A. Van Reine. 2005. Taxonomic survey of seaweeds from Eritrea. *Blumea* 50, no. 1: 65–111. http://doi.org/10.3767/000651905X623292.

Bahartan, K., Zibdah, M., Ahmed, Y., Israel, A., Brickner, I., and A. Abelson. 2010. Macroalgae in the coral reefs of Eilat (Gulf of Aqaba, Red Sea) as a possible indicator of reef degradation. *Mar Pollut Bull* 60: 759–764. http://doi.org/10.1016/j.marpolbul.2009.11.017.

Banat, K.M., Howari, F.M., and K.A. Kadi. 2005. Water chemical characteristics of Red Sea coastal sabkhas and associate evaporite and carbonate minerals. *J Coast Res* 21: 1068–1081. http://doi.org/10.2112/04-0181.1.

Barratt, L., Dawson-Shepherd, A., Ormond, R., and R. McDowell. 1987. Yemen Arab Republic Marine Conservation Survey, Volume 1. Distribution of habitats, species along the YAR coastline. *Unpublished Report.* International Union for Conservation of Nature and Natural Resources.

Basyoni, M.H., and M.A. Aref. 2016. Composition and origin of sabkha brine and their environmental impact of infrastructure in Jizan, Red Sea coast, Saudi Arabia. *Environ Earth Sci* 75: 105.

Bawazir, G.M. 2009. Marine biodiversity of Aden wetland protected areas. Aden wetland conservation project. *Unpublished Report.*

Bird, E. 2010a. Saudi Arabia, Red Sea Coast. In: *Encyclopedia of the World's Coastal Landforms.* Ed. E.C.F. Bird. Dordrecht: Springer.

Bird, E. 2010b. Eritrea. In: *Encyclopedia of the World's Coastal Landforms.* Ed. E.C.F. Bird. Dordrecht: Springer.

Bontognali, T.R.R., Vasconcelos, C., Warthmann, R.J., Bernasconi, S.M., Dupraz, C., Strohmenger, C.J., and J.A. Mckenzie. 2010. Dolomite formation within microbial mats in the coastal sabkha of Abu Dhabi (United Arab Emirates). *Sedimentology* 57, no. 3: 824–844. http://doi.org/10.1111/J.1365-3091.2009.01121.X.

Bui, E.N. 2019. Evidence for the role of salinity and alkalinity in plant diversification in Australia. In: *Sabkha Ecosystems. Tasks for Vegetation Science*, vol. 49. Ed. B. Gul, B. Böer, M. Khan, M. Clüsener-Godt and A. Hameed. Cham: Springer. http://doi.org/10.1007/978-3-030-04417-6_2.

Cantin, N.E., Cohen, A.L., Karnauskas, K.B., et al. 2010. Ocean warming slows coral growth in the central Red Sea. *Science* 329: 322–325.

Carreira, C., Lønborg, C., Kühl, M., Lillebø, A., Sandaa, R., Villanueva, L., and S. Cruz. 2020. Fungi and viruses as important players in microbial mats. *FEMS Microbiol Ecol* 96, no. 11. http://doi.org/10.1093/femsec/fiaa187.

Chai, W.-W., Wang, W.Y., Ma, Q., Yin, H.-J., Hepworth, S.R., and S.-M. Wang. 2019. Comparative transcriptome analysis reveals unique genetic adaptations conferring salt tolerance in a xerohalophyte. *Functional Plant Biol* 46: 670–683. http://doi.org/10.1071/FP18295.

Dar, B.A., Assaeed, A.M., Al-Rowaily, S.L., Al-Doss, A.A., and A.M. Abd-ElGawad. 2022. Vegetation composition of the halophytic grass *Aeluropus lagopoides* communities within coastal and inland sabkhas of Saudi Arabia. *Plants* 11: 666. http://doi.org/10.3390/plants11050666.

DiBattista, J.D., Choat, J.H., Gaither, M.R., Hobbs, J.P., Lozano-Cort, D.F. Myers, R.F., Paulay, G., Rocha, L.A., Toonen, R.J., Westneat, M.W., and M.L. Berumen. 2016. On the origin of endemic species in the Red Sea. *J Biogeogr* 43: 13–30. http://doi.org/10.1111/jbi.12631.

Di Carlo, G., and M. Tombolahy. 2011. Seagrasses and algae of north-eastern Madagascar. In: *A Rapid Marine Biodiversity Assessment of the Coral Reefs of Northeast Madagascar*, vol. 61. Eds. D. Obura, G. Di Carlo, A. Rabearisoa and T. Oliver, 44–52. Arlington, VA: RAP Bull Biol Assessment Conservation International. http://doi.org/10.1896/054.061.0106.

Diloreto, Z.A., Garg, S., Bontognali, T.R.R., and M. Dittrich. 2021. Modern dolomite formation caused by seasonal cycling of oxygenic phototrophs and anoxygenic phototrophs in a hypersaline sabkha. *Sci Rep* 11: 4170. http://doi.org/10.1038/s41598-021-83676-1.

Einav, R., Guiry, M.D., and A. Israel. 2021. A revised list of seaweeds from the Red Sea (1756–2020). *Israel J Plant Sci* 67: 1–73.

El-Magsodi, M.O., and D.A. Haddoud. 2010. The salt marsh (sabkha) in the western part of Libya. In: *Sabkha Ecosystems. Tasks for Vegetation Science*, vol. 46. Eds. M. Öztürk, B. Böer, H.J. Barth, M. Clüsener-Godt, M. Khan and S.W. Breckle. Dordrecht: Springer. http://doi.org/10.1007/978-90-481-9673-9_9.

El-Manawy, I.M. 2001. Floristic composition and zonation of seaweeds on Zabargad Reef (Red Sea, Egypt). *Taeckholmia* 21, no. 1: 115–134.

El-Manawy, I.M., and S.H. Rashedy. 2022. The ecology and physiology of seaweeds: An overview. In: *Sustainable Global Resources of Seaweeds*, vol. 1. Eds. A.R. Rao and G.A. Ravishankar, 3–16. Cham, Switzerland: Springer. http://doi.org/10.1007/978-3-030-91955-9_1.

El-Omla, M.M., and H.A. Abou-Lela. 2012. Environmental and mineralogical studies of sabkhas soil at Ismailia-roadbed, southern of Suez Canal district, Egypt. *Open J Geol* 2: 165–181.

El Shaer, H.M. 2011. Potential role of salt marshes in the sabkhas of Egypt. In: *Sabkha Ecosystems, Tasks for Vegetation Science 46*. Ed. M. Öztürk, et al., 95–98. Cham: Springer. http://doi.org/10.1007/978-90-481-9673-9_11.

El Shaffai, A. 2016. *Field Guide to Seagrasses of the Red Sea*, 2nd ed. Eds. A. Rouphael and A. Abdulla. Gland, Switzerland: IUCN and Courbevoie, France: Total Foundation, 56 p.

El Shaffai, A., Hanafy, M.H., and A.A.F.A. Gab-Alla. 2014. Distribution, abundance and species composition of seagrasses in Wadi El Gemal National Park, Red Sea, Egypt. *Indian J Appl Res* 4, no. 3: 1–9.

El-Sharouny, H.M., El-Tayeb, M.A., and M.S. Ismail. 2001. Macroalgae associated with mangroves at Hurghada and Safaga of the Egyptian Red Sea coast. *J King Abdulaziz Univ Mar Sci* 12 (Special Issue): 241–251. http://doi.org/10.4197/mar.12-1.18.

Engel, M., Strohmenger, C.J., Peis, K.T., Pint, A., Brill, D., and H. Brückner. 2021. High-resolution facies analysis of a coastal sabkha in the eastern Gulf of Salwa (Qatar): A spatio-temporal reconstruction. *Sedimentology* 69, no. 3: 1119–1150. http://doi.org/10.1111/sed.12938.

Fahmy, M.A., Fattah, L.M.A., Abdel-Halim, A.M., Aly-Eldeen, M.A., Abo-El-Khair, E.M., Ahdy, H.H., Hemeilly, A., El-Soud, A.A., and M.A. Shreadah. 2016. Evaluation of the quality of the Egyptian Red Sea coastal waters during 2011–2013. *J Environ Prot* 7: 1810–1834. http://doi.org/10.4236/jep.2016.712145.

FAO. 2008. *AQUASTAT Country Profile – Israel.* Food and Agriculture Organization of the United Nations (FAO), Rome.

Farag, I., and M.A.Y. Kordofani. 2012. *Field Guide to the Wild Plants of the Sudanese Coastal Plain.* UNESCO Chair on Marine Biology and Oceanography, 80 p.

Gaib Alla, A.K. 2005. *Impacts of Coastal Development Activities on Seagrasses Communities Along the Sudanese Red Sea Coast.* Unpublished thesis, University of Khartoum, Sudan.

Galil, S.B., Danovaro, R., Rothman, S.B.S., Gevili, R., and M. Goren. 2019. Invasive biota in the deep-sea Mediterranean: An emerging issue in marine conservation and management. *Biol Invasions* 21, no. 2: 281–288.

Geneid, Y.A.H. 2009. Distribution of seagrass species along the Egyptian Red Sea coast. *Egypt J Aquat Res* 35, no. 1: 58–68.

Ghazanfar, S.A., Böer, B., Al Khulaidi, A.W., El-Keblawy, A., and S. Alateeqi. 2019. Plants of sabkha ecosystems of the Arabian Peninsula. In: *Sabkha Ecosystems, Tasks for Vegetation Science*, vol. VI. Eds. B. Gul et al., 55–80. Cham, Switzerland: Springer Nature. http://doi.org/10.1007/978-3-030-04417-6_5.

Ghebretinsae, M., Adhanom, A., Ghebremedhin, H., and B. Tsegay. 2019. A review on bioactive secondary metabolites of seagrass of the southern Red Sea, Eritrea. *Int J Adv Res* 7, no. 3: 306–314.

González, M.B. 2019. Adaptation of halophytes to different habitats. In: *Seed Dormancy and Germination*. Ed: J.C. Jimenez-Lopez. IntechOpen. http://doi.org/10.5772/intechopen.87056.

Green, E.P., and F.T. Short. 2003. World Atlas of Seagrasses. *Prepared by the UIMEP World Conservation Monitoring Centre*. Berkeley: University of California Press.

Greve, T.M., and T. Binzer. 2004. Which factors regulate seagrass growth and distribution? In: *European Seagrasses: An Introduction to Monitoring and Management*, 19–23. EU Project Monitoring and Managing of European Seagrasses.

Grigore, M.N., and C. Toma. 2017. Definition and classification of halophytes. In: *Anatomical Adaptations of Halophytes*. Cham, Switzerland: Springer. http://doi.org/10.1007/978-3-319-66480-4_1.

Guiry, M.D., and G.M. Guiry. 2022. *AlgaeBase*. World-wide Electronic Publication, National University of Ireland, Galway. www.algaebase.org. Accessed July 2022.

Gunatilaka, A. 1990. Anhydrite diagenesis in a vegetated sabkha, Al-Khiran, Kuwait, Arabian Gulf. *Sediment Geol* 69: 95–116.[36]

Guvensen, A., Guven, G., and M. Ozturk. 2006. An overview of the halophytes in Turkey. In: *Sabkha Ecosystems. Tasks for Vegetation Science*, vol. 42. Eds. M.A. Khan, B. Boer, G.S. Kust and H.J. Barth, 9–30. Dordrecht: Springer. http://doi.org/10.1007/978-1-4020-5072-5_2.

Hafeez, M.B., Raza, A., Zahra, N., Shaukat, K., Akram, M.Z., Iqbal, S., and A.M.A. Basra. 2021. Gene regulation in halophytes in conferring salt tolerance. In: *Handbook of Bioremediation*. Eds. M. Hasanuzzaman and M.N.V. Prasad, 341–370. Academic Press. http://doi.org/10.1016/b978-0-12-819382-2.00022-3.

Hasanuzzaman, M., Nahar, K., Alam, M.M., Bhowmik, P.C., Hossain, M.A., Rahman, M.M., Prasad, M.N.V., Ozturk, M., and M. Fujita. 2014. Potential use of halophytes to remediate saline soil. *BioMed Res Int* 2014: 589341. http://doi.org/10.1155\2014589341.

Hegazy, A.K, Lovett-Doust, J., Hammouda, O., and N.H. Gomaa. 2007. Vegetation distribution along the altitudinal gradient in the northwestern Red Sea region. *Community Ecol* 8, no. 2: 151–162.

Husain, T., and A. Khalil. 2013. Environment and sustainable development in the Kingdom of Saudi Arabia: Current status and future strategy. *J Sustain Dev* 6, no. 12.

Issa, A., Hifney, A., Abdel-Gawad, K., and M. Gomaa. 2014. Spatio temporal and environmental factors influencing macroalgal β diversity in the Red Sea, Egypt. *Bot Mar* 57: 99–110. http://doi.org/10.1515/bot-2013-0107.

Kearney, W.S., and S. Fagherazzi. 2016. Saltmarsh vegetation promotes efficient tidal channels network. *Nat Comm* 7: 12287.

Khalaf, M.A., Al-Rousan, S., and F.A. Al-Hourani. 2012. Fish assemblages in seagrass habitat along the Jordanian coast of the Gulf of Aqaba. *Nat Sci* 4, no. 8: 51–525. http://doi.org/10.4236/ns.2012.48069.

Krokos, G., Papadopoulos, V.P., Sofianos, S.S., Ombao, H., Dybczak, P., and I. Hoteit. 2019. Natural climate oscillations may counteract Red Sea warming over the coming decades. *Geophys Res Lett* 46, no. 6: 3454–3461. http://doi.org/10.1029/2018GL081397.

Krumbein, W.E., Gorbushina, A.A., and E. Holtkamp-Tacken. 2004. Hypersaline microbial systems of sabkhas: Examples of life's survival in "extreme" conditions. *Astrobiology* 4, no. 4: 450–460.

Kürten, B., Khomayis, H.S., Devassy, R., Audritz, S., Sommer, U., Struck, U., El-Sherbiny, M.M., and A.M. Al-Aidaroos. 2014. Ecohydrographic constraints on biodiversity and distribution of phytoplankton and zooplankton in coral reefs of the Red Sea, Saudi Arabia. *Mar Ecol* 36, no. 4. http://doi.org/10.1111/maec.12224.

Le Houerou, H.N. 1993. Salt-tolerant plants for the arid regions of the Mediterranean isoclimatic zone. In: *Towards the Rational Use of High Salinity Tolerant Plants*. Eds. H. Leith and A. Al Masoom, 403–422. Dordrecht: Kluwer Academic Publishers.

Le Houerou, H.N. 2003. Bioclimatology and phytogeography of the Red Sea and Aden Gulf basins: A monograph (with a particular reference to the highland evergreen sclerophylls and lowland halophytes). *Arid Land Res Manag* 17, no. 3: 177–256. http://doi.org/10.1080/15324980301601.

Lei, L., and H. Xiaoping. 2012. Three tropical seagrasses as potential bio-indicators to trace metals in Xincun Bay, Hainan Island, South China. *Chin J Oceanol Limnol* 30, no. 2: 212–224.

Lokier, S.W., Andrade, L.L., Court, W.M., Dutton, K.E., Head, I.M., Der Land, C.V., Paul, A., and A. Sherry. 2017. A new model for the formation of microbial polygons in a coastal sabkha setting. *Depos Rec* 3, no. 2: 201–208. http://doi.org/10.1002/dep2.33.

Lugendo, B. 2016. Mangroves, salt Marshes, and seagrass beds. In: *Regional State of the Coast Report*. https://policycommons.net/artifacts/1008929/mangroves-salt-marshes-and-seagrass-beds/. Accessed 29 March 2022. CID: 20.500. 12592/61fc8h.

Manaa, A.A., and M.A. Aref. 2022. Microbial mats and evaporite facies variation in a supralittoral, ephemeral lake, Red Sea coast, Saudi Arabia. *Facies* 68, no. 1. http://doi.org/10.1007/s10347-021-00641-0.

Mansour, A.M., Mohamed, A.W., Eldien, K.M., and S.M. Sayed. 2007. Sedimentological and environmental impacts of development projects along the coast of Hurghada, Red Sea, Egypt. *Egypt J Aquat Res* 339, no. 1: 59–84.

Mohamed, T., Abel Kareem, M.S.H., Aldoushy, M., Mohamed, Z.N., and N.F. Abo-Dahab. 2022. Status of seaweeds community in the mangrove forest and sandy shore ecosystems, Red Sea, Egypt. *Egypt J Aquat Biol Fish* 26, no. 2: 567–585. www.ejabf.journals.ekb.eg.

Mohamedzein, Y.E.A., and A.A. Al-Rawas. 2011. Cement-stabilization of sabkha soils from Al-Auzayba, Sultanate of Oman. *Geotech Geol Eng* 29: 999–1008.

Mohorjy, A.M., and A.M. Khan. 2006. Preliminary assessment of water quality along the Red Sea coast near Jeddah, Saudi Arabia. *Water Int* 31, no. 1: 109–115.

Moon, J.-S., Jung, H.S., Lee, S., and S.-T. Kang. 2019. Ground improvement using dynamic compaction in sabkha deposit. *Appl Sci* 9: 2506. http://doi.org/10.3390/app9122506.

Moreira-Saporiti, A., Bejarano, S., Viana, I.G., Belshe, E.F., Mtolera, M.S.P., and M. Teichberg. 2021. Local victory: Assessing interspecific competition in seagrass from a trait-based perspective. *Front Plant Sci* 12: 709257. http://doi.org/10.3389/fpls.2021.709257.

Nasr, D., Shawky, A.M., and P. Vine. 2019. Status of Red Sea dugongs. In: *Oceanographic and Biological Aspects of the Red Sea*. Eds. N.M.A. Rasul and I.C.F. Stewart, 327–354. Cham, Switzerland: Springer Nature.

Omer, A.O. 2001. *Geology and Coastal Plain Evolution of Towaratit, South of Port Sudan, Red Sea, Sudan*. Unpublished Thesis, Alneelain University, Sudan.

Oren, A. 2009. Microbial diversity and microbial abundance in salt-saturated brines: Why are the waters of hypersaline lakes red? *Nat Resour Environ Issues* 15: 49. https://digitalcommons.usu.edu/nrei/vol15/iss1/49.

Osman, N.A., and S.E. Mohammed. 2016. A compiled checklist of seaweeds of Sudanese Red Sea coast. *J Coastal Life Med* 4, no. 2: 114–120. http://doi.org/10.12980/jclm.4.2016j5-231.

Ouni, Y., Lakhdar, A., Rabhi, M., Smaoui, A., Maria, R.A., and A. Chedly. 2013. Effects of the halophytes Tecticornia indica and Suaeda fruticosa on soil enzyme activities in a Mediterranean sabkha. *Int J Phytoremediation* 15, no. 2: 188–197. http://doi.org/10.1080/15226514.2012.687194.

Papenfuss, G.F. 1968. A history, catalogue and bibliography of benthic Red Sea algae. *Israel J Botany* 17: 1–118.

Pena-Garcia, D., Ladwig, N., Turki, A.J., and M.S. Mudarris. 2014. Input and dispersion of nutrients from the Jeddah Metropolitan Area, Red Sea. *Mar Pollut Bull* 80, no. 1–2: 41–51.

PERSGA. 2003. Standard survey methods: Seagrasses and seaweeds. *PERSGA Technical Series No. 10.*

PERSGA. 2006. State of the marine environment. *Report for the Red Sea and Gulf of Aden*. PERSGA, Jeddah.

Piñeiro-Corbeira, C., Barreiro, R., Cremades, J., and F. Arenas. 2018. Seaweed assemblages under a climate change scenario: Functional responses to temperature of eight intertidal seaweeds match recent abundance shifts. *Sci Rep* 8: 12978. http://doi.org/10.1038/s41598-018-31357-x.

The Plant List (v. 1.1). www.theplantlist.org. Accessed 29 February 2022.

Price, A.R.G., Crossland, C.J., Shepherd, A.R., McDowall, R.J., Medley, P.A.H., Stafford, M.G., Smith, R.F.G., and T.J. Wrathall. 1988. Aspects of seagrass ecology along the Eastern coast of the Red Sea. *Bot Mar* 31: 83–92.

Priyashree, S., Jha, S., and S.P. Pattanayak. 2010. A review on Cressa cretica Linn.: A halophytic plant. *Pharmacogn Rev* 4, no. 8: 161–166. http://doi.org/10.4103/0973-7847.70910.

Qurban, M.A.B., Karuppasamy, M., Krishnakumar, P.K., Garcias-Bonet, N., and C.M. Duarte. 2019. Seagrass distribution, composition and abundance along the Saudi Arabian coast of Red Sea. In: *Oceanographic and Biological Aspects of the Red Sea*. Eds. N.M.A. Rasul and I.C.F. Stewart, 367–385. Cham, Switzerland: Springer Nature. http://doi.org/10.1007/978-3-319-99417-8_20.

Rasul, N.M.A., and I.C.F. Stewart (eds.). 2019. *Oceanographic and Biological Aspects of the Red Sea*. Cham, Switzerland: Springer Nature, 550 p. http://doi.org/10.1007/978-3-319-99417-8.

Sabtan, A., Al-Saify, M., and A. Kazi. 1995. Moisture retention characteristics of sabkha. *Quart J Eng Geol* 28: 37–46.

Saeed, W., Shouakar-Stash, O., Wood, W., and P. Parker. 2020. Groundwater and solute budget (a caste study from Sabkha Matti, Saudi Arabia). *Hydrology* 7: 94. http://doi.org/10.3390/hydrology7040094.

Scheffers, A., and T. Browne. 2010. Egypt, Red Sea Coast. In: *Encyclopedia of the World's Coastal Landforms*. Ed. E.C.F. Bird. Dordrecht: Springer.

Shawky, A.M. 2019. Evidence of the occurrence of a large dugong in the Red Sea, Egypt. *Egypt J Aquat Res* 45, no. 3: 247–250.

Sheppard, C., Price, A., and C. Roberts. 1992. *Marine Ecology of the Arabian Region*. London: Academic Press, 359 p.

Short, F., Polidoro, B., Livingstone, S.R., Carpenter, K.E., Bandeira, S., Bujang, J.S., Calumpong, H.P., et al. 2011. Extinction risk assessment of the world's seagrass species. *Biol Conserv* 144: 1961–1971.

Short, F., Short, C., and A. Novak. 2016. Seagrasses. In: *The Wetland Book*. Eds. C.M. Finlayson, G.R. Milton, R.C. Prentice and N.C. Davidson. Springer Science. http://doi.org/10.1007/978-94-007-6173-5_262-1.

Sonnewald, M., and M.M. El-Sherbiny. 2017. Editorial: Red Sea biodiversity. *Mar Biodiv* 47: 991–993. http://doi.org/10.1007/s12526-017-0648-9.

Taher, A.G. 2014. Microbially induced sedimentary structures in evaporite – siliciclastic sediments of Ras Gemsa sabkha, Red Sea Coast, Egypt. *J Adv Res* 5: 577–586. http://doi.org/10.1016/j.jare.2013.07.009.

Tregarot, E., Touron-Gardic, G., Cornet, C.C., and P. Failler. 2020. Valuation of coastal ecosystem services in the Large Marine Ecosystems of Africa. *Environ Dev* 36: 100584. http://doi.org/10.1016/j.envdev.2020.100584.

Turschwell, M.P., Connolly, R.M., Dunic, J.C., Sievers, M., Buelow, C.A., Pearson, R.M., Tulloch, V.D.J., Cote, I.M., Unsworth, R.K.F., Collier, C.J., and C.J. Brown. 2021. Anthropogenic pressures and life history predict trajectories of seagrass meadow extent at a global scale. *PNAS* 118, no. 45: e2110802118. http://doi.org/10.1073/pnas.2110802118.

UNEP. 2015. Marine resources in the Arab Region. *Regional Coordination Mechanism (RCM) Issues Brief for the Arab Sustainable Development Report.*

Vieira, V.M.N.C.S., Lopes, I.E., and J.C. Creed. 2018. The biomass-density relationship in seagrasses and its use as an ecological indicator. *BMC Ecol* 18: 44. http://doi.org/10.1186/s12898-018-0200.

Walker, D.I. 1987. Benthic algae. In: *Key Environment, Red Sea*. Pergamon Press, 441 p. www.worldfloraonline.org/taxon/wfo-0000777168. Accessed February 2022.

Winters, G., Beer, S., Willette, D.A., Viana, I.G., Chiquillo, K.L., Beca-Carretero, P., Villamayor, B., Azcárate-García, T., Shem-Tov, R., Mwabvu, B., Migliore, L., Rotini, A., Oscar, M.A., Belmaker, J., Gamliel, I., Alexandre, A., Engelen, A.H., Procaccini, G., and G. Rilov. 2020. The tropical seagrass Halophila stipulacea: Reviewing what we know from its native and invasive habitats, alongside identifying knowledge gaps. *Front Mar Sci* 7: 300. http://doi.org/10.3389/fmars.2020.00300.

Winters, G., Tov, R., Elmaliach, T., Edelist, D., and G. Rilov. 2015. Assessment of seagrass communities in the northern Gulf of Aqaba: Mapping their distribution. http://doi.org/10.7287/PEERJ.PREPRINTS.1175V1.

Yechieli, Y., and W. Wood. 2002. Hydrogeologic processes in saline systems: Playas, sabkhas, and saline lakes. *Earth-Sci Rev* 58: 343–365. http://doi.org/10.1016/S0012-8252(02)00067-3.

Zahran, M.A. 1982. Ecology of the halophytic vegetation of Egypt. In: *Tasks for Vegetation Science*, vol. 2. Eds. D.N. Sen and K.S. Rajpurohit, 3–20. Dordrecht: Springer.

The grazing effects of dugongs on seagrass meadows

25

A field experiment at Wadi El Gemal National Park, Red Sea, Egypt

Ahmed M. Shawky and Soha Shabaka

25.1 INTRODUCTION

Dugongs (*Dugong dugon*), or sea cows, are herbivorous marine mammals that feed on seagrass (Marsh et al. 2011). Dugongs are found in the tropical to warm temperate areas across a latitudinal range from 0–10° to 30–40° (Marsh 2008; Heck et al. 2021). Dugongs are listed as vulnerable to extinction on the IUCN Red List and Appendix 1 of CITES (Marsh and Sobtzick 2015), because it has a low reproductive rate and feeding within constricted habitats in coastal waters (Marsh et al. 2002).

The dugong as a megagrazer on seagrass has a significant role in intertidal habitats (Heck and Valentine 2006; Ebrahim et al. 2014; Budiarsa et al. 2021). Dugongs are able to graze up to 40 kg of seagrass per day (Aragones 1996; Marsh et al. 2011) and have digestion times from six to eight days (Lanyon and Marsh 1995). Herbivores have been demonstrated to have a major impact on the species composition of seagrass communities (Preen 1995; Carruthers et al. 2002; Stapel 2012). Some plant species require disturbance such as grazing to allow them to successfully compete (Owen-Smith 1988). In the absence of grazing, tall grasses can overgrow and shade shorter species to the point of exclusion (Edroma 1989). This increases the role of dugongs and manatees in encouraging resilience in seagrass ecosystems (Christianen et al. 2012). These herbivores may reduce the accumulation of organic matter in sediments and decrease sediment hypoxia (Valentine and Duffy 2006). They may also raise the income of leaves, decreasing the possibility of algal overgrowth (Valentine and Duffy 2006), therefore increasing the resistance of the seagrass ecosystem to stressors (Unsworth et al. 2015). High grazing by macro grazers can eliminate extensive biomass, cause seagrass meadows to fragment (Fourqurean et al. 2010) and affect the function of the seagrass ecosystem (Christianen et al. 2023) while low grazing can encourage growth and productivity (Christianen et al. 2012), and even lesser grazers extensively affect the structure of seagrass meadows (Ebrahim et al. 2014; Smulders et al. 2023). While feeding on seagrass, dugongs prefer species with high nitrogen and low fibre contents like *Halophila ovalis* and *Halodule uninervis* (Preen 1992, 1995). We report that dugongs graze on the same site for periods of up to a month or more, which might have an effect on the seagrasses.

In the Egyptian Red Sea, the density of feeding trails is different between sites (Shawky 2019a, 2019b). In Wadi El-Gemal National Park (WGNP), the Shams Alam site was the most dominant for feeding trails (3.3±0.3 trails/m^2), with a maximum of six trails/m^2, as opposed to only two and three trails for other sites (Nasr et al. 2019; Shawky 2024). The aim of this study is to examine the impact of dugong grazing on seagrass structure by monitoring the recovery of each species for one year through a field experiment simulation of different intensities of dugong grazing.

25.2 MATERIALS AND METHODS

25.2.1 Study site

A manipulation experiment by Preen (1992) was adopted and carried out in January 2016 at a shallow depth of 1.5 m to 2 m in Shams Alam in WGNP (Figure 25.1). This site was selected due to a high density of feeding trails and a combination of

DOI: 10.1201/9781003321408-32

several species of seagrass, which are found in a sheltered area away from human disturbance (Figure 25.2). Four seagrass species were recorded: *Halophila stipulacea*, *Halophila ovalis*, *Halodule uninervis* and *Cymodocea rotundata* (Shawky 2024).

25.2.2 Methodology

Following Preen (1992), the enclosure was designed to deter dugong and turtle grazing without affecting light and water flow regimes. Each treatment consisted of steel stakes projecting 25 cm above the sediment and connected by a clear rope (4 mm) around the boundary and across the top (Figure 25.3), forming a 25 cm-high fence around the enclosed area with an open net (1 m^2 mesh) over the top. The enclosure would not stop a dugong from feeding through the top mesh, but the boundary would interrupt the dugong feeding motion and so prevent feeding trails from crossing the experimental areas. The enclosure was an obstacle rather than a barrier to grazing. Each 9 m^2 was marked with four thin ropes and a small buoy which floated 50 cm above the bottom to be clearly identified underwater during diving. The ropes of the enclosure were cleaned of epiphytic algae every two to four weeks. All ropes were replaced once during the study period.

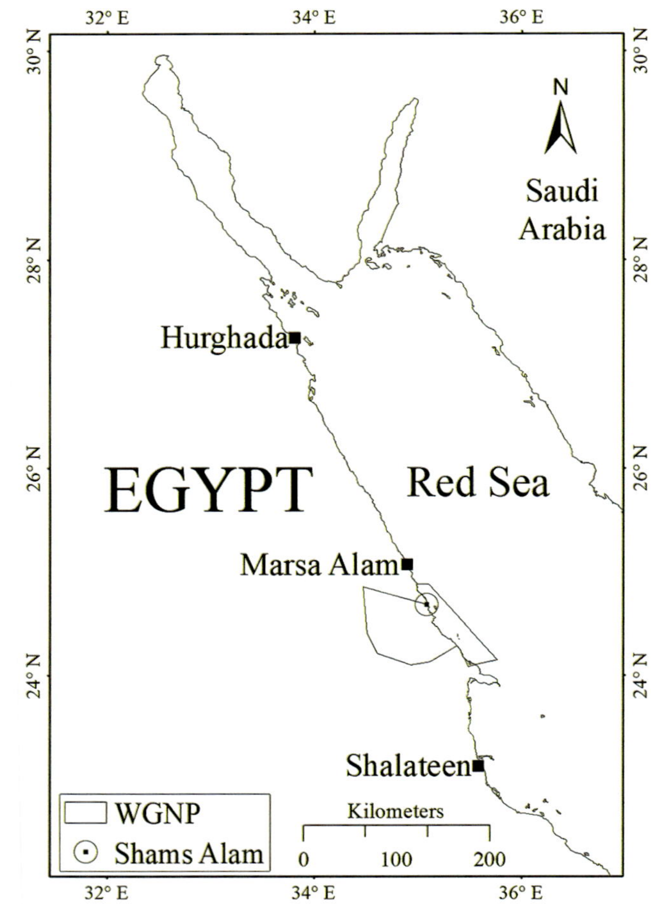

FIGURE 25.1 Map of the Egyptian Red Sea showing the study site of Shams Alam in Wadi El Gemal National Park (WGNP).

FIGURE 25.2 Dugong grazing on a seagrass bed in Shams Alam in Wadi El Gemal National Park.

Three simulated treatments were performed: no grazing, low grazing and high-intensity grazing. The no-grazing treatment (control) was to monitor changes in the abundance of each seagrass species in the absence of grazing by dugongs. At low-intensity grazing, three feeding trails were simulated by removing seagrass (shoots and rhizomes). The parallel "feeding trails" were 1 m apart, 3 m long, 18 cm wide and about 6 cm deep. The edges were determined by ropes to distinguish them from seagrass growth. In the high-intensity treatment, intensive grazing was simulated over 9 m^2 areas until the site resembled a heavily grazed meadow.

The shoot density of the seagrasses was counted in 0.0225 m^2 quadrats (15 cm × 15 cm) systematically located in each treatment. Thirty quadrats (six quadrats along every five transects) were monitored within each treatment except in the low-intensity grazing treatment (six quadrats along each of three simulated trails). In order to repeat counts in the same locations, the position of each quadrat was strictly controlled based on the marked corner for no- and high-grazing treatments and the limiting ropes for the simulated trails of the low-grazing treatment. The abundance of seagrass in no-, low- and high-intensity treatments was measured immediately after the seagrass manipulations. The abundance of seagrass was monitored for one year on six occasions (average 63±23 days apart) between 15 January and 11 November 2016 by SCUBA divers. Each quadrat was documented by video images using an HD digital camera (GoPro Hero 4 Silver Edition set in 4K mode). Images were displayed later on a computer screen and zoomed to count the seagrass shoots and identify them, following El Shaffai (2016).

25.2.3 Statistical analysis

Statistical analysis was performed using Statistica (Version 7, StatSoft Inc.). Before conducting the statistical analysis, a Shapiro-Wilk test was applied to test if the data was significantly different from a normal distribution ($p < 0.05$). However, the data showed mixed normal to non-normal distributions after $\log_{10}(x + 1)$ transformation that was applied. Accordingly, both parametric factorial ANOVA and non-parametric Kruskal-Wallis tests were applied. According to McNeish et al. (2012, 2018), presentation of both non-parametric and parametric analyses balances the interpretation of the lower power non-parametric test when datasets are a mix of both normal and non-normal Gaussian distributions. The analysis was followed by post-hoc pairwise Kruskal-Wallis comparisons to identify which treatment (no grazing, low grazing, high grazing) had a significant effect on seagrass recovery. Tukey's post-hoc test was conducted to test for significant differences in means of seagrass shoot densities under a combined effect of months and treatments.

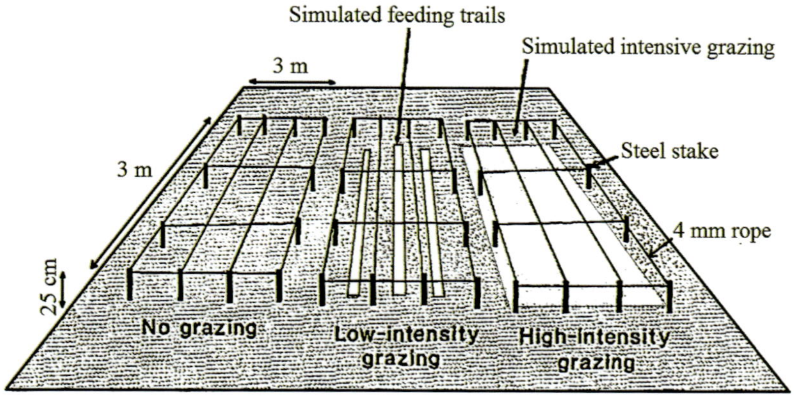

FIGURE 25.3 Design of the enclosure and the set of seagrass treatments.

Source: Preen (1992)

25.3 RESULTS

Changes in shoot density (shoots/m²) of the four seagrass species under the three treatments are shown in Table 25.1. Factorial ANOVA results showed that seagrass densities varied significantly between treatments throughout the duration of the experiment and under the combined effect of treatments and through the experiment duration (Table 25.2). Seagrass densities showed different recovery strategies among the three treatments during the experiment. In addition, this strategy varied among seagrass species. A Kruskal-Wallis test showed that only *H. stipulaceae* ($\chi 2 = 89.4$, $H = 104.1$, $p = 0.00$) and *C. rotundata* ($\chi 2 = 27.07$, $H = 46.1$, $p = 0.00$) showed a significant recovery among the three treatments.

In the control experiment (no grazing), *H. stipulaceae* showed a clear temporal trend ($p < 0.05$), increasing from January to June and declining afterward (Figure 25.4). *H. Ovalis* had significantly higher densities in January, decreased through March and increased in April before it declined during the remaining course of the experiment (Figure 25.4). *H. uninervis* and *C. rotundata* did not show significant temporal variations under the no-grazing treatment during the course of the experiment (Figure 25.4).

In the low-grazing and high-grazing treatments, the sea grass recolonized the feeding trails primarily through ingrowths from the edges by vegetative reproduction. *H. stipulacea* attained the pretreatment level in June under the low-grazing treatment and in July under the high-grazing treatment (Figure 25.4). The same was observed for *C. rotundata*; in addition, it showed a significant recovery under the high-grazing treatment by the end of the experiment (Figure 25.4).

The growth of seagrasses was significantly different in the three treatments all year round ($P = <0.05$). Changes in shoot density (shoots/m²) of four species of seagrass under three treatments were recorded (Table 25.1). In the no-grazing treatment, *H. stipulacea* and *H. ovalis* increased in April and decreased

TABLE 25.1 Changes in shoot density (shoots/m²) of four species of seagrass under three treatment regimes. Values compare species within time intervals.

TREATMENT	*DURATION OF MONITORING (DAYS)*	*HALOPHILA STIPULACEA*		*HALOPHILA OVALIS*		*HALODULE UNINERVIS*		*CYMODOCEA ROTUNDATA*	
		MEAN	*SE*	*MEAN*	*SE*	*MEAN*	*SE*	*MEAN*	*SE*
No grazing	1	1,013	103	1,633	216	1,013	103	1,633	216
	46	1,633	106	261	67	1,633	106	261	67
	94	2,490	215	599	137	2,490	215	599	137
	157	2,415	249	135	34	2,415	249	135	34
	230	1,579	162	12	7	1,579	162	12	7
	334	1,809	267	47	13	1,809	267	47	13
Low grazing	1	240	30	74	20	10	8	35	11
	46	299	57	111	38	72	30	94	20
	94	9.0	1.0	4.0	1.0	–	–	2	0
	157	2,412	192	316	55	47	18	351	48
	230	1,081	211	141	52	67	26	232	36
	334	1,840	175	178	45	170	38	346	54
High grazing	1	207	38	74	22	27	8	127	14
	46	2	0	61	20	84	38	6	1
	94	92	22	533	85	18	7	284	39
	157	1,071	151	378	97	59	21	10	2
	230	881	136	76	25	74	23	721	105
	334	1,721	202	234	64	305	53	559	82

in July, *H. uninervis* increased in February and November but decreased in April while *C. rotundata* increased in July and decreased in February (Figure 25.4). The abundance of *H. stipulacea* was significantly less ($P = <0.05$) than *H. ovalis* at the start of the experiment and increased considerably over the monitoring period while that of *H. ovalis* decreased. The abundance of *C. rotundata* ($P = 0.9$) showed no trend when compared to *H. ovalis* while *H. uninervis* ($P = 0.001$) differed from *H. ovalis* in April and November.

TABLE 25.2 Factorial ANOVA test of significance for $\log_{10}(x+1)$ transformed data of seagrass density among months and treatments.

	EFFECT	*ERROR*	*F*	*P*
Treatments	8	896.0	25.85	0.00
Months	20	1,486.8	10.58	0.00
*Grazing*Months*	40	1,700.6	6.11	0.00

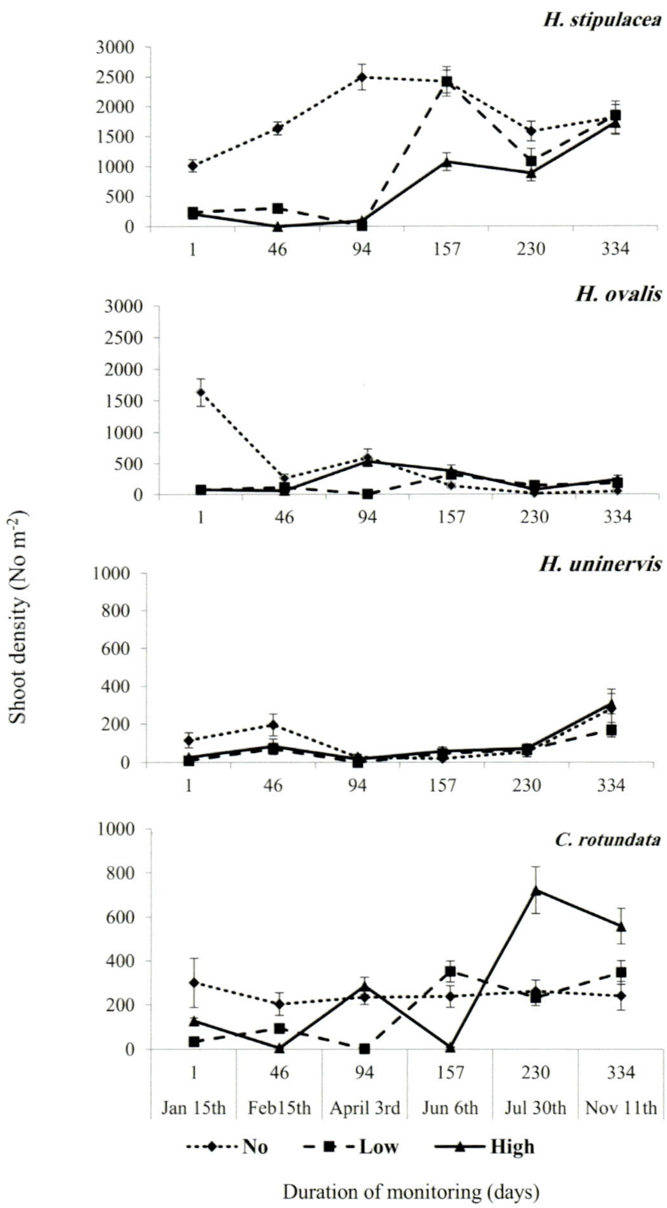

FIGURE 25.4 Trend in recovery of seagrass species.

In low grazing, 91% of the shoots were removed along the simulated feeding trails. This reduction was in accordance with the level of shoot removal from real feeding trails (87%–92%). The seagrass recolonized the feeding trails primarily through ingrowth from the edges. All seagrass species decreased in April and increased in June, except *H. uninervis*, which increased in February and November. *H. ovalis*, however, was significantly less abundant (4±1 shoots/m^2, $P = <0.05$), and *H. uninervis* completely disappeared. The first 100 days of the experiment corresponded with winter and spring. Hence, *H. stipulacea*, which has a winter-spring growth period, colonized the feeding trails first. *H. ovalis* and *C. rotundata* have a summer-autumn growth period. By the end of the experiment, *H. stipulacea* and *C. rotundata* had significantly recovered to pretreatment levels (both $P = <0.05$).

During high grazing, as an average across species, 87% of shoots were removed from the plots designed to simulate intensive grazing by dugongs. The winter-spring conditions at the start of the experiment delayed any recovery for the first 46 days (in contrast to the low-intensity grazing treatments), especially the *H. stipulacea*, which showed the lowest recovery (2±0 shoots/m^2). *H. ovalis* showed significantly greater recovery than *H. stipulacea* or *C. rotundata*, increasing its relative and absolute abundance during the 80-to-100-day period. *H. ovalis* maintained its relative and significantly greater abundance during the 50 days period than *H. uninervice* at the end of the experiment (265 days).

25.4 DISCUSSION

In this study, the enclosure experiment showed that disturbance (i.e., dugong grazing) can change the relative abundance of seagrasses. The recovery of the seagrass communities varied between the low and high intensities of simulated grazing (Scott et al. 2021). In low-intensity grazing, all seagrass species recovered beyond one month, but in high-intensity grazing, recovery occurred after one month for *H. uninervis*, three months for *H. ovalis* and *C. rotundata* while *H. stipulacea took* eleven months. The disturbance encouraged the growth of *H. ovalis* while it slowed down the expansion of *H. stipulacea*. At the high-grazing treatment (after three months), the enclosure was dominated by *H. ovalis* while after an additional five months, it was dominated by *H. stipulacea*. Aragones and Marsh (2000) reported that the recovery of seagrasses after grazing disruption depended on the timing and intensity of the grazing disturbance, the species structure and locality. They recorded that the time period taken between low and high grazing varied between three to seven months, respectively. *H. ovalis* recovered most quickly, followed by *H. uninervis* (Preen 1995). *H. ovalis* is a pioneer species, and its presence is a good indication of dugong habitats with their low-fibre and high-nitrogen content (Birch and Birch 1984; Preen 1995; De Iongh et al. 2007). The high production level of *H. ovalis* could be responsible for its fast recovery after dugong grazing (Dos Santos et al. 2012). During the experiment, *H. uninervis* grew faster than *H. ovalis* and dominated the treatment area after eleven months. Burkholder (2013) recorded that *H. uninervis* in enclosures grew higher than at control sites. De Iongh et al. (1995) estimated that the biomass of *H. uninervis* in Indonesia took five months to recover from dugong grazing in the rainy season and did not recover during the dry season. Anderson (1998) and Masini et al. (2001) concluded that dugongs favourably feed on *H. uninervis* to maximize energy intake as its rhizomes are rich in starch.

Aragones et al. (2006) mention that the variable periods for seagrasses to recover from sirenians feeding show that the timing for dugongs and manatees to return to sites varies significantly in both time and space. Consequently, it is not surprising that there is considerable variability in the feeding patterns of sirenians and the return times to particular locations (e.g., De Iongh et al. 1995, 2007; Masini et al. 2001; McMahon 2005). There are many variables linked with environmental and anthropogenic changes to seagrass beds that make estimates of seagrass recovery and sirenian return times difficult, including the carrying capacity (Marsh et al. 2011; Budiarsa et al. 2021).

This experiment supports the hypothesis that disturbance, such as that caused by the extensive grazing of dugongs, can alter the recovery of three different species in a seagrass meadow from one to three months and at least eleven months for *H. stipulaces*. Fast-growing species such as *H. ovalis* have an apparent advantage over high-abundance species such as *H. stipulacea*.

25.5 CONCLUSION

In a one-year experiment, simulated grazing levels were shown to impact competition and recovery between the seagrass species compared to those in a control area with no grazing. In a no-grazing experiment, the abundance of *H. stipulacea* increased over the monitoring period while that of *H. ovalis* decreased. In the case of low grazing, *H. stipulacea* colonized the "feeding trails" first and at the end of the experiment; *H. stipulacea* and *C. rotundata* had recovered to pretreatment levels. At high grazing, *H. ovalis* showed significantly greater recovery than *H. stipulacea* or *C. rotundata*, increasing its relative and absolute abundance within 80- and 100-day periods, respectively. The experiment confirmed that intensive grazing by dugongs changes the relative abundance of different species of seagrass, and the high density of feeding trails has a role in controlling the seagrass structure. This study demonstrated that seagrass dispersion assists in the recovery of coastal marine habitats and alters the distribution, species structure, composition, nutrient status and productivity of specific seagrass beds on which dugongs graze for a month or more. Feeding trails are created through an excavating process, reducing the density of the seagrass species and improving its recovery.

ACKNOWLEDGEMENTS

This study would not have been possible without the financial assistance of Rufford Small Grant (RSG: 17553–1), which is gratefully acknowledged. We extend our thanks to our friends in the Egyptian dugong team (EDT), Ayman Nasr, Ahmed Gad and Kriollos Fayez, for helping me in the field work. Many thanks go to Professor Helene Marsh for her advice. Special thanks to Farag Saad, the ranger of WGNP, for his logistical assistance.

REFERENCES

Anderson, P.K. 1998. Shark Bay dugongs (*Dugong dugon*) in summer: II. Foragers in a Halodule-dominated community. *Mammalia* 62: 409–425.

Aragones, L.V. 1996. *Dugongs and Green Turtles: Grazers in the Tropical Seagrass Ecosystem*. Ph.D. Thesis, James Cook University, Australia.

Aragones, L.V., Lawler, I.R., Foley, W.J., and H. Marsh. 2006. Dugong grazing and turtle cropping: Grazing optimization in tropical seagrass systems? *Oecologia* 149: 635–647.

Aragones, L.V., and H. Marsh. 2000. Impact of dugong grazing and turtle cropping on tropical seagrass communities. *Pac Conserv Biol* 5: 278–288.

Birch, W.R., and M. Birch. 1984. Succession and pattern of tropical intertidal seagrasses in Cockle Bay, Queensland, Australia: A decade of observations. *Aquat Bot* 19: 343–336.

Budiarsa, A.A., De Iongh, H.H., Kustiawan, W., et al. 2021. Dugong foraging behavior on tropical intertidal seagrass meadows: The influence of climatic drivers and anthropogenic disturbance. *Hydrobiologia* 848: 4153–4166. http://doi.org/10.1007/s10750-021-04583-0.

Burkholder, D.A. 2013. Patterns of top-down control in a seagrass ecosystem: Could a roving apex predator induce a behaviour-mediated trophic cascade? *J Animal Ecol* 82: 1192–1202.

Carruthers, T.J.B., Dennison, W.C., Longsta, B.J., Waycott, M., Abal, E.G., and L.J. McKenzie. 2002. Seagrass habitats of Northeast Australia: Models of key processes and controls. *Bull Mar Sci* 71: 1153–1169.

Christianen, M.J.A., Govers, L.L., Bouma, T.J., Kiswara, W., Roelofs, J.G.M., and L.P.M. Lamers. 2012. Marine megaherbivore grazing may increase seagrass tolerance to high nutrient loads. *J Ecol* 100: 546–560.

Christianen, M.J.A., Smulders, F.O.H., Vonk, J.A., Becking, L.E., Bouma, T.J., Engel, S.M., James, R.K., Nava, M.I., de Smit, J.C., van der Zee, J.P., Palsbøll, P.J., and E.S. Bakker. 2023. Seagrass ecosystem multifunctionality under the rise of a flagship marine megaherbivore. *Glob Change Biol* 29. 215–230. http://doi.org/10.1111/gcb.16464.

De Iongh, H.H., Kiswara, W., Kustiawan, W., and P.E. Loth. 2007. A review of research on the interactions between dugongs (Dugong dugon Müller 1776) and intertidal seagrass beds in Indonesia. *Hydrobiologia* 591: 73–83.

De Iongh, H.H., Wenno, B.J., and E. Meelis. 1995. Seagrass distribution and seasonal biomass changes in relation to dugong grazing in the Moluccas, East Indonesia. *Aquat Bot* 50: 1–19.

Dos Santos, V.M., Matheson, F.E., Pilditch, C.A., and A. Elger. 2012. Is black swan grazing a threat to seagrass? Indications from an observational study in New Zealand. *Aquat Bot* 100: 41–50.

Ebrahim, A., Olds, A.D., Maxwell, P.S., Pitt, K.A., Burfeind, D.D., and R.M. Connolly. 2014. Herbivory in a subtropical seagrass ecosystem: Separating the functional role of different grazers. *Mar Ecol Prog Ser* 511: 83–91.

Edroma, E.L. 1989. The response of tropical vegetation to grazing and browsing in Queen Elizabeth National Park, Uganda. *Symposia Zool Soc London* 61: 1–13.

El Shaffai, A. 2016. *Field Guide to Seagrasses of the Red Sea*, 2nd ed. Eds. A. Rouphael and A. Abdulla. Gland, Switzerland: IUCN and Courbevoie, France: Total Foundation, viii+56 p.

Fourqurean, J.W., Manuel, S., Coates, K.A., Kenworthy, W., and S.R. Smith. 2010. Effects of excluding sea turtle herbivores from a seagrass bed: Overgrazing may have led to loss of seagrass meadows in Bermuda. *Mar Ecol Prog Ser* 419: 223–232.

Heck, K.L., Samsonova, M., Poore, A.G.B., et al. 2021. Global patterns in seagrass herbivory: Why, despite existing evidence, there are solid arguments in favor of latitudinal gradients in seagrass herbivory. *Estuaries Coasts* 44: 481–490. http://doi.org/10.1007/s12237-020-00833-x.

Heck, K.L., and J.F. Valentine. 2006. Plant-herbivore interactions in seagrass meadows. *J Exp Mar Biol Ecol* 330: 420–436.

Lanyon, J., and H. Marsh. 1995. Digesta passage times in the dugong. *Aust J Zool* 43: 119–127. http://doi.org/10.1071/ZO9950119.

Marsh, H. 2008. Dugong dugon. In: *IUCN Red List of Threatened Species v. 2012.2*. www.iucnredlist.org. Accessed 10 November 2012.

Marsh, H., Penrose, H., Eros, C., and J. Hugues. 2002. *Dugong: Status Report and Action Plans for Countries and Territories*. UNEP Early Warning Assessment Report Series, vol. 1.

Marsh, H., O'Shea, T., and J. Reynolds. 2011. *Ecology and Conservation of the Sirenia: Dugongs and Manatees*. New York: Cambridge University Press.

Marsh, H., and S. Sobtzick. 2015. Dugong dugon. *IUCN Red List of Threatened Species*. http://doi.org/10.2305/iucn.uk.2015-4.rlts.t6909a43792211.en.

Masini, R.J., Anderson, P.K., and A.J. McComb. 2001. A Halodule-dominated community in a subtropical embayment: Physical environment productivity, biomass, and impact of dugong grazing. *Aquat Bot* 71: 179–197.

McMahon, K. 2005. *Recovery of Subtropical Seagrasses from Natural Disturbance*. PhD Thesis, University of Queensland, Brisbane, Qld.

McNeish, R.E., Benbow, M.E., and R.W. McEwan. 2012. Riparian forest invasion by a terrestrial shrub (Lonicera maackii) impacts aquatic biota and organic matter processing in headwater streams. *Biol Invasions* 14: 1881–1893.

McNeish, R.E., Kim, L.H., Barrett, H.A., Mason, S.A., Kelly, J.J., and T.J. Hoellein. 2018. Microplastic in riverine fish is connected to species traits. *Sci Rep* 8, no. 1: 11639. http://doi.org/10.1038/s41598-018-29980-9.

Nasr, D., Shawky, A.M., and P. Vine. 2019. Status of Red Sea dugongs. In: *Oceanographic and Biological Aspects of the Red Sea*. Eds. N.M.A. Rasul and I.C.F. Stewart, 327–354. Cham, Switzerland: Springer Nature.

Owen-Smith, R.N. 1988. *Megaherbivores: The Influence of Very Long Body Size on Ecology*. Cambridge: Cambridge University Press.

Preen, A.R. 1992. *Interactions Between Dugongs and Seagrasses in a Subtropical Environment*. Ph.D. Thesis, James Cook University, Townsville, Queensland.

Preen, A.R. 1995. Impacts of dugong foraging on seagrass habitats: Observational and experimental evidence for cultivation grazing. *Mar Ecol Prog Ser* 124: 201–213.

Scott, A.L, York, P.H, Macreadie, P.I., and M.A. Rasheed. 2021. Spatial and temporal variability of green turtle and dugong herbivory in seagrass meadows of the southern Great Barrier Reef (GBR). *Mar Ecol Prog Ser* 667: 225–231. http://doi.org/10.3354/meps13703.

Shawky, A.M. 2019a. Evidence of the occurrence of a large dugong in the Red Sea, Egypt. *Egypt J Aquat Res* 45: 247–250. http://doi.org/10.1016/j.ejar.2019.08.001.

Shawky, A.M. 2019b. Analysis of feeding trails provides evidence of the number of Dugongs excavating seagrass at key habitats seasonally in the Egyptian Red Sea. *Final Report to Rufford Foundation*. http://doi.org/10.13140/RG.2.2.15151.97449.

Shawky, A.M. 2024. Dugong feeding ecology along the Egyptian Red Sea. In: *Coral Reefs and Associated Marine Fauna around the Arabian Peninsula*. Eds. N.M.A. Rasul and I.C.F. Stewart. Taylor and Francis.

Smulders, F.O.H., Bakker, E.S., O'Shea, O.R., Campbell, J.E., Rhoades, O.K., and M.J.A. Christianen. 2023. Green turtles shape the seascape through grazing patch formation around habitat features: Experimental evidence. *Ecology* 104, no. 2: e3902. http://doi.org/10.1002/ecy.3902.

Stapel, J. 2012. *Nutrient Dynamics in Indonesian Seagrass Beds: Factors Determining Conservation and Loss of Nitrogen and Phosphorous*. Amsterdam: Radboud University Nijmegen.

Unsworth, R.K.F., Collier, C.J., Waycott, M., Mckenzie, L.J., and L.C. Cullen-Unsworth. 2015. A framework for the resilience of seagrass ecosystems. *Mar Pollut Bull* 100: 34–46.

Valentine, J.F., and E. Duffy. 2006. The central role of grazing in seagrass ecology. In: *Seagrass Biology*. Eds. A.W. Larkum, C.M. Duarte and R. Orth, 463–501. Dordrecht: Springer.

Changes in key habitats on the southern Sudanese Red Sea coast using remote sensing and GIS (1972–2013)

26

Awatif Alnair, Dirar H. Nasr, Amna A. Hamid and Sumaya A. Zakieldeen

26.1 INTRODUCTION

Coastal systems are particularly sensitive to three key drivers related to climate change: sea level, ocean temperature and ocean acidity. Coastal systems and low-lying areas will increasingly experience adverse impacts such as submergence, coastal flooding and coastal erosion due to relative sea-level rise (IPCC 2007). The predicted temperature rises of surface waters and a change in the ocean circulation would substantially impact the structure and location of marine habitats (Brierley and Kingsford 2009). Ocean warming is the primary cause of mass coral bleaching and mortality (De Vantier and Al-Hariri 2000; De Vantier et al. 2000a, 2000b; Glynn 1996; Hoegh-Guldberg 1999; Kemp 1998; Kemp and Benzoni 2000; Lough 2000; PERSGA/GEF 2003a, 2003b, 2004; Sheppard and Sheppard 1991; Sheppard et al. 1992; Wilkinson 1998).

The entire area represented by the Red Sea state in Sudan covers an area of about 218,887 km². The Regional Organization for the Conservation of the Environment of the Red Sea and Gulf of Aden (PERSGA) is currently encouraging Sudan to assess and map the vulnerability of coastal and marine environments, promote capacities, raise public awareness, facilitate information dissemination and develop efficient observation systems. The coastal ecosystems are to be monitored periodically for better management plans. Satellite-based sensors provide valuable information useful in the assessment, monitoring and management of coastal ecosystems. Optical remote-sensing data is very useful for mapping coral reef, mangrove and lagoon ecosystems. The information thus derived can be very useful in coastal ecosystem management, which is essential for the sustainable use, development and protection of the coastal and marine areas (Teodoro 2016).

Remote-sensing (RS) and geographic information system (GIS) techniques are providing new tools for advanced ecosystem management. The collection of remotely sensed data facilitates the synoptic analyses of the earth-system function, patterns and change at local, regional and global scales over time; such data also provide an important link between intensive, localized ecological research and the regional, national and international conservation and management of biological diversity (Wilkie and Finn 1996).

The key habitats in the Sudanese Red Sea coast may face the implications of global climate change, particularly the impact of sea-level rise. The overall objectives of this work are to assess the impact of climate change on the southern Sudanese Red Sea coast; the study area extends from Suakin to the south along the coast to just north of the Eritrean border in an area which extends up to 2 km into the Red Sea and 50 km inland. This zone is characterized by flat sandy areas up to 20 km in width yet is traversed by three large and poorly defined rocky canyons running from north to south.

Fusing remote-sensing and GIS techniques, the study concentrates on the following:

- Inundation and displacement of wetlands and lowlands
- Saltwater intrusion into estuaries and freshwater aquifers (especially at the Tokar Delta)
- Assessing the impacts of sea-level rise on the land cover in the study area
- Assessing the effect of sea-level rise on potable water reservoirs (underground water) in the Tokar Delta area
- Developing recommendations based on the results for the enhancement of coastal zone management

DOI: 10.1201/9781003321408-33

26.2 THE STUDY AREA

Since the coastal habitats are confined to a narrow area, and the fresh water is located in the Tokar Delta, the current study has been conducted in a small area comprising the Tokar Delta and the coastal zone, as shown in Figure 26.1.

26.2.1 Key habitats

The coastal and marine environments of the Sudanese Red Sea coast consist of a range of ecosystems: an arid coastal zone, coastal wetlands, mangroves, seagrass beds and coral reefs. They form the basis of biodiversity, fish production and recreational values. Figure 26.2 shows the land cover classification map for the study area in 1972 while Figure 26.3 shows the land cover classification map for the study area in 2013.

26.2.2 Mapping and monitoring land cover in the coastal zone

Geo-corrected cloud-free Landsat MSS 1972 data was obtained from the Remote Sensing Authority, and ETM+ (2013) images were downloaded from the USGS website. The following procedures were used:

- **Image segmentation:** Segmentation process carried out using eCognition serves to reduce the heterogeneity of objects at a specified scale.
- **Image interpretation:** Classification, legend and the generation of land use classes.
- **MadCat (Mapping Device – Change Analysis Tool):** The labeling process of the clustered polygons follows the FAO land-use Land Cover Classification System (LULC).
- **Change detection analysis method:** Post-classification comparison is conceptually one of the most simple change detection methods. It involves an initial independent classification of each image. The area covered by each class in the years 1972 and 2013 was calculated.
- **Accuracy assessment method:** In parallel to the remote-sensing work, field work is carried out to collect data for accuracy assessment.

26.3 CLIMATE VARIABILITY/CHANGE

The analysis of meteorological information shows the variability in the three climate elements (average temperature, average relative humidity and rainfall) during the period 1984–2013 (about 30 years) in order to detect the climate change. Changes in annual average temperature and humidity show an increase from 1984

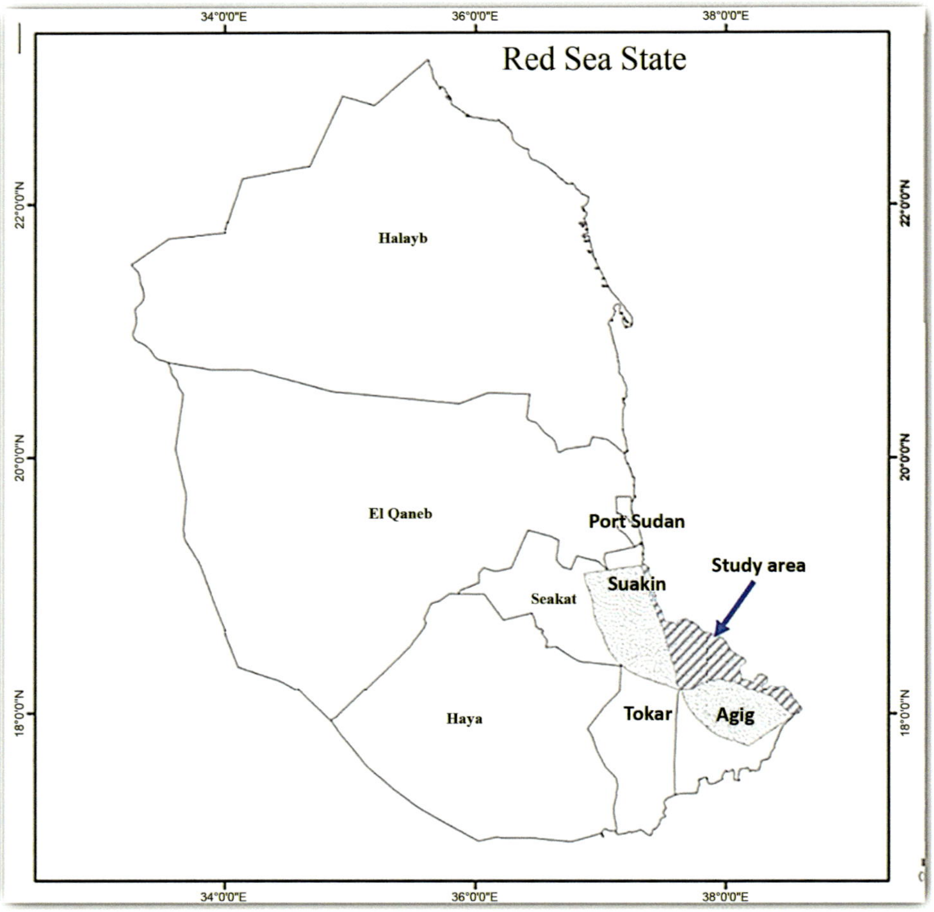

FIGURE 26.1 Location map of the study area in the Tokar Delta and the coastal zone.

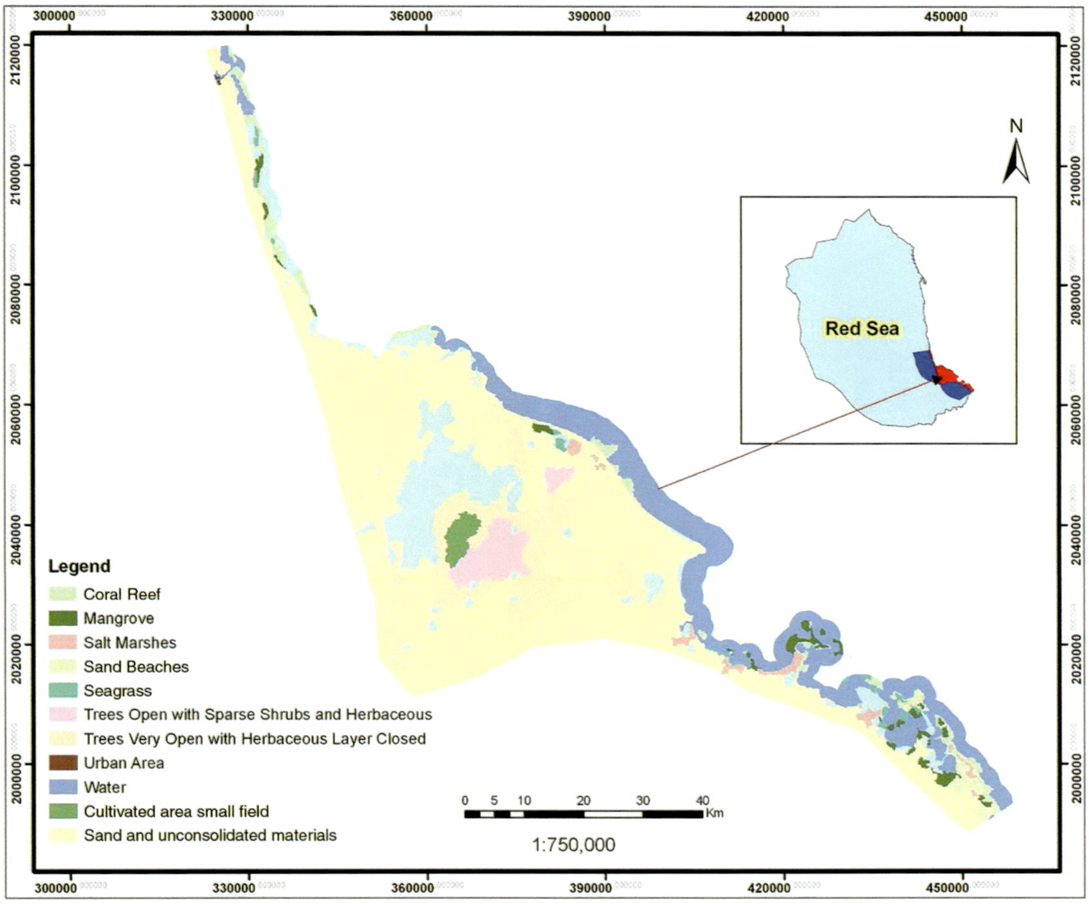

FIGURE 26.2 Land cover classification map for the study area in 1972.

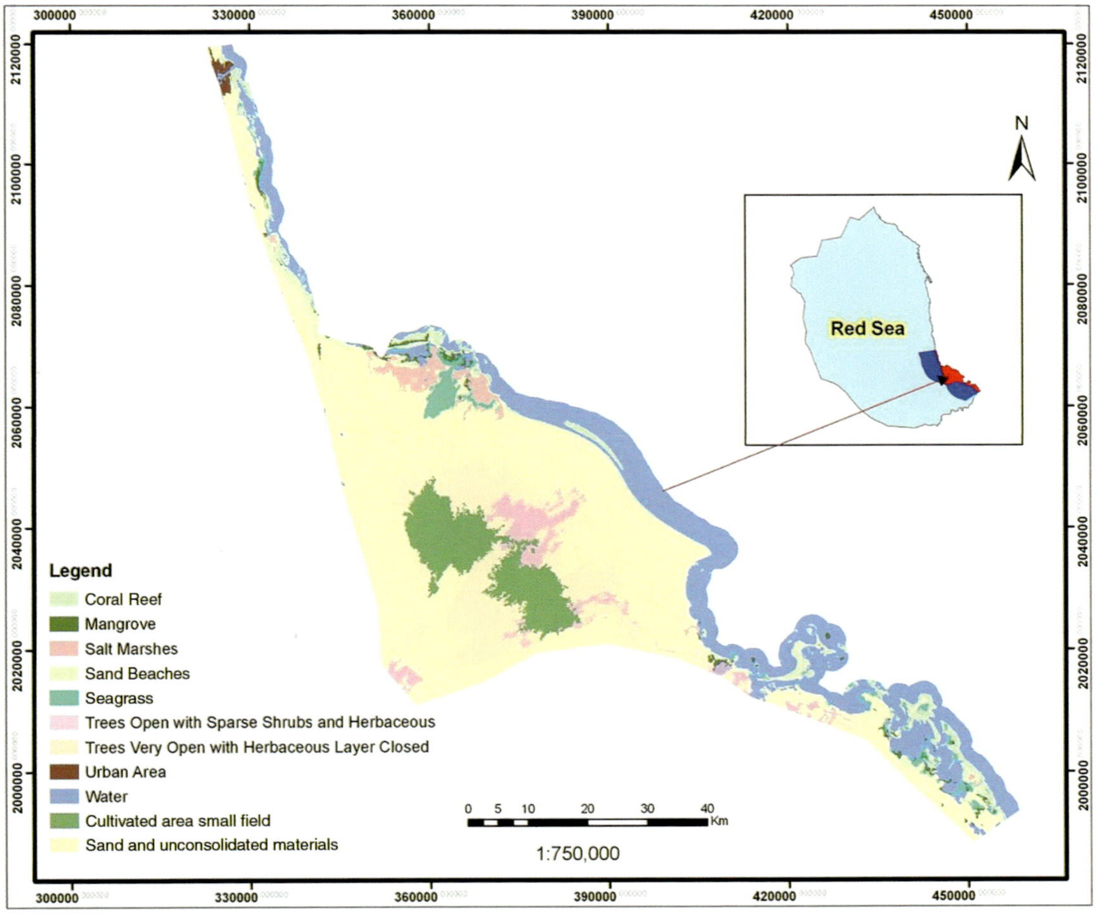

FIGURE 26.3 Land cover classification map for the study area in 2013.

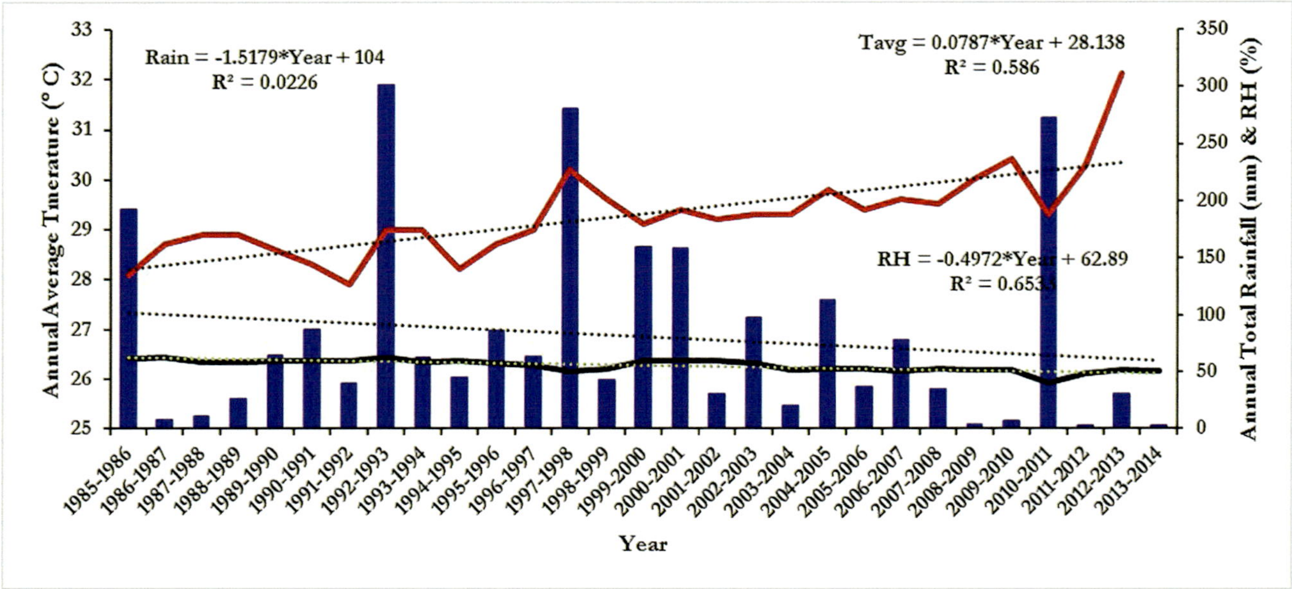

FIGURE 26.4 Annual total rainfall (mm – blue bars), annual relative humidity (solid black line) and annual average temperature (°C – solid red line) at Port Sudan station for the period 1984–2013.

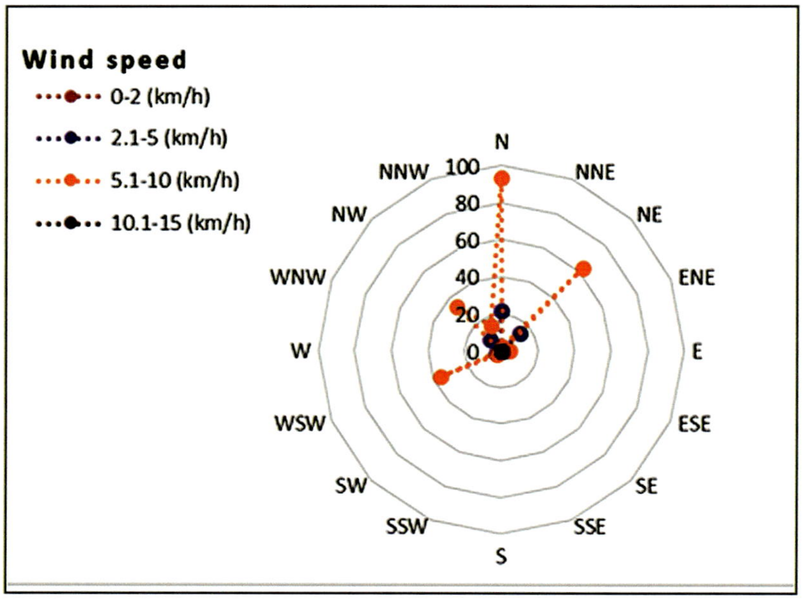

FIGURE 26.5 Wind speed and direction at Port Sudan Station (1985–2013).

to 2013 (Figure 26.4), with corresponding decreases in rainfall in the same period and variations in wind speed and direction, shown in Figure 26.5.

26.4 MEAN SEA LEVEL AT PORT SUDAN

The Sudan tide gauge during the period 1986–1995, as retrieved from the Permanent Service for Mean Sea Level (PSMSL), indicates a gradual increase in sea level (Nasr et al. 2013). However, the maximum tidal range is about 55 cm (Schroeder 1982; Sheppard at al. 1988). Sudan's coastal zone is likely to be a vulnerable area, given the increased certainty of sea-level rise. The increasing concentration of infrastructure, industrial activity and population renders these areas vulnerable to impacts such as inundation, erosion, flooding and changes in sea surface temperature and salinity.

26.5 SEA-LEVEL RISE AND INUNDATION

From sea-level rise modeling of the Red Sea State in Sudan's Second National Communication under the United Nations Framework Convention on Climate Change report and for

TABLE 26.1 Sea-level rise scenarios for Red Sea state.

	SEA-LEVEL RISE OVER MEAN SEA LEVEL IN 2010 (METRES)		
Scenario	2030	2060	2100
Scenario #1: Lower bound SLR rate of 2.0 mm/year at high tide	0.32	0.38	0.46
Scenario #2: Upper bound SLR rate of 6.0 mm/year at high tide	0.40	0.58	0.82

TABLE 26.2 Study area land cover classes, inundation by area in 2030.

	NAME	*AREA (KM²)*	*PERCENTAGE OF INUNDATED LAND COVER*
1	Coral reef	2.663074	2.2%
2	Cultivated area, small fields	0.000377	1.4×10^{-4}%
3	Mangrove	0.306853	0.88%
4	Salt marshes	0.651417	0.89%
5	Sand beaches	0.624929	0.6%
6	Seagrass	0.367415	0.6%
7	Trees, open with sparse shrubs and herbaceous	0.025115	4.6×10^{-3}%
8	Trees, very open with closed herbaceous layer	0.27996	0.2%
9	Urban area	0.063437	0.5%
10	Sand and unconsolidated materials	14.362331	1%

TABLE 26.3 Study area land cover classes, inundation by area in 2100.

	NAME	*AREA (KM²)*	*PERCENTAGE OF INUNDATED LAND COVER*
1	Coral reef	4.691055	4.03%
2	Cultivated area small field	0.183353	0.067%
3	Mangrove	3.443184	9.9%
4	Salt marshes	5.909121	8.1%
5	Sand beaches	10.175362	10.4%
6	Seagrass	6.766935	11.9%
7	Trees, open with sparse shrubs and herbaceous	0.679443	0.5%
8	Trees, very open with closed herbaceous layer	5.627229	1%
9	Urban area	6.643158	56.8%
10	Sand and unconsolidated materials	86.697176	6.35%

the purpose of the assessment of the impact on the built environment, two plausible future sea-level rise scenarios were developed. The first scenario is based on the upper bound of the observed sea-level rise rates over the past century of about 2.0 mm/year. This is consistent with an extrapolation of past trends evident in the area. The second scenario assumed a mean sea-level rise rate of 6.0 mm/year (Table 26.1). Simulation of the second scenario (SLR rate of 6.0 mm/year) in the study area (coastline of South Suakin) showed that there is inundation in 2030 in all environments with different proportions and influence, as shown in Table 26.2. The results of immersion for the land cover in each study area in 2100 are shown in Table 26.3.

The areas affected by inundation as a result of sea-level rise in 2030 are illustrated by overlaying the inundation map on a high-resolution image (Figure 26.6); it shows that there is immersion in the salt marshes and an increase in depth in areas where there are coral reefs. Similarly, Figure 26.7 shows that there is an increase in inundation in the salt marshes and a full immersion of the area covered by the mangroves in 2100.

26.6 SEAWATER INTRUSION IN TOKAR DELTA AREA

Saltwater intrusion, as defined by the Southwest Florida Water Management District (SWFWMD), is the movement of saline water into regions of an aquifer previously occupied by fresher and less saline groundwater and can be the result of natural hydrologic processes or dramatically enhanced by overpumping groundwater (SWFWMD 2003). The contamination of groundwater due to saltwater intrusion has become a major concern for coastal communities, which rely on groundwater as their principal source of drinking water. Thirteen water samples were collected during the wet season from wells distributed in the Tokar Delta. The remaining 12 wells out of 25 recommended for checking through the field work were closed by timber and other material to prevent their destruction in the Khor Baraka flood. Based on the Sudanese standards for drinking water properties, the tool Map Algebra, included in the Spatial Analyst Toolbox, was then used to combine five conditional layers (CL × Na × NO_3 × TDS × SO_4) to produce the suitability map of the drinking water.

Hydro-chemical evaluations of groundwater flow systems are generally based on the availability of information on groundwater chemistry (Shuxian 2013). Groundwater wells were sampled and analyzed for chloride (Cl), total dissolved solids (TDS) and other geochemical parameters to determine the extent of the saltwater intrusion. In this research, samples of groundwater were collected to assess its contamination.

Sodium/chloride ratios can be used to differentiate between seawater intrusion and other sources of saltwater. Jones et al. (1999) suggested that sodium/chloride ratios in advance of a seawater intrusion front will be below 0.86 (molar ratio). Such an assumption is an indicator of seawater intrusion that clearly shows the increase of chloride, sodium, TDS and low Na/Cl ratio concentrations from the Tokar Delta toward the sea. This is due to overpumping of coastal wells as a result of the increase of the cultivated area in the period from 1972 (31.9 km²) to 2013 (272.6 km²). The Tokar Delta is one of the important agricultural areas of Sudan, having the most fertile soil and the largest cultivated area in the Red Sea state (RSS), and

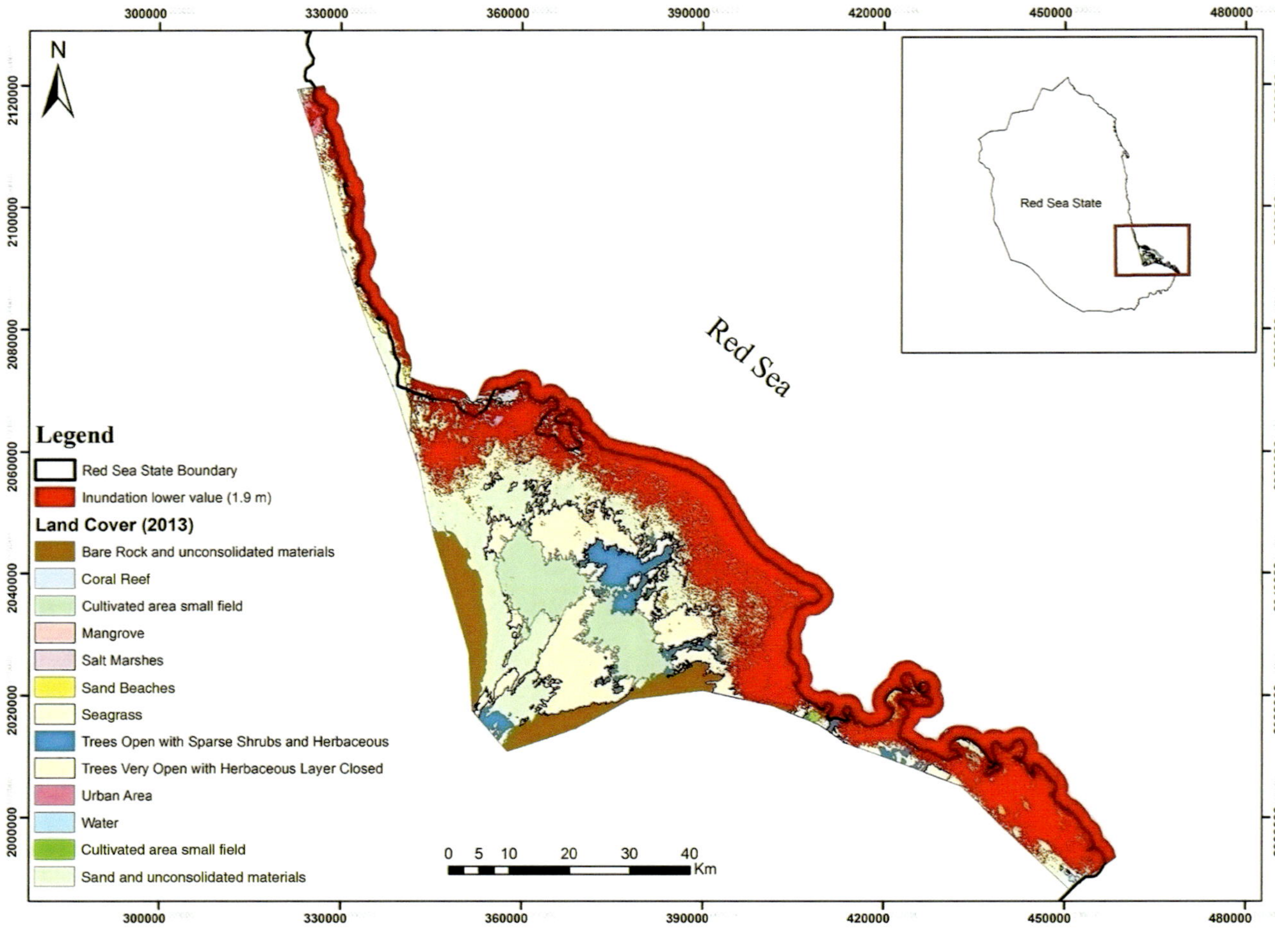

FIGURE 26.6 Land cover classes and anticipated inundation by 2030.

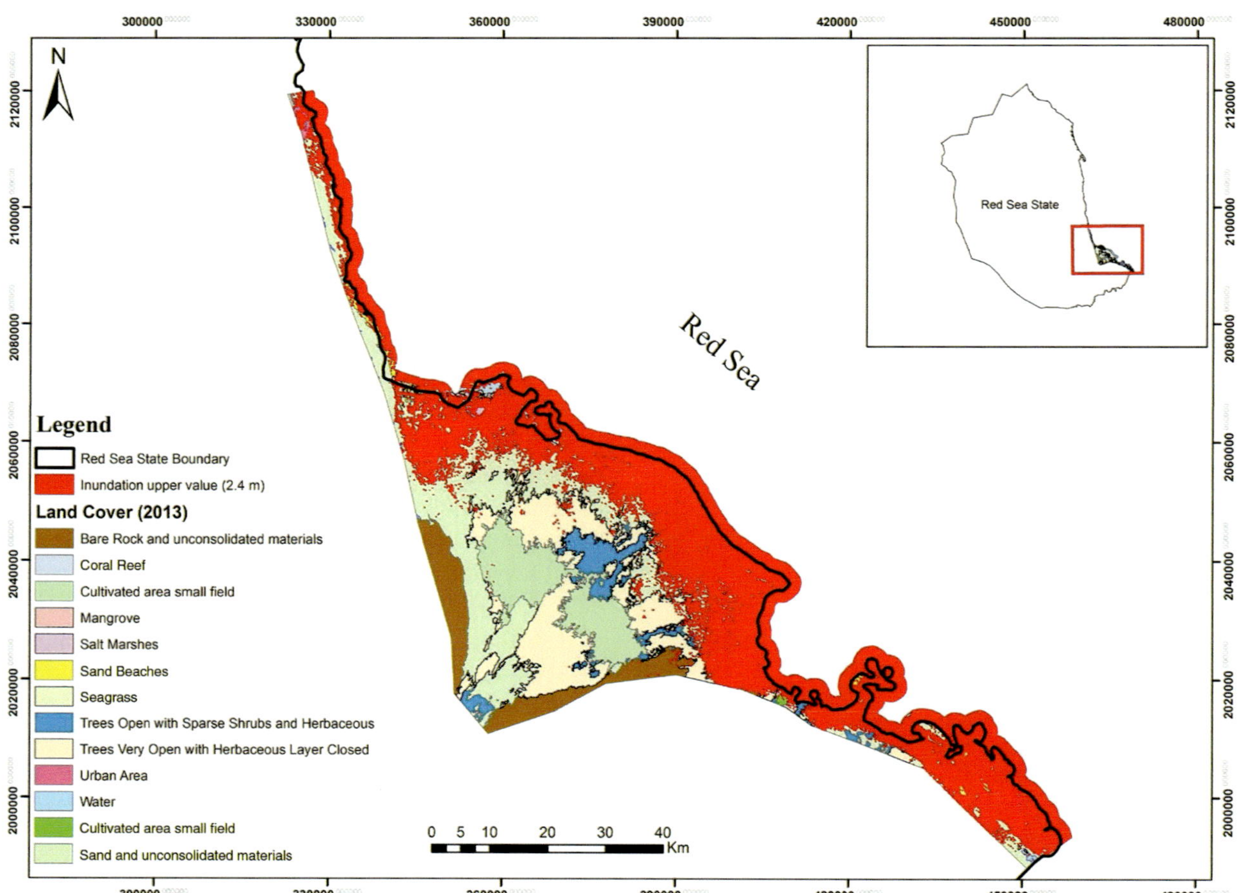

FIGURE 26.7 Land cover classes and anticipated inundation by 2100.

this large area is irrigated from groundwater wells. Also, the increase of mesquite, which was confirmed in field surveys in 2015, has caused a reduction in fresh water. When groundwater is pumped from the freshwater aquifers above the saline zone faster than it can be replenished, upward or lateral movements of saline water toward the pumped wells occur.

Under natural conditions, there is a hydraulic gradient in the fresh groundwater system toward the sea coast, and seepage of fresh groundwater out into the sea occurs at and offshore of the coastal margin. At greater depths, the location and shape of the interface between the fresh groundwater and the seawater are largely determined by the hydraulic gradients and the mixing that occurs between these two water types, recognizing that due to the greater density of seawater, it has a greater total head than simply the elevation of sea level. This can allow the interface to occur at inland locations in the deeper parts of an aquifer that has a permeable hydraulic connection to the sea under natural conditions (Callander et al. 2011). In order to evaluate the state of groundwater salination by seawater intrusion in the Tokar Delta, several geochemical criteria were used to identify the origin of the salinity, especially the detection of seawater intrusion as opposed to other salinity sources in coastal aquifers; these included total dissolved solids (TDS), chloride, electrical conductivity (EC) and sodium and the Na/Cl ratio.

Total dissolved solids (TDS) include all solid material in solution, whether ionized or not. They do not include suspended sediments, colloids or dissolved gasses (Norelmadina 1995). TDS assessment was conducted in order to investigate and monitor the evolution of seawater intrusion over time and space in 1982 and 1995 (Figure 26.8). The TDS appear to increase as we move toward the sea in the years 1982 to 1995, and it is estimated that the contour value of 2,000 in 1982 reached 3,159 mg/l in 1995.

The results of the water analysis for the year 2015 (Figure 26.9) show that TDS in high concentrations (3,630 mg/l) were found in Tokar wells, 2,070 mg/l in the area of the Marafit well and low concentrations up to 629 mg/l in the Dealbyie 2 well. To determine the status of the waters of the Tokar Delta wells based on TDS concentrations, Figure 26.9 shows that fresh water is found in four wells (Doalbyie1, Doalbyie2, Kirimbet and Tablyanie), and most of the wells have brackish water. Seawater intrusion in the Tokar Delta is shown in Figure 26.10; there is an indication of seawater intrusion in most of the Tokar Delta wells.

26.7 ASSESSMENT OF WATER SUITABILITY FOR DRINKING

The most important elements studied in 2015 to determine the levels (mg/l) likely to give rise to consumer complaints are nitrate, chloride, sodium and sulfate. The level of sulfate in the drinking water shows that none of the wells adjacent to the sea is suitable. For chloride, most of the wells are not suitable while nitrate does not have much effect. Sodium is also an important factor in the southeast Delta.

From consumer complaints about drinking water, Figure 26.11 shows only two wells in this study are suitable for drinking. A map showing the locations likely to give rise to consumer complaints about drinking water for all parameters is shown in Figure 26.12. Irrigation water, whether pumped from groundwater wells or containing appreciable quantities of chemical substances in solution, reduces crop yields and deteriorates soil fertility.

26.8 CONCLUSIONS

The general conclusions can be summarized as follows:

- The inundation model as a GIS sampling tool for water was carried out in 13 wells. The ArcGIS9.3 inverse distance weighting interpolation tool (available in the Geostatistical Analyst toolbar) was used as a tool for the properties of well water and for datasets to produce contour maps for each property; it showed increased seawater intrusion into the Tokar Delta, which will also seriously affect the ecosystem of this basin.
- The natural occurrence and movement of saltwater has been changed by the utilization of groundwater resources for human uses and irrigation over recent decades. Worldwide, increasing water demands have often lowered groundwater levels and caused saltwater intrusion into aquifers. Recent cultivated area increases along coastal zones suggest that demands for groundwater resources will steadily rise in the future. The need for water will pose a number of challenges (like the optimal exploitation of fresh groundwater resources and the control of seawater intrusion) to engineers and decision makers.
- The combination of tools used for the purpose of this study provides an opportunity to assemble, standardize and analyze scientific data that comes from a variety of sources, making it available to others who may be able to continue with further research and/or use the information for improved coastal management strategies in the area.
- The preservation and sustainable development of the marine resources of Sudan will require an integrated approach.

26.9 RECOMMENDATIONS

- Climate change consequences on coastal zone ecosystems should be integrated into coastal strategy and planning. Management activities should be practised to cope with long-term trends in the key habitats to adapt to climate change.
- Proper habitat protection and land-use management strategies are needed to protect the important coastal zone and land cover types before they disappear.
- Strengthening of an observation system, including meteorological and oceanographic stations along the Sudanese Red Sea, is required.

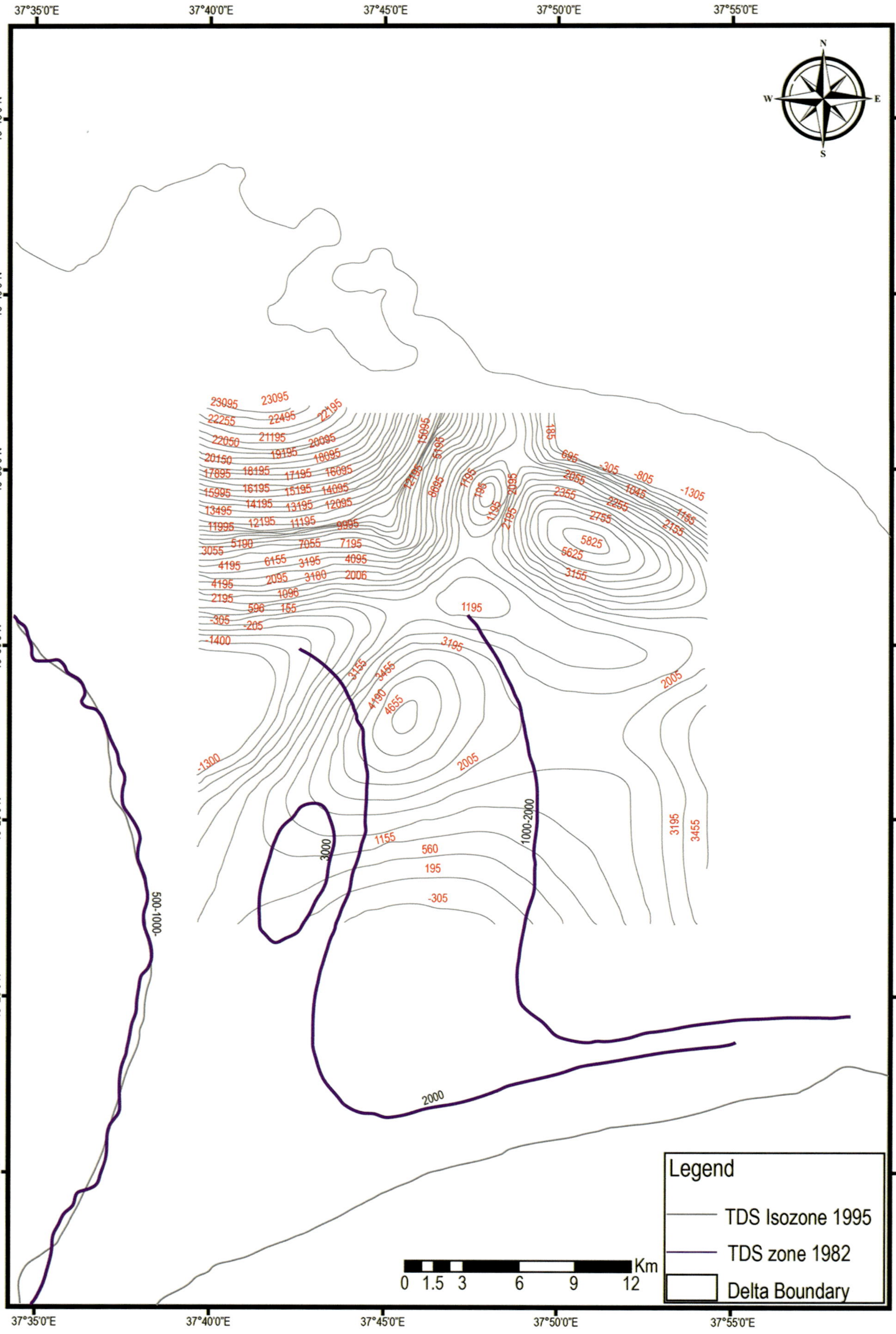

FIGURE 26.8 TDS (mg/l) in the Tokar Delta in 1982 and 1995.

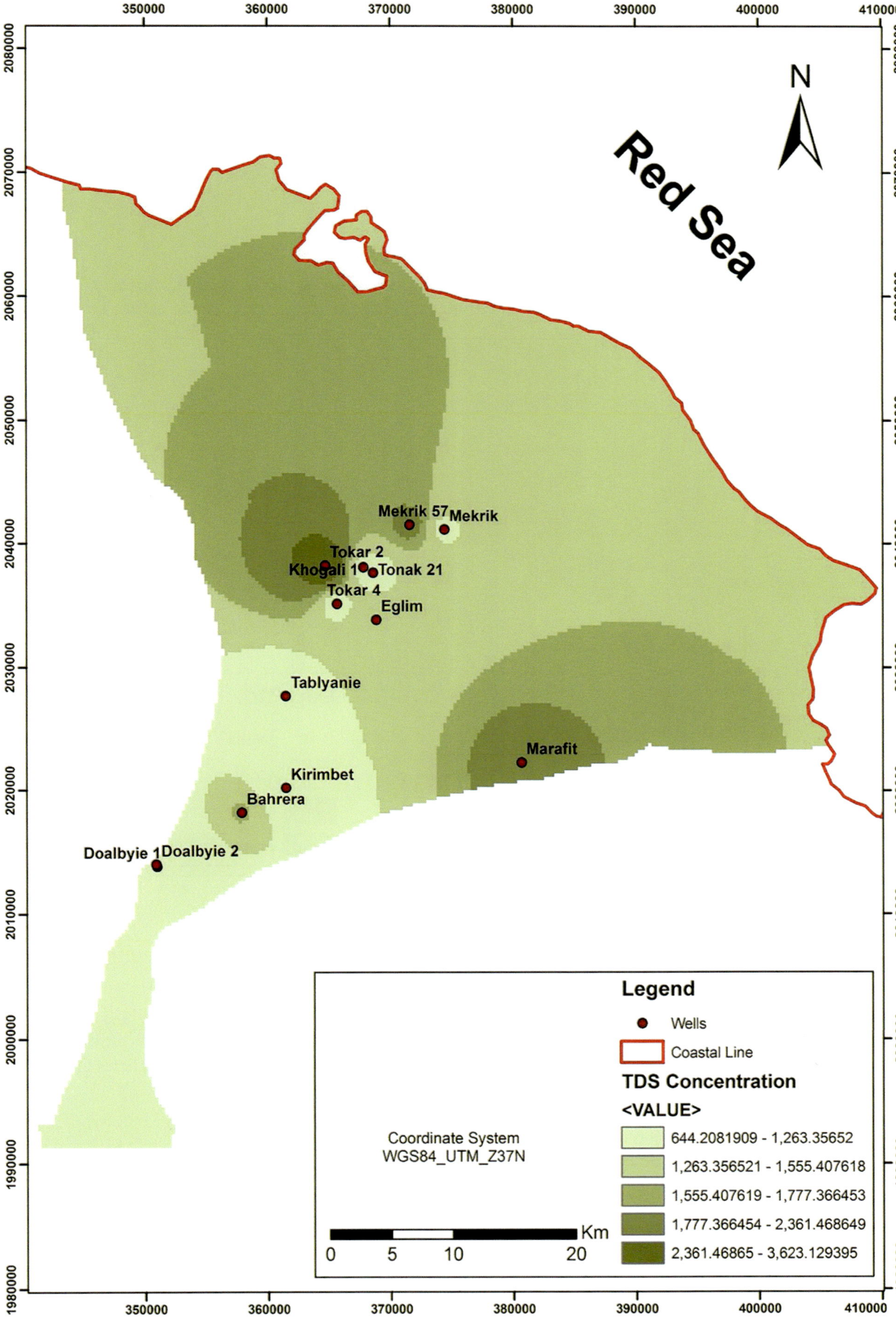

FIGURE 26.9 TDS (mg/l) in the Tokar Delta in 2015.

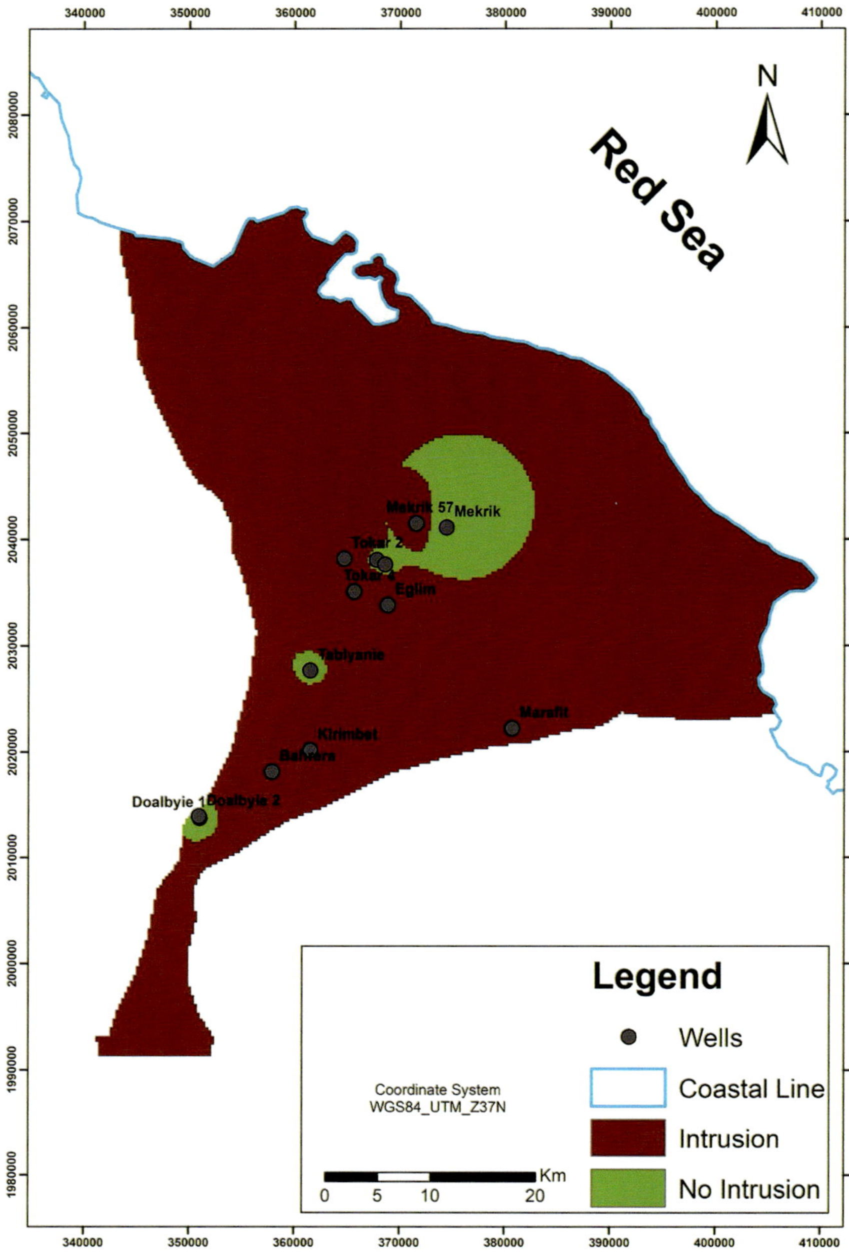

FIGURE 26.10 Seawater intrusion map in the Tokar Delta.

- Construction of surface dams and improved water harvesting techniques to reduce dependence on groundwater in the region are needed.
- Adoption of integrated water resource management and application of numerical models for groundwater flow and transport are required to improve early prediction of changes in the groundwater and hydrogeological system as a whole for an efficient, reliable response to different pumping scenarios, land use and climate change.
- In-depth quantitative analysis and assessment of the marine habitat are needed to provide a scientific basis for the formulation of policies and conservation strategies.

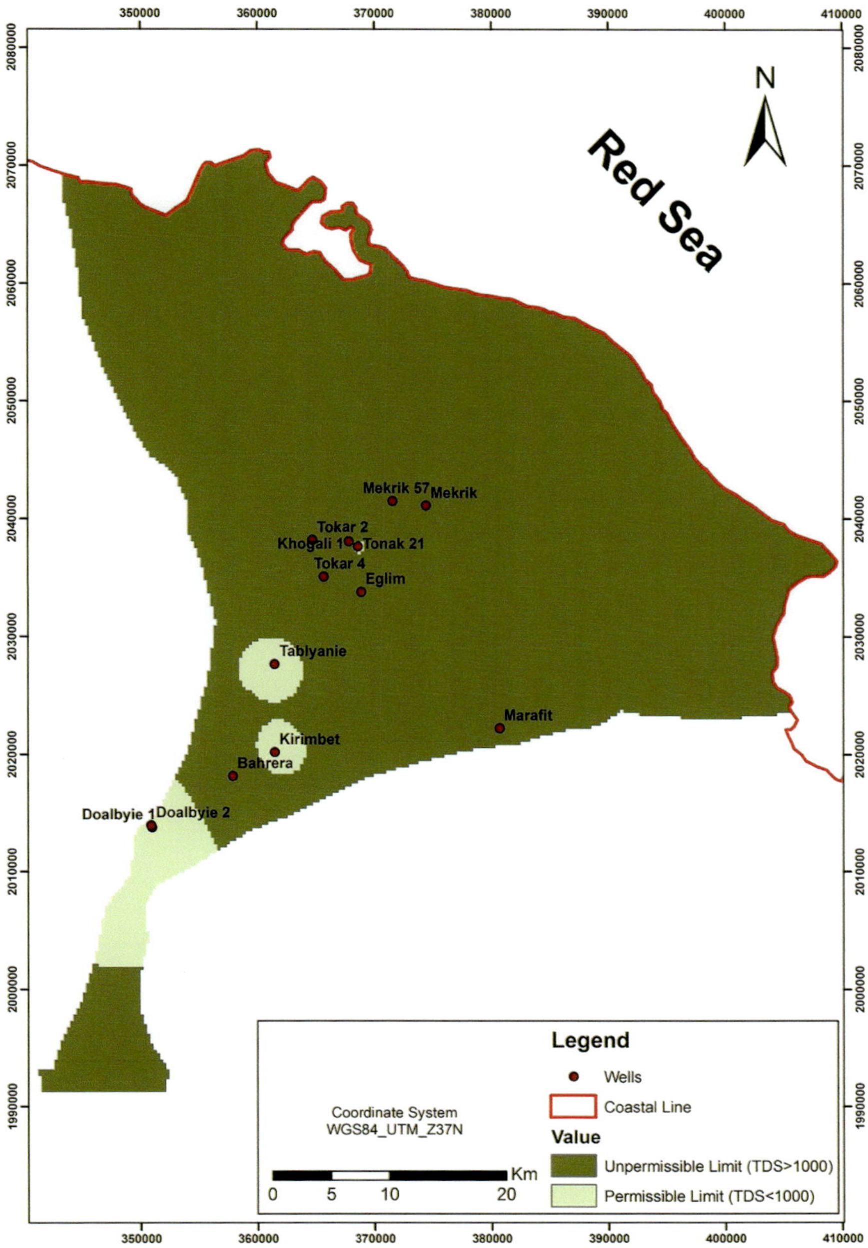

FIGURE 26.11 Classification of well water based on TDS concentration in 2015.

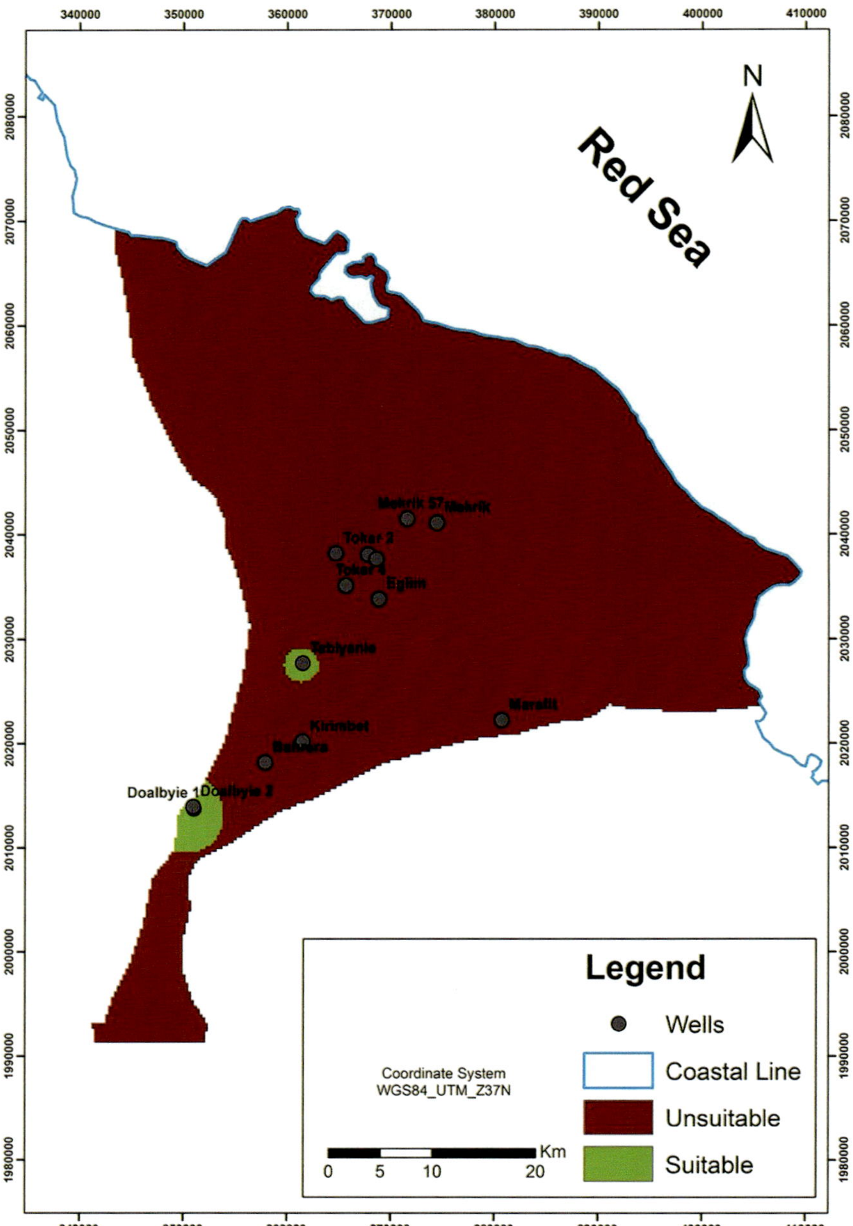

FIGURE 26.12 Map showing the locations likely to give rise to consumer complaints about drinking water for all parameters.

REFERENCES

Brierley, A.S., and M.J. Kingsford. 2009. Impacts of climate change on marine organisms and ecosystems. *Curr Biol* 19, no. 14: R602–R614.

Callander, P., Lough, H., and C. Steffens. 2011. New Zealand guidelines for the monitoring and management of sea water intrusion risks on groundwater. *Report Prepared for Envirolink (Project 420-NRLC50)*. Pattle Delamore Partners Ltd., Christchurch, New Zealand. http://envirolink.govt.nz/assets/Envirolink/420-NLRC50-Guidelinesfor-the-monitoring-and-management-of-sea-water-intrusion-risks-on-groundwater.pdf.

De Vantier, L., and K. Al-Hariri. 2000. Preliminary ecological assessment of coral communities of the north-east Gulf of Aden, Yemen with reference to the 1998 coral bleaching event. In: *Proc International Symposium on the Extent of Coral Bleaching in the Arabian Region, Riyadh (5–9 February 2000), Kingdom of Saudi Arabia*. Eds. A.H. Abuzinada, E. Joubert and F. Krupp, 110–127. Jeddah: PERSGA; Kuwait: ROPME; Riyadh: NCWCD.

De Vantier, L., Turak, E., and K. Al-Shaikh. 2000b. Coral bleaching in the central-northern Saudi Arabian Red Sea – September 1998. In: *Proc International Symposium on the Extent of Coral Bleaching in the Arabian Region, Riyadh (5–9 February 2000), Kingdom of Saudi Arabia*. Eds. A.H. Abuzinada, E. Joubert and F. Krupp, 75–90. Jeddah: PERSGA; Kuwait: ROPME; Riyadh: NCWCD.

De Vantier, L., Turak, E., Al-Shaikh, K., and G. De'ath. 2000a. Coral communities of the central-northern Saudi Arabian Red Sea. *Fauna Arabia* 18: 23–66.

Glynn, P.W. 1996. Coral reef bleaching: Facts, hypotheses and implications. *Glob Change Biol* 2: 495–509.

Hoegh-Guldberg, O. 1999. Climate change, coral bleaching and the future of the world's coral reefs. *Mar Freshw Res* 50: 839–866.

IPCC. 2007. Climate change 2007: The physical science basis. In *Contribution of Working Group I to the Fourth Assessment Report of the Intergovernmental Panel on Climate Change.*

Eds. S. Solomon, D. Qin, M. Manning, Z. Chen, M. Marquis, K.B. Averyt, M. Tignor and H.L. Miller, 996 p. Cambridge and New York: Cambridge University Press.

Jones, B.F., Vengosh, A., Rosenthal, E., and Y. Yechieli. 1999. Geochemical investigations. In: *Seawater Intrusion in Coastal Aquifers – Concepts, Methods and Practices*. Eds. J. Bear, A.H.-D. Cheng, S. Sorek, D. Ouazar, and I. Herrera, 51–71. Dordrecht, The Netherlands: Kluwer Academic Publishers.

Kemp, J.M. 1998. The occurrence of *Nizamuddinia zanardinii* (Schiffner) P.C. Silva (Phaeophyta: Fucales) at the Socotra Archipelago. *Bot Mar* 41: 345–348.

Kemp, J.M., and F. Benzoni. 2000. A preliminary study of coral communities in the northern Gulf of Aden. *Fauna Arabia* 18: 67–86.

Lough, J.M. 2000. 1997–98: Unprecedented thermal stress to coral reefs? *Geophys Res Lett* 27: 3901–3904.

Nasr, D.H., Hamza, M.A., and M. Eltayeb. 2013. Assessment of Sudanese coastal zone vulnerability to climate change. *Report Submitted to the Higher Council for Environment and Natural Resources*, Khartoum.

Norelmadina, E.K. 1995. *Assessment of the Groundwater Resources of the Tokar Delta with Special Emphasis on the Impact of the Expected Sea Level Rise on the Shallow Groundwater Aquifer.* Sudan: Port Sudan University.

PERSGA/GEF. 2003a. Regional action plan for the conservation of coral reefs in the Red Sea and Gulf of Aden. *PERSGA Technical Series No. 3*. PERSGA, Jeddah.

PERSGA/GEF. 2003b. Coral reefs in the Red Sea and Gulf of Aden. Surveys 1990 to 2000 summary and recommendations. *PERSGA Technical Series No. 7.* PERSGA, Jeddah.

PERSGA/GEF. 2004. Survey of the proposed marine protected area at Dungonab Bay and Mukawwar Island, Sudan. *Report.* PERSGA, Jeddah.

Schroeder, J. 1982. Aspects of coastal management in the Sudanese Red Sea. *J Faculty Mar Sci* 2: 45–68.

Sheppard, C.R.C., Price, A., and C. Roberts. 1992. *Marine Ecology of the Arabian Region: Patterns and Processes in Extreme Tropical Environments.* London: Academic Press.

Sheppard, C.R.C., and A.L.S. Sheppard. 1991. Corals and coral communities of Arabia. *Fauna Saudi Arabia* 12: 3–170.

Sheppard, C.R.C., Wells, S.M., and M.D. Jenkins. 1988. *Coral Reefs of the World. Vol. 2: Indian Ocean, Red Sea and Gulf.* Gland, Switzerland: IUCN, Conservation Monitoring Centre.

Shuxian, W. 2013. Groundwater quality and its suitability for drinking and agricultural use in the Yanqi Basin of Xinjiang Province, Northwest China. *Environ Monit Assess* 185, no. 9: 7469–7484. http://doi.org/10.1007/s10661-013-3113-7.

SWFWMD. 2003. District data on-line. *Southwest Florida Water Management District.* www.swfwmd.state.fl.us/data/dataonline.htm.

Teodoro, A.C. 2016. Optical satellite remote sensing of the coastal zone environment – an overview. In: *Environment Applications of Remote Sensing.* Ed. M. Marghany, 165–196. London: InTechOpen.

Wilkie, D.S., and J.T. Finn. 1996. *Remote Sensing Imagery for Natural Resources Monitoring: A Guide for First-Time Users.* New York: Columbia University Press.

Wilkinson, C.R. 1998. *The 1997–1998 Mass Bleaching Event Around the World.* Australian Institute of Marine Science, Townsville. www.aims.gov.au/pages/research/coral-bleaching/1997-98-mbe/mbe-00.html.

PART 8

Climate variability and its effects

Climate change and risk assessment of the Red Sea region, Egypt

27

Abbas M. Mansour and Mohamed Abdelkareem

27.1 INTRODUCTION

Changes in climate events and the unpredictability of weather patterns will have substantial impact on human and environmental systems. Heat stress, droughts and floods are expected to occur more frequently throughout the rest of this century. In addition to the negative consequences brought on by changes in mean variables alone, these occurrences are projected to have a wide range of unfavourable effects on food, human and health systems (IPCC 2012; Thornton et al. 2014). The amount of land between 60°S and 75°N that is classified as a dry zone has expanded significantly since the 1970s, rising from 17% in the 1950s to around 27% in the 2000s (Dai 2011; Thornton et al. 2014). People are mainly impacted by drought because of its effects on food systems, and while floods tend to be more localized, they can still have a significant global impact.

In the context of climate change and growing development, the Red Sea region of Egypt is expected to be exposed to a range of potential risks that can affect environmental quality and resource sustainability and, consequently, impact Egypt's economy significantly. The Red Sea, like other marine ecosystems, experiences global warming and ocean acidification (Steiner et al. 2018). Climatic models predict increasing frequency and intensity of heatwaves (Osman et al. 2018; Genevier et al. 2019), which may bring about extreme warm weather in the Red Sea region sooner than is predicted by the measured warming trend (Kleinhaus et al. 2020). Rapid ocean warming due to climate change leads to coral bleaching and poses a serious risk to the survival of coral reefs. If the global warming trend continues, the coral of the Red Sea could cease growing by 2070 (Cantin et al. 2010). The region already suffers from increasing habitat loss due to growing unplanned urbanization, pollution, coastal land filling, flash flooding and the negative effects of tourism. Oil spillage from pipelines and offshore storage facilities also threatens the reefs. Deterioration of reefs and other resources along the Red Sea coast of Egypt will impact millions of people through damage to fisheries and loss of income from tourism and put them at an immediate risk. The assessment indicates that Quseir City, valley mouths, bays, pocket beaches and the marine ecosystem face climate changes and are under pressure and high risk (Mansour and Abd El-Sadek 2021). In order to expand our knowledge and reduce the danger to the Red Sea ecosystem, it is vital to adapt to extreme weather occurrences. This requires a thorough risk assessment and management strategy.

Coastal habitats are already among the most vulnerable ecosystems in the world, and the intensive use of coastal areas for man-made activities (living, tourism, economic activities) places significant pressure on them (Barbier et al. 2011; Cunha et al. 2021). The protection and stabilization of ecosystems against coastal natural hazards including flooding, coastal erosion, storms and surges depend heavily on coastal and marine resources like beaches, coastal dunes, mangroves and associated wetlands (Cunha et al. 2021). Global warming is expected to lead to higher heat waves, and an increase of evaporation, flooding (Serrano et al. 2019) and the intensity and frequency of sand and dust storms as the Red Sea region has a prevailing arid climate (Zittis et al. 2022). The response of the coast to sea-level rise can be severe, such as inundation of valley mouths and lowlands of downstream areas, erosion of exposed raised coral reefs, shoreline changes and loss of a sizable proportion of the coast and settlement areas. In addition to increased sedimentation and contamination and smothering of seagrass beds and coral reefs are mangrove deterioration and retreat (Mansour and Abd El-Sadek 2021).

An essential part of monitoring the earth's climate is satellite remote sensing, which collects data from sensors on satellites (including geodetic satellites) about the planet's surface, interior and atmosphere (Yang et al. 2013). Integration of remote sensing in combination with GIS and GPS is considered to be the best solution for detecting regional changes, assessing land cover change (Müller and Zeller 2002), monitoring temporal changes and assessing the hazards (El-Asmar et al. 2015). Satellite remote sensing has the potential to provide accurate and timely geospatial information

DOI: 10.1201/9781003321408-35

describing changes in land use and land cover (LU/LC) (Foody 2003; Herold et al. 2002, 2003; Yuan et al. 2005). Generally, change detection involves the application of multi-temporal datasets to quantitative analysis of the temporal change in the phenomena over the earth's surface (Lu et al. 2004; Srivastava et al. 2012).

The coastal region of the Egyptian Red Sea is characterized by highly productive and diverse coastal and marine ecosystems such as coral reefs, seagrasses, mangroves, beaches and coastal sand that are important for tourism. Therefore, this chapter describes the prevailing natural, anthropogenic and environmental conditions in Egypt's coastal zones of the Red Sea in order to 1) evaluate the current vulnerability in terms of climate change and sustainable development; 2) assess the risk from natural events to coral reefs, mangrove stands and shoreline change; and 3) predict areas vulnerable to flash flood hazards and their impact on the coastal ecosystem.

27.2 THE RED SEA COAST OF EGYPT

The Egyptian Red Sea coast is known for its highly productive and diversified marine and coastal ecosystems, which include coral reefs, seagrass beds, mangrove forests, beaches, elevated coral reefs and coastal dunes. Thus, the area is one of the most coveted for long-term sustainable development, and it represents a very significant zone for Egypt as the coastal and marine natural resources are vital for social, economic and ecological development.

27.2.1 Physical conditions and risks

The Red Sea region lies in the arid zone between two great deserts, the Sahara of North Africa and the Arabian Peninsula in Asia. It is mainly regulated by the distribution of winds and changes in atmospheric pressure over a very wide area. These pressures undergo variations during the course of the year, causing widespread seasonal changes (Mansour and Abd El-Sadek 2021). The Red Sea receives a very small amount of water through precipitation and runoff. However, with global warming and the expected increase of evaporation, the area could be exposed to intense flooding (Zittis et al. 2022). The topography and wadis along the Egyptian Red Sea coast are shown in Figure 27.1.

27.2.1.1 Temperature

Temperatures in Egypt have increased at a rate of 0.1°C per decade on average between 1901 and 2013. However, substantially stronger warming was observed over the past 30 years, with average annual temperatures increasing by 0.53°C per decade. Warming has been more pronounced during the summer than the winter with 0.31°C and 0.07°C per decade increase in average temperatures, respectively, since 1960 (World Bank Group 2021). According to an analysis by the German Climate Service Center (GERICS 2019) of 32 global climate models (GCMs), Egypt is expected to experience a change in annual mean temperature from 1.8°C to 5.2°C by the 2080s.

The air temperature at Hurghada, obtained in 2001 by the meteorological station of the National Institute of Oceanography and Fisheries (NIOF 2019), varies between 28.77 and 30.84°C (average of ~29.83°C) in summer and 15.72 and 21.77 °C (average 16.1°C) in winter. The climate of Ras Banas in the south shows annual mean minimum and maximum temperatures of 19.1°C and 32.4°C, with an annual mean of 17.4 mm/year rainfall (Abdel-Hamid et al. 2018). In addition, the water temperature ranges between 28–29°C in summer and 22–23°C in winter, but the extreme shallow areas record higher values due to solar radiation (Mansour and Abd El-Sadek 2021). Piller and Pervesler (1989) showed that the water temperature in Safaga Bay clearly reflects seasonal changes, except in the shallow areas, where higher values were reached due to solar radiation. Therefore, in shallow waters of normal marine salinity, temperature plays the most significant role in determining the distribution and diversity of marine invertebrates.

27.2.1.2 Winds

The Hurghada meteorological station recorded a mean annual wind speed of about 5.8 m/s, which represents the highest wind speed along the Red Sea coast; the highest recorded peak wind speed was 13.8 m/s, although about 98% of the recorded wind-speed records are in the range of 3 to 10m/s (Essa and Mubarak 2006). The results obtained from the meteorological station of the NIOF at the Red Sea branch in 2001 show that the prevailing wind is NE, E in summer and N, NE and NW in the winter. An increase in peak wind intensities is expected as a result of climate change (IPCC 2007b), which may lead to increased storm surge heights in some regions (Nicholls et al. 2008). Egypt also experiences hot wind storms, known as *khamasin*, which carry sand and dust. These storms typically occur between March and May and can increase the temperature by 20°C in two hours, and last for several days (USAID 2018).

27.2.1.3 Precipitation

Egypt is an arid/hyper-arid country that receives very little annual precipitation. Egypt has observed a statistically significant reduction of approximately 22% in total annual precipitation over the past 30 years (World Bank Group 2021). While overall annual mean precipitation is expected to decrease, the intensity of heavy rain events is expected to increase by the 2080s (GERICS 2019). Reduced precipitation and increased temperature are expected to impact evaporation and the water balance as well as drought conditions. Water routing, storage and other management options can be highly varied, depending on whether the precipitation input comes frequently or with long periods of aridity between rainfalls (GERICS 2019).

Precipitation decreases from the north along the Mediterranean coast southward, and Cairo receives little more than 1 cm of precipitation each year. Areas south of Cairo receive less intense rainfall yet can suddenly experience extreme precipitation events, resulting in flash floods (World Bank Group 2021).The Eastern Desert of Egypt suffers from a hyper-arid climate that lacks regular precipitation. Precipitation occasionally occurs in the region in the form of short-lived, intense

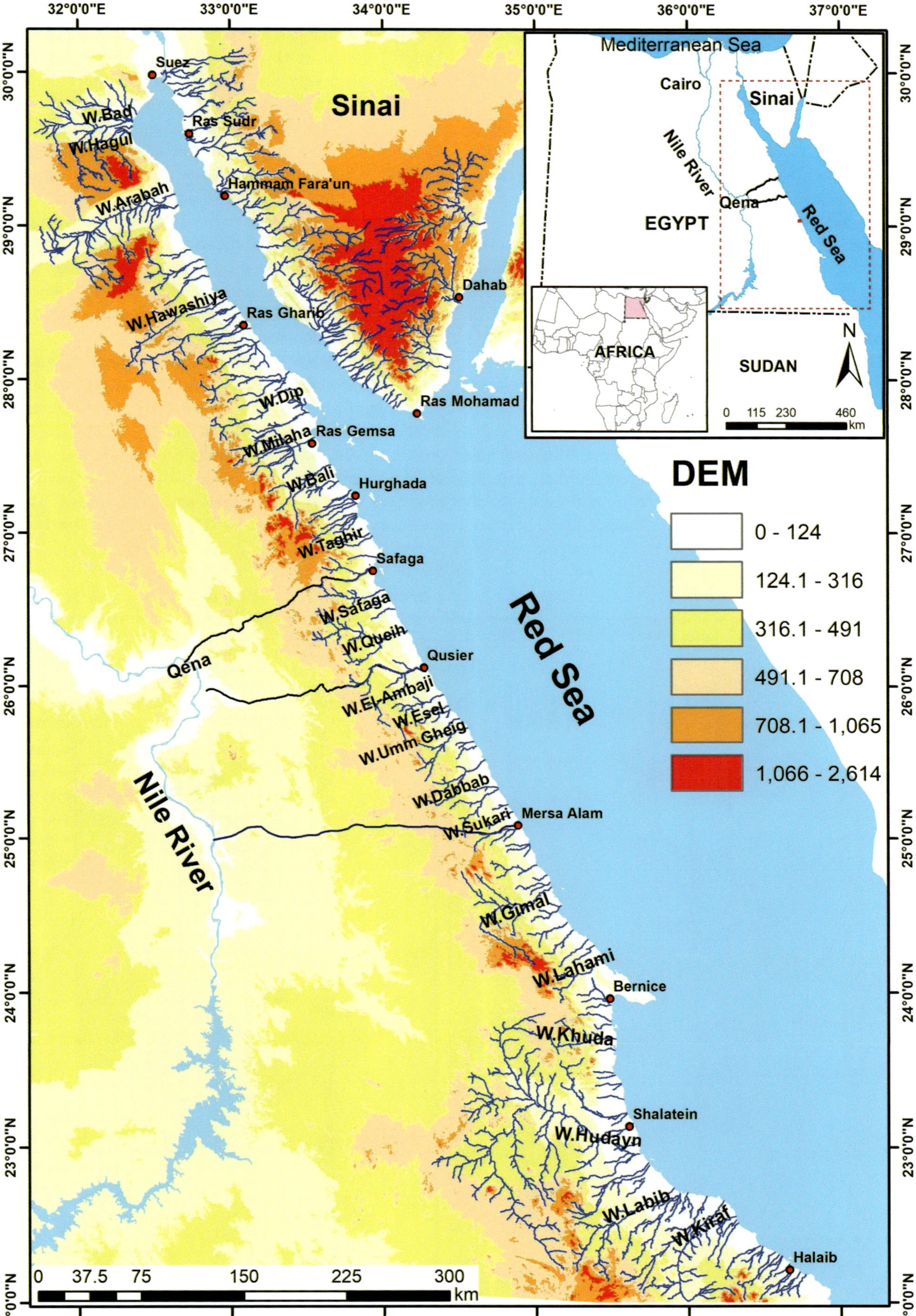

FIGURE 27.1 Elevation map derived from the SRTM DEM overlaid by the main wadis that drain into the Red Sea.

rainstorm events. Understanding the hydrological properties of an area is the first step in identifying sites at risk of flash flood damage. Based on the morphometric characteristic of the drainage systems, wadis catch rain that can result in a surge of water inundating towns within a few hours. Within 14 hours at a maximum peak flow of 875 $m^3 s^{-1}$, a 60-mm rain event at Wadi El-Ambagy would result in a surge of water inundating the town of Quseir (Mashaly and Ghoneim 2018).

In remote arid areas with scarce hydrological records and field measurements, these hydrological properties can be derived from remotely sensed data. Based on tropical rainfall measuring mission (TRMM) data, the rainfall distribution during the interval from 1998 to 2015 along the Red Sea reveals a high intensity at the Gulf of Suez and the Gulf of Aqaba and along Lake Nasser (Abdelkareem and El-Baz 2015) (Figure 27.2a).

27.2.1.4 Tides, waves and currents

The tide is one of the important influencing forces on the coastal zone. Tides are oscillatory and semi-diurnal, with an average spring range of 0.5 m and a mean sea level up to 1.0 m higher in winter. A six-hour time difference exists between tide times in the southern and northern areas (Morcos 1970). The spring range is about 0.6 m near Hurghada and 0.7 m at Quseir (Moussa et al. 1986). Tides in the Red Sea are semi-diurnal with a very low tidal range, and water currents follow the NNW-SSE direction throughout the year (Churchill et al. 2014). The maximum wave height is 1.3 m, and the wave direction is toward the south (Frihy et al. 2006). Winds create a mainly NE-SW-oriented distinct wave motion and also drive long-shore currents affecting sedimentation. In Sharm El Bahari and Sharm El Qibli, waves redistribute terrigenous sediments on the rocky tidal flat around the sharm (Mansour 1995). The currents in the Red Sea are predominantly weak, and the average flow at any season tends to be in alignment with the sea itself, NW-SE (David 1987). The coastal current along the Egyptian side is generally in a southerly direction.

The Inter-governmental Panel on Climate Change (IPCC 2007b) reported that the global sea level will rise from 0.18 to 0.59 m by the end of this century. The gradual increase in sea level can exacerbate the erosion and slowly inundate low-lying coastal regions. This will contribute to the worsening of the area of inundation and erosion with already-existing dynamic coastal processes such as waves, tides and storm surges. The global mean sea level increased by 0.20 m between 1901 and 2018

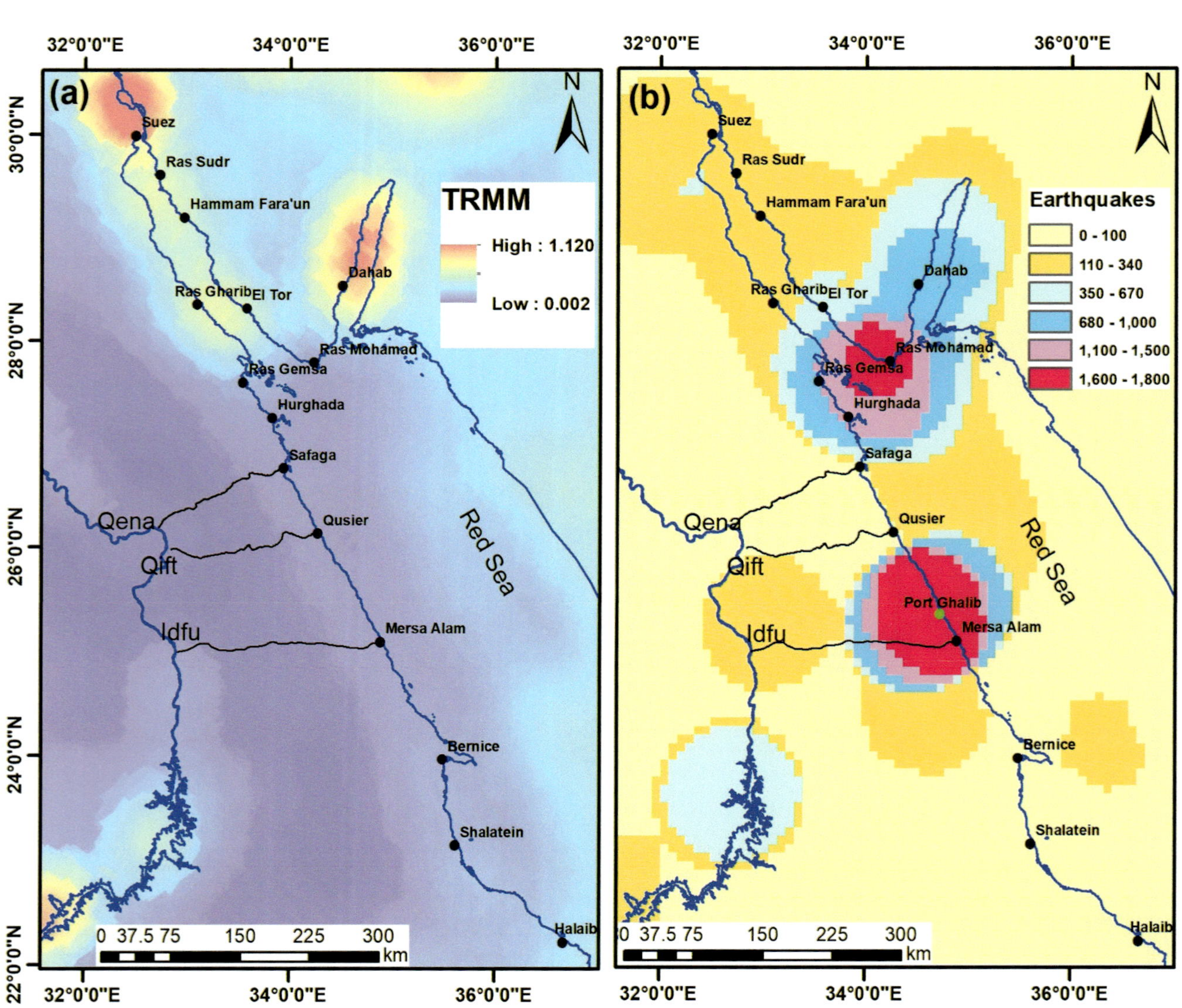

FIGURE 27.2 The Red Sea: (a) rainfall precipitation, mm/day (1998–2015); (b) earthquake density map, number of events per km^2; the magnitudes of these events range from 0.1 to 8 during the period of 1976 to 2003.

(IPCC 2021), leading to an increase in flood hazards in coastal regions (IPCC 2007a). Climate change will also lead to a possible increase in peak wind intensities (IPCC 2007b), which may lead to increased storm surge heights in some regions (Nicholls et al. 2008). The vulnerability of coastal cities will also increase due to socioeconomic trends such as the continued demographic and economic expansion in these areas (Bouwer et al. 2007). Estimates of the vulnerability of cities are essential in the planning of adaptation measures (Aerts et al. 2009).

27.2.1.5 Earthquake activity

The Gulf of Suez, the Red Sea, the Gulf of Aqaba and their triple junction are among the many locations in the study region where earthquakes occur frequently. Local areas of higher activity exist around Abo-Dabab and Port Ghalib, located north of Mersa Alam, along with areas southeast of Esna and at Lake Nasser (Figure 27.2b). Earthquakes frequently occur where the Gulf of Suez, Red Sea and Gulf of Aqaba converge (Abdelkareem and El-Baz 2016), due to continuing movement along the sinistral strike-slip faulting in the Gulf of Aqaba and the sinistral oblique-slip faulting in the Gulf of Suez (e.g., Morgan 1990). It is noteworthy that earthquakes have impacted some tourism areas north of Mersa Alam at Port Ghalib, initiating a crack about 320 m long, with a 7 m width and about 8 m depth (Figure 27.3a, b). The fissure is filled with water and is evidence of the risk of earthquakes in residential and tourism areas along the Red Sea (Figure 27.3d).

27.2.2 The Red Sea resources and risks

The Egyptian coasts extend over 3,500 km, with two-thirds along the Red Sea and the coasts of the Sinai Peninsula; for example, the Red Sea coast from Suez in the north to the Egypt/Sudan border in the south (latitude 22°N) (Figure 27.1) covers about 1,200 km. The coastal zone of the Red Sea on the Egyptian side is generally narrow because a high mountain range runs nearly parallel and close to the shoreline. It varies from a few metres in width at the Gulf of Suez to more than 35 km near the Egypt/Sudan border. Pleistocene raised coral reef terraces represent one of the major geomorphic features of the coastal Red Sea region and extend from the intertidal zone to a few metres above the level of the high tide. Many wadis fan out from these hills, with the raised terraces being dry most of the time, with only sporadic and abrupt fluvial activity. The coast is composed of a large number of protected shallow bays, sharms, lagoons and small beaches. Fragmented and extended coral reef communities, with their rich marine life, extend over large areas of the coast. These protected shallow environments provide favourable habitats for the growth of mangroves.

Several world-renowned diving destinations, such as Hurghada, Marsa Alam, Sharm El Sheikh, Dhahab and Nuweiba, significantly benefit the Egyptian economy via the profits from tourism. There are also trade and transport networks and a large number

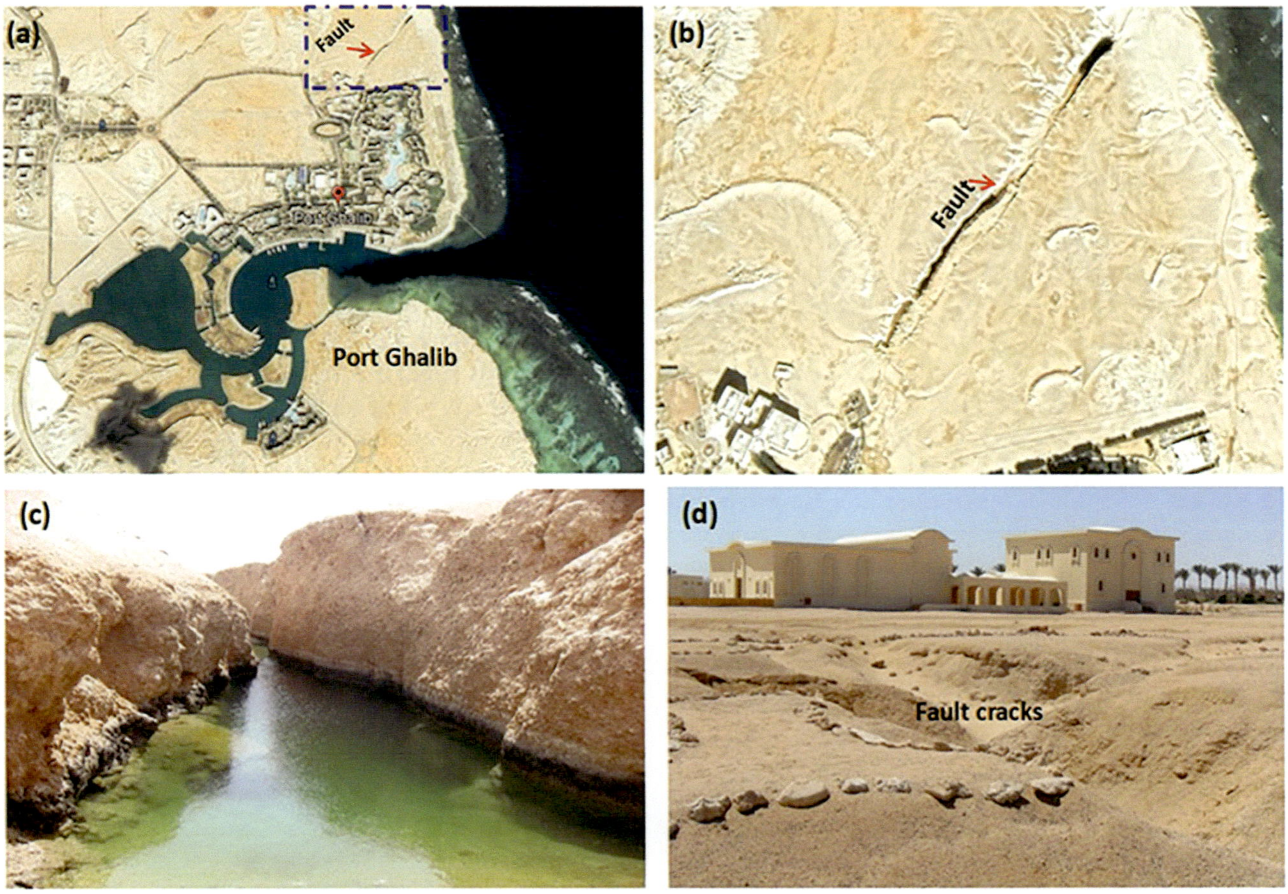

FIGURE 27.3 (a), (b) Impact of earthquake activity, fracture in land surface at Port Ghalib area north of Mersa Alam; (c) field photo of the crack that filled with water after earthquake activity; (d) fault cracks as a result of earthquake activity near residential areas (HEPCA center).

of ports on the coast of the Red Sea. The coast is an important source of fisheries, providing income and food security.

Climate change, especially global warming, sea-level rise and acidification, will pose a great risk to the marine ecosystems. Also, any increase in the water temperature can lead to the deterioration of the symbiotic relationship between coral and algae and lead to coral bleaching or even death. Increasing the rate of carbon dioxide in the atmosphere and change in water chemistry greatly impacts coral calcification and formation. Flash floods impact the coastal developments and bring detrital sediments and contaminants from the wadis into the sea (Mansour and Abd El-Sadek 2021). Global sea-level rise will lead to an increase in flood hazards in coastal regions (IPCC 2007a). Bleaching has occurred in the Red Sea as far north as Hurghada (Furby et al. 2013; El-Askary et al. 2014; Monroe et al. 2018; Khaled 2019).

Human activities and developments along the coast are growing very fast, and most land is occupied by large hotels, tourist villages and extensions of cities. Harbours and ports are widely distributed in different areas. Phosphate mines and some industrial activities, including phosphate and other mineral shipments, are hosted by the coast. The natural resources along the coast are exposed to many stressors from growing development, tourism and climate changes (Mansour and Abd El-Sadek 2021). Landsat data collected from 1973 to 2015 reflects a loss of 6.21 square kilometres of coral cover and an increase of 13.4 square kilometres of coastal development along the reef in the Hurghada region (Khaled 2019). Oil leaks, damage from vessels and potential spillage from oil pipelines and offshore storage facilities threaten the reefs and put millions of people at immediate risk through damage to fisheries and loss of income from tourism. To monitor and control these hazards, strong institutions and legislation are needed. Research and conservation of the Red Sea coral reefs are also required due to the increased demands on the resources and sustainable development in Egypt.

Egypt initiated one of the earliest conservation efforts in the Red Sea in 1983, with the establishment of the Ras Mohamad National Park (Ras Mohammed National Park 2019). The Northern Red Sea Islands were announced as protectorates by the resolution of the prime minister of Egypt, no. 1618 for the year 2006. The protectorate comprises sensitive coastal ecosystems such as scattered clusters of seagrass, mangroves and more than a thousand species of living coral, echinoderms, molluscs, crustaceans, fish, reptiles, birds and mammals (EEAA 2008).

Egypt has also an annual monitoring programme for marine resources and water quality along its coastline, conducted by the Red Sea National Park Protectorate and the National Institute of Oceanography and Fisheries (NIOF 2019). The Hurghada Environmental Protection and Conservation Association (HEPCA) is an NGO specializing in the field of marine and land conservation in the Egyptian Red Sea. PERSGA is the Regional Organization for the Conservation of the Environment of the Red Sea and Gulf of Aden. Both organizations conduct monitoring for the Red Sea coast. However, reliably tracking and detecting anthropogenic impacts and rapid climate changes requires a frequent and spatially detailed assessment and monitoring, and conservation initiatives should be conducted in each country and with other countries along the Red Sea, with regular publication of the collected data.

27.2.2.1 Marine environments

Coral reef ecosystems, mangroves, seagrass and other marine communities characterize the Red Sea coast. These ecosystems are highly sensitive to human stressors, global warming, sea-level rise and flash floods and can be classified as highly vulnerable. The marine environment of the Red Sea coast receives inputs of hazardous substances through wadis, harbours, ports and atmospheric deposition. Pollution sources range from oil spills to dumping of treated wastewater from tourist resorts and urban areas into the sea. Changes in the nature of the sedimentary environment (i.e., pollution) have been caused by land filling, along with urbanization and building of tourist resorts (Attia et al. 2010). Eighteen oil accidents from tankers and other causes in the period from 1989 to 1994 were recorded in the marine environment of the northern Red Sea and the Gulfs of Suez and Aqaba (Abdel-Maksoud 1997). Pollution has adverse effects on the marine environment, society and the economy. Heavy metals and pollution elements are detected at high levels in the sediments and waters in several areas along the coast (Mansour et al. 2011, 2013; Abdelkader et al. 2018) and, consequently, impact the organisms living in these ecosystems. Unfortunately, little is known about the transfer of these chemicals to and their effects on the different components of coastal/marine ecosystems. This hampers an ecologically sound risk assessment and/or derivation of science-based environmental quality criteria for these compounds (DeFur et al. 1999). The links between observed effects and the concentrations of the hazardous substances are still lacking.

27.3 DATA AND METHODS

The data used in this study were collected from different sources, including the Shuttle Radar Topography Mission (SRTM) with 30 m resolution, and optical images such as Landsat-8 data, Sentinel-2 Google Earth images and the Tropical Rainfall Measuring Mission (TRMM). The SRTM DEM data is selected to map the topographic characteristics and compute the catchments' parameters. The optical images were used for land use and land cover mapping and change detection. All data were processed using ArcGIS software packages. Model parameters were calibrated, and results were validated using field data. Combining the optical and radar remotely sensed data allowed mapping flood hazards and natural resources, including land, water and biological and climate resources, which impacted their persistence as well as expanding human activities.

The SRTM DEM data was used to extract watersheds and stream networks using the D8 approach (O'Callaghan and Mark 1984). Following Strahler (1957, 1964), the stream orders were determined and used to calculate other characteristics. This also provides information about elevation, slope, stream power index (SPI) and drainage density, as well as the sediment transport index (STI) and the terrain roughness index (TRI), which are used to define the area.

The optical sensor of Landsat-8 with nine spectral bands, the Landsat Operational Land Imager (OLI), was released on 11 February 2013. The collected scenes were mosaicked and corrected to the UTM Zone 37N coordinate system. Then, image transformation and enhancement processes were carried out using the OLI bands 2, 3, 4, 5, 6 and 7.

Sentinel-2 is an optical satellite platform from the ESA. These two land-monitoring satellites cover a bigger portion of the earth's surface than earlier Landsat missions and produce large optical images more reliably. Future investigations will greatly benefit from Sentinel-2 satellites' ten- and five-day temporal resolution. The Sentinel-2 scene captured was processed using the ENVI and SNAP software packages.

The areas vulnerable to flood hazards were mapped in order to characterize the catchment features and produce the output map. Complex datasets are required for more data that is needed for flash flood hazard (FFHz) modeling. The elevation parameters, drainage systems, rainfall, land use and land cover are the main foundations of the GIS-based model that is produced. The need for such a strategy arises from the threat to residents of infrastructure destruction. Global warming puts a strain on the hydrologic cycle; hence, it is generally believed that flash flood hazards are increasing as a result of global climate change.

The occurrence of mangroves and seagrasses were determined using Sentinel-2 data and in the field and was verified in Google Earth images. Field validation was conducted by selecting random points for each habitat class along the entire area of study. Shoreline orientation relative to the dominant wind direction (NW) was measured to determine the degree of risk from wave action.

27.4 RISK ASSESSMENT OF CLIMATE CHANGE ON THE COASTAL AREA

27.4.1 Risk assessment of flood hazards and their impact on coastal areas

The elevation is widely used to describe regions that are susceptible to FFHz (Das 2018). The elevation of the study area ranges from 0 to 2614 m (a.s.l). In comparison to areas with a higher elevation, which are less susceptible, the areas of lower elevation are more susceptible to flash floods. This is because lowland areas collect a lot of rainwater (Waqas et al. 2021) and may flood faster as water flows more quickly from high elevations to low elevations (Das 2018). As water flows from the mountain uplands and connects with streams at lower elevations, flooding is caused. As a result, most of the coast is susceptible to flash floods (Figure 27.4a). In addition, the slope is the most essential index for characterizing surface runoff and designating areas vulnerable to FFHz (Abdelkareem 2017; Zhu and Abdelkareem 2021). Since areas of higher slope are more vulnerable to runoff and the impact of floods, the area is classified into four ranges from 0 to 76.05 degrees. Most of the coastal areas range from flat to gently sloped (Figure 27.4b).

Drainage density (Dd) is calculated by dividing the total length of the stream network by its area. Infiltration and drainage density interact negatively because locations with a high Dd are more closely related to surface runoff (Hong and Abdelkareem 2022); thus, flooding and drainage density have a positive relationship. Higher drainage density, which denotes a high surface runoff, is directly related to higher flood susceptibility (Sharma et al. 2021). Therefore, the Dd of the studied area is classified into four classes: low (0–11), moderate (12–17), high (18–23) and very high (24–37) (Figure 27.4c).

The potential for flow erosion at a certain point on the topographic surface is measured by the stream power index (SPI) (Tehrany et al. 2014). The SPI is a gauge of how much water flow can erode land. The stream power index was calculated in equation 1 as follows (Moore et al. 1991):

$$\mathrm{SPI} = \mathrm{Ac}^{*} \tan S \qquad (27.1)$$

where Ac is the specific catchment area (m^2m^{-1}), and S is the slope gradient in degrees (Regmi et al. 2010). A higher SPI and erosion risk are brought on by an increased slope gradient and catchment area, which produces a large volume of water and increases the velocity of water flow from upslope zones. Slope and flow accumulation characteristics were used in the Arc GIS programme version 10.5 to calculate the SPI. The SPI is positively related to the impact of FFHz (Cao et al. 2016). The SPI map is divided into two classes (Figure 27.5a). In addition to SPI, the sediment transport index (STI) illustrates the eroding capacity of overland streams as related to two variables: alluvium flow transport concentration and basin evolution. Consequently, it is an important criterion in the analysis of flash flood hazards (Bui et al. 2019). The STI was calculated in equation 2 as follows:

$$STI = \left(\frac{\mathrm{As}}{22.13}\right)0.6\left(\frac{\sin\beta}{0.0896}\right)1.3 \qquad (27.2)$$

where As is the specific catchment area (m^2m^{-1}), and β (radian) is the slope gradient (Moore and Burch 1986; Regmi et al. 2010). The obtained STI map was grouped into four classes: 0–2, 2–50, 50–500 and 500–1300 (Figure 27.5b).

A flash flood hazard map also benefited from applying the terrain roughness index (TRI), as it is one of the important morphological factors connected to flash flooding (Tehrany et al. 2015). The TRI is positively related to flash flood hazard; thus, the area is classified into four classes: low (0.11–0.39), moderate (0.4–0.49), high (0.5–0.59) and very high (0.6–0.89) (Figure 27.6a).

The rainfall conditions and amount of rainfall within a short period of time play an important role in the risk of flash floods (Brooks and Stensrud 2000). FFHz and rainfall are closely related (Waqas et al. 2021). Kriging interpolation was used to illustrate the distribution of precipitation in the study area using average daily precipitation data during the interval from 1 January 1998 to 30 November 2013 (Figure 27.6b). Note that the northern part of the study area is characterized by high rainfall, particularly along the Gulf of Suez, Sinai, and in south Aswan and Lake Nasser in southern Egypt (Abdelkareem and El-Baz 2015).

Nine predictor maps were used to build the final FFHz map, which was created using an integrated model with input from remote-sensing data through a GIS analysis tool. Elevation, slope, STI, SPI, drainage density, TRI and rainfall are input predictors used to generate FFHz maps. The thematic layers'

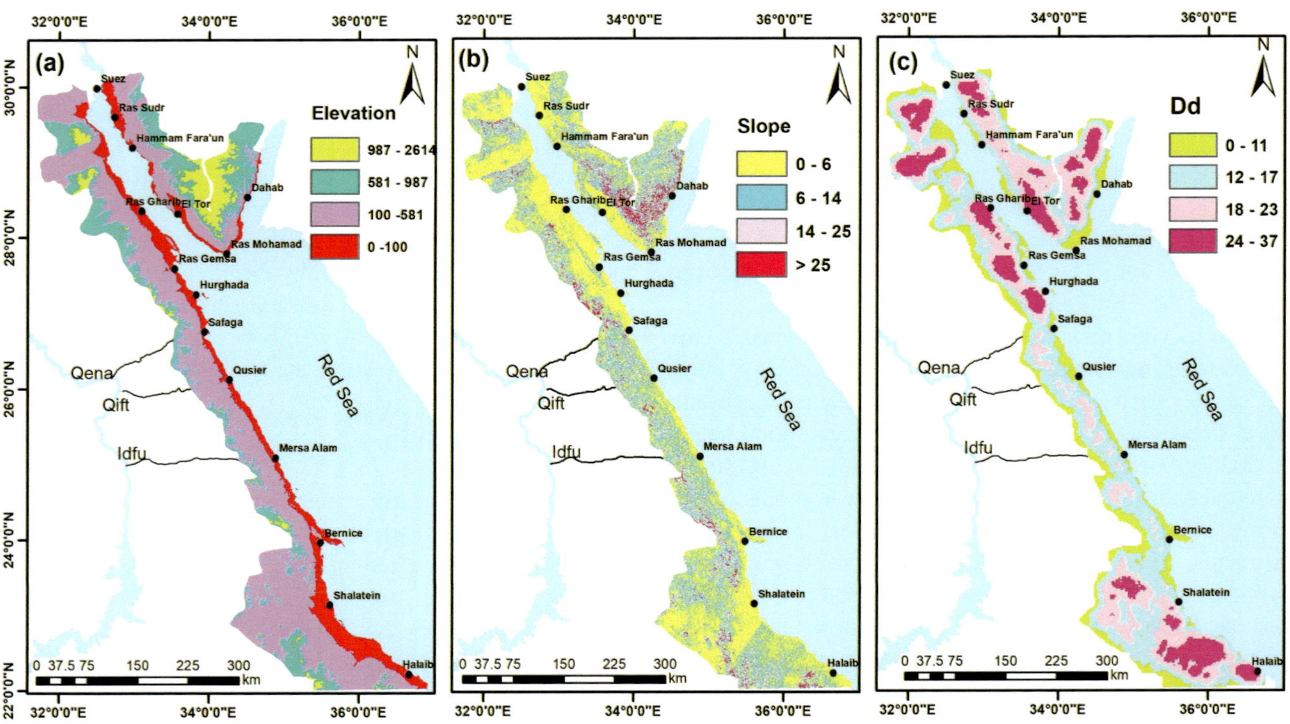

FIGURE 27.4 The Red Sea coast: (a) elevation in m, (b) slope in degrees, (c) drainage density (Dd) (km/km²).

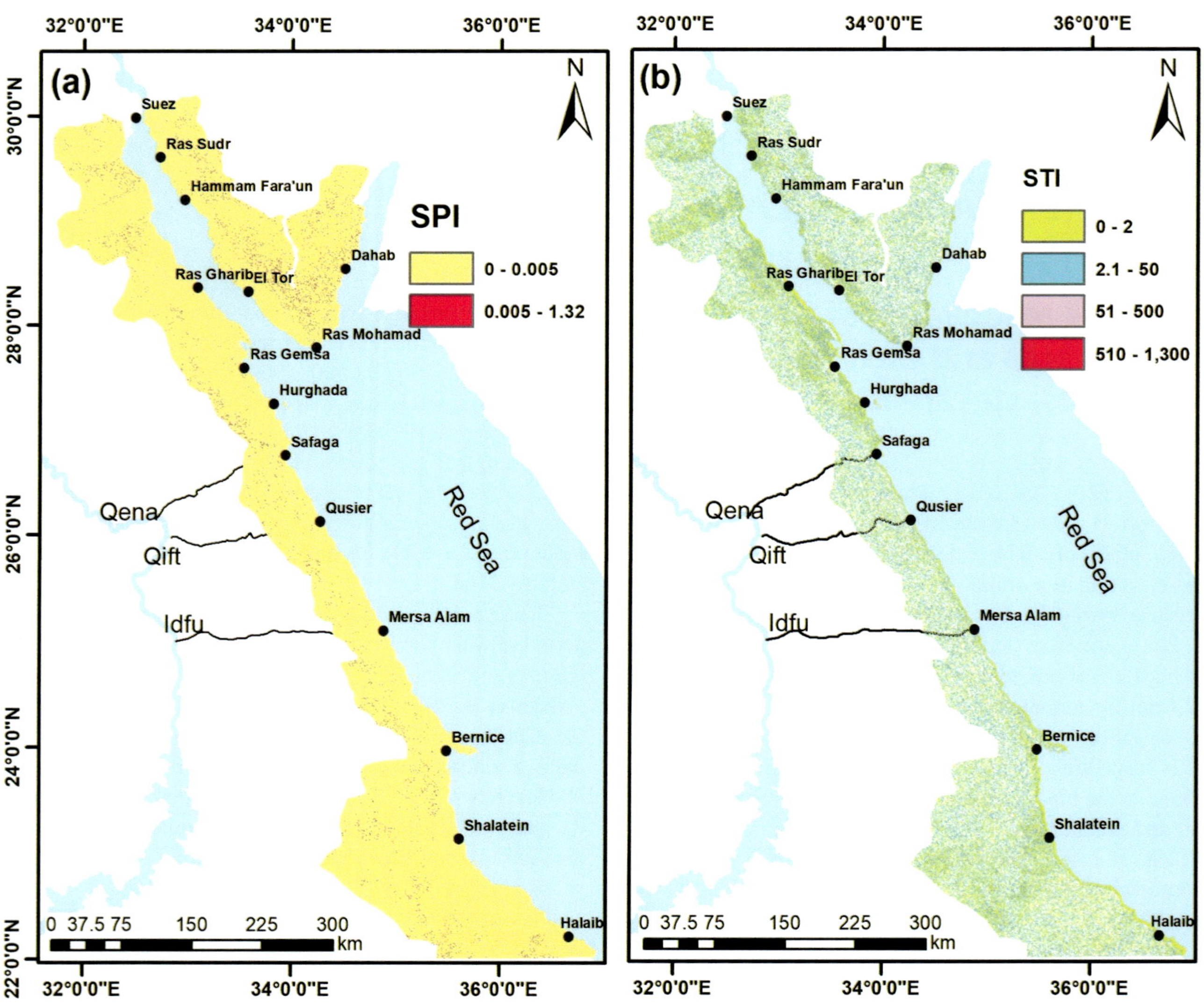

FIGURE 27.5 The Red Sea coast: (a) SPI, (b) STI.

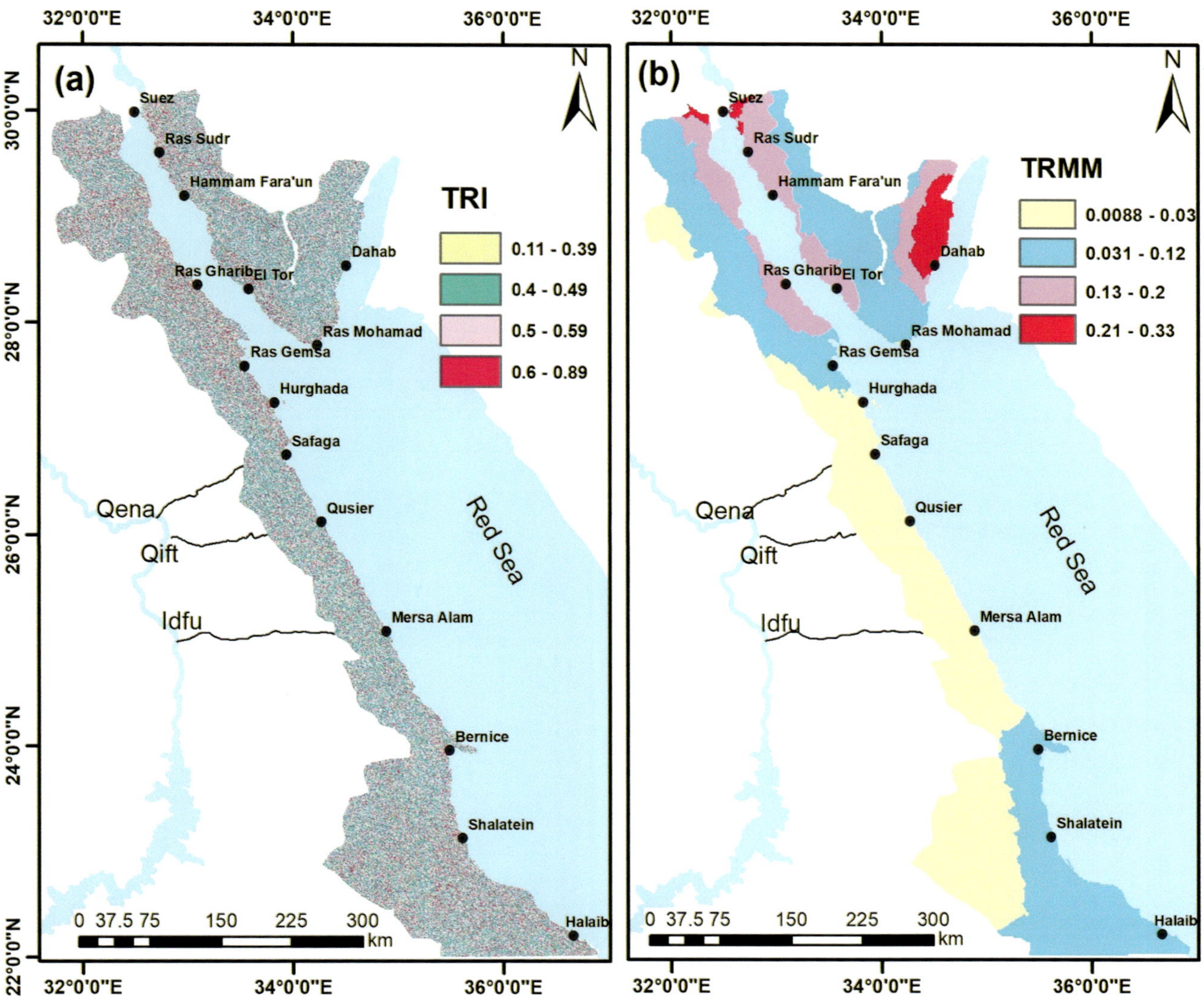

FIGURE 27.6 The Red Sea coast: (a) TRI, (b) TRMM.

FFHz were produced via a multi-criteria GIS-based overlay technique, with each GIS cell's layers fitting to the same pixel location. This method calculates the multiplied predictor layers' average rank. As a result, the sub-class characteristics that were normalized before combining were given grades from 1 to 5, and the predictor maps were classified into five classes – very low, low, moderate, high and very high risk – covering 17.42, 26.79, 29.19, 14.74 and 8.6% of the entire area, respectively (Figure 27.7).

Based on the FFHz map, the northern part of the study area that includes Hurghada, Ras Gemsa, Ras Gharib, Ain Sokhna, Ras Sudr, Hammam Fara'un, El Tor and Dahab is susceptible to flash flood hazards. In addition, the southern part of the study area (e.g., Bernice, Shalatein and Halaib) is characterized by high to very high risk. On the other hand, the areas of Mersa Alam to Safaga are susceptible to moderate flash flood hazards (Figure 27.7). Note that the majority of the tourism area in the northern part is susceptible to flood hazards, such as the area of Ain Sokhna, which has an impressive coastal zone and coral reef (Figures 27.7b-e).

Cities along the coast are expanding: in particular, along the courses of wadis, where potential flood damage is high. To minimize this damage, houses and other construction can be built on the top of the banks of these wadis, but these locations may still be sensitive to lateral erosion from flood waters and hazards associated with flash flood events. However, the hyper-arid climate makes implementing mitigation efforts essential to stopping or slowing the flow of flood waters and, ideally, storing this fresh water for future use and recharging the local aquifers. A few protective measures have been undertaken recently in some wadis to alleviate the flood damage in several cities along the coast, especially Ras Gharib, Hurghada, Safaga, Quseir and Marsa Alam; the highways along the coast; and those connecting these cities to the Nile Valley. Amongst these mitigation measures are the construction of small retention dams across the downstream area to reduce the speed of the flood water and the creation of an artificial lake behind the dams to reduce flood power. Many segments of the highway have also been lowered to the level of the wadi floors in order to avoid water blocking and, hence, erosion along the highway sides during flash flood events.

27.4.2 Coral reefs

A significant proportion and considerable range of the coral reefs occur in the Red Sea of Egypt, with about 3,800 km^2 of

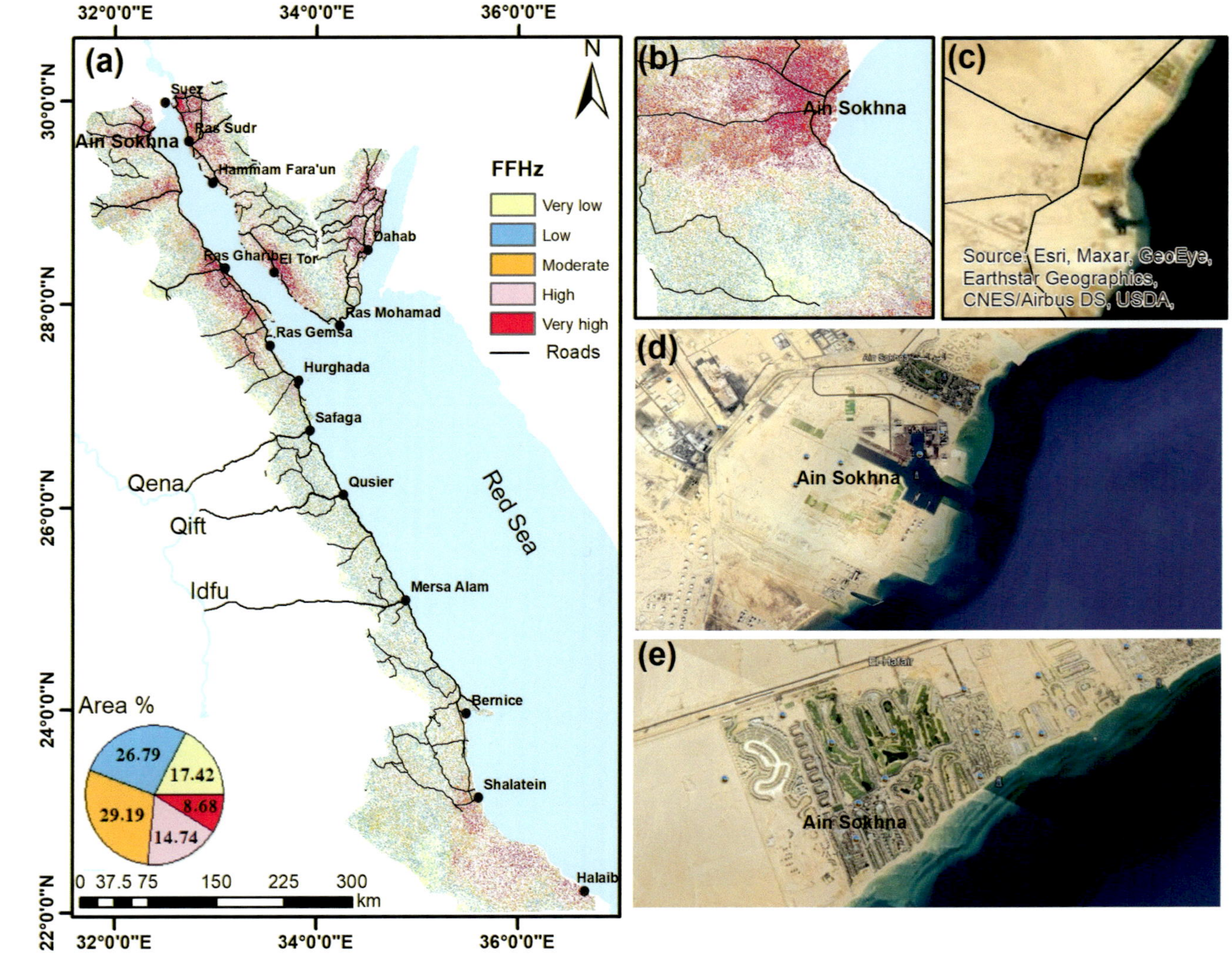

FIGURE 27.7 The Red Sea coast: (a) flash flood hazards; (b) subset of FFHz; (c), (d), (e) view of Ain Sokhna and the coastal zone.

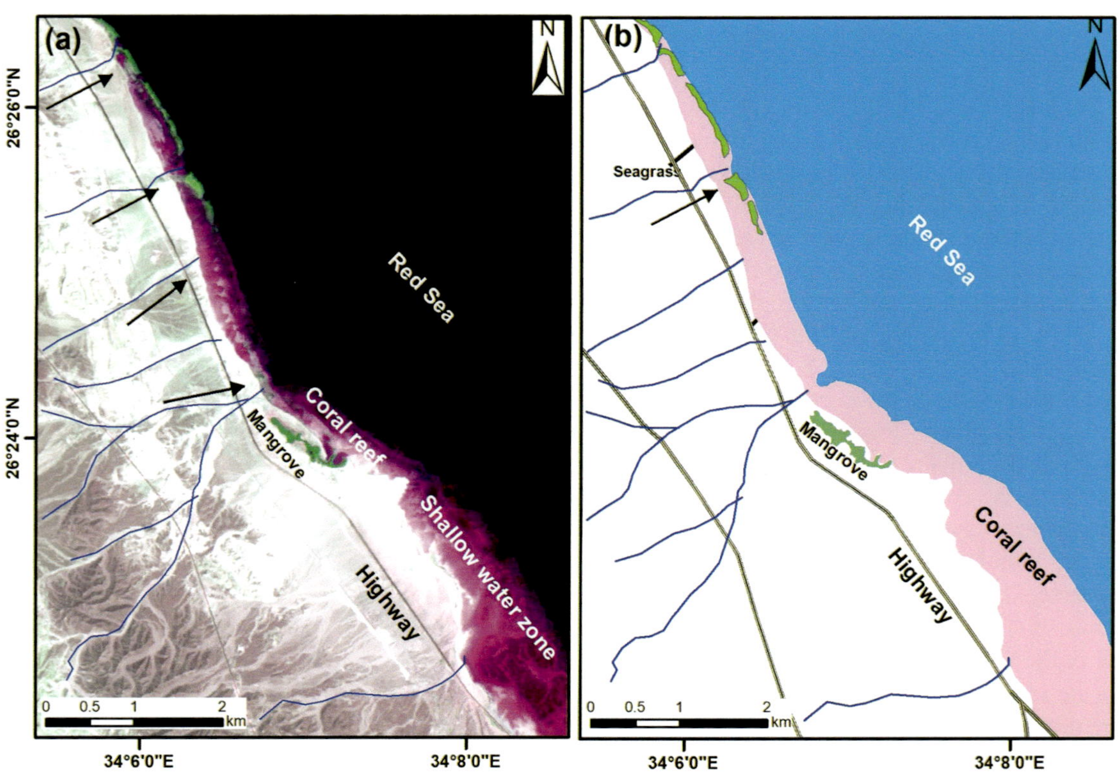

FIGURE 27.8 (a) Sentinel-2 image (bands 2, 8 and 12) shows the impact of flash flood erosion on the coastal and marine resources at the area between Quseir and Safaga, (b) digitizing of coral reefs, seagrass and mangroves along the Red Sea.

reef area (Spalding et al. 2001). About two-thirds of the hard corals of the Red Sea are found in the Egyptian reefs, including some endemic species (Kotb et al. 2004). Fringing coral reefs (Figure 27.8a, b) along the Egyptian Red Sea coastline are not continuous because periodic flooding from wadis creates gaps within the reef system. Uncontrolled development activities constitute a major threat to coral reefs through direct damage caused by tourist use of the reefs and anthropogenic impacts. Artificial beaches on rocky shores increase sedimentation and water turbidity. Sewage runoff, sedimentation following urban construction, dredging, coastal alteration and overfishing are other anthropogenic impacts. The flash floods during rainstorms impact the coral reef by causing erosion and deterioration (Figure 27.9a, b).

Coral bleaching and mortality will increase as a result of climate change. Two bleaching events that occurred in 2007 reflect the impact on Egyptian coral reefs (Eladawy et al. 2015). Threat levels due to thermal stress and ocean acidification are projected to increase in the Middle East at nearly 90% of the reefs by 2030. Climate change impacts combined with current local impacts will push all Egyptian reefs to a threatened status by 2050 (Burke et al. 2011). Bleaching events in 2011 and 2012 impacted only coral reefs from south of Qusier to the far south and were found to be in conformity with the water circulation pattern on the Egyptian coast of the Red Sea (Hanafy and Ismail 2014). In comparison, bleaching on the coral reefs of the Gulfs of Aqaba and Suez and the area between Hurghada and Safaga was not recorded (Hanafy and Ismail 2014). The increase in water temperature affected certain reef-building coral genera, including Montipora, Porites, Acropora, Stylophora and Pocilliopora; some non-reef-building corals, including Millipora; and some of the soft corals and sea anemones (Hanafy and Ismail 2014).

Sea level is largely responsible for the balance of erosive and constructive processes that shape a reef, the dynamics of waves and ocean currents, and the distribution of nutrients within reefs. Sea-level rise may lead to increases in sedimentation for reefs located near land-based sources of sediment. Sedimentation runoff can lead to the smothering of coral (NOAA 2021). Waves effectively transfer sediment from the near shore and beach face to the reef crest. Coral reefs can keep up with a gradual rise in sea level by growing upward toward the surface. However, if sea level rises more rapidly than a reef can grow, the reef will eventually drown and be overgrown by deep-water coralline algae (Webster et al. 2004). Turbidity on fringing reefs from 0.2 m of sea-level rise can be increased through resuspension of fine sediment on reef flats (Blanchon et al. 2009). However, coral reefs of the Red Sea coast of Egypt can be resilient to sea-level rise due to tectonic movements. Pleistocene coral reef terraces that today exist above present sea level are evidence of tectonic movement (Mansour 2000).

27.4.3 Mangroves

Two halophytic types of vegetation have been recorded along the Red Sea coast of Egypt: mangrove and salt marsh (Madkour et al. 2013). The mangrove vegetation is represented by a single community dominated by *Avicennia marina*. Twelve communities are distinguished in salt marshes, occupying the salt ground and the dunes behind it. Human-induced stresses, including overgrazing, solid waste disposal, overfishing, urbanization and changes in hydrological patterns, now threaten the survival of mangroves in Egypt (Madkour et al. 2013; Mansour and Abd El-Sadek 2021). Heavy grazing by camels is recorded at the mangrove stand of Sharm el-Qibli, especially at its southern edge. The mangrove stand in the Erier area is now separated from the sea by a newly formed sandy, beach which has cut off tidal flow. The landward fringes of the mangrove stands of Hamata and other areas like Wadi El-Gimal and Qulan are also impacted by camel grazing. Loss of mangroves will not only deplete the resources available within their boundaries but also affect the productivity of offshore waters. Therefore, all mangrove forests in Egypt are protected according to Laws No. 102/83 and 4/94 and by Prime Ministerial Decrees No. 1067/1983 and 642/1995.

The dependence of mangrove occurrence on the sea level and climate change is an important topic. Despite of their economic and ecological importance and protection by law, mangroves along the Red Sea coast experience high yearly loss rates. Therefore, for conservation and sustainable management, the mapping and assessing of changes in mangrove extent and condition are in growing demand, especially in the context of climate change, land use change and related threats to coastal ecosystems (Abdel-Hamid et al. 2018). Remote sensing is the most appropriate tool for mapping and assessment of mangroves, due to its

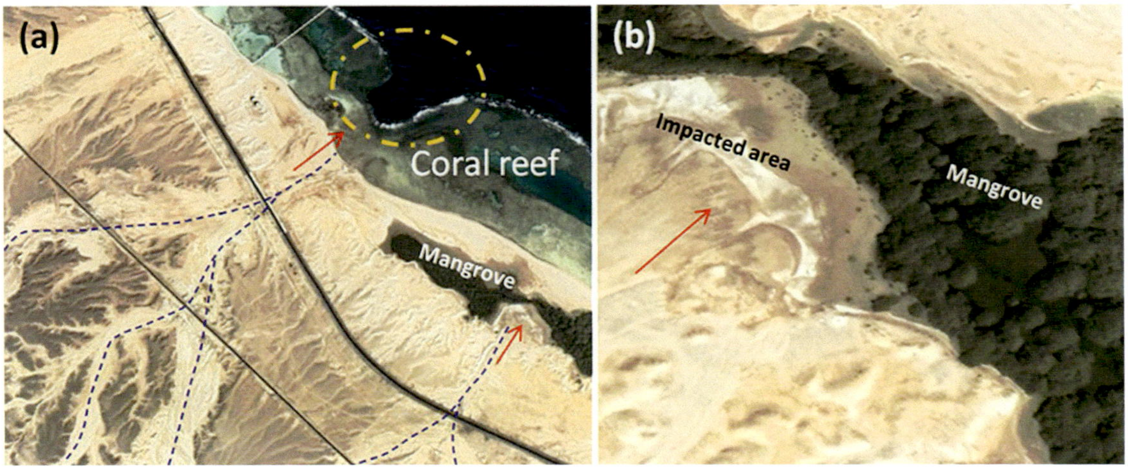

FIGURE 27.9 (a), (b) Impact of the wadi flooding on coral reef and mangroves.

ability to capture high spatio-temporal variability over large geographical scales (Archibald and Scholes 2007). Sentinel-2 band composite 2, 8 and 12 revealed mangrove areas that were clearly impacted by the floods of the wadis during rainstorms in green (Figures 27.8, 27.9, 27.10, 27.11).

In terms of climate change, a rise in sea level is expected to be the greatest factor influencing the future distribution of mangroves, having the greatest negative impact on mangrove forests (Field 1995; Gilman et al. 2008). Sea-level rise will result in increased flooding and potential redistribution of mangrove species (Gilman et al. 2008). However, mangroves are very resilient ecosystems that have survived extreme fluctuations in sea-level rise over the last 65 million years because below-ground root production coupled with sediment loads are allowing these forests to keep up with sea-level rise (Mackenzie and Friess 2015).

Mangrove vulnerability and resilience to relative sea-level rise are largely dependent on mangrove sediment surfaces, species composition and the ability of different species to colonize new habitats. Moreover, the slope of the adjacent land relative to that of the land that the mangroves currently occupy; the presence of obstacles that can impede landward migration; and the effects of other factors, such as overharvesting of trees or pollution (Mackenzie and Friess 2015), all impact the mangroves. Sea-level rise is the climate change impact that is expected to have the greatest negative impact on mangrove forests. Increased sea level will result in a shift in the distribution of mangrove species or the ultimate loss of mangrove species (Mackenzie and Friess 2015).

Mangroves provide habitats for both fish and invertebrates and other wildlife, as well as a filtration function for land-based sediment and pollutants at ambient concentrations and shoreline stabilization functions along coastal shores. These functions may be impaired if sea-level rise or other exposure pathways are modified such that one or more of the influential factors impact mangroves and associated wetlands in significant or chronic ways.

27.4.4 Shoreline changes

Along the Red Sea coasts of Egypt, raised coral reef terraces with their rocky nature extend from the intertidal zone to few metres above the seashore, and therefore, the shoreline is stable, with the exception of the areas facing strong waves, which suffer from erosion (Mansour and Abd El-Sadek 2021). If the shoreline is directly subjected to waves, the risk of inundation and erosion is much greater than the risk to sheltered shorelines. Shores directly facing the northwest are the most vulnerable to wave action. Lowland areas and valley mouths are subject to inundation during high tide, and, consequently, some erosion occurs during water retreat and drifting currents. Accretions occur rarely in some sheltered areas. Unfortunately, due to the rocky nature of tidal flats, landfilling and dredging processes are widely observed along the coast, especially in Hurghada. These processes are widely used to gain land for construction, prepare swimming pools, and construct marinas; they highly

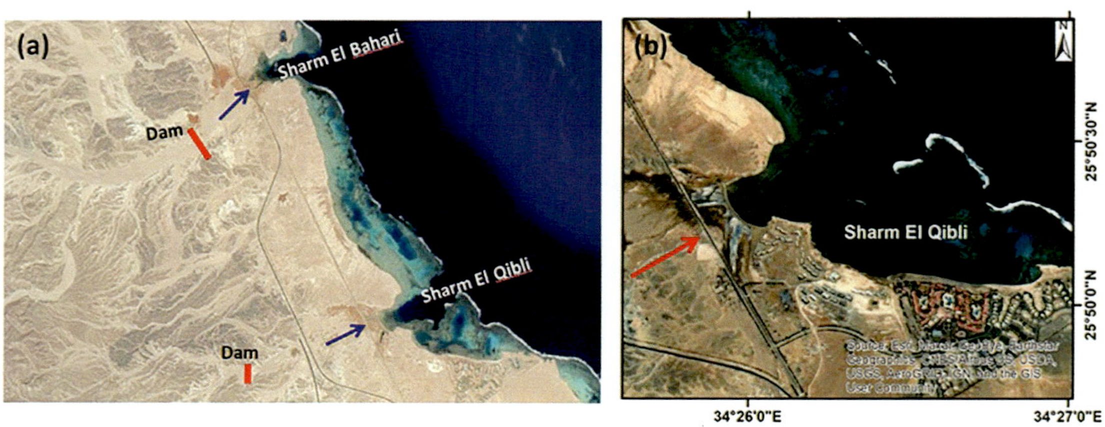

FIGURE 27.10 (a) Impact of drainage system on destruction of coral reefs and mangroves of Sharm El Bahari and Sharm El Qibli areas, (b) close-up view of Sharm El Qibli area.

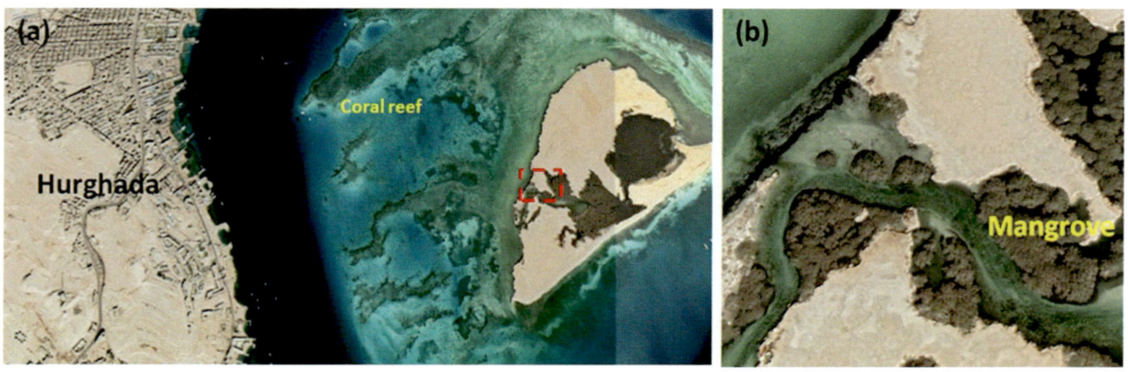

FIGURE 27.11 (a) Irregular shoreline at Hurghada and degradation of coral reefs, (b) close-up view of mangroves at Abu Monqar Island east of Hurghada.

damage the living organisms of tidal flat areas, change the shoreline and impact the whole marine ecosystem, especially coral reef communities. Such processes produce erosion and consequently accretion patterns. The satellite image from 1984 shows the coast before development activities with no evidence of anthropogenic interference while, in a recent year (2020), images showed that the situation was completely changed, with patterns of dredging and landfilling detected at different locations of the coast (Figure 27.10). Urban communities and resort areas at the valley mouths are usually under the threat of flash flooding and are have been determined to be highly vulnerable areas.

The orientation of the shoreline was measured in a clockwise direction (0–360° from N), and the degree of risk was determined in relation to the dominant wind direction (i.e., NW). Therefore, shores directly facing the northwest are the most vulnerable to wave action. Shoreline exposure was classified into four categories (Hereher 2015): very highly exposed (351°–80°), highly exposed (81°–170°), moderately exposed (171°–260°) and low-exposed (261°–350°) shorelines.

27.5 RISK ASSESSMENT AND COASTAL ZONE MANAGEMENT

It is important to understand the environmental risks associated with climate change and the Red Sea coastal development of Egypt and the potential ramifications of their effects on the quality of both environmental and human health and the economy. Ecological risk assessment operates in a framework that identifies either individual or specific sets of environmental threats and analyzes the available information. These are often contaminants (such as specific chemicals), but any threat to environmental quality, such as litter along the shoreline or excessive sediment, can be classified as a potentially influential factor. The relative vulnerability of different coastal environments along the Egyptian Red Sea coast to long-term sea-level rise were quantified based on variables including coastal geomorphology, regional coastal slope, rate of sea-level rise, wave and tide characteristics and historical shoreline change rates. The rankings for each of the six variables at any particular location can be averaged to produce an overall coastal vulnerability index from very low to very high (Figure 27.7). This index provides an overview of how different regions of the coast are likely to change in response to erosion and deterioration related to flood hazards.

The high sensitivity of the Red Sea coastal region to climate change, extreme events and anthropogenic activities requires the establishment of an environmental emergency plan, including warning systems, monitoring fluctuations of the sea level and water temperature, guides and instructions for local people and public awareness about land use and the expected impacted sites. Therefore, the future plan for protecting the Red Sea ecosystems should promote continuous coastal studies and monitoring of the region. Coastal protection measures should also be implemented in a provisional manner to protect the environment and adapt to climate change and the development of the region.

27.6 CONCLUSION

Understanding global changes, anthropogenic impacts and potential future risks in the Red Sea coastal region allows for more informed decision making and, ultimately, safer and more resilient development. This chapter discusses the importance of the Red Sea ecosystem for sustainable development. A detailed flood hazard map is given to show safe areas for settlement. Based on this map, five zones of hazard are estimated: very low, low, moderate, high and very high risk that cover 17.42, 26.79, 29.19, 14.74 and 8.6% of the total area, respectively.

Sea-level rise along the Red Sea coast of Egypt would cause significant coastal retreat due to wave erosion, terrestrial processes (e.g., rockfalls, floods) and anthropogenic activities. Areas of lower topography retreat more than resistant bedrock cliffs and headlands. Coastal sand dunes and beaches, which consist primarily of unconsolidated sand, provide little resistance to wave and tidal attack, especially at times of elevated sea level. The future retreat of beaches will depend on the rate of sea-level rise and, to a lesser extent, the amount of sediment input and loss.

Some of the coastal damage expected from sea-level rise may be mitigated in some areas by mangrove vegetation through their high roughness, stem height and density, which reduce wave height and energy. An adequate supply of sediment and/or organic matter accumulation produced by flash floods supports mangrove stands against the risk of high waves and protects inland areas from inundation. Mangroves may migrate inland if the area is unobstructed, but unless they maintain their elevation under sea-level rise, the vegetation will be limited by the slope of the land surface and the tidal range. A detailed assessment of the response of mangroves to sea-level rise along the Red Sea coast is required. The assessment depends on the geological, hydrological and biological factors and comparable long-term data.

This study is a guide for coastal planners and conservation managers as well as official agencies in order to encourage a comprehensive plan for coastal zone management. Local coral bleaching monitoring and prediction programmes should be established. The programmes should locate potential stressors, such as warm water, as well as determine the coral's reaction to those stressors. Future studies should determine the areas at greatest risk and map the best areas for regeneration of coral reefs in terms of physical conditions.

In summary, we recommend 1) continuous coastal studies and monitoring of the region to protect the Red Sea ecosystems; and 2) building dams at the mouths of major wadis to manage flood hazards to avoid loss of a sizable proportion of the coast and settlement/tourism areas and human activities.

ACKNOWLEDGEMENT

The authors would like to extend their sincere appreciation to the Academy of Scientific Research and Technology (ASRT) for support through project number 19476.

REFERENCES

Abdel-Hamid, A., Dubovyk, O., Abou El-Magd, I., and G. Menz. 2018. Mapping mangroves extents on the Red Sea coastline in Egypt using polarimetric SAR and high resolution optical remote sensing data. *Sustainability* 10, no. 3: 646. http://doi.org/10.3390/su10030646.

Abdelkader, A.I., Abuelregal, M., El-Metwally, M.E, Hassaan, M.A., and E.F. Sanad. 2018. Ecological risk assessment of heavy metals in Hurghada coastal sediment, Red Sea, Egypt. *Pollution* 4, no. 4: 759–774. http://doi.org/10.22059/poll.2018.253426.444.

Abdelkareem, M. 2017. Targeting flash flood potential areas using remotely sensed data and GIS techniques. *Nat Hazards* 85: 19–37.

Abdelkareem, M., and F. El-Baz. 2015. Regional view of a Trans-African drainage system. *J Adv Res* 6, no. 3: 433–439.

Abdelkareem, M., and F. El-Baz. 2016. Mode of formation of the Nile Gorge in northern Egypt by DEM-SRTM data and GIS analysis. *Geol J* 51, no. 5: 760–778.

Abdel-Maksoud, M. 1997. *The Accounting Measurement of Removal Costs of Marine Pollution by Petroleum Resulting from Ships Under Arm's Length Standard.* Suez Canal University, Report.

Aerts, J.C.J.H., Major, D.C., Bowman, M.J., Dircke, P., Marfai, M.A., Abidin, H.Z., Ward, P.J., Botzen, W., Bannink, B.A., Nickson, A., and T. Reeder. 2009. *Connecting Delta Cities. Coastal Adaptation, Flood Risk Management and Adaptation to Climate Change.* Amsterdam: VU University Press.

Archibald, S., and R.J. Scholes. 2007. Leaf green-up in a semi-arid African savanna – separating tree and grass responses to environmental cues. *J Veg Sci* 181: 583–594.

Attia, O.E.A., Khadra, A.M.A., Nawwar, A.H., and G.E. Radwan. 2010. Impacts of human activities on the sedimentological and geochemical characteristics of Mabahiss Bay, North Hurghada, Red Sea, Egypt. *Arab J Geosci* 5, no. 3: 481–499. http://doi.org/10.1007/s12517-010-0193-3.

Barbier, E.B., Hacker, S.D., Kennedy, C., Koch, E.W., Stier, A.C., and S.R. Silliman. 2011. The value of estuarine and coastal ecosystem services. *Ecol Monogr* 81, no. 2: 169–193. http://doi.org/10.1890/10-1510.1.

Blanchon, P., Eisenhauer, A., Fietzke, J., and V. Liebetrau. 2009. Rapid sea-level rise and reef back-stepping at the close of the last interglacial highstand. *Nature* 458: 881–884.

Bouwer, L.M., Crompton, R.P., Faust, E., Höppe, P., and R.A. Pielke. 2007. Confronting disaster losses. *Science* 318: 753.

Brooks, H.E., and D.J. Stensrud. 2000. Climatology of heavy rain events in the United States from hourly precipitation observations. *Mon Weather Rev* 128: 1194–1201.

Bui, D.T., Tsangaratos, P., Ngo, P.-T.T., Pham, T.D., and B.T. Pham. 2019. Flash flood susceptibility modeling using an optimized fuzzy rule based feature selection technique and tree based ensemble methods. *Sci Total Environ* 668: 1038–1054.

Burke, L., Reytar, K., Spalding, M., and A. Perry. 2011. *Reefs at Risk Revisited.* Washington, DC: World Resources Institute.

Cantin, N.E., Cohen, A.L., Karnauskas, K.B., Tarrant, A.M., and D.C. McCorkle. 2010. Ocean warming slows coral growth in the central Red Sea. *Science* 329, no. 5989: 322–325.

Cao, C., Xu, P., Wang, Y., Chen, J., Zheng, L., and C. Niu. 2016. Flash flood hazard susceptibility mapping using frequency ratio and statistical index methods in coal mine subsidence areas. *Sustainability* 8, no. 9: 1–18. http://doi.org/10.3390/su8090948.

Churchill, J., Lentz, S., Farrar, J., and Y. Abualnaja. 2014. Properties of Red Sea coastal currents. *Cont Shelf Res* 78: 51–61.

Cunha, J., Cardona, F.S., Bio, A., and S. Ramos. 2021. Importance of protection service against erosion and storm events provided by coastal ecosystems under climate change scenarios. *Front Mar Sci* 8: 726145. http://doi.org/10.3389/fmars.2021.726145.

Dai, A. 2011. Drought under global warming: A review. *WIREs Clim Change* 2, no. 1: 45–65.

Das, S. 2018. Geographic information system and AHP-based flood hazard zonation of Vaitarna basin, Maharashtra, India. *Arab J Geosci* 11, no. 19: 576. http://doi.org/10.1007/s12517-018-3933-4.

David, A.C.F. 1987. Red Sea and Gulf of Aden Pilot: Suez Canal, Gulf of Suez and Gulf of Aqaba, Red Sea, Gulf of Aden, southeast coast of Arabia from Ra's Ba̅ Ghashwah to Ra'sJunayz, coast of Africa from Raas Caseyr to Raas Binna, Suquṭrá and adjacent islands. UK Hydrographic Office, *Hydrographer of the Navy.* In English – 12 ed., 1987, rev. 313 p.

DeFur, P.L., Crane, M., Ingershold, C., and L. Tattersfield. 1999. *Endocrine Disruption in Invertebrates: Endocrinology, Testing and Assessment.* Pensacola, FL: Soc Environ Toxicology and Chemistry, 303 p.

Eladawy, A., Negm, A.M., and K. Nadaoka. 2015. Environmental characteristics and future challenges of the Egyptian-side of Red Sea: State of the art review. *Eighteenth International Water Technology Conference*, IWTC18 Sharm El Sheikh, 12–14 March 2015. http://doi.org/10.13140/RG.2.1.2458.1287.

El-Askary, H., Abd El-Mawla, S.H., Li, J., El-Hattab, M.M., and M. El-Raey. 2014. Change detection of coral reef habitat using LANDSAT-5 TM, Landsat 7 ETM+ and Landsat 8 OLI data in the Red Sea (Hurghada, Egypt). *Int J Remote Sens* 35: 2327–2346. http://doi.org/10.1080/01431161.2014.894656.

El Asmar, H.M., Ahmed, M.H., El Kafrawy, S.B., Outbids-Allah, M.T.A., and M.A. Khaled. 2015. Monitoring and assessing the coastal ecosystem at Hurghada, Red Sea coast, Egypt. *J Environ Earth Sci* 5, no. 6: 144–160.

Environmental Affairs Agency (EEAA). 2008. Strategy for implementing ICZM process in Egypt. *Draft Report 2.* Cairo, Egypt.

Essa, K.S.M., and F. Mubarak. 2006. Survey and assessment of wind-speed and wind power in Egypt, including air density variation. *Wind Eng* 30, no. 2: 95–106. http://doi.org/10.1260/030952406778055081.

Field, C.D. 1995. Impact of expected climate change on mangroves. *Hydrobiologia* 295: 75–81.

Foody, G. 2003. Remote sensing of tropical forest environments: Towards the monitoring of environmental resources for sustainable development. *Int J Remote Sens* 24: 4035–4046.

Frihy, O., Hassan, A., El Sayed, W., Iskandera, M., and M. Sherif. 2006. A review of methods for constructing coastal recreational facilities in Egypt (Red Sea). *Ecol Eng* 27: 1–12.

Furby, K.A., Bouwmeester, J., and M. Berumen. 2013. Susceptibility of central Red Sea corals during a major bleaching event. *Coral Reefs* 32: 505–513. http://doi.org/10.1007/s00338-012-0998-5.

Genevier, L.G.C., Jamil, T., Raitsos, D.E., Krokos, G., and I. Hoteit. 2019. Marine heatwaves reveal coral reef zones susceptible to bleaching in the Red Sea. *Glob Change Biol* 25: 2338–2351. http://doi.org/10.1111/gcb.14652.

GERICS. 2019. *Climate Fact Sheet – Egypt.* www.climate-service-center.de/products_and_publications/fact_sheets/climate_fact_sheets/index.php.en.

Gilman, E.L., Ellison, J., Duke, N.C., and C.B. Field. 2008. Threats to mangroves from climate change and adaptation options: A review. *Aquat Bot* 89: 237–250.

Hanafy, M., and M. Ismail. 2014. *Climate Change and Its Effect on the Egyptian Coast of the Red Sea: The First Recorded Coral Mass Bleaching Event.* Hurghada Environmental Protection and Conservation Assoc (HEPCA), Egypt.

Hereher, M.E. 2015. Assessment of Egypt's Red Sea coastal sensitivity to climate change. *Environ Earth Sci* 74: 2831–2843.

Herold, M., Goldstein, N., and K. Clarke. 2003. The spatiotemporal form of urban growth: Measurement, analysis and modelling. *Remote Sens Environ* 86: 286–302.

Herold, M., Scepan, J., and K.C. Clarke. 2002. The use of remote sensing and landscape metrics to describe structures and changes in urban land uses. *Environ Plan* 34: 1443–1458.

Hong, Y., and M. Abdelkareem. 2022. Integration of remote sensing and a GIS-based method for revealing prone areas to flood hazards and predicting optimum areas of groundwater resources. *Arab J Geosci* 15, no. 1: 1–14.

IPCC. 2007a. Climate Change 2007: impacts, adaptation and vulnerability. *Contribution of Working Group II to the Fourth Assessment Report of the Intergovernmental Panel on Climate Change*. Cambridge: Cambridge University Press.

IPCC. 2007b. Climate change 2007: the physical science basis. *Contribution of Working Group I to the Fourth Assessment Report of the Intergovernmental Panel on Climate Change*. Cambridge: Cambridge University Press.

IPCC. 2012. Managing the risks of extreme events and disasters to advance climate change adaptation. In: *A Special Report of Working Groups I and II of the Intergovernmental Panel on Climate Change*. Eds. C.B. Field, V. Barros, T.F. Stocker, et al. Cambridge and New York: Cambridge University Press.

IPCC. 2021. Climate Change 2021, the physical science basis, summary for policymakers. In: *Working Group I Contribution to the Sixth Assessment Report of the Intergovernmental Panel on Climate Change*. Cambridge: Cambridge University Press. http://doi.org/10.1017/9781009157896.001.

Khaled, M. 2019. Using Landsat data to assess the status of coral reefs cover along the Red Sea Coast, Egypt. *Int J Ecotoxicol Ecobiol* 4: 17–31.

Kleinhaus, K., Al-Sawalmih, A., Barshis, D.J., Genin, A., Grace, L.N., Hoegh-Guldberg, O., Loya, Y., Meibom, A., Osman, E.O., Ruch, J-D., Shaked, Y., Voolstra, C.R., Zvuloni, A., and M. Fine. 2020. Science, diplomacy, and the Red Sea's unique coral reef: It's time for action. *Front Mar Sci* 7: 90. http://doi.org/10.3389/fmars.2020.00090.

Kotb, M., Abdulaziz, M., Al-Agwan, Z., Alshaikh, K., Al-Yami, H., Banajah, A., Devantier, L., Eisinger, M., Eltayeb, M., Hassan, M., et al. 2004. Status of coral reefs in the Red Sea and Gulf of Aden in 2004. In: *Status of Coral Reefs of the World, 2004*. Ed. C. Wilkinson, 137–151. Australia: Global Coral Reef Monitoring Network and Australian Institute of Marine Science , Townsville, Queensland. http://hdl.handle.net/1834/389

Lu, D., Mausel, P., Brondizio, E., and E. Moran. 2004. Change detection techniques. *Int J Remote Sens* 25: 2365–2407.

Mackenzie, R., and D. Friess. 2015. Mangrove and sea-level rise (PowerPoint presentation). In: *SWAMP Toolbox: Theme E Section E1*. www.cifor.org/SWAMP-toolbox. Accessed 15 August 2022.

Madkour, H., Mansour, A., Ahmed, A., and A. El-Taher. 2013. Environmental texture and geochemistry of the sediments of a subtropical mangrove ecosystem and surrounding areas, Red Sea Coast, Egypt. *Arab J Geosci* 7, no. 9. http://doi.org/10.1007/s12517-013-0977-3.

Mansour, A.M. 1995. Sedimentary facies and carbonate – siliciclastic transition of Sharm El Bahari and Sharm El Qibli, Red Sea, Egypt. *Egypt J Geol* 39, no. 1: 57–76.

Mansour, A.M. 2000. Quaternary reef terraces and their relation to tectonics/eustatics of the Red Sea region, Egypt. *Sedimentology of Egypt* 8: 19–33.

Mansour, A.M., and M.S. Abd El-Sadek. 2021. Risk assessment of climate change on the coastal area of Quseir, Red Sea, Egypt. In: *Climate Change Management Through Adaptation and Mitigation*. Eds. R. Privitera, D. La Rosa, V. Pappalardo and F. Martinico, 64–69. Maggioli Editore March 2021. Italy: Open Access Creative Commons License CC BY-NC-ND 4.0 International Attribution. Maggioli Editore, Santarcangelo di Romagna, Italy.

Mansour, A.M., Askalany, M.S., Madkour, H.A., and B.B. Assran. 2013. Assessment and comparison of heavy-metal concentrations in marine sediments in view of tourism activities in Hurghada area, northern Red Sea, Egypt. *Egypt J Aquat Res* 39: 91–103.

Mansour, A.M., Nawar, A.H., and H.A. Madkour. 2011. Metal pollution in marine sediments of selected harbours and industrial areas along the Red Sea coast of Egypt. *Ann Naturhist Mus Wien, Serie A* 113: 225–244.

Mashaly, J., and E. Ghoneim. 2018. Flash flood hazard using optical, radar, and stereo-pair derived DEM: Eastern Desert, Egypt. *Remote Sens* 10, no. 8: 1204. http://doi.org/10.3390/rs10081204.

Monroe, A.A., Ziegler, M., Roik, A., Röthig, T., Hardenstine, R.S., Emms, M.A., and M.L. Berumen. 2018. In situ observations of coral bleaching in the central Saudi Arabian Red Sea during the 2015/2016 global coral bleaching event. *PLoS ONE* 13: e0195814. http://doi.org/10.1371/journal.pone.0195814.

Moore, I.D., and G.J. Burch. 1986. Sediment transport capacity of sheet and rill flow: Application of unit stream power theory. *Water Res* 22: 1350–1360.

Moore, I.D., Grayson, R.B., and A.R. Ladson. 1991. Digital terrain modelling: A review of hydrological, geomorphological, and biological applications. *Hydrol Process* 5: 3–30.

Morcos, S.A. 1970. Physical and chemical oceanography of the Red Sea. *Oceanogr Mar Biol Ann Rev* 8: 73–202.

Morgan, P. 1990. Egypt in the tectonic framework of global tectonics. In: *The Geology of Egypt*. Ed. R. Said, 91–111. Rotterdam: Balkema.

Moussa, A.A., Moussa, K.A., and A.H. El-Mamony. 1986. Beach sediments and littoral processes along the Red Sea coast of Egypt. *Bull Inst Oceanogr Fish Egypt* 12: 301–313.

Müller, D., and M. Zeller. 2002. Land use dynamics in the central highlands of Vietnam: A spatial model combining village survey data with satellite imagery interpretation. *Agric Econ* 27: 333–354.

National Institute of Oceanography and Fisheries. 2019. www.niof.sci.eg/. Accessed 30 October 2019.

National Oceanic and Atmospheric Administration (NOAA). 2021. www.noaa.gov.

Nicholls, R.J., Hanson, S., Herweijer, C., Patmore, N., Hallegatte, S., Corfee-Morlot, J., Château, J., and R. Muir-Wood. 2008. Ranking port cities with high exposure and vulnerability to climate extremes: Exposure estimates. *OECD Environment Working Papers No. 1, ENV/WKP (2007)*. OECD, Paris, France. www.olis.oecd.org/olis/2007doc.nsf/linkto/env-wkp(2007)1. Accessed 25 December 2009.

O'Callaghan, J.F., and D.M. Mark. 1984. The extraction of drainage networks from digital elevation data. *Comput Vis Graph Image Process* 28: 323–3444.

Osman, E.O., Smith, D.J., Ziegler, M., Kürten, B., Conrad, C., El-Haddad, K.M., and D.J. Suggett. 2018. Thermal refugia against coral bleaching throughout the northern Red Sea. *Glob Change Biol* 24: e474–e484. http://doi.org/10.1111/gcb.13895.

Piller, W.E., and P. Pervesler. 1989. An Actuopalaeontological approach, the northern Bay of Safaga, (Red Sea, Egypt). *Beitr Paleont Oesterr Wien* 15: 103–147.

Ras Mohammed National Park. 2019. *Atlas of Marine Protection*. www.mpatlas.org/mpa/sites/758/. Accessed 22 October 2019.

Regmi, N.R., Giardino, J.R., and J.D. Vitek. 2010. Modelling susceptibility to landslides using the weight of evidence approach: Western Colorado, USA. *Geomorphology* 115: 172–187.

Serrano, O., Lovelock, C.E., Atwood, T.B., et al. 2019. Australian vegetated coastal ecosystems as global hotspots for climate change mitigation. *Nat Commun* 10: 4313. http://doi.org/10.1038/s41467-019-12176-8.

Sharma, T.P.P., Zhang, J., Raj Khanal, N., Prodhan, F.A., Nanzad, L., Zhang, D., and P. Nepal. 2021. A geomorphic approach for identifying flash flood potential areas in the East Rapti River Basin of Nepal. *ISPRS Int J Geo-Inf* 10: 247.

Spalding, M.D., Ravilious, C., and E.P. Green. 2001. *World Atlas of Coral Reefs*. Berkeley: University of California Press, 424 p.

Srivastava, P.K., Han, D., Rico-Ramirez, M.A., Bray, M., and T. Islam. 2012. Selection of classification techniques for land use/land cover change investigation. *Adv Space Res* 50, no. 9: 1250–126.

Steiner, F.M., Csősz, S., Markó, B., Gamisch, A., Rinnhofer, L., Folterbauer, C., Hammerle, S., Stauffer, C., Arthofer, W., and B.C. Schlick-Steiner. 2018. Turning one into five: Integrative taxonomy uncovers complex evolution of cryptic species in the harvesterant Messor "structor". *Mol Phylogenetics Evol* 127: 387–404.

Strahler, A.N. 1957. Quantitative analysis of watershed geomorphology. *Eos Trans Am Geophys Union* 38: 913–920.

Strahler, A.N. 1964. Quantitative geomorphology of drainage basins and channel networks. In: *Handbook of Applied Hydrology*. Ed. V.T. Chow, Section 4II. New York: McGraw Hill Book Company.

Tehrany, M.S., Pradhan, B., and M.N. Jebur. 2014. Flood susceptibility mapping using a novel ensemble weights-of-evidence and support vector machine models in GIS. *J Hydrol* 512: 332–343.

Tehrany, M.S., Pradhan, B., Mansor, S., Ahmad, N., et al. 2015. Flood susceptibility assessment using GIS-based support vector machine model with different kernel types. *Catena* 125: 91–101.

Thornton, P.K., Ericksen, P.J., Herrero, M., and A.J. Challinor. 2014. Climate variability and vulnerability to climate change: A review. *Glob Change Biol* 20: 3313–3328. http://doi.org/10.1111/gcb.12581.

USAID. 2018. *Climate Risk Profile – Egypt*. Fact Sheet. www.climatelinks.org/sites/default/files/asset/document/2018_USAID-ATLAS-Project_Climate-Risk-Profile-Egypt.pdf.

Waqas, H., Lu, L., Tariq, A., Li, Q., Baqa, M.F., Xing, J., and A. Sajjad. 2021. Flash flood susceptibility assessment and zonation using an integrating analytic hierarchy process and frequency ratio model for the Chitral District, Khyber Pakhtunkhwa, Pakistan. *Water* 13: 1650.

Webster, J.R., Clague, D.A., Riker-Coleman, K., Gallup, C., Braga, J.C., Potts, D., Moore, J.G., Winterer, E.L., and C.K. Paull. 2004. Drowning of the -150 m reef off Hawaii: A casualty of global meltwater pulse 1A? *Geology* 32, no. 3: 249–252. http://doi.org/10.1130/G20170.1.

World Bank Group. 2021. *Climate Risk Country Profile: EGYPT*. 23 p. https://climateknowledgeportal.worldbank.org. Accessed 6 September 2022.

Yang, J., Gong, P., Fu, R., Zhang, M., Chen, J., Liang, S., Xu, B., Shi, J., and R. Dickinson. 2013. The role of satellite remote sensing in climate change studies. *Nat Clim Change* 3: 875–883.

Yuan, F., Sawaya, K.E., Loeffelholz, B.C., and M.E. Bauer. 2005. Land cover classification and change analysis of the Twin Cities (Minnesota) Metropolitan Area by multi-temporal Landsat remote sensing. *Remote Sens Environ* 98: 317–328.

Zhu, Q., and M. Abdelkareem. 2021. Mapping groundwater potential zones using a knowledge-driven approach and GIS analysis. *Water* 13: 579.

Zittis, G., Almazroui, M., Alpert, P., Ciais, P., Cramer, W., Dahdal, Y., et al. 2022. Climate change and weather extremes in the Eastern Mediterranean and Middle East. *Rev Geophys* 60: e2021RG000762. http://doi.org/10.1029/2021RG000762.

PART 9

New developments

28 New research opportunities in extreme saline environments of the Arabian Gulf and the Red Sea

From biotechnology to astrobiology

Yang He, Júnia Schultz, Marta F. Simões, Vigneshwaran Krishnamoorthi, Alexandre S. Rosado, Dimitra Atri and André Antunes

28.1 INTRODUCTION

The Arabian Peninsula is located between 13° and 32°N and 35° and 60°E. Due to its low latitude, low annual precipitation and high evaporation rate, it has semi-arid to extreme hyper-arid conditions and includes some of the saltiest and driest locations on Earth (De Pauw 2002). The peninsula also hosts a minor tectonic plate (the Arabian Plate), whose tectonic divergence and down-faulting with the African Plate originated the Red Sea basin while convergence and subduction with the Asian Plate originated the Arabian Gulf basin (Vita-Finzi 2001). The plate is surrounded by the marginal seas of the Indian Ocean: Arabian Gulf, Gulf of Oman, Gulf of Aden and the Red Sea.

28.2 ENVIRONMENT AND ECOLOGY

28.2.1 Environment and ecology of sabkhas in the Arabian Peninsula

Salt flats or bodies of saline water are ubiquitous features in Earth's arid and semi-arid areas (Yechieli and Wood 2002). The term "sabkha" is adopted from the Arabic word for typical geological features that consist of salt-encrusted flat surfaces. While originally used only for the description of salt flats in coastal desert settings, the term is now also widely used to describe inland salt flats (Mountney 2005). Sabkhas are a common feature of the Arabian Peninsula, being present in the Red Sea area and, more abundantly, in the Arabian Gulf (Figure 28.1).

Sabkhas are evaporite deposits, hydrologically similar to pans, playas, salinas, saline lakes and salt flats, differing only in their boundary conditions (Yechieli and Wood 2002). Based on location, we can differentiate between 1) inland sabkhas, which are more mature; and 2) coastal sabkhas (or supratidal flats), which are more widespread (Ashour et al. 2013). The former are composed of continental or earlier marine sediments (with associated evaporites deriving from evaporated continental waters) while the latter are supratidal surfaces formed by depositional offlap of marine sediments (with associated evaporites deriving from seawater) (Bahafzullah et al. 1993). Coastal sabkhas can be further divided into open supratidal environments with an unrestricted and continuous hydrodynamic flow regime and sea marginal brine pans (pools) with a restricted and uneven marine water supply (Gavish 1980). In open supratidal sabkhas, brines form saline groundwater tables that extend continuously from the sea edge toward land. Brines get into the sediments by continuous subsurface seepage and occasional floods.

Besides evaporation, the most important physical processes affecting sabkhas include those operative under 1) marine-, 2) fluvial-lacustrine- and 3) eolian-dominated conditions. Dominance of one or more of these in the proper settings gives rise to marine coastal sabkhas, continental playas and interdune sabkhas (Handford 1981).

DOI: 10.1201/9781003321408-37

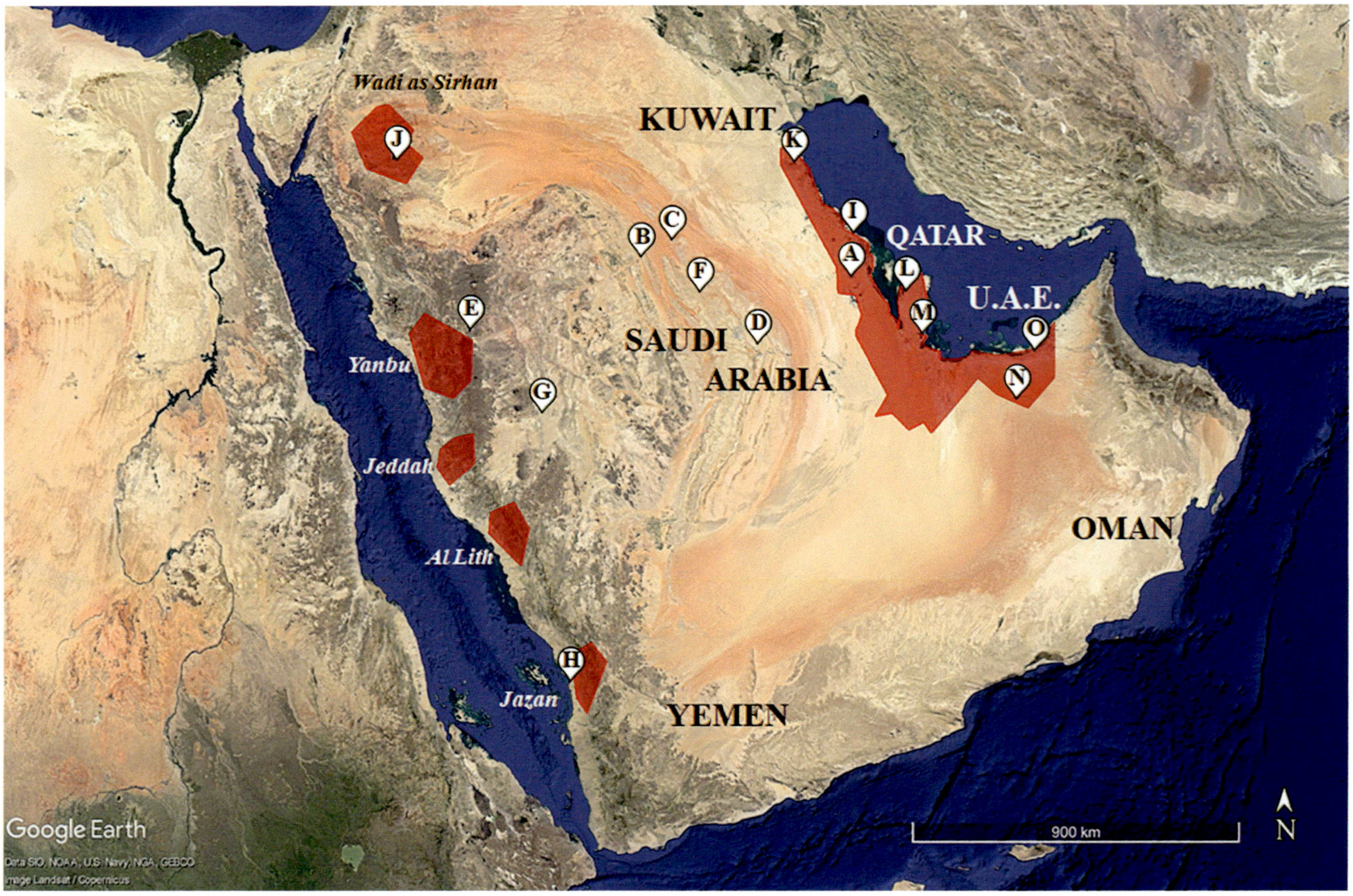

FIGURE 28.1 Distribution of main active sabkhas in the Arabian Peninsula (red areas), based on Al-Amoudi (2008). Locations highlighted in the current study: Saudi Arabia (A, Abqaiq; B, Al-Aushazia; C, Al-Kasr; D, Al-Kharj; E, Al-Madina; F, Al-Qasab; G, Al-Taif; H, Jazan; I, Jubail; J, Tabuk), Kuwait (K), Qatar (L, northwest; M, southeast), and United Arab Emirates (N, Abu Dhabi; O, Empty Quarter).

Source: Google Earth

28.2.2 Environment and ecology of the deep-sea brines of the Red Sea

The deep-sea brines of the Red Sea represent a unique type of environment, combining multiple extremes. The topographical depressions that host the brines formed due to the tectonic splitting of the Arabian and African plates, a process that also exposed evaporite deposits from the Miocene, which underlie the Red Sea basin. The exposure of these deposits and local hydrothermal activity resulting from the tectonic splitting allow seawater to be enriched in salt and heavy metals while suffering an increase in temperature and depletion of oxygen. The brines formed in this process accumulate in the deeps due to their high density and, due to their typical stability, usually form sharp transition zones with seawater with a drastic change in conditions (Antunes et al. 2011).

After the discovery of the first deep-sea brine in the 1950s, the following decades brought the discovery of multiple additional sites, with close to 30 deeps now identified (and even similar types of environments in the Mediterranean and the Gulf of Mexico; Antunes et al. 2011). Despite their similar origin, deep-sea brines of the Red Sea have diverse physical-chemical conditions, making each brine unique. Detailed information on the physical-chemical conditions of each brine can be found in previous reviews (e.g., Antunes et al. 2011, 2019).

28.3 MICROBIAL DIVERSITY

28.3.1 Microbial diversity in sabkhas of the Arabian Peninsula

Sabkhas are very complex ecosystems with high concentrations of salts and heavy metals, high light intensity, low nutrient levels and fluctuations in temperature, pH and water (Krumbein 1985). These environments are subject to an intricate interplay of physical, chemical and hydrodynamic factors, influencing the presence of organisms and their diversity.

Sabkhas of the Arabian Peninsula have been the target of a few studies looking into their microbial diversity, mostly centered on coastal sabkhas and microbial mats (Tables 28.1, 28.2 and 28.3). Representatives of Bacteria, Archaea and Eukarya can be found within the main layers of the mats. Typically, the upper layer is dominated by oxygenic photosynthetic microorganisms, whereas the deeper layers are characterized by anoxygenic sulfur bacteria, heterotrophs and sulfate-reducing bacteria (Al-Thani et al. 2014). Communities are dominated by bacteria, mostly associated with the Pseudomonadota, Bacillota and Actinomycetota (DiLoreto et al. 2019). There is limited data for the archaea, but there is a clear predominance of members of the *Halobacteria* (Table 28.2), explained by the high salinity values

TABLE 28.1 Reported bacterial diversity of sabkhas in the Arabian Peninsula.

TAXONOMY			*LOCATION**													
			SAUDI ARABIA							*OTHER COUNTRIES*						
PHYLUM:	*CLASS:*	*GENUS:*	*B*	*C*	*D*	*F*	*G*	*H*	*I*	*K*	*L*	*M*	*N*	*O*	*SAMPLE TYPE***	*REF*
Actinomycetota	Actinomycetes	*Actinopolyspora*						×							S3	1
		Rhodococcus								×					S1	2
	Rubrobacteria	*Rubrobacter*												×	S2	3
Bacteroidota	Cytophagia	*Flexibacter*										×			S2	4
		Hymenobacter										×			S2	4
		Marivirga												×	S4	3
	Flavobacteriia	*Lutibacter*										×			S2	4
		Salegentibacter								×					S4	5
	Saprospiria	*Phaeodactylibacter*										×			S2	4
Balneolota	Balneolia	*Gracilimonas*												×	S4	3
Chloroflexota	Anaerolineae	*Thermomarinilinea*										×			S2	4
	Chloroflexia	*Heliothrix*										×			S2	4
		Oscillochloris										×			S2	4
Cyanobacteria	Cyanophyceae	*Aphanocapsa*											×		S2	6
		Aphanothece											×		S2	6
		Chroococcus											×		S2	6
		Chroococcidiopsis											×		S2	6
		Cyanothece												×	S1	3
		Entophysalis											×		S2	6
		"Euhalothece"												×	S1	3
		Gomphosphaeria											×		S2	6
		Halomicronema											×		S2	6
		Leptolyngbya											×		S2	6
		Lyngbya											×		S2	6
		Microcoleus											×		S2	6
		Oscillatoria											×		S2	6
		Phormidium											×		S2	6
		Rhabdoderma											×		S2	6
		Schizothrix											×		S2	6
		Spirulina											×		S2	6
Bacillota	Bacilli	*Bacillus*	×	×	×	×	×	×	×	×	×				S1	2; 5; 7-8
		Exiguobacterium								×					S1	2
		Halobacillus						×		×					S1	1; 5
		Lactobacillus	×	×	×	×	×		×						S5	8
		Oceanobacillus								×					S1	5
		Planococcus								×					S3	5
		Pontibacillus								×					S4	5
		Staphylococcus									×				S1	7
		Virgibacillus						×			×				S1	1; 7
	Clostridia	*Moorella*										×			S2	4
		Thermaerobacter												×	S2	3
Pseudomonadota	Alphaproteobacteria	*Afifella*										×			S2	4
		Azospirillum												×	S2	3
		Bradyrhizobium										×			S2	4
		Erythrobacter												×	S1	3
		Limimonas												×	S2	3
		Ochrobactrum								×					S5	2
		Parvibaculum										×			S2	4
		Rhodoligotrophos										×			S2	4
		Rhodovibrio										×			S2	4
		Sulfitobacter								×					S4	5

(Continued)

TABLE 28.1 (Continued)

TAXONOMY			*LOCATION**													
			SAUDI ARABIA							*OTHER COUNTRIES*						
PHYLUM:	*CLASS:*	*GENUS:*	*B*	*C*	*D*	*F*	*G*	*H*	*I*	*K*	*L*	*M*	*N*	*O*	*SAMPLE TYPE***	*REF*
	Gammaproteobacteria	*Aquisalimonas*												×	S1	3
		Arhodomonas						×						×	S1	1;3
		Chromohalobacter								×					S1	2
		Halomonas						×		×		×		×	S1	1-5
		Halospina												×	S4	3
		Idiomarina								×					S1	5
		Marinobacter								×					S1	2; 5
		Methylohalomonas												×	S4	3
		Pseudoalteromonas								×					S6	5
		Pseudomonas								×		×			S1	2; 4
		Salinivibrio								×	×				S1	2; 7
		Stenotrophomonas								×					S1	2
		Tamilnaduibacter						×							S3	1
		Thioalkalivibrio												×	S1	3
		Thiohalorhabdus												×	S2	3
		Thiothrix										×			S2	4
	Deltaproteobacteria	*Desulfococcus*										×			S2	4
		Desulfohalobium										×			S2	4
		Desulfomonile										×			S2	4
		Desulfovibrio											×		S2	9
Rhodotherota	Rhodothermia	*Salinibacter*												×	S1	3
Spirochaetota	Spirochaetia	*Spirochaeta*										×			S2	4
Verrucomicrobiota	Opitutae	*Coraliomargarita*												×	S4	3

*Location: Saudi Arabia (B, Al-Aushazia; C, Al-Kasr; D, Al-Kharj; F, Al-Qasab; G, Al-Taif; H, Jazan; I, Jubail; J, Tabuk), Kuwait (K), Qatar (L, northwest; M, southeast), and United Arab Emirates (N, Abu Dhabi; O, Empty Quarter);
**Sample type: S1 = mixed samples, S2 = mat, S3 = salt, S4 = sediments, S5 = soil, S6 = water; Ref= Reference (1. Albokari et al., 2017; 2. Al-Mailem et al., 2014; 3. McKay et al., 2016; 4. DiLoreto et al., 2019; 5. Alagarsamy et al., 2021; 6. Abed et al., 2008; 7.Al Disi et al., 2017; 8. Alotaibi et al., 2020; 9. Bontognali et al., 2012).

in these environments. Eukaryotes also tend to be understudied but are important components of the microbial communities of sabkhas. Here, we observe a prevalence of fungi of the phylum Ascomycota (particularly for the genera *Alternaria*, *Aspergillus*, *Chaetomium* and *Fusarium*) (Table 28.3).

28.3.2 Microbial diversity in deep-sea brines of the Red Sea

While the first studies on the microbiology of the deep-sea brines of the Red Sea relied on traditional cultivation methods and led researchers to believe that they were sterile (Watson and Waterbury 1969), later studies using more up-to-date approaches quickly changed this perspective. The first molecular-based studies on these sites uncovered unexpectedly high diversities, with the discovery of several new, deeply-branching sequence groups (Eder et al. 1999, 2002). Surprisingly, some of these new phylogenetic groups had a widespread distribution, being detected in multiple deep-sea brines, despite variations in their conditions and location (including detection in similar environments outside the Red Sea) (Antunes et al. 2011) while others had a much narrower distribution. Both these cases point to an adaptation to this specific niche environment. Molecular-based studies also revealed that the microbial communities present in the brine-seawater interface were stratified (Antunes et al. 2019). Further exploration of metagenomic datasets later allowed us to obtain the first glimpse into the viral communities present at some of these sites, which revealed that these mirrored microbial communities in their stratification (Antunes et al. 2015).

Cultivation-based approaches have received less attention but have been moderately successful in isolating and describing several new taxa from the deeps. These included a relatively high percentage of representatives of new, higher-ranked taxa (orders and phyla), as well as a few genera and several new species (Table 28.4 and references cited therein). The description of these new taxa is centered mostly on samples from the brine-seawater interface and a relatively limited number of less extreme deep-sea brines. However, the descriptions of some of the most unusual microbes and novel higher-ranked taxa are associated with either sediment samples or more extreme conditions. Further efforts are clearly necessary, particularly as the isolation of new strains is essential for downstream applications (Antunes et al. 2016).

TABLE 28.2 Reported archaeal diversity of sabkhas in the Arabian Peninsula.

TAXONOMY			*LOCATION**			*SAMPLE TYPE***	*REF*
			SAUDI ARABIA	*KUWAIT*	*QATAR*		
PHYLUM:	*CLASS:*	*GENUS:*	*H*	*K*	*M*		
Euryarchaeota	Halobacteria	*"Candidatus* Nitrosocaldus"			×	S2	1
		Haloarcula	×			S3	2
		Halobacterium	×	×		S1	2-3
		Halobaculum	×			S3	2
		Halobellus	×			S3	
		Haloferax	×	×		S1	2-3
		Halogeometricum	×			S3	2
		Halomicroarcula	×			S3	
		"Halomicrobium"		×		S1	3
		Haloplanus	×			S3	2
		Halorhabdus	×			S3	
		Halorientalis		×		S1	3
		Halorientalis	×			S3	2
		Halorussus		×		S1	3
		Halosimplex	×			S3	2
		Natrinema	×			S3	
		Salarchaeum	×			S3	
		Salinigranum	×			S3	
		Salinirubrum	×			S3	
		Several undefined species			×	S2	1
	Methanomicrobia	Several undefined species			×	S2	
	Thermoplasmata	Several undefined species			×	S2	
Thermoproteota	Thermoprotei	*"Candidatus Crenarchaeum"*	×			S3	2

*Location: Saudi Arabia (H, Jazan), Kuwait (K), and Qatar (M, southeast); **Sample type: S1 = mixed samples, S2 = mat, S3 = salt; Ref= Reference (1. Al-Thani et al., 2014; 2. Albokari et al., 2017; 3. Al-Mailem et al., 2014).

28.4 BIOTECHNOLOGICAL APPLICATIONS

Humankind has benefited from halophilic microbes for over 5,000 years, namely by using them for increased salt crystallization (due to the red colouration seen in many salterns and associated with high densities of halophiles) and for the production of sauces and fermented foods (e.g., Antunes et al. 2017). However, the last decades have brought an exponential increase in applications of halophiles, which now extend to a wide range of fields (Antunes et al. 2017). Pigments were one of the first indirect applications of halophiles and are still in high demand. They are currently used as natural food colourants and antioxidants, as additives in cosmetics and in multivitamin and health food products (nutraceuticals) (Margesin and Schinner 2001).

Enzymes from extremophiles (called extremozymes) are the most common biotechnological applications of these organisms, being relevant and valuable products for a myriad of industrial applications. This is due to their innate stability and activity under hostile conditions, similar to those found in many industrial processes (Dumorné et al. 2017). A more recent focus for bioprospection of extremophiles is the search for bioactive molecules and antimicrobial or antitumor compounds. For example, microbes from extreme marine environments are considered a new frontier for natural product and drug discovery (Rotter et al. 2021).

28.4.1 Bioprospection and biotechnology of sabkhas on the Arabian Peninsula

Sabkhas on the Arabian Peninsula are a promising source of microbes for different applications but remain vastly underexplored. Most bioprospection studies are narrowly focused, looking at a specific application or a limited number of sites, yet they have some relevant results. Next, we highlight some examples of such studies, yet we must note that bioprospection of sabkhas elsewhere points to an even wider potential. Such studies have frequently identified similar taxa but tested and confirmed a wider range of applications (Selama et al. 2014).

Regarding pigment production, carotenoid-producing microbes were reported in cyanobacterial mats in the Khor Al-Adaid sabkha, Qatar (Paulo and Dittrich 2013; Al-Thani and Yasseen 2021), although they seem not to have been subjected to further testing for biotechnological purposes.

TABLE 28.3 Reported fungal diversity of sabkhas in the Arabian Peninsula.

TAXONOMY			*LOCATION**										*SAMPLE TYPE***	*REF*
			SAUDI ARABIA									*KUWAIT*		
PHYLUM:	*CLASS:*	*GENUS:*	*A*	*B*	*C*	*D*	*E*	*F*	*G*	*I*	*J*	*K*		
Ascomycota	Dothideomycetes	*Alternaria*		×			×	×	×	×	×	×	S1	1-2
		Ascochyta										×	S1	1
		Aureobasidium			×							×	S5	1-2
		Cladosporium			×					×		×	S1	
		Cochliobolus		×							×	×	S1	
		Drechslera										×	S1	1
		Embellisia								×		×	S1	1-2
		Epicoccum										×	S1	1
		Monodictys										×	S6	
		Paraphaeosphaeria						×					S5	2
		Phoma						×				×	S1	1-2
		Pleosporales				×							S5	2
		Preussia						×					S5	
		Sirodesmium						×					S5	
		Sporomia						×					S5	
		Sporormiella										×	S1	1
		Ulocladium									×	×	S1	1-2
	Eurotiomycetes	*Amauroascopsis*										×	S1	1
		Aspergillus		×				×	×	×		×	S5	1-2
		Emericella								×			S5	
		Eurotium										×	S1	1
		Gymnoascus		×				×				×	S1	1-2
		Lithothelium						×					S5	2
		Penicillium						×		×		×	S1	1-2
	Leotiomycetes	*Botrytis*										×	S1	1
	Pezizomycetes	*Ascobolus*						×					S5	2
	Sordariomycetes	*Acremonium*										×	S1	1
		Chaetomium			×	×		×	×		×	×	S1	1-2
		Fusarium		×		×			×	×	×	×	S5	
		Hypocreales						×					S5	2
		Madurella						×					S5	
		Microascus							×				S5	
		Myceliophthora						×					S5	
		Nigrospora										×	S5	1
		Stachybotrys										×	S5	
		Thielavia	×									×	S1	1-2
		Trichoderma										×	S6	1
Glomeromycota		*Claroideoglomus*		×									S5	3
		Funneliformis		×									S5	
		Gigaspora		×									S5	
		Scutellospora		×									S5	
Mucoromycota	Mucoromycetes	*Mucor*							×				S5	2

*Location: Saudi Arabia (A, Abqaiq; B, Al-Aushazia; C, Al-Kasr; D, Al-Kharj; E, Al-Madina; F, Al-Qasab; G, Al-Taif; I, Jubail; J, Tabuk) and Kuwait (K, Ahmadi); **Sample type: S1 = mixed samples, S5 = soil, S6 = water; Ref= Reference (1. Al-Musallam et al., 2011; 2. Alotaibi et al., 2020; 3. Malik et al., 2022).

Enzyme bioprospection has also been sparsely explored, although hydrolytic activity has been reported in many strains from sabkhas in Kuwait (Al-Zour and Mina Abdulla). These included strains of the genera *Halobacillus*, *Bacillus*, *Planococcus* and *Idiomarina* with proteolytic activity and strains of the genera *Bacillus* and *Halobacillus* with amylolytic, DNase and lipolytic activity (Alagarsamy et al. 2021). The same study also found strains of the genera *Bacillus*, *Pontibacillus* and *Pseudoalteromonas* with antimicrobial activity and strains of *Halomonas shengliensis* and *Idiomarina piscisalsi*, which harboured genes coding for antitumor compounds.

TABLE 28.4 Overview of new taxa described from the deep-sea brines of the Red Sea.

DOMAIN	*CLASS*	*FAMILY*	*SPECIES*	*NOVELTY*	*ORIGIN**	*REF*
Bacteria	Alphaproteobacteria	*Propionibacteriaceae*	*Ponticoccus marisrubri*	new species	Erba (bsi)	1
		Rhodobacteraceae	*Ruegeria profundi*	new species	Erba (bsi)	2
			Ruegeria marisrubri	new species	Erba (bsi)	
	Gammaproteobacteria	*Alteromonadaceae*	*Marinobacter salsuginis*	new species	Shaban (bsi)	3
		Salinisphaeraceae	*Salinisphaera shabanensis*	new order	Shaban (bsi)	4
	Deferribacteres	*Deferribacteraceaee*	*Flexistipes sinusarabici*	new phylum	Atlantis II (bsi)	5
	N/A	*Haloplasmataceae*	*Haloplasma contractile*	new order/ phylum?	Shaban (sed)	6
Archaea	Halobacteria	*Haloarculaceae*	*Halorhabdus tiamatea*	new species	Shaban (sed)	7
		Haloferacaceae	*Haloferax profundi*	new species	Discovery (bsi)	8
			Haloferax marisrubri	new species	Discovery (bsi)	
			Haloprofundus marisrubri	new genus	Discovery (bsi)	9

*Origin: bsi = brine-seawater interface; sed = sediment; N/A = Not applied. Ref= Reference (1. Zhang et al. 2017b; 2. Zhang et al. 2017c; 3. Antunes et al. 2007; 4. Antunes et al. 2003; 5. Fiala et al. 1990; 6. Antunes et al. 2008a; 7. Antunes et al. 2008b; 8. Zhang et al. 2020; 9. Zhang et al. 2017a).

A stronger emphasis on the bioprospection of the sabkhas on the Arabian Peninsula has been placed on bioremediation. Hypersaline environments are frequently contaminated with petroleum and derived compounds (Fowler et al. 1993; Al-Mailem et al. 2013; Al-Mueini et al. 2007). Given that conventional processes have limitations at higher salinities, strategies to enhance bioremediation of such locations should make use of halophilic or halotolerant microbes capable of degrading such compounds (Fathepure 2014). Due to the high levels of local contamination observed as part of the first Gulf War, a few studies on sabkhas from Kuwait have focused on this application and targeted the isolation of strains with such capabilities. Identified strains were assigned to the archaeal genera *Halobacterium*, *Halococcus* and *Haloferax*, or the bacterial genera *Exiguobacterium*, *Halomonas*, *Marinobacter*, *Marinococcus* and *Salinococcus* (Al-Mailem et al. 2010, 2012, 2013).

A very different path for bioremediation is the use of microbially induced calcite precipitation (MICP), an effective and eco-friendly technology that can be applied for removal of heavy metals and radionuclides but is also used in bioconsolidation, biocement, CO_2 sequestration and a few other applications (Dhami et al. 2013). Sabkhas in the Arabian Peninsula are also a potential source of microbes with such capabilities. Some strains of the genera *Virgibacillus* and *Salinivibrio*, isolated in Dohat Faishakh (NW Qatar), precipitated carbonate minerals with varying magnesium content, including high magnesium calcites (Al Disi et al. 2017).

A less traditional bioproduct is squalene, an industrially important compound and a vital ingredient in skin cosmetics due to its photoprotective role. The controversial primary origin of this bioactive isoprenoid (liver oil of deep-sea sharks and whales) has increased the drive to find alternative microbial-based sources. A recent wider-scoped bioprospection survey in sabkhas on the coast of Kuwait reported the isolation of a few squalene-producing strains assigned to the species *Bacillus vietnamensis*, *Oceanobacillus iheyensis*, *Oceanobacillus picturae*, *Planococcus rifietoensis* and *Planococcus maritimus* (Alagarsamy et al. 2021).

28.4.2 Bioprospection and biotechnology of deep-sea brines of the Red Sea

Brine pools exhibit a unique combination of extreme conditions reflected in a myriad of unique microbes inhabiting these salty, anoxic and high-pressure environments (Antunes et al. 2011) and making them hotspots of extremophilic microbial life with numerous metabolic activities and applications. Despite their potential for significant novel discoveries, most microbes from these brines have never been described and/or explored for biotechnology.

The search for new extremozymes from the deep-sea brines of the Red Sea has been the most prolific, with a current total of 12 experimentally characterized enzymes from Atlantis II, Discovery and Kebrit. Most of these enzymes belong to 1) oxidoreductases, 2) transferases or 3) hydrolases, including aldehyde dehydrogenase, proteases, cellulases, esterases, ferredoxin oxidoreductase, agarase, amylases, κ-Carragenases, ketoreductases and cyclodextrin glycosyltransferase (Varrella et al. 2020; Renn et al. 2021). Grötzinger et al. (2018) reported two alcohol dehydrogenases (oxidoreductases) exhibiting thermostability and halophily, withstanding organic solvents and accepting primary long-chain and aromatic alcohols as substrates. Badiea et al. (2019) observed the production of a halophilic, thermophilic thioredoxin-reductase lacking selenocysteine, produced by a polyextremophilic bacteria isolated from Atlantis II. Also from this brine, a new DNA polymerase (transferase) was identified and characterized: it was active at 55°C, unusually tolerant to high salt and metal ion concentrations and had a unique ability to use Zn^{2+} as a cofactor (Takahashi et al. 2018). This team bioengineered the polymerase in *Thermococcus kodakarensis*, resulting in a patented halophilic chimera used for the development of new detection essays. Further microbial strains from Atlantis II produced esterases with high activity under a wide range of temperatures (30–80°C) and high salinities (2–4.5 M NaCl) while not being affected by heavy metals (Mohamed et al. 2013). Also, *Pontibacillus chungwhensis*, *Halomonas meridiana* and *Zunongwangia profunda* (commonly reported in the Red Sea brines) have been shown to produce new thermoactive and alkali-tolerant α-amylases.

Microbes from the Red Sea brines are also promising sources of bioactive compounds, as originally uncovered by Sagar et al. (2013a, 2013b) and Esau et al. (2019). These studies, focusing on strains from Atlantis II, Discovery, Erba, Kebrit and Nereus, showed high cytotoxic and apoptotic activity of most bacterial strain extracts against different human cancer cell lines, highlighting strains assigned to the genera *Halomonas*, *Sulfitobacter*, *Idiomaria*, *Virgibacillus*, *Sediminimonas*, *Staphylococcus*, *Chromohalobacter*, *Halobacillus*, *Marinobacter* and *Zunongwangua*. While analyzing metabolome studies of extremophiles from Atlantis II, Kebrit and Discovery, Ziko et al. (2019) also observed 2,751 specialized metabolism gene clusters and predicted several new compounds, including potential new antimicrobials.

A final, less explored aspect is the potential use of microbes from these deep-sea brines in the degradation of oil-derived compounds and bioremediation. Many strains isolated from different brines belong to genera reported as good candidates for bioremediation, being able to grow on different polyaromatic hydrocarbons sources, including naphthalene, dibenzothiophene, pyrene and phenanthrene (Budiyanto et al. 2018; Thompson et al. 2020). These genera include *Marinobacter*, *Alcanivorax*, *Pseudomonas*, *Vibrio*, *Arthrobacter*, *Ponticoccus*, *Pseudoalteromonas*, *Idiomarina*, *Halomonas*, *Ruegeria*, *Chromohalobacter*, *Thalassospira* and *Salinisphaera* (Antunes et al. 2003, 2007; Sagar et al. 2013a, 2013b; Zhang et al. 2017a, 2017b; Esau et al. 2019). This potential application is further supported by the observed presence of degradation pathways of 13 aromatic compounds in metagenomes from Atlantis II and Discovery (Wang et al. 2013).

28.5 ASTROBIOLOGICAL RELEVANCE

One of the main pillars of astrobiology research is the study of terrestrial analogue sites – locations that share some relevant similarities with other parts of the solar system. As outlined by Wu et al. (2022), the study of such sites is seen as vital for 1) studying the limits of life, 2) retrieving new microbes for astrobiological exposure experiments, 3) analyzing long-term preservation and viability of biomolecules and microbes, 4) life-detection technology development and testing for space missions and 5) defining and refining planetary protection measures.

In general, high salinity locations are becoming increasingly relevant terrestrial analogue sites, given that there is growing evidence that water bodies on Mars and on the exo-oceans of the icy moons of the outer solar system are associated with high salinity levels. Salt would play a vital role in ensuring the stability of liquid water at low temperatures and pressures. Hence, microbes living in high salinities are rightfully recognized as a rich source of information for astrobiology and of possible model organisms (e.g., Wu et al. 2022).

28.5.1 Astrobiological relevance of sabkhas of the Arabian Peninsula

Sabkhas are generally recognized as valuable terrestrial analogue sites, relevant for the study of evaporite-bearing Martian deposits and possible life on Mars (Stivaletta et al. 2009). Part of this link between Mars and sabkhas has been strengthened by the detection of sulfate-rich deposits on Mars, thought to have been formed under evaporitic settings analogous to those taking place in sabkhas (Gendrin et al. 2005). The study of sabkhas and the active role of their microbial communities in shaping surface geomorphologies have also been highlighted as relevant to searching for signs of life beyond our planet (e.g., Barbieri and Cavalazzi 2018).

Another relevant aspect is that salt crusts in sabkhas can trap or cover and protect biological material, eventually preserving microbes, biomarkers or other evidence of life (McKay et al. 2016; Barbieri and Cavalazzi 2018). Therefore, several studies have investigated this effect with the detection of preserved biomarkers, such as pigments and DNA molecules (in, e.g., endolithic communities in gypsum mounds of continental sabkhas in the Sahara Desert) (Stivaletta et al. 2010; Barbieri and Stivaletta 2011).

As pointed out by Edwards et al. (2006), the successful adoption of specific equipment for detection of extant life or biomolecular traces of extinct life on missions to Mars and beyond depends on the interpretation of data from terrestrial Mars analogues such as sabkhas. These authors were the first to report on Raman spectroscopic biosignatures of cyanobacteria (collected from sabkhas near Abu Dhabi, UAE). Further studies were made with samples from Dhukan Sabkha in Qatar (Edwards et al. 2010) and a wider study with mixed samples from a coastal sabkha in the UAE analyzed visible, near-infrared, thermal infrared reflectance and thermal infrared transmission spectra, as well as XRD and SEM/EDX (Howari 2006).

28.5.2 Astrobiological relevance of the deep-sea brines of the Red Sea

The relevance of deep-sea brines for astrobiology has only recently been recognized but is a very promising research thread. These unusual brine pools have a unique combination of deep-sea conditions with high salinities and hydrothermal input, together with variability in local physical-chemical conditions, which provide some of the best candidates for terrestrial analogues for the exo-oceans found in the icy moons of Jupiter and Saturn (as proposed by Antunes et al. 2020).

Microbes from deep-sea brines have gained recent relevance as targets for astrobiology exposure experiments. Within all the new taxa isolated from the deep-sea brines of the Red Sea, *Salinisphaera shabanensis* seems to be the most widely used: it was one of the microbes studied as part of a launch and exposure to stratospheric conditions (Cortesão et al. 2021) and was also included in a set of strains of the Mars exposed extremophile mixture (MExEM) project that were approved to be tested in the International Space Station within the next two years. This species is recognized as a new model organism for astrobiology (Beblo-Vranesevic et al. 2022).

The exploration of the deep-sea brines under the scope of astrobiology might still be in its early days, but it is increasingly relevant (e.g., Antunes et al. 2020; Fisher et al. 2021; McClain et al. 2022). It is worth noting that they have been flagged as high-priority sites by the reports on planetary protection of the outer solar system (Rettberg et al. 2019), so further studies and developments are expected in the near future.

ACKNOWLEDGEMENTS

We acknowledge the support of the Science and Technology Development Fund and funding from the Macau University of Science and Technology and the State Key Laboratory of Lunar and Planetary Science (FRG, AA: grant No. FRG-22–079-LPS; MFS: grant No. FRG-22- RG-22–080-LPS; OPEN Project, DA and AA: ABAA Project).

REFERENCES

Abed, R.M., Kohls, K., Schoon, R., Scherf, A.K., Schacht, M., Palinska, K.A., Al-Hassani, H., Hamza, W., Rullkötter, J., and S. Golubic. 2008. Lipid biomarkers, pigments and cyanobacterial diversity of microbial mats across intertidal flats of the arid coast of the Arabian Gulf (Abu Dhabi, UAE). *FEMS Microbiol Ecol* 65: 449–462.

Alagarsamy, S., Habeebullah, S.F.K., and F. Al-Yamani. 2021. Bioprospecting potentials of moderately halophilic bacteria and the isolation of squalene producers from Kuwait sabkha. *Int Microbiol* 24: 373–384.

Al-Amoudi, O.S.B. 2008. Testing and stabilization of saline sabkha soils: A review. *Sixth Int Conf Case Histor Geotech Eng* 45.

Albokari, M., Cinar, S., and M. Mutlu. 2017. Microbial characterization of Jazan sabkha in Saudi Arabia. *Appl Ecol Environ Res* 15: 1069–1077.

Al Disi, Z.A., Jaoua, S., Bontognali, T.R., Attia, E.S., Al-Kuwari, H.A.A.S., and N. Zouari. 2017. Evidence of a role for aerobic bacteria in high magnesium carbonate formation in the evaporitic environment of Dohat Faishakh Sabkha in Qatar. *Front Env Sci-Switz* 5: 1.

Al-Mailem, D.M., Eliyas, M., Khanafer, M., and S. Radwan. 2014. Culture-dependent and culture-independent analysis of hydrocarbonoclastic microorganisms indigenous to hypersaline environments in Kuwait. *Microb Ecol* 67: 857–865.

Al-Mailem, D.M., Eliyas, M., and S.S. Radwan. 2012. Enhanced haloarchaeal oil removal in hypersaline environments via organic nitrogen fertilization and illumination. *Extremophiles* 16 :751–758.

Al-Mailem, D.M., Eliyas, M., and S.S. Radwan. 2013. Oil-bioremediation potential of two hydrocarbonoclastic, diazotrophic *Marinobacter* strains from hypersaline areas along the Arabian Gulf coasts. *Extremophiles* 17: 463–470.

Al-Mailem, D.M., Sorkhoh, N.A., Al-Awadhi, H., Eliyas, M., and S.S. Radwan. 2010. Biodegradation of crude oil and pure hydrocarbons by extreme halophilic archaea from hypersaline coasts of the Arabian Gulf. *Extremophiles* 14: 321–328.

Al-Mueini, R., Al-Dalali, M., Al-Amri, I.S., and H. Patzelt. 2007. Hydrocarbon degradation at high salinity by a novel extremely halophilic actinomycete. *Environ Chem* 4: 5–7.

Al-Musallam, A.A.S., Al-Sammar, H.F., and N.A. Al-Sané. 2011. Diversity and dominance of fungi inhabiting the sabkha area in Kuwait. *Bot Mar* 54: 83–94.

Alotaibi, M.O., Sonbol, H.S., Alwakeel, S.S., Suliman, R.S., Fodah, R.A., Jaffal, A.S.A., AlOthman, N.I., and A.E. Mohammed. 2020. Microbial diversity of some sabkha and desert sites in Saudi Arabia. *Saudi J Biol Sci* 27: 2778–2789.

Al-Thani, R., Al-Najjar, M.A., Al-Raei, A.M., Ferdelman, T., Thang, N.M., Al-Shaikh, I., Al-Ansi, M., and D. de Beer. 2014. Community structure and activity of a highly dynamic and nutrient-limited hypersaline microbial mat in Um Alhool Sabkha, Qatar. *PLoS ONE* 9: e92405.

Al-Thani, R.F., and B.T. Yasseen. 2021. Microbial ecology of Qatar, the Arabian Gulf: Possible roles of microorganisms. *Front Mar Sci* 8: 697269.

Antunes, A., Alam, I., Simões, M.F., Daniels, C., Siam, R., El-Dorry, H., and V.B. Bajic. 2015. First insights on the viral communities of the deep-sea anoxic brines of the Red Sea. *Genom Proteom Bioinform* 13: 304–309.

Antunes, A., Eder, W., Fareleira, P., Santos, H., and R. Huber. 2003. *Salinisphaera shabanensis* gen. nov., sp. nov., a novel, moderately halophilic bacterium from the brine-seawater interface of the Shaban Deep, Red Sea. *Extremophiles* 7: 29–34.

Antunes, A., França, L., Rainey, F.A., Huber, R., Nobre, M.F., Edwards, K.J., and M.S. da Costa. 2007. *Marinobacter salsuginis* sp. nov., isolated from the brine-seawater interface of the Shaban Deep, Red Sea. *Int J Syst Evol Microbiol* 57: 1035–1040.

Antunes, A., Kaartvedt, S., and M. Schmidt. 2019. Geochemistry and life at the interfaces of brine-filled deeps in the Red Sea. In: *Oceanographic and Biological Aspects of the Red Sea*. Eds. N.M.A. Rasul and I.C.F. Stewart, 185–194. Cham, Switzerland: Springer Nature.

Antunes, A., Ngugi, D.K., and U. Stingl. 2011. Microbiology of the Red Sea (and other) deep-sea anoxic brine lakes. *Environ Microbiol Rep* 3: 416–433.

Antunes, A., Olsson-Francis, K., and T.J. McGennity. 2020. Exploring deep-sea brines as potential terrestrial analogues of oceans in the icy moons of the outer solar system. *Curr Issues Mol Biol* 38: 123–162.

Antunes, A., Rainey, F., Wanner, G., Taborda, M., Pätzold, J., Nobre, M.F., da Costa, M.S., and R. Huber. 2008b. A new lineage of halophilic, wall-less, contractile bacteria from a brine-filled Deep of the Red Sea. *J Bacteriol* 190: 3580–3587.

Antunes, A., Simões, M.F., Grötzinger, S.W., Eppinger, J., Bragança, J., and V.B. Bajic. 2017. Bioprospecting archaea: Focus on extreme halophiles. In: *Bioprospecting*. Eds. R. Paterson and N. Lima, 81–112. Cham, Switzerland: Springer.

Antunes, A., Stackebrandt, E., and N. Lima. 2016. Fueling the bioeconomy: European culture collections and microbiology education and training. *Trends Microbiol* 24: 77–79.

Antunes, A., Taborda, M., Huber, R., Moissl, C., Nobre, M.F., and M.S. da Costa. 2008a. *Halorhabdus tiamatea* sp. nov., a non-pigmented, extremely halophilic archaeon from a deep-sea hypersaline anoxic basin of the Red Sea, and emended description of the genus *Halorhabdus*. *Int J Syst Evol Microbiol* 58: 215–220.

Ashour, M.M., Mogheeth, S.A., Metwali, A., Al-Ghazali, G., Ghafour, S.A., Shaksipi, R., and A. Ali. 2013. Sabkhas in Qatar peninsula. *Landsc Geodivers* 1: 10–35.

Badiea, E.A., Sayed, A.A., Maged, M., Fouad, W.M., Said, M.M., and A.Y. Esmat. 2019. A novel thermostable and halophilic thioredoxin reductase from the Red Sea Atlantis II hot brine pool. *PLoS ONE* 14: e0217565.

Bahafzullah, A., Fayed, L.A., Kazi, A., and M. Al-Saify. 1993. Classification and distribution of the Red Sea coastal sabkhas near Jeddah – Saudi Arabia. *Carbonate Evaporite* 8: 23–38.

Barbieri, R., and B. Cavalazzi. 2018. Microterracettes in Sabkha Oum Dba (Western Sahara, Morocco): Physical and biological interactions in the formation of a surface micromorphology. *Astrobiology* 18: 1351–1367.

Barbieri, R., and N. Stivaletta. 2011. Continental evaporites and the search for evidence of life on Mars. *Geol J* 46: 513–524.

Beblo-Vranesevic, K., Piepjohn, J., Antunes, A., and P. Rettberg. 2022. Surviving Mars: New insights into the persistence of facultative anaerobic microbes from analogue sites. *Int J Astrobiol* 21: 110–127.

Bontognali, T.R.R., Vasconcelos, C., Warthmann, R.J., Lundberg, R., and J.A. McKenzie. 2012. Dolomite-mediating bacterium isolated from the sabkha of Abu Dhabi (UAE). *Terra Nova* 24: 248–254.

Budiyanto, F., Thukair, A., Al-Momani, M., Musa, M.M., and A. Nzila. 2018. Characterization of halophilic bacteria capable of efficiently biodegrading the high-molecular-weight polycyclic aromatic hydrocarbon pyrene. *Environ Eng Sci* 35: 616–626.

Cortesão, M., Siems, K., Koch, S., Beblo-Vranesevic, K., Rabbow, E., Berger, T., Lane, M., James, L., Johnson, P., Waters, S.M., Verma, S.D., Smith, D.J., and R. Moeller. 2021. MARSBOx: Fungal and bacterial endurance from a balloon-flown analog mission in the stratosphere. *Front Microbiol* 12: 601713.

De Pauw, E. 2002. *An Agroecological Exploration of the Arabian Peninsula*. Eds. Icarda (International Center for Agricultural Research in the Dry Areas). Aleppo, Syria, 77 p.

Dhami, N.K., Reddy, M.S., and A. Mukherjee. 2013. Biomineralization of calcium carbonates and their engineered applications: A review. *Front Microbiol* 4: 314.

DiLoreto, Z.A., Bontognali, T.R., Al Disi, Z.A., Al-Kuwari, H.A.S., Williford, K.H., Strohmenger, C.J., Sadooni, F., Palermo, C., Rivers, J.M., McKenzie, J.A., Tuite, M., and M. Dittrich. 2019. Microbial community composition and dolomite formation in the hypersaline microbial mats of the Khor Al-Adaid sabkhas, Qatar. *Extremophiles* 23: 201–218.

Dumorné, K., Córdova, D.C., Eló, M.A., and P. Renganathan. 2017. Extremozymes: A potential source for industrial applications. *J Microbiol Biotechnol* 27: 649–659.

Eder, W., Ludwig, W., and R. Huber. 1999. Novel 16S rRNA gene sequences retrieved from highly saline brine sediments of Kebrit Deep, Red Sea. *Arch Microbiol* 172: 213–218.

Eder, W., Schmidt, M., Koch, M., Garbe-Schönberg, D., and R. Huber. 2002. Prokaryotic phylogenetic diversity and corresponding geochemical data of the brine-seawater interface of the Shaban Deep, Red Sea. *Environ Microbiol* 4: 758–763.

Edwards, H.G., Mohsin, M.A., Sadooni, F.N., Nik Hassan, N.F., and T. Munshi. 2006. Life in the sabkha: Raman spectroscopy of halotrophic extremophiles of relevance to planetary exploration. *Anal Bioanal Chem* 385: 46–56.

Edwards, H.G., Sadooni, F., Vítek, P., and J. Jehlička. 2010. Raman spectroscopy of the Dukhan sabkha: Identification of geological and biogeological molecules in an extreme environment. *Philos Trans Roy Soc A* 368: 3099–3107.

Esau, L., Zhang, G., Sagar, S., Stingl, U., Bajic, V.B., and M. Kaur. 2019. Mining the deep Red-Sea brine pool microbial community for anticancer therapeutics. *BMC Complement Altern Med* 19: 142.

Fathepure, B.Z. 2014. Recent studies in microbial degradation of petroleum hydrocarbons in hypersaline environments. *Front Microbiol* 5: 173.

Fiala, G., Woese, C.R., Langworthy, T.A., and K.O. Stetter. 1990. *Flexistipes sinusarabici*, a novel genus and species of eubacteria occurring in the Atlantis II Deep brines of the Red Sea. *Arch Microbiol* 154: 120–126.

Fisher, L.A., Pontefract, A., Som, S.M., Carr, C.E., Klempay, B., Schmidt, B.E., Bowman, J.S., and D.H. Bartlett. 2021. Current state of athalassohaline deep-sea hypersaline anoxic basin research – recommendations for future work and relevance to astrobiology. *Environ Microbiol* 23: 3360–3369.

Fowler, S.W., Readman, J.W., Oregioni, B.J.P.V., Villeneuve, J.P., and K. McKay. 1993. Petroleum hydrocarbons and trace metals in nearshore Gulf sediments and biota before and after the 1991 war: An assessment of temporal and spatial trends. *Mar Pollut Bull* 27: 171–182.

Gavish, E. 1980. Recent sabkhas marginal to the southern coasts of Sinai, Red Sea. In: *Developments in Sedimentology*. Ed. A. Nissenbaum, 233–251. Amsterdam: Elsevier.

Gendrin, A., Mangold, N., Bibring, J.P., Langevin, Y., Gondet, B., Poulet, F., Bonello, G., Quantin, C., Mustard, J., Arvidson, R., and S. LeMouélic. 2005. Sulfates in Martian layered terrains: The OMEGA/Mars Express view. *Science* 307: 1587–1591.

Grötzinger, S.W., Karan, R., Strillinger, E., Bader, S., Frank, A., Al Rowaihi, I.S., Akal, A., Wackerow, W., Archer, J.A., Rueping, M., Weuster-Botz, D., Groll, M., Eppinger, J., and S.T. Arold. 2018. Identification and experimental characterization of an extremophilic brine pool alcohol dehydrogenase from single amplified genomes. *ACS Chem Biol* 13: 161–170.

Handford, C.R. 1981. A process-sedimentary framework for characterizing recent and ancient sabkhas. *Sediment Geol* 30: 255–265.

Howari, F. 2006. Spectral analyses of sabkha sediments with implications for remote sensing on Mars. *Int J Astrobiol* 5: 47–56.

Krumbein, W.E. 1985. Introduction and definitions. In: *Hypersaline Ecosystems*. Eds. G.M. Friedman and W.E. Krumbein, 13–17. Berlin and Heidelberg: Springer.

Malik, J.A., AlQarawi, A.A., Dar, B.A., Hashem, A, Alshahrani, T.S., AlZain, M.N., Habib, M.M., Javed, M.M., and E.F. Abd-Allah. 2022. Arbuscular mycorrhizal fungi isolated from highly saline "Sabkha Habitat" soil alleviated the NaCl-induced stress and improved *Lasiurus scindicus* Henr. growth. *Agriculture* 12: 337.

Margesin, R., and F. Schinner. 2001. Potential of halotolerant and halophilic microorganisms for biotechnology. *Extremophiles* 5: 73–83.

McClain, C.R., Bryant, S.R., Hanks, G., and M.W. Bowles. 2022. Extremophiles in Earth's deep seas: A view toward life in exo-oceans. *Astrobiology* 22: 1009–1028.

McKay, C.P., Rask, J.C., Detweiler, A.M., Bebout, B.M., Everroad, R.C., Lee, J.Z., Chanton, J.P., Mayer, M.H., Caraballo, A.A.L., Kapili, B., Al-Awar, M., and A. Al-Farraj. 2016. An unusual inverted saline microbial mat community in an interdune sabkha in the Rub' al Khali (the Empty Quarter), United Arab Emirates. *PLoS ONE* 11: e0150342.

Mohamed, Y.M., Ghazy, M.A., Sayed, A., Ouf, A., El-Dorry, H., and R. Siam. 2013. Isolation and characterization of a heavy metal-resistant, thermophilic esterase from a Red Sea brine pool. *Sci Rep* 3: 3358.

Mountney, N.P. 2005. Sedimentary environments | Deserts. In: *Encyclopedia of Geology*. Eds. R.C. Selley, L.R.M. Cocks and I.R. Plimer, 539–549. Oxford: Elsevier.

Paulo, C., and M. Dittrich. 2013. 2D Raman spectroscopy study of dolomite and cyanobacterial extracellular polymeric substances from Khor Al-Adaid sabkha (Qatar). *J Raman Spectrosc* 44: 1563–1569.

Renn, D., Shepard, L., Vancea, A., Karan, R., Arold, S.T., and M. Rueping. 2021. Novel enzymes from the Red Sea brine pools: Current state and potential. *Front Microbiol* 12: 732856.

Rettberg, P., Antunes, A., Brucato, J., Cabezas, P., Collins, G., Haddaji, A., Kminek, G., Leuko, S., McKenna-Lawlor, S., Moissl-Eichinger, C., Fellous, J.L., Olsson-Francis, K., Pearce, D.,

Rabbow, E., Royle, S., Saunders, M., Sephton, M., Spry, A., Walter, N., Schweingruber, R.W., and J.C. Treuet. 2019. Biological contamination prevention for Outer Solar System moons of astrobiological interest – What do we need to know? *Astrobiology* 19: 1–24.

Rotter, A., Barbier, M., Bertoni, F., Bones, A.M., Cancela, M.L., Carlsson, J., Carvalho, M.F., Cegłowska, M., Chirivella-Martorell, J., Dalay, M.C., Cueto, M., Dailianis, T., Deniz, I., Díaz-Marrero, A.R., Drakulovic, D., Dubnika, A., Edwards, C., Einarsson, H., Erdoğan, A., Erdoğan, O.T., Ezra, D., Fazi, S., FitzGerald, R.J., Gargan, L.M., Gaudêncio, S.P., Udovič, M.G., DeNardis, N.I., Jónsdóttir, R., Kataržytė, M., Klun, K., Kotta, J., Ktari, L., Ljubešić, Z., Bilela, L.L., Mandalakis, M., Massa-Gallucci, A., Matijošytė, I., Mazur-Marzec, H., Mehiri, M., Nielsen, S.L., Novoveská, L., Overlingė, D., Perale, G., Ramasamy, P., Rebours, C., Reinsch, T., Reyes, F., Rinkevich, B., Robbens, J., Röttinger, E., Rudovica, V., Sabotič, J., Safarik, I., Talve, S., Tasdemir, D., Schneider, X.T., Thomas, O.P., Torunska-Sitarz, A., Varese, G.C., and M.I. Vasquez. 2021. The essentials of marine biotechnology. *Front Mar Sci* 8: 629629.

Sagar, S., Esau, L., Hikmawan, T., Antunes, A., Holtermann, K., Stingl, U., Bajic, V.B., and M. Kaur. 2013b. Cytotoxic and apoptotic evaluations of marine bacteria isolated from brine-seawater interface of the Red Sea. *BMC Complement Altern Med* 13: 29.

Sagar, S., Esau, L., Holtermann, K., Hikmawan, T., Zhang, G., Stingl, U., Bajic, V.B., and M. Kaur. 2013a. Induction of apoptosis in cancer cell lines by the Red Sea brine pool bacterial extracts. *BMC Complement Altern Med* 13: 344.

Selama, O., Amos, G.C.A., Djenane, Z., Borsetto, C., Laidi, R.F., Porter, D., Nateche, F., Wellington, E.M.H., and H. Hacène. 2014. Screening for genes coding for putative antitumor compounds, antimicrobial and enzymatic activities from haloalkalitolerant and haloalkaliphilic bacteria strains of Algerian Sahara soils. *Biomed Res Int* 2014: 317524.

Stivaletta, N., Barbieri, R., Picard, C., and M. Bosco. 2009. Astrobiological significance of the sabkha life and environments of southern Tunisia. *Planet Space Sci* 57: 597–605.

Stivaletta, N., López-García, P., Boihem, L., Millie, D.F., and R. Barbieri. 2010. Biomarkers of endolithic communities within gypsum crusts (southern Tunisia). *Geomicrobiol J* 27: 101–110.

Takahashi, M., Takahashi, E., Joudeh, L.I., Marini, M., Das, G., Elshenawy, M.M., Akal, A., Sakashita, K., Alam, I., Tehseen, M., Sobhy, M.A., Stingl, U., Merzaban, J.S., Di Fabrizio, E., and S.M. Hamdan. 2018. Dynamic structure mediates halophilic adaptation of a DNA polymerase from the deep-sea brines of the Red Sea. *FASEB J* 32: 3346–3360.

Thompson, H.F., Summers, S., Yuecel, R., and T. Gutierrez. 2020. Hydrocarbon-degrading bacteria found tightly associated with the 50–70 μm cell-size population of eukaryotic phytoplankton in surface waters of a northeast Atlantic region. *Microorganisms* 8: 1955.

Varrella, S., Tangherlini, M., and C. Corinaldesi. 2020. Deep hypersaline anoxic basins as untapped reservoir of polyextremophilic prokaryotes of biotechnological interest. *Mar Drugs* 18: 91.

Vita-Finzi, C. 2001. Neotectonics at the Arabian plate margins. *J Struct Geol* 23: 521–530.

Wang, Y., Cao, H., Zhang, G., Bougouffa, S., Lee, O.O., Al-Suwailem, A., and P.Y. Qian. 2013. Autotrophic microbe metagenomes and metabolic pathways differentiate adjacent Red Sea brine pools. *Sci Rep* 3: 1748.

Watson, S.W., and J.B. Waterbury. 1969. The sterile hot brines of the Red Sea. In: *Hot Brines and Recent Heavy Metal Deposits in the Red Sea*. Eds. E.T. Degens and D.A. Ross, 272–281. New York: Springer-Verlag.

Wu, J.-H., McGenity, T.J., Rettberg, P., Simões, M.F., Li, W.-J., and A. Antunes. 2022. The archaeal class *Halobacteria* and Astrobiology: Knowledge gaps and research opportunities. *Front Microbiol* (in press).

Yechieli, Y., and W.W. Wood. 2002. Hydrogeologic processes in saline systems: Playas, sabkhas, and saline lakes. *Earth Sci Rev* 58: 343–365.

Zhang, G., Dong, X., Sun, Y., Antunes, A., Hikmawan, T., Haroon, M.F., Wang, J., and U. Stingl. 2020. *Haloferax profundi* sp. nov. and *Haloferax marisrubri* sp. nov., isolated from the Discovery Deep brine-seawater interface in the Red Sea. *Microorganisms* 8: 1475.

Zhang, G., Gu, J., Zhang, R., Rashid, M., Haroon, M.F., Xun, W., Ruan, Z., Dong, X., and U. Stingl. 2017c. *Haloprofundus marisrubri* gen. nov., sp. nov., an extremely halophilic archaeon isolated from a brine-seawater interface. *Int J Syst Evol Microbiol* 67: 9–16.

Zhang, G., Haroon, M.F., Zhang, R., Dong, X., Liu, D., Xiong, Q., Xun, W., Dong, X., and U. Stingl. 2017a. *Ponticoccus marisrubri* sp. nov., a moderately halophilic marine bacterium of the family Rhodobacteraceae. *Int J Syst Evol Microbiol* 67: 4358–4364.

Zhang, G., Haroon, M.F., Zhang, R., Dong, X., Wang, D., Liu, Y., Xun, W., Dong, X., and U. Stingl. 2017b. *Ruegeria profundi* sp. nov. and *Ruegeria marisrubri* sp. nov., isolated from the brine-seawater interface at Erba Deep in the Red Sea. *Int J Syst Evol Microbiol* 67: 4624–4631.

Ziko, L., Adel, M., Malash, M.N., and R. Sia. 2019. Insights into Red Sea brine pool specialized metabolism gene clusters encoding potential metabolites for biotechnological applications and extremophile survival. *Mar Drugs* 17: 273.

Index

For Product Safety Concerns and Information please contact our EU representative GPSR@taylorandfrancis.com Taylor & Francis Verlag GmbH, Kaufingerstraße 24, 80331 München, Germany

Batch number: 10733477

Printed by Printforce, the Netherlands